# *Study Guide*

*to accompany*

Principles of

SECOND EDITION

Hillis • Sadava • Hill • Price

Kristine F. Nowak
*Kennesaw State University*

Joseph A. Bruseo
*Holyoke Community College*

Betty McGuire
*Cornell University*

Shannon L. Compton
*University of Massachusetts, Amherst*

Meredith G. Safford
*Johns Hopkins University*

Mary S. Tyler
*University of Maine, Orono*

Cover photograph © Ch'ien Lee/Minden Pictures.

**Study Guide** to accompany ***Principles of Life*, Second Edition**



**Address editorial correspondence to:**
Sinauer Associates, Inc.
P.O. Box 407
Sunderland, MA 01375 U.S.A.
Internet: www.sinauer.com; publish@sinauer.com

**Address orders to:**
MPS / W.H. Freeman & Co. Order Department
16365 James Madison Highway, U.S. Route 15
Gordonsville, VA 22942 U.S.A. or call 1-888-330-8477
Examination copy information: www.whfreeman.com/request or 1-800-446-8923

This SFI label applies to the text paper.

ISBN 978-1-4641-8475-8
Printed in U.S.A.

4 3 2 1

# To the Student

Biology is an incredibly exciting field of study, but in order to appreciate new discoveries and discussions, it is necessary to have a firm grasp of the underlying concepts and ideas. Your textbook is designed to give you a comprehensive overview of important biological phenomena. It will also serve as a resource to you in future studies. Together with your instructor, your textbook will provide you with invaluable information for beginning your study of biology.

This Study Guide is designed to supplement, not replace, your textbook and your instructor. It was written for you, the student, in language that you can understand, but it does emphasize proper usage of biological terminology. Important concepts and ideas have been synthesized into easy-to-read text that provide an overview of the biological concepts discussed in your textbook. Each Study Guide chapter includes three review elements: (1) The Big Picture, an introductory overview of the topics covered in the chapter; (2) Study Strategies, tips for effective ways to study the material and master common problem areas; and (3) Key Concept Review, a detailed outline of each numbered concept in the chapter interspersed with short answer and diagram questions. These will help you preview a chapter before reading it, check your understanding, and review the chapter later.

Each Study Guide chapter also includes a series of questions that have been grouped into two categories: Key Concept Review questions and Test Yourself questions. Key Concept Review questions are integrated into the summary section, allowing you to test your factual and conceptual knowledge of each chapter while reading the pertinent review text. In some cases, diagrams are provided and you are asked to label them or answer questions based on them, while in other chapters, you are asked to create your own diagrams based on a question or series of questions. Key Concept Review questions ask you to apply the knowledge you have gleaned from a chapter to answer questions that are more open-ended. These questions require you to have assimilated several concepts and to think beyond what you have just read. Your instructor may ask questions similar to these on exams, or may use an entirely different approach to assess your knowledge, but these questions will be a good check of how well you understand the material. Test Yourself questions are multiple choice style questions, designed to determine if you have retained information from a chapter, and if you can put together various concepts in order to answer questions. Answers to both types of questions are provided at the end of each chapter of the Study Guide. These answers are not exhaustive, but instead are designed to point you to the correct concepts in the textbook. Because of the nature of many of the Key Concept Review questions, your answers should be more expansive than the short explanations given in this Study Guide.

## Strategies for Studying Biology

Each individual has his or her own unique study pattern. However, there are some successful study strategies that are universal. We recommend that you first preview a chapter in your textbook. In this initial preview, it is important to note the organization of the chapter and the main points, and to go over the chapter summary at the end. By referring to the Study Guide at this point, you will further understand the organization of the material. If you follow these steps, you might have some specific questions in your mind that you will expect your reading to answer.

Reading a textbook is an active process. We recommend that you always have a pencil and paper available. Jotting notes in margins serves as an excellent mental trigger when it comes time to review. As you read each section of your textbook, see if you can summarize it in your own words. Compare your summaries to those provided at the end of each section and at the end of the chapter. You want to assure yourself that the main points you are noting match those that the author has selected. Refer back to those questions that came up as you previewed the chapter. You should be able to answer your own questions by the time you have finished your reading.

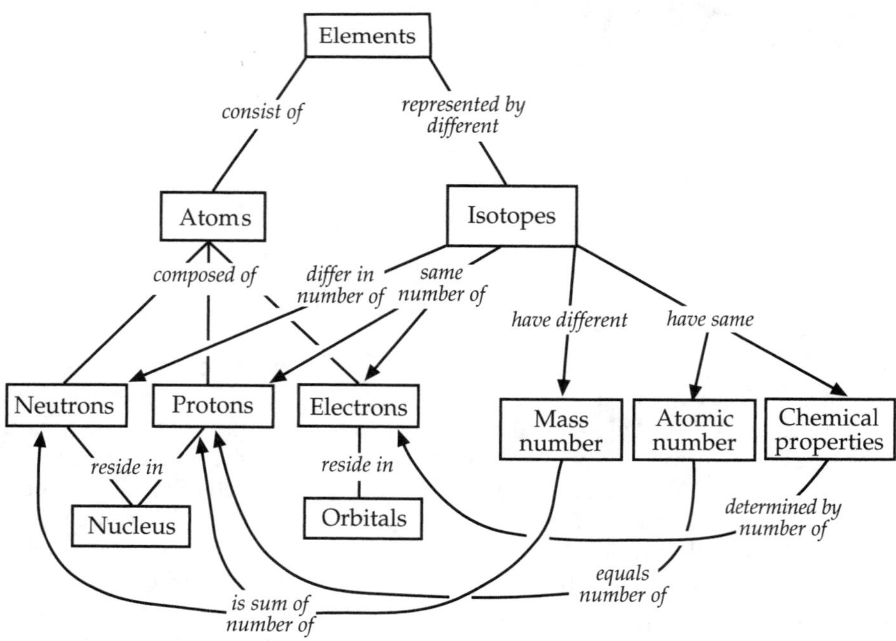


As you are reading, take time to review all the figures and tables. Your textbook makes use of figures to illustrate points, describe pathways, and give visual representation to complex topics. Often these figures are more helpful than the paragraphs of written explanation. In places where figures might be beneficial to you but are not provided in the textbook, try to draw them yourself. This is especially important when attempting to understand structures and pathways. You will find that the Study Guide also refers you to specific animated tutorials, activities, analyze the data exercises, and apply the concept exercises, all of which are available within LaunchPad, at macmillanhighered.com/launchpad. Each of these tools will aid in your understanding of the material. In addition, LaunchPad includes many other study and review resources, some of which your instructor may assign.

Once you have read the material, summarized it in your own words, and reviewed all of the figures and tables, it is time to do another quick review. Scan the summary in your textbook again and read over the Key Concept Review section of this Study Guide. Additionally, the textbook includes many features to help you learn the material as you read and review, including Checkpoint questions, Apply the Concept questions, Analyze the Data questions, and Chapter Summaries. (Refer to the inside front cover of the textbook for more about these features.) We recommend further that you construct a "concept map" of the main points of a chapter. This will assure you that you see how concepts are interrelated and that you understand the organization of the material. An example of a concept map is shown above. Main concepts are in boxes, and arrows are used to connect concepts and show relationships.

Finally, when you are comfortable with chapter content, work on the questions section of this Study Guide (both from the Key Concept Review and Test Yourself sections). We recommend working through all of the questions before checking your answers. Any questions that you do not understand or that you answer incorrectly should be flagged as material you need to review. Remember that your goal is to understand the material completely, not just to answer these particular questions.

Experiments have suggested that the average attention span of a reader is approximately 12 minutes. You cannot expect to sit down and master an entire chapter in one sitting. Break up your study into short segments of approximately 30 minutes each. This will give you time to get down to business, maximize your attention, and learn without becoming frustrated or drained. At the end of this time, move to some other activity or area of study. When you return to biology, you will find your mind is ready to absorb more.

This entire process should take place well in advance of your exams. You cannot learn even one chapter adequately the day before an exam. Mastering biology requires daily study and review. If you keep up with learning the concepts as your course proceeds, you will find that reviewing the textbook summaries and the Study Guide, along with the online review tools, is adequate preparation. The key to learning biology well is slow and diligent daily work.

## Doing Well on Biology Exams

Your instructors are your best resource for doing well on exams. They are there to assist you in learning the material that is presented in your textbook. They are experts in their fields and understand how the concepts presented are vital to your biology education. Follow their lead in preparing for exams. Exam performance is directly linked to classroom attendance and daily study. Sit front and center in your lecture hall and pay close attention to what your instructor writes and provides in presentations or overhead displays. Your instructor will give you guidance about the most important points to study. You should take careful notes and ask for clarification when necessary.

You should always compare your classroom and lecture notes to your textbook and the Study Guide. Take note of the points your instructor emphasizes. Were there specific questions that the instructor brought up in class? Do these correspond to the questions in the Study Guide? If so, chances are you will see them again on an exam. This practice should be part of your daily study regimen.

Approach the exam itself with confidence. Read the directions carefully and read the questions completely. One of the biggest mistakes students make on exams has to do with not following instructions. We recommend that you read carefully through the entire exam even before you begin to answer the first question. You may find that early questions are answered partially in later ones, and that an overview of the whole exam helps to trigger your memory of the material.

If your exam contains multiple choice questions, treat each of the answer options as if it were a separate true/false question. In other words, mentally fill in the question with each answer and ask yourself if the resulting statement is true or false. Be sure to read each of the options, even if the first one strikes you as the correct one. You may be dealing with a question that has multiple correct answers or asks you to select "all of the above" or "none of the above." If you are unsure about how to answer a question, begin by ruling out answers that you know are incorrect. By eliminating some possibilities, your chance of selecting the right one is greatly improved! If you narrow the choices down to two possible answers and just can't decide between them, go with your gut response. You just may be right. And when you go over your answers, change your response only if you know you answered a question incorrectly. If you aren't sure, it may be best to leave the answer as-is. Your first instincts are often correct.

Essay questions require you not only to think through the concepts, but also to decide how you will present them. Organized, concise essay answers are always preferred to rambling answers with little direction. Take a look at the point value of each question. This is often an indication of how many "points" you need to make in your discussion. A 10-point question rarely can be answered with a single sentence. Look carefully at what your instructor is asking. If you are required to "discuss" a concept or a problem, do not present a list or some scattered phrases. Write in complete sentences and paragraphs unless you are asked specifically to list or itemize the material. Be sure that you address all points in the question. Essay questions frequently have multiple parts to test your understanding of the links between various concepts.

Always be aware of the time you have to complete your exam. Answer the questions you know swiftly, and allow yourself time to go back and really think about those you are struggling with. Be sure to concentrate on the questions with the highest point value. Those are the questions testing the most critical concepts and they are the ones that will have the most influence on your grade.

Once you think you have completed the exam, take a few minutes to go back over it. If there are any questions about which you are still unclear, be sure to ask your instructor for clarification.

## Learn from Your Mistakes

After your instructor returns your exam or posts the answers, review the mistakes you have made. Take this opportunity to clarify misunderstandings and re-learn material when necessary. You will find this helpful not only for the final exam, but also as you approach upper-level courses. Material reviewed, corrected, and re-learned is more likely to be retained in the long run.

## Get Help before It Is Too Late

Many students come to college without adequate study habits and/or are ill prepared for the rigor of biology at the college level. But this is a problem that can be solved fairly easily. Many colleges and universities have academic centers that assist with locating tutors, teaching study habits, and acquainting you with a variety of other resources. Make good use of these facilities. They are there for you. Also, for many additional study tips and strategies, see the *Survival Skills* document that is available within LaunchPad.

*Good luck in your study of introductory biology!*

# Table of Contents

# Table of Contents (continued)

# 1 Principles of Life

## The Big Picture

- Biology is the study of living things. Living things are composed of cells, which contain genetic material, require nutrients for energy, maintain a constant internal environment, and interact with one another. All available evidence points to an origin of life some 4 billion years ago. Natural selection and evolution have led to the current host of organisms inhabiting the planet. Life has specific characteristics, and the evolutionary roots of these characteristics can be traced through time.
- All living things, and their components, have places in an organizational hierarchy of life, from the smallest molecules up to the biosphere of the entire Earth. Biological systems are found at every level of biological organization.
- Based on molecular evidence, all living things on Earth are organized as a tree of life. When scientists study organisms from an evolutionary perspective, they are concerned with evolutionary relatedness and common ancestors.
- Scientific inquiry is pursued according to a specific hypothesis–prediction method, which sets it apart from other methods of inquiry. This method requires a testable and falsifiable hypothesis, control of variables, and repeated testing.

## Study Strategies

- The tenets of evolutionary biology may conflict with some of your personal beliefs. However, the textbook is concerned with science and scientific evidence, which ideally are impersonal, objective, and value neutral.
- The hypothesis–prediction method should not be thought of or memorized merely as a series of steps. Think about how each step in the process follows from the preceding step. Two characteristics of scientific investigation that set it apart from other modes of inquiry are the setting of hypotheses and the continuous nature of the process. With each conclusion leading to a new set of hypotheses, science has the capacity to alter our overall base of knowledge.
- Familiarity with the hypothesis–prediction method will help you understand the evidence behind evolution. It may make sense to turn to Section 1.5 ("Science Is Based on Quantitative Observations, Experiments, and Reasoning") before studying the rest of the chapter.
- Drawing a timeline or examining the calendar model of Figure 1.1 will help you understand the chronology of the major evolutionary events discussed in the textbook.
- Go to **LaunchPad** (or use the Web addresses listed) to review the following additional resources:

  Animated Tutorial 1.1 System Simulation (PoL2e.com/at1.1)

  Animated Tutorial 1.2 Using Scientific Methodology (PoL2e.com/at1.2)

  Activity 1.1 The Hierarchy of Life (PoL2e.com/ac1.1)

## Key Concept Review

### 1.1 Living Organisms Share Common Aspects of Structure, Function, and Energy Flow

Life as we know it had a single origin

Major steps in the history of life are compatible with known physical and chemical processes

Biologists can trace the evolutionary tree of life

Life's unity allows discoveries in biology to be generalized

All species on Earth share a common ancestor. Evidence for this theory includes the fact that living organisms share several characteristics, such as a similar set of chemical components, a common code of genetic information, and a set of basic genes that operate in all organisms. Life arose on Earth approximately 4 billion years ago. A random aggregation of complex chemicals led to the existence of the first biological molecules. These initial simple molecules led to molecules that could reproduce themselves and act as templates for larger, more complex molecules. The next step in the origin of life probably involved the enclosure of these molecules within a membrane, a development that created an environment favorable for the necessary biological molecules to interact in more complete ways. Early cells were simple, ocean-living prokaryotes with no internal membrane-enclosed compartments. The ability to carry out photosynthesis arose approximately 2.7 billion years ago. Photosynthesis involves converting light energy from the sun into usable chemical energy with oxygen ($O_2$) as a by-product. Large numbers of photosynthetic organisms increased atmospheric $O_2$ levels, allowing for the evolution of the much more efficient aerobic metabolism. Photosynthetic organisms also contributed to the insulating ozone ($O_3$) layer that shields Earth from radiation and modifies temperatures. This shield eventually made life on land possible.

Eukaryotic cells have intracellular membrane-enclosed compartments, known as organelles, which carry out specific cellular functions. These compartments are thought to have arisen when larger prokaryotic cells engulfed and assimilated smaller cells. Multicellularity, which evolved about 1 billion years ago, made it possible for cells to specialize and for organism size to increase. Different species come about when two groups of organisms in one species are isolated from each other so that they can no longer mate and produce viable offspring. As mutations in the separated genomes accumulate, these two groups become new species.

Organisms are referred to by their binomial genus and species names (e.g., humans are *Homo sapiens*). By examining gene sequences, biologists can compile phylogenetic trees that show the evolutionary relationships among species. All life is thereby placed into one of three major domains: the prokaryotic Archaea and Bacteria and the eukaryotic Eukarya. Because all life is related, scientists work with model organisms to help them study general patterns in living things.

**Question 1.** Scientists interested in human biology typically perform experiments on other species. Many individuals think it is wrong to experiment with animals. Defend the idea of working with model systems in this way.

**Question 2.** What is the hypothesis for the evolution of eukaryotic cells?

## 1.2 Life Depends on Organization and Energy

Organization is apparent in a hierarchy of levels from molecules to ecosystems

Each level of biological organization consists of systems

Biological systems are highly dynamic even as they maintain their essential organization

Positive and negative feedback are common in biological systems

Systems analysis is a conceptual tool for understanding all levels of organization

Organization is critical for living cells, since the tendency to become disorganized can be harmful. Maintaining organization requires energy. Cells synthesize the chemical components they require by organizing smaller chemical components. Organization is necessary for multicellular organisms to function. There are many levels of hierarchical organization in a multicellular organism: small molecules, large molecules, cells, tissues, organs, and organ systems. Organisms themselves interact in ecosystems in a hierarchical fashion: populations, communities, landscapes, and the biosphere (the entire Earth). The internal hierarchies of an organism are governed by its genome, whereas the external hierarchies involve complex interactions among species. A system includes the components and processes by which the components interact. Biological systems are dynamic and involve the flow of matter and energy; biological systems constantly exchange matter and energy with the environment.

Feedback occurs when one component affects the rate of an earlier process in the system. Feedback can be positive, speeding up that process, or negative, slowing it down. Regulatory systems, which tend to stabilize amounts or concentrations of a molecule, are typically controlled via negative feedback. Systems analysis is used to understand how biological systems function. This often requires expressing the processes of a system mathematically. These mathematical models can be used to make predictions about biological systems.

**Question 3.** The study of biology can be organized from the most basic unit, the molecule, up to the biosphere. Create a flow chart describing how each level is connected with the level below it.

**Question 4.** Positive feedback is often seen as detrimental, since it can create a situation in which a process cannot be stopped. Describe an instance in which positive feedback might have some benefit in a living system.

## 1.3 Genetic Systems Control the Flow, Exchange, Storage, and Use of Information

Genomes encode the proteins that govern an organism's structure

Genomes provide insights into all aspects of an organism's biology

The genome of a cell contains all the genetic information in the form of deoxyribonucleic acid (DNA). DNA contains the code to make the proteins required for a cell to function. Segments of DNA make up genes, which are used to produce the proteins necessary for life. Proteins regulate the chemical reactions occurring within the cell. All of the cells in an organism contain the same genome, but they express different genes. Mutations, either spontaneous or caused by chemicals or radiation, occur in the DNA sequence and can alter protein function; these mutations are the basis for evolution. Sequencing the genome of organisms allows biologists to look for the genetic basis for disease; it also allows them to compare the evolution of genes among species. A relatively new field, bioinformatics, is involved in studying the large amount of data generated from genome sequencing.

**Question 5.** Your genome has over 3 billion nucleotides. Does each cell use every one of these nucleotides?

**Question 6.** Assume that you have sequenced the genomes of 15 different species of turtle. What can this vast amount of information tell you?

### 1.4 Evolution Explains the Diversity as Well as the Unity of Life

Natural selection is an important process of evolution

Evolution is a fact, as well as the basis for broader theory

Evolution, the change in populations over time, is the unifying principle of biology. Charles Darwin proposed that all living things are related to one another and that species have evolved through the process of natural selection. Natural selection is the differential survival and reproduction among individuals in a population. Natural selection can produce adaptations, which are structural, physiological, or behavioral traits that enhance the chance of survival of an organism in a given environment. Proximate and ultimate explanations of adaptations explain why and how the adaptations evolved. Evolution is a theory, which is a body of scientific work in which rigorously tested and well-established facts and principles are used to make predictions about the natural world. Scientists can directly measure and observe past and present evolution.

**Question 7.** Create a scenario for how natural selection might act upon a population of rabbits with fur colors that vary from white to brown.

**Question 8.** Many individuals do not accept the theory of evolution and fail to see its importance in biological studies. Defend the idea that the theory of evolution is the unifying theory of biology, and thus absolutely necessary for an understanding of the living world.

### 1.5 Science Is Based on Quantitative Observations, Experiments, and Reasoning

Observing and quantifying are important skills

Scientific methods combine observation, experimentation, and logic

Getting from questions to answers

Well-designed experiments have the potential to falsify hypotheses

Statistical methods are essential scientific tools

Not all forms of inquiry into nature are scientific

Consider the big themes of biology as you read this book

Biologists study life by means of observation and experimentation. Biologists observe organisms in order to understand their natural history, and they try to quantify these observations. The scientific method, or the hypothesis–prediction method, is the basis of most scientific investigations. This method involves making observations and asking questions that lead to the formation of hypotheses, which are provisional answers to the questions. Predictions are made in regard to the hypotheses by means of deductive logic, and then they are tested.

The predictions derived from a hypothesis can be tested by comparative or controlled experiments. A controlled experiment involves isolating variables of interest (independent variables) while keeping other variables that may influence the outcome as steady as possible. Observations are then made of the dependent variables to test the hypothesis. A comparative experiment involves gathering data from multiple groups or conditions to examine the patterns found in nature. Statistical methods are used to determine if the results of the comparison are significant, or if the differences are greater than random chance alone could have produced. Typically, these statistical tests start with a null hypothesis stating that there are no differences.

Science depends on a hypothesis that is testable and that can be rejected by direct observation and experiments. Observations must also be reproducible and quantifiable for a conclusion to be considered scientific.

**Question 9.** The scientific method advances the study of biology. Create a flow chart showing the steps involved in the scientific method.

**Question 10.** Look over a newspaper to see how many articles are directly related to biology. Select one article and discuss how the researchers followed or did not follow the hypothesis–prediction method.

## Test Yourself

1. Life arose on Earth approximately _______ years ago.
   a. 4 billion
   b. 4 million
   c. 4,000
   d. 1.5 billion
   e. 400,000
   ***Textbook Reference:*** *Concept 1.1 Living Organisms Share Common Aspects of Structure, Function, and Energy Flow; Major steps in the history of life are compatible with known physical and chemical processes*

2. Populations of organisms have been able to inhabit a wide variety of environments on Earth because they
   a. have a genome.
   b. contain organelles.
   c. carry out photosynthesis.
   d. adapt through evolution.
   e. are similar to model organisms.
   ***Textbook Reference:*** *Concept 1.4 Evolution Explains the Diversity as Well as the Unity of Life; Natural selection is an important process of evolution*

3. Which of the following is a characteristic that is *not* found in most living organisms?
   a. Regulation of the internal environment
   b. One or more cells
   c. Ability to produce biological molecules

d. Ability to reproduce
e. Ability to create energy
***Textbook Reference:*** *Concept 1.1 Living Organisms Share Common Aspects of Structure, Function, and Energy Flow; Life as we know it had a single origin*

4. Photosynthesis was a major evolutionary milestone because
a. photosynthetic organisms contributed oxygen to the environment, which led to the evolution of aerobic organisms.
b. it arose after organisms began to live on land.
c. it is the only metabolic process that can provide food for organisms.
d. it arose after the ozone layer in the atmosphere began to protect the planet surface.
e. its appearance was followed immediately by the appearance of multicellular animals.
***Textbook Reference:*** *Concept 1.1 Living Organisms Share Common Aspects of Structure, Function, and Energy Flow; Major steps in the history of life are compatible with known physical and chemical processes*

5. Which of the following is *not* an example of a biological system?
a. A coral reef
b. An immune system
c. The synthesis of a biological molecule
d. An enzyme
e. A population of rabbits in a forest
***Textbook Reference:*** *Concept 1.2 Life Depends on Organization and Energy; Each level of biological organization consists of systems*

6. A group of cells that work together to carry out a similar function is known as a(n)
a. tissue.
b. organ system.
c. unicellular organism.
d. protein.
e. gene.
***Textbook Reference:*** *Concept 1.2 Life Depends on Organization and Energy; Organization is apparent in a hierarchy of levels from molecules to ecosystems*

7. In order for natural selection to occur,
a. certain traits must provide greater chances for survival and reproduction than other traits.
b. random survival of all organisms must occur.
c. sexual selection must occur.
d. no genetic mutations can occur.
e. organisms must get larger and stronger over time.
***Textbook Reference:*** *Concept 1.4 Evolution Explains the Diversity as Well as the Unity of Life; Natural selection is an important process of evolution*

8. Which of the following is a domain on the tree of life?
a. Archaea
b. Plantae
c. Animalia
d. Prokaryota
e. Protista
***Textbook Reference:*** *Concept 1.1 Living Organisms Share Common Aspects of Structure, Function, and Energy Flow; Biologists can trace the evolutionary tree of life*

9. The information needed to produce proteins is contained in
a. nutrients.
b. tissues.
c. evolution.
d. organs.
e. genes.
***Textbook Reference:*** *Concept 1.3 Genetic Systems Control the Flow, Exchange, Storage, and Use of Information; Genomes encode the proteins that govern an organism's structure*

10. Evolution is
a. relevant only to the study of biology.
b. the change in the genetic makeup of a population through time.
c. the change in protein expression of a population through time.
d. not influenced by natural selection.
e. seen only in fossil evidence.
***Textbook Reference:*** *Concept 1.4 Evolution Explains the Diversity as Well as the Unity of Life*

11. In a model experiment, researchers subjected frogs to various levels of atrazine while keeping all other variables constant. This is an example of a _______ experiment.
a. controlled
b. repeated
c. laboratory
d. comparative
e. variable
***Textbook Reference:*** *Concept 1.5 Science Is Based on Quantitative Observations, Experiments, and Reasoning; Well-designed experiments have the potential to falsify hypotheses*

12. For a hypothesis to be scientifically valid, it must be _______ and it must be possible to _______ it.
a. testable; prove
b. testable; reject
c. controlled; prove
d. controlled; reject
e. testable; control
***Textbook Reference:*** *Concept 1.5 Science Is Based on Quantitative Observations, Experiments, and Reasoning; Well-designed experiments have the potential to falsify hypotheses*

13. Eukaryotic cells differ from prokaryotic cells in that eukaryotic cells have

a. genes.
b. proteins.
c. organelles.
d. membranes.
e. DNA.
***Textbook Reference:*** *Concept 1.1 Living Organisms Share Common Aspects of Structure, Function, and Energy Flow; Major steps in the history of life are compatible with known physical and chemical processes*

14. In the scientific names of organisms, the _______ is placed first and the _______ is placed second.
a. species; genus
b. genus; domain
c. domain; genus
d. genus; species
e. domain; species
***Textbook Reference:*** *Concept 1.1 Living Organisms Share Common Aspects of Structure, Function, and Energy Flow; Biologists can trace the evolutionary tree of life*

15. A record of the numerical value of an observation is termed _______ while a descriptive observation is termed _______.
a. mathematical; verbal
b. qualitative; quantitative
c. quantitative; qualitative
d. subjective; objective
e. a valuation; a generalization
***Textbook Reference:*** *Concept 1.5 Science Is Based on Quantitative Observations, Experiments, and Reasoning; Observing and quantifying are important skills*

## Answers

### *Key Concept Review*

1. Model systems are useful in the study of biology because all organisms evolved from a common ancestor. Cellular pathways in a zebrafish, for example are very similar to those found in the human. Model systems are also valuable because in many cases they can be manipulated experimentally. Without this type of work, it would have been very difficult to make the dramatic progress in medical science that has been made.

2. Eukaryotic cells contain a number of membrane-bound organelles. It is hypothesized that organelles such as mitochondria and chloroplasts evolved from engulfed prokaryotic organisms that were not digested, but began a mutual relationship with the new host cell.

3. The study of biology involves the smallest molecules to the entire biosphere. See Figure 1.6.

4. There are a number of examples of positive feedback in living systems. During childbirth, uterine contractions cause the production of more oxytocin, which increases the amplitude and frequency of contractions. Blood clotting is initiated when injured tissue provides a signal that activates platelets. The so-called evolutionary arms races, whereby organisms acquire adaptations that overcome the abilities of other organisms in an escalating fashion (as in predator–prey relationships, or pathogenic evasion of immune function) are also positive feedback loops. Positive feedback also functions in the enhancement of B lymphocyte action in the immune system.

5. In your genome, genes are made of specific DNA nucleotide sequences. Each cell uses these genes to code for the proteins necessary for that cell type to function. Not all cell types need all the proteins that can be coded for by your genome, and therefore, they do not use all of genome.

6. With the advent of new sequencing techniques, the genomes of different species can be sequenced more easily than in the past. The sequences from the 15 turtles could be used to help determine the evolutionary relationship among these species. They might also allow you to look at specific genes of interest to see how they differ among the species and how this relates to differences in the turtles' physiology or morphology.

7. Natural selection refers to the differential survival and reproduction among individuals in a population according to their differing traits. In one possible scenario, these rabbits could be living in high mountains that receive a great deal of snow. The white rabbits would be most successful in hiding from predators, and thus the white coloration would increase an animal's likelihood of living and contributing offspring to the next generation.

8. It is only in light of the theory of evolution that we can use model organisms to study how all life on Earth functions. It is the hypothesis of evolutionary relationships, and thus functional similarities, between mice and humans that allow us to do biomedical research in mice. Without this overarching concept of biology, many studies of biological systems would be essentially meaningless, because we would not be able to use the relatedness of all living things to help us understand the natural world.

9.

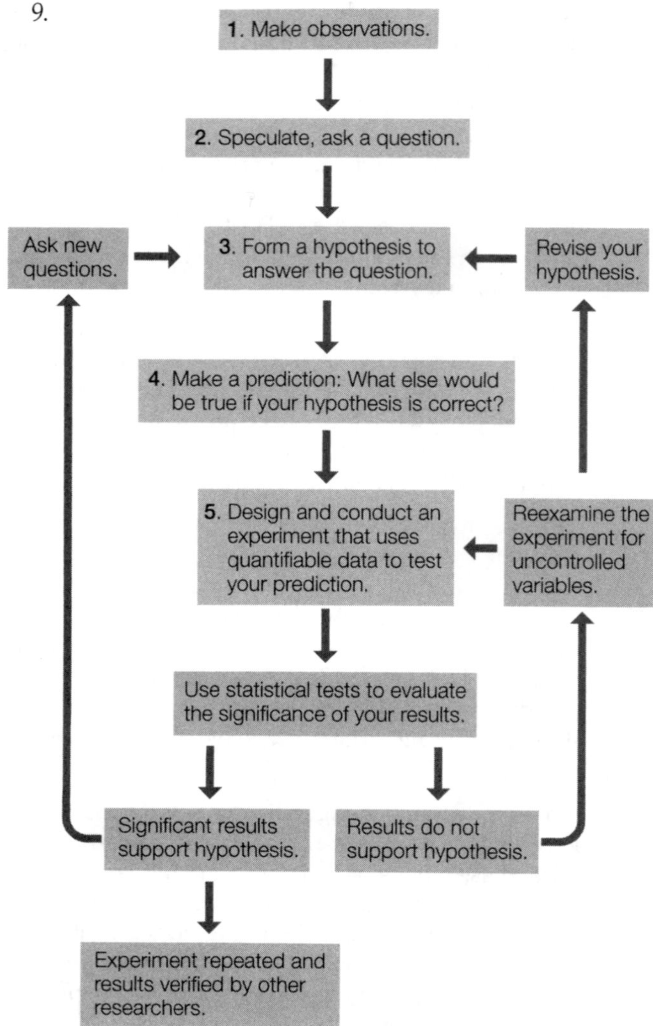


10. Often an article reporting on science news does not provide enough information for the reader to know whether a particular study followed the hypothesis–prediction approach. It pays to be a smart reader and consumer in this respect.

### *Test Yourself*

1. **a.** Available evidence places the beginning of life at about 3.8 billion years ago.
2. **d.** Adaptations are the differences found in organisms that allow them to live in an environment.
3. **e.** Most living organisms are composed of cells, have the ability to make biological molecules, can regulate their internal environment, and reproduce. Living organisms cannot create energy, but they can obtain it from their environment.
4. **a.** Photosynthesis caused the accumulation of oxygen in the atmosphere and contributed to the formation of an ozone layer. The presence of oxygen made the evolution of aerobic organisms possible, and the ozone layer shielded Earth from harmful radiation. These two phenomena contributed to the evolution of terrestrial life. Because photosynthesis is the only metabolic process that can convert light energy into chemical energy, it provides food for many organisms, but it is not the only way this can occur. It was many millions of years after the evolution of photosynthesis before multicellular life arose.
5. **d.** An enzyme by itself is not a system, although it can be a component of a biological system. The other examples are all groups of components that interact in one way or another, making them biological systems.
6. **a.** Multicellular organisms have tissues that are formed from many similar cells.
7. **a.** Natural selection relies on the fact that living organisms vary in their traits and that this variability provides some individuals with an advantage in terms of survival and reproduction.
8. **a.** Plantae and Animalia are kingdoms found in the domain Eukarya. Protista is an archaic term for a group of organisms that are generally described as "not animals, plants, or fungi." Prokaryota is not a real domain; both domains Archaea and Bacteria are prokaryotic.
9. **e.** Genes are specific sequences of DNA that contain the information used to make proteins.
10. **b.** Evolution is a change in the frequency of genes within a population over time and can occur through natural selection. This is the major unifying principle of biology.
11. **a.** For experiments to be scientifically valid, they must be controlled.
12. **b.** Scientific hypotheses are set apart from mere conjecture by being testable and falsifiable.
13. **c.** Both eukaryotic and prokaryotic cells contain genes (which are made of DNA), proteins, and some sort of membrane. Only the eukaryotes have organelles.
14. **d.** An example is *Homo sapiens.*
15. **c.** Numerical values are quantities, and thus this constitutes quantitative data. General descriptions are qualitative (record the "qualities" of something).

# 2 The Chemistry and Energy of Life

## The Big Picture

- Understanding the chemical building blocks of all matter is essential to understanding the biology of organisms. Atoms, which contain protons, neutrons, and electrons, form elements or combine to form molecules. Molecules may be held together with covalent or ionic bonds and stabilized in three-dimensional conformations by hydrogen bonds. Hydrogen bonds are important for determining the properties of water such as its high heat capacity and cohesion.
- Four groups of macromolecules are biologically most important: carbohydrates, lipids, proteins, and nucleic acids. Carbohydrates are used for energy, storage, structure, and signaling. Polysaccharides are complex sugars made up of multiple simple monosaccharides linked by glycosidic linkages. Lipids are insoluble in water and make up a large component of the cell membrane phospholipid bilayer. Their properties depend on the length and number of double bonds of the hydrocarbon chains.
- Chemical reactions change reactants to products without creating or destroying matter. Metabolism is a series of energy-transferring reactions that fuel the processes of the cell. All reactions either give off energy (exergonic) or require energy to proceed (endergonic). Energy may be stored in chemical bonds (potential energy) and released when the bonds are broken to do work (kinetic energy).

## Study Strategies

- It is easy to become overwhelmed by the different types of chemical bonds. Consult Table 2.1 in the textbook to make sure you understand the components and how they interact. Keep it handy as a reference.
- Use the figures in your book to help you visualize what atoms and molecules look like and how they interact.
- Be careful not to get lost in terminology. Think about what each term means, and focus on understanding the concept while memorizing the term.
- Go to **LaunchPad** (or use the Web addresses listed) to review the following additional resources:

  Animated Tutorial 2.1 Chemical Bond Formation (PoL2e.com/at2.1)

  Animated Tutorial 2.2 Macromolecules: Carbohydrates and Lipids (PoL2e.com/at2.2)

  Animated Tutorial 2.3 Synthesis of Prebiotic Molecules (PoL2e.com/at2.3)

  Activity 2.1 Functional Groups (PoL2e.com/ac2.1)

  Activity 2.2 Forms of Glucose (PoL2e.com/ac2.2)

  Analyze the Data in textbook Figure 2.16

  Apply the Concept in textbook Sections 2.2 and 2.5

## Key Concept Review

### 2.1 Atomic Structure Is the Basis for Life's Chemistry

An element consists of only one kind of atom

Electrons determine how an atom will react

Atoms combine to form all matter. The nucleus of an atom contains a defined number of positively charged protons and neutral neutrons. Negatively charged electrons move in their shells, which surround the nucleus. When an atom has equal numbers of electrons and protons it is electrically neutral. The mass of one proton is defined as a dalton. An element is made up of only one type of atom. There are 94 natural elements and 24 or more man-made elements. Six elements compose the majority of every living organism: carbon, hydrogen, oxygen, nitrogen, phosphorus, and sulfur. The number of protons in an element determines its type and is known as its atomic number. Neutrons are found in every element except hydrogen. The total number of protons and neutrons make up the mass number of the atom. An isotope is a variant of an element that has a different number of neutrons.

The Bohr model shows the interaction between electrons in orbits around the central nucleus of the atom. Interactions between atoms involve the associations of their electrons. Electrons continuously orbit the nucleus of an atom in a defined space. The orbit of an atom is the space in which an electron is found. These patterns of orbits compose a series of electron shells, or energy levels, each with a specific number of electrons. The first shell can contain up to two electrons. The second shell can have as many as eight, third shells up to eighteen, and the fourth and subsequent shells up to thirty-two electrons. The number of electrons in the outermost shell determines how the atom interacts with other atoms. Two or more linked atoms compose a molecule, resulting in stabilization of the outer electron shell of the atoms. Many biologically important atoms, such as carbon and nitrogen, are stable when they have eight electrons in the outermost shell. This phenomenon is referred to as the octet rule. Molecules form when atoms share electrons.

**Question 1.** For the following elements, draw the electron shells and place electrons in the appropriate shells based on their atomic numbers: carbon ($_6C$), nitrogen ($_7N$), sodium ($_{11}Na$), chlorine ($_{17}Cl$), argon ($_{18}Ar$), and oxygen ($_8O$).

**Question 2.** Compare and contrast the atomic number and the mass number of an element.

## 2.2 Atoms Interact and Form Molecules

Covalent bonds consist of shared pairs of electrons

Hydrogen bonds may form within or between molecules with polar covalent bonds

Polar and nonpolar substances: Each interacts best with its own kind

Ionic attractions form between anions and cations

Functional groups confer specific properties to biological molecules

Macromolecules are formed by the polymerization of smaller molecules

Chemical bonds link two atoms together to make a molecule. The bonds include covalent bonds, hydrogen bonds, and ionic bonds.

Atoms share pairs of electrons to stabilize their outer shells in covalent bonding. This type of bond is very stable and strong and can be broken only with a great deal of energy. All molecules have a three-dimensional shape, and the interactions between a given pair of atoms always have the same length, angle, and direction. A molecule's shape affects how it behaves. Multiple covalent bonds may exist. Although a single covalent bond involves one pair of shared electrons, double bonds involve two pairs of shared electrons. Triple bonds share three pairs, but are rare.

Covalent bonding between atoms of the same element results in equal sharing of electrons. The attractive force that an atom exerts on electrons is known as electronegativity. Two atoms that have the same electronegativity share electrons equally and form nonpolar covalent bonds. Bonds between different elements generally result in polar covalent bonds with unequal sharing of electrons. Unequal sharing of electrons results in partial ($\delta$) charges in molecules because one nucleus is more electronegative than the other, attracting the electrons more strongly. One atom of a molecule may be partially negative, whereas the other is partially positive. This balance of partial charges results in a polar molecule with a $\delta^-$ pole and $\delta^+$ pole.

Hydrogen bonds form when the positively charged hydrogen atom on one polar molecule attracts a negatively charged atom on another polar molecule. This occurs between molecules of water where the positive hydrogen is attracted to the negative oxygen. This is a weak bond that is easily broken because no electrons are shared, but this type of bond is important in stabilizing the three-dimensional shape of large molecules such as proteins and DNA.

Hydrogen bonds in water provide for its heat capacity and cohesion. An average of 3.4 hydrogen bonds are formed by each water molecule. Water is an ideal solvent for many biological molecules. Water has a high heat of vaporization, so a great deal of heat is required to change it from a liquid to a gaseous state. Much of the energy is used to break hydrogen bonds. This makes evaporating water an effective coolant. The polar nature of water and the formation of hydrogen bonds contribute to the cohesion of water, as well as its adhesion to surrounding solid surfaces and its surface tension. Polar molecules are hydrophilic, or "water loving," because of the partial charges and the hydrogen bonds they form. Nonpolar molecules are hydrophobic, or "water hating," and they have hydrophobic interactions. Polar molecules tend to aggregate with other polar molecules and nonpolar molecules aggregate with other nonpolar molecules.

Ions are formed when atoms lose or gain electrons, resulting in a net positive or negative charge. Cations are positively charged and have fewer electrons than protons. Anions are negatively charged and have more electrons than protons. Ionic attractions form between ions of opposite charge and are often called salts. When in solution, these bonds are weak.

Functional groups are small groups of atoms that provide specific properties to the molecules to which they are attached. In chemical structures, the functional group is often diagramed as bonded to an "R" to indicate that the functional group can be attached to a wide variety of carbon skeletons. Biological molecules typically contain hydrophobic, polar, and charged functional groups that determine shape and function.

Macromolecules form from the covalent bonding of smaller monomers. All organisms are composed of four major biological macromolecules: proteins, nucleic acids, carbohydrates, and lipids. In condensation, the removal of water links monomers together (see Figure 2.8A). Polymers may be broken down into their constituent monomers through hydrolysis (see Figure 2.8B). In all life forms, the specific types of macromolecules share similar roles that are dependent on the chemical properties of their component monomers.

**Question 3.** Calcium has an atomic number of 20. Draw structures for Ca and $Ca^{2+}$. What is the difference between these structures?

**Question 4.** Water is a polar molecule. This property contributes to cohesion and surface tension. Draw six water molecules. In your drawing, indicate how hydrogen bonding between molecules contributes to cohesion and surface tension. Be sure to include appropriate covalent bonds in each molecule.

## 2.3 Carbohydrates Consist of Sugar Molecules

### Monosaccharides are simple sugars

### Glycosidic linkages bond monosaccharides

### Polysaccharides store energy and provide structural materials

Carbohydrates have the general formula $C_n(H_2O)_n$. They are used to store and transport energy, in carbon skeletons, and as signaling molecules in organisms. They can be small monomers or large polymers. Monosaccharides are simple sugar monomers that are used in the synthesis of complex carbohydrates. Examples include six-carbon hexoses such as glucose, which is used as an energy source, or five-carbon pentoses such as ribose and deoxyribose, which make up the structural backbones of RNA and DNA, respectively.

Disaccharides, oligosaccharides, and polysaccharides are constructed from monosaccharides covalently bonded by condensation reactions that form glycosidic linkages. Disaccharides are formed by two linked monosaccharides and oligosaccharides are formed by several linked monosaccharides. Polysaccharides are very large polymers of monosaccharides that provide energy storage or structural support. The specific structure of polysaccharides contributes to their function. For instance, starch, glycogen, and cellulose are all chains of glucose, but the glycosidic bonds are in different orientations and yield very different chemical properties.

**Question 5.** Complex carbohydrates should be a mainstay of one's diet. What properties of carbohydrates make them excellent food sources?

**Question 6.** Examine the structures of glucose polymers in Figure 2.10. Hypothesize why cellulose is a better structural polysaccharide than either starch or glycogen.

## 2.4 Lipids Are Hydrophobic Molecules

### Fats and oils are triglycerides

### Phospholipids form biological membranes

Lipids are insoluble in water due to many nonpolar covalent bonds. They aggregate because of their nonpolar nature and are held together by van der Waals interactions. They play a variety of roles in the biology of organisms, including energy storage, structure in cell membranes and on body surfaces, and thermal insulation. Fats and oils are also known as triglycerides, or simple lipids. These molecules, composed of fatty acids and glycerol, function primarily in the storage of energy. At room temperature, fats are solid and oils are liquid. Each triglyceride contains a glycerol bound to three fatty acids with carbon atom chains that are all single-bonded (saturated fatty acids) or contain double bonds (unsaturated fatty acids). During condensation, three fatty acids are covalently linked to the glycerol. The characteristics of the carbon bonds of fatty acids (i.e., whether they are saturated or unsaturated) influence the shape of the molecule, affect how densely it can pack with other fatty acid molecules, and determine its melting point. The greater the saturation of the fatty acid chains, the higher the melting point and the denser the packing of the triglyceride will be.

Phospholipids form cellular membranes. In phospholipids, one of the hydrophobic fatty acids of a typical lipid is replaced with a hydrophilic phosphate group. This allows phospholipids to be amphipathic: they have a hydrophilic "water-loving" head and hydrophobic "water-hating" tails. In an aqueous environment, the hydrophobic tails of the phospholipids tend to aggregate, with the phosphate heads facing out. This bilayer effect allows for the establishment of a hydrophobic inside surrounded by an aqueous environment. Biological membranes are made of such phospholipid bilayers.

**Question 7.** Consider the following triglyceride:

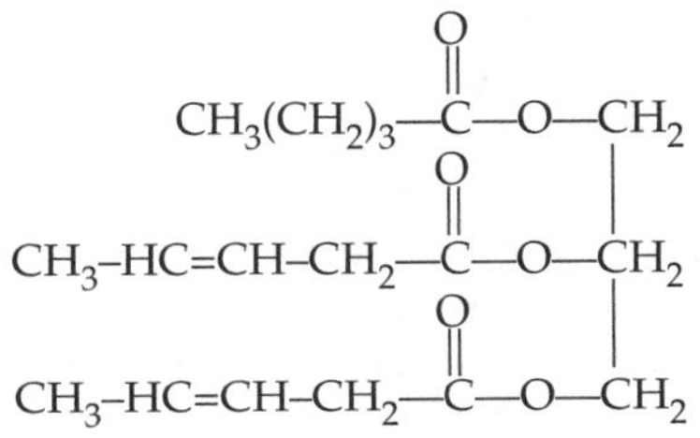

$$CH_3(CH_2)_3{-}\overset{\overset{\large O}{\|}}{C}{-}O{-}CH_2$$
$$CH_3{-}HC{=}CH{-}CH_2{-}\overset{\overset{\large O}{\|}}{C}{-}O{-}CH$$
$$CH_3{-}HC{=}CH{-}CH_2{-}\overset{\overset{\large O}{\|}}{C}{-}O{-}CH_2$$

a. Circle the remnant of the glycerol portion of the triglyceride.

b. How many water molecules result from the formation of this triglyceride from glycerol and three fatty acids?

**Question 8.** Which triglyceride (*A* or *B*) is probably a solid at room temperature? Explain your answer.

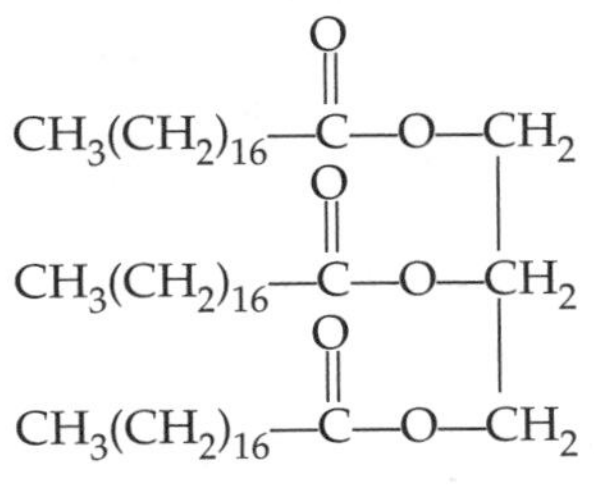

*A*

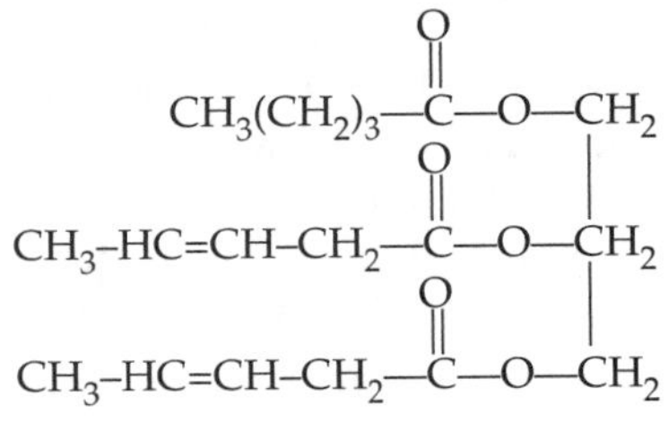

$$CH_3(CH_2)_{16}{-}\overset{\overset{\large O}{\|}}{C}{-}O{-}CH_2$$
$$CH_3(CH_2)_{16}{-}\overset{\overset{\large O}{\|}}{C}{-}O{-}CH$$
$$CH_3(CH_2)_{16}{-}\overset{\overset{\large O}{\|}}{C}{-}O{-}CH_2$$

*B*

$$CH_3(CH_2)_3{-}\overset{\overset{\large O}{\|}}{C}{-}O{-}CH_2$$
$$CH_3{-}HC{=}CH{-}CH_2{-}\overset{\overset{\large O}{\|}}{C}{-}O{-}CH$$
$$CH_3{-}HC{=}CH{-}CH_2{-}\overset{\overset{\large O}{\|}}{C}{-}O{-}CH_2$$

**Question 9.** Draw a phospholipid and a bilayer. What characteristics of phospholipids make them perfectly suited for membranes? What do you think might happen if phospholipids did not form a bilayer? How might they arrange themselves in an aqueous environment?

**Question 10.** Dietary guidelines encourage people to stay away from saturated fats. What is meant by the term "saturated fat"? Why is this type of fat of more concern than unsaturated fats in the diet? What is the structural difference between saturated and unsaturated fats?

### 2.5 Biochemical Changes Involve Energy

#### Metabolism involves reactions that store and release energy

#### Biochemical changes obey physical laws

Chemical reactions involve reactants that combine or change bonding partners to produce products. Metabolism is all of the chemical reactions occurring in a living organism. Kinetic energy is energy of movement and does work. Potential energy is stored energy. Energy can be stored biologically in chemical bonds of fatty acids and other molecules. Breaking the bonds converts the energy to kinetic energy. Anabolic reactions, or anabolism, link together simple molecules to create complex molecules and to store energy in the resulting bonds. Catabolic reactions, or catabolism, break down complex molecules and release stored energy.

The laws of thermodynamics apply to all matter and energy transformations. The first law of thermodynamics states that energy is neither created nor destroyed. Energy can be converted from one form to another. The second law of thermodynamics states that not all energy can be used to do work in a given reaction. When energy is converted from one form to another, some becomes unusable and there is an increase in entropy. The change in available energy that occurs during a reaction is called free energy (*G*). Reactions are spontaneous when the reactants have a higher *G* than the products.

**Question 11.** It is estimated that approximately 90 percent of energy that passes between levels in a food web is "lost" at each level. Explain the first law of thermodynamics and discuss why this apparent loss of energy does not contradict the law.

**Question 12.** You decide to purchase a new water heater and start looking at the energy efficiency ratings. You find one unit that is labeled as 100 percent energy efficient, and the salesperson says that the more efficient the appliance is, the more money you will save. However, you don't trust that the store is providing accurate information, and you do not buy the product. Was this decision correct?

## Test Yourself

1. The stability of the three-dimensional shape of many large molecules is dependent on
   a. covalent bonds.
   b. ionic bonds.
   c. hydrogen bonds.
   d. van der Waals attractions.
   e. hydrophobic interactions.
   ***Textbook Reference:*** *Concept 2.2 Atoms Interact and Form Molecules; Covalent bonds consist of shared pairs of electrons*

2. Which statement about water is true?
   a. Water has a low heat of vaporization.
   b. Water has a high heat capacity.
   c. Evaporation of water provides a warming effect.
   d. Water molecules can only form hydrogen bonds with other water molecules.
   e. Water can never support the weight of an object.
   ***Textbook Reference:*** *Concept 2.2 Atoms Interact and Form Molecules; Hydrogen bonds may form within or between molecules with polar covalent bonds*

3. Which statement about chemical reactions is true?
   a. The bonding partners of atoms remain constant.
   b. All reactions release energy as they proceed.
   c. The bonding partners of atoms change.
   d. All reactions consume energy as they proceed.
   e. All reactions occur spontaneously.
   ***Textbook Reference:*** *Concept 2.5 Biochemical Changes Involve Energy*

4. Which of the following is *not* a polymer?
   a. A protein
   b. A nucleic acid, such as DNA
   c. A polysaccharide carbohydrate
   d. An oligosaccharide carbohydrate
   e. A lipid
   ***Textbook Reference:*** *Concept 2.3 Carbohydrates Consist of Sugar Molecules; Glycosidic linkages bond monosaccharides*

5. The atomic number of an atom is determined by the number of
   a. protons and neutrons.
   b. electrons.
   c. neutrons.
   d. protons.
   e. protons, neutrons, and electrons.
   ***Textbook Reference:*** *Concept 2.1 Atomic Structure Is the Basis for Life's Chemistry; An element consists of only one kind of atom*

6. Cellulose and starch have structural and functional differences. Which of the following is the characteristic that accounts for those differences?
   a. Different types of glycosidic linkages
   b. Different numbers of glucose monomers
   c. Different types of bonds holding them together
   d. Different types of sugar monomers
   e. A linear shape in one versus a ring shape in the other

   ***Textbook Reference:*** *Concept 2.3 Carbohydrates Consist of Sugar Molecules; Polysaccharides store energy and provide structural materials*

7. Triglycerides are synthesized from _______ and _______.
   a. glycerol; amino acids
   b. amino acids; cellulose
   c. steroid precursors; starch
   d. cholesterol; glycerol
   e. fatty acids; glycerol

   ***Textbook Reference:*** *Concept 2.4 Lipids Are Hydrophobic Molecules; Fats and oils are triglycerides*

8. Which characteristic distinguishes carbohydrates from other macromolecule types?
   a. Carbohydrates are constructed of monomers that always have a ring structure.
   b. Carbohydrates never contain nitrogen.
   c. Carbohydrates consist of a carbon bonded to hydrogen and a hydroxyl group.
   d. Carbohydrates contain glycerol.
   e. Carbohydrates always contain the same number of carbon molecules in the monomeric component.

   ***Textbook Reference:*** *Concept 2.3 Carbohydrates Consist of Sugar Molecules*

9. One characteristic of phospholipids that allows them to form a bilayer is their
   a. hydrophilic fatty acid tail.
   b. hydrophobic head.
   c. hydrophobic fatty acid tail.
   d. hydrophilic glycogen acid tail.
   e. hydrogen bonding between fatty acid tails.

   ***Textbook Reference:*** *Concept 2.4 Lipids Are Hydrophobic Molecules; Phospholipids form biological membranes*

10. A five-carbon sugar is known as a
    a. glutamine.
    b. glucose.
    c. hexose.
    d. pentose.
    e. mannose.

    ***Textbook Reference:*** *Concept 2.3 Carbohydrates Consist of Sugar Molecules; Monosaccharides are simple sugars*

11. Covalent bonds form when
    a. atoms of opposite charge are attracted to each other.
    b. hydrogen and oxygen interact.
    c. hydrophilic molecules bind hydrophobic molecules.
    d. electrons of nonpolar substances interact.
    e. atoms share electrons.

    ***Textbook Reference:*** *Concept 2.2 Atoms Interact and Form Molecules; Covalent bonds consist of shared pairs of electrons*

12. _______ energy is the energy of movement.
    a. Potential
    b. Kinetic
    c. Entropic
    d. Enthalpic
    e. Heat

    ***Textbook Reference:*** *Concept 2.5 Biochemical Changes Involve Energy*

13. _______ energy is the energy of state or position.
    a. Potential
    b. Kinetic
    c. Entropic
    d. Enthalpic
    e. Physical

    ***Textbook Reference:*** *Concept 2.5 Biochemical Changes Involve Energy*

14. Which statement concerning energy transformations is true?
    a. Increases in entropy reduce usable energy.
    b. Energy may be created during energy transformations.
    c. Potential energy increases with each transformation.
    d. Increases in thermal energy decrease total amount of energy available.
    e. Decreases in entropy reduce usable energy.

    ***Textbook Reference:*** *Concept 2.5 Biochemical Changes Involve Energy; Biochemical changes obey physical laws*

15. A reaction with a $\Delta G$ of –20 kcal/mol is
    a. endergonic, and equilibrium is far toward completion.
    b. exergonic, and equilibrium is far toward completion.
    c. endergonic, and the forward reaction occurs at the same rate as the reverse reaction.
    d. exergonic, and the forward reaction occurs at the same rate as the reverse reaction.
    e. of an indeterminate nature, according to the information supplied.

    ***Textbook Reference:*** *Concept 2.5 Biochemical Changes Involve Energy; Biochemical changes obey physical laws*

# Answers

## *Key Concept Review*

1.

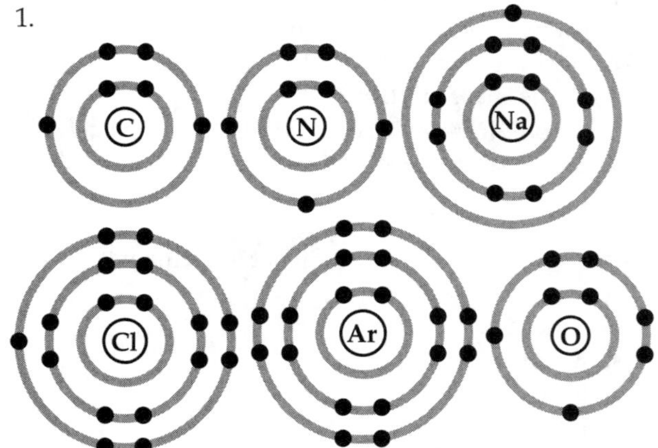


2. The atomic number of an element reflects the number of protons in the nucleus, and thus is unique for each element. The mass number is the mass of an atom of that element, and thus includes the number of protons and neutrons present in a single atom. The total mass of an atom would also take into account the electrons, but since their mass is negligible, they are often omitted in calculating atomic or molecular masses.

3.

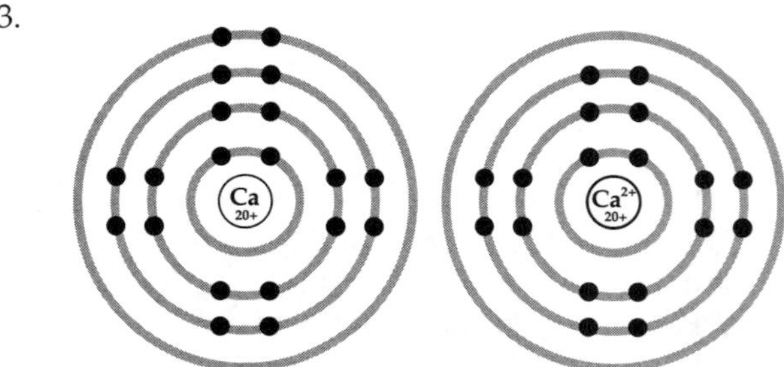


The difference between these structures is that calcium (Ca) has two electrons in its outer shell. Calcium ion ($Ca^{2+}$) has lost its two outer electrons and therefore has a positive charge of 2 because it has two more protons than electrons.

4. The partially positive hydrogen atoms of one water molecule are attracted to the partially negative oxygen atoms of another molecule of water. This attraction tends to cause water molecules to "stick" together, creating surface tension.

*Hydrogen bond*

5. Complex carbohydrates are easily broken down into glucose monomers, which provide nearly all cellular energy. By storing glucose monomers in large carbohydrates, the osmotic strain on any given cell is reduced without sacrificing availability of energy.

6. Cellulose has an unbranched, linear structure. Both starch and glycogen are branched molecules. The individual cellulose fibers can pack together very tightly, and the hydrogen bonds between the different fibers also provide stability. The branched starch and glycogen do not pack as tightly as cellulose does, and thus are not as favorable for providing structure to a cell.

7.

a.

$CH_3(CH_2)_3$–C(=O)–O–$CH_2$
$CH_3$–HC=CH–$CH_2$–C(=O)–O–$CH_2$
$CH_3$–HC=CH–$CH_2$–C(=O)–O–$CH_2$

b. Three water molecules will result; one for each of the three fatty acids added to glycerol by a condensation reaction.

8. Triglyceride *A* is probably solid at room temperature. Its fatty acid chains are saturated (no double bonds) and relatively long, both of which are characteristics of solid, animal-derived triglycerides.

9. The hydrophilic head and the hydrophobic tail of phospholipids allow them to have an "inside" that resists an aqueous environment and an "outside" that can reside in such an environment. When they exist as a bilayer, the hydrophobic tails aggregate. If they did not exist in two layers, the tails would still try to aggregate.

   This would result in a spherical aggregation of phospholipids called a micelle, in which the tails are arranged toward the center of the sphere, away from the aqueous environment, and the heads are immersed in the aqueous environment.

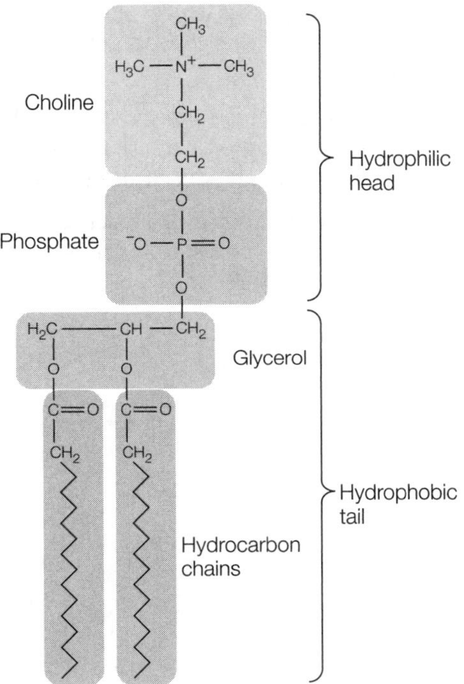


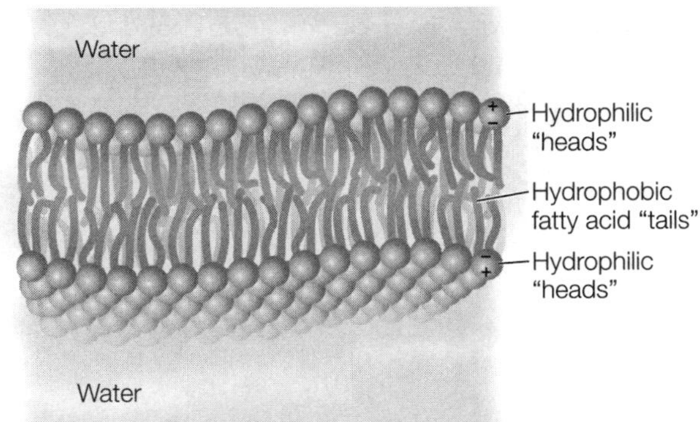


10. Saturated fats contain only single bonds and are "saturated" with hydrogen, which allows fat molecules to pack together densely. This is the reason that most saturated fats are solid at room temperature. Unsaturated fats contain double bonds that affect the shape of the molecule and are not "saturated" with hydrogen. This characteristic keeps them from packing together tightly, and they tend to be liquid at room temperature. Saturated fats are dangerous because of this "packing" ability, which can affect membrane function.

11. The first law of thermodynamics states that energy cannot be created or destroyed, but that it may be converted from one form to another. In the transfer between levels of a food web, approximately 90 percent of the energy is converted to unusable heat energy. There is a net loss of usable energy during each conversion, but the total amount of energy (usable and unusable) remains the same.

12. The decision was correct. An appliance with 100 percent energy efficiency is not possible. The second law of thermodynamics indicates that every time energy is transformed, some is lost in the form of entropy. An appliance with no energy lost to entropy therefore does not exist.

### *Test Yourself*

1. **c.** Hydrogen bonds, though weak individually, are quite effective in large numbers and are responsible for maintaining the structural integrity of many large molecules, such as proteins and DNA.

2. **b.** Water has both a high heat of vaporization and a high heat capacity. Evaporation provides a cooling effect because the water absorbs some of the heat from

the adjacent surface. Adhesion is the hydrogen bonding of water to surrounding surfaces. Surface tension provided by cohesion and adhesion allows small organisms such as insects to walk on the surface of water.

3. **c.** Chemical reactions involve the combining or changing of bonding partners of the reactants to produce the products. During this process matter is neither created nor destroyed. These reactions can either release energy or require energy. Reactions that require energy will not be spontaneous.
4. **e.** Proteins, polysaccharides, oligosaccharides, and nucleic acids are all polymers made by condensation to link monomers together. Lipids are monomers that do not form polymers when they interact.
5. **d.** The atomic number of an atom is the number of protons in the nucleus.
6. **a.** Starch and cellulose have different types of glycosidic linkages. This difference accounts for structural and functional differences between the two macromolecules.
7. **e.** Triglycerides are formed from one glycerol and three fatty acid molecules.
8. **c.** Carbohydrates always have carbon atoms bonded to hydrogen atoms and hydroxyl groups. They may have a variety of other associated molecules in addition to these.
9. **c.** Phospholipids are composed of a hydrophilic head and a hydrophobic tail. When they are placed in water, the hydrophobic tails come together in the interior of the bilayer, surrounded by the hydrophilic heads facing outward.
10. **d.** Five-carbon sugars are known as pentoses and some types form the backbones of RNA and DNA.
11. **e.** Covalent bonds form when atoms share electron pairs.
12. **b.** The released energy is available to do work, therefore, it is kinetic energy. Some energy is lost in the form of entropy, but it will not be used by the cell to do work.
13. **a.** Potential energy is energy held within chemical bonds that may be converted to working kinetic energy.
14. **a.** Total energy = Free energy + Unusable energy. Any increase in entropy, or increase in unusable energy, is necessarily going to reduce free energy.
15. **b.** A negative $\Delta G$ indicates a reaction in which energy is liberated; it will be a spontaneous reaction, and will tend to go in the direction from reactants to products.

# Nucleic Acids, Proteins, and Enzymes

## The Big Picture

- Sequences of nucleic acids are either single-stranded RNA or double-stranded DNA and contain the genetic information needed to make the proteins required for life. Because DNA holds essential information, it must be reproduced exactly, so that each new cell or organism receives a complete set of DNA. The information stored in DNA is transcribed into RNA and then translated into specific proteins. Monomer amino acids are joined together to produce proteins. Each amino acid contains an R group that provides it with specific chemical characteristics. The sequence of amino acids in a protein determines its shape and ultimate function within the cell. Many proteins act as enzymes within the cell.
- Enzymes aid biological reactions by lowering the activation energy required to start the reaction. Each enzyme has a specific three-dimensional conformation that interacts specifically with the substrate. Interaction between the enzyme and substrate results in an optimal orientation for a reaction to take place. Enzymes are affected by temperature, pH, reactants, activators, and inhibitors.
- Metabolism is regulated by enzymes. Most metabolic pathways are under allosteric control. Enzyme complexes allow for interactions and regulation of adjacent active sites. Often the final product of a specific pathway regulates the commitment step of the pathway by feedback inhibition.

## Study Strategies

- Do not get overwhelmed by the different types of macromolecules found in biological systems. Remember that each macromolecule is synthesized from smaller components that give each class of molecule its specific function.
- Make a table of macromolecule characteristics. This will help you see the patterns of similarity and difference. Avoid the temptation merely to memorize structures. If you understand condensation and hydrolysis and know the basic components of the macromolecules, memorizing the structures of the large macromolecules is unnecessary.
- Allosteric regulation can be confusing because the terms "allosteric regulation" and "allosteric enzyme" are related but slightly different. Remember that an allosteric enzyme is an enzyme with more than one binding site: an active site (for binding and acting on substrates) and one or more allosteric regulatory sites. When allosteric regulators are bound to the allosteric site, they control whether or not the enzyme can bind substrate.
- With this chapter, the best strategy is to take "small bites." You may find the concepts to be very unfamiliar; therefore, make sure you understand each section before proceeding. This chapter contains a large amount of terminology. Making a terminology/vocabulary list may be helpful.
- When thinking of enzyme-mediated reactions, be sure to consider that alterations on the left side of the equation lead to changes in the amount of product formed. Remember that enzymes are always conserved across the equation.
- Go to **LaunchPad** (or use the Web addresses listed) to review the following additional resources:

  Animated Tutorial 3.1 Macromolecules: Nucleic Acids and Proteins (PoL2e.com/at3.1)

  Animated Tutorial 3.2 Enzyme Catalysis (PoL2e.com/at3.2)

  Animated Tutorial 3.3 Allosteric Regulation of Enzymes (PoL2e.com/at3.3)

  Activity 3.1 Nucleic Acid Building Blocks (PoL2e.com/ac3.1)

  Activity 3.2 DNA Structure (PoL2e.com/ac3.2)

  Activity 3.3 Features of Amino Acids (PoL2e.com/ac3.3)

  Activity 3.4 Free Energy Changes (PoL2e.com/ac3.4)

  Analyze the Data in textbook Figure 3.10

  Apply the Concept in textbook Sections 3.2 and 3.4

## Key Concept Review

### 3.1 Nucleic Acids Are Informational Macromolecules

Nucleotides are the building blocks of nucleic acids

Base pairing occurs in both DNA and RNA

DNA carries information and is expressed through RNA

The DNA base sequence reveals evolutionary relationships

The primary role of nucleic acids is to store and transmit hereditary information. DNA (deoxyribonucleic acid) and RNA (ribonucleic acid) are the two different types of nucleic acids. Nucleic acids are made up of nucleotide monomers. Each nucleotide monomer consists of a pentose (five-carbon) sugar (either dexoyribose or ribose), a phosphate group, and a nitrogenous base. The bases of nucleotides contain either a single ring with six members (pyrimidine) or a fused double-ring (purine). The backbone of DNA consists of alternating sugars and phosphates with the bases projecting outward.

In both RNA and DNA, monomers are joined by phosphodiester linkages between one nucleotide and the phosphate group of the next nucleotide, growing from the 5′ to the 3′ direction. Nucleic acids are either short oligonucleotides (such as RNA molecules used for DNA replication or regulation of gene expression) or longer polynucleotides (such as DNA). Four types of nucleotides are found in DNA: two purines (adenine and guanine) and two pyrimidines (thymine and cytosine).

In RNA, thymine is replaced by uracil. In DNA, bases are capable of complementary base pairing, with adenine pairing with thymine and cytosine pairing with guanine. Complementary base pairing is facilitated by the size and shape of the molecules, the geometry of the sugar–phosphate backbone, and hydrogen bonding. DNA is typically found as a double helix with two complementary strands paired and twisted together. The two strands in DNA are stabilized by hydrogen bonds and run in opposite directions. Most RNA is a single strand of nucleotides that in some cases can, through hydrogen bonding, fold in on itself or interact with other RNA molecules to produce different three-dimensional structures.

The sequences of bases in DNA provide a means of storing information in the cell. To ensure the correct use of this information, DNA is replicated precisely and can be transcribed into RNA, which in turn is translated into specific polypeptides. Replication and transcription rely on properties of the five bases. During transcription, the bases in DNA are paired with their appropriate RNA counterparts following base pairing rules (A-U, T-A, G-C, and C-G). When a molecule of DNA is replicated, the entire DNA molecule is copied. The genome contains a complete set of an organism's DNA. The specific portions of the DNA that encode for proteins are genes; they are transcribed only when the cell requires those specific proteins.

Similarities in base sequences of DNA can help determine evolutionary relationships among organisms because it is a conserved and consistent molecule. Closely related species tend to have greater similarity in base sequences compared to distantly related species. The increased ability of scientists to sequence DNA is leading to insights into the evolutionary relationships among organisms that were not available based on anatomical and behavioral comparisons.

**Question 1.** If you were to transcribe both strands of the DNA sequence below, what two RNA sequences would result?

5′-AAGCGTC-3′
3′-TTCGCAG-5′

**Question 2.** In the diagram below, use base-pairing rules to label the bases on the strand of RNA (right) that is complementary to the single strand of DNA (left), where C = cytosine, G = guanine, A = adenine, T = thymine, and U = uracil. Then circle and label an example of a nucleotide and a nucleoside. Based on the orientations of the sugar molecules, label the four ends of the molecule as 3′ or 5′.

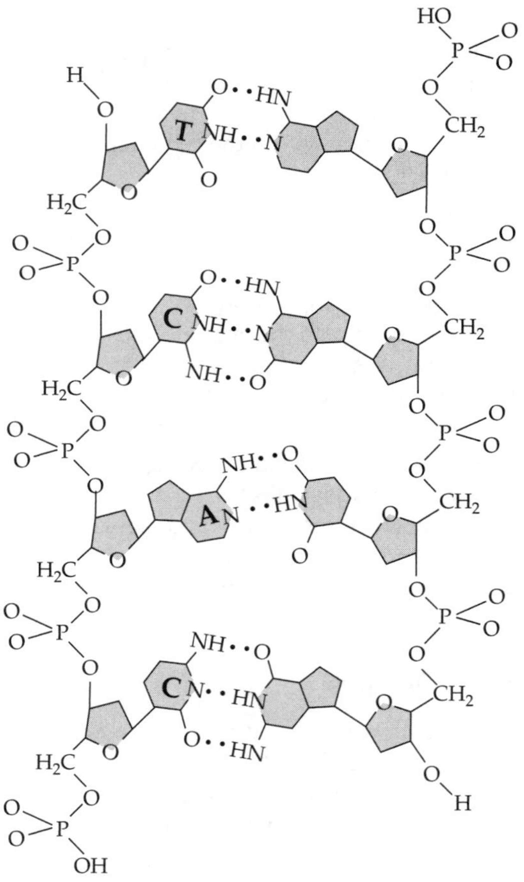

H
O
T
C
A
C
$H_2C$
O
P
OH
HO
P
$CH_2$
O
H
O··HN
NH··N
NH··O
NH··O
N··HN
O··HN

**Question 3.** What properties of a DNA structure make it well suited for its function as an informational molecule?

## 3.2 Proteins Are Polymers with Important Structural and Metabolic Roles

Amino acids are the building blocks of proteins

Amino acids are linked together by peptide bonds

Higher-level protein structure is determined by primary structure

Protein structure can change

Proteins have many functions in an organism, including enzymatic activity, defense, hormonal regulation, signaling, storage, structural support, transport, and genetic regulation. Proteins are polymers composed of amino acid monomers, each of which has a central carbon atom with a hydrogen atom, a basic amino group, an acidic carboxyl group, and a unique side chain (or R group) attached to it. The shape and structure of proteins is influenced by the properties of the side chains of the constituent amino acids. Proteins contain only 20 amino acids, which are grouped based on the characteristics of their side chains. Five amino acids have charged side chains, which attract water; five have polar, hydrophilic side chains, which form hydrogen bonds; seven have hydrophobic, nonpolar side chains; and three have special hydrophobic functions.

The side chain of cysteine, —SH, can form disulfide bridges with other cysteines and influences protein chain folding; the two —SH groups lose their hydrogens and convert from a reduced state to an oxidized state. Glycine is small and fits into folds of proteins. Proline has limited bonding ability and often contributes to looping in proteins due to its ring shape. Proteins range in size from small oligopeptides (also called peptides) to large polypeptides (also called proteins). Amino acids are linked together during condensation. The resulting bonds between amino and carboxyl groups are called peptide bonds. Peptide chains always grow directionally from the N terminus to the C terminus. The subsequent sequence of amino acids is known as the primary structure of the protein.

The primary structure of a protein is held together by covalent bonds; however, all higher levels of structure are determined largely by the specific amino acid sequence in the primary structure of the protein. Secondary structure refers to regular, repeated spatial patterns of the amino acid chain stabilized by hydrogen bonds between the N—H and C=O groups on the protein backbone. The α helix consists of a right-handed coiling of a single polypeptide chain. The β pleated sheet is formed from two or more hydrogen-bonded chains. Tertiary structure refers to an amino acid chain that is bent and folded into a more complex pattern. This structure is stabilized by covalent disulfide bridges, hydrophobic interactions, van der Waals forces, ionic interactions, and hydrogen bonds between the R groups. The sequence of R groups in the primary structure is responsible for this level of folding. Because a protein's three-dimensional structure is stabilized by relatively "weak" bonds, environmental conditions such as temperature, pH ($H^+$ concentration), and the presence of high concentrations of polar or nonpolar substances can change the shape of the protein into an inactive form. This is called denaturation, and it can be reversible or irreversible depending on the protein. Quaternary structure refers to the three-dimensional interactions of multiple protein subunits. Hydrophobic interactions, hydrogen bonds, and ionic interactions are all involved in maintaining the quaternary structure. Chemical alterations to the side chains of amino acids can change the three-dimensional shape of the protein: for instance, by changing the hydrophobicity of a region of the protein.

**Question 4.** Differentiate between primary, secondary, tertiary, and quaternary structures as they relate to proteins. If a protein is immersed in an unfavorable pH solution, which structures are most likely to disassociate first, and why?

**Question 5.** How do the different chemical properties of amino acid R groups contribute to the final three-dimensional shape of the molecule?

**Question 6.** Suppose that you have isolated a protein with the following amino acid sequence: RSCFLA. Referring to Table 3.2 of the textbook, draw this protein. In your drawing, label the N terminus and the C terminus and show all peptide linkages. How many water molecules are generated in the synthesis of this protein?

## 3.3 Some Proteins Act as Enzymes to Speed up Biochemical Reactions

An energy barrier must be overcome to speed up a reaction

Enzymes bind specific reactants at their active sites

Energy barriers between reactants and products slow down reactions. Activation energy ($E_a$) must be added to get past this barrier; it may be thought of as a little "shove" to get the reaction going. The activation energy is used to destabilize the reactants slightly and change them into a transition state that has higher free energy (available energy) than either the reactants or the products. This is accomplished with the help of a catalyst, which serves to lower the amount of activation energy needed to initiate a chemical reaction. Most biological catalysts are proteins called enzymes. Enzymes have specific binding areas on their surfaces, called active sites, where reactants (substrates) can bind. The enzyme's specificity comes from the shape of its active site. The names for enzymes typically end in the suffix *-ase*. Only specific substrates can fit in and bind with an enzyme's active site. Once bound, the site is referred to as an enzyme–substrate complex (ES). The enzyme alters the conformation of the substrate, lowering the activation energy for the reaction.

The enzyme–substrate complex gives rise to the product and the enzyme remains unchanged. The reaction can be represented as follows:

$$E + S \rightarrow ES \rightarrow E + P$$

Enzymes speed up reactions by orienting substrates for maximum chemical interactions, by inducing strain on the substrate, or by adding chemical groups to substrates. Binding of small molecules to the active sites depends on hydrogen bonds, charge interactions, and hydrophobic interactions. Binding of substrate by an enzyme can change the enzyme's shape to produce an "induced fit." This is made possible by the large size of an enzyme, which helps it position the correct amino acids at the active site and regulate the site's shape. Cofactors, coenzymes, and prosthetic groups all contribute to an enzyme's activity. Cofactors are inorganic ions such as zinc, copper, or iron. Coenzymes are organic molecules that bind in the active site and can be thought of as a co-substrate. Organic prosthetic groups are permanently bound to the enzyme. The rate of a reaction increases as the substrate concentration increases, until all enzyme active sites are occupied. At that point, no amount of additional substrate will increase the reaction rate because all of the active sites are saturated and no more enzyme is available for catalysis. Enzyme efficiency is measured in turnover number, or how fast an enzyme can convert substrate to product and free up its active site.

**Question 7.** Discuss how a protein's three-dimensional structure makes it perfect for acting as a carrier and receptor molecule. In what ways are proteins uniquely suited for this function compared to other macromolecules?

**Question 8.** Explain how substrate concentration affects the rate of an enzyme-mediated reaction.

**Question 9.** Label the graph below with the following: Activation energy ($E_a$) for catalyzed reaction, activation energy ($E_a$) for uncatalyzed reaction, $\Delta G$ for catalyzed reaction, $\Delta G$ for uncatalyzed reaction, free energy of reactants, free energy of products, least stable state on graph, most stable state on graph.

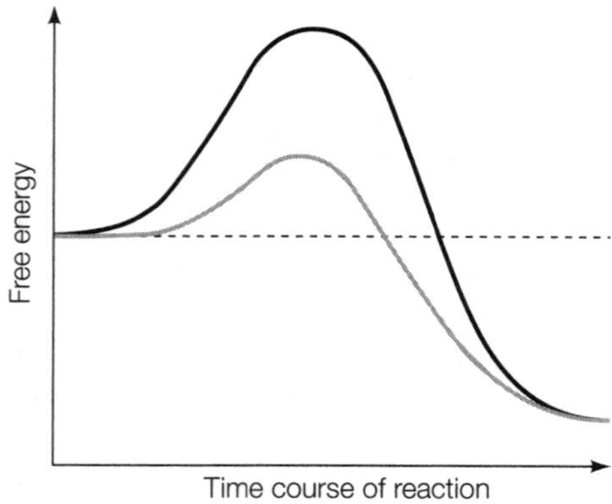


### 3.4 Regulation of Metabolism Occurs by Regulation of Enzymes

- Enzymes can be regulated by inhibitors
- An allosteric enzyme is regulated by changes in its shape
- Some metabolic pathways can be controlled by feedback inhibition
- Enzymes are affected by their environment

All life must maintain stable internal conditions. The chemical reactions occurring in an organism are organized into specific metabolic pathways that are catalyzed by specific enzymes at each step. Study of these pathways is the goal of systems biology. Enzyme function within a cell is regulated by both the amount and the activity of the enzyme. Inhibitors are substances that bind with enzymes to inhibit or slow their function. Irreversible inhibition occurs when an inhibitor forms a covalent bond with an enzyme and permanently destroys the active site. In reversible inhibition, an inhibitor binds noncovalently to the active site and alters it, but the inhibition is reversible, depending on the concentration of the inhibitor and substrate. Some reversible inhibitors are called competitive inhibitors because they compete with the substrate for the active site. Others are called noncompetitive inhibitors because they bind elsewhere in the enzyme but alter the active site so that the substrate cannot bind, or its rate of binding is reduced.

Allosteric regulation occurs when a molecule other than the substrate binds to the enzyme at a site other than the active site, resulting in a change in the enzyme shape. Noncompetitive inhibitors are an example of inhibitors that work allosterically. Most allosteric enzymes exist naturally in an active form and an inactive form, and can switch back and forth between the two. When an allosteric regulator binds to the enzyme, the enzyme becomes locked in either its active form or inactive form. Enzymes' allosteric sites are modified either covalently or noncovalently, causing a change in enzyme shape. Phosphorylation is an important regulator of enzyme activity. Protein kinases covalently attach phosphate groups to amino acids in the enzyme; phosphatases can remove the phosphate group. Both kinases and phosphatases are important enzymes that regulate other metabolic enzymes.

The first step in an enzyme-mediated pathway is referred to as the commitment step. Once initiated, the pathway is followed to completion. Frequently, the final product allosterically inhibits the enzyme of the commitment step, preventing overproduction of the final product. This process is called end-product inhibition or feedback inhibition.

Enzyme activity is dependent upon the surrounding environment. Changes in pH influence charges of amino and other groups on the enzyme or substrate. This may change the folding of the enzyme or its ability to interact with the substrate and drastically alter its catalytic ability. Enzymes typically have an ideal pH at which they function best. Increases in temperature may aid in reduction of activation energy by adding kinetic energy; however, this may also result in the denaturing of enzymes. Each enzyme has an optimal temperature. Organisms may produce different forms of the same enzyme, called isozymes, which have different optimal temperatures.

**Question 10.** You discover a fish that possesses two versions of an enzyme. One version functions best at a high temperature, around 37°C, and the other at a lower temperature, around 30°C. Draw a graph showing how temperature affects enzyme activity of these two enzymes. Why would a fish have versions of an enzyme that function optimally at two different temperatures?

**Question 11.** Amylase is a digestive enzyme that breaks down starch and is secreted in the human mouth. Although it functions well in the mouth, it ceases to function once it contacts the acidic environment of the stomach. Explain why amylase does not function in the stomach.

**Question 12.** Figure 3.20a shows the behavior of an allosteric enzyme that has binding sites for a positive regulator. Describe the behavior of an enzyme with binding sites for a negative regulator, as opposed to a positive regulator.

**Question 13.** Enzyme X has an optimal pH of 9.0 and is completely inactive at pH 7.0 and pH 11.0. Draw the predicted enzyme activity and determine if this enzyme prefers an acidic, neutral, or basic environment.

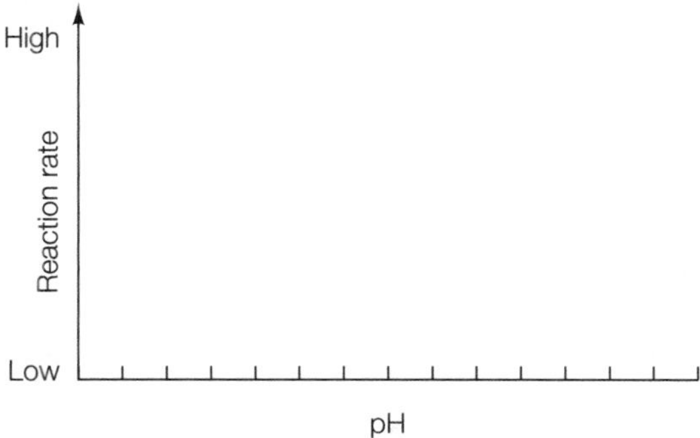


## Test Yourself

1. You are investigating a newly discovered species. This organism lives in acidic pools in volcanic craters where temperatures often reach 100°C and normally stay above 90°C. You determine that it has a surface enzyme that catalyzes a reaction leading to its protective coating, and you decide to study this enzyme in the laboratory. At which temperature would you most likely find optimal activity of this enzyme?
   a. 0°C
   b. 37°C
   c. 55°C
   d. 95°C
   e. 105°C
   ***Textbook Reference:*** *Concept 3.4 Regulation of Metabolism Occurs by Regulation of Enzymes; Enzymes are affected by their environment*

2. Which equation represents an enzyme-catalyzed reaction? (E = enzyme, P = product, S = substrate)
   a. $E + P \rightarrow E + S$
   b. $E + S \rightarrow E + P$
   c. $E + S \rightarrow P$
   d. $E + S \rightarrow E$
   e. $P + S \rightarrow E$
   ***Textbook Reference:*** *Concept 3.3 Some Proteins Act as Enzymes to Speed up Biochemical Reactions; Enzymes bind specific reactants at their active sites*

3. A disulfide bridge is formed by
   a. two cysteine side chains.
   b. two glycerol linkages.
   c. two proline side chains.
   d. condensation.
   e. hydrolysis.
   ***Textbook Reference:*** *Concept 3.2 Proteins Are Polymers with Important Structural and Metabolic Roles; Amino acids are the building blocks of proteins*

4. In DNA and RNA, nucleotides in a chain are joined by
   a. phosphodiester linkages.
   b. hydrogen bonds.
   c. peptide linkages.
   d. glycosidic linkages.
   e. ionic attractions.
   ***Textbook Reference:*** *Concept 3.1 Nucleic Acids Are Informational Macromolecules; Nucleotides are the building blocks of nucleic acids*

5.–6. Ascorbic acid, which is found in citrus fruits, acts as an inhibitor to catecholase, the enzyme responsible for the browning reaction in fruits such as apples, peaches, and pears.

5. One explanation for the inhibiting function of ascorbic acid could be its similarity, in terms of size and shape, to catechol, the substrate of the browning reaction. If this explanation is correct, then this inhibition is most likely an example of _______ inhibition.
   a. competitive
   b. indirect
   c. noncompetitive
   d. allosteric
   e. feedback
   ***Textbook Reference:*** *Concept 3.4 Regulation of Metabolism Occurs by Regulation of Enzymes; Enzymes can be regulated by inhibitors*

6. Suppose further studies indicate that ascorbic acid is not similar to catechol in size and shape but that the pH of the ascorbic acid solution alters the protein folding of catecholase. If this is true, then this inhibition is most likely an example of
   a. competitive inhibition.
   b. enzyme denaturation.
   c. noncompetitive inhibition.
   d. allosteric regulation.
   e. feedback inhibition.
   ***Textbook Reference:*** *Concept 3.4 Regulation of Metabolism Occurs by Regulation of Enzymes; Enzymes are affected by their environment*

7. The end product of transcription is (are) _______ and the end product of translation is (are) _______.
   a. proteins; DNA
   b. proteins; RNA
   c. RNA; DNA
   d. DNA; RNA
   e. RNA; proteins
   ***Textbook Reference:*** *Concept 3.1 Nucleic Acids Are Informational Macromolecules; DNA carries information and is expressed through RNA*

8. You are given an unlabeled enzyme in a container and told to add a compound that will bind irreversibly to the enzyme and increase its function. You ask for information about the enzyme, but your instructor simply hands you a list of possible compounds. Based on what you have learned about the enzyme partners below, which one is the best choice?
   a. Coenzyme A
   b. Zinc ($Zn^{2+}$)
   c. Flavin
   d. ATP
   e. NAD
   ***Textbook Reference:*** *Concept 3.4 Regulation of Metabolism Occurs by Regulation of Enzymes; Enzymes bind specific reactants at their active sites*

9. Which of the following statements about proteins is *false?*
   a. Enzymes are proteins.
   b. Proteins are part of the phospholipid bilayer.
   c. Some hormones are proteins.
   d. Proteins are structural components of the cell.
   e. Proteins can move substances across a cell membrane.
   ***Textbook Reference:*** *Concept 3.2 Proteins Are Polymers with Important Structural and Metabolic Roles*

10. Metabolism is organized into pathways that are linked in which of the following ways?
    a. All cellular functions feed into a central pathway.
    b. All steps in the pathway are catalyzed by the same enzyme.
    c. The product of one step in the pathway functions as the substrate in the next step.
    d. Products of the pathway accumulate and are secreted from the cell.
    e. Different substrates are acted on by the same enzyme.
    ***Textbook Reference:*** *Concept 3.4 Regulation of Metabolism Occurs by Regulation of Enzymes*

11. Which of the following is a factor that contributes to the specificity of enzymes?
    a. Each enzyme is active over a wide range of temperatures.
    b. Each enzyme is active over a wide range of pH conditions.
    c. Each enzyme has an active site that interacts with many substrates.
    d. Substrates themselves may alter the active site slightly for optimum catalysis.
    e. Enzymes are more active at higher temperatures than at lower temperatures.
    ***Textbook Reference:*** *Concept 3.3 Some Proteins Act as Enzymes to Speed up Biochemical Reactions; Enzymes bind specific reactants at their active sites*

12. Which statement about enzymes is true?
    a. They are consumed by the enzyme-mediated reaction.
    b. They are not altered by the enzyme-mediated reaction.
    c. They raise activation energy.
    d. They can be composed of RNA or proteins.
    e. They are rarely regulated.
    ***Textbook Reference:*** *Concept 3.3 Some Proteins Act as Enzymes to Speed up Biochemical Reactions; Enzymes bind specific reactants at their active sites*

13. DNA utilizes the bases guanine, cytosine, thymine, and adenine. In RNA, _______ is replaced by _______.
    a. adenine; arginine
    b. thymine; uracil
    c. cytosine; uracil
    d. cytosine; arginine
    e. cytosine; thymine
    ***Textbook Reference:*** *Concept 3.1 Nucleic Acids Are Informational Macromolecules; Base pairing occurs in both DNA and RNA*

14. Proteins consist of amino acids linked by
    a. noncovalent bonds.
    b. peptide bonds.
    c. phosphodiester bonds.
    d. van der Waals forces.
    e. ionic attraction.
    ***Textbook Reference:*** *Concept 3.2 Proteins Are Polymers with Important Structural and Metabolic Roles; Amino acids are the building blocks of proteins*

15. You fill two containers with identical amounts of reactants A and B and enzymes 1–4. In the reactions shown below, if product D inhibits enzyme 2 and product F is an allosteric stimulator of enzyme 1, what will be the final result if you add extra product D to the second container? (Assume that both containers are given enough time for the reactions to go to completion.)

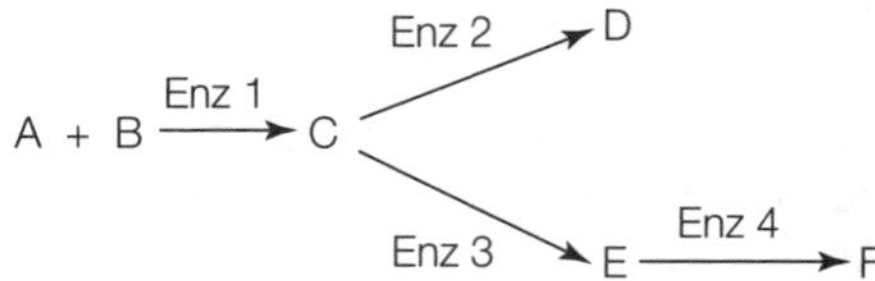


    a. The concentration of product C will increase in the second container and there will be no change in the concentration of product F.
    b. The concentration of reactants A and B will increase in the second container.
    c. The concentration of product F will increase in the second container because more of D is converted back to C.
    d. The concentration of products E and F will both increase in the second container, since D inhibits enzyme 2.

e. The concentration of product F will increase in the second container, since enzyme 2 will have been inhibited from converting C into D.

***Textbook Reference:*** *Concept 3.4 Regulation of Metabolism Occurs by Regulation of Enzymes; Some metabolic pathways can be controlled by feedback inhibition*

16. The double-helix formation of DNA is caused by
    a. ionic bonds.
    b. covalent bonds.
    c. hydrogen bonds.
    d. hydrophobic side chains.
    e. glycosidic linkages.

***Textbook Reference:*** *Concept 3.1 Nucleic Acids Are Informational Macromolecules; Base pairing occurs in both DNA and RNA*

17. Enzymes are biological catalysts that function by
    a. increasing free energy in a system.
    b. lowering activation energy of a reaction.
    c. lowering entropy in a system.
    d. increasing temperature near a reaction.
    e. altering the equilibrium of the reaction.

***Textbook Reference:*** *Concept 3.3 Some Proteins Act as Enzymes to Speed up Biochemical Reactions; An energy barrier must be overcome to speed up a reaction*

18. The R groups of amino acids located on the surface of protein molecules in the interior of biological membranes are
    a. hydrophobic.
    b. hydrophilic.
    c. polar.
    d. able to form disulfide bridges.
    e. electrically charged.

***Textbook Reference:*** *Concept 3.2 Proteins Are Polymers with Important Structural and Metabolic Roles; Amino acids are the building blocks of proteins*

19. In the pathway A + B → C + D, enzyme X facilitates the reaction. If compound D inhibits enzyme X, you could conclude that
    a. enzyme X is an allosteric inhibitor of the reaction.
    b. compound D is an allosteric stimulator of the reaction.
    c. compound D is a competitive inhibitor of the reaction.
    d. enzyme X is subject to feedback stimulation.
    e. compound D is a coenzyme in the reaction.

***Textbook Reference:*** *Concept 3.4 Regulation of Metabolism Occurs by Regulation of Enzymes; Some metabolic pathways can be controlled by feedback inhibition*

20. The pairing of purines with pyrimidines to create a double-stranded DNA molecule is called
    a. complementary base pairing.
    b. phosphodiester linkage.
    c. antiparallel synthesis.
    d. dehydration.
    e. the genome.

***Textbook Reference:*** *Concept 3.1 Nucleic Acids Are Informational Macromolecules; Base pairing occurs in both DNA and RNA*

21. Enzymes alter the _______ of the reaction.
    a. $\Delta G$ value
    b. activation energy
    c. equilibrium
    d. rate
    e. substrates

***Textbook Reference:*** *Concept 3.4 Regulation of Metabolism Occurs by Regulation of Enzymes; An energy barrier must be overcome to speed up a reaction*

## Answers

### *Key Concept Review*

1. RNA sequence for top: 3′-UUCGCAG-5′; RNA sequence for bottom: 5′-AAGCGUC-3′.
2. nucleotide 5′ 3′ nucleoside 3′ 5′
3. DNA structure is highly regular and conserved, which makes it usable by all cells, while the different bases allow it to encode specific genetic information.

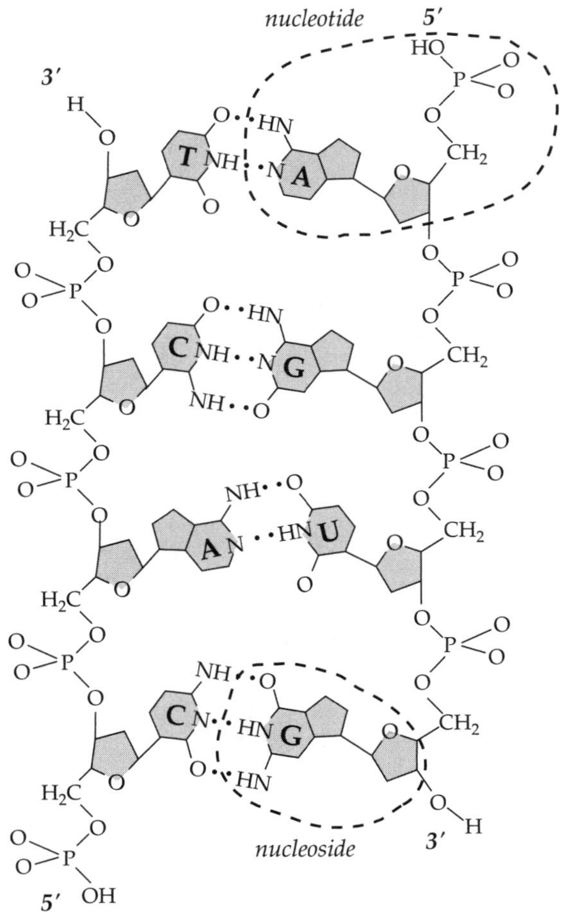

4. See Figure 3.7 in the textbook for diagrams of primary, secondary, tertiary, and quaternary structures. The quaternary structure is the least stable and breaks down first in unfavorable conditions. Protein structures continue to denature by unfolding the tertiary structures, then the secondary structures. Primary structure is maintained by covalent bonds and is the last to break down. Disruptions in tertiary and quaternary structures are often reversible. Disruptions in primary or secondary structures are generally irreversible.

5. The size of the R group, the charge of the R group, and any special binding properties all contribute to the final orientation of a protein molecule.

6. Five water molecules are generated.

```
N terminus                                                          C terminus
 H    H O H  H O H  H O H  H O H  H O H  H O H                        O
  \   | ‖ |  | ‖ |  | ‖ |  | ‖ |  | ‖ |  | ‖ |                       //
H—N—C—C—N—C—C—N—C—C—N—C—C—N—C—C—N—C—C
  /   |         |        |        |           |            |          \
 H   CH2      H2O-OH    CH2      CH2         CH2          CH3          O-
      |                  |        |           |
     CH2                SH    (benzene)       CH
      |                                      /  \
     CH2                                  H3C    CH3
      |
     N-H
      |
     C=NH2+
      |
     NH2
```

7. The three-dimensional nature of proteins allows them to form binding sites. These binding sites are uniquely shaped to interact with other molecules. In the case of an enzyme, the binding site for reactants is called the active site

8. Increasing substrate concentration will result in an increased rate of reaction until all available active sites are occupied. At that point, no amount of substrate increase will increase the rate of reaction (see Figure 3.15).

9.

Least stable state
$E_a$
Uncatalyzed reaction
$E_a$
Free energy
Reactants
Catalyzed reaction
$\Delta G$
Products
Most stable state
Time course of reaction

10. It is possible that this fish lives in an environment that changes temperature over some time period, (e.g., daily or seasonally). The two different versions of the enzyme allow it to function well over a range of body temperatures.

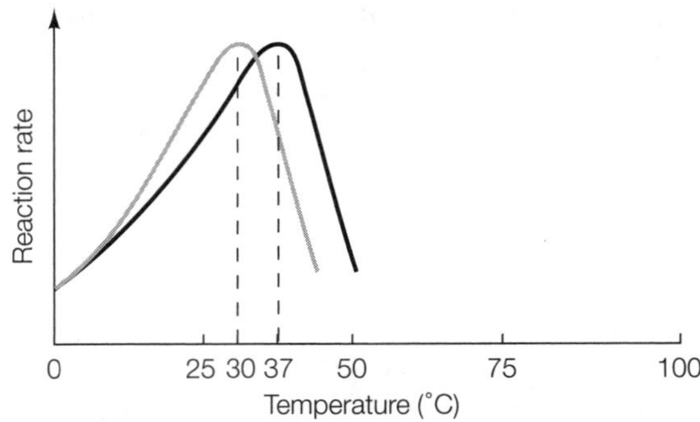


11. The pH optimum of amylase is approximately 7. At this pH, the protein has a three-dimensional shape that allows starch to bind to its active site and catalyze its hydrolysis. When it is in the stomach (pH approximately 2), the protein is denatured, and its three-dimensional shape and active site are lost; therefore, it can no longer catalyze the reaction.

12. A negative regulator would stabilize the enzyme in its inactive form. In the absence of the regulator, the enzyme alternates between its inactive and active forms. When it encounters substrate, it is able to bind only if it happens to be in its active form. When bound to the negative regulator, the enzyme is fixed in its inactive form, and is not able to bind substrate.

13.

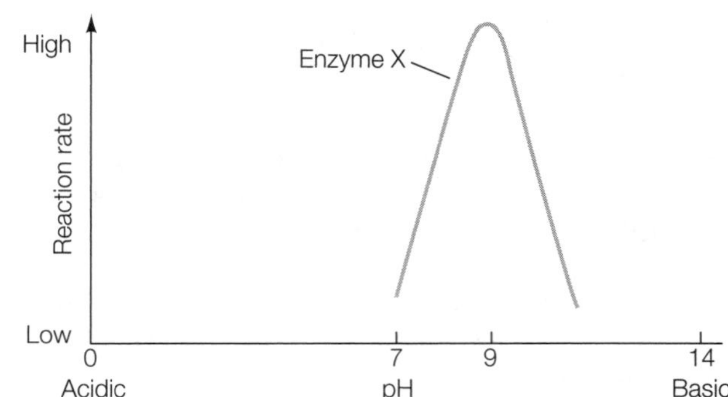


The enzyme prefers a basic environment. Your graph should be symmetrical around a peak at pH 9.0.

## *Test Yourself*

1. **d.** Enzymes typically work at a maximal rate at a particular temperature or within a range of temperatures. This optimal temperature tends to be correlated with the body temperature of the organism.

2. **b.** Substrate is converted to product and the enzyme is unchanged.

3. **a.** Disulfide bridges are formed when the—SH groups of two cysteines interact.

4. **a.** The sugar of one nucleotide and the phosphate of the next are joined by phosphodiester linkages in both DNA and RNA.

5. **a.** Competitive inhibitors compete for the active site with the substrate.

6. **b.** Destroying the three-dimensional structure of an enzyme, or denaturing it, is usually irreversible.

7. **e.** Transcription involves making RNA from DNA, and translation involves making proteins from RNA.

8. **c.** Of the five compounds listed, only one is a prosthetic group, which is irreversibly bound to its target enzyme.

9. **b.** Proteins do not make up the phospholipid bilayer, which is composed of phospholipids. Proteins do perform all of the other functions.

10. **c.** Within a given pathway, the products of the preceding step act as substrates for subsequent steps.

11. **d.** Enzymes are specific to particular substrates that may actually "adjust" the fit of the active site. They also function in specific narrow optimum ranges of pH and temperature.

12. **b.** Enzymes are not consumed or altered in any way during an enzyme-mediated reaction, and they function to lower the activation energy of a reaction. Ribozymes (composed of RNA) are catalysts, but they are not true enzymes.

13. **b.** RNA and DNA differ by one oxygen molecule in their ribose sugar and in the substitution of uracil (RNA) for thymine (DNA).

14. **b.** The peptide bond that links amino acids to form proteins is a type of covalent bond.

15. **e.** The addition of more D to the second container will reduce the activity of enzyme 2. The pathway from A + B going all the way to F will be the predominant reaction that takes place, leading to a greater final concentration of F in the second container. The first container will convert some of C into D, and thus the level of intermediate E and final product F will be lower than in the second container. (Hint: Draw a diagram of the experiment.)

16. **c.** The double-helix structure of DNA is due to the hydrogen bonding.

17. **b.** Enzymes reduce activation energy and speed up reactions.

18. **a.** The interior of the membrane is hydrophobic; therefore, an embedded protein would have hydrophobic residues on its exposed surfaces.

19. **c.** Products of a reaction that inhibit the enzymes via feedback inhibition can be competitive or allosteric inhibitors (not stimulators).

20. **a.** Complementary base pairing results from the attraction of charges and the ability of purines to pair with pyrimidines through hydrogen bond formation.

21. **b.** The free energy levels of the reactants and products are not changed by enzymes. Only the activation energy is altered. See Figure 3.13.

# Cells: The Working Units of Life

## The Big Picture

- Cell theory states that all living things are composed of cells and all cells come from preexisting cells. Living things have one of two basic cell forms; the small prokaryotic cell has no organelles, and the eukaryotic cell has many membrane-enclosed organelles.
- The organelles of eukaryotic cells have a variety of functions, including information storage and transmission, modification and packaging of cellular products, and energy transformation. Figure 4.7 shows the organization, structure, and function of each component.
- The cytoskeleton of the cell provides support, structure, and protection. Components of the cytoskeleton also aid in movement of the entire cell or cell components. Cilia and flagella provide cell mobility.
- Extracellular matrix surrounds eukaryotic cells and has many roles, including support and protection. Cell junctions serve to attach cells and allow them to share and move cell components among their cytoplasms.

## Study Strategies

- It is easy to become overwhelmed by the number of organelles and their diverse functions in a eukaryotic cell. Remember that each organelle has a distinct location within the cell and that an organelle's structure is related to its function.
- Terminology is the most difficult part of this material. The terminology is easier to learn if you understand the concepts before you memorize the terms.
- Refer to Figure 4.7 while you are studying the organelles and their components in order to visualize how their functions relate to their structure.
- Go to **LaunchPad** (or use the Web addresses listed) to review the following additional resources:

  Animated Tutorial 4.1 Eukaryotic Cell Tour (PoL2e.com/at4.1)

  Animated Tutorial 4.2 The Golgi Apparatus (PoL2e.com/at4.2)

  Activity 4.1 The Scale of Life (PoL2e.com/ac4.1)

  Activity 4.2 Lysosomal Digestion (PoL2e.com/ac4.2)

  Activity 4.3 Animal Cell Junctions (PoL2e.com/ac4.3)

  Analyze the Data in textbook Figure 4.14

  Apply the Concept in textbook Section 4.3

## Key Concept Review

### 4.1 Cells Provide Compartments for Biochemical Reactions

Cell size can be limited by the surface area-to-volume ratio

Cells can be studied structurally and chemically

The cell membrane forms the outer surface of every cell

Cells are classified as either prokaryotic or eukaryotic

The cell theory states that the cell is the basic unit of life, that all organisms are composed of cells, and that all cells come from preexisting cells. This means that when we study cell biology we are studying life. An important concept to remember is that life is continuous; each organism and cell could, in theory, be traced back through its ancestors to the first living cell. Cell size is limited by surface area-to-volume ratio. Very large cells are not feasible because as volume increases, surface area increases at a slower rate. A cell's capacity for chemical activity is related to the cell volume. However, cells with large volumes are unable to take up enough material across their surfaces to support the metabolic activity of the volume. Because cells are small, either light microscopes or electron microscopes must be used to visualize them. Light microscopes, which use lenses and light, can resolve objects as small as 0.2 μm. Electron microscopes, which use electron beams, can resolve objects as small as 0.1 nm. Staining techniques assist in visualizing cellular components. Cells can also be studied through chemical analysis that involves breaking open the cells to make a cell-free extract and examining the contents (e.g., enzyme activity).

All cells are surrounded by membranes composed of a phospholipid bilayer with many embedded and protruding proteins. The cell membrane is selectively permeable, permitting some small molecules to pass through but preventing others from doing so. The membrane allows for maintenance of the internal cellular environment and is responsible for communication and interactions with other cells. Prokaryotes, which include members of the domains Archaea and Bacteria, lack a nucleus and other membrane-bound internal compartments. Eukaryotes, which are member of the domain Eukarya, have organelles and a nucleus containing the cell's DNA. The organelles compartmentalize the different functions occurring in the cell.

**Question 1.** What is the primary function of a cell membrane? What characteristics of membranes allow them to contribute to metabolic activity?

**Question 2.** What is the primary benefit of studying a cell-free extract? Name one drawback to using this method.

## 4.2 Prokaryotic Cells Do Not Have a Nucleus

Prokaryotic cells share certain features

Specialized features are found in some prokaryotes

In terms of numbers and diversity, prokaryotes are the most successful cell type on Earth. Prokaryotes are one-celled organisms found only in domains Archaea and Bacteria. They do not have internal membrane compartments and are generally smaller than eukaryotic cells. The single cells have a cell membrane and a nucleoid containing the genetic material, which is not membrane-enclosed. The cytoplasm of the cell consists of the liquid cytosol and insoluble particles that act as subcellular machinery (i.e., ribosomes for protein synthesis). The cytoplasm is in constant motion to ensure that reactants come together to bring about biochemical reactions.

Most prokaryotes have a rigid cell wall located outside the cell membrane that supports the cell and determines its structure. The cell walls of bacteria contain peptidoglycan, which is relatively permeable; cell wall type is frequently used as a characteristic to identify bacteria. A relatively permeable outer membrane surrounds the cell wall in some bacteria. Some prokaryotes produce an outer slimy layer, or capsule, rich in polysaccharides, that encloses the cell wall. Though it provides protection, it is not necessary for the cell's survival. Some prokaryotes have an internal membrane that provides some compartmentalization. Photosynthetic prokaryotes have an internal membrane system that contains the molecules necessary for photosynthesis. Other prokaryotes have membrane folds that function in energy reactions and cell division. Some prokaryotes have structures to help with movement or support. Flagella are tiny protein machines containing the protein flagellin that facilitate motion of the cell. The cytoskeleton is a system of protein filaments that provide structure in the cell; one such prokaryotic protein is similar to eukaryotic actin.

**Question 3.** In your biology course, your professor gives you a sample of cells and tells you that they can only be prokaryotic cells. What characteristics identify them definitively as prokaryotic cells?

**Question 4.** Some prokaryotes have internal membranes that allow some limited compartmentalization. Compare and contrast these regions with the defined organelles found in eukaryotic cells.

## 4.3 Eukaryotic Cells Have a Nucleus and Other Membrane-Bound Compartments

Compartmentalization is the key to eukaryotic cell function

Ribosomes are factories for protein synthesis

The nucleus contains most of the cell's DNA

The endomembrane system is a group of interrelated organelles

Some organelles transform energy

Several other membrane-enclosed organelles perform specialized functions

Eukaryotic cells are larger than prokaryotic cells and are distinguished by their membrane-enclosed internal organelles. Each organelle has its own internal environment uniquely suited to its function. Animal and plant cells share common organelles, while other organelles are found only in plant cells. Ribosomes, which consist of RNA and protein molecules and are found in the cytoplasm of both eukaryotic and prokaryotic cells, act as information transcription centers and guide the synthesis of proteins from the messenger RNA nucleic acid blueprints. The nucleus stores DNA and is also the site of DNA duplication and regulation of DNA transcription into RNA. The nucleolus region of the nucleus is involved in ribosome assembly and RNA synthesis. The nucleus is bounded by a double lipid bilayer called the nuclear envelope. Small openings in the nuclear envelope, called nuclear pores, allow passage of RNA and ribosomes to the cell cytoplasm. Inside the nucleus, DNA and proteins combine in chromosomes (also called chromatin). At cell division, chromatin condenses to make the chromosomes more compact. The outer membrane of the nuclear envelope is continuous with the endoplasmic reticulum (ER).

The endomembrane system consists of the nuclear envelope, the ER, the Golgi apparatus, and lysosomes. Membrane-bound vesicles move various substances within the endomembrane system. The endoplasmic reticulum is a complex of membrane sacs throughout the cell that is continuous with the nuclear membrane. It is classified into two types based on the presence or absence of attached ribosomes. Rough endoplasmic reticulum (RER) is studded with active ribosomes and is involved in the synthesis, storage, transport, and modification of new proteins. Many of the cell's membrane-bound proteins are produced in the RER; a ribosome is directed to the RER by a short sequence of amino acids in the polypeptide that is forming. Proteins are modified, such as by addition of carbohydrate groups to make glycoproteins, in the ER. These carbohydrate groups help, for example, in identifying proteins and ensuring that they reach the correct destinations within the cell. Smooth endoplasmic reticulum (SER) has a role in protein modification and transport. Its more important functions are in the modification of chemicals taken in by the cell, such as drugs and pesticides, and as the site of hydrolysis of glycogen and synthesis of steroids and lipids. The SER can also store calcium ions, which are released to cause certain effects in the cell (such as muscle contraction). The Golgi apparatus is an organelle composed of flattened membrane stacks called cisternae and membrane-enclosed vesicles. It contributes to further modification, packaging, and concentration of proteins. In plants, it is also the site of some polysaccharide synthesis. The three different regions of the Golgi have different enzymes and functions. The *cis* region is closest to the nucleus or RER, the *trans* region is closest to the cell surface, and the medial region lies in between. Proteins are released from the ER in a vesicle and transported to the *cis* region, where the vesicle membrane fuses with the Golgi membrane and the contents are released. Vesicles containing proteins pinch off of the *trans* Golgi for transport to the plasma membrane or lysosomes. Lysosomes are "digestion centers" within a cell that break down proteins, polysaccharides, nucleic acids, and lipids into their monomer components. Primary lysosomes released from the Golgi contain a host of powerful enzymes in a slightly acidic environment; these break down engulfed molecules and cellular waste. In the process of phagocytosis, the cell takes up material in a phagosome, which fuses with a primary lysosome to make a secondary lysosome. Additionally, through the process of autophagy, lysosomes digest organelles such as mitochondria, breaking them down to monomers for reuse in new organelles. Phagocytes are specialized cells that take in and digest materials. Lysosomal storage diseases occur when lysosomes fail to digest internal components. Plant cells do not contain lysosomes.

Two organelles, mitochondria and chloroplasts, are involved in harvesting energy. During cellular respiration, mitochondria convert potential chemical energy stored in glucose and other molecules into adenosine triphosphate (ATP), a form of energy readily usable by the cell. Cells can contain anywhere from one to a few hundred thousand mitochondria. Mitochondria have two membranes—an outer smooth membrane with pores and a highly folded internal membrane. The folds are called cristae, and the remaining internal space is called the matrix. The matrix holds ribosomes and DNA. Protein complexes used during cellular respiration are embedded in the cristae. Plastids are found in plants and protists, but not in animal cells. Plant cells have several different types of plastids, each with unique functions. Chloroplasts containing the photosynthetic pigment chlorophyll are the site of photosynthesis, where light energy is converted to chemical energy. Chloroplasts have two membranes. The innermost membrane contains compartments called thylakoids which are folded into stacks called grana and hold chlorophyll and enzymes for photosynthesis. The liquid content of the chloroplast outside the thylakoids is called the stroma and contains ribosomes and DNA. Other plastid types include chromoplasts, which contain pigments involved in flower color, and starch-storing leucoplasts. Other specialized organelles are found in some, but not all, cells. Peroxisomes function to break down harmful peroxide by-products. Glyoxysomes, which are found in plants, convert lipids to carbohydrates. Vacuoles, particularly found in plants and protists, have various functions depending on the organism, including storage of waste products, structural support, reproduction, food storage, and water regulation.

**Question 5.** Eukaryotic cells possess organelles where specific metabolic functions occur. What are the benefits of compartmentalization to a cell?

**Question 6.** If we can assume that form follows function, what would be the explanation for the structural similarities between mitochondria and chloroplasts?

**Question 7.** The role of a certain cell in an organism is to secrete a protein. Create a flow chart in which you trace the production of that protein from the nucleus through all necessary organelles to the point of release from the cell.

## 4.4 The Cytoskeleton Provides Strength and Movement

Microfilaments are made of actin
Intermediate filaments are diverse and stable
Microtubules are the thickest elements of the cytoskeleton
Cilia and flagella provide mobility
Biologists manipulate living systems to establish cause and effect

The cytoskeleton is involved in cell support, position and movement of organelles, cytoplasmic streaming, and anchoring of the cell. Microfilaments, composed of the protein actin, assist with contraction of the cell. These filaments are involved in cell division, cytoplasmic streaming, cell movement, and stabilizing cell shape. The cytoskeleton can lengthen or shorten due to the dynamic instability of actin. In muscle cells, the motor protein myosin interacts with actin to produce muscle contraction. Intermediate filaments are found only in multicellular organisms and function in stabilizing structure and resisting tension. They also help to anchor the nucleus and maintain rigidity with desmosomes. Hair and fingernails are composed of the intermediate filament keratin.

Microtubules are long hollow tubes of the dimer protein tubulin that contribute to the rigidity of the cell and act as a framework for the movement of motor proteins. They have a very specific structure that can be quickly added to or reduced, showing dynamic instability.

Cilia and flagella of eukaryotes are powered by microtubules. Though cilia and flagella differ in size and function, they have the same basic structure: a "9 + 2" arrangement of microtubules. Movement of cilia and flagella occurs when the microtubules slide past one another. The sliding is caused by an ATP-driven shape change in molecules of dynein bound to the microtubules. Dynein binds two microtubule doublets. When nexin cross-links the doublets, the cilium bends. Dynein and kinesin, both motor proteins, deliver vesicles or organelles to various locations in the cell by moving directionally along microtubule "tracks."

Biologists can test their hypotheses by manipulating living systems to establish cause and effect relationships. It is important to understand which simultaneous observations are indicative of a correlative rather than causative relationship. Two approaches to testing these hypotheses are inhibiting a function and mutating a gene that codes for a particular protein that performs that function.

**Question 8.** Explain how microtubules and dynein function to make cilia and flagella move.

**Question 9.** As you examine a cell under the microscope, you notice what appear to be small, membrane-bound units moving along a path within the cell. Based on your knowledge of filaments in a cell, describe what you are observing.

### 4.5 Extracellular Structures Provide Support and Protection for Cells and Tissues

The plant cell wall is an extracellular structure

The extracellular matrix supports tissue functions in animals

Cell junctions connect adjacent cells

Extracellular matrix has two components: a fibrous macromolecule and a gel-like medium. Plants have semirigid cell walls composed of fibrous cellulose and a gel-like polysaccharide and protein matrix. The cell wall functions to support and protect the cell. Plasmodesmata are small holes in the cell walls that allow connections between plant cells. The extracellular matrix of some animal cells, which is composed of collagen and proteoglycans, functions to hold cells together, filter materials, orient cell movement, and assist with chemical signaling. Some cells, such as bone cells, secrete an elaborate and rigid matrix. The extracellular matrix and cell membrane are connected by proteins such as integrin.

Cell junctions help to hold the cells of multicellular animals together. The three types are tight junctions, desmosomes, and gap junctions. Tight junctions, which are found in the epithelium of the bladder, do not allow for movement of substances and prevent urine from leaking out into the body. Desmosomes are involved in joining cells strongly while allowing for some movement through the extracellular matrix. Gap junctions, which are structurally similar to plasmodesmata of plants, connect cells and act as channels between the interiors of the adjoining cells.

**Question 10.** You are looking at a group of cells from an animal and notice that the cell junctions look like channels. What type of junctions are these, what do they do, and what type of tissue are you most likely examining?

**Question 11.** You are examining a cross-section of intestinal lining under the microscope. What type(s) of cell junction would you expect to find holding these cells together? Explain why.

## Test Yourself

1. A mass of cells is found in the sediment surrounding a thermal vent in the ocean floor. The salinity in the area is quite high. Microscopic examination of one of the cells reveals no evidence of membrane-enclosed organelles. This cell would be classified as a(n) _______ cell.
   a. eukaryotic
   b. prokaryotic
   c. plant
   d. animal
   e. fungal
   ***Textbook Reference:*** *Concept 4.1 Cells Provide Compartments for Biochemical Reactions; Cells are classified as either prokaryotic or eukaryotic*

2. Centrifugation of a cell results in the rupture of the cell membrane and the compacting of the contents into a pellet in the bottom of the centrifuge tube. Bathing this pellet with a glucose solution yields metabolic activity, including the production of ATP. Which of the following is most likely one of the contents of this pellet?
   a. Cytosol
   b. Mitochondria
   c. Lysosomes
   d. Golgi bodies
   e. Thylakoids
   ***Textbook Reference:*** *Concept 4.3 Eukaryotic Cells Have a Nucleus and Other Membrane-Bound Compartments; Some organelles transform energy*

3. The cell theory states or implies all of the following *except*
   a. all living things are composed of cells.
   b. cells are the fundamental units of life.
   c. all cells come from preexisting cells.
   d. all cells contain mitochondria.
   e. all living things belong to a continuous chain of life dating back to the first cell.
   ***Textbook Reference:*** *Concept 4.1 Cells Provide Compartments for Biochemical Reactions*

4. Though science fiction has produced stories like "The Blob," we do not see very many large single-celled organisms. Which of the following tends to limit cell size?
   a. The difficulty in maintaining a continuous large membrane
   b. The difficulty of reproduction in a large cell
   c. Surface area-to-volume ratios
   d. The ability to make enough proteins to fill a large cell volume
   e. The amount of energy required to maintain a large cell
   ***Textbook Reference:*** *Concept 4.1 Cells Provide Compartments for Biochemical Reactions; Cell size can be limited by the surface area-to-volume ratio*

5. Microscopes are used to resolve images that cannot be seen with the unaided eye. Electron microscopes use _______ to resolve images, whereas light microscopes use _______ to resolve images.
   a. light and lenses; diffraction of electron beams
   b. diffraction of electron beams; light and lenses
   c. lasers; light and lenses
   d. light and lenses; lasers
   e. diffraction of electron beams; lasers
   ***Textbook Reference:*** *Concept 4.1 Cells Provide Compartments for Biochemical Reactions; Cells can be studied structurally and chemically*

6. The cellular function of the RER is
   a. DNA synthesis.
   b. photosynthesis.
   c. cellular respiration.
   d. protein synthesis.
   e. mRNA degradation.
   ***Textbook Reference:*** *Concept 4.3 Eukaryotic Cells Have a Nucleus and Other Membrane-Bound Compartments; The endomembrane system is a group of interrelated organelles*

7. Photosynthesis occurs in the
   a. chloroplast.
   b. mitochondria.
   c. Golgi apparatus.
   d. nucleus.
   e. RER.
   ***Textbook Reference:*** *Concept 4.3 Eukaryotic Cells Have a Nucleus and Other Membrane-Bound Compartments; Some organelles transform energy*

8. Lysosomes are involved in
   a. DNA synthesis.
   b. the breakdown of phagocytized material.
   c. protein folding.
   d. pigment production.
   e. cell membrane production.
   ***Textbook Reference:*** *Concept 4.3 Eukaryotic Cells Have a Nucleus and Other Membrane-Bound Compartments; The endomembrane system is a group of interrelated organelles*

9. The packaging of proteins to be used outside the cell occurs in the
   a. nucleus.
   b. SER.
   c. Golgi apparatus.
   d. chromoplast.
   e. nuclear pore.
   ***Textbook Reference:*** *Concept 4.3 Eukaryotic Cells Have a Nucleus and Other Membrane-Bound Compartments; The endomembrane system is a group of interrelated organelles*

10. Which of the following organelles is *not* enclosed in two membranes?
    a. Nucleus
    b. Chloroplast
    c. Mitochondrion
    d. RER
    e. Leucoplast
    ***Textbook Reference:*** *Concept 4.3 Eukaryotic Cells Have a Nucleus and Other Membrane-Bound Compartments; Some organelles transform energy*

11. Movement of cells in both prokaryotes and eukaryotes is accomplished by which of the following structures?
    a. Cilia
    b. Pili
    c. Dynein
    d. Cell membranes
    e. Flagella
    ***Textbook Reference:*** *Concept 4.4 The Cytoskeleton Provides Strength and Movement; Cilia and flagella provide mobility*

12. Which statement about mitochondria and chloroplasts is true?
    a. Animal cells produce chloroplasts.
    b. Both mitochondria and chloroplasts may be found in the same cell.
    c. Mitochondria and chloroplasts are not found in the same cell.
    d. In certain conditions, chloroplasts can revert to mitochondria.
    e. Chloroplasts evolved from mitochondria.
    ***Textbook Reference:*** *Concept 4.3 Eukaryotic Cells Have a Nucleus and Other Membrane-Bound Compartments; Some organelles transform energy*

13. Which statement about ribosomes is true?
    a. Ribosomes guide protein synthesis.
    b. Ribosomes are found only in the nucleus or on the RER.
    c. There are no ribosomes in the mitochondria.
    d. Ribosomes are the site of photosynthesis.
    e. Ribosomes are responsible for moving secreted proteins to the cell membrane for export.
    ***Textbook Reference:*** *Concept 4.3 Eukaryotic Cells Have a Nucleus and Other Membrane-Bound Compartments; Ribosomes are factories for protein synthesis*

14. Nuclear DNA exists in a complex with proteins, together called _______; the proteins attached to the DNA help to condense the _______ during cellular division.
    a. chromosomes; chromatin
    b. chromatids; chromosomes
    c. chromophores; chromatin
    d. chromatin; chromosomes
    e. chromatids; chromatin
    ***Textbook Reference:*** *Concept 4.3 Eukaryotic Cells Have a Nucleus and Other Membrane-Bound Compartments; The nucleus contains most of the cell's DNA*

15. Rough endoplasmic reticulum and smooth endoplasmic reticulum differ
    a. only in the presence (RER) or absence (SER) of ribosomes.
    b. only in their function.
    c. only in their microscopic appearance.
    d. both in their function and in terms of the presence (RER) or absence (SER) of ribosomes.
    e. in unknown ways.
    ***Textbook Reference:*** *Concept 4.3 Eukaryotic Cells Have a Nucleus and Other Membrane-Bound Compartments; The endomembrane system is a group of interrelated organelles*

16. Which prokaryotic component can be responsible for causing bacterial disease in humans?
    a. Folded internal membrane
    b. Flagellum
    c. Nucleoid
    d. Cell wall
    e. Capsule
    ***Textbook Reference:*** *Concept 4.2 Prokaryotic Cells Do Not Have a Nucleus; Specialized features are found in some prokaryotes*

17. Which of the following is *not* a type of cell junction?
    a. Tight junction
    b. Desmosome
    c. Plasmodesmata
    d. Proteoglycan
    e. Gap junction
    ***Textbook Reference:*** *Concept 4.5 Extracellular Structures Provide Support and Protection for Cells and Tissues; Cell junctions connect adjacent cells*

## Answers

### *Key Concept Review*

1. The cell membrane allows for the enclosure of biochemical functions within a membrane and acts as a selectively permeable barrier. It also allows a cell to maintain homeostasis and is important in communication with adjacent cells and in receiving signals from the environment. In addition, the cell membrane plays an important structural role and contributes to cell shape

2. The major advantage of studying a cell-free extract is that all of the cell contents are exposed to any type of measuring or recording method, and thus the function of the cellular components can be studied. Some drawbacks are that small organelles may not be opened, and that it can be hard to isolate which molecules are responsible for the results of an experiment when the cell contents are pooled.

3. Prokaryotic cells have a number of features that are not shared with eukaryotic cells, such as a nucleoid instead of a nucleus. It is also likely that the cells will have a cell wall. Unlike eukaryotic cells, they will not contain any organelles, and will be small.

4. In prokaryotes, the defined regions formed by the internal folds do not exist separately within the cytoplasm, but lie at the periphery inside the cell membrane.

5. Organelles allow different metabolic environments to exist in the same cell. This partitioning of jobs allows for greater specialization.

6. Both mitochondria and chloroplasts are involved in energy-transformation activities that require many enzymes and other molecules that are found embedded in membranes. The stacking or folding of membranes provides enzymatic activity centers for these reactions.

7.

Protein synthesis

In nucleus | In cytosol | On RER

Transported to RER

Packaged and transported to *cis* region of Golgi

Encased in vesicle and released from *trans* region of Golgi

Target location?

Inside cell → Vesicles become lysosomes

Outside cell → Vesicle fuses with plasma membrane and proteins released

8. Dynein molecules bind to pairs of microtubules in the flagella or cilia. With the addition of cellular energy, the dynein molecules undergo a conformational change that causes the microtubules to slide past one another, resulting in a whiplike motion of the flagella.

9. The moving units are vesicles that are travelling along microtubules within the cell. The vesicles are attached to the microtubules by kinesin and dynein motor proteins. The kinesin and dynein "walk" along the microtubule by detaching and reattaching to it.

10. The junctions between these cells are gap junctions. Gap junctions function as channels allowing two adjoining cells to communicate with each other. Heart muscle is a classic example of a cell type that relies on gap junctions to pass along information.

11. Intestinal lining is an epithelial tissue, like skin. It undergoes significant mechanical stress, and thus the cells are anchored together with desmosomes to provide strength. At the same time, there are tight junctions in order to prevent leakage of body fluids into the intestine and of intestinal contents into the body cavity.

### *Test Yourself*

1. **b.** Several characteristics suggest that this is a prokaryote, such as its ability to survive in high salinity and high heat. The sure indication is that it contains no membrane-enclosed organelles. Prokaryotes are in the domain Archaea and Bacteria.
2. **b.** The pellet is undergoing cellular respiration, a function that occurs in the mitochondria. You can also assume that if the single membrane of the cell itself is ruptured, other organelles enclosed in single membranes will be ruptured as well.
3. **d.** The cell theory states that cells are the fundamental units of life, that all living things are composed of cells, and that all cells come from preexisting cells. It also implies that life is continuous and traceable back to the very first cell.
4. **c.** As volume increases, the surface area available for exchange does not increase proportionally. Eventually the surface is not large enough for maintenance of the metabolic activity of the cell.
5. **b.** In electron microscopy, a concentrated beam of electrons is focused on an object, allowing resolution of structures as small as 0.1 nm. Light microscopy, using lights and lenses, can only resolve objects to approximately 0.2 μm.
6. **d.** The RER is a site of protein synthesis, for membrane proteins or exported proteins.
7. **a.** The chloroplasts are the organelles involved in photosynthesis.
8. **b.** Lysosomes are organelles that contain digestive enzymes used to break down macromolecules taken in by phagocytosis.
9. **c.** The Golgi apparatus packages proteins for both internal and external use.
10. **d.** The nucleus, mitochondria, and plastids are the only organelles enclosed in two membranes.
11. **e.** Though the flagella have different structures, they serve the same role in prokaryotes and eukaryotes.
12. **b.** Mitochondria and chloroplasts may be found in the same cell. Almost all eukaryotic cells contain mitochondria. It is thought that mitochondria and chloroplasts evolved from separate endosymbiotic events.
13. **a.** Ribosomes, found in the nucleus, the cytosol, and in organelles such as the mitochondria, RER, and chloroplasts, complex with RNA to guide protein production. Vesicles move secreted proteins to the cell membrane for export.
14. **d.** The complex of proteins and DNA is called chromatin. Chromosomes condense and become visible under a light microscope during cell division.
15. **d.** Both the structure and the function of RER and SER differ.
16. **e.** The capsule is found outside the cell wall of some bacteria, and is implicated in the ability of some strains to cause disease in humans.
17. **d.** Proteoglycans are proteins with attached carbohydrate side chains. The other four answers are all cell junctions.

# 5 Cell Membranes and Signaling

## The Big Picture

- Cellular membranes are a dynamic composition of phospholipid bilayers, integral and peripheral proteins, and carbohydrates. The nature of the constituent phospholipids allows for the formation of a barrier that is semipermeable. Small hydrophobic molecules can traverse the membrane via simple diffusion, water can cross the membrane through aquaporins by osmosis, small charged ions can pass through protein channels, and some other molecules can be shepherded through by carrier proteins. Transport of molecules against their concentration gradient is active and requires energy input, either directly from ATP or coupled to ATP-driven transport. Larger substances depend on endocytosis and exocytosis for transport into and out of the cell.
- Signal transduction is the means by which cells receive information from the environment or other cells and react to those signals. Transduction is a highly regulated series of events that depends on the binding of a signal ligand to a receptor protein. The signal binding causes a change in the shape of the receptor protein, which causes a responder protein to initiate events in the cell that change its function. The effects of signals are often mediated by second messengers.

## Study Strategies

- In the study of membranes, the processes of diffusion and osmosis are the most difficult to understand. It is very easy to get the terminology confused, especially the terms "hypertonic" and "hypotonic." A consideration of the Latin roots of the words is helpful. "Hyper-" generally means excess, and "hypo-" generally means "less than." "Tonic" refers to solute concentration. Therefore, "hypertonic" means excess solutes, and "hypotonic" means fewer solutes.
- For the study of diffusion and osmosis, draw diagrams of the movement of water and solutes. Diagrams will help you visualize what is happening across a membrane.
- Secondary active transport also tends to be confusing. Remember that secondary active transport does not use ATP directly, but is tightly coupled to ion transport that does require ATP.
- When studying the fluid mosaic model, think about the properties of the constituent molecules. This will make understanding the membrane's structure much easier. Draw a cartoon of the cell membrane and all the potential components.
- Create a flow chart or a diagram of the signal transduction pathways.
- Make a list of the different secondary signals and provide an example of how each signal is activated and what it affects.
- Review the chapter figures to clarify the different examples of signal transduction.
- It is easy to become overwhelmed by the different examples of signal transduction. Try to focus on particular details of the systems. Compare cell membrane receptors with intracellular receptors. List the three kinds of second messengers for signal transduction. Then, pick one of the signal transduction examples and create a table that includes the signal, the receptor, the transduction (responders and amplification), and the effect in the cell. Expand your table to include other examples of signal transduction.
- G protein action is a difficult concept. Remember that the G protein interacts with the receptor, binds GTP, and then interacts with the effector protein.
- Go to **LaunchPad** (or use the Web addresses listed) to review the following additional resources:

  Animated Tutorial 5.1 Lipid Bilayer Composition (PoL2.com/at5.1)

  Animated Tutorial 5.2 Passive Transport (PoL2.com/at5.2)

  Animated Tutorial 5.3 Active Transport (PoL2.com/at5.3)

  Animated Tutorial 5.4 Endocytosis and Exocytosis (PoL2.com/at5.4)

  Animated Tutorial 5.5 G Protein–Linked Signal Transduction and Cancer (PoL2.com/at5.5)

Animated Tutorial 5.6 Signal Transduction Pathway (PoL2.com/at5.6)

Activity 5.1 Membrane Molecular Structure (PoL2.com/ac5.1)

Activity 5.2 Concept Matching (PoL2.com/ac5.2)

Analyze the Data in textbook Figures 5.2 and 5.5

Apply the Concept in textbook Sections 5.1 and 5.4

## Key Concept Review

### 5.1 Biological Membranes Have a Common Structure and Are Fluid

Lipids form the hydrophobic core of the membrane

Proteins are important components of membranes

Cell membrane carbohydrates are recognition sites

Membranes are constantly changing

Biological membranes evolved to provide a barrier for cellular life. A biological membrane is mostly composed of lipids, with phospholipids as the most abundant component. Phospholipids have both hydrophilic ("water-loving") polar phosphate heads and hydrophobic ("water-hating") nonpolar fatty tails. They arrange themselves into a bilayer with the hydrophobic tails touching and the heads extending into the aqueous environment inside and outside the cell. This arrangement allows for the fluid movement of the two layers on top of each other and the sealing of any disruptions in the membrane. This membrane is about 8 nm thick. Though the basic structure of a bilayer is always the same, the inner and outer halves can differ in lipid composition and thus have slightly different properties. Phospholipids differ in their length, degree of unsaturation, and degree of polarity. Phospholipids that are saturated can be packed together very closely in the membrane, while unsaturated phospholipids have kinks in their fatty acids that make them pack less densely. Cholesterol can constitute up to 25 percent of the lipid content in the membrane of animals. The amount of cholesterol present and the degree of fatty acid saturation (and thus membrane kinks) influences the fluidity of the membrane. Increases in cholesterol, length of hydrophobic chains, and fatty acid saturation make the membrane less fluid. Decreases in temperature also make the membrane less fluid. Organisms have the ability to change the lipid composition of their membranes in response to a change in temperature.

The typical ratio of proteins to phospholipid molecules in a cell membrane is 1 to 25. This ratio can vary: in mitochondria the ratio is 1 protein to 5 lipids, while in the myelin sheath of neurons it is 1 protein to 70 lipids. Proteins may be embedded in or extend completely through membranes. Regions or domains of a membrane protein with hydrophobic amino acid side chains tend to be found in the hydrophobic environment within the membrane. The regions of a membrane protein containing hydrophilic amino acid side chains extend out from the membrane. The phospholipids and proteins interact noncovalently, allowing for movement throughout the membrane. Anchored membrane proteins have lipid groups that essentially tether them within the bilayer. Proteins that penetrate the phospholipid bilayer are called integral proteins. Special types of integral proteins called transmembrane proteins span the entire bilayer. Those not embedded are referred to as peripheral proteins, and make noncovalent attachments to the hydrophilic lipid heads. Proteins are not necessarily distributed evenly, and the numbers of proteins and their placement vary greatly based on cell type. The "inside" and "outside" of a membrane often have different properties due to the different characteristics of the peripheral proteins on the two sides. Some membrane proteins may be anchored to cytoskeletal components.

Membrane carbohydrates serve as recognition sites for other cells and molecules. Carbohydrates are frequently bound to lipids or proteins, forming glycolipids or glycoproteins, respectively. A proteoglycan contains more oligosaccharides than a glycoprotein with greater numbers of monosaccharide units per oligosaccharide. Because of the large diversity of oligosaccharide conformations and contents, they contribute to differential cell communication and adhesion. Cell adhesion is due to interactions between carbohydrate groups on two different cells, between a carbohydrate on one cell and a protein on another cell, or between two proteins on two different cells. Membranes are in a constant state of flux. Phospholipids from the membrane of the smooth endoplasmic reticulum are distributed to the Golgi via the vesicles that transport other molecules. From the Golgi they move to the cell membrane. The membranes of the different organelles differ in their phospholipid content.

**Question 1.** Diagram a cell membrane and label the phospholipid bilayer, integral proteins, peripheral membrane proteins, and carbohydrates. Describe the fluid mosaic model with reference to your diagram.

**Question 2.** You are examining the membrane of a cell and notice many carbohydrates in the membrane. What role do these carbohydrates play in the cell?

### 5.2 Passive Transport across Membranes Requires No Input of Energy

Simple diffusion takes place through the phospholipid bilayer

Osmosis is the diffusion of water across membranes

Diffusion may be aided by channel proteins

Carrier proteins aid diffusion by binding substances

Cell membranes have selective permeability, meaning that certain substances can pass through the bilayer via simple diffusion. Passive transport is the movement of a substance across a lipid bilayer down its concentration gradient and does not require an input of energy. Active transport is the movement of molecules against a concentration gradient, which does require energy input. Diffusion is the process of random movement toward a state of equilibrium, and can be described as the net movement of a substance from an area of greater concentration to an area of lesser concentration. A membrane is said to be permeable to those substances that can pass through it by diffusion and impermeable to those that cannot. If a substance can pass through a membrane, it will diffuse until concentrations on either side of the membrane are equal. At equilibrium, the molecules of the substance continue to move across the membrane, but the net movement of molecules is zero. The rate of diffusion depends on the size of the diffusing substance, the temperature of the solution, and the concentration gradient. Diffusion within small areas such as single cells may occur rapidly, but diffusion occurs more slowly with increasing distance. In simple diffusion, small molecules pass through a membrane independently. The more lipid-soluble the molecules are, the faster they diffuse across the membrane. In contrast, charged and polar molecules do not readily pass through a membrane due to the formation of many hydrogen bonds with water and the hydrophobic nature of the internal layer of the membrane.

Osmosis is the diffusion of water across a membrane, which typically occurs through specialized membrane channels. Water will move across a membrane from areas of low solute concentration (high water concentration) to areas of high solute concentration (low water concentration) in order to equalize solute concentrations on either side of the membrane. Osmotic pressure is the amount of pressure that needs to be applied to prevent flow across a membrane by osmosis; it is defined mathematically as $cRT$, where $c$ is total solute concentration, $R$ is the gas constant, and $T$ is the absolute temperature. Solute concentrations in a membrane compartment are classified as isotonic, hypertonic, or hypotonic. An isotonic solution has the same solute concentration on both sides of a membrane. A hypertonic solution has a solute concentration that is higher than the concentration on the other side of the membrane, and a hypotonic solution has a concentration that is lower. Environmental and cellular solute concentrations dictate the direction of osmosis in living cells. The pressure within cells, called turgor pressure, changes according to the amount of water taken up by osmosis. Turgor pressure can increase only if there is a cell wall to limit cell expansion and prevent bursting of the cell membrane.

Facilitated diffusion is the passive movement of a substance across a membrane with the help of specialized membrane-bound proteins. A substance can cross the membrane by facilitated diffusion through protein channels inserted in the cell membrane. The pores of these proteins have polar amino acids that allow polar molecules and ions to cross the membrane, bypassing the hydrophobic environment of the membrane interior. The best-studied channels are ion channels. Most ion channels are either ligand-gated or voltage-gated, allowing the passage of ions to be controlled depending on the cellular environment. Specific channels called aquaporins allow water to cross the membrane by osmosis. Water can also diffuse through ion channels. Facilitated diffusion may be aided by carrier proteins that bind a substance and transport it across the membrane. Diffusion in this case is limited not only by the concentration gradient, but also by the number of available membrane carrier proteins. When all of the carrier proteins are bound to the substance, they are saturated, limiting the rate of diffusion.

**Question 3.** A marathon runner has just arrived in the emergency room with severe dehydration, and the physician must decide which type of solution to pump into his veins: pure water, 0.9 percent saline, or 1.5 percent saline. In order to be certain, blood samples are treated with each solution and observed under a microscope. Describe what is likely to happen to the blood cells when they are exposed to each solution. (Hint: Blood cells are approximately 0.9 percent saline.) Which solution should the physician choose for rehydrating the runner?

**Question 4.** The cell membrane is a good barrier to the movement of water. How, therefore, does water typically cross the cell membrane into a cell?

## 5.3 Active Transport Moves Solutes against Their Concentration Gradients

### Different energy sources distinguish different active transport systems

Primary active transport requires the energy stored in ATP to move ions against their concentration gradient. This is a directional process in which a molecule is moved either into or out of a cell. Primary active transport makes direct use of energy from the hydrolysis of ATP to drive the transporter. The sodium–potassium pump is an integral membrane glycoprotein found in animal cells that uses ATP to transport two $K^+$ ions into the cell and three $Na^+$ ions out of the cell. Secondary active transport uses ATP indirectly by coupling solute transport with an ion concentration gradient established by primary transport. It typically moves amino acids and sugars across the membrane with the help of an ion like $Na^+$. The ion and transported molecule may move in the same direction across the membrane or in opposite directions.

**Question 5.** You are examining a membrane that contains a number of proteins that appear to be involved in transport of an ion across that membrane. Design an experiment that will allow you to determine if these transporters are passive transporters or active transporters.

**Question 6.** The sodium–potassium pump is important for maintaining gradients of $Na^+$ and $K^+$ between the inside and outside of the cell. Describe how these gradients function in secondary active transport to move another molecule, such as glucose.

## 5.4 Large Molecules Cross Membranes via Vesicles

### Exocytosis moves materials out of the cell

### Macromolecules and particles enter the cell by endocytosis

### Receptor endocytosis often involves coated vesicles

Some macromolecules are unable to cross the cell membrane because of their size or charge, or because they are polar. Cell products to be secreted need to move across the cell membrane; this is accomplished using vesicles via exocytosis (see Figure 5.8B). Specific proteins found on the surface of the vesicles bind with receptor proteins on the cytoplasmic side of the cell membrane. The vesicle membrane fuses with the cell membrane, and the contents of the vesicle are released outside the cell.

Endocytosis is the process by which a cell brings large molecules inside. This is accomplished with receptors embedded in the cell membrane, which folds around the substance to form an endocytotic vesicle (see Figure 5.8A). Large substances and even entire cells are engulfed in the process of phagocytosis. Once inside the cell, the vesicle can fuse with a lysosome for digestion. The cell takes up liquids from the outside in small vesicles during pinocytosis, which is relatively nonspecific. Animal cells use receptor-mediated endocytosis to capture specific macromolecules such as cholesterol from the environment. Receptors for specific macromolecules cluster together on the cell surface in coated pits containing the protein clathrin. Upon binding of the specific molecule to the receptors, the coated pit invaginates to form a vesicle. The resulting vesicle is clathrin-coated until it is well inside the cell, where it loses its coat and fuses with an endosome. In this new fused compartment the vesicle contents are separated and sorted.

**Question 7.** Cells have the ability to take in large molecules by endocytosis and secrete them to the environment by exocytosis. Describe each process and explain why both are important for the cell.

**Question 8.** Compare and contrast phagocytosis and receptor-mediated endocytosis.

## 5.5 The Membrane Plays a Key Role in a Cell's Response to Environmental Signals

Cells are exposed to many signals and may have different responses

Receptors can be classified by location and function

Many receptors are associated with the cell membrane

The process of cell signaling allows cells to process information from their environments. A signal may be a physical stimulus (e.g., light or heat) or a chemical (such as a hormone). Another name for a chemical signal is a ligand, a molecule that binds to a receptor protein. The cell response to this signal is achieved through a signal transduction pathway. Cells in multicellular organisms receive signals from the outside environment, from other cells, or from extracellular fluid. Autocrine signals are local signals that affect the cells that make them. Paracrine signals are local signals that affect nearby cells. Juxtracrine signals require contact between two cells. This usually does not require a ligand, but instead involves direct interaction between two signaling molecules on the surfaces of the two cells. Hormones are circulatory signals that travel through the circulatory system (of plants as well as animals) and affect distant cells. Only those cells that have the correct receptor will respond to the chemical signal.

A cell's response to a signal is dependent upon the presence of membrane proteins that function as receptors. Commonly, these have an effect through allosteric regulation. The response may occur through a short-acting enzyme or longer-term gene expression. There are two classes of receptors: intracellular receptors, which bind small nonpolar ligands that can diffuse across the cell membrane, and surface membrane receptors, which bind large and/or polar ligands that cannot cross the cell membrane. Binding to the receptor requires a fit of the ligand to the receptor binding site. Often the catalytic domain of the protein is on the cytoplasmic side of the membrane, and is activated through allosteric regulation by binding of the ligand. The ligand is not changed in the binding process, and binding of the signal to the receptor is reversible. This reversibility is important in order to prevent the receptor from being permanently stimulated.

An inhibitor can bind to the signal site on a receptor, preventing the binding of the usual ligand. The inhibitors can be natural or artificial. There are three main types of cell membrane receptors: ion channel receptors, protein kinase receptors, and G protein–linked receptors. In the case of ion channel receptors, ligand binding causes a conformational change to open "gates" that allow ions (e.g., $Na^+$, $K^+$, $Ca^+$, or $Cl^-$) to pass through. An example is the acetylcholine receptor found on muscle cells. When acetylcholine binds to the receptor, the channel in the receptor opens and sodium flows through. In the case of protein kinase receptors, ligand binding stimulates the transfer of a phosphate group from ATP to a target protein. Insulin works through a protein kinase receptor. The term "G protein–linked receptor" applies to a group of receptors for which ligand binding changes its shape so that a G protein on the cytoplasmic side can bind. When the G protein is activated (by binding to the cytoplasmic side of the G protein–linked receptor), it exchanges a bound GDP molecule for a GTP and then activates a third effector protein, leading to signal amplification in the cell. G protein–linked receptors are important in animal sensory systems.

**Question 9.** Acetylcholine is a polar ligand that initiates a series of events. Based on the properties of this ligand, what types of receptors would you predict that it could interact with? Why?

**Question 10.** Describe the steps involved in a signaling cascade acting through a G protein–linked receptor.

## 5.6 Signal Transduction Allows the Cell to Respond to Its Environment

Cell functions change in response to environmental signals

Second messengers can stimulate signal transduction

A signaling cascade involves enzyme regulation and signal amplification

Signal transduction is highly regulated

Activation of cell receptors leads to a series of events inside the cell to distribute and amplify a signal. These signal transduction pathways bring about the response in the cell in a number of different ways. Some signal transduction pathways open ion channels. Others alter gene expression, by either increasing or decreasing transcription of particular genes. Signal transduction pathways also influence enzyme function, either through inhibition or activation (as with epinephrine stimulation).

Second messengers, such as cyclic AMP, help to transduce the message from an activated receptor to the associated events in the cell. The advantage of second messengers is that they distribute and amplify the initial signal. Often other protein kinases are involved by phosphorylating other enzymes in the cell that participate in the pathway. Phosphorylation can either inhibit or activate an enzyme. Signal transduction pathways are highly regulated to ensure their proper function. Function in a pathway is dependent on the synthesis and breakdown of the enzymes involved and their activation or inhibition. A balance is achieved among all of these to create an effect in a cell at the appropriate time.

**Question 11.** Discuss the role of second messengers in a signaling pathway. How are they different from ligands and receptors? What roles do second messengers and signals have in common?

**Question 12.** Describe how G protein–linked receptors and protein kinases interact in a signal transduction cascade.

**Question 13.** Why are signal transduction pathways highly regulated? What would happen if they were not?

## Test Yourself

1. Which statement regarding cellular membranes is *false?*
   a. The hydrophobic nature of the phospholipid tails limits the migration of polar molecules across the membrane.
   b. Integral proteins and phospholipids move fluidly throughout the membrane.
   c. Membrane phospholipids flip back and forth from one side of the bilayer to the other.
   d. Glycolipids and glycoproteins serve as recognition sites on the cell membrane.
   e. Hydrophobic regions of proteins become embedded in the membrane.

   ***Textbook Reference:*** *Concept 5.1 Biological Membranes Have a Common Structure and Are Fluid*

2. Paracrine signals
   a. act on the cells that made them.
   b. move through the blood and act on cells far from their source.
   c. act on cells that are near to those that secrete them.
   d. do not act through receptors.
   e. require large concentrations of the signaling molecule to function.

   ***Textbook Reference:*** *Concept 5.5 The Membrane Plays a Key Role in a Cell's Response to Environmental Signals; Cells are exposed to many signals and may have different responses*

3. You are monitoring the diffusion of a molecule across a membrane. An internal concentration of _______ and an external concentration of _______ will result in the fastest rate of diffusion.
   a. 5; 60
   b. 20; 50
   c. 35; 40
   d. 50; 50
   e. 60; 20

   ***Textbook Reference:*** *Concept 5.2 Passive Transport across Membranes Requires No Input of Energy; Simple diffusion takes place through the phospholipid bilayer*

4. Caffeine is a stimulant that works because it acts as a(n) _______ to the adenosine receptors in a person's brain, and stimulates _______ in that person's heart and liver that increases blood flow and blood glucose.
   a. effector; a pathway
   b. inhibitor; a cascade pathway
   c. inhibitor; a ligand
   d. signal; inhibitors
   e. pathway; a ligand

   ***Textbook Reference:*** *Concept 5.6 Signal Transduction Allows the Cell to Respond to Its Environment; Signal transduction is highly regulated*

5. Which statement about osmosis is *false?*
   a. Osmosis refers to the movement of water along a concentration gradient.
   b. In osmosis, water moves to equalize solute concentrations on either side of the membrane.
   c. The movement of water across a membrane can affect the turgor pressure of some cells.
   d. If osmosis occurs across a membrane, then diffusion is not occurring.
   e. During osmosis, water is moving through membrane channels.

   ***Textbook Reference:*** *Concept 5.2 Passive Transport across Membranes Requires No Input of Energy; Diffusion may be aided by channel proteins*

6. Channel proteins allow ions that would not normally pass through the cell membrane to pass through via the channel. What property of the channel proteins makes this possible?
   a. A pore of polar amino acid groups
   b. A pore of hydrophobic amino acid groups
   c. A pore of $Ca^{2+}$
   d. A pore lined with clathrin
   e. Use of ATP to carry the ions through

   ***Textbook Reference:*** *Concept 5.2 Passive Transport across Membranes Requires No Input of Energy; Diffusion may be aided by channel proteins*

7. Active transport differs from passive transport in that active transport
   a. requires energy.
   b. never requires direct input of ATP.

c. moves molecules via a concentration gradient.
d. always requires ATP as an energy source.
e. occurs spontaneously.
***Textbook Reference:*** *Concept 5.3 Active Transport Moves Solutes against Their Concentration Gradients*

8. Single-celled animals, such as amoebas, engulf entire cells for food. This manner of "eating" is called
a. exocytosis.
b. endocytosis.
c. facilitative transport.
d. active transport.
e. osmosis.
***Textbook Reference:*** *Concept 5.4 Large Molecules Cross Membranes via Vesicles; Macromolecules and particles enter the cell by endocytosis*

9. Many cells have a sodium–potassium pump. In order to function, sodium–potassium pumps require
a. ATP.
b. a channel protein.
c. the absence of a concentration gradient.
d. ADP.
e. glycolipids.
***Textbook Reference:*** *Concept 5.3 Active Transport Moves Solutes against Their Concentration Gradients; Different energy sources distinguish different active transport systems*

10. Bacterial cells are often found in very hypotonic environments. Which of the following characteristics keeps them from taking in too much water from their environment?
a. The presence of a cell wall, which allows for a buildup of turgor pressure, preventing additional water from entering the cell
b. The presence of a cell wall, which allows for a buildup of tonic pressure, preventing additional water from entering the cell
c. The capacity of the cell to expel water as quickly as it takes it up
d. The presence of an active water pump
e. The ability to expand as they take on water
***Textbook Reference:*** *Concept 5.2 Passive Transport across Membranes Requires No Input of Energy; Osmosis is the diffusion of water across membranes*

11. The rate of diffusion of a particular solute is *not* affected by
a. temperature.
b. molecule size.
c. the concentration gradient.
d. the electrical charge.
e. concentration gradients of other solutes.
***Textbook Reference:*** *Concept 5.2 Passive Transport across Membranes Requires No Input of Energy; Simple diffusion takes place through the phospholipid bilayer*

12. Which of the following actions does *not* occur during signal transduction?
a. Binding of ligand to receptor
b. Conformational change to the receptor protein
c. Conformational change of the signal
d. Alteration of cellular activity
e. Allosteric regulation
***Textbook Reference:*** *Concept 5.5 The Membrane Plays a Key Role in a Cell's Response to Environmental Signals; Many receptors are associated with the cell membrane*

13. Which statement about second messengers is true?
a. They amplify the signal.
b. They bind to the active site of the receptor.
c. They exit the cell to bind to another membrane receptor.
d. They act via their enzymatic activity.
e. They only bind to and affect one downstream molecule in a given signal transduction pathway.
***Textbook Reference:*** *Concept 5.6 Signal Transduction Allows the Cell to Respond to Its Environment; Second messengers can stimulate signal transduction*

14. If a red blood cell with an internal salt concentration of about 0.85 percent is placed in a saline solution that is 4 percent, the
a. cell will lose water and shrivel.
b. cell will gain water and burst.
c. turgor pressure in the cell will increase greatly.
d. turgor pressure in the cell will decrease greatly.
e. cell will remain unchanged.
***Textbook Reference:*** *Concept 5.2 Passive Transport across Membranes Requires No Input of Energy; Osmosis is the diffusion of water across membranes*

15. Intracellular receptors bind
a. small signals that can diffuse through the cell membrane.
b. second messengers such as cAMP.
c. hydrophilic molecules.
d. cell membrane receptors.
e. protein hormones.
***Textbook Reference:*** *Concept 5.5 The Membrane Plays a Key Role in a Cell's Response to Environmental Signals; Receptors can be classified by location and function*

16. Which statement about receptors is true?
a. Receptors are found only on the surface of cells.
b. Receptors are specific to the signal ligand.
c. Most receptors can bind with many different types of ligands.
d. All receptors can act as ion channels.
e. All receptors have protein kinase activity.
***Textbook Reference:*** *Concept 5.5 The Membrane Plays a Key Role in a Cell's Response to Environmental Signals; Many receptors are associated with the cell membrane*

17. Which cell membrane component serves as a recognition site for interactions between cells?
    a. Cholesterol
    b. Glycolipids or glycoproteins
    c. Phospholipids
    d. Carrier proteins
    e. Cytoskeleton

    ***Textbook Reference:*** *Concept 5.1 Biological Membranes Have a Common Structure and Are Fluid; Cell membrane carbohydrates are recognition sites*

18. Which of the following represents the correct order of steps in a signal transduction pathway?
    a. Binding of signal, release of second messenger, alteration of receptor conformation, alteration of cellular function
    b. Binding of signal, release of second messenger, alteration of receptor conformation, transcription of gene
    c. Binding of signal, activation of target protein by responder, alteration of receptor conformation, release of second messenger; transcription of gene
    d. Binding of signal, alteration of receptor conformation, alteration of cellular function, release of second messenger
    e. Binding of signal, alteration of receptor conformation, activation of target protein by responder, alteration of cellular function

    ***Textbook Reference:*** *Concept 5.5 The Membrane Plays a Key Role in a Cell's Response to Environmental Signals; Many receptors are associated with the cell membrane*

## Answers

### *Key Concept Review*

1.

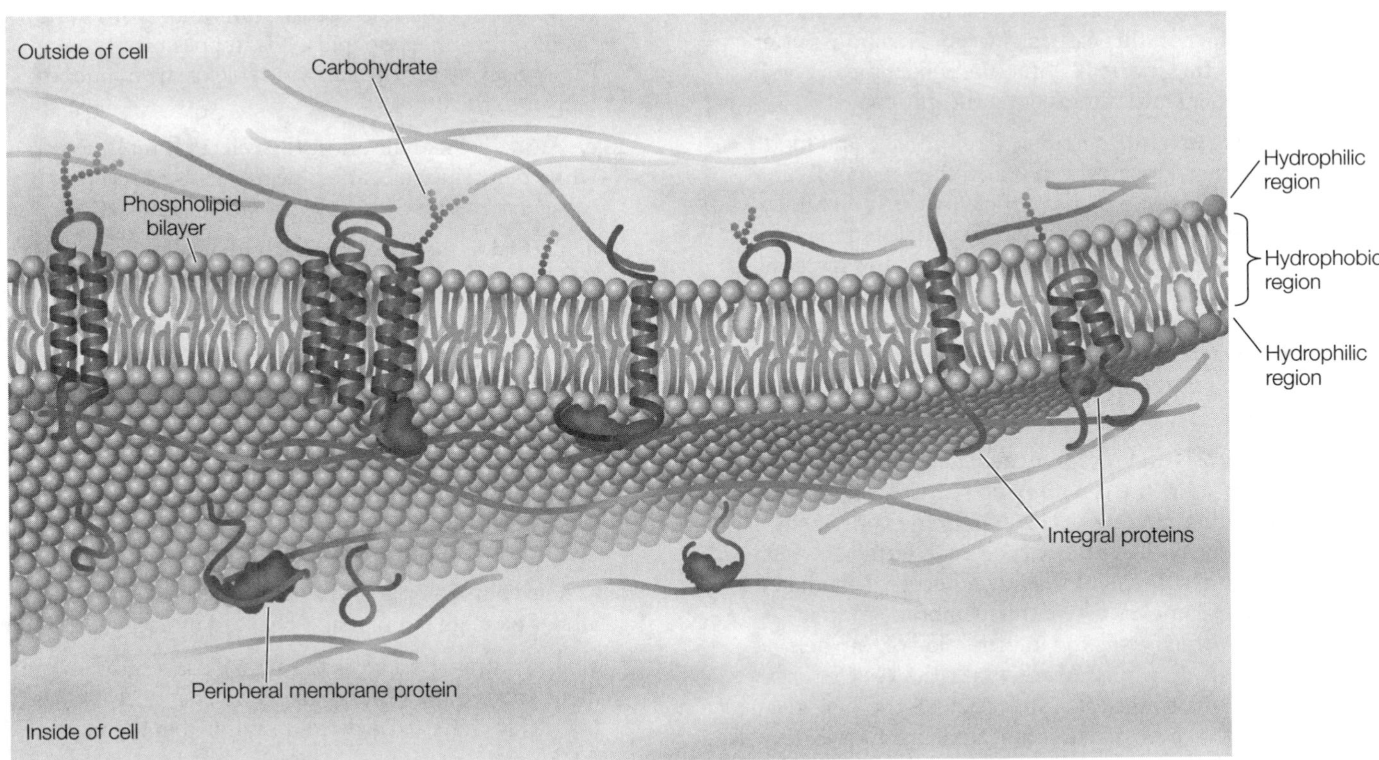


The phospholipid bilayer allows the embedded proteins to float freely through the membrane. This general design is known as the fluid mosaic model.

2. Carbohydrates located on the surface of the cell membrane are typically involved in recognition. This can be recognition between cells or as sites responsible for cell communication.

3. In pure water, blood cells will take on water through osmosis, swell, and eventually rupture. In 0.9 percent saline, the cells should neither gain nor lose a significant amount of water. In a 1.5 percent saline solution,

the cells should lose water and shrivel. In order to rehydrate the runner, a solution isotonic to the patient's blood cells, 0.9 percent, should be infused into his bloodstream. A hypotonic solution would end up rupturing the patient's cells.

4. Some cell membranes contain special protein channels known as aquaporins, which allow for the movement of water by osmosis into and out of the cell.

5. The main difference between active and passive transport is that active transport goes against a concentration gradient and requires energy, whereas passive transport diffuses passively and does not require energy. To test this, you would want to measure the movement of the ions across a membrane. One side of the membrane should have a high concentration of the ion and the other a low concentration. If ions move from the area of high concentration to the area of low concentration, then the transporter is a passive transporter. If, instead, the addition of ATP is needed for movement to occur, it is an active transporter.

6. The sodium–potassium pump sets up a gradient of $Na^+$ and $K^+$, with more $Na^+$ outside the cell and more $K^+$ inside the cell. A secondary active transport mechanism has a transport protein for the specific molecule, which in this case is glucose. This transporter passively transports $Na^+$ into the cell by means of the $Na^+$ gradient and brings along the glucose.

7. Cells take up large particles, foreign cells, and food sources by endocytosis, in which the cell membrane of the cell surrounds the particle to form an endocytotic vesicle. Substances such as undigested material, digestive enzymes, neurotransmitters, and material for plant wall construction are secreted by the cell by means of exocytosis. During exocytosis, the membrane of a secretory vesicle fuses with the cell membrane and the contents are released to the outside of the cell.

8. Phagocytosis and receptor-mediated endocytosis are both ways that the cell takes up large substances by surrounding them with cell membrane and pinching off vesicles. Both are fairly specific in terms of which items they will target for uptake. Receptor-mediated endocytosis makes use of the membrane protein clathrin, which coats the internal side of the membrane in shallow pits. After uptake the vesicle fuses with an endosome; after phagocytosis the vesicle fuses with a lysosome in order to digest the vesicle contents.

9. Polar molecules such as acetylcholine are unable to cross the cell membrane and thus cannot enter the cell freely. Therefore, they require a receptor that is located on the outside of the cell membrane. This receptor must span the membrane and initiate a signaling cascade upon the binding of acetylcholine.

10. A G protein–linked receptor is stimulated by hormone binding to its external binding site. This in turn activates the G protein, which activates an effector protein. The effector protein converts reactants into signaling products, amplifying the signal from the single hormone.

11. Second messengers function to amplify and distribute signals. They do not bind to receptors and do not act like signals in the cascade. Signals result in a change in cell function, and second messengers are part of the pathway that results in this change.

12. G protein–linked receptors frequently expose the protein kinase activities of effector molecules. Often several protein kinases will participate in a particular signal transduction pathway.

13. Signal transduction pathways are highly regulated because they are temporary events within a cell that help to regulate the function of the cell. If they were not highly regulated, the response of the cell would lag behind the changes in the environment. This could potentially lead to cell death.

### *Test Yourself*

1. **c.** Because of the hydrophobic tails and hydrophilic heads of the phospholipids, it is impossible for them to flip back and forth from one side of the membrane bilayer to the other.

2. **c.** Autocrine signals affect the cells that make them, while paracrine cells affect nearby cells. A receptor in the cell membrane binds hydrophilic signals.

3. **a.** Diffusion may take place in either direction across a membrane and always follows a concentration gradient. The larger the gradient, the faster the diffusion will occur.

4. **b.** Caffeine and adenosine bind to the same receptor protein. The binding of caffeine prevents adenosine from binding in the brain. In other organs, a series of cascades begin in response to caffeine binding.

5. **d.** Diffusion and osmosis are not mutually exclusive and may take place at the same time.

6. **a.** The charged or polar lining of the channel proteins allows passage of polar and charged molecules.

7. **a.** Active transport works against a concentration gradient and requires energy to do so. That energy does not always have to be directly supplied in the form of ATP.

8. **b.** Cells carry out cellular eating by phagocytosis, which is a type of endocytosis.

9. **a.** Sodium–potassium pumps are forms of primary active transport and require energy in the form of ATP.

10. **a.** Turgor pressure limits osmosis, and once a cell is turgid, no more water may be taken on.

11. **e.** Temperature, molecule size, molecule charge, and concentration gradients all affect the rate at which diffusion takes place. The concentration gradients of other solutes have no effect on the rate of diffusion.

12. **c.** In order for signal transduction to take place, the ligand must bind to the receptor, the receptor must undergo a conformational change, and the activity of the cell must be altered. The signal itself is not changed.

13. **a.** Second messengers amplify the signal, interacting with many downstream molecules. They do not have enzymatic activity, and they do not leave the cell.

14. **a.** The cell will lose water as solute concentrations on both sides of the membrane equalize.

15. **a.** To bind to a cytoplasmic receptor, the ligand must be able to pass through the cell membrane. Protein hormones tend to be too big to enter the cell, and interact with membrane receptors.

16. **b.** Receptors interact only with specific ligands to bring about a response in the cell.

17. **b.** Both glycolipids and glycoproteins serve as recognition sites.

18. **e.** Signals must bind their receptors, causing change in the conformation of the receptor. A responder then activates a target protein, finally causing change in cell function.

# Pathways that Harvest and Store Chemical Energy

## The Big Picture

- Cells require energy to carry out their functions. Metabolism is a series of energy-transferring reactions that fuel the processes of the cell. All reactions either give off energy (exergonic) or require energy to proceed (endergonic). Energy may be stored in chemical bonds (potential energy) and released when the bonds are broken to do work (kinetic energy). ATP serves as the energy shuttle for many metabolic processes. Redox reactions are important in these energy conversions.
- Metabolism is regulated by enzymes. Most metabolic pathways are under allosteric control. Enzyme complexes allow for interactions and regulation of adjacent active sites. Often the final product of a specific pathway regulates the commitment step of the pathway itself.
- Glucose provides cellular energy and may be metabolized in the absence of oxygen through glycolysis and fermentation. In the presence of oxygen, it is metabolized through glycolysis and cellular respiration in the form of the citric acid cycle, electron transport, and oxidative phosphorylation. Compared to aerobic respiration, anaerobic respiration produces significantly less cellular energy (in the form of ATP) from the same amount of glucose.
- Photosynthesis allows organisms with the appropriate pigments and metabolic processes to convert light energy from the sun into chemical energy that can be used in the cell or stored. The process of photosynthesis occurs in two steps. The first step utilizes light energy to produce ATP and NADPH with the help of the electron transport system. The second step uses the products of the light reactions to fix $CO_2$ in the Calvin cycle. Photosynthesis forms the basis of all food webs and is vital to life on Earth.

## Study Strategies

- The chemistry presented in this chapter can be difficult, and you may question why you are studying it in a biology course. Remember that biological processes are based on physical and chemical properties. To understand the function of organisms, it is necessary to understand the basics of energy transfer. If you do not have a firm understanding of topics covered in previous chapters—potential energy, entropy, and enzyme activity—it will be helpful to review these before learning more about them in the context of cellular metabolism.
- When you begin your study, do not focus on the pathways themselves; instead, focus on the beginning and end products and why each pathway does what it does. Once you understand conceptually why the pathway is present, move on to studying the steps in the pathway itself. You may find the concepts to be very unfamiliar; make sure you understand each section before proceeding. Spend a significant amount of time on the diagrams so you can visualize the pathways' purposes and interactions.
- This chapter contains a great deal of new terminology. Making a vocabulary list may be helpful. You might also benefit from writing the various reactants, products, key intermediates, and electron carriers on index cards. Use the cards to arrange the relevant molecules into the proper sequence for the pathway you are studying (i.e., glycolysis, cellular respiration, photosynthesis). Pay attention to multiple pathways that use the same molecules but in different ways, and also note which molecules are limited to a single pathway.
- Remember that ATP and electron carriers such as $NAD^+$ are energy currencies; they move energy from pathway to pathway. In order to understand the pathways completely, think about the roles of ATP and electron carriers, like $NAD^+$, in redox reactions. Other electron carriers mentioned in this chapter are $NADP^+$ and FAD. It may help to mentally link the P in $NADP^+$ with photosynthesis so you do not confuse it with the electron carriers used in glycolysis ($NAD^+$) and cellular respiration ($NAD^+$ and FAD).
- The properties of chlorophyll and light itself greatly affect how well a plant carries out photosynthesis. Take some time to understand the properties of light and how pigments interact with it.

- A common mistake is to think that the light reactions occur in the light and the Calvin cycle occurs in the dark. This is not the case! The light reactions must occur simultaneously with the Calvin cycle to supply the needed energy in the form of ATP and NADPH + $H^+$.
- When you are studying energy pathways in the cell, it is easy to forget that all living plant cells use cellular respiration. Cellular respiration takes the energy stored in photosynthesis and makes it available to drive other cellular processes. Photosynthesis is limited to photosynthetic cells in the plant; cellular respiration is more widespread in the organism.
- Go to **LaunchPad** (or use the Web addresses listed) to review the following additional resources:

  Animated Tutorial 6.1 Electron Transport and ATP Synthesis (PoL2e.com/at6.1)

  Animated Tutorial 6.2 Two Experiments Demonstrate the Chemiosmotic Mechanism (PoL2e.com/at6.2)

  Animated Tutorial 6.3 Photophosphorylation (PoL2e.com/at6.3)

  Animated Tutorial 6.4 The Source of the Oxygen Produced by Photosynthesis (PoL2e.com/at6.4)

  Animated Tutorial 6.5 Tracing the Pathway of $CO_2$ (PoL2e.com/at6.5)

  Activity 6.1 ATP and Coupled Reactions (PoL2e.com/ac6.1)

  Activity 6.2 The Citric Acid Cycle (PoL2e.com/ac6.2)

  Activity 6.3 Respiratory Chain (PoL2e.com/ac6.3)

  Activity 6.4 Glycolysis and Fermentation (PoL2e.com/ac6.4)

  Activity 6.5 Energy Levels (PoL2e.com/ac6.5)

  Activity 6.6 Regulation of Energy Pathways (PoL2e.com/ac6.6)

  Activity 6.7 The Calvin Cycle (PoL2e.com/ac6.7)

  Analyze the Data in textbook Figure 6.11

  Apply the Concept in textbook Sections 6.2 and 6.5

## Key Concept Review

### 6.1 ATP and Reduced Coenzymes Play Important Roles in Biological Energy Metabolism

ATP hydrolysis releases energy

Redox reactions transfer electrons and energy

The processes of NADH oxidation and ATP production are coupled

This chapter builds on your understanding of energy, enzymes, and metabolism to explain how they all function in living cells. Energy is stored in chemical bonds and can be released and transformed in metabolic pathways. Metabolic pathways which are similar (and thus evolutionarily conserved) among all organisms comprise separate reactions, with each catalyzed by a separate enzyme. In eukaryotes the pathways are compartmentalized in order to increase efficiency; each pathway is controlled by activation and inhibition of key enzymes. In cells, energy-transforming reactions are often coupled. Two of the most widely used coupling molecules are ATP and NADH.

ATP is the "energy currency" of the cell. When the bond between the second and third phosphate groups is hydrolyzed, energy is released. The linkage of ATP hydrolysis with an endergonic reaction is called reaction coupling. Redox reactions are coupled reduction and oxidation reactions. Reduction is the gain of one or more electrons by an atom, and oxidation is the loss of one or more electrons. Oxidation and reduction reactions always occur together, with one molecule donating an electron to the other. When the electrons are transferred, protons are also transferred; for this reason, some prefer to think of redox reactions as transferring hydrogen atoms.

A common electron carrier in redox reactions is nicotinamide adenine dinucleotide ($NAD^+$/NADH). Reduction of $NAD^+$ to NADH is an endergonic reaction. While $NAD^+$ is a common electron carrier in cells, there are other molecules with the same role. Flavin adenine dinucleotide (FAD) plays a role in the citric acid cycle and nicotinamide adenine dinucleotide phosphate ($NADP^+$) has a function in photosynthesis. An understanding of redox reactions, electron carriers, and ATP synthesis is necessary for an understanding of the process of oxidative phosphorylation, by which NADH is oxidized and ATP is synthesized.

**Question 1.** What is the result of the hydrolysis of ATP? Is there an alternative pathway that results in the release of more energy? Draw the products for both reactions using the reactant structures given in the diagram below as a guide.

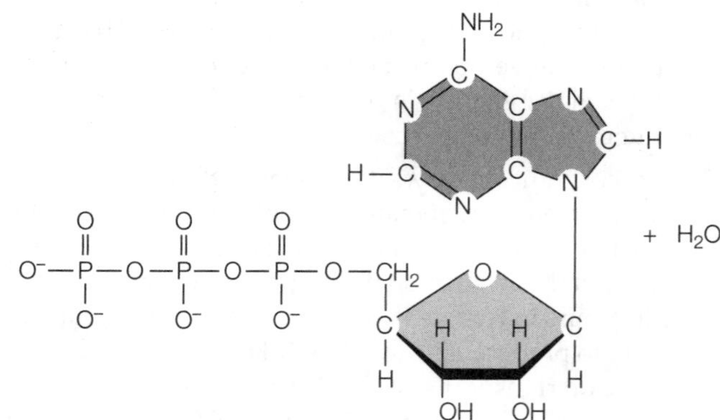


**Question 2.** It is estimated that approximately 90 percent of the energy that passes between levels in a food web is "lost" at each level. Explain this in the context of ATP usage in the cell.

**Question 3.** In the diagram below, label each compound as either oxidized or reduced. Circle the two lowest energy compounds. Which is the oxidizing agent and which is the reducing agent?

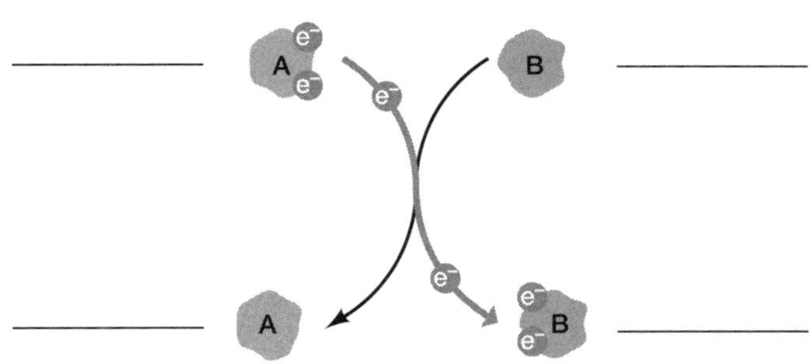


## 6.2 Carbohydrate Catabolism in the Presence of Oxygen Releases a Large Amount of Energy

In glycolysis, glucose is partially oxidized and some energy is released

Pyruvate oxidation links glycolysis and the citric acid cycle

The citric acid cycle completes the oxidation of glucose to $CO_2$

Energy is transferred from NADH to ATP by oxidative phosphorylation

Chemiosmosis uses the proton gradient to generate ATP

Chemiosmosis can be demonstrated experimentally

Oxidative phosphorylation and chemiosmosis yield a lot of ATP

Cellular respiration is the set of metabolic reactions used by the cell to release energy from food. Glucose is one type of molecule that is oxidized to $CO_2$ by the reactions of respiration. The overall process is highly efficient. The oxidation of glucose under aerobic conditions occurs through several steps that are grouped into three linked pathways: glycolysis, pyruvate oxidation, and the citric acid cycle. Glycolysis occurs in the cytosol and oxidizes one molecule of glucose to form two molecules of pyruvate; the other products are two molecules of ATP and two molecules of NADH. Two stereotypical reactions that occur repeatedly in glycolysis and many other metabolic pathways are oxidation–reduction (creating NADH) and substrate-level phosphorylation (forming ATP from ADP and an inorganic phosphate).

Pyruvate oxidation generates an acetate molecule that is then bound to coenzyme A (CoA), creating acetyl-CoA, which is used in various reactions. Acetyl CoA enters the citric acid cycle and is oxidized to two molecules of $CO_2$. The citric acid cycle operates twice for each glucose (once for each of the two pyruvate generated in glycolysis). Each turn of the citric acid cycle results in energy being captured by one molecule of ATP, three molecules of NADH, and one molecule of $FADH_2$. Oxidative phosphorylation uses NADH oxidation to transport $H^+$ ions (protons) across the inner mitochondrial membrane to generate a proton gradient. The respiratory chain is a series of redox electron carriers embedded in the inner mitochondrial membrane. Electron transport passes electrons from oxidation of NADH and $FADH_2$ from one carrier to the next, and the energy is used to pump protons across the membrane. ATP is generated when the $H^+$ ions flow back across the membrane down their concentration gradient through ATP synthase. The process is called chemiosmosis. ATP synthase is highly conserved across all living organisms. Artificial proton gradients will also drive ATP synthesis. The net result of aerobic cellular respiration is 32 molecules of ATP per fully oxidized glucose. Thus, $O_2$ is essential, acting as an electron receptor, allowing highly efficient oxidative phosphorylation to generate a proton gradient to be used to synthesize ATP.

**Question 4.** Explain how a proton motive force drives chemiosmosis.

**Question 5.** In the diagram below, fill in the numbers of reactants and products on each side of the equation.

1 Glucose { + ____ ADP; + ____ $NAD^+ + H^+$; + ____ $P_i$; + ____ ATP } → ____ Pyruvate { + ____ ATP; + ____ NADH; + ____ $H_2O$; + ____ ADP; + ____ $P_i$ }

Cross out anything appearing on both sides of the equation to get the net reaction shown below.

1 Glucose { + ____ ADP; + ____ $NAD^+ + H^+$; + ____ $P_i$ } → ____ Pyruvate { + ____ ATP; + ____ NADH; + ____ $H_2O$ }

**Question 6.** In the diagram below, label all of the structures and indicate the necessary reactants and products for the different steps of the pathway.

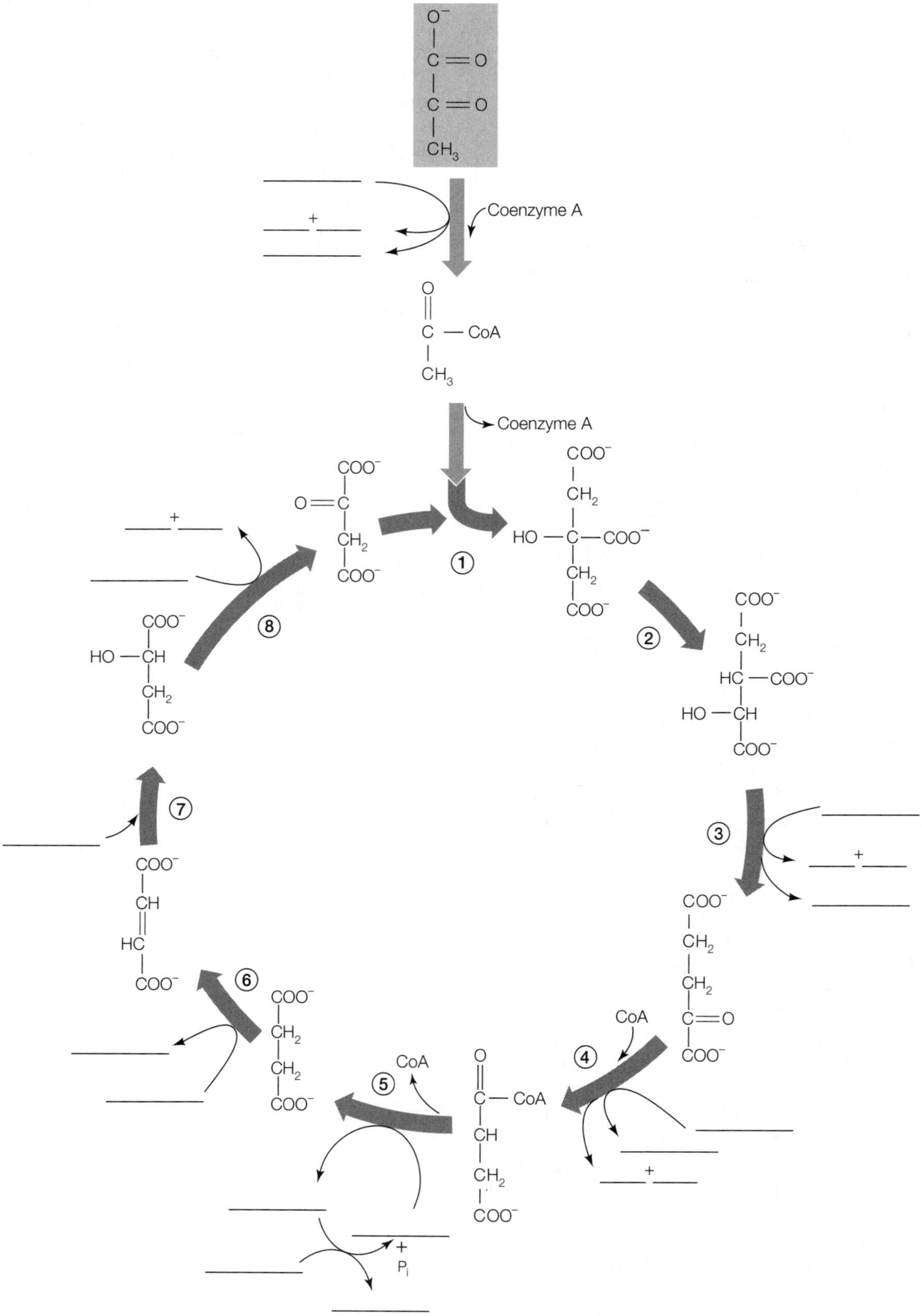

Coenzyme A

C — CoA

$CH_3$

Coenzyme A

1 2 3 4 5 6 7 8

CoA

CoA

+ $P_i$

**Question 7.** Label the diagram below showing the respiratory chain and chemiosmosis in the mitochondria. Be sure to label the mitochondrial parts as well.

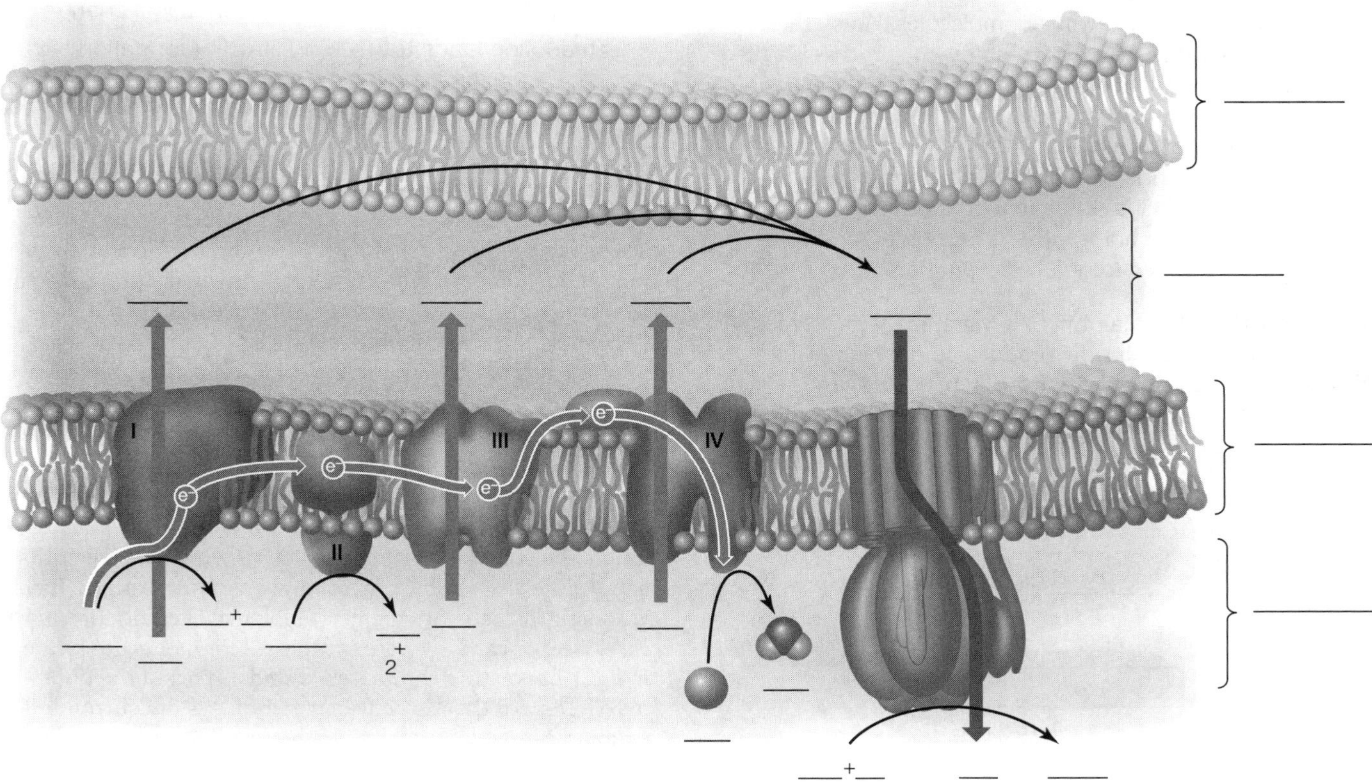


**Question 8.** One of the by-products of aerobic cellular respiration is carbon dioxide. Assume that you are working with labeled glucose, and trace the fate of that molecule until carbon dioxide is released.

## 6.3 Carbohydrate Catabolism in the Absence of Oxygen Releases a Small Amount of Energy

In the absence of oxygen (in an anaerobic environment), NADH cannot be oxidized back to $NAD^+$; at this point, glycolysis ceases because of a lack of electron acceptors. To solve this problem, organisms use fermentation. There are two fermentation pathways—lactic acid fermentation and alcoholic fermentation—that regenerate $NAD^+$ so that glycolysis can continue. There is no additional ATP obtained from the fermentation reactions, but the overall glycolysis-to-fermentation pathway does yield two ATPs per glucose molecule as a product of substrate-level phosphorylation. The fermentation pathways oxidize pyruvate and convert it into lactic acid and ethanol, respectively. Both reactions are essentially reversible, and pyruvate is generated when $O_2$ becomes available again.

**Question 9.** Why is oxygen necessary for aerobic respiration?

**Question 10.** Cyanide is poisonous to humans because it inhibits cytochrome oxidase in the mitochondria so that oxygen can no longer be utilized and the electron transport chain is halted. However, many cells in the human body are capable of lactic acid fermentation. Since cyanide does not inhibit glycolysis and fermentation, what could explain cyanide's lethal effect?

**Question 11.** What is responsible for the bubbles in beer?

**Question 12.** Glycolysis yields two molecules of pyruvate, two ATP, and two NADH + $H^+$, regardless of whether oxygen is present or not. What are the fates of these molecules in the absence of oxygen? What would happen if NADH + $H^+$ were not recycled?

**Question 13.** Compare and contrast energy yields from aerobic respiration and fermentation.

## 6.4 Catabolic and Anabolic Pathways Are Integrated

Catabolism and anabolism are linked

Catabolism and anabolism are integrated into a system

ATP and reduced coenzymes link catabolism, anabolism, and photosynthesis

A cell needs raw materials both to build new macromolecules and for catabolism. The same molecules can be used for both purposes depending on the needs of the cell at a particular time. For this reason, the different pathways all connect to one another. As Figure 6.13 shows, all four groups of macromolecules (lipids, carbohydrates, nucleic acids, and proteins) can feed into the glycolysis and cellular respiration pathways. Likewise, intermediates of glycolysis and cellular respiration can be diverted from energy production and used to build new macromolecules. When these intermediates are used to form glucose, the process is called gluconeogenesis. The metabolic pathways are integrated into a single system that "decides" how to use different molecules based on the needs of the cell or cells. Regulatory mechanisms include allostery and feedback inhibition. Since the energy that is stored in food molecules ultimately came from the sun, photosynthesis is another metabolic pathway that is linked to cellular respiration.

**Question 14.** In the diagram below, name the processes that are bracketed, enter the products in the blanks indicated, and indicate how many ATP molecules are generated.

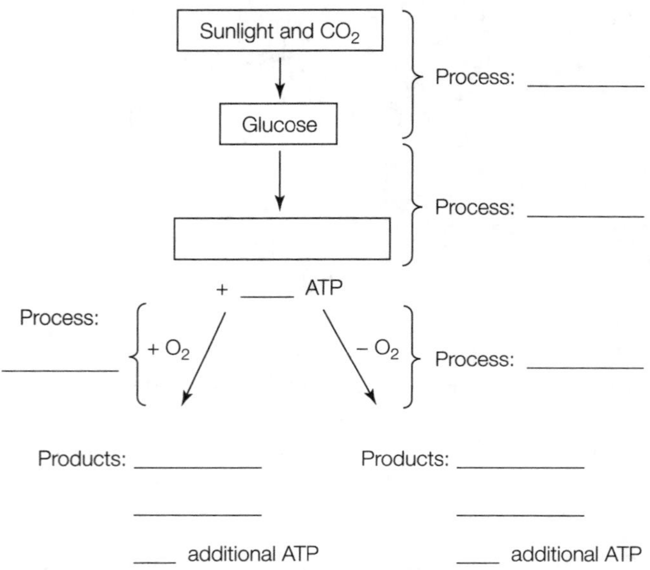


**Question 15.** The fate of acetyl CoA differs according to how much ATP is present in the cell. Explain what happens to acetyl CoA when ATP is limited, and compare this to what happens when ATP is abundant. How do these processes help regulate metabolism?

**Question 16.** When a person consumes a packet of pure sugar and burns it for energy, where does the carbon in the sugar ultimately go? Is the same true of the carbons in fat molecules when a person loses weight?

## 6.5 During Photosynthesis, Light Energy Is Converted to Chemical Energy

Light energy is absorbed by chlorophyll and other pigments

Light absorption results in photochemical change

Reduction leads to ATP and NADPH formation

Photosynthesis, the conversion of light energy into carbohydrate bonds, is the ultimate source of energy for most living things. Two pathways constitute photosynthesis: the light reactions that generate ATP and NADPH, and the carbon-fixation reactions that use the products of the light reactions to produce carbohydrates. Light is a form of electromagnetic radiation, whose energy is inversely proportional to wavelength. Light behaves as particles called photons. Receptor molecules in photosynthetic organisms absorb photons at a particular energy level (wavelength). The energy of the photon excites an electron in the receptor, which then gives off the energy and returns to its ground state.

Pigments are molecules that absorb light in the visible range. The pigment chlorophyll absorbs light in the blue and red portions of the light spectrum and thus appears green. An absorption spectrum traces which wavelengths a molecule absorbs best. An action spectrum traces which wavelengths generate the most biological activity (in this case, ATP or carbohydrate synthesis). Pigments are located in light-harvesting complexes, found as part of the photosystems. All the molecules in a photosystem surround a reaction center. Chlorophyll at the reaction center absorbs a photon and becomes excited, and then the excited electron is transferred to an electron acceptor in a redox reaction.

In green plants there are two photosystems, photosystem I and photosystem II. An excited chlorophyll in photosystem II loses its electron and becomes unstable; the chlorophyll is reduced by electrons that come from breaking bonds in water. The electron is passed through a series of electron carriers to a final acceptor, generating a proton gradient that is used by the enzyme complex ATP synthase to synthesize ATP. When the chlorophyll in photosystem I is oxidized following excitation, it replaces its electron with one from the final acceptor in the electron transport system. Cyclic electron transport uses only photosystem I and produces only ATP to provide the extra ATP needed for generating carbohydrates in the Calvin cycle; electrons are handed off to the electron transport chain and the final electron acceptor is used to reduce the chlorophyll at the photosystem I reaction center, making this a cyclic pathway.

**Question 17.** Label the photosystems, pathways, reactants, and products in the diagram below. Include both electron pathways in your answer.

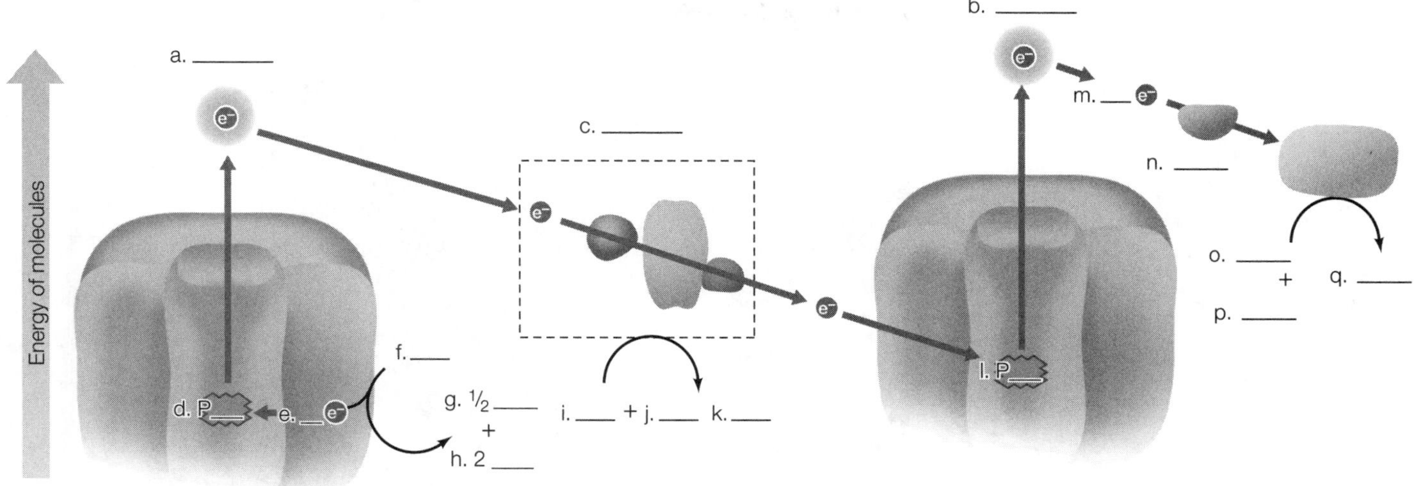


**Question 18.** Plants consume $CO_2$ and give off $O_2$. How is this possible if plants must also undergo cellular respiration?

**Question 19.** Explain the differences between cyclic and noncyclic electron flow. Why are both processes necessary?

**Question 20.** How do accessory pigments enhance photosynthetic activity in plants?

**Question 21.** Why are plants green?

## 6.6 Photosynthetic Organisms Use Chemical Energy to Convert $CO_2$ to Carbohydrates

The Calvin cycle takes place in the chloroplast stroma and has three stages: carbon fixation, reduction of sugars, and RuBP regeneration. In the first stage, an enzyme called ribulose bisphosphate carboxylase/oxygenase (rubisco) uses energy in the 5-carbon RuBP molecule to incorporate a molecule of $CO_2$. This reaction yields two 3-carbon molecules. The second stage reduces the 3-carbon molecules, yielding a sugar precursor called glyceraldehyde 3-phosphate (G3P). Approximately one-sixth of the G3P product is used by the cell; either to synthesize other molecules or catabolized as an energy source. The remaining G3P is recycled into RuBP. Autotrophs (including photosynthetic organisms) provide chemical energy for heterotrophs. In this way, the light energy harvested in photosynthesis is used to feed most organisms living on the planet.

**Question 22.** Label the stages, reactants, and products in the diagram below. Which enzyme is most important in this system?

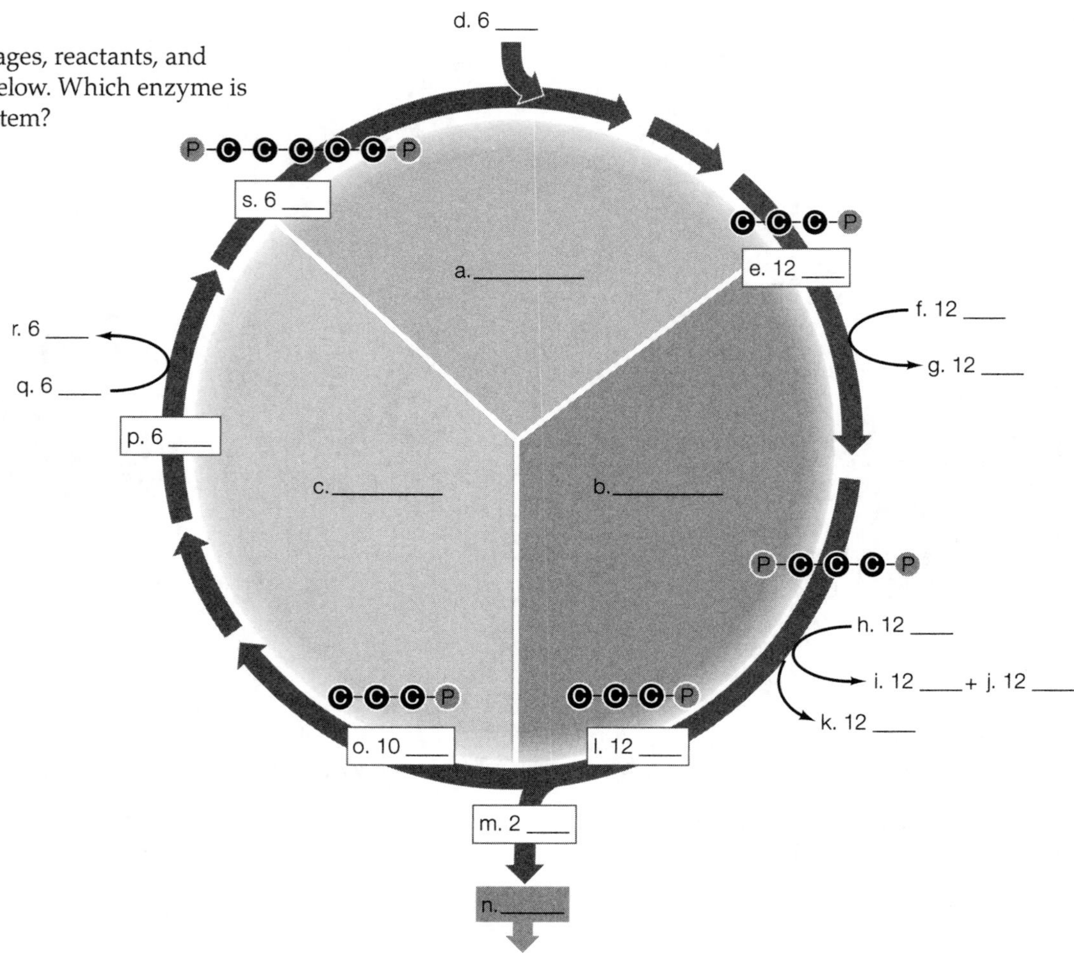


**Question 23.** Label the cycles, pathways, reactants, and products in the diagram below. As you work, focus on the connections between the two systems. Note how the products of one system provide the raw materials for the other system.

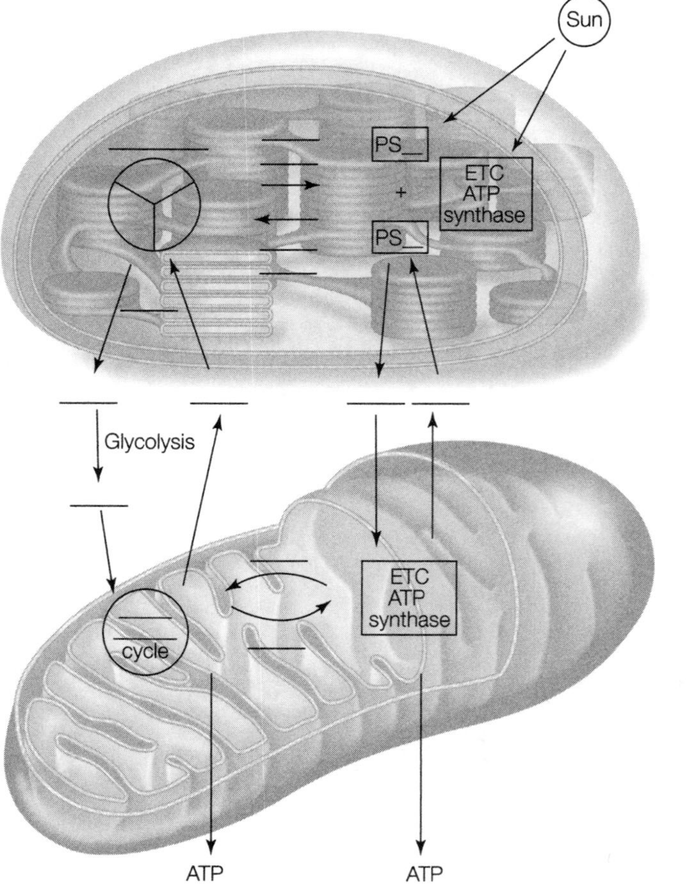

**Question 24.** Cellular respiration occurs simultaneously with many other cellular processes. Describe, in general, how cellular respiration interacts with other cellular metabolic events in plant and animal cells.

**Question 25.** Why do plants undergo both photosynthesis and cellular respiration, even during the daytime? Why don't they simply use the ATP produced in the light reactions of photosynthesis to drive cellular processes?

**Question 26.** The Calvin cycle was once referred to as the "dark" reactions of photosynthesis. Why is this a misnomer?

## Test Yourself

1. Before ATP is split into ADP and $P_i$, it holds what type of energy?
   a. Potential
   b. Kinetic
   c. Entropic
   d. Enthalpic
   e. Physical
   ***Textbook Reference:*** *Concept 6.1 ATP and Reduced Coenzymes Play Important Roles in Biological Energy Metabolism*

2. ATP hydrolysis is
   a. endergonic.
   b. exergonic.
   c. chemoautotrophic.
   d. anabolic.
   e. endothermic.
   ***Textbook Reference:*** *Concept 6.1 ATP and Reduced Coenzymes Play Important Roles in Biological Energy Metabolism; ATP hydrolysis releases energy*

3. Which of the following cellular metabolic processes is active in all cells, regardless of the presence or the absence of oxygen?
   a. The citric acid cycle
   b. Electron transport
   c. Glycolysis
   d. Fermentation
   e. Pyruvate oxidation
   ***Textbook Reference:*** *Concept 6.3 Carbohydrate Catabolism in the Absence of Oxygen Releases a Small Amount of Energy*

4. Which statement regarding glycolysis is *false*?
   a. A 6-C sugar is broken down to two 3-C molecules.
   b. Two ATP molecules are consumed.
   c. Glycolysis requires oxygen.
   d. A net sum of two ATP molecules is generated.
   e. Glycolysis occurs in the cytosol.
   ***Textbook Reference:*** *Concept 6.2 Carbohydrate Catabolism in the Presence of Oxygen Releases a Large Amount of Energy; In glycolysis, glucose is partially oxidized and some energy is released*

5. During which process is most ATP generated in the cell?
   a. Glycolysis
   b. The citric acid cycle
   c. Electron transport coupled with chemiosmosis
   d. Fermentation
   e. Pyruvate oxidation
   ***Textbook Reference:*** *Concept 6.2 Carbohydrate Catabolism in the Presence of Oxygen Releases a Large Amount of Energy; Oxidative phosphorylation and chemiosmosis yield a lot of ATP*

6. Which of the following is a function of the electron transport chain?
   a. Cycling NADH back to $NAD^+$
   b. Using the intermediates from the citric acid cycle
   c. Breaking down pyruvate
   d. Increasing the number of protons in the mitochondrial matrix
   e. Consuming excess ATP
   ***Textbook Reference:*** *Concept 6.2 Carbohydrate Catabolism in the Presence of Oxygen Releases a Large Amount of Energy; Energy is transferred from NADH to ATP by oxidative phosphorylation*

7. Which statement best describes the role of the inner mitochondrial membrane?
   a. It acts as an anchor for the membrane-associated enzymes of cellular respiration.
   b. It allows for the establishment of a proton gradient.
   c. It separates the mitochondria's environment from that of the cytosol.
   d. It anchors enzymes and allows for the establishment of the proton gradient, but it is not involved in separating the contents of the mitochondria from the cytosol.
   e. It anchors enzymes, allows for the establishment of the proton gradient, and is involved in separating the contents of the mitochondria from the cytosol.
   ***Textbook Reference:*** *Concept 6.2 Carbohydrate Catabolism in the Presence of Oxygen Releases a Large Amount of Energy; Energy is transferred from NADH to ATP by oxidative phosphorylation*

8. In the following redox reaction, _______ is oxidized and _______ is reduced.

   Glyceraldehyde 3-phosphate (G3P) + $P_i$ → 1,3-Bisphosphoglycerate (BPG)
   a. G3P; $NAD^+$
   b. BPG; NADH
   c. G3P; NADH
   d. $NAD^+$; NADH
   e. BPG; $P_i$
   ***Textbook Reference:*** *Concept 6.1 ATP and Reduced Coenzymes Play Important Roles in Biological Energy Metabolism; Redox reactions transfer electrons and energy*

9. Which statement about redox reactions is true?
   a. Oxidizing agents accept electrons, reducing another molecule.
   b. Oxidizing agents donate electrons.
   c. A molecule that accepts electrons is said to be oxidized.
   d. A molecule that donates electrons is said to be reduced.
   e. Oxidizing agents accept electrons and are reduced in the process.

   ***Textbook Reference:*** *Concept 6.1 ATP and Reduced Coenzymes Play Important Roles in Biological Energy Metabolism; Redox reactions transfer electrons and energy*

10. Which of the following is matched correctly with its catabolic product?
    a. Polysaccharides – amino acids
    b. Lipids – glycerol and fatty acids
    c. Proteins – glucose
    d. Polysaccharides – glycerol and fatty acids
    e. Nucleic acids – monosaccharides

    ***Textbook Reference:*** *Concept 6.4 Catabolic and Anabolic Pathways Are Integrated; Catabolism and anabolism are linked*

11. Plants give off $O_2$ because
    a. $O_2$ results from the incorporation of $CO_2$ into sugars.
    b. they do not respire; they photosynthesize.
    c. water is the initial electron donor, leaving $O_2$ as a photosynthetic by-product.
    d. electrons moving down the electron chain bind to water, releasing $O_2$.
    e. $O_2$ is synthesized in the Calvin cycle.

    ***Textbook Reference:*** *Concept 6.5 During Photosynthesis, Light Energy Is Converted to Chemical Energy; Reduction leads to ATP and NADPH formation*

12. Which statement concerning the synthesis of ATP in the mitochondria is *false*?
    a. ATP synthesis cannot occur without the presence of ATP synthase.
    b. The proton motive force is the establishment of a charge and concentration gradient across the mitochondrial membrane.
    c. The proton motive force drives protons back across the membrane through channels established by the ATP synthase channel protein.
    d. The ATP synthase protein is composed of two units.
    e. The intermembrane space is more acidic than the mitochondrial matrix.

    ***Textbook Reference:*** *Concept 6.2 Carbohydrate Catabolism in the Presence of Oxygen Releases a Large Amount of Energy; Chemiosmosis uses the proton gradient to generate ATP*

13. The Calvin cycle results in the production of
    a. glucose.
    b. starch.
    c. rubisco.
    d. G3P.
    e. ATP.

    ***Textbook Reference:*** *Concept 6.6 Photosynthetic Organisms Use Chemical Energy to Convert $CO_2$ to Carbohydrates*

14. Which of the following molecules is *not* recycled and reused in cellular metabolism?
    a. ADP
    b. $NAD^+$
    c. FAD
    d. $P_i$
    e. Glucose

    ***Textbook Reference:*** *Concept 6.2 Carbohydrate Catabolism in the Presence of Oxygen Releases a Large Amount of Energy*

15. In fermentation, for each molecule of glucose, how many ATPs are synthesized?
    a. 0
    b. 1
    c. 2
    d. 3
    e. 4

    ***Textbook Reference:*** *Concept 6.3 Carbohydrate Catabolism in the Absence of Oxygen Releases a Small Amount of Energy*

16. The main function of photosynthesis is the
    a. consumption of $CO_2$.
    b. production of ATP.
    c. conversion of light energy to chemical energy.
    d. production of starch.
    e. production of $O_2$.

    ***Textbook Reference:*** *Concept 6.5 During Photosynthesis, Light Energy Is Converted to Chemical Energy*

17. Which of the following best represents the components necessary for photosynthesis?
    a. Mitochondria, accessory pigments, visible light, water, and $CO_2$
    b. Chloroplasts, accessory pigments, visible light, water, and $O_2$
    c. Mitochondria, chlorophyll, visible light, water, and $O_2$
    d. Chloroplasts, chlorophyll, visible light, water, and $CO_2$
    e. Chlorophyll, accessory pigments, visible light, water, and $O_2$

    ***Textbook Reference:*** *Concept 6.5 During Photosynthesis, Light Energy Is Converted to Chemical Energy; Light energy is absorbed by chlorophyll and other pigments*

18. Which of the following is *not* a characteristic of chlorophyll that makes it well-suited for the capture of light energy?

a. It is raised to an excited state by certain wavelengths of light.
b. In its excited state it gives off electrons.
c. Its structure allows it to attach to thylakoid membranes.
d. It can transfer absorbed energy to another molecule.
e. It is degraded when plants become dormant, leaving accessory pigments behind.

***Textbook Reference:*** *Concept 6.5 During Photosynthesis, Light Energy Is Converted to Chemical Energy; Light absorption results in photochemical change*

19. The main function of cellular respiration is the
a. conversion of energy stored in the chemical bonds of glucose to an energy form that the cell can use.
b. recovery of $NAD^+$ from NADPH.
c. conversion of kinetic to potential energy.
d. creation of energy in the cell.
e. elimination of excess glucose from the cell.

***Textbook Reference:*** *Concept 6.2 Carbohydrate Catabolism in the Presence of Oxygen Releases a Large Amount of Energy*

20. In plants, cyclic electron flow and noncyclic electron flow serve to
a. meet the ATP demands of the Calvin cycle.
b. produce excess NADPH.
c. synthesize equal amounts of ATP and NADPH in the chloroplast.
d. consume the products of the Calvin cycle.
e. produce $O_2$ for the atmosphere.

***Textbook Reference:*** *Concept 6.5 During Photosynthesis, Light Energy Is Converted to Chemical Energy; Reduction leads to ATP and NADPH formation*

21. Which statement about the light reactions of photosynthesis is true?
a. Photosystem I cannot operate independently of photosystem II.
b. Photosystems I and II are activated by different wavelengths of light.
c. Photosystems I and II transfer electrons and create proton equilibrium across the thylakoid membrane.
d. Photosystem I is more significant than photosystem II.
e. Oxygen gas is a product of photosystem I.

***Textbook Reference:*** *Concept 6.5 During Photosynthesis, Light Energy Is Converted to Chemical Energy; Reduction leads to ATP and NADPH formation*

22. ATP is produced during the light reactions via
a. $CO_2$ fixation.
b. chemiosmosis.
c. reduction of water.
d. glycolysis.
e. noncyclic electron flow from photosystem I.

***Textbook Reference:*** *Concept 6.5 During Photosynthesis, Light Energy Is Converted to Chemical Energy; Reduction leads to ATP and NADPH formation*

23. Because of the properties of chlorophyll, plants need adequate _______ light to grow properly.
a. green
b. blue and red
c. infrared
d. ultraviolet
e. blue and blue-green

***Textbook Reference:*** *Concept 6.5 During Photosynthesis, Light Energy Is Converted to Chemical Energy; Light energy is absorbed by chlorophyll and other pigments*

24. Which statement concerning the Calvin cycle is *false*?
a. Light energy is not required for the cycle to proceed.
b. $CO_2$ is assimilated into sugars.
c. RuBP is regenerated.
d. It makes use of energy stored in ATP and NADPH.
e. It takes place in the stroma of the chloroplast.

***Textbook Reference:*** *Concept 6.6 Photosynthetic Organisms Use Chemical Energy to Convert $CO_2$ to Carbohydrates*

25. Which of the following events initiates the Calvin cycle and results in the entire pathway being carried out under environmental conditions?
a. 3PG is reduced to G3P using ATP and NADPH + $H^+$.
b. RuBP is regenerated.
c. $CO_2$ and RuBP join, forming 3PG.
d. G3P is converted into glucose and fructose.
e. Any of the above; since it is a cycle, it can start at any point.

***Textbook Reference:*** *Concept 6.6 Photosynthetic Organisms Use Chemical Energy to Convert $CO_2$ to Carbohydrates*

26. Which of the following processes does *not* occur in the mitochondria of eukaryotic cells?
a. Fermentation
b. Oxidative phosphorylation
c. Citric acid cycle
d. Electron transport chain
e. Creation of a proton gradient

***Textbook Reference:*** *Concept 6.3 Carbohydrate Catabolism in the Absence of Oxygen Releases a Small Amount of Energy*

27. Which statement regarding the relationship between photosynthesis and cellular respiration in plants is true?
a. Photosynthesis occurs in specialized photosynthetic cells.
b. Cellular respiration occurs in specialized respiratory cells.
c. Both cellular respiration and photosynthesis occur in all cells of every plant.

d. Photosynthesis is limited to specialized plant cells and cellular respiration does not occur.
e. Cellular respiration does not occur in specialized photosynthetic cells.
***Textbook Reference:*** *Concept 6.4 Catabolic and Anabolic Pathways Are Integrated; ATP and reduced coenzymes link catabolism, anabolism, and photosynthesis*

28. Photosynthesis occurs
a. in all plant cells.
b. only in photosynthetic plant cells.
c. only in plant cells lacking mitochondria.
d. only in the stroma.
e. only on the thylakoid membrane.
***Textbook Reference:*** *Concept 6.6 Photosynthetic Organisms Use Chemical Energy to Convert $CO_2$ to Carbohydrates*

29. Activities such as amino acid synthesis and active transport in plant cells are powered by
a. the light reactions and carbon-fixation reactions of photosynthesis.
b. ATP from the light reactions of photosynthesis.
c. ATP from alcoholic fermentation.
d. ATP from glycolysis and cellular respiration.
e. ATP from lactic acid fermentation.
***Textbook Reference:*** *Concept 6.4 Catabolic and Anabolic Pathways Are Integrated; ATP and reduced coenzymes link catabolism, anabolism, and photosynthesis*

## Answers

### *Key Concept Review*

1. Products of reaction 1:

+ HO–P–O⁻ (Pi)

Products of reaction 2:

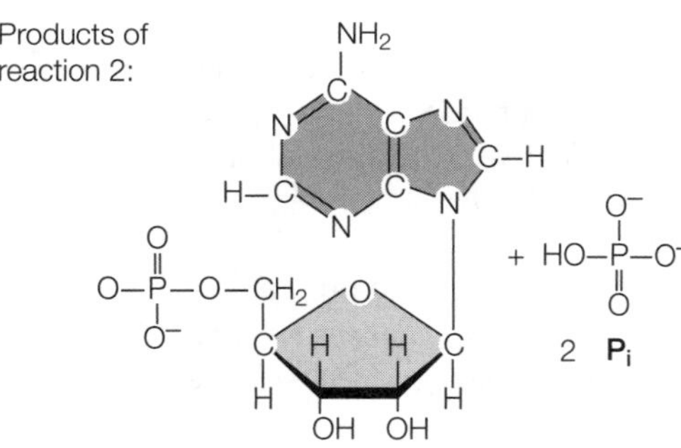


2. Recall that energy cannot be created or destroyed, but that it may be converted from one form to another. Many of these energy transformations use ATP as the energy carrier. As each molecule of ATP is hydrolyzed, some energy is converted to unusable heat energy. This phenomenon is not limited to ATP hydrolysis, but applies to every energy conversion reaction in the cell. Thus, there is a net loss of usable energy between different members of the food web.

3.

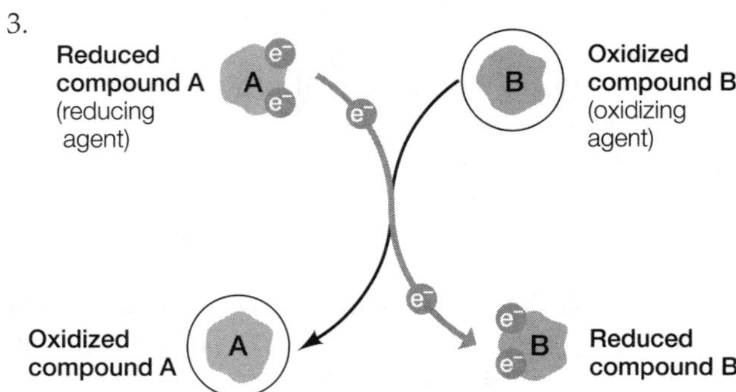


4. The proton motive force results in a concentration and charge gradient across the mitochondrial membrane. For that gradient to equalize, the protons must flow through a channel protein. If this channel protein has an associated ATP synthase, ATP is generated as protons flow through.

5.

1 Glucose + ~~4~~ ADP (becomes 2 ADP) + 2 $NAD^+$ + $H^+$ + 4 $P_i$ (becomes 2 $P_i$) + ~~2 ATP~~ → 2 Pyruvate + ~~4~~ ATP (becomes 2 ATP) + 2 NADH + 2 $H_2O$ + ~~2 ADP~~ + ~~2 $P_i$~~

Net Reaction:

1 Glucose + 2 ADP + 2 $NAD^+$ + $H^+$ + 2 $P_i$ → 2 Pyruvate + 2 ATP + 2 NADH + 2 $H_2O$

6.

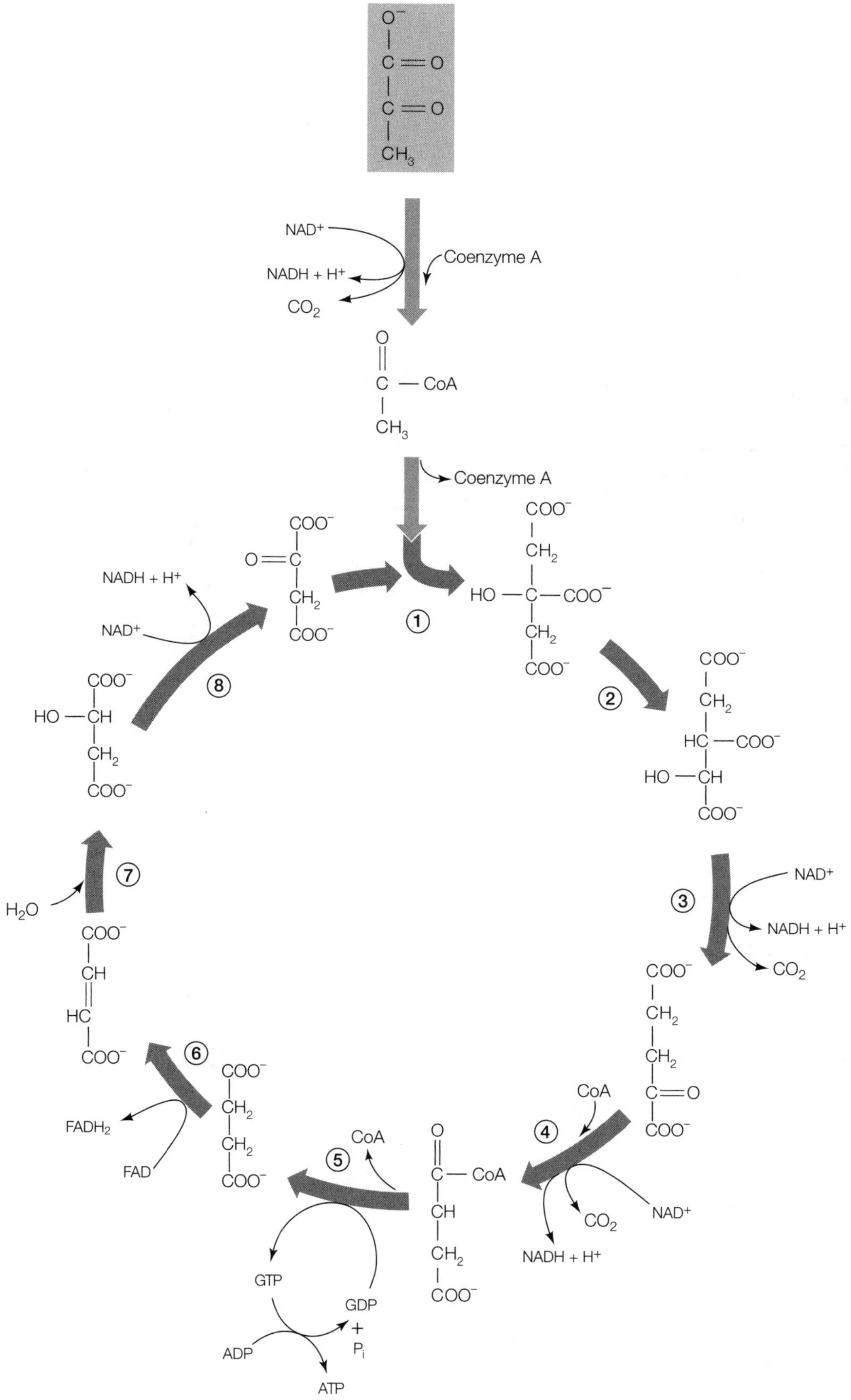

O⁻
C=O
C=O
CH3
NAD+
NADH + H+
CO2
Coenzyme A
O
C — CoA
CH3
Coenzyme A
COO⁻
O=C
CH2
COO⁻
1
COO⁻
CH2
HO — C — COO⁻
CH2
COO⁻
2
COO⁻
CH2
HC — COO⁻
HO — CH
COO⁻
3
NAD+
NADH + H+
CO2
COO⁻
CH2
CH2
C=O
COO⁻
4
CoA
NAD+
CO2
NADH + H+
O
C — CoA
CH2
CH2
COO⁻
5
CoA
GTP
GDP
+
Pi
ADP
ATP
COO⁻
CH2
CH2
COO⁻
6
FADH2
FAD
COO⁻
CH
HC
COO⁻
7
H2O
COO⁻
HO — CH
CH2
COO⁻
8
NADH + H+
NAD+

7.

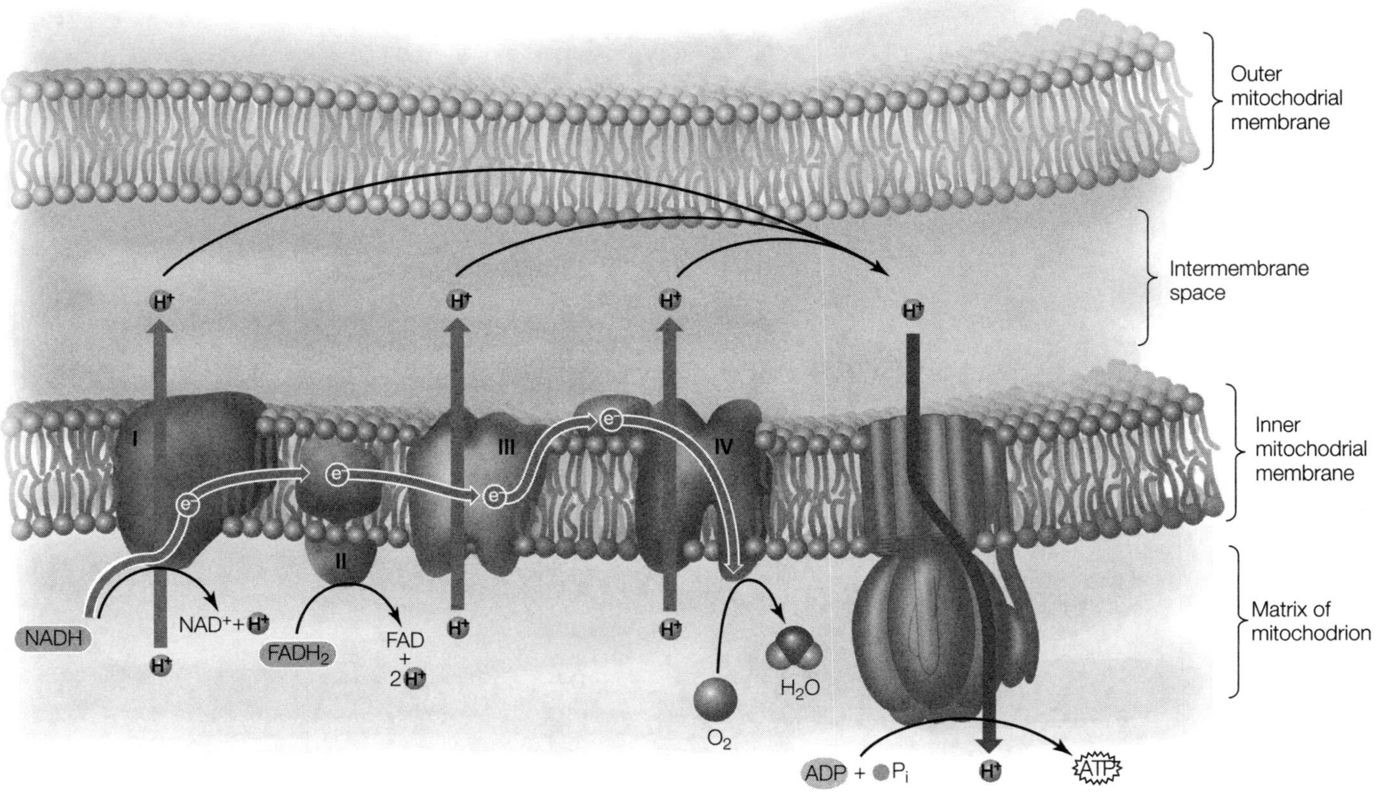


8. Carbon dioxide is liberated when pyruvate is oxidized to acetate (see In-text Art, p. 110) and in the citric acid cycle (Figure 6.8).

9. Oxygen acts as the terminal electron acceptor in the electron transport pathway. Without it, NADH + H$^+$ cannot be cycled back to NAD$^+$. The accumulated NADH + H$^+$ acts as an inhibitor to the citric acid cycle and effectively shuts it down. Therefore, in the absence of oxygen, a cell can undergo glycolysis only.

10. The reduced efficiency of glycolysis and fermentation for ATP synthesis is one reason that oxygen deprivation is so deadly to humans. (And, as later chapters will show, the resulting lactic acid buildup also causes problems.) In addition, not all cells are capable of carrying out the reactions of fermentation. Brain cells, for example, will simply die in the absence of oxygen.

11. The bubbles in beer are bubbles of $CO_2$ released during the fermentation of pyruvate into ethyl alcohol (see Figure 6.12).

12. In the absence of oxygen, pyruvate is either reduced to lactate (in lactic acid fermentation) or metabolized, and its metabolites are reduced to ethyl alcohol (in alcoholic fermentation). In either case, NADH + H$^+$ is the reducing agent, and it is oxidized back to NAD$^+$ in the process. The two molecules of ATP are used to power cellular activities. If NADH + H$^+$ were not oxidized to NAD$^+$, there would eventually be no NAD$^+$ available for glycolysis.

13. Fermentation yields only 2 ATP per glucose. Cellular respiration yields 32 ATP per glucose.

14.

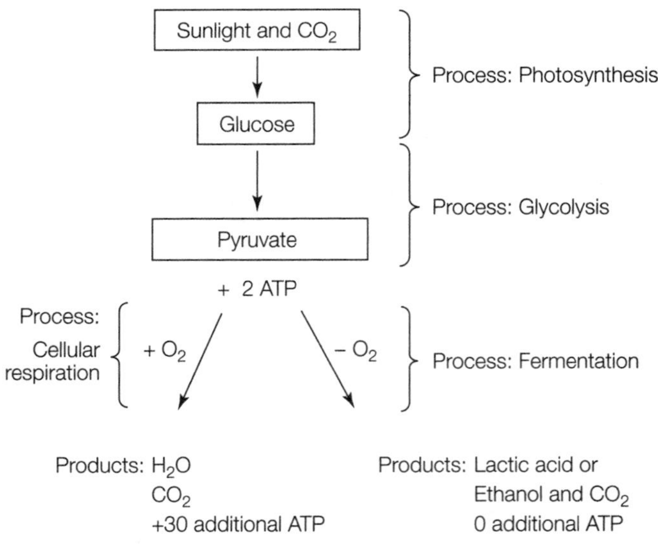


15. If ATP is limited, acetyl CoA enters the citric acid cycle and cellular respiration utilizes it to produce ATP. If ATP is abundant, acetyl CoA is shunted to fatty acid

synthesis, thus storing the energy in chemical bonds. This enables the cell to store energy for times when energy sources are abundant and consume the stored energy when external sources are limited.

16. The carbon in the sugar is exhaled in the form of $CO_2$. When a person loses weight, the carbon from the fat molecules is also exhaled in the form of $CO_2$.

17.
  a. Photosystem II
  b. Photosystem I
  c. Electron transport
  d. 680
  e. 2
  f. $H_2O$
  g. $O_2$
  h. $H^+$
  i. ADP
  j. $P_i$
  k. ATP
  l. 700
  m. 2
  n. Electron carrier
  o. $NADP^+$
  p. $H^+$
  q. NADPH

18. Cellular respiration occurs in all living plant cells. Therefore, all living cells consume $O_2$. Photosynthesis occurs in specialized cells that consume both $CO_2$ and $O_2$. Because atmospheric $O_2$ levels are high, excess $O_2$ is available for plants to utilize; therefore, $O_2$ continues to be emitted from plants.

19. Noncyclic electron flow involves both photosystems I and II. It results in the synthesis of equal amounts of ATP and NADPH. However, more ATP than NADPH is required for the Calvin cycle. To provide the additional ATP, photosystem I sends electrons to the electron carrier in the electron transport chain driving ATP synthesis. This cyclic pathway provides the necessary ATP for the Calvin cycle to regenerate RuBP.

20. Accessory pigments allow utilization of light in many wavelengths of the visible spectrum that could not be used by chlorophyll alone. The energy absorbed is channeled to the reaction centers of the photosystems.

21. The primary pigments in plants are chlorophylls. Chlorophylls absorb blue and orange-red wavelengths of light and reflect green light, thus making plants appear green. See the absorption and action spectra of chlorophyll in Figure 6.17.

22. The most important enzyme in this system is rubisco, which is active in the first stage.
  a. Carbon fixation
  b. Reduction and sugar production
  c. Regeneration of RuBP
  d. $CO_2$
  e. 3PG
  f. ATP
  g. ADP
  h. NADPH
  i. $NADP^+$
  j. $H^+$
  k. $P_i$
  l. G3P
  m. G3P
  n. Sugars
  o. G3P
  p. RuMP
  q. ATP
  r. ADP
  s. RuBP

23.

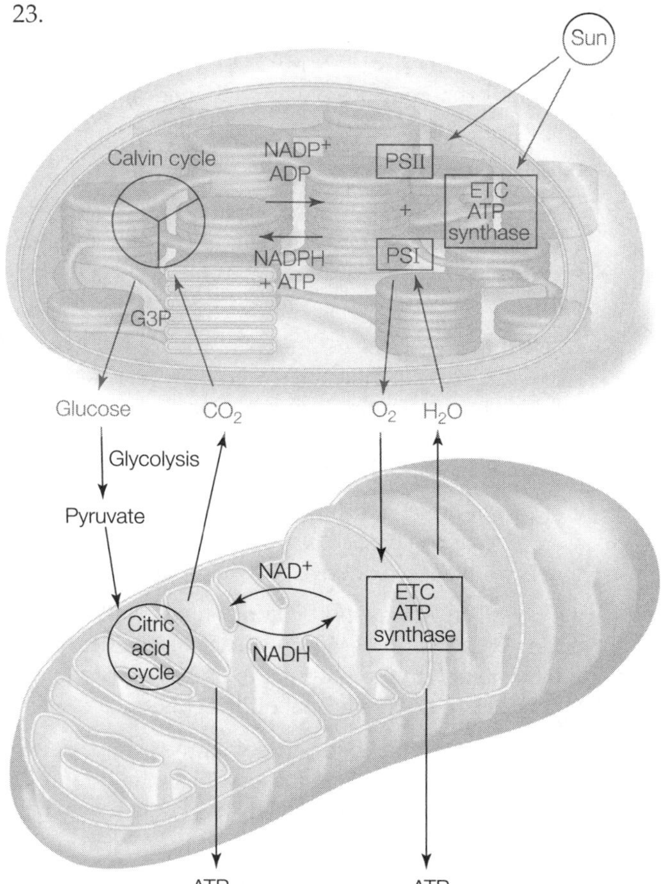


24. Consult Figure 6.14 to see where different metabolic pathways in the cell interact. It is crucial to remember that photosynthetic plant cells contain both chloroplasts and mitochondria. Living nonphotosynthetic plant cells have only mitochondria and lack chloroplasts. Animals, in contrast, have only mitochondria. The pathways seen in Figure 6.14 are common to organisms in both kingdoms.

25. The light reactions of photosynthesis produce ATP. ATP cannot be stored for use later (such as when light is not available) or transported to nonphotosynthetic cells. Therefore, there has to be a mechanism for this energy to be stored in a more stable form. The Calvin cycle stores it in the chemical bonds of G3P, which can be incorporated into carbohydrates for longer-term storage. When the cell needs energy for cellular processes, the carbohydrate is processed through the catabolic pathways (glycolysis, cellular respiration) previously discussed.

26. Light is required for both the light reactions of photosynthesis and the Calvin cycle. The Calvin cycle depends on the ATP generated during the light-dependent reactions.

### *Test Yourself*

1. **a.** Potential energy is energy held within chemical bonds that may be converted to working kinetic energy.
2. **b.** ATP hydrolysis is exergonic.
3. **c.** Glycolysis proceeds during both fermentation and cellular respiration. Only in cellular respiration is oxygen needed as the terminal electron acceptor of the pathway.
4. **c.** During glycolysis, 6-C glucose is broken down into two 3-C pyruvate molecules. In the process, four total ATP molecules are produced, but two are consumed, leaving a net production of two ATP molecules. No oxygen is required in glycolysis.
5. **c.** Most of the ATP produced during cellular respiration is produced by electron transport and chemiosmosis coupled in oxidative phosphorylation.
6. **a.** The electron transport chain is responsible for oxidizing NADH + $H^+$ + ½ $O_2$ → $NAD^+$ + $H_2O$. Protons are pumped into the intermembrane space of the mitochondrion, not the mitochondrial matrix.
7. **e.** The inner mitochondrial membrane is necessary for the anchoring of proteins as well as the establishment of a barrier across which a gradient can be established.
8. **a.** A molecule is oxidized when it loses electrons or protons and is reduced when it gains electrons or protons. In this reaction, G3P donates electrons and therefore is oxidized, while $NAD^+$ accepts them and thus is reduced.
9. **e.** Oxidizing agents accept electrons and cause oxidation of another molecule. Reducing agents donate electrons and cause the reduction of another molecule.
10. **b.** Lipids are broken down into glycerol and fatty acids; polysaccharides are broken down into glucose; proteins are broken down into amino acids. Nitrogenous bases from nucleic acids are converted into some amino acids and fed into the citric acid cycle.
11. **c.** Water is split at photosystem II to donate electrons to the reaction center. The resulting protons are moved across the membrane to establish the proton gradient, and $O_2$ is given off as a by-product.
12. **a.** Substrate-level phosphorylation occurs in step 5 of the citric acid cycle in the mitochondria, also producing ATP.
13. **d.** The Calvin cycle produces only G3P, which can then be metabolized into storage products, such as sugars and starch.
14. **e.** ADP, $NAD^+$, FAD, and $P_i$ are all recycled and reused in the process of cellular respiration.
15. **a.** Fermentation regenerates $NAD^+$ so that glycolysis can continue.
16. **c.** Photosynthetic organisms, including but not limited to plants, are the only life forms capable of trapping light energy and converting it to chemical energy by synthesizing simple carbohydrates. Because of this they form the basis of many of Earth's food webs.
17. **d.** Chloroplasts are the site of the photosynthetic reactions; chlorophyll is excited by photons of light and serves as a reaction center for the photosystems; visible light is necessary to excite chlorophyll and accessory pigments; water is the initial electron donor for the pathway; and $CO_2$ is necessary to make precursor molecules for energy storage.
18. **e.** The "tails" of chlorophyll molecules are associated with the thylakoid membranes of the chloroplasts. This close membrane association assists with establishing the proton motive force that will drive ATP synthesis. When excited by light, the chlorophyll moves into an excited state and passes electrons to acceptor molecules. This begins to set up the proton gradient across the membrane that will drive ATP synthesis.
19. **a.** Cellular respiration is the cell's way of converting potential energy in the chemical bonds of glucose to potential energy that the cell can use to perform other reactions.
20. **a.** More ATP than NADPH + $H^+$ is required in the Calvin cycle; therefore there must be a mechanism for producing more of it. Cyclic electron flow provides that mechanism. If noncyclic electron flow were to be sped up to meet ATP needs, an excess of NADPH would result. Shifting between cyclic and noncyclic flow balances ATP/NADPH ratios. Oxygen gas is a by-product of the light reactions, but its production is not the purpose of the reactions.
21. **b.** Photosystems I and II operate depending on whether electron flow is cyclic or noncyclic. The ATP levels in the chloroplast control activity. Compared to photosystem I, photosystem II is activated by light of a higher energy level. Both photosystems transfer electrons and create proton gradients across the thylakoid membranes; photosystem I does this via the cyclic pathway. Water is split by a structure embedded in the photosystem II complex.
22. **b.** In the light reactions, ATP synthesis occurs when protons flow through an ATP synthase channel protein in the thylakoid membrane. This is a chemiosmotic process. Photosystem II always generates a proton gradient (and thus ATP), while photosystem I generates a proton gradient only through cyclic electron flow.
23. **b.** Chlorophyll and accessory pigments absorb light in the blue and red wavelengths of visible light. Green light is reflected; therefore, plants appear green. (Accessory pigments allow energy from additional wavelengths to be absorbed as well.)

24. **a.** Light energy is required for the Calvin cycle to proceed. ATP synthesis is dependent on light energy, and the Calvin cycle is dependent on ATP. The Calvin cycle takes place in the chloroplast stroma.

25. **c.** The first step of the Calvin cycle is the fixation of $CO_2$ into 3PG. This is the regulatory step, and it requires ATP and NADPH. While it is true that the Calvin cycle is a cycle, there is a net consumption of $CO_2$ for the purpose of building carbohydrates.

26. **a.** Fermentation occurs in the cytosol, whereas all the other processes occur in the mitochondria of eukaryotic cells.

27. **a.** Photosynthesis occurs only in plant cells that have the necessary structures, while cellular respiration occurs in every living plant cell that has mitochondria and $O_2$.

28. **b.** Photosynthesis is limited to photosynthetic plant cells. There are many plant cells that are not exposed to light or that lack chloroplasts; these cells rely on cellular respiration.

29. **d.** Plant cells have mitochondria and rely on the processes of glycolysis and cellular respiration to provide ATP for cellular activities. Photosynthesis converts light energy into potential energy stored in chemical form, but then the cells must make the energy usable. Plant cells release this stored energy via glycolysis and cellular respiration.

# 7 The Cell Cycle and Cell Division

## The Big Picture

- Cellular reproduction is a controlled process that ensures that the daughter cells receive a complete set of genetic instructions. Prokaryotic cells reproduce by the asexual process of binary fission. Single-celled eukaryotes and some multicellular organisms reproduce by the asexual process of mitosis. In mitosis, replicated chromosomes are segregated by one cell division into two daughter cells that are genetically identical to the parent cell. Mitosis is also used to build tissues and organs. Multicellular organisms can also sexually reproduce via meiosis. In meiosis, the replicated chromosomes are segregated by two cell divisions to generate four haploid cells (gametes). Two gametes fuse to form a new diploid organism. During meiosis, crossing over between homologous chromatids and random segregation of those chromatids create genetically diverse gametes.
- Cell reproduction is tightly controlled by regulatory molecules. The cells of multicellular organisms undergo programmed cell death (apoptosis) when they are no longer needed or have accumulated DNA damage.

## Study Strategies

- Be sure that you understand the similarities and differences between mitosis and meiosis. Compare, side-by-side, mitosis, meiosis I, and meiosis II.
- It is helpful to work through the processes of mitosis and meiosis with various props. For instance, yarn of different colors, different flavors of licorice, or Play-Doh "snakes" can represent cells with at least two different chromosome pairs. The hands-on work will allow you to see more clearly what is taking place in the drawings in the book. After you understand the overall process, you can record it in sequential diagrams. Alternatively, you can use a digital camera to photograph your models in different stages, print them out, and then shuffle them. Practice putting the images into the proper order and explain the purpose of each phase of mitosis and meiosis. Once you understand how cell division in mitosis and meiosis works, explore the processes of crossing over and nondisjunction in order to see how these influence genetic composition of daughter cells.
- It also might be helpful to work with a study partner. For instance, you can diagram the starting cell, some of the intermediate steps, and the resulting cells and then challenge your partner to fill in the missing steps. You and your study partner can also take turns introducing different errors in cell division and challenging each other to explain how these errors will influence the genetic composition of the daughter cells.
- Draw a picture of the cell cycle and indicate the points at which the cyclin–CDK complexes act during that cycle. Use paper cups to represent CDK. Replicate the synthesis of cyclin by adding candies to the cups. Draw a large circle and label the stages of the cell cycle. Then place the cyclin–CDK complexes in their proper places on the circle. What would happen if you were unable to degrade cyclin (cup stays full) or unable to synthesize cyclin (cup stays empty)? Repeat these activities with a study partner so you can practice breaking the cell cycle in different ways while you challenge each other to explain the outcomes.
- Always keep in mind that the "stages" of the cell cycle, like the stages of mitosis and meiosis, are artificial labels used to help us understand the overall process. The cells do not lurch from stage to stage in incremental jumps! But thinking in terms of phases of the cycle allows us to identify key features of the overall process as it flows from beginning to end.
- Remember that programmed cell death is a beneficial process. List examples of how the sacrifice of individual cells aids in the development and/or survival of the multicellular organism.
- Go to **LaunchPad** (or use the Web addresses listed) to review the following additional resources:

  Animated Tutorial 7.1 Mitosis (PoL2e.com/at7.1)

  Animated Tutorial 7.2 Meiosis (PoL2e.com/at7.2)

  Activity 7.1 Sexual Life Cycle (PoL2e.com/ac7.1)

Activity 7.2 The Mitotic Spindle (PoL2e.com/ac7.2)
Activity 7.3 Images of Mitosis (PoL2e.com/ac7.3)
Activity 7.4 Images of Meiosis (PoL2e.com/ac7.4)
Analyze the Data in textbook Figure 7.9
Apply the Concept in textbook Sections 7.4 and 7.5

## Key Concept Review

### 7.1 Different Life Cycles Use Different Modes of Cell Reproduction

Asexual reproduction by binary fission or mitosis results in genetic constancy

Sexual reproduction by meiosis results in genetic diversity

Sexual life cycles are diverse

Asexual reproduction produces daughter cells that have the same genetic make-up as that of the parent cell. Prokaryotic cells reproduce by the asexual cell division process of binary fission. Single-celled eukaryotes and some multicellular organisms, including fungi and plants, can reproduce asexually through the similar process of mitosis. Mitosis gives rise to diploid (2n) daughter cells. Multicellular organisms also use mitosis to build tissues and organs. Any alterations in the DNA are due to rare occurrences of replication errors (mutations) that occur during DNA synthesis or due to environmental factors.

Multicellular eukaryotic organisms can also reproduce by the sexual process of meiosis. In meiosis, the replicated chromosomes undergo two cell divisions, resulting in four cells (gametes), each of which has only one set of chromosomes (haploid). Gametes do not share the identical DNA makeup of their parental cells, due to the combined processes of homologous recombination and independent assortment that occur during meiosis. Genetic diversity in offspring increases the probability of survival and provides the raw material for natural selection and evolution.

There are multiple versions of the sexual life cycle. Animals reproduce by means of a diplontic life cycle. In this cycle, the organism is diploid and the gametes are the only haploid stage. The gametes fuse to form a diploid, mature (multicellular) organism. Organisms that have a haplontic life cycle are haploid in their mature form and diploid only in their zygote stage. In a third system, called alternation of generations, both the haploid and diploid generations have multicellular stages.

**Question 1.** How does binary fission permit the evolution of new organisms when the benefit of genetic diversity from sexual reproduction is not possible?

**Question 2.** Consider cell division by a somatic cell and cell division by a cell that will produce gametes. Which of these cells will undergo mitotic cell division? Which cell will undergo meiotic cell division? Explain why each cell type undergoes its respective form of cell division.

**Question 3.** For each diagram below, identify the sexual life cycle strategy that is represented (diplontic, haplontic, or alternation of generations). Provide one example of an organism that demonstrates that life cycle strategy.

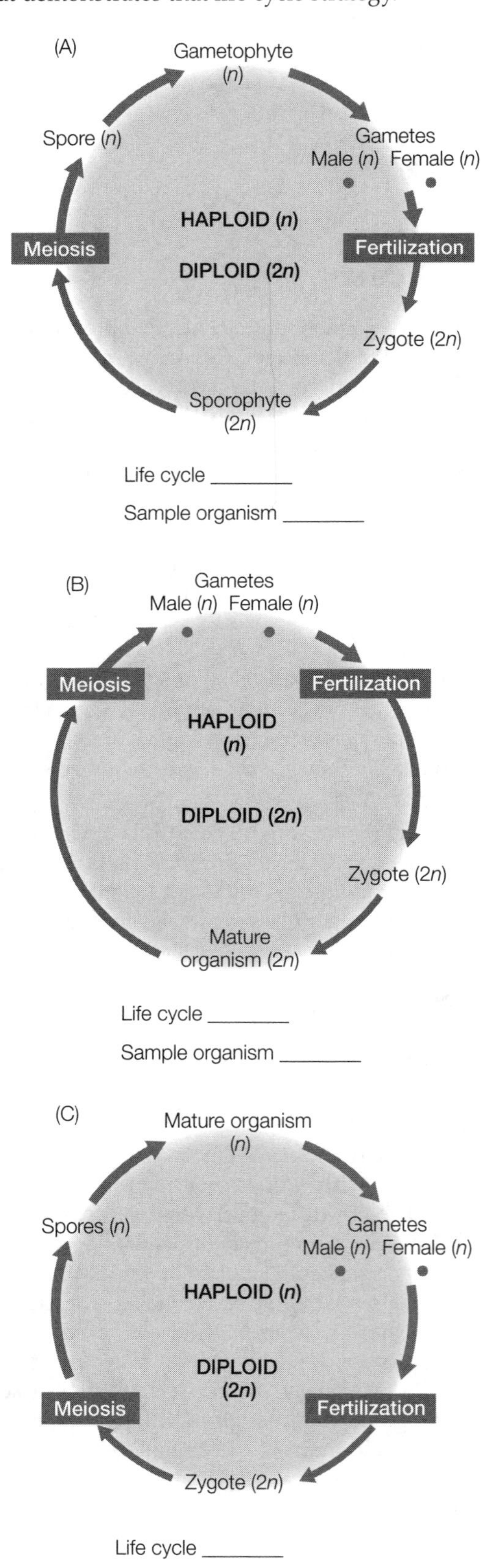

### 7.2 Both Binary Fission and Mitosis Produce Genetically Identical Cells

Prokaryotes divide by binary fission

Eukaryotic cells divide by mitosis followed by cytokinesis

Prophase sets the stage for DNA segregation

Chromosome separation and movement are highly organized

Cytokinesis is the division of the cytoplasm

Cell division in both prokaryotic and eukaryotic cells consists of three stages: DNA replication, DNA segregation, and cytokinesis. Prokaryotic cells divide by means of binary fission. Since the cell has only a single circular chromosome, there is no need to have a system to organize multiple chromosomes prior to cell division. Eukaryotic cells, by contrast, have multiple chromosomes, each of which needs to be replicated and distributed to the daughter cells. This is the reason why you see chromosomes aligning at the equatorial plate in mitosis (and meiosis). Mitosis is a more complex process than binary fission. It can be conceptualized in terms of stages, with each stage flowing directly into the next. In the interphase stage of the cell cycle most metabolic functions operate, and DNA is replicated in the S phase.

By the time the cell enters mitosis, the DNA is condensed, and replication would no longer even be possible. In mitosis, the division of the nucleus takes place in a number of stages: prophase, prometaphase, metaphase, anaphase, and telophase. In prophase, the DNA is packaged into a very condensed state. The kinetochores assemble on the centromeres and the spindle fibers form. In prometaphase, the nuclear envelope breaks down and the chromosomes attach to the spindles extending to each pole. In metaphase, the chromosomes line up at the midline in preparation for moving away from each other in anaphase. In telophase, the nuclear envelope re-forms around the nucleus and the process of nuclear division is terminated.

Telophase is followed by cytokinesis, which is the division of the cytoplasm and the organelles it contains. In animal cells, a contractile ring composed of microfilaments "pinches" the cell in half. In plant cells, vesicles form the boundary between the two daughter nuclei and generate both the cell membrane and the cell wall.

**Question 4.** How does the process of cell division differ in prokaryotic cells and animal cells? Which elements are the same?

**Question 5.** Diagram normal mitosis in a diploid cell ($n = 2$). Use either colors or sizes to differentiate between homologous and nonhomologous chromosomes.

**Question 6.** Describe how two meters of DNA in a typical human cell can fit into the nucleus, which is 5 μm in diameter.

**Question 7.** How does cytokinesis differ in animal and plant cells?

### 7.3 Cell Reproduction Is Under Precise Control

The eukaryotic cell division cycle is regulated internally

The cell cycle is controlled by cyclin-dependent kinases

Cell reproduction is a process that is carefully controlled and includes multiple regulatory molecules and checkpoints. Growth factors are required for a cell to pass from G1 into S phase, and proteins within the cell, such as the cyclin-dependent kinases (CDKs), regulate transitions from one phase to the next in the cell cycle.

CDKs are allosterically activated when they are bound by a cyclin protein. Different CDK molecules regulate different checkpoints of the cell cycle. For example, the G1-S cyclin–CDK catalyzes the phosphorylation of the retinoblastoma protein (RB). This inactivates RB's function as an inhibitor of the transition from G2 phase into S phase, thereby allowing DNA replication to proceed.

**Question 8.** What would happen to a cell if it were unable to synthesize cyclin (as described in the textbook)?

**Question 9.** In the cell cycle shown in the diagram below, where is the checkpoint (restriction point) that determines whether the cell will proceed from G1 to S? Label G1, S, and G2 phases and the restriction point. Then add CDK and cyclin to the diagram in the appropriate places and indicate if they are present in an intact or degraded form.

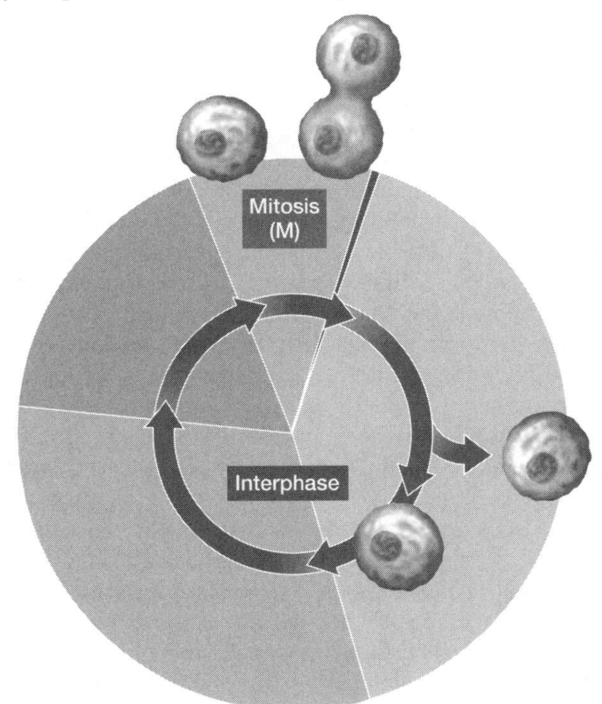

### 7.4 Meiosis Halves the Nuclear Chromosome Content and Generates Diversity

Meiotic division reduces the chromosome number

Crossing over and independent assortment generate diversity

Meiotic errors lead to abnormal chromosome structures and numbers

To generate haploid gametes, two sequential nuclear divisions (referred to as meiosis I and II) must occur during meiosis. The first round of nuclear divisions separates homologous chromosomes, while the second round separates sister chromatids. As in mitosis, DNA replication occurs in the S phase of interphase and the DNA is replicated a single time.

In meiosis I, the replicated homologous chromosomes pair up at the equator. The pairing of homologous chromosomes results in a tetrad. The chromosomes in the tetrads are aligned along their lengths in a process referred to as synapsis. They exchange material via crossing over, which results in the formation of recombinant chromatids. This is one source of genetic diversity in the next generation. A second source of genetic diversity during meiosis is the phenomenon known as independent assortment. In essence, each tetrad separates independently of the other tetrads. Because of crossing over, in each gamete no chromosome is a match to the parent chromosome, and because of independent assortment, no gamete has a chromosome assortment identical to a parent. These are key sources of genetic diversity in sexually reproducing organisms.

Meiosis, like mitosis, is tightly regulated, but errors are possible. One type of error is a result of chromosome breakage and incorrect repair, resulting in the movement of a genetic sequence to a different location on a chromosome or even to a different chromosome altogether. This is called translocation. A second error can occur when homologous chromosomes fail to separate (nondisjunction) in meiosis I or II; the gametes will have either two copies or zero copies of that chromosome. Many of these irregularities are fatal, but there are examples in which the zygote survives with the extra or missing chromosome (aneuploidy). Trisomy 21 in humans is not lethal like many other aneuploidy cases, but it does result in phenotypic changes compared to diploid 21 individuals; trisomy 21 is referred to as Down syndrome.

Polyploid organisms may have three, four, or even more sets of homologous chromosomes. As long as the ploidy number is even, meiosis can continue normally and the organism is fertile. However, a triploid nucleus cannot undergo normal meiosis because one-third of the chromosomes would lack partners; the organism would be sterile. In plants, evenly polyploid crops are more robust and may produce larger fruits or flowers, while oddly polyploid crops are seedless.

**Question 10.** A cell at the start of meiosis has two $2n$ chromosomes, representing the replicated DNA of a diploid cell. If the cell has 1 microgram of DNA in G1, it would have 2 micrograms of DNA at the start of meiosis. At the end of meiosis I, there is 1 microgram of DNA in a cell, but the cell is not diploid. Why not?

**Question 11.** Starting with a diploid cell ($n = 2$), diagram normal meiosis. How is this process different from mitosis?

**Question 12.** Starting with a diploid cell ($n = 2$), diagram meiosis with a nondisjunction in meiosis I, and then meiosis with a nondisjunction in meiosis II. Label the gametes that would result from each meiotic event. Describe the types of zygotes that would be produced from these gametes when they are fertilized by a normal gamete.

**Question 13.** Describe two ways in which the genetic diversity of organisms is increased during meiosis.

**Question 14.** Suppose that by using a chemical that inhibits cytokinesis, you have created peaches that are tetraploid. How many sets of chromosomes do these peaches have? What will be the ploidy of the gametes?

**Question 15.** Are the tetraploid peaches described in Question 14 fertile? Would triploid peaches be fertile? Explain.

### 7.5 Programmed Cell Death Is a Necessary Process in Living Organisms

Cell death can occur in cells in two ways: by necrosis (cells are damaged by toxins or starved of oxygen or essential nutrients) or by apoptosis (programmed cell death). Apoptosis is a normal process during development that eliminates unneeded tissue, such as the webbing between the fingers of a developing human fetus. Apoptosis also occurs in cells that have acquired genetic damage, as this can negatively affect function and possibly lead to the development of cancer.

Signals for cell death include a lack of mitotic signals, recognition of DNA damage, or changes in a receptor protein that activate signal transduction. This in turn leads to the activation of caspases, which are proteases that hydrolyze target molecules leading to cell death.

**Question 16.** Describe a situation in which apoptosis plays an important role during normal human embryogenesis.

**Question 17.** Describe how a cell exposed to radiation for a prolonged period of time might trigger and proceed with apoptosis.

## Test Yourself

1. Which statement about mitosis is *false*?
   a. It is followed by cytokinesis.
   b. DNA replication is completed prior to the beginning of this phase.
   c. The chromosome number of the resulting cells is the same as that of the parent cell.
   d. The daughter cells are usually genetically identical to the parental cell.
   e. Homologous chromosomes align at the equatorial plate.

   ***Textbook Reference:*** *Concept 7.2 Both Binary Fission and Mitosis Produce Genetically Identical Cells; Eukaryotic cells divide by mitosis followed by cytokinesis*

2. Which statement about meiosis is true?
   a. The chromosome number in the resulting cells is halved.
   b. DNA replication occurs before meiosis I and again before meiosis II.
   c. The homologs pair during prophase II.
   d. The daughter cells are genetically identical to the parent cell.
   e. The chromosome number of the resulting cells is the same as that of the parent cell.

   ***Textbook Reference:*** *Concept 7.4 Meiosis Halves the Nuclear Chromosome Content and Generates Diversity; Meiotic division reduces the chromosome number*

3. Which statement about kinetochores on mitotic chromosomes is *false?*
   a. They are located at the centromere of each chromosome.
   b. They are the sites where microtubules attach to separate the chromosomes.
   c. They are organized so that there are two per sister chromatid.
   d. Kinetochore microtubules from opposite poles attach to each sister chromatid.
   e. They contain molecular motor proteins.

   ***Textbook Reference:*** *Concept 7.2 Both Binary Fission and Mitosis Produce Genetically Identical Cells; Prophase sets the stage for DNA segregation*

4. Which statement about the mitotic spindle is true?
   a. It is composed of polar and kinetochore microtubules, both of which attach to chromosomes.
   b. It is composed of actin and myosin microfilaments.
   c. It is composed of kinetochores at the metaphase plate.
   d. It is composed of microtubules, which help separate the chromosomes to opposite poles of the cell.
   e. It originates only at the centrioles in the centrosomes.

   ***Textbook Reference:*** *Concept 7.2 Both Binary Fission and Mitosis Produce Genetically Identical Cells; Prophase sets the stage for DNA segregation*

5. Imagine that there is a mutation in the cyclin gene such that its gene product is nonfunctional. What kind of effect would this mutation have on a skin cell in the area of a cut?
   a. CDK would not be synthesized.
   b. There would be no effect, because skin cells do not replicate.
   c. The cell would be stuck in S phase and unable to replicate.
   d. The cell would not be able to enter G1.
   e. The cell would be unable to reproduce itself.

   ***Textbook Reference:*** *Concept 7.3 Cell Reproduction Is Under Precise Control; The cell cycle is controlled by cyclin-dependent kinases*

6. Imagine that there is a mutation in a CDK gene such that its gene product is present but nonfunctional. Based on what you know about CDK and protein RB interactions, what kind of effect would this mutation have on a mammalian cell that has received a growth signal?
   a. The cell would replicate its DNA and then fail to enter G2.
   b. The cell would remain in G1.
   c. The cell would skip S phase and enter G2.
   d. The cell would not be able to phosphorylate its associated cyclin.
   e. The cell would enter the cell cycle but not be able to undergo cytokinesis.

   ***Textbook Reference:*** *Concept 7.3 Cell Reproduction Is Under Precise Control; The cell cycle is controlled by cyclin-dependent kinases*

7. Which statement about DNA replication and cytokinesis in *E. coli* is true?
   a. DNA replication occurs in the nucleus.
   b. Cytokinesis is facilitated by microfilaments of actin and myosin.
   c. DNA replication occurs during the S phase of the cell cycle.
   d. Cell reproduction is initiated by reproductive signals, which lead to DNA replication, DNA segregation, and cytokinesis.
   e. The *E. coli* chromosome is linear.

   ***Textbook Reference:*** *Concept 7.2 Both Binary Fission and Mitosis Produce Genetically Identical Cells; Eukaryotic cells divide by mitosis followed by cytokinesis*

8. Which statement about chromatids is *false?*
   a. They are replicated chromosomes still joined together at the centromere.
   b. They are identical in mitotic chromosomes.
   c. They share one kinetochore.
   d. They separate during anaphase.
   e. They form during S phase.

   ***Textbook Reference:*** *Concept 7.2 Both Binary Fission and Mitosis Produce Genetically Identical Cells; Prophase sets the stage for DNA segregation*

9. Chromosome movement during anaphase is the result of all of the following *except*
   a. the hydrolysis of ATP by dynein.
   b. molecular motors at the kinetochores that move the chromosomes toward the poles.
   c. molecular motors at the centrosome that pull the microtubules toward the poles.
   d. shortening of the microtubules at the centrosome that pull the chromosomes toward the poles.
   e. the hydrolysis of ATP by kinesin.

   ***Textbook Reference:*** *Concept 7.2 Both Binary Fission and Mitosis Produce Genetically Identical Cells; Chromosome separation and movement are highly organized*

10. Programmed cell death (apoptosis)
    a. occurs in cells that have been deprived of essential nutrients.
    b. occurs only in cells that have damaged DNA.
    c. is signaled by the initiation of mitosis.
    d. is a natural process during development.
    e. is well controlled in cancer cells.

    ***Textbook Reference:*** *Concept 7.5 Programmed Cell Death Is a Necessary Process in Living Organisms*

11. If the *ori* site on the *E. coli* chromosome is deleted,
    a. nothing will happen.
    b. replication will not start.
    c. replication will start but not be able to continue.
    d. replication will be initiated at another *ori* site on the chromosome.
    e. the chromosome will be replicated but the cell will not be able to divide.

    ***Textbook Reference:*** *Concept 7.2 Both Binary Fission and Mitosis Produce Genetically Identical Cells; Prokaryotes divide by binary fission*

12. Which statement about chiasmata is true?
    a. They are sites where nonsister chromatids can exchange genetic material during meiosis and increase genetic variation in gametes.
    b. They are sites where sister chromatids can exchange genetic material during meiosis and increase genetic variation in gametes.
    c. They increase genetic variation among the gametes due to exchange between all members of the tetrad during meiosis.
    d. They increase genetic variation among the gametes due to exchange between all members of the tetrad during mitosis.
    e. They are sites where nonsister chromatids can exchange genetic material during mitosis and increase genetic variation in gametes.

    ***Textbook Reference:*** *Concept 7.4 Meiosis Halves the Nuclear Chromosome Content and Generates Diversity; Crossing over and independent assortment generate diversity*

13. The difference between asexual and sexual reproduction is that
    a. asexual reproduction results in an organism that is identical to the parent, whereas sexual reproduction results in an organism that is not identical to either parent.
    b. asexual reproduction results from meiosis, whereas sexual reproduction results from mitosis.
    c. asexual reproduction occurs only in bacteria, whereas sexual reproduction occurs in plants and animals.
    d. asexual reproduction results from the fusion of two gametes, whereas sexual reproduction produces clones of the parent organism.
    e. asexual reproduction occurs only in haplontic organisms, whereas sexual reproduction occurs only in diplontic organisms.

    ***Textbook Reference:*** *Concept 7.1 Different Life Cycles Use Different Modes of Cell Reproduction*

14. A chromatid is
    a. a chromosome before it has undergone DNA replication.
    b. one of the pairs of homologous chromosomes.
    c. a homologous chromosome.
    d. a newly replicated bacterial chromosome.
    e. one-half of a newly replicated eukaryotic chromosome.

    ***Textbook Reference:*** *Concept 7.2 Both Binary Fission and Mitosis Produce Genetically Identical Cells; Prophase sets the stage for DNA segregation*

15. A diploid cell in G1 has _______ number of micrograms of DNA as one of its daughter cells at the end of meiosis I. At the end of meiosis II it has _______ number of chromosomes as one of its daughter cells.
    a. twice the; the same
    b. one-half the; twice the
    c. the same; twice the
    d. one-half the; the same
    e. the same; one-half the

    ***Textbook Reference:*** *Concept 7.4 Meiosis Halves the Nuclear Chromosome Content and Generates Diversity; Meiotic division reduces the chromosome number*

## Answers

### *Key Concept Review*

1. Binary fission typically results in two daughter cells that are genetically identical to the parent cell. Errors in DNA replication, while rare, can introduce changes to the genetic code. These changes are then passed on in subsequent rounds of cell division.
2. Somatic cells undergo mitotic cell division for the growth and repair of tissues and organs. The daughter cells should contain identical genetic information, as they will have a tissue function in the multicellular organism. The cell that produces gametes will undergo meiosis to produce haploid daughter cells that can fuse with other haploid gametes to form diploid zygotes. Meiosis also provides genetic diversity so that the genetic constitution of the offspring is not identical to that of either parent. Genetic diversity increases the probability of survival.
3. 

   (A) Life cycle: Alternation of generations; Sample organism: Most plants, some fungi

   (B) Life cycle: Diplontic; Sample organism: Animals, brown algae, some fungi

(C) Life cycle: Haplontic; Sample organism: Most protists, fungi, some green algae

4. All cells divide by replicating their DNA, segregating the DNA, and then splitting the cytoplasm by cytokinesis. In most prokaryotic cells there is only one circular chromosome. As the cell enlarges to prepare for division, the newly replicated daughter chromosomes are separated at opposite sides of the cell. During fission, the cell membrane pinches in, and cell wall components are synthesized between the daughter cells. In animal cells, there are more chromosomes, and they are linear. The cell undergoes a sequential set of steps called the cell cycle, in which the chromosomes are replicated and then separated to opposite poles of the cell. The chromosomes are segregated equally into the daughter cells by means of microtubules, and actin filaments and myosin cause the cell membrane to form a contractile ring and separate to form two daughter cells.

5.

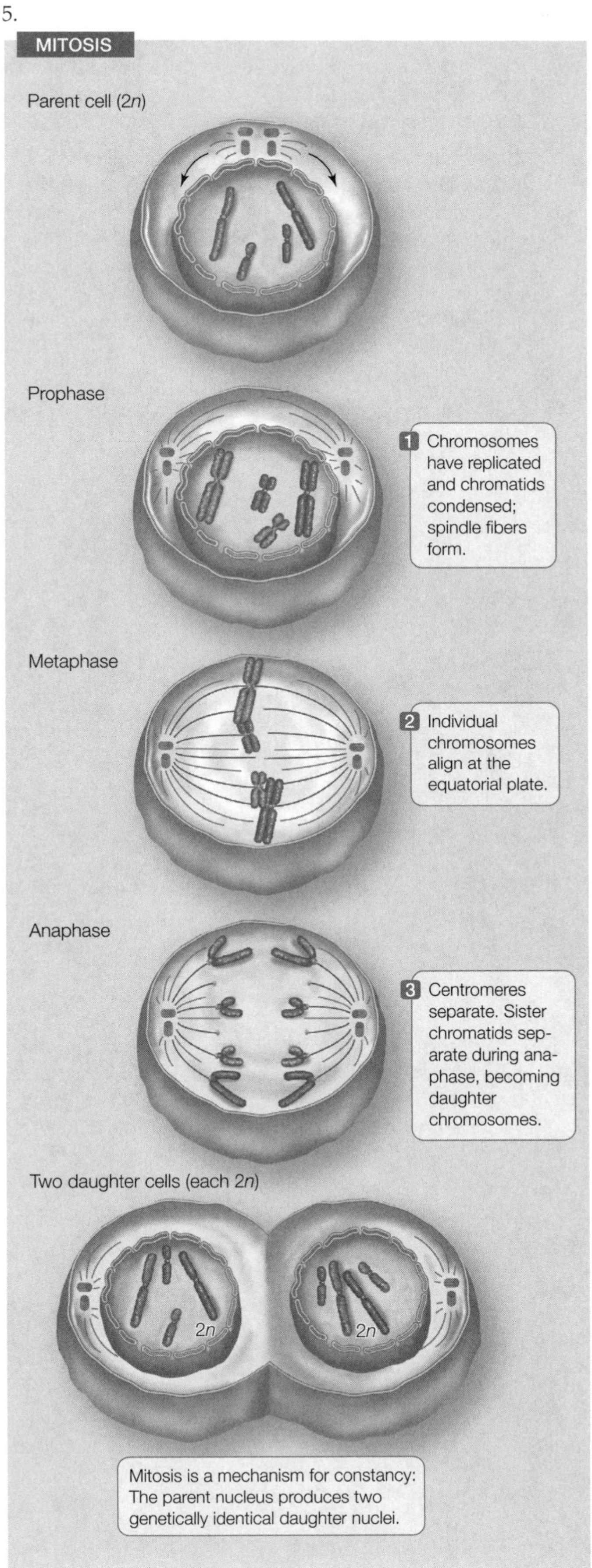

6. The DNA is very tightly condensed because it is packaged with proteins that stabilize the double helix in organized DNA coils. (See Figure 11.14 for an illustration of the first level of DNA packaging that results in the condensed chromosome seen in Figure 7.5).
7. In animal cells, cytokinesis results from the interaction of actin filaments and myosin, which causes the cell membrane to pinch in and divide the cytoplasm into two cells. In plant cells, a cell plate forms between the newly segregated chromosomes, and Golgi vesicles fuse at that site to form the new cell membranes. Cell wall components are then secreted between the plasma membranes to complete cytokinesis.
8. The cell would never be able to activate CDK. The active CDK is required for the cell to move past the G1/S checkpoint.
9.

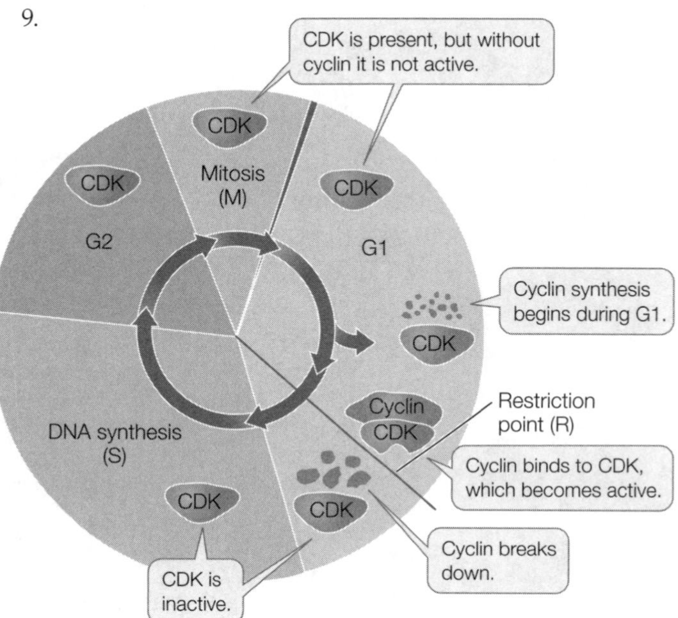


10. At the start of meiosis, the 2 micrograms of DNA are approximately divided among a total of $2n = 2$ (4 total, or 0.5 micrograms each) chromosomes. At the end of meiosis I, each of the cells has $n = 2$ chromosomes (2 total) because the homologous pairs have separated but the sister chromatids have not. Therefore, each cell has one-half the amount of DNA (1 microgram), but it is haploid, not diploid.
11. The difference between mitosis and meiosis is the number and ploidy of the daughter cells. In mitosis, cell division results in two diploid cells, whereas meiosis results in four haploid cells.

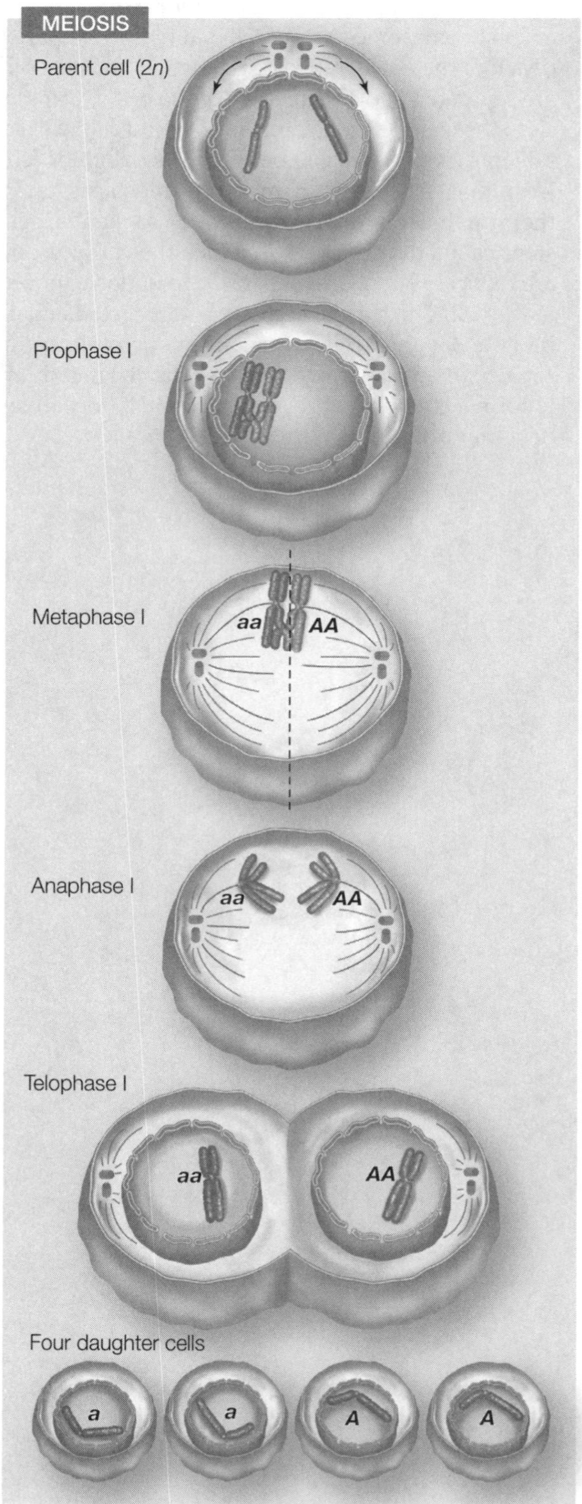

12.

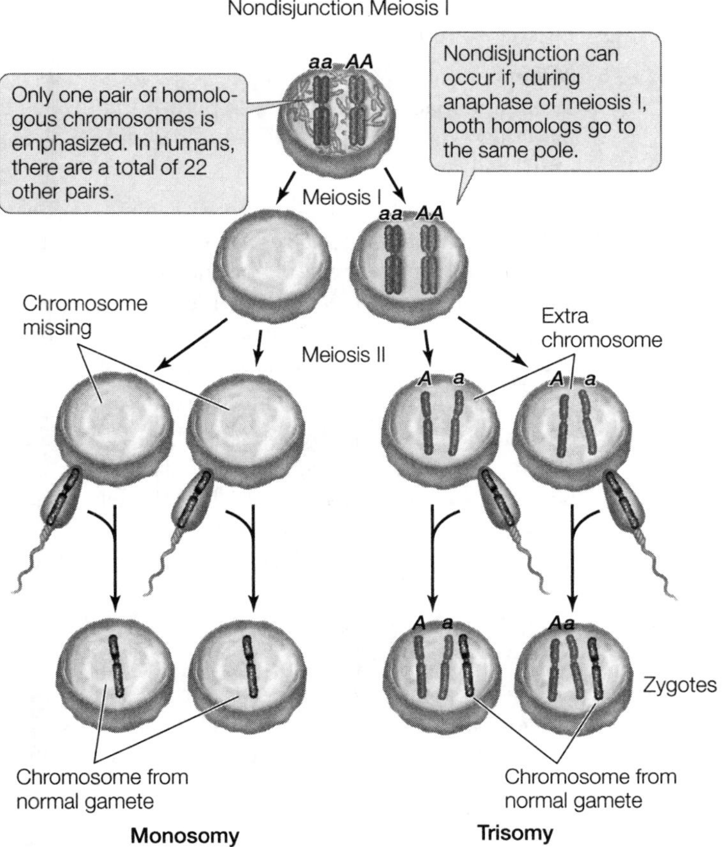


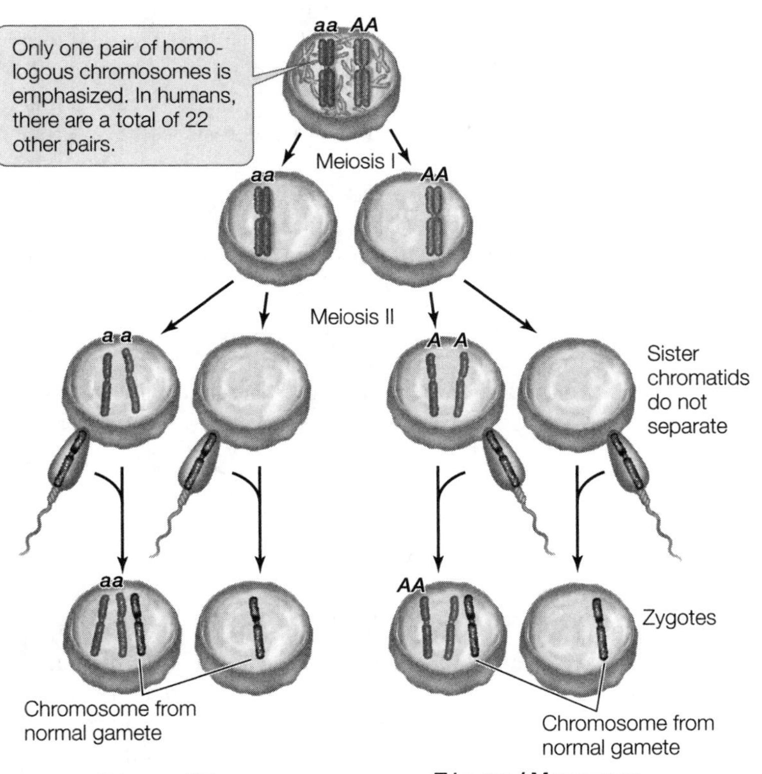


If a chromosome undergoes nondisjunction during meiosis I, the pairs of homologous chromosomes (homologs *A* and *a*) will fail to separate during anaphase. Consequently, one of the two nuclei will have both homologous pairs, and the other nuclei will have none. Following meiosis II, the cell with neither chromosomal pair will produce two gametes lacking the chromosome, whereas the cell containing both homologous pairs will produce two gametes that contain a copy of both homologs (*Aa*).

If the same chromosome separates properly during meiosis I but undergoes nondisjunction during meiosis II (in both cells), each cell will produce a gamete that has both members of the homologous pair (*Aa*), and a gamete that has neither member of the homologous pair.

For both scenarios, following fertilization of these aneuploidy gametes by a normal gamete, two zygotes will have only one copy of the chromosome (monosomy *A*/*a*), whereas the other two zygotes will have three copies of the chromosome (trisomy *A*/*a*). However, note that the zygotes from nondisjunction during meiosis I will have both homologs (*Aa*), whereas the zygotes from nondisjunction during meiosis II will have inherited two of the same homologs (*AA* or *aa*).

13. Genetic diversity is increased during crossing over of prophase I of meiosis so that each gamete has chromosomes with combinations of alleles that are different from those of the parents. During meiosis, each homologous chromosome is randomly sorted to one of the four daughter cells.
14. Peaches that are tetraploid have four sets of chromosomes. Because there is an even number of chromosomes ($4n$), each replicated homologous chromosome will be able to find a replicated homolog to pair with at meiosis and fertile gametes will result. These gametes will be diploid ($2n$).
15. Because these gametes are diploid, they will be fertile. Triploid peaches would not be fertile because one of the three homologs would not find its pair during prophase of meiosis I, and the single homologs would be sorted randomly into the daughter cells.
16. Apoptosis is important during the development of human fingers. In the human fetus, the fingers initially develop with connecting tissue similar to the tissue that creates the webbed toes of a duck's foot. In humans this tissue eventually disappears (during development) as the result of the cells undergoing apoptosis.
17. A cell that has been damaged by radiation likely has DNA damage. This is the internal signal for a signal transduction pathway that leads to the activation of caspases. When the caspases degrade the nuclear envelope, nucleosomes, and plasma membrane, the cell dies.

### *Test Yourself*

1. **e.** Individual chromosomes, rather than homologous chromosomes, align at the equatorial plate during mitosis.
2. **a.** Meiosis occurs after one round of DNA replication. Homologous chromosomes pair during prophase I of meiosis, and after meiosis II the resulting cells have half the number of chromosomes as the parent cell. Those chromosomes are not genetically identical to the parent cells.
3. **c.** Kinetochores, *one* per sister chromatid, are assembled at the centromere of each chromosome and are the sites at which microtubules from opposite poles attach to segregate the chromosomes. They contain molecular motor proteins, including dynein and kinesin.
4. **d.** The mitotic spindle is composed of microtubules, not actin and myosin filaments. The spindle originates from the centrosome, which may or may not have centrioles, and only the kinetochore microtubules attach to the chromosomes.
5. **e.** While many cells lose the ability to divide when they mature, you know from experience that skin wounds heal. Without the presence of cyclin, however, there would be no signal to the cell to replicate, and that particular skin cell would not participate in healing the wound.
6. **b.** Functional CDK is required to complex with its cyclin and inactivate protein RB, permitting the cell to pass the restriction point in G1 and pass into S phase of the cell cycle.
7. **d.** *E. coli* is a prokaryote. It lacks a nucleus and does not undergo the cell cycle seen in eukaryotes. It has a circular chromosome, and does not synthesize actin or myosin proteins. Cytokinesis in *E. coli* is a result of a reproductive signal that causes the DNA to be replicated and segregated, and finally causes the cell to divide.
8. **c.** Chromatids are highly condensed, newly replicated chromosomes that form during S phase and that will be segregated to the daughter cells. After DNA replication, chromatids are still attached to each other at the centromere. A kinetochore assembles on the centromere of each chromatid and is involved in separating the chromatids during anaphase. Mitotic sister chromatids are identical.
9. **e.** Chromosomes are attached to the microtubules at their kinetochores. Dynein and kinesin molecular motors at the kinetochores (but not the centrosome) hydrolyze ATP and help move the chromosomes to opposite poles. Chromosomes are also pulled toward the poles by the shortening of the kinetochore microtubules.
10. **d.** Programmed cell death occurs during the development of many organisms (for instance, tadpoles lose their tails to become adult frogs). One of the stimuli for programmed cell death is DNA damage, but it is not the only cause of death. Necrosis (cell death that is not programmed) occurs when cells have been deprived of essential nutrients. The initiation of mitosis is part of the cell cycle in which cells reproduce and is not a step in programmed cell death. Apoptosis is not well controlled in cancer cells.
11. **b.** Without the origin of replication (there is only one in *E. coli*), there is no site for the replication proteins to bind in order to initiate DNA replication, so DNA synthesis will not start.
12. **c.** Chiasmata are sites where both homologous and sister chromatids exchange genetic material during meiosis (not mitosis). The exchange between homologous chromosomes increases genetic variation in the gametes. The exchange between sister chromatids has no effect, even though it does occur.
13. **a.** Asexual reproduction, which can occur in plants, results from mitosis. It produces cells that are identical to the parent cells. Sexual reproduction can occur in haplontic organisms (such as fungi) and results in an organism that is not genetically identical to either parent.

14. **e.** A chromatid is one-half of a newly replicated eukaryotic chromosome and is connected to the other (sister) chromatid at the centromere.

15. **c.** The cell in G1 has not yet replicated the DNA, so it has one copy of the 2*n* genome. At the end of meiosis I, the DNA replicated in S phase is still present, but the homologous chromosomes have been separated. Thus, there are two copies of the 1*n* genome in each daughter cell, or twice the original DNA but half the number of chromosomes. Thus, the cell in G1 has the same number of micrograms and DNA as the daughter cell after meiosis I. (In meiosis II, the result is one copy of the 1*n* genome compared to the original cell, or half the total DNA and half the number of chromosomes in each daughter cell). The cell in G1 has twice the number of chromosomes as the daughter cells after meiosis II.

# Inheritance, Genes, and Chromosomes

## The Big Picture

- Genes are units of inheritance that are passed down to progeny. Phenotypes are observable traits that result from gene expression. Diploid organisms have two sets of genes, which may be different forms (alleles), but only a single phenotype results. Meiosis generates gametes that contain only one set of genes. Copies of different genes are segregated independently, unless they are expressed on the same chromosome (i.e., linked). Gametes fuse to form a new diploid organism.
- Alleles arise by mutation. By studying the expression of particular phenotypes in successive generations, we can begin to determine, according to Mendel's laws, which alleles control those phenotypes and patterns of inheritance (i.e., which are rare recessive, rare dominant, or sex-linked). Exceptions to Mendel's laws also help explain the interactions of gene products that lead to an observed phenotype (i.e., incomplete dominance, codominance, epistasis, quantitative traits, linkage).
- Bacteria do not undergo meiosis. Instead, they transfer genes to other recipient bacterial cells through conjugation.

## Study Strategies

- It is easy to be overwhelmed by all the exceptions to Mendel's laws. Initially, make sure you understand the laws of segregation and independent assortment. The law of independent assortment focuses on different genes, while the law of segregation applies to different alleles of the same gene. In studying these laws, think about the molecular role of these alleles. For example, what is the explanation for a codominant phenotype at the molecular level, such as the AB blood type in humans? Both gene products are synthesized by the cell and modify the cell surface glycoproteins, producing both A and B antigens. What is the molecular explanation for epistasis? The enzyme required to produce an initial substrate necessary for pigmentation is missing, even though the other gene products (farther down the biochemical pathway) for producing a particular color are present in the cell.
- Review the figures of meiosis in Chapter 7 to see how Mendel's laws function in meiosis. Draw the chromosomes with different alleles (*R* on one chromosome, *r* on the other) and follow how those alleles are segregated to the gametes during meiosis.
- Use a Punnett square to predict the outcome of different monohybrid and dihybrid crosses. Use the probability rules to predict the genotypes and phenotypes of these same crosses and compare the two methods.
- After reviewing Mendel's laws of independent assortment and allele segregation, make a list of the exceptions to those laws with specific examples for each exception.
- Go to **LaunchPad** (or use the Web addresses listed) to review the following additional resources:

  Animated Tutorial 8.1 Independent Assortment of Alleles (PoL2e.com/at8.1)

  Animated Tutorial 8.2 Pedigree Analysis Simulation (PoL2e.com/at8.2)

  Animated Tutorial 8.3 Alleles That Do Not Assort Independently (PoL2e.com/at8.3)

  Activity 8.1 Homozygous or Heterozygous? (PoL2e.com/ac8.1)

  Activity 8.2 Concept Matching I (PoL2e.com/ac8.2)

  Activity 8.3 Concept Matching II (PoL2e.com/ac8.3)

  Analyze the Data in textbook Figures 8.1 and 8.4

  Apply the Concept in textbook Sections 8.2 and 8.3

## Key Concept Review

### 8.1 Genes Are Particulate and Are Inherited According to Mendel's Laws

Mendel used the scientific method to test his hypotheses

Mendel's first experiments involved monohybrid crosses

Mendel's first law states that the two copies of a gene segregate

Mendel verified his hypotheses by performing test crosses

Mendel's second law states that copies of different genes assort independently

Probability is used to predict inheritance

Mendel's laws can be observed in human pedigrees

The scientific method was used by Mendel and by those who came after him, who analyzed his results in the context of new evidence. Mendel's work was conducted long before the structure of DNA was known or the process of meiosis understood. As we have gained an understanding of cell division, genes, and chromosomes, we have been able to further verify Mendel's original hypotheses and refine his original ideas.

Mendel's experiments with peas revealed the way multiple alleles for a trait are expressed in an individual and passed on to the next generation, often in a new combination and resulting in a phenotype that is different from that of the parent. Mendel demonstrated that different alleles of the same gene are sorted into different gametes in meiosis (the law of segregation) and that different genes for different traits move to gametes independently of one another (the law of independent assortment). We know that the latter law is affected by the physical placement of genes on a single chromosome, and is referred to as linkage.

These two laws allow us to predict the probability of a certain progeny's having a particular phenotype and genotype when the parents' genotypes are known. This ability to make predictions can be particularly useful when potential parents are counseled about the probability that a certain trait will be passed on to their children.

**Question 1.** You have two plants of the same species with the phenotype of red and white, where red (*R*) is dominant and white (*r*) is recessive. Using a Punnett Square, show how you would determine the genotype of the red plant.

**Question 2.** Diagram two separate pairs of chromosomes, each bearing a different allele; for instance, *Rr* and *Yy*, as seen in the dihybrid cross with seed shape (*Rr*) and seed color (*Yy*). Show how these alleles assort independently during meiosis to produce haploid gametes.

**Question 3.** Suppose you are a genetics counselor and you are working with a 21-year-old pregnant woman who has just discovered that her father has Huntington's chorea, a rare dominant autosomal trait. This disease usually develops in middle age, so without genetic testing people carrying this trait are not aware of this until midlife. What are the chances that the child she is carrying will develop the disease? (Assume that her husband's family has no history of the disease.) What is the chance that she has Huntington's chorea?

## 8.2 Alleles and Genes Interact to Produce Phenotypes

New alleles arise by mutation

Dominance is not always complete

Genes interact when they are expressed

The environment affects gene action

A genetic code is much like a blueprint; it provides directions but must be translated into a product in order to be useful. As alleles for traits are expressed, they may interact with each other. This is often true if one allele does not code for a functional product and the other does. In other cases, both alleles code for a product and neither product can mask the presence of the other, leading to a blending of traits we call codominance. Alternately, the products of both alleles may be visible as separate traits, as seen in human blood types. The environment may also influence the expression of the genes. The final outcome is the phenotype, and this is what we are able to observe and measure, much like a finished building.

Not all traits are the result of a single gene product. In many cases, there are multiple genes that influence the final phenotype. Examples of this include epistasis, in which one set of genes may affect the expression of other genes (as in Labrador retriever coat color), and quantitative traits, in which multiple genes all contribute to a measurable outcome, such as height or grain yield.

In all cases, is it important to remember that genes are not expressed in a vacuum. The environment plays a very important role in determining the final phenotype of an individual. In order to meet one's full genetic potential, the environment must have sufficient resources. A plant without water will not produce a large crop any more than a child with insufficient food will reach his or her maximum height. We now know that phenotype is not determined by either nature or nurture alone, but by how they work together.

**Question 4.** A woman who has O type blood has a son with A type blood and a daughter with O type blood. What is the genotype of each parent? If the children in the family both have A type blood, would this change your answer? If so, how?

**Question 5.** One gene in rabbits has been directly linked to four phenotypes: dark gray, chinchilla (a lighter gray), Himalayan (white with dark ears, paws, nose, and tail), and albino (no color). How can one gene be linked to more than two phenotypes?

**Question 6.** In mice, agouti (fur with dark and light coloration) and full color are both dominant traits, while black and albino are recessive traits. We know that coat color is an epistatic trait. If you cross a population of true-breeding albino mice with a population of true-breeding agouti mice, you get progeny that are all agouti. If you then cross members of the agouti population (the $F_1$ generation) with each other, you get the following results for 16 progeny: 9 agouti, 4 albino, 3 black. How do you explain the emergence of black mice? What are the original genotypes of the parents? Diagram the agouti ($F_1$) crosses to determine the original genotypes of the parents and the genotypes associated with the phenotypes of agouti, black, and albino. Use *X* to represent unknown alleles in the $F_1$ genotypes. Agouti = *A*, Black = *a*, Color = *C*, Albino = *c*.

### 8.3 Genes Are Carried on Chromosomes

Genes on the same chromosome are linked, but can be separated by crossing over in meiosis

Linkage is also revealed by studies of the X and Y chromosomes

Some genes are carried on chromosomes in organelles

The DNA sequences coding for particular products are located on chromosomes in both prokaryotes and eukaryotes. Since chromosomes are much longer than individual genes, some traits are linked and tend to travel together to gametes in meiosis. The farther apart the genes are on the chromosome, the more likely it is that crossing over in meiosis I will separate them to different gametes.

The sex chromosomes also carry linked genes, and due to the different sizes of the X and Y chromosomes, an allele is sometimes found on one sex chromosome and not on the other. Since a single allele on the X chromosome has no homologue on the Y chromosome, males are said to be hemizygous for these traits. Sex-linked traits are inherited according to Mendel's laws, but the phenotypes are more likely to be observed in males.

Chromosomes in the organelles are not subject to the recombination seen in meiosis and are passed virtually unchanged from one parent to the offspring. In humans, cytoplasmic chromosomes are inherited from the mother.

**Question 7.** Draw a pedigree for three generations in which the father has red–green color blindness, his daughter is a carrier, and this daughter has four sons. Predict how many of the grandsons of the man with red–green color blindness will be color blind themselves.

**Question 8.** Draw a sample pedigree with three generations in which the paternal grandfather has a rare dominant autosomal trait. What is the probability that one of his children will have the disease? What is the probability that one of his grandchildren will have the disease?

**Question 9.** Draw a sample pedigree with three generations in which the maternal grandmother and paternal grandfather are carriers of a rare recessive autosomal trait. What is the probability that one of their children will be carriers of this trait? What is the probability that a grandchild will have the disease?

**Question 10.** Cytoplasmic traits in certain species of trees are passed from the male plant to all of its progeny. Compare this phenomenon to that of cytoplasmic inheritance in humans.

### 8.4 Prokaryotes Can Exchange Genetic Material

Bacteria exchange genes by conjugation

Plasmids transfer genes between bacteria

The evolution of drug-resistant bacteria is a major public health problem

Prokaryotic organisms have single, haploid chromosomes but divide by means of binary fission. This deprives prokaryotes of the benefits of sexual reproduction: namely the guarantee of genetic diversity from one generation to the next. While mutations do occur, their occurrence is relatively rare.

The solution to this problem is that many prokaryotes have developed ways of sharing genetic information. The first is though conjugation, by which a sequence of chromosomal DNA is transferred to a cell, recombines with the host chromosome, and results in a chromosome with a genotype different from that of either parent cell. The second is through the transfer of plasmid DNA. Plasmids do not become incorporated into the bacterial chromosome, but they are replicated during cell division and transmitted to the next generation. Plasmids also move through the conjugation tube and transfer new genes to the recipient cell.

**Question 11.** Two strains of *E. coli*, each having alleles *ABCdef* or *abcDEF*, are grown together in the laboratory for several weeks. Most of the resulting cells are genetically identical to the parent strains, but some have the genotypes *ABCDef*, *ABCDEf*, *abcdEF*, or *abcdeF*. How is this likely to have occurred?

**Question 12.** Many genes for antibiotic resistance are located on plasmids, and many antibiotics are given to healthy livestock in order to prevent illness and increase productivity. Explain why the use of antibiotics in livestock has raised concern about use of antibiotics in humans.

## Test Yourself

1. Hemophilia is a trait carried by the mother and passed to her sons. The allele for hemophilia, therefore,
   a. is carried on one of the mother's autosomal chromosomes.
   b. is carried on the Y chromosome.
   c. can be carried on the X or Y chromosome.
   d. is on the X chromosome and can be inherited by the son only if the mother is a carrier (heterozygous).
   e. is carried in the mitochondrial genome because a son inherits this allele from his mother.

   ***Textbook Reference:*** *Concept 8.3 Genes Are Carried on Chromosomes; Linkage is also revealed by studies of the X and Y chromosomes*

2. Before Mendel, genetic inheritance was thought to be a function of the blending of traits from the two parents. Which exception to Mendel's laws is in fact an example of blending?
   a. X linkage
   b. Polygenic inheritance
   c. Incomplete dominance
   d. Codominance
   e. Pleiotropism

   ***Textbook Reference:*** *Concept 8.2 Alleles and Genes Interact to Produce Phenotypes; Dominance is not always complete*

3. Which statement about true-breeding plants is true?
   a. When a true-breeding plant with a particular trait is crossed with another plant of the same variety, all of their offspring produce plants with that same trait.
   b. When a true-breeding plant with a particular trait is crossed with another plant of the same variety, all of their offspring will be sterile.
   c. They result from a monohybrid cross.
   d. They result from a dihybrid cross.
   e. They result from crossing over during prophase I of meiosis.

   ***Textbook Reference:*** *Concept 8.1 Genes Are Particulate and Are Inherited According to Mendel's Laws; Mendel used the scientific method to test his hypotheses*

4. What is the probability that a cross between a true-breeding pea plant with round seeds and a true-breeding pea plant with wrinkled seeds will produce $F_1$ progeny with round seeds?
   a. 0
   b. ⅛
   c. ¼
   d. ½
   e. 1

   ***Textbook Reference:*** *Concept 8.1 Genes Are Particulate and Are Inherited According to Mendel's Laws; Mendel's first experiments involved monohybrid crosses*

5. Which statement about the pattern of inheritance for a rare recessive allele is true?
   a. Every affected person has an affected parent.
   b. Unaffected parents can produce children who are affected.
   c. Affected parents do not produce affected children.
   d. Unaffected mothers have affected sons and daughters who are carriers.
   e. If a person is affected, both parents must also be affected.

   ***Textbook Reference:*** *Concept 8.1 Genes Are Particulate and Are Inherited According to Mendel's Laws; Mendel's laws can be observed in human pedigrees*

6. Which statement about the pattern of inheritance for a rare dominant allele is true?
   a. Every affected person has an affected parent.
   b. Unaffected parents can produce children who are affected.
   c. Affected parents do not produce affected children.
   d. Unaffected mothers have sons who are affected and daughters who are carriers.
   e. Unaffected fathers have sons who are affected and daughters who are carriers.

   ***Textbook Reference:*** *Concept 8.1 Genes Are Particulate and Are Inherited According to Mendel's Laws; Mendel's laws can be observed in human pedigrees*

7. Which statement about the pattern of inheritance for a recessive X-linked allele is true?
   a. Every affected person has an affected parent.
   b. Unaffected parents can produce daughters who are affected.
   c. Affected parents do not produce affected children.
   d. Unaffected parents can have sons who are affected and daughters who are carriers.
   e. Unaffected fathers can be carriers.

   ***Textbook Reference:*** *Concept 8.3 Genes Are Carried on Chromosomes; Linkage is also revealed by studies of the X and Y chromosomes*

8. The terms "penetrance" and "expressivity" refer to
   a. the increased expression of a particular trait when a hybrid species is formed.
   b. quantitative traits that diminish or intensify a particular phenotype.
   c. the influence of environment on the expression of a particular genotype.
   d. the expression of one gene masking the effects of another gene.
   e. the expression of a dominant phenotype in a heterozygote.

   ***Textbook Reference:*** *Concept 8.2 Alleles and Genes Interact to Produce Phenotypes; The environment affects gene action*

9. In what way is sex determination similar in humans and *Drosophila?*
   a. In both species females are hemizygous.
   b. In both species males have one X chromosome and females have two X chromosomes.
   c. In both species all males have one Y chromosome and two X chromosomes.
   d. In both species secondary sex characteristics are determined by genes on the X chromosome.
   e. In both species the ratio of X chromosomes to sets of autosomes determines maleness or femaleness.

   ***Textbook Reference:*** *Concept 8.3 Genes Are Carried on Chromosomes; Linkage is also revealed by studies of the X and Y chromosomes*

10. Linked genes are genes that
    a. assort independently.
    b. segregate equally in the gametes during meiosis.
    c. always contribute the same trait to the zygote.
    d. are found on the same chromosome.
    e. recombine during mitosis.

    ***Textbook Reference:*** *Concept 8.3 Genes Are Carried on Chromosomes; Genes on the same chromosome are linked, but can be separated by crossing over in meiosis*

11. Cytoplasmic inheritance
    a. results from polygenic nuclear traits.
    b. is determined by nuclear genes.
    c. is the result of the gametes' contributions of equal amounts of cytoplasm to the zygote.

d. is determined by genes on DNA molecules in mitochondria and chloroplasts.
e. follows Mendel's law of segregation.

***Textbook Reference:*** *Concept 8.3 Genes Are Carried on Chromosomes; Some genes are carried on chromosomes in organelles*

12. Epistasis is
a. the degree to which a particular genotype is expressed in an individual.
b. the proportion of individuals within a group that have a particular genotype and show the expected phenotype.
c. a situation in which a heterozygotic individual expresses phenotypic traits that are intermediate between those of the parents.
d. a situation in which one gene masks the expression of another gene.
e. a situation in which both alleles are expressed equally.

***Textbook Reference:*** *Concept 8.2 Alleles and Genes Interact to Produce Phenotypes; Genes interact when they are expressed*

13. Which of the following statements about quantitative traits is *false*?
a. They are affected by the environment.
b. They affect the same physical characteristic.
c. Each allele intensifies or diminishes the phenotype.
d. They are controlled by multiple genes.
e. They are controlled by one gene.

***Textbook Reference:*** *Concept 8.2 Alleles and Genes Interact to Produce Phenotypes*

14. A bacterial cell's genotype can be altered by
a. the introduction of a plasmid that carries some of the bacterium's genes.
b. mating with a bacterial cell with the same genotype.
c. homologous recombination with human DNA.
d. the forming of a conjugation tube with another bacterial cell.
e. the transferring of genetic material from a different strain of bacteria.

***Textbook Reference:*** *Concept 8.4 Prokaryotes Can Exchange Genetic Material; Bacteria exchange genes by conjugation*

15. A test cross
a. is used to determine if an organism that is displaying a dominant trait is heterozygous or homozygous for that trait.
b. is used to determine if an organism that is displaying a recessive trait is heterozygous or homozygous for that trait.
c. causes the loss of hybrid vigor.
d. results in an $F_2$ generation with a phenotypic ratio of ¾ dominant to ¼ recessive.
e. results in the transfer of the same alleles from generation to generation.

***Textbook Reference:*** *Concept 8.1 Genes Are Particulate and Are Inherited According to Mendel's Laws; Mendel verified his hypotheses by performing test crosses*

## Answers

### *Key Concept Review*

1. To determine the genotype of the red plant, you would do a cross of the red and white plants. The outcome of the cross would determine if the red plant is homozygous (*RR*), or heterozygous (*Rr*). If the red plant is *RR*, then all progeny will be red. If the red plant is *Rr*, then the red and white seed phenotypes will be 1:1.

| | *R* | *R* |
|---|---|---|
| *r* | *Rr* | *RR* |
| *r* | *Rr* | *Rr* |

| | *R* | *r* |
|---|---|---|
| *r* | *Rr* | *rr* |
| *r* | *Rr* | *rr* |

2. Refer to figure 8.6
3. There is a 50 percent chance that she will develop Huntington's chorea. Because the trait is an autosomal dominant allele, half of her father's gametes will contain the homologous chromosome carrying that allele and half of his gametes will contain the homologous chromosome that carries the wild-type allele. If she received the Huntington's allele, her child has a 50 percent chance of receiving this allele from her. The product rule is used to predict the probability that her child will inherit the Huntington's allele: ½ (the probability that she has the Huntington's allele) × ½ (the probability that her child will inherit this allele from her) = ¼ (the probability that her child has the allele). Her child has a 25 percent chance of carrying the Huntington's chorea allele and thus of developing the disease.
4. In the first case, the mother and the child with O type blood must have genotypes of $I^OI^O$, since $I^O$ is recessive to all other alleles. The second child, with A type blood, must have a genotype of $I^AI^O$ and had to have inherited the $I^A$ allele from the father, which means that the father has a genotype of $I^AI^O$. In the second scenario, if both of the children have A type blood, there is a greater probability that the father has an $I^AI^A$

genotype. However, since the sample size is small, it is also possible that the father has an $I^AI^O$, or even an $I^AI^B$, genotype.

5. The *C* gene that determines coat color in rabbits has multiple alleles. These alleles show a hierarchy of dominance when present in heterozygous rabbits. For example, dark gray (*C*) is dominant to the other three alleles—chinchilla ($c^{chd}$), Himalayan ($c^h$), and albino (*c*)—whereas chinchilla is dominant to albino.
6. Agouti = *A*, Black = *a*, Color = *C*, Albino = *c*. The true-breeding agouti parent must be homozygous for agouti (*AA*) and for color expression (*CC*). The true-breeding albino parent must be homozygous for albinism: *cc*. Because all of the $F_1$ progeny are agouti, you do not know from the $F_1$ generation if the parent albino mouse carried *aa*, *Aa*, or *AA* alleles. Therefore, write the genotype of the albino parent as *ccXX* for now, using *X's* as placeholders for the unknown alleles. So *ccXX* = true-breeding albino parent, *CCAA* = true-breeding agouti parent. Therefore, all $F_1$ progeny are *CcAX*. Continue to use *X* as a placeholder for the $F_1$ cross:

| | CA | cA | CX | cX |
|---|---|---|---|---|
| CA | CCAA | CcAA | CCAX | CcAX |
| cA | CcAA | ccAA | CcAX | ccAX |
| CX | CCAX | CcAX | CCXX | CcXX |
| cX | CcAX | ccAX | CcXX | ccXX |

Resulting phenotypes: albino *cc* (white), agouti *AA* and *Aa* (light gray), three are unknown (black).

Now go back to the original problem and see what else you can determine from the filled-out Punnett square. You are told that the progeny ratio is 9 agouti, 4 albino, 3 black. The black progeny must be the unknown phenotypes above and *XX* must correspond to the genotype *aa*, since black is recessive (*a*). Fill in those three squares. This information can then be used to figure out the genotypes of the parents. Below is the Punnett square with all of the alleles filled in.

| | CA | cA | Ca | ca |
|---|---|---|---|---|
| CA | CCAA | CcAA | CCAa | CcAa |
| cA | CcAA | ccAA | CcAa | ccAa |
| Ca | CCAa | CcAa | CCaa | Ccaa |
| ca | CcAa | ccAa | Ccaa | ccaa |

Resulting phenotypes and required alleles: albino *cc* (white), agouti *AA* or *Aa* (light gray), black *aa* (dark gray).

Now you know that the $F_1$ parents were actually all *CcAa* because it is the only way to get three black mice.

Going back to the original albino parents, the $F_1$ mice had to get the *A* allele from the true-breeding agouti parent (*CCAA*), so the *a* allele must have come from the albino parent (*ccaa*).

7. One-half of the grandsons could be color-blind.

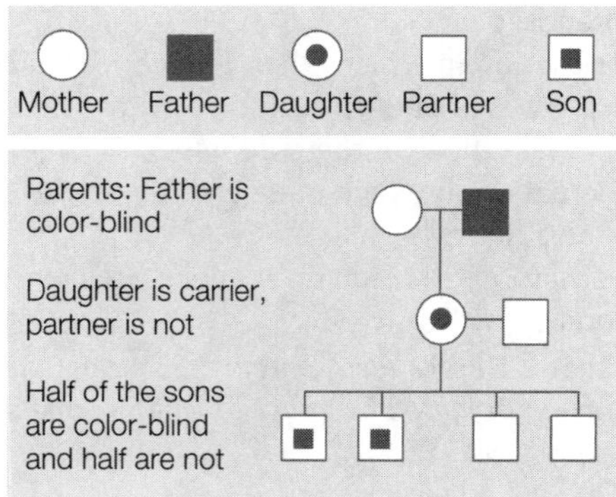

8. One-half of his children could get the disease; one-quarter of the grandchildren could get the disease.

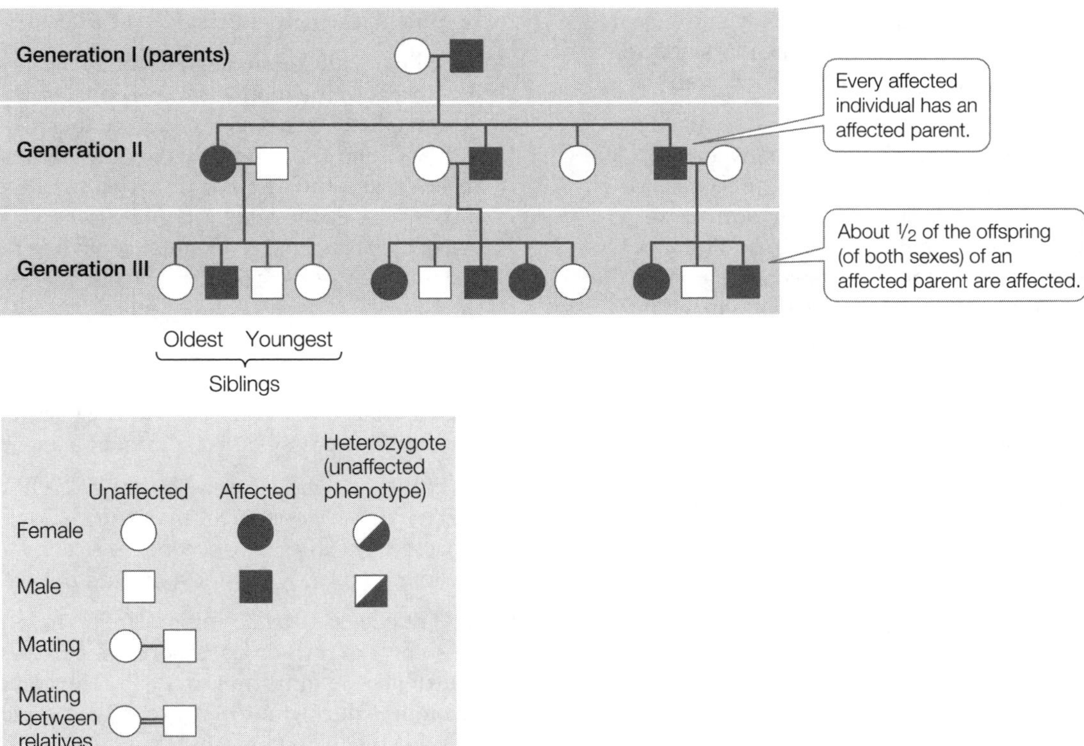


9. One-half of the children of these grandparents could be carriers. One-sixteenth of the grandchildren could have the disease.

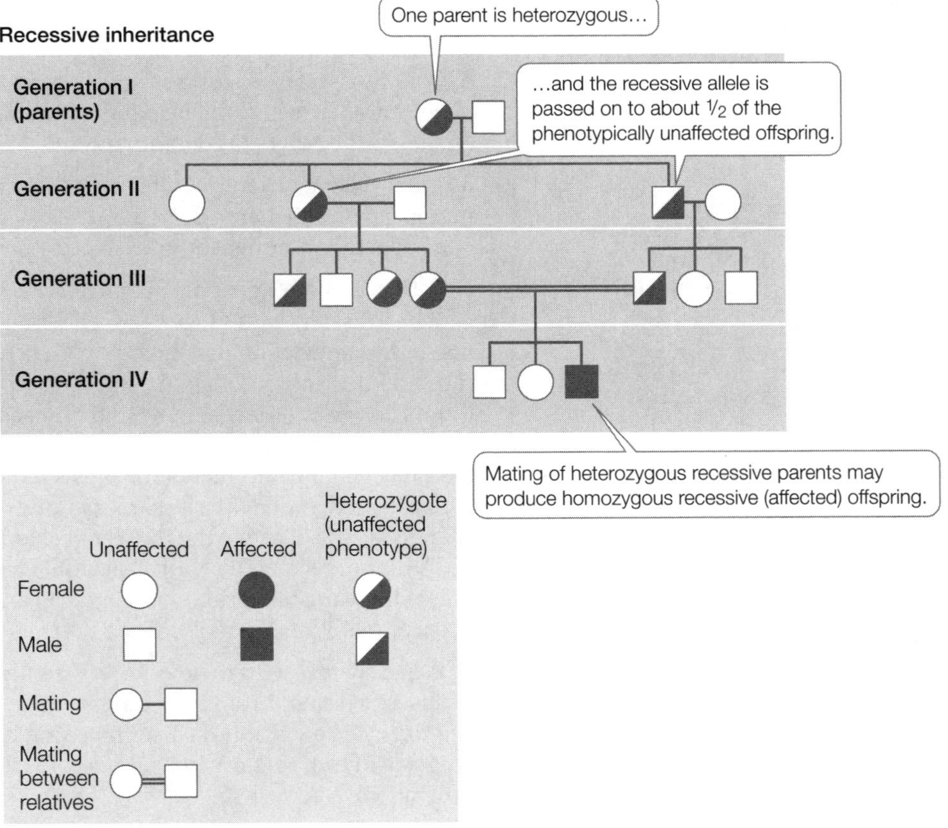

10. In humans, the gamete with the largest cytoplasmic contribution is the egg, so cytoplasmic inheritance is passed from the female parent to all her children. In certain tree species, the male gamete contributes most of the cytoplasm to the zygote, so all the mitochondria and chloroplasts in the zygote are inherited from the male parent.
11. Conjugation allows for recombination between the DNA transferred though the pilus and the recipient chromosome. The process of conjugation can be disrupted if the cells are pulled apart. In addition, because crossing over is random, there is no guarantee that all of the transferred DNA will be exchanged into the new chromosome. Either or both of these events could result in the genotypes seen in the results.
12. We know that plasmids are transferred between bacteria, including bacteria of different strains. Antibiotics in the livestock feed select for those bacteria containing antibiotic resistance genes. This makes it more likely that they will be transferred to strains of bacteria that cause disease in humans and be resistant to the antibiotics normally used to treat the infection. (Alternatively, the bacteria themselves might cause diseases in humans that are untreatable with the antibiotics normally used.)

### *Test Yourself*

1. **d.** Hemophilia is an X-linked trait and can only be inherited by the son from his mother's X chromosome (and not her mitochondrial chromosome). The father contributes the Y chromosome to his son (not his X chromosome) and thus cannot pass any of his X-linked alleles to his son.
2. **c.** Incomplete dominance results in the progeny's expressing an intermediate form of the two parental alleles. In a cross between red-flowered plants and white-flowered plants, for example, the expression of pink-flowered plants would be a "blend" of the parental traits. Codominance is not an example of blending because both alleles are fully expressed in the individual.
3. **a.** True-breeding individuals are always homozygous. Monohybrid and dihybrid crosses produce heterozygous individuals.
4. **e.** This is an example of a monohybrid cross. The $F_1$ generation would all have the genotype *Rr*, producing the phenotype of round seeds, because the round allele (*R*) is dominant to the wrinkled allele (*r*). All of the $F_1$ progeny, therefore, would have round seeds.
5. **b.** Rare recessive alleles can be carried by both parents but not expressed in those parents. If the parents are heterozygous for this allele (*Aa*), their children will have a ¼ probability of expressing that recessive allele (*aa*). If both parents are affected (*aa*), their children will also be affected (*aa*).
6. **a.** If an allele is dominant, every affected individual has at least one dominant allele and must have received that allele from a parent. Because the allele is dominant, the parent must also be affected.
7. **d.** X-linked alleles are passed from a mother to her son (because the mother always donates one of her X chromosomes to her son, and the father always donates a Y chromosome to his son). Daughters can also receive the X-linked allele from their mothers, but the father donates the other X chromosome, so daughters are usually carriers. A daughter could be affected if the mother was a carrier and the father was affected.
8. **c.** "Penetrance" and "expressivity" refer to the effects of the environment on a particular phenotype. Answer a. refers to hybrid vigor, answer b. refers to quantitative traits, answer d. refers to epistasis, and answer e. refers to expression of a dominant allele.
9. **b.** In these species, females have two X chromosomes and males have one X chromosome and one Y chromosome. The alleles for secondary sex characteristics are found on the X chromosome and the autosomes.
10. **d.** Linked genes, by definition, are on the same chromosome and thus do not assort independently, do not contribute the same trait to the zygote, and do not recombine during mitosis or segregate equally to the gametes during meiosis.
11. **d.** The genes on the mitochondria and chloroplast chromosomes, which are cytoplasmically inherited (unlike nuclear genes), are passed on to all of the progeny from the gamete that contributes most of the cytoplasm.
12. **d.** Answer a. refers to expressivity, answer b. refers to penetrance, answer c. refers to incomplete dominance, and answer e. refers to codominance.
13. **e.** Quantitative traits are traits that are conferred by multiple genes that are affected by the environment and can either diminish or intensify one phenotype.
14. **e.** A bacterial cell's genotype changes when new genetic information is introduced, either on a plasmid or by conjugation (which requires a conjugation tube and the transfer of DNA). If conjugation occurs, those genes need to recombine into the recipient cell's chromosome in order to be maintained. Transferring the same genetic information on a plasmid or mating with a bacterial cell with the same genotype will not change the genotype of the recipient bacteria. Human DNA cannot recombine with a bacterial chromosome unless there are identical (homologous) DNA sequences in both DNA molecules.
15. **a.** A test cross is used to determine if an organism that is expressing a dominant trait is homozygous or heterozygous for that trait. True-breeding individuals continue to express the same alleles generation after generation.

# DNA and Its Role in Heredity

## The Big Picture

- The identification of DNA as the genetic material and the elucidation of its structure have revealed how hereditary information is passed on at the molecular level. Advances in our knowledge of DNA replication have led to technologies, such as DNA sequencing and PCR, which facilitate further studies of gene function, expression, and their relatedness in different organisms.
- Accuracy in DNA replication is necessary for the survival of an organism, yet this hereditary information is altered by point mutations and chromosomal mutations that occur spontaneously or are induced by mutagens. Alterations are often silent, but sometimes they can be harmful to the organism or, if occurring in germline cells, harmful to the organism's offspring. Thus cells have developed various mechanisms to protect against and repair DNA damage. Alterations in the genetic material may also provide evolutionary benefits to an organism, as they provide the genetic diversity that makes natural selection possible.

## Study Strategies

- Be careful to keep track of the DNA's orientation. The 5′ and 3′ ends are often confused when parental (template) strands and daughter (newly synthesized) strands are compared. Remember that on both strands, the 3′ end corresponds to the site (or former site) of a hydroxyl group (—OH), and the 5′ end corresponds to the site (or former site) of a phosphate tail.
- Study the figures in the textbook to visualize what is occurring during DNA replication. Make your own diagrams of these processes or arrange simple manipulatives. For instance, to visualize the differences between leading and lagging strands, lay out colored candies on a strip of paper. Separate the strips, and "synthesize" a new strand of DNA using the complementary colors of candy (set up a pairing scheme of your choosing). Make sure you work from a 5′-to-3′ direction.
- Take advantage of laboratory activities that simulate or allow you to experience sequencing or PCR. Nearly all molecular research labs utilize one or both of these techniques.
- Draw a picture of the replication fork. Position the primers on the fork, and then draw in the leading and lagging strands. Label all of the 3′ and 5′ ends. Draw a box to indicate where ligase will seal up the ends.
- Many different mutations can occur in a gene that will alter the function of the gene product. Some of those mutations may have more deleterious effects than others. Make a table listing the different types of mutations, the scale of the changes, and where appropriate, a simple illustration of the change compared to a template DNA sequence you design or to a template chromosome.
- Go to **LaunchPad** (or use the Web addresses listed) to review the following additional resources:

  Animated Tutorial 9.1 The Hershey–Chase Experiment (PoL2e.com/at9.1)

  Animated Tutorial 9.2 Experimental Evidence for Semiconservative DNA Replication (PoL2e.com/at9.2)

  Animated Tutorial 9.3 DNA Replication, Part 1: Replication of a Chromosome and DNA Polymerization (PoL2e.com/at9.3)

  Animated Tutorial 9.4 DNA Replication, Part 2: Coordination of Leading and Lagging Strand Synthesis (PoL2e.com/at9.4)

  Animated Tutorial 9.5 Polymerase Chain Reaction Simulation (PoL2e.com/at9.5)

  Activity 9.1 DNA Polymerase (PoL2e.com/ac9.1)

  Analyze the Data in textbook Figure 9.3

  Apply the Concept in textbook Section 9.2 and 9.3

## Key Concept Review

### 9.1 DNA Structure Reflects Its Role as the Genetic Material

Circumstantial evidence suggested that DNA is the genetic material

Experimental evidence confirmed that DNA is the genetic material

Four key features define DNA structure

### The double-helical structure of DNA is essential to its function

Today we take for granted our ability to sequence the genome of a given species. But the role of DNA in a cell was not always understood and the structure of the DNA molecule was also a mystery. Before the mid-twentieth century scientists could study the fluctuations of DNA content during mitosis and meiosis in eukaryotes using DNA dyes, but the available technology made it difficult to directly verify that DNA was the genetic material. The issue was finally resolved when bacteria were observed to acquire viral DNA, but not protein, when infected by bacteriophage T2. Once DNA was established as the genetic material, it was important to identify its structure. It had already been observed that the ratios of adenine to thymine and cytosine to guanine are always fixed in the DNA polymer. Watson and Crick made use of Franklin's work in X-ray crystallography as well as models to determine that the DNA molecule is a double helix with antiparallel strands. This structure proved to be an elegant solution to the question of how DNA stores genetic information in a form that can be efficiently copied and "read" by the cell machinery.

**Question 1.** Diagram the double helix. Be sure to label those properties that make it most suited as the genetic material. Think about DNA replication and encoding of information.

**Question 2.** Explain why in a DNA molecule of any species the amount of adenine is equal to thymine (A = T) and cytosine is equal to guanine (C = G) but (A + T) does not equal (C + G).

**Question 3.** Given the following DNA sequence, what would the sequence of the complementary strand look like?

5′-GCTAACTGTGATCGTATAAGCTGA-3′

## 9.2 DNA Replicates Semiconservatively

### DNA polymerases add nucleotides to the growing chain

### Telomeres are not fully replicated in most eukaryotic cells

### Errors in DNA replication can be repaired

### The basic mechanisms of DNA replication can be used to amplify DNA in a test tube

A key feature of DNA is that it replicates semiconservatively. This means that each "half" of the double helix serves as the template for a new, complementary DNA strand. The process of replication requires several proteins as well as an abundant supply of energy and nucleotides. Starting at an origin of replication, the hydrogen bonds between the complementary bases are broken and the strands are pulled apart. Replication begins with the synthesis of an RNA primer at the origin of replication. Then, DNA polymerase synthesizes the new strand of DNA by adding nucleotides to the 3′ end of the primer. One of the strands (the leading strand) is in the proper orientation to allow for continuous synthesis from an origin of replication toward the replication fork. The other strand (the lagging strand) is antiparallel and must be synthesized away from the replication fork. This necessitates a discontinuous form of DNA synthesis in which short stretches of DNA (Okazaki fragments) are synthesized from a series RNA primers laid down on the single-stranded template that becomes available as the replication fork progresses along the DNA molecule. Later, the cell removes the RNA primers, fills in the resulting gaps in the sequence with DNA, and ligates the backbone so the lagging strand is complete.

In linear chromosomes, the directional nature of DNA results in a continued shortening of the lagging strand with each round of DNA replication due to the requirement of an RNA primer to initiate DNA synthesis. To ensure that this shortening does not affect coding regions, and to prevent chromosomal ends from joining, telomeres are located at the ends of chromosomes. These stretches of repeating sequences, which are capped by a protein shelterin complex, can be lengthened with the action of the enzyme telomerase, which is not expressed in most somatic cells. The absence of telomerase acts as a "countdown clock" to tell a cell when to stop dividing. Cells that should continue to divide, such as stem cells, contain active telomerase that keeps the telomeres from being lost over time.

DNA replication needs to be accurate. To help ensure the accuracy, the cell has several strategies to monitor the process. DNA polymerase itself has a proofreading capability and will remove an incorrectly selected nucleotide if it makes a mistake. There are other proteins that monitor the DNA strand for mismatched base pairs and will remove the incorrect nucleotides. (DNA polymerase will then replace them.)

DNA replication is possible in a test tube. By selecting primers for a DNA region of interest, scientists can amplify a target sequence by means of PCR (the polymerase chain reaction). This is a powerful tool that has revolutionized molecular biology research.

**Question 4.** Diagram the replication complex and label the following: DNA polymerase, RNA primer, Okazaki fragment, the leading and lagging strands, and the leading and lagging template strands. Be sure to label the 5′ and 3′ ends of the parental and newly replicated DNA.

**Question 5.** Explain the role of Okazaki fragments in the synthesis of the lagging strand.

**Question 6.** Differentiate between proofreading and mismatch repair.

**Question 7.** Explain how PCR amplifies a particular sequence of DNA. Is the replication process linear or exponential? Explain your answer.

**Question 8.** Refer to the following DNA sequence. Note that only the "top" strand is provided here.

TATCGTCAGA TTTCAATCTA TTGGCGTTGT TAAAAAACTA

TGGTTATACT AACGGCAAAA ACGCTCTGAA ACTAGATCCT

AATGAAGTCT TCAACGTGAC TTTTGACCGT TCAATGTTCA

The nucleotides can be referred to by number (i.e., the first one is number 1 and the last is 120). Design PCR primers that would amplify a PCR product that includes nucleotides 31–80. Your primers should be beyond the region you are amplifying and should be 15 nucleotides long. What do you need in the test tube to make the reaction work?

### 9.3 Mutations Are Heritable Changes in DNA

Mutations can have various phenotypic effects

Point mutations are changes in single nucleotides

Chromosomal mutations are extensive changes in the genetic material

Mutations can be spontaneous or induced

Some base pairs are more vulnerable than others to mutation

Mutagens can be natural or artificial

Mutations have both benefits and costs

We attempt to minimize our exposure to mutagens

Mutations are alterations to the genetic code. Mutations in somatic cells are confined to the individual, whereas mutations in germline cells are passed to the next generation. Regardless of the cell type, mutations may be limited to a single nucleotide change or they may be chromosomal mutations in which entire chromosome fragments are moved, lost, duplicated, or flipped. Whether or not the mutation is beneficial, harmful, or neutral to the organism is dependent upon how the mutation affects the phenotype.

The causes of mutations are numerous and may be natural or artificial. DNA replication itself can introduce changes to the genetic code (spontaneous mutation), and this happens naturally in a cell. Chromosomes may break or fail to separate, or they can be repaired incorrectly. Exposure to radiation and mutagenic chemicals can lead to induced mutations. While an individual cannot prevent spontaneous mutations, limiting one's exposure to mutagens can reduce the incidence of induced mutations.

**Question 9**. Two individuals have a mutation in gene *X* but at different sites. The mutation affects the first individual adversely, and the second individual experiences no effect. Explain this observation.

**Question 10**. Are all mutations inherently bad? What about those in somatic cells?

**Question 11.** Starting with chromosome ABCDEFG and the nonhomologous chromosome LMNOPQR, illustrate the following:

a. Deletion of segment C
b. A duplication of segment C
c. A duplication of segment D and a deletion of segment E
d. An inversion of a segment, going from A to C
e. A reciprocal translocation between CDEFG and LMNO

## Test Yourself

1. Which of the following chromosomal mutations would unlikely alter the functional activity of protein X ?
   a. Deletion of the last 100 codons of gene *X*
   b. A translocation of 100 codons of another gene into the middle of gene *X*
   c. An inversion of the last 100 codons of gene *X*
   d. A translocation of the last 100 codons of gene *X* to another chromosome
   e. A duplication of gene *X*

   ***Textbook Reference:*** *Concept 9.3 Mutations Are Heritable Changes in DNA; Chromosomal mutations are extensive changes in the genetic material*

2. Watson and Crick's model allowed them to visualize all of the following *except*
   a. the molecular bonds of DNA.
   b. the sugar and phosphate component of the DNA molecule's surface.
   c. how the purines and pyrimidines fit together in a double helix.
   d. the antiparallel design of two strands of the DNA double helix.
   e. that DNA replication was semiconservative.

   ***Textbook Reference:*** *Concept 9.1 DNA Structure Reflects Its Role as the Genetic Material; Experimental evidence confirmed that DNA is the genetic material*

3. A fundamental requirement for the functioning of genetic material is that it must be
   a. conserved among all organisms with very little variation.
   b. passed intact from one species to another.
   c. replicated accurately and passed from a parent to its offspring or from a cell to its daughter cells.
   d. found outside the nucleus.
   e. replicated accurately without any errors over many millions of years.

   ***Textbook Reference:*** *Concept 9.2 DNA Replicates Semiconservatively*

4. The primary function of DNA polymerase is to
   a. add nucleotides to the growing daughter strand.
   b. seal nicks along the sugar–phosphate backbone of the daughter strand.
   c. unwind the parent DNA double helix.
   d. generate primers to initiate DNA synthesis.
   e. prevent reassociation of the denatured parental DNA strands.

   ***Textbook Reference:*** *Concept 9.2 DNA Replicates Semiconservatively; DNA polymerases add nucleotides to the growing chain*

5. When the lagging daughter strand of DNA is synthesized, unreplicated gaps are formed on the parental

DNA. Lagging strand synthesis fills these gaps by all of the following mechanisms *except* for
a. synthesis of short Okazaki fragments in a 5′-to-3′ direction.
b. synthesis of multiple short RNA primers to initiate DNA replication.
c. the use of DNA polymerase I to remove RNA primers from Okazaki fragments.
d. the use of DNA polymerase I to add nucleotides in a 3′-to-5′ direction.
e. the filling of the gaps with new strands of complementary DNA as the replication fork proceeds.
***Textbook Reference:*** *Concept 9.2 DNA Replicates Semiconservatively; DNA polymerases add nucleotides to the growing chain*

6. RNA primers are necessary in DNA synthesis because
a. DNA polymerase is unable to initiate replication without an origin.
b. the DNA polymerase enzyme can catalyze the addition of deoxyribonucleotides only onto the 3′ (—OH) end of an existing strand.
c. RNA primase is the first enzyme in the replication complex.
d. primers mark the sites where helicase has to unwind the DNA.
e. they are the source of nucleotides for DNA polymerases.
***Textbook Reference:*** *Concept 9.2 DNA Replicates Semiconservatively; DNA polymerases add nucleotides to the growing chain*

7. Proofreading and repair occur
a. at any time during or after synthesis of DNA.
b. only before mitosis.
c. only in the presence of DNA polymerase.
d. only after replication is complete.
e. only during replication.
***Textbook Reference:*** *Concept 9.2 DNA Replicates Semiconservatively; Errors in DNA replication can be repaired*

8. If 30 percent of the bases in a sample of DNA extracted from eukaryotic cells are adenine, what percentage of the bases in this DNA are cytosine?
a. 10 percent
b. 20 percent
c. 30 percent
d. 40 percent
e. 50 percent
***Textbook Reference:*** *Concept 9.1 DNA Structure Reflects Its Role as the Genetic Material; Experimental evidence confirmed that DNA is the genetic material*

9. Gain-of-function mutations
a. are dominant mutations that are expressed in wild-type cells.
b. are dominant mutations that are expressed in mutant cells.
c. are the cause of continuous division in cancer cells.
d. can be analyzed only under restrictive conditions.
e. are expressed only in response to the appropriate environmental signals.
***Textbook Reference:*** *Concept 9.3 Mutations Are Heritable Changes in DNA; Mutations can have various phenotypic effects*

10. The PCR technique
a can amplify only very small samples of DNA.
b. amplifies several random DNA sequences within a genome.
c. requires synthetic primers to flank the regions of interest.
d. is accomplished in three sequential steps: binding, denaturation, and replication.
e. generates DNA molecules that all have variable sequences.
***Textbook Reference:*** *Concept 9.2 DNA Replicates Semiconservatively; The basic mechanisms of DNA replication can be used to amplify DNA in a test tube*

11. Which of the following represents a bond between a purine and a pyrimidine, respectively?
a. C–T
b. G–A
c. G–C
d. T–A
e. A–G
***Textbook Reference:*** *Concept 9.1 DNA Structure Reflects Its Role as the Genetic Material; Experimental evidence confirmed that DNA is the genetic material*

12. Which of the following would *not* be found in a DNA molecule?
a. Purines
b. Ribose sugars
c. Phosphates
d. Sulfur
e. Nitrogenous bases
***Textbook Reference:*** *Concept 9.1 DNA Structure Reflects Its Role as the Genetic Material; Four key features define DNA structure*

13. If a high concentration of a particular nucleotide lacking a hydroxyl group at the 3′ end is added to a PCR reaction,
a. no additional nucleotides will be added to a growing strand containing that nucleotide.
b. strand elongation will proceed as normal.
c. nucleotides will be added only at the 5′ end.
d. DNA polymerase will become nonfunctional.
e. the primer will be unable to bind with the DNA template.
***Textbook Reference:*** *Concept 9.2 DNA Replicates Semiconservatively; The basic mechanisms of DNA replication can be used to amplify DNA in a test tube*

14. The telomeres at the ends of linear chromosomes allow
    a. the 5′ ends of the chromosomes to undergo recombination.
    b. the gaps left by primer removal of lagging strands to be repaired by telomerase.
    c. DNA repair enzymes to recognize those ends and remove them.
    d. normal cells to divide continuously.
    e. DNA breaks to be examined at cell division checkpoints.

    ***Textbook Reference:*** *Concept 9.2 DNA Replicates Semiconservatively; Telomeres are not fully replicated in most eukaryotic cells*

15. Chargaff observed that the amount of _______ was *not* equal to the amount of _______ in all tested organisms.
    a. purines; pyrimidines
    b. A; T
    c. A + T; G + C
    d. A + G; T + C
    e. G; C

    ***Textbook Reference:*** *Concept 9.1 DNA Structure Reflects Its Role as the Genetic Material; Experimental evidence confirmed that DNA is the genetic material*

16. Which of the following statements about DNA replication is *false?*
    a. Okazaki fragments are synthesized as part of the leading strand.
    b. Replication forks represent areas of active DNA synthesis on the chromosomes.
    c. Error rates for DNA replication are reduced by proofreading of the DNA polymerase.
    d. Ligases and polymerases function in the vicinity of replication forks.
    e. The sliding clamp protein increases the rate of DNA synthesis.

    ***Textbook Reference:*** *Concept 9.2 DNA Replicates Semiconservatively; DNA polymerases add nucleotides to the growing chain*

## Answers

### *Key Concept Review*

1. See Figure 9.5 in the textbook. Be sure that the diagram includes the following features: (1) the capacity of nucleotides to form hydrogen bonds only with their complement (C and G, A and T); (2) the antiparallel structure of double-stranded DNA and the addition of new nucleotides only to the 3′ end; (3) the possible appearance of nucleotides in any order on a strand of DNA, allowing the DNA sequence to be unique when the nucleotides are "read" by cellular machinery.

2. The purine adenine (A) always pairs with the pyrimidine thymine (T), and the purine guanine (G) always pairs with the pyrimidine cytosine (C). Therefore, in a double-stranded molecule of DNA, the total amount of A is equal to T , and the total amount of G is equal to C. The numbers of As and Gs in a strand can be very different, however, so the total abundance of (A + T) is not equal to the total abundance of (G + C).

3. The complementary strand sequence would be:

   3′-CGATTGACACTAGCATATTCGACT-5′

4. See Figure 9.11.

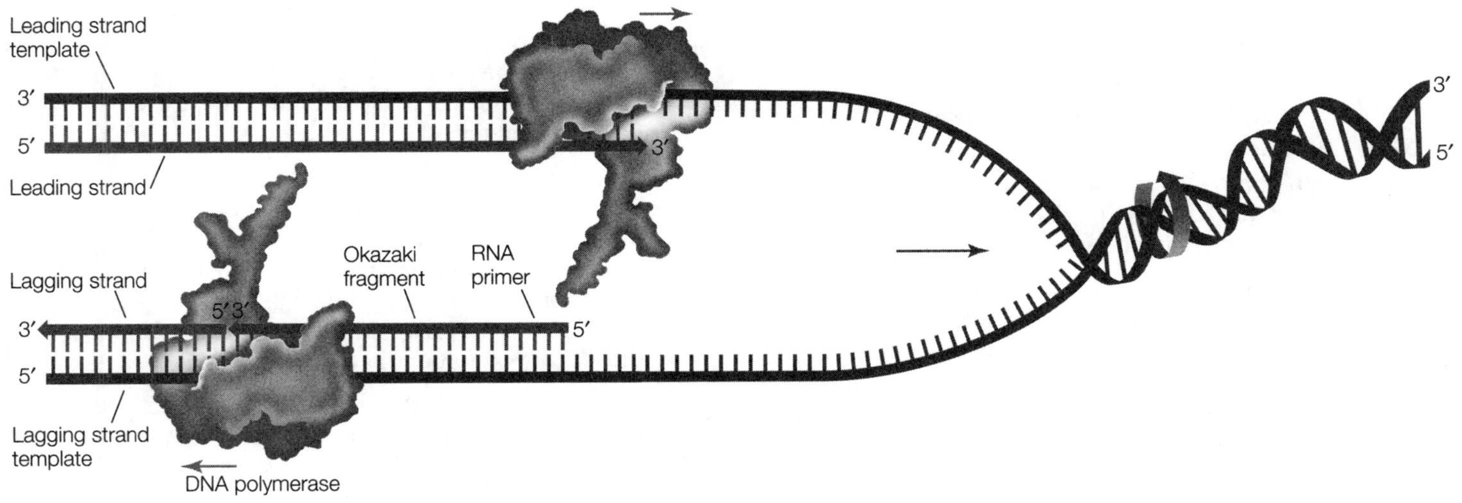

5. Okazaki fragments are short, discontinuous stretches of sequence that are formed when the lagging strand is replicated. Because only a portion of the DNA is opened up at the replication fork at any given time, and because DNA polymerase adds nucleotides to the 3′ end of the previous nucleotide, continuous growth is not possible on the lagging strand. Therefore, replication of the lagging strand produces short, discontinuous stretches of DNA that are synthesized in the 5′-to-3′ direction—the opposite direction from the movement of the replication fork.

6. Proofreading occurs during synthesis of the DNA molecule, whereas mismatch repair occurs after synthesis.

7. Refer to Figure 9.15 in the textbook. PCR amplifies a sequence of DNA by means of a laboratory technique that uses a sample of double-stranded DNA, two artificially synthesized primers that are complementary to the ends of the sequence to be amplified, the four dNTPs, and a DNA polymerase. During each cycle of a PCR reaction, the double-stranded DNA is separated and replicated, producing two double-stranded copies. These copies are themselves copied in the next cycle of the PCR reaction. Since every cycle of the reaction results in a doubling of the DNA, the amount of DNA increases exponentially.

8. Any 15 nucleotide sequence identical to the sequence shown between 1 and 30 would work for one of the primers. For example,

   5′-ATCTATTGGCGTTGT-3′ goes from 16 to 30. It will bind to the complementary sequence and allow DNA to be synthesized from position 31 to 120. To amplify the double-stranded DNA sequence, a primer is also needed to bind to the other strand of DNA and synthesize a strand complementary to the one shown. For example, the sequence between 81 and 95 is

   5′-AATGAAGTCTTCAAC-3′

   3′-TTACTTCAGAAGTTG-5′

   with the bottom strand showing the primer sequence. Convention has us write single-stranded DNA from 5′ to 3′ (as this is how they are synthesized), so the two primers will be

   5′-AATGAAGTCTTCAAC-3′ (top), and

   5′-GTTGAAGACTTCATT-3′ (bottom).

   The test tube must also contain the template DNA, both primers, DNA polymerase that is heat resistant, dATP, dTTP, dCTP, dGTP nucleotides, salts, and a buffer solution.

9. The mutation in gene *X* in the first individual must have occurred in an essential region of the gene that is required for its function. The mutation in gene *X* in the second individual may be a silent mutation, or it may be in a region that is nonessential for the function of that protein.

10. Mutations may be beneficial, neutral, or deleterious to an organism. Somatic cell mutations will only affect the individual, whereas germ cell mutations will be passed to the next generation. If a mutation results in the loss of function, the phenotype may be altered if the homologous gene's expression is insufficient. In the case of a germ cell loss-of-function mutation, the mutation may be deleterious in the next generation if two nonfunctional genes are inherited. Some mutations do not alter the function of genes and are called silent. Still other mutations may result in gene products with altered function. Depending on the environment, this altered function may provide some benefit for survival. It is also possible that this altered function is deleterious. On a species level, mutations in germ cells are one source of genetic diversity. In organisms that replicate asexually, all cell mutations are a potential source of genetic diversity. Genetic diversity helps ensure that at least some organisms will have a chance of surviving in spite of changing environmental conditions.

11. Refer to Figure 9.17.

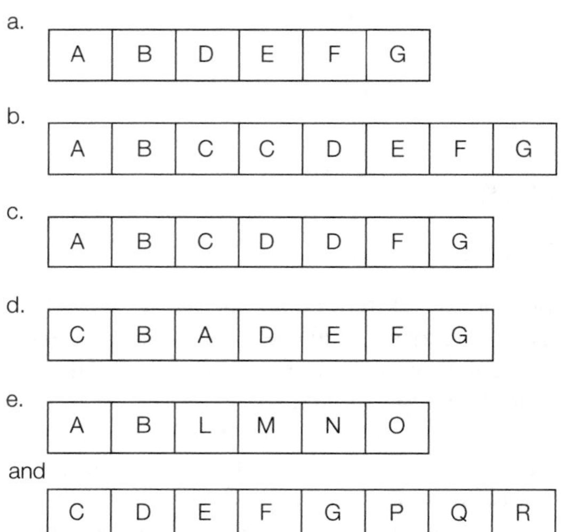


### *Test Yourself*

1. **e.** A duplication of gene *X* might allow the protein to be functional. The other chromosomal mutations would remove large regions of the coding sequence or would position part of that sequence in the opposite direction (inversion), which would produce a nonfunctional protein X.

2. **e.** Model-building by Watson and Crick created a three-dimensional model of the size, bond angles, base pairings, and overall structure of the DNA molecule. The information that contributed to their model came from a variety of experiments performed by other scientists. The model did not prove or disprove that DNA replication was semiconservative.

3. **c.** Replication of DNA is a fundamental requirement for its function as the genetic material. DNA must be correctly replicated in each cell of an organism

and passed from parent to offspring or from cell to daughter cell. DNA is not passed from one species to another. There are variations in DNA sequences from different organisms. DNA is found outside the nucleus in eukaryotic cells (specifically in the mitochondria, and in the chloroplasts of plants), but this DNA is inherited in a non-Mendelian fashion (see Chapter 8). For evolution to occur there must be some mistakes in DNA replication over time.

4. **a.** DNA polymerase adds nucleotides to an existing nucleotide strand, ligase seals nicks, and other proteins unwind the DNA, separate the strands, and prevent reassociation of the parental strands.

5. **d.** Okazaki fragments are small segments of newly synthesized DNA that have been added to short RNA primers. The RNA is removed from these small fragments by DNA polymerase I and replaced with DNA. The remaining DNA fragments are ligated (covalently bound) to form a continuous, newly synthesized DNA strand. DNA polymerase uses free nucleotide bases for DNA synthesis.

6. **b.** DNA polymerase cannot initiate synthesis of a nucleotide strand; it can only add to an existing strand of RNA primer or DNA. Primers provide DNA polymerase with the 3′ (—OH) end required for the addition of deoxyribonucleotides. The origin is also required for DNA synthesis, but it is the site where the replication complex is initially assembled. Primase is not the first enzyme in the replication complex.

7. **a.** The mismatch repair mechanism operates before the newly synthesized DNA strand is methylated, but for the integrity of DNA to be maintained, repair mechanisms must be active during synthesis, modification, and utilization of DNA.

8. **b.** If 30 percent of DNA is adenine, then according to Chargaff's rule, 30 percent must be thymine. The remaining 40 percent of the DNA is cytosine and guanine. Because the ratio of cytosine to guanine must be equal, the percentage of cytosine in this DNA must be 20 percent.

9. **b.** Gain-of-function mutations are dominant and are expressed in mutant cells. They do not respond to the appropriate environmental signals and do not have a restrictive condition under which they are expressed. Some gain-of-function mutations are seen in cancer cells, but mutations in tumor suppressors are also responsible for unregulated cell division in cancer cells.

10. **c.** PCR amplifies specific DNA sequences from small or large samples of DNA and requires three sequential steps: denaturation, binding, and replication. The amplified DNA molecules do not have variable DNA sequences.

11. **c.** Guanine is a purine, and its paired pyrimidine is cytosine.

12. **d.** Sulfur is a constituent of many protein molecules, but it is not found in DNA.

13. **a.** A hydroxyl group at the 3′ position of a nucleotide is necessary for the binding of any additional nucleotides. If this hydroxyl group is absent, no other nucleotides can be added to a growing strand.

14. **b.** Telomerase binds the telomeres at the ends of the chromosomes, protects them from being degraded or recombined with other chromosomes, and ensures that these ends will be replicated after primer removal. Telomeres are not associated with cell division checkpoints or continuous cell division.

15. **c.** Chargaff found that in most DNA sampled, the amount of A equaled the amount of T and the amount of G equaled the amount of C. It follows that the amount of purines (A + G) equals the amount of pyrimidines (T + C). The same does not hold for the amount of (A + T) versus the amount of (G + C).

16. **a.** Okazaki fragments are involved in synthesis of the lagging strand.

# 10 From DNA to Protein: Gene Expression

## The Big Picture

- Every protein and RNA in the cell has a DNA blueprint (the gene sequence) that specifies the amino acid or nucleotide sequence of that gene product. The sequences determine how that gene product will fold three-dimensionally and ultimately function in the cell.
- Alterations in the gene products, which are caused by mutations in the genes, help us understand how those proteins and RNAs function in the cell.
- Many proteins are modified after translation or combined with other peptides to generate a complete, functional product.

## Study Strategies

- Don't try to understand every detail of gene expression without first comprehending the larger picture: the way genetic information is accessed in the DNA and expressed in the cell so that the cell can function. Familiarize yourself with the details of the central dogma, but then take a step back and understand that all of these details describe the steps required to synthesize gene products.
- It is easy to confuse transcription and translation. Draw diagrams of the processes of both, including initiation, elongation, and termination. Carefully review the template, the product, and the sites of initiation and termination. Label the components that play a role in each process. Be sure to orient the 5′ and the 3′ ends of the nucleic acid, and the N and C terminus of the protein.
- Choose a hypothetical gene that encodes a peptide of four amino acids and draw a flowchart that shows how that gene in the chromosome is made into protein. Next, follow the details of gene expression: the synthesis of the RNA from the DNA, and the synthesis of the protein from the mRNA. Be sure to include the start and stop signals in each of these processes. Next, alter one of those nucleotides in the DNA sequence and transcribe and translate the gene. Is the protein sequence changed? Repeat this process using a different mutation in the DNA.
- Review the experiments that revealed the presence of introns in the primary RNA transcript and describe how introns are removed during splicing. Develop analogies to explain the editing process to a non-biology student.
- Introns and exons are easily confused. It may help to remember that "Introns Interrupt" and "Exons are Expressed."
- When taking notes in class and from the book, you will find you will be writing the words "transcription" and "translation" again and again. Many scientists use "txn" as shorthand for "transcription" and "tln" as shorthand for "translation." The abbreviations for ribosomal RNA (rRNA), transfer RNA (tRNA), and messenger RNA (mRNA) can also be added to your repertoire.
- Go to **LaunchPad** (or use the Web addresses listed) to review the following additional resources:

  Animated Tutorial 10.1 Transcription (PoL2e.com/at10.1)

  Animated Tutorial 10.2 RNA Splicing (PoL2e.com/at10.2)

  Animated Tutorial 10.3 Deciphering the Genetic Code (PoL2e.com/at10.3)

  Animated Tutorial 10.4 Genetic Mutations Simulation (PoL2e.com/at10.4)

  Animated Tutorial 10.5 Protein Synthesis (PoL2e.com/at10.5)

  Activity 10.1 Eukaryotic Gene Expression (PoL2e.com/ac10.1)

  Activity 10.2 The Genetic Code (PoL2e.com/ac10.2)

  Analyze the Data in textbook Figure 10.10

  Apply the Concept in textbook Sections 10.1 and 10.3

## Key Concept Review

### 10.1 Genetics Shows That Genes Code for Proteins

Observations in humans led to the proposal that genes determine enzymes

The concept of the gene has changed over time

Genes are expressed via transcription and translation

In metabolic pathways, the product of one reaction becomes the substrate for the next reaction. If we understand how the larger pathways work, we can figure out which step failed in a metabolic disorder such as alkaptonuria or phenylketonuria. In both of these disorders, the mutation in genes coding for enzymes leads to the buildup of an undesirable intermediate. Study of this type of disorder led to the idea of one gene correlating to one enzyme.

Further work revealed that some proteins comprise multiple peptide subunits. This led to the refinement of the idea that a single gene coded for a single protein and the relationship was modified to one gene–one polypeptide. While still useful for many gene products, this relationship oversimplifies how genes work. For example, RNA is a gene product that may or may not be translated into a protein. Even so, the transcription of the genome into an RNA sequence followed by translation into a polypeptide chain is true in many cases in a cell and in all instances of protein synthesis.

**Question 1.** Suppose that two different mutant strains of a bacterium are unable to grow on a minimal medium without the addition of the amino acid lysine, even though the "normal" strain does not need a lysine supplement. Explain how different mutations in each strain, either in the same gene or in two different genes, might have resulted in this lysine-requiring phenotype.

**Question 2.** Imagine that you are trying to explain the process of translation and transcription to a non-biologist, and you decide to use the analogy of a builder's construction of a new house. How would you create this analogy? What are the biological equivalents of the architect's blueprints, the construction site, the builder's photocopies of the blueprints, the actual building materials, the Bobcat loader, and the finished building? Be sure to detail the roles of DNA, mRNA, rRNA, tRNA, amino acids, and protein.

### 10.2 DNA Expression Begins with Its Transcription to RNA

RNA polymerases share common features

Transcription occurs in three steps

Eukaryotic coding regions are often interrupted by introns

Eukaryotic gene transcripts are processed before translation

In both eukaryotes and prokaryotes, transcription occurs in three steps: initiation, elongation, and termination. The process of initiation begins with a promoter sequence, which orients RNA polymerase and indicates which strand of DNA is to be transcribed. RNA polymerase does not bind all by itself; rather, it is one of many proteins that are recruited to the promoter region in order to initiate transcription. In the second stage, elongation, RNA polymerase unwinds the DNA double helix and synthesizes a single strand of mRNA in a 5′-to-3′ direction while reading the DNA template in the 3′-to-5′ direction. The complementary triphosphate nucleotides are added using energy contained in the bonds between the three phosphates. When RNA polymerase encounters the termination sequence in the DNA, either the RNA polymerase falls off the DNA template or additional proteins remove it from the DNA template, and transcription ends.

In eukaryotes, the coding sequences (exons) often have noncoding "interruptions" called introns. The product of transcription includes all of these sequences, and before translation begins the cell works to remove the introns by a process called RNA splicing. Splicing is completed by an RNA protein complex called a spliceosome, which reads consensus sequences at the boundaries of introns and exons in order to splice the mRNA properly. Once splicing is complete, a GTP cap is added to the 5′ end of the pre-mRNA and a poly A tail is added to the 3′ end. These modifications help mRNA bind to the ribosome, protect the mRNA from premature degradation, and assist in the transport of mRNA from the nucleus to the cytosol.

**Question 3.** Suppose that you place a double-stranded DNA sequence in a test tube along with the following components: RNA polymerase, dATP, dCTP, dGTP, dUTP, and the required transcription factors. A coworker uses the same solution with a different DNA template sequence and finds that mRNA is the result. Your procedure, however, does not result in mRNA, even though you are sure that the reagents are sound. What went wrong?

**Question 4.** On the DNA and mRNA fragments below, label the following: 5′ GPT cap, 3′ poly A tail, introns, exons, promoter, terminator, and splice sites.

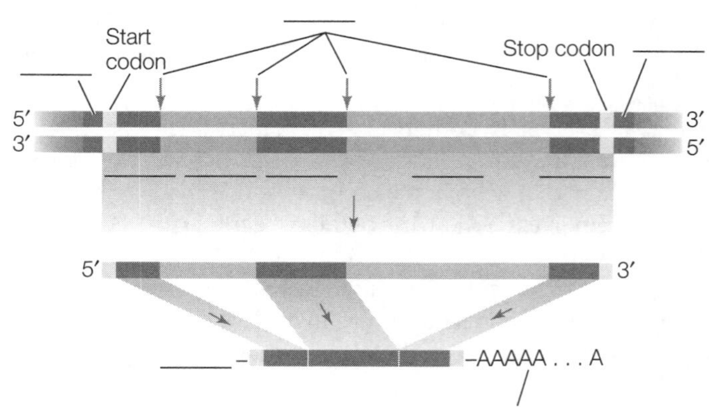

## 10.3 The Genetic Code in RNA Is Translated into the Amino Acid Sequences of Proteins

The information for protein synthesis lies in the genetic code

Point mutations confirm the genetic code

Trios of nucleotides (called a codon) code for individual amino acids. There is more than one codon for many amino acids, as well as a start codon and three stop codons. The genetic code is very highly conserved in living things, indicating that it arose in a common ancestor early in the evolution of life. Alterations to the genetic code may not alter the amino acid, especially if the change is in the third nucleotide of the codon (a silent mutation). In other cases, a single nucleotide change may change the codon to that of a different amino acid (a missense mutation) or a premature stop codon (a nonsense mutation). If an extra nucleotide is added to a DNA sequence, or one is lost, the resulting shift in reading frame will result in the production of a different polypeptide (a frame-shift mutation).

**Question 5.** Starting with the following mature mRNA sequence, identify the start codon, stop codon, and amino acid sequence.

```
         10         20         30         40         50         60
5'–UUAGCTAUCC UAAAGUAUGC GUCAUUCUCA AAUCGUUUGG GGUUGUUAAU GUAAACGUCA–3'
```

**Question 6.** Using the mature mRNA sequence shown in Question 5 as a starting point, determine the amino acid sequences that would result from the following changes and classify them as silent, missense, nonsense, or frame-shift mutations:

a. Nucleotide 29 is replaced with guanine.
b. Nucleotide 29 is replaced with uracil.
c. Nucleotide 37 is replaced with adenine.
d. One uracil is inserted between nucleotides 10 and 11.
e. The nucleotide at position 20 is deleted.
f. One guanine is inserted between nucleotides 17 and 18.

## 10.4 Translation of the Genetic Code Is Mediated by tRNAs and Ribosomes

Transfer RNAs carry specific amino acids and bind to specific codons

Each tRNA is specifically attached to an amino acid

Translation occurs at the ribosome

Translation takes place in three steps

Polysome formation increases the rate of protein synthesis

Translation is the process of reading the codons in the mRNA to properly link amino acids in a peptide chain. The job of tRNA is to transport specific amino acids to the ribosome. By reading the codon, the tRNA is able to bind to the mRNA at the appropriate time, allowing the attached amino acid to be linked to the peptide chain. The other main structure involved is rRNA, which is composed of two subunits. These subunits position the tRNAs along the mRNA sequence and catalyze peptide bond formation.

Like transcription, translation has three steps: initiation, elongation, and termination. The initiation process involves binding of the small ribosomal subunit and a charged tRNA to the start codon on the mRNA. The large subunit of the ribosome then joins the complex. Elongation breaks the bond between the tRNA and the amino acids while forming peptide bonds between amino acids already present at the P site and those delivered to the A site. The ribosome and mRNA shift relative to each other by one codon and the process is repeated until a termination codon is reached. At this point, a protein release factor frees the peptide chain and translation is complete.

**Question 7.** Draw a diagram of a eukaryotic cell and indicate where in the cell the gene is transcribed and translated. In your diagram, indicate how this particular gene product is targeted to a compartment in the cell such as the endoplasmic reticulum. Create a stepwise list of all the important proteins and enzymes required for this process.

**Question 8.** What would happen if the tRNA synthase for tryptophan added a phenylalanine to the tryptophan tRNAs instead of tryptophan?

### 10.5 Proteins Are Modified after Translation

Signal sequences in proteins direct them to their cellular destinations

Many proteins are modified after translation

The polypeptide released from a ribosome is often not a complete, functional protein. Proteins are sent to different organelles and they may be modified after translation is complete.

Proteins are sent to the proper destination by use of signal sequences at the N terminus of the peptide chain. They may be modified with the addition of phosphates or sugars. They may be synthesized in an inactive form and become active only after being cleaved by proteases. The combination of signal sequences and posttranslational modifications may confer additional functions to the proteins.

**Question 9.** Suppose that a protein is supposed to go to the endoplasmic reticulum and the DNA encoding the signal sequence for that gene product is deleted. What would be the result?

**Question 10.** What would happen if you put an NLS sequence on a cytoplasmic protein?

## Test Yourself

1. Transcription in prokaryotic cells
   a. occurs in the nucleus, whereas translation occurs in the cytoplasm.
   b. is initiated at a start codon with the help of initiation factors and the small subunit of the ribosome.
   c. is initiated at a promoter and uses only one strand of DNA (the template strand) to synthesize a complementary RNA strand.
   d. is terminated at a stop codon.
   e. does not need a promoter.
   ***Textbook Reference:*** *Concept 10.2 DNA Expression Begins with Its Transcription to RNA; Transcription occurs in three steps*

2. Which of the following statements about RNA polymerase is *false*?
   a. It synthesizes mRNA in a 5′-to-3′ direction, reading the DNA strand 3′ to 5′.
   b. It synthesizes mRNA in a 3′-to-5′ direction, reading the DNA strand 5′ to 3′.
   c. It binds at the promoter and unwinds the DNA.
   d. It does not require a primer to initiate transcription.
   e. It uses only one strand of DNA as a template for synthesizing RNA.
   ***Textbook Reference:*** *Concept 10.2 DNA Expression Begins with Its Transcription to RNA; Transcription occurs in three steps*

3. Translation of messenger RNA into protein occurs in a _______ direction and from the _______ terminus to the _______ terminus.
   a. 3′-to-5′; N; C
   b. 5′-to-3′; N; C
   c. 3′-to-5′; C; N
   d. 5′-to-3′; C; N
   e. 3′-to-5′; C; C
   ***Textbook Reference:*** *Concept 10.4 Translation of the Genetic Code Is Mediated by tRNAs and Ribosomes; Translation takes place in three steps*

4. If codons were read two bases at a time instead of three bases at a time, how many different possible amino acids could be specified?
   a. 8
   b. 16
   c. 32
   d. 64
   e. 128
   ***Textbook Reference:*** *Concept 10.3 The Genetic Code in RNA Is Translated into the Amino Acid Sequences of Proteins; The information for protein synthesis lies in the genetic code*

5. Peptidyl transferase is an
   a. enzyme found in the nucleus of the cell that assists in the transfer of mRNA to the cytoplasm.
   b. enzyme that adds the amino acid to the 3′ end of the tRNA.
   c. enzyme found in the large subunit of the ribosome that catalyzes the formation of the peptide bond in the growing polypeptide.
   d. RNA molecule that is catalytic.
   e. Both c and d
   ***Textbook Reference:*** *Concept 10.4 Translation of the Genetic Code Is Mediated by tRNAs and Ribosomes; Translation takes place in three steps*

6. What would happen if a mutation occurred in DNA such that the second codon of the resulting mRNA were changed from UGG to UAG?
   a. Translation would continue and the second amino acid would be the same.
   b. Nothing. The ribosome would skip that codon and translation would continue.
   c. Translation would continue, but the reading frame of the ribosome would be shifted.
   d. Translation would stop at the second codon, and no functional protein would be made.
   e. Translation would continue, but the second amino acid in the protein would be different.
   ***Textbook Reference:*** *Concept 10.3 The Genetic Code in RNA Is Translated into the Amino Acid Sequences of Proteins; The information for protein synthesis lies in the genetic code*

7. If the following synthetic RNA were added to a test tube containing all the components necessary for protein translation to occur, what would the amino acid sequence be?

5′-AUAUAUAUAUAU-3′

a. Polyphenylalanine
b. Isoleucine–Tyrosine–Isoleucine–Tyrosine
c. Isoleucine–Isoleucine–Isoleucine–Isoleucine
d. Tyrosine–Tyrosine–Tyrosine–Tyrosine
e. Aspargine–Aspargine–Aspargine–Aspargine

***Textbook Reference:*** *Concept 10.3 The Genetic Code in RNA Is Translated into the Amino Acid Sequences of Proteins; The information for protein synthesis lies in the genetic code*

8. Which part of the tRNA base-pairs with the codon in the mRNA?
a. The 3′ end, where the amino acid is covalently attached
b. The 5′ end
c. The anticodon
d. The start codon
e. The promoter

***Textbook Reference:*** *Concept 10.4 Translation of the Genetic Code Is Mediated by tRNAs and Ribosomes; Transfer RNAs carry specific amino acids and bind to specific codons*

9. Translate the following mRNA:
3′-GAUGGUUUUAAAGUA-5′
a. $NH_2$ Met—Lys—Phe—Leu—Stop COOH
b. $NH_2$ Met—Lys—Phe—Trp—Stop COOH
c. $NH_2$ Asp—Gly—Phe—Lys—Val COOH
d. $NH_2$ Met—Gly—Phe—Lys—Val COOH
e. $NH_2$ Asp—Gly—Phe—Lys—Stop COOH

***Textbook Reference:*** *Concept 10.3 The Genetic Code in RNA Is Translated into the Amino Acid Sequences of Proteins; The information for protein synthesis lies in the genetic code*

10. Termination of translation requires
a. a termination signal, RNA polymerase, and a release factor.
b. a release factor, initiator tRNA, and ribosomes.
c. initiation factors, the small subunit of the ribosome, and mRNA.
d. elongation factors and charged tRNAs.
e. a stop codon positioned at the A site of the ribosome and a release factor.

***Textbook Reference:*** *Concept 10.4 Translation of the Genetic Code Is Mediated by tRNAs and Ribosomes; Translation takes place in three steps*

11. If the DNA encoding a nuclear signal sequence were placed in the gene for a cytoplasmic protein, the protein would
a. be modified in the Golgi.
b. be directed to the lysosomes.
c. be directed to the nucleus.
d. be directed to the cytoplasm.
e. stay in the endoplasmic reticulum.

***Textbook Reference:*** *Concept 10.5 Proteins Are Modified after Translation; Signal sequences in proteins direct them to their cellular destinations*

12. The process of gene expression involves two steps in which _______ is transcribed into _______, which is (are) translated into _______.
a. a gene; polypeptides; a gene product
b. protein; DNA; RNA
c. DNA; mRNA; tRNA
d. DNA; RNA; protein
e. RNA; DNA; protein

***Textbook Reference:*** *Concept 10.2 DNA Expression Begins with Its Transcription to RNA; Transcription occurs in three steps*

13. If the third nucleotide in the codon 5′-AAA-3′ is changed from A to G, a _______ mutation will occur and the result will be _______.
a. silent; no effect
b. nonsense; no effect
c. missense; formation of a stop codon
d. frame-shift; no effect
e. nonsense; formation of a stop codon

***Textbook Reference:*** *Concept 10.3 The Genetic Code in RNA Is Translated into the Amino Acid Sequences of Proteins; The information for protein synthesis lies in the genetic code*

14. The enzyme that catalyzes the synthesis of RNA is
a. peptidyl transferase.
b. DNA polymerase.
c. tRNA synthase.
d. ribosomal RNA.
e. RNA polymerase.

***Textbook Reference:*** *Concept 10.2 DNA Expression Begins with Its Transcription to RNA; RNA polymerases share common features*

15. A mutation occurs such that a spliceosome cannot remove one of the introns in a gene. What effect will this have on that gene?
a. It will have no effect; the gene will be transcribed and translated into protein.
b. Transcription will terminate early and the protein will not be made.
c. Transcription will proceed, but translation will stop at the site where the intron remains.
d. Translation will continue, but a nonfunctional protein will be made.
e. Translation will continue and will skip the intron sequence.

***Textbook Reference:*** *Concept 10.2 DNA Expression Begins with Its Transcription to RNA; Eukaryotic gene transcripts are processed before translation*

## Answers

### *Key Concept Review*

1. The mutations in these two strains of bacteria apparently interfere with lysine synthesis. Both mutations might be in the same gene coding for an enzyme necessary for lysine synthesis, but one could be a nonsense mutation in the fifth codon, for example, and the other could be a frame-shift mutation in the twenty-third codon (the number of mutations that can disable a gene is enormous). If lysine synthesis in this bacterium requires more than one enzyme (as is likely), the two mutations could be in different genes coding for different enzymes. In this case, the phenotypes would not be strictly identical; it would be possible to distinguish between the two by trying to grow them on minimal media to which different intermediates in the synthesis of lysine were added.
2. The master blueprint is DNA, which is too precious to be taken to the actual worksite and possibly damaged. Instead, the DNA is photocopied into mRNA. Multiple copies of the same blueprint (mRNA) may be generated to allow the same blueprint to be read multiple times in multiple manufacturing sites in the cell. Ribosomal RNA serves as the main workbench where all of the raw materials are assembled. It brings together the mRNA template along with the amino acids in order to build a protein. The amino acids can be thought of as the actual building materials. They are brought to the workbench in the appropriate order by the tRNA, which serves as a kind of Bobcat loader, fetching the appropriate materials and delivering them at the proper time and in the proper amount. The mRNA blueprint is read carefully as each building block is put into place on the ribosome. Finally, the building (the protein) is constructed according to the plans (mRNA), and the construction crew (RNAs, energy, amino acids) can leave the scene and go to a new site.
3. Your DNA sequence does not have a promoter. The process of transcription depends on the promoter sequence to recruit RNA polymerase to the DNA, orient the RNA polymerase, and indicate which DNA strand is the template strand. Your coworker's DNA template included a promoter sequence.
4. 

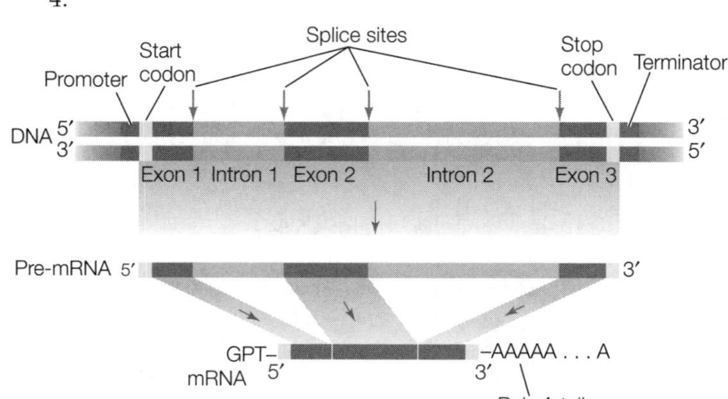


5. The amino acid sequence is coded from nucleotides 17–46. The stop codon is from nucleotides 47–49. The amino acid sequence is: Met–Arg–His–Ala–Gln–Ile–Val–Trp–Gly–Cys

```
              10          20          30          40          50          60
5′–UUAGCTAUCC UAAAGUAUGC GUCAUGCUCA AAUCGUUUGG GGUUGUUAAU GUAAACGUCA–3′
                     Met  Arg  His  Ala  Gln  Ile  Val  Trp  Gly  Cys  Stop
```

6. 
   a. Met–Arg-His-Ala-Glu-Ile–Val–Trp–Gly–Cys (Missense, amino acid 5)
   b. Met–Arg–His–Ala (Nonsense; premature stop after serine)
   c. Met–Arg–His–Ala–Gln–Ile–Val–Trp–Gly–Cys (Silent)
   d. Met–Arg–His–Ala–Gln–Ile–Val–Trp–Gly–Cys (Silent; the insertion was before the start codon, so the reading frame was not changed.)
   e. Met–Val–Met–Leu–Lys–Ser–Phe–Gly–Val–Val–Asn–Val–Asn–Val (Frame-shift; there is no stop codon in this sequence in frame with the new reading frame.)
   f. Met–Leu–Lys–Ser–Phe–Gly–Val–Val–Asn–Val–Asn–Val (Frame-shift; the first start codon was disrupted, so the second Met in answer e. is the new start codon here. There is no stop codon in this sequence in frame with the new reading frame.)

7.

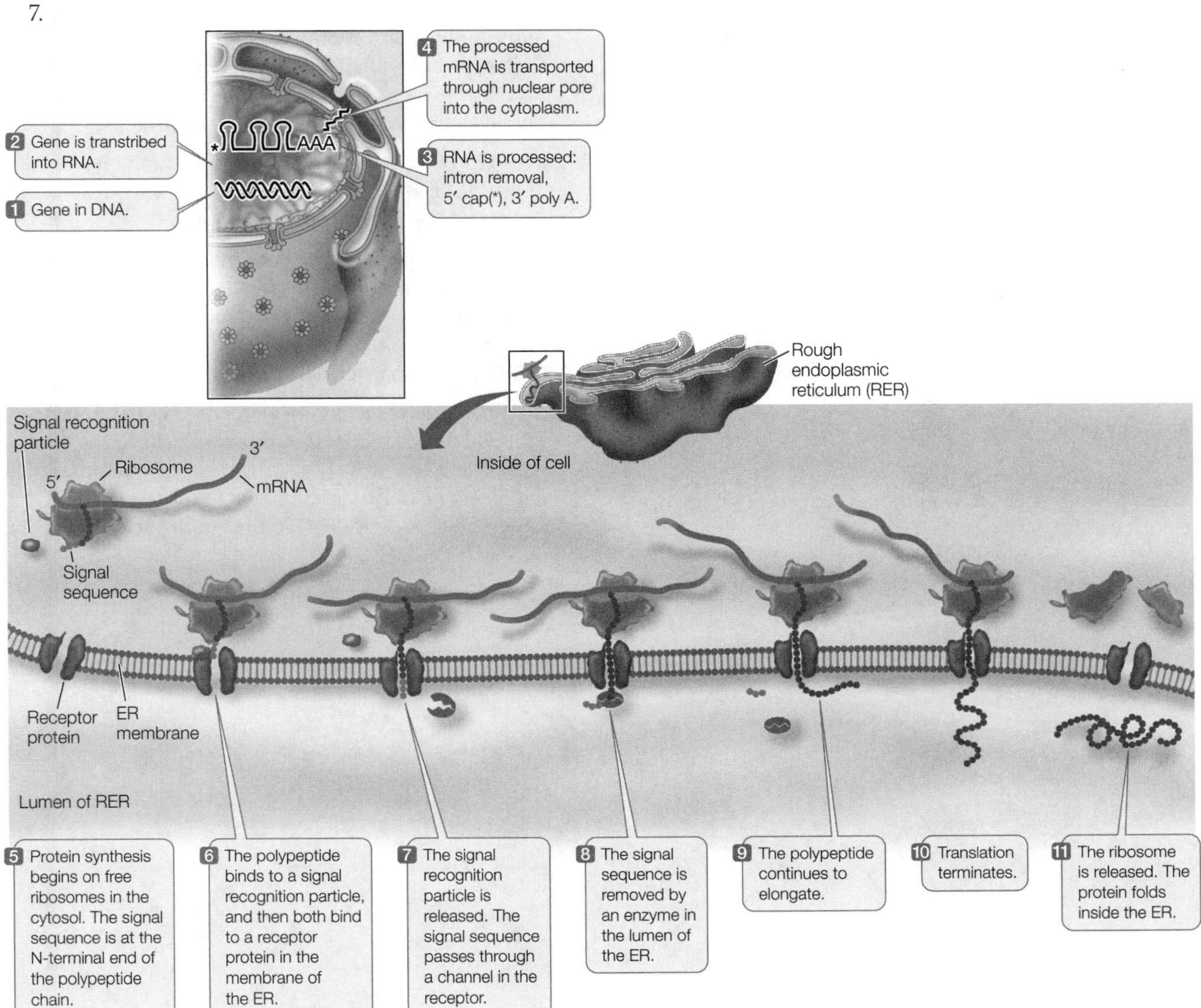


8. If the tRNA synthetase for tryptophan added phenylalanine to the tryptophan tRNAs, phenylalanine would be added to the polypeptide whenever a tryptophan codon was read by these tryptophan tRNAs. This would create proteins that were nonfunctional, and the cell would die.

9. Deletion of the signal sequence would not affect the transcription or translation of this gene, but it would affect the targeting of the gene product. During translation of the mRNA, no signal peptide would be made, and the signal recognition particle would be unable to bind. The mRNA would continue to be translated in the cytoplasm and remain there.

10. Adding an NLS sequence to a protein targets the protein to the nucleus. Therefore, the cytoplasmic protein would be targeted to the nucleus rather than remaining in the cytoplasm.

### *Test Yourself*

1. **c.** Transcription occurs in the nucleus and translation occurs in the cytoplasm of eukaryotic cells, not of prokaryotic cells. Translation, rather than transcription, is initiated at start codons and terminates at a stop codon.

2. **a.** RNA polymerase binds at a promoter, unwinds the DNA, synthesizes mRNA in a 5′-to-3′ (not 3′-to-5′) direction, and does not require a primer to synthesize the RNA.

3. **b.** Translation of messenger RNA occurs 5′ to 3′, and the polypeptide is synthesized from the N terminus to the C terminus.

4. **b.** Four possible bases read two at a time would yield $4^2$, or 16, different codons.

5. **e.** Peptidyl transferase is the enzyme that catalyzes the formation of the peptide bond, and it is located in

the large subunit of the ribosome. Its catalytic activity is due to ribosomal RNA found in the large subunit of the ribosome.

6. **d.** UAG is a stop codon, so translation would terminate at that site.
7. **b.** See the codon table (Figure 10.11).
8. **c.** Neither the 3′ end nor the 5′ end of the tRNA is part of the anticodon. The promoter is a DNA sequence, to which RNA polymerase binds to initiate transcription. The start codon is found in the mRNA.
9. **c.** See the codon table (Figure 10.11). Recall that translation occurs in the 5′-to-3′ direction.
10. **e.** Termination of translation requires a stop codon positioned at the A site of the ribosome, and a release factor.
11. **c.** The nuclear sequence would direct this protein to the nucleus.
12. **d.** Genes are not transcribed into polypeptides, protein is not used to synthesize DNA, and messenger RNAs are not translated into tRNAs. RNA can be used to synthesize DNA using reverse transcriptase, but DNA cannot be utilized to make protein.
13. **a.** A G to C mutation in the third position results in an mRNA sequence of UUC, which also codes for phenylalanine, thus a silent mutation.
14. **e.** DNA polymerase catalyzes the synthesis of DNA, tRNA synthase covalently attaches amino acids to tRNAs, and ribosomal RNA (peptidyl transferase in the large subunit) catalyzes the formation of the peptide bond during translation.
15. **d.** When an intron fails to be removed, that noncoding sequence is retained in the RNA within the coding sequence. When this RNA is translated, the protein will likely be nonfunctional due to the insertion of a noncoding sequence within the coding sequence.

# 11 Regulation of Gene Expression

## The Big Picture

- Gene expression is highly regulated, allowing an organism to respond to external and internal cues. Genes can be expressed in response to specific environmental or developmental conditions (inducible) or all the time (constitutive). Regulation of genes can potentially occur at many points. Gene expression is controlled at transcription by activators or repressors.
- Viruses and bacteriophage are acellular and depend on host cell metabolism to carry out their gene expression and reproduction. Viruses can have a DNA or RNA genome, and inject it into the host cell when they infect it. They may enter a lytic or lysogenic phase once in the cell.
- Prokaryotes regulate gene expression transcriptionally at promoters and posttranscriptionally by mRNA degradation, translational regulation, and protein degradation. Prokaryotic cells use operons to coordinately regulate the expression of several genes.
- Eukaryotes regulate gene expression at the transcriptional level with general transcription factors as well as specific repressors and activators, and they coordinate regulation by placing regulatory sequences in front of different genes. Epigenetic modifications also affect gene expression. Posttranscriptional regulation includes alternate splicing, modification of the 5′ cap, control of translational initiation, and protein degradation.

## Study Strategies

- It is easy to confuse prokaryotic inducible operons (the *lac* operon) with repressible operons (the *trp* operon) because they both use repressors to regulate gene expression. Review the environmental conditions that must exist for each repressor to bind its operator site and what causes each repressor to release the operator site.
- It may at first seem puzzling that some viruses insert their chromosome into the host chromosome, because the goal of viral infection seems to be rapid multiplication and infection of adjacent cells. However, lysogeny allows the viral genome to persist while local environmental resources are abundant. When local resources are depleted or cell damage occurs, the virus can enter the lytic cycle to escape and infect other cells.
- Gene expression in eukaryotes can be regulated epigenetically by DNA methylation and chromatin remodeling. RNA polymerase and transcription factors require access to the DNA sequences before transcription can begin. Diagram the process of transcriptional activation for a gene that can be modified by methylation, and for a gene that undergoes histone modification. When is the gene active? When is it silent?
- Gene expression often occurs in response to environmental signals. Make a list of the environmental signals that a prokaryote receives and then outline the steps initiated by the cell in response to those signals. Include the following signals: lactose in the cell, glucose and lactose in the cell, high levels of glucose in the cell, high or low levels of tryptophan in the cell, bacteriophage infection under poor or rich environmental conditions.
- Outline the different viral life cycles, from the start of infection through multiplication of virus to viral release (see Figure 11.7). Compare the lytic and lysogenic life cycles of bacteriophage.
- Gene regulation can occur at many steps in a eukaryotic cell. Starting with chromatin remodeling, list the steps that must occur for a gene to be made into functional protein in the cytoplasm of the cell.
- Go to **LaunchPad** (or use the Web addresses listed) to review the following additional resources:

  Animated Tutorial 11.1 The *lac* Operon (PoL2e.com/at11.1)

  Animated Tutorial 11.2 The *trp* Operon (PoL2e.com/at11.2)

  Animated Tutorial 11.3 Initiation of Transcription (PoL2e.com/at11.3)

  Activity 11.1 Eukaryotic Gene Expression Control Points (PoL2e.com/ac11.1)

Activity 11.2 Concept Review (PoL2e.com/ac11.2)
Analyze the Data in textbook Figure 11.10
Apply the Concept in textbook Sections 11.1 and 11.4

## Key Concept Review

### 11.1 Many Prokaryotic Genes Are Regulated in Operons

Genes are subject to positive and negative regulation

Regulating gene transcription is a system that conserves energy

Operons are units of transcriptional regulation in prokaryotes

Operator–repressor interactions regulate transcription in the *lac* and *trp* operons

RNA polymerase can be directed to a class of promoters

Viruses use gene regulation strategies to hijack host cells

Genes can be negatively or positively regulated at many points: before transcription, during transcription, after transcription but before translation, at translation, and after translation. At transcription, gene expression can be negatively regulated by a repressor that inhibits transcription, or positively regulated by an activator protein that stimulates transcription.

Prokaryotes conserve energy by making proteins only when they need them. The most efficient means of regulating gene expression is at the level of transcription, because protein synthesis is energetically expensive. Another method of gene regulation (allosteric regulation) allows fine tuning of metabolism.

An example of transcription-level gene regulation is the response to food availability of the gut bacteria *E. coli*. Lactose uptake by *E. coli* involves three proteins: β-galactoside permease, which transports lactose into the cell; β-galactosidase, which breaks down lactose into glucose and lactose; and β-galactoside transacetylase, which transfers acetyl groups to certain β-galactosides.

The genes that encode these three enzymes are called structural genes, and they reside close to each other in the *E. coli* genome. A cluster of genes regulated by a single promoter is called an operon. *E. coli* responds to changes between glucose and lactose availability using the *lac* operon. The *lac* operon includes the sequences for the promoter, the operator, and the structural genes for the enzymes involved in lactose metabolism (see Figure 11.3). The operator controls transcription of the structural genes.

The *lac* operon is an inducible operon and is transcribed only in the presence of the inducer, β-galactoside (e.g., lactose; see Figure 11.4). Other operons (e.g., *trp* operon) are repressible operons that are turned off only in response to a repressor.

In the inducible *lac* operon, a repressor protein is normally bound to the operator, preventing transcription. When the inducer is abundant, it binds to the repressor, causing it to change shape so it can no longer bind to the DNA and block transcription.

In summary, inducible operons have the following features: In the absence of the inducer, the operator is bound by a regulatory protein (the repressor) to prevent transcription. In the presence of the inducer, the repressor binds to the inducer so that it changes shape and no longer binds with the DNA, so transcription of the operon occurs.

The *trp* operon is a repressible operon. Like an inducible operon, the repressible operon is switched off when the repressor is bound to the operator. However, in the case of the repressible operon, the repressor is not bound to the operator unless another molecule, the co-repressor (tryptophan, in this case), first binds to the repressor. When this occurs, the repressor becomes active and binds to the operator, preventing transcription (see Figure 11.5).

In repressible systems (like that of the *trp* operon), the product of the metabolic pathway (the co-repressor) interacts with the regulatory protein, enabling it to bind the promoter and block transcription. The presence of the product thus turns off the transcriptional machinery necessary to produce more of that product.

In general, inducible systems like the *lac* operon regulate catabolic pathways that are turned on only when the substrate is available, while repressible systems like the *trp* operon regulate anabolic pathways that are always turned on unless the concentration of the product becomes excessive.

Other proteins in prokaryotes called sigma factors regulate transcription by binding to RNA polymerase and guiding it to specific promoters.

Viruses are acellular and perform no metabolic functions. They develop and reproduce only inside the cells of specific hosts. Outside the host cell, viruses exist as virions, which consist of nucleic acid and proteins. By means of gene regulation, they manipulate the host cell to replicate the virus particles (virions). Some viruses are lytic (they immediately activate the host cell to replicate the virus and lyse, releasing the virions), and some alternate between lytic and lysogenic periods. The latter is a dormant period during which the virus genome is replicated with the cell genome until it is induced to begin the lytic stage.

Many bacteriophages are lytic. Lytic viral cycles have two stages, early and late. During the early phase, the phage injects its nucleic acid into the host cytoplasm, and viral genes are transcribed and translated. These early gene products shut down host transcription, degrade host DNA, and stimulate viral genome replication and viral gene transcription. Late gene products include viral capsid proteins and proteins that lyse the cell at the end of the lytic cycle. The lytic cycle takes about 30 minutes and produces hundreds of bacteriophage per cell (see Figure 11.7).

**Question 1.** Which type of gene regulation do prokaryote genomes employ: positive, negative, or a combination of both?

**Question 2.** Suppose that a cell has a mutation that deletes the gene encoding the repressor for a certain operon, and a plasmid is introduced into the host cell that carries a wild-type copy of the gene for the repressor. Is normal regulation of this operon restored in the presence of this plasmid?

### 11.2 Eukaryotic Genes Are Regulated by Transcription Factors

Transcription factors act at eukaryotic promoters

The expression of transcription factors underlies cell differentiation

Transcription factors can coordinate the expression of sets of genes

Eukaryotic viruses can have complex life cycles

Eukaryotic promoters have three important sequences: a regulatory binding site, a transcription binding site that often has a TATA box, and an RNA polymerase binding site. Initiation of transcription by RNA polymerase II in eukaryotic cells requires regulatory proteins called general transcription factors (TFIID). These are proteins that bind to the promoter and form a transcription complex to which RNA polymerase II can bind in order to initiate transcription. General transcription factors are different from transcription factors that are specific to promoters or classes of promoters.

TFIID binds the TATA box, changing the shape of the DNA and the transcription factor itself. Other transcription factors then bind the complex on the promoter, and RNA polymerase II binds to initiate transcription (see Figure 11.10).

Other short DNA sequences bind regulatory proteins that affect transcription positively (enhancers, which bind activator proteins) and negatively (silencers, which bind repressors). Regulatory DNA sequences can be located close to or far away from the gene they affect. Often, many binding factors are involved and transcription of any eukaryotic gene is determined by the combination of transcription factors, repressors, and activators.

Recognition of specific nucleotide sequences by transcription factors involves available sites for hydrogen bonding, hydrophobic interactions, and induced fit.

Coordinated gene regulation is achieved when genes share regulatory sequences that bind the same transcription factors. An example of coordinated gene regulation is seen in plant drought response genes. Regulatory elements known as dehydration response elements (DREs) are found near the promoters of genes that respond to drought (see Figure 11.11).

Human immunodeficiency virus (HIV) is a retrovirus (RNA genome) that infects only immune system cells and alternates between lytic and lysogenic stages. HIV is an enveloped virus, enclosed in a plasma membrane derived from the previous host cell. HIV copies its RNA genome into the host DNA using reverse transcriptase. The DNA copy of the viral genome is then integrated into a host cell chromosome where it becomes a provirus (see Figure 11.12). The provirus may remain dormant in the host virus for years.

When activated, the provirus is transcribed as mRNA and translated into protein by the host cell's translation machinery. Usually, the cell has regulatory mechanisms to terminate expression of invader virus genes. The HIV viral protein Tat (transactivation of transcription) prevents termination, allowing viral gene transcription by the host RNA polymerase (see Figure 11.13).

**Question 3.** Some people who are exposed to the HIV virus do not become infected. Describe two possible ways their cells might resist the HIV invasion.

**Question 4.** Suppose you are engineering gene *Y*, such that when it is inserted into a eukaryotic chromosome it will be expressed continuously. Which specific sequences must be part of this gene so that it will be expressed?

**Question 5.** Suppose that you are engineering a new plant in which gene *Y* will be activated under drought conditions. What kinds of DNA sequences must be present to ensure activation of the gene under these conditions?

**Question 6.** Diagram a gene in both a prokaryotic cell and a eukaryotic cell. Include the types of sequences that are important for transcriptional regulation, the location of those sequences, and the proteins that bind those regions.

### 11.3 Gene Expression Can Be Regulated via Epigenetic Changes to Chromatin

Modification of histone proteins affects chromatin structure and transcription

DNA methylation affects transcription

Epigenetic changes can be induced by the environment

DNA methylation can result in genomic imprinting

The term "epigenetics" refers to reversible changes in the DNA structure of a gene or genes that occurs without any change in the gene sequence (no mutation). These are heritable. Epigenetic processes include DNA methylation and chromatin remodeling.

One mechanism for gene regulation is chromatin remodeling. The DNA in eukaryotic chromosomes is attracted to positively charged histone proteins of the nucleosomes that can inhibit the initiation and elongation steps of transcription. Chromatin remodeling must occur to make the DNA accessible to transcription complexes (see Figure 11.14).

In methylation, cytosine residues are methlyated by DNA methyl transferase and the methlyated state can be inherited. In mammals, this usually happens in regions rich in CG-adjacent pairs called CpG islands. Maintenance methylases catalyze the formation of 5-methylcytosine on the replicated DNA strand. Repressor proteins bind methylated regions on DNA, so methylated genes are usually silenced. Reversal of methylation is accomplished when demethylases remove the methyl group from the cytosine (see Figure 11.15).

DNA methylation occurs in male and female genomes upon fertilization to silence duplicate genes on the extra X chromosome, and in tumor suppressor genes of cancer cells. When a long stretch or entire chromosome is methylated, it appears as heterochromatin (which stains darkly). The X chromosome in female animals is an example of heterochromatin. The genotype for the sex chromosomes in mammals is XY in males and XX in females. However, the expression of X-linked genes (the gene dosage) is the same in both sexes, due to X-inactivation in females. Early in embryonic development, one of the two X chromosomes is randomly inactivated, producing a highly condensed heterochromatic chromosome called a Barr body (see In-Text Art, p. 228 [2]).

Epigenetic changes that occur in germline cells can be inherited but later modified by environmental factors, including stress experienced by offspring. Monozygotic twins at the age of 3 have almost exactly the same methylation pattern. However, by age 50 their patterns are significantly different.

Genomic imprinting is the heritable difference between the methylation patterns in males versus females. This means that for a small subset of genes, methylated genes are silenced such that the genes are either expressed *only* from the nonmethylated genes inherited from the mother, or in other instances from the nonmethylated genes inherited from the father. Further, this pattern is necessary for viable offspring to be produced.

**Question 7.** Suppose you want to study the effects of silencing a gene. What is a heritable but nonpermanent mechanism by which this gene could be silenced?

**Question 8.** Recently Sally was diagnosed with breast cancer. Examination of her tumor showed that she was not expressing the *BRCA1* gene, a DNA repair gene whose loss is associated with an increased risk of the disease. However, a biopsy of breast tissue from her identical twin sister, Suzie, showed expression of the *BRCA1* gene. What could explain this difference in gene expression in identical twins?

### 11.4 Eukaryotic Gene Expression Can Be Regulated after Transcription

Different mRNAs can be made from the same gene by alternative splicing

MicroRNAs are important regulators of gene expression

Translation of mRNA can be regulated

Protein stability can be regulated

Alternate splicing of mRNA is a mechanism for generating different proteins from the same DNA (see Figure 11.16). There are only about 24,000 protein-coding genes in the human genome, but there are many more human mRNAs than human genes, and about 80 percent of human genes are alternatively spliced.

Some "noncoding" regions of DNA code microRNAs (miRNAs about 22 bp long) that bind specific mRNAs and block their translation. Each miRNA has dozens of target mRNAs. MiRNAs are transcribed as longer precursors that can fold into double-stranded hairpin structures. The folded miRNAs are then cleaved by dicer protein to produce short double-stranded miRNAs (see Figure 11.17). These miRNAs are converted to single-stranded RNA by another protein complex and guided to a target, where they bind to and inhibit translation of mRNAs and target them for degradation.

The amount of protein in a cell does not always correlate with the amount of mRNA, so translation of mRNAs in the cytoplasm must sometimes be regulated, or some other process must regulate how long the proteins persist in the cell.

Two other ways that translation of mRNA can be regulated is by modification of the 5′ cap and binding of translational repressor proteins. If the mRNA 5′ cap is not modified, the mRNA will not be translated. These unprocessed mRNAs can be stored until needed, and their 5′ cap will then be modified by GTP. Translational repressor proteins block translation by binding to mRNAs and preventing them from binding to ribosomes (see Figure 11.18).

Protein content of cell is a balance of synthesis and degradation. Proteins are targeted for degradation when an enzyme attaches ubiquitin to a lysine residue of the protein to be destroyed. Subsequently, more ubiquitin chains attach, forming a polyubiquitin chain. This polyubiquitin complex binds to a proteasome complex.

When a protein–ubiquitin complex enters the proteasome, ubiquitin is removed using ATP, the protein is unfolded, and three proteases digest the protein (see Figure 11.19).

**Question 9.** Although miRNA appear to be evolutionarily ancient and biologically important, what is a drawback of this form of gene regulation? What might favor its persistence?

**Question 10.** Describe how a translational repressor functions.

## Test Yourself

1. Which of the following about miRNAs is *false*?
   a. They are usually about 22 nucleotides long.
   b. They are double-stranded RNAs.
   c. They are complementary to their target mRNAs.
   d. They are translation inhibitors.
   e. They each have dozens of mRNA targets.
   ***Textbook Reference:*** *Concept 11.4 Eukaryotic Gene Expression Can Be Regulated after Transcription; MicroRNAs are important regulators of gene expression*

2. Lytic bacterial viruses
   a. infect the cell, replicate their genomes, and lyse the cell.
   b. infect the cell, replicate their genomes, transcribe and translate their genes, and lyse the cell.

c. infect the cell, replicate their genomes, transcribe and translate their genes, package those genomes into viral capsids, and lyse the cell.
d. infect the cell, transcribe and translate their RNA, replicate their genomes, package those genomes into viral capsids, and lyse the cell.
e. insert their chromosome into the host chromosome.

***Textbook Reference:*** *Concept 11.1 Many Prokaryotic Genes Are Regulated in Operons; Viruses use gene regulation to hijack host cells*

3. Retroviruses such as HIV
a. have DNA as their genome.
b. are prophages.
c. copy their RNA genome into DNA using reverse transcriptase.
d. replicate their genome using RNA polymerase.
e. can only undergo a lytic infection cycle.

***Textbook Reference:*** *Concept 11.2 Eukaryotic Genes Are Regulated by Transcription Factors; Eukaryotic viruses can have complex life cycles*

4. If the TATA box for gene *X* becomes highly methylated, how will this affect the expression of gene *X*?
a. There will be no effect.
b. Gene X will be transcribed but not translated.
c. Gene X will be transcribed if the transcription factors receive the appropriate environmental signal.
d. Gene X will not be transcribed or translated.
e. Gene X will be transcribed if the histones become acetylated.

***Textbook Reference:*** *Concept 11.3 Gene Expression Can Be Regulated via Epigenetic Changes to Chromatin; DNA methylation affects transcription*

5. If the gene encoding the *lac* repressor is mutated so that the repressor can no longer bind the operator, will transcription of that operon occur?
a. Yes, because the repressor transcriptionally activates the *lac* genes.
b. Yes, but only when lactose is present.
c. No, because RNA polymerase is needed to transcribe the genes.
d. Yes, because RNA polymerase will be able to bind the promoter and transcribe the operon.
e. No, because cAMP levels are low when the repressor is nonfunctional.

***Textbook Reference:*** *Concept 11.1 Many Prokaryotic Genes Are Regulated in Operons; Operator–repressor interactions regulate transcription in the* lac *and* trp *operons*

6. If the gene encoding the *trp* repressor is mutated such that it can no longer bind tryptophan but can still bind the operon, will transcription of the *trp* operon occur?
a. Yes, because the *trp* repressor can bind the *trp* operon and block transcription only when it is bound to tryptophan.
b. No, because this mutation does not affect the part of the repressor that can bind the operator.
c. No, because the *trp* operon is repressed only when tryptophan levels are high.
d. Yes, because the *trp* operon can allosterically regulate the enzymes needed to synthesize the amino acid tryptophan.
e. No, because the repressor will be continuously bound to the operator.

***Textbook Reference:*** *Concept 11.1 Many Prokaryotic Genes Are Regulated in Operons; Operator–repressor interactions regulate transcription in the* lac *and* trp *operons*

7. Transcriptional regulation in prokaryotes can occur by
a. a repressor binding an operator and preventing transcription.
b. proteins that direct the RNA polymerase to specific promoters.
c. activator proteins that bind to DNA elements near the promoter and promote transcription.
d. the control of promoter efficiency.
e. All of the above

***Textbook Reference:*** *Concept 11.1 Many Prokaryotic Genes Are Regulated in Operons; Genes are subject to positive and negative regulation*

8. The Tat protein in HIV disables the host cell's negative regulatory system by binding to the
a. RNA polymerase.
b. host mRNA to block the host terminator genes.
c. host terminator protein complex.
d. viral mRNA to block the terminator protein.
e. viral mRNA and to proteins associated with RNA polymerase.

***Textbook Reference:*** *Concept 11.3 Gene Expression Can Be Regulated via Epigenetic Changes to Chromatin; Eukaryotic viruses can have complex life cycles*

9. An operon
a. is regulated by a repressor binding at the promoter.
b. has structural genes that are all transcribed from same promoter.
c. has several promoters, but all of the structural genes are related biochemically.
d. is a set of structural genes that are all under the same translational regulation.
e. is transcribed when RNA polymerase binds the operator.

***Textbook Reference:*** *Concept 11.1 Many Prokaryotic Genes Are Regulated in Operons; Operons are units of transcriptional regulation in prokaryotes*

10. How will the expression of gene *X* be affected if it is moved to a part of the chromosome where the histones are highly acetylated?
    a. There will be no effect.
    b. It will be transcribed but not translated.
    c. It will be transcribed if the transcription factors receive the appropriate environmental signal.
    d. It will not be transcribed or translated.
    e. It will be transcribed if the histones become deacetylated.
    ***Textbook Reference:*** *Concept 11.3 Gene Expression Can Be Regulated via Epigenetic Changes to Chromatin; Modification of histone proteins affects chromatin structure and transcription*

11. Which of the following is an example of regulation of eukaryotic transcription?
    a. Iron binding the repressor protein for the ferritin mRNA and increasing ferritin expression
    b. Proteostome breakdown of protein–ubiquitin complexes
    c. MicroRNAs binding their target mRNA and causing their degradation
    d. Alternate splicing of an mRNA transcript
    e. Activator proteins binding an enhancer
    ***Textbook Reference:*** *Concept 11.2 Eukaryotic Genes Are Regulated by Transcription Factors; Transcription factors act at eukaryotic promoters*

12. Which of the following about histone modifications is *false*?
    a. They cause some genes to be transcriptionally activated.
    b. They can result in the repression of gene transcription.
    c. They are reversible.
    d. They cause Barr bodies to form.
    e. All of the above are false, none is true.
    ***Textbook Reference:*** *Concept 11.3 Gene Expression Can Be Regulated via Epigenetic Changes to Chromatin; Modification of histone proteins affects chromatin structure and transcription*

13. What would happen initially to cells that lack a functional ubiquitin?
    a. Translation of proteins would be more efficient.
    b. Transcriptional initiation would increase.
    c. Protein degradation would decrease.
    d. Histone modifications would increase.
    e. There would be no effect.
    ***Textbook Reference:*** *Concept 11.4 Eukaryotic Gene Expression Can Be Regulated after Transcription; Protein stability can be regulated*

14. Which of the following is *not* a means by which translation can be regulated in a eukaryotic cell?
    a. Inhibition of translation with miRNAs
    b. Modification of the 5′ cap
    c. Translational repressor protein translation
    d. Production of ferritin mRNA
    e. Alternative splicing
    ***Textbook Reference:*** *Concept 11.4 Eukaryotic Gene Expression Can Be Regulated after Transcription; Different mRNAs can be made from the same gene by alternative splicing*

15. Viral genomes can be
    a. double-stranded DNA.
    b. single-stranded DNA.
    c. double-stranded RNA.
    d. single-stranded RNA.
    e. All of the above
    ***Textbook Reference:*** *Concept 11.1 Many Prokaryotic Genes Are Regulated in Operons; Viruses use gene regulation to hijack host cells*

## Answers

### *Key Concept Review*

1. Viral genomes use positive regulation of transcription including a promoter that binds host RNA polymerase. They also act at the posttranscription level by using virally encoded enzymes to break down host mRNA before it can be translated.
2. Yes. The repressor gene can be transcribed and translated from the plasmid DNA, and normal regulation will be restored.
3. The surface receptor on their immune system cells may possess a variant of the CD4 receptor that the virus does not recognize. The terminator protein of the host human may not be recognizable to the viral Tat and thus may be able to terminate the transcription of viral DNA.
4. In order to be expressed, the gene would need a promoter that binds transcription factors (such as a TATA box), RNA polymerase, and regulatory binding sites that bind activator proteins. It would also have to be placed in a chromosomal region that had not been silenced by condensed nucleosomes.
5. The plant will require stress response elements (SREs) in front of the promoter for gene *Y*. SREs are bound by transcription factors that are sensitive to drought, and genes with SRE sequences in front of their promoters can be coordinately regulated.

6.

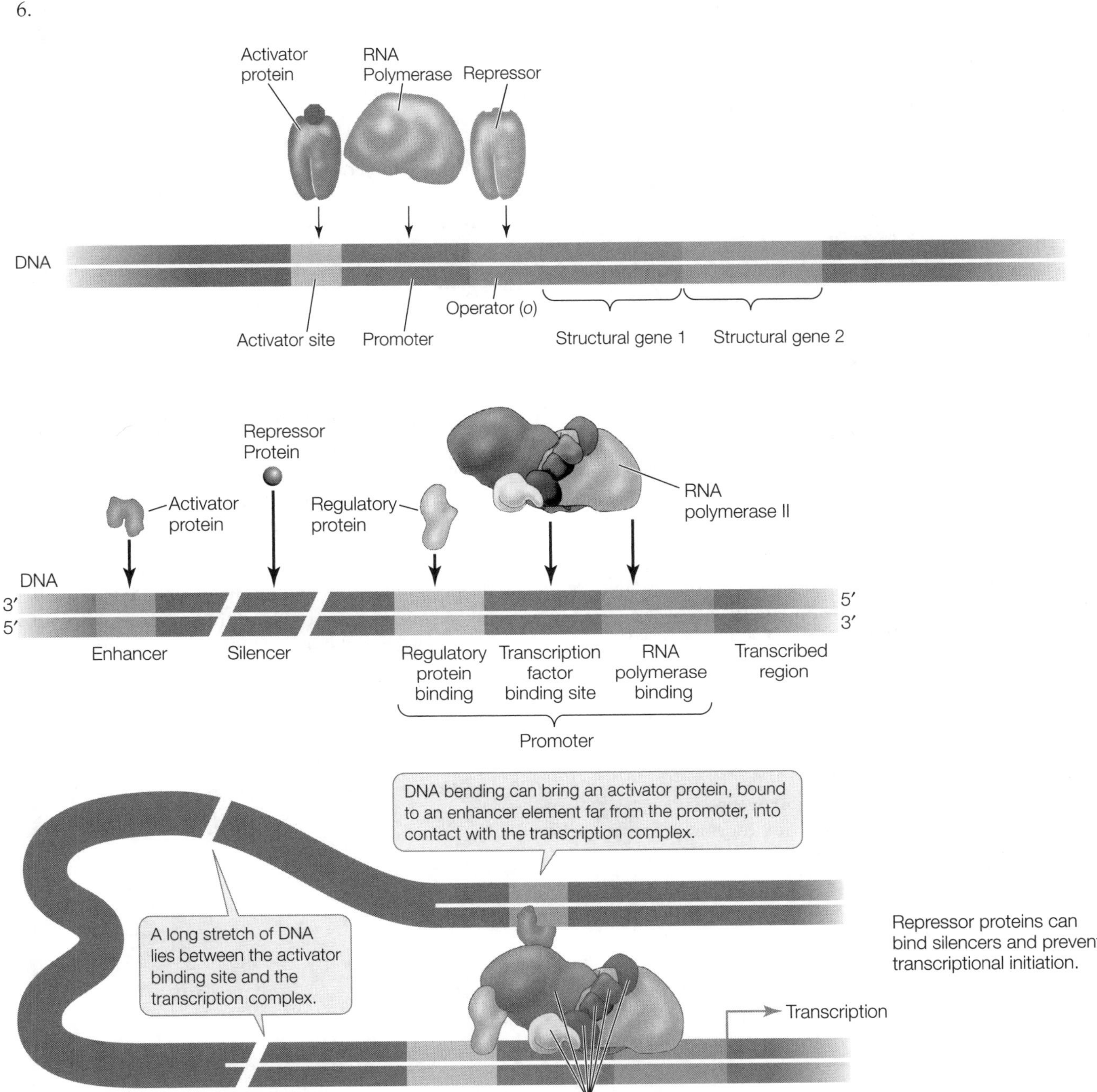


7. DNA methyltransferase could be used to methylate the gene. Methylated DNA binds specific proteins that are involved in the repression of transcription, thus heavily methylated genes tend to be inactive (silenced).

8. Although methylation patterns are considered stable and heritable, the environment can induce changes in the patterns. Since methylation silences gene expression, it is likely that Sally was exposed to an environmental stressor that caused methylation of her *BRCA1* gene whereas her sister was not.

9. Since miRNA degrades mRNA, this means that transcription is already complete, so the cell has wasted energy. However, because of its small size and the potential for many combinations of miRNA and protein complex guides, miRNA provides a system for fine-tuning regulation at the posttranscription level.

10. A translational repressor binds to the mRNA and prevents it from attaching to the ribosome. The repressor can be removed via allosteric regulation. For example, free iron ions that are necessary in low concentrations but toxic in high concentrations will bind to the

repressor when in high concentrations, thereby causing it to release the mRNA and permitting translation of ferritin. Ferritin binds and sequesters excess iron ions. Once they are in low concentrations there will be few to bind the repressor and translation will cease.

### *Test Yourself*

1. **b.** They are single-stranded RNAs that are complementary to their target RNAs.
2. **d.** This sequence includes the most complete details of the viral life cycle. Viral transcription and translation have to occur first so that viral gene products needed for viral replication will be synthesized.
3. **c.** Answer b. describes a provirus, which is a bacterial virus that has inserted its genome into a host chromosome. Animal viruses replicate their RNA genomes using reverse transcriptase and do not lyse the cells they infect.
4. **d.** Gene *X* will not be transcribed or translated, since methylation sites on DNA are transcriptionally inactive.
5. **d.** If the *lac* repressor is nonfunctional, it cannot bind the operator site, and transcription of the *lac* operon will occur at all times, whether or not lactose is present.
6. **a.** If the repressor can no longer bind tryptophan, then it cannot bind the operator, and transcription of the *trp* operon will always be on, whether tryptophan levels in the cell are high or low.
7. **e.** Answer a. refers to the *lac* and *trp* repressors, answer b. to the sigma factors, and answer c. to promoters proteins. Answer d. refers to the *lac* operon.
8. **c.** The Tat complex binds the terminator proteins so they cannot interfere with the RNA polymerase and it can complete transcription of the HIV DNA.
9. **b.** An operon is a set of genes that are all transcribed from the same promoter, which is the site where RNA polymerase binds. The repressor binds at the operator site, which overlaps the promoter. Answer d. is not correct because the operon is regulated transcriptionally, not translationally.
10. **c.** Acetylation leads to loosening of the nucleosomes, resulting in more DNA sequence being accessible for transcription. If transcription factors receive the appropriate environmental signal, transcription will occur. Deacetylation leads to tighter packing of the nucleosomes.
11. **e.** Answer a. refers to translational regulation; answer b. refers to the regulation of protein longevity; answers c. and d. refer to posttranscriptional regulation.
12. **d.** The Barr body is heterochromatin that is heavily methylated DNA. All of the other statements about histone modifications are true.
13. **c.** In the absence of ubiquitin, protein degradation would decrease, since ubiquitin targets proteins for degradation in the proteasome.
14. **e.** Alternative splicing occurs after primary transcription but before mature messenger RNA is ready to be translated. All of the other processes are means by which translation is regulated.
15. **e.** Viral genomes can be single- or double-stranded and can be DNA or RNA.

# 12 Genomes

## The Big Picture

- The science of genomics is the study and comparison of genomes from different organisms and is used to identify organisms, genes, potential genes, and their regulatory sequences. Information from these studies has been used to design new medicines to combat pathogens and to understand the minimal requirements for a cell to sustain life. Knowing the DNA sequence of an organism and the genes it contains permits comparisons of normal and mutant genes and the study of gene product variants (usually proteins).
- The science of proteomics focuses on the regulation of protein production and the consequences of changes to the proteins expressed by different genes and alternative splicing.
- Metabolomics is the study of the active metabolites in the cell (primary and secondary metabolites) that control cell functions, and how metabolite pools change under different environmental and developmental conditions.
- These subdisciplines have many goals, including a better understanding of diseases and disabilities as well as improvements in agriculture, pharmacology, and environmental reclamation. Some diseases in humans can be directly related to a change in the genome that causes the production of an abnormal gene product. Other molecular diseases are affected by the individual's environment. In agriculture, the production of disease resistance compounds and how to enhance expression of inserted genes are areas of research. An understanding of the molecular alterations in a particular gene product and their effects may lead to therapeutic approaches, new diagnostic tools, and preventative measures in health care, agriculture, and many other areas.

## Study Strategies

- It may seem confusing that so little of the eukaryotic genome codes for proteins. Remember that noncoding regions (transposons, introns, and repetitive regions of the chromosome) may have played a role in creating new functional genes over the course of evolution.
- There are many methods for analyzing genomes and proteomes, including those of pharmagenomics and metagenomics. Review what kinds of sequences are analyzed by each of these methods and think of examples of how these analyses have furthered our understanding of the functions of genes, proteins, and cells in different organisms.
- Describe the steps that must be followed in order to generate a genome sequence from a particular organism.
- Comparing genomes among different organisms has allowed us to determine which genes are essential for all cells and which genes are designated for specific adaptive functions. Compare prokaryotic and eukaryotic genomes and list classes of genes that are essential for both types of organisms and classes of genes that are specialized for each type.
- Compare the information in and applications of genome and proteome data.
- List the agricultural, medical, and environmental benefits of genome sequencing.
- Go to **LaunchPad** (or use the Web addresses listed) to review the following additional resources:

  Animated Tutorial 12.1 Sequencing the Genome (PoL2e.com/at12.1)

  Animated Tutorial 12.2 High-Throughput Sequencing (PoL2e.com/at12.2)

  Activity 12.1 Concept Review (PoL2e.com/ac12.1)

  Analyze the Data in textbook Figure 12.8

  Apply the Concept in textbook Sections 12.3 and 12.4

## Key Concept Review

### 12.1 There Are Powerful Methods for Sequencing Genomes and Analyzing Gene Products

Methods have been developed to rapidly sequence DNA

Genome sequences yield several kinds of information

Phenotypes can be analyzed using proteomics and metabolomics

The Human Genome Project, completed in 2003, sequenced all 3.3 billion base pairs in haploid cells. Simultaneously, genomes from model organisms across all domains of life have been sequenced, allowing for general comparisons.

To sequence genomic DNA using high-throughput methods, it must first be cut into smaller (100 bp) fragments. DNA fragments are denatured using heat, creating single-strand DNA templates. These fragments are attached to short adapter sequences that are mounted on a solid surface. Each fragment is then amplified by PCR. Synthesis of DNA complementary to each tethered, amplified fragment is carried out one nucleotide at a time using universal primers, DNA polymerase, and the four nucleotides. Each nucleotide is labeled with a different color fluorescent tag. Unused nucleotides are removed after each nucleotide addition. A camera records the color of the fluorescent tag identifying the nucleotide after each addition, building a color-coded sequence of the DNA fragment. This powerful method is fully automated, can process millions of fragments simultaneously (is massively parallel), and is fast and economical. Determining the entire genome sequence from these fragments is possible because the fragments are overlapping.

Bioinformatics was developed to reconstruct whole genome sequences from DNA fragment sequences and to analyze sequence information using complex mathematics and computer programs.

Two fields of research utilize genome sequence data. Functional genomics involves identification of functional areas of the genome. In comparative genomics, the genomes of different organisms are compared to understand evolutionary relationships among the organisms and to discover their similar areas of function.

Proteomics is the study of the variation and interactions of proteins translated in an organism. The total amount of proteins produced by an organism is called its proteome. Two common methods used together to study the proteome are separation through gel-electrophoresis, and mass spectrometry, which determines mass and structure of proteins. Both methods are used to isolate, analyze, and characterize expressed proteins.

Metabolomics is the study of the metabolites in a cell or organism, focusing on their function in variable cellular environments. Primary metabolites are involved in normal cell processes, like glycolysis and hormone signaling, while secondary metabolites are involved in special responses to the environment and can be unique to the organism. These include antibiotics made by bacteria and disease resistance and other defense compounds made by plants.

Together, genomics, proteomics, and metabolomics increase our understanding of the genotype reaction to the environment that produces the phenotype.

**Question 1.** Diagram the steps required to sequence a DNA molecule using the high-throughput sequencing method described in the textbook. In your diagram be sure to include all of the necessary components and technologies that are utilized for DNA sequencing.

**Question 2.** Suppose you possess the sequenced genome for a prokaryote and you are looking for the gene(s) that code(s) for a given protein. How would you begin your search?

## 12.2 Prokaryotic Genomes Are Small, Compact, and Diverse

- Prokaryotic genomes are compact
- Metagenomics reveals the diversity of viruses and prokaryotic organisms
- Some sequences of DNA can move about the genome
- Will defining the genes required for cellular life lead to artificial life?

The genomes of prokaryotes and archaea are generally small (160,000–2 million bp). They are organized in a single circular chromosome and are compact (85 percent of the DNA codes proteins or RNA). Smaller circular DNA molecules, called plasmids, are often present in addition to the main chromosome. Most prokaryote genes do not possess introns, but archaeal rRNA and tRNA genes do. Other than these general similarities, these groups possess varied genomes, reflecting the broad range of environments they inhabit.

Functional genomic analysis has been used to assign functions to gene products. Genes can be identified that are involved in a prokaryote's metabolism, transport, and infectious properties. Prokaryotes with smaller genomes tend to have fewer types of proteins dedicated to a given function (see Table 12.1).

Comparative genomics is used to examine the differences among the genomes of prokaryotes and correlate that information with functional differences between the species. This provides insight about requirements for gene regulation and function, and ultimately how these organisms adapt to different environments.

Metagenomics is used to analyze gene sequences from complex samples without isolating individual organisms. Using metagenomics, thousands of new viruses and bacteria have been found in sea water, marine sediment, and mine water runoff (see Figure 12.6). It is estimated that 90 percent of the microbial world was invisible to microbiologists prior to the development of metagenomic sequence analysis methods.

Genome sequencing has enabled scientists to better understand transposons (transposable elements), which are small (1,000–2,000 bp) mobile genetic elements that can move from one site in the genome to another. Transposons can affect the phenotype of the organism when they insert themselves into a gene's coding region, altering the coding sequence. Some transposons carry the genes for antibiotic resistance (see Figure 12.7).

Some genes or gene segments appear universal, or nearly so, among all genomes compared to date. An example is the sequence that codes for ATP binding sites on proteins. Comparative genomics can be used to determine the minimal number of genes needed for life. The *Mycoplasma genitalium* genome has been mutated to determine the smallest number of genes (382) needed for survival, and experiments are now under way to make synthetic genomes based on that of *M. discoides.* This new knowledge has the potential to help us create new microbes to degrade oil spills, reduce tooth decay, or convert cellulose to ethanol for use as fuel.

**Question 3.** Genomics has revealed that the bacterium that causes tuberculosis has more than 250 genes that metabolize lipids. What does this finding suggest about the bacterium and about medical approaches to fighting this disease?

**Question 4.** Some transposons carry the genes for antibiotic resistance when they move. By what mechanism could this happen?

**Question 5.** You have isolated a new strain of bacteria from a contaminated piece of Swiss cheese. Diagram how you would use transposon mutagenesis to find out the minimal number of genes in the bacterium's genome that are necessary for its survival on Swiss cheese. Include the results you would expect for an essential gene and for gene that is not essential.

## 12.3 Eukaryotic Genomes Are Large and Complex

Model organisms reveal many characteristics of eukaryotic genomes

Gene families exist within individual eukaryotic organisms

Eukaryotic genomes contain many repetitive sequences

Eukaryotic genomes are larger and have more protein-coding genes and regulatory sequences than prokaryotic genomes, reflecting their greater cellular and organismal complexity. Eukaryotes have multiple chromosomes. Much of eukaryotic DNA is noncoding; noncoding DNA may be introns, regulatory sequences, or repetitive sequences.

Model eukaryotic organisms include *Saccharomyces cerevisiae, Caenorhabditis elegans, Drosophila melanogaster, Arabidopsis thaliana,* and *Oryza sativa* (see genome comparisons in Table 12.2).

*Saccharomyces cerevisiae* (budding yeast) is a single-celled eukaryote that has 16 linear chromosomes. The genomes of *S. cerevisiae* and *E. coli* have the same number of genes for the basic functions of cell survival, but *S. cerevisiae* has 5,770 total genes while *E. coli* has 4,377. Many of the additional genes possessed by *S. cerevisiae* code for products involved in protein targeting and organelle function required for eukaryotic compartmentalization.

The soil-dwelling nematode *Caenorhabditis elegans* is a simple multicellular organism used to study development. The worm's body is transparent, and its growth from a fertilized egg to a differentiated adult with 959 cells takes just three days. The genome of *C. elegans* is eight times larger than that of yeast and has 3.5 times the number of protein-coding genes (19,427 proteins). Gene inactivation studies have revealed that the worm can survive in laboratory culture with only 10 percent of those genes.

The fruit fly *Drosophila melanogaster* has ten times as many cells as *C. elegans.* Its genome has more DNA but it contains fewer genes than *C. elegans. D. melanogaster* has many genes that code for transcription factors needed to regulate its more complex development, which includes egg, larva, pupa, and adult stages (see Figure 12.9 for distribution of gene functions in *D. melanogaster*).

The genome of the *Arabidopsis thaliana* plant has 28,000 protein-coding genes, many of which are duplicates of each other. When the duplicates are removed, only 15,000 unique genes remain, and many of these are very similar to genes found in nematodes and fruit flies. *Arabidopsis* contains genes unique to plants, including those involved in photosynthesis, water transport, cell-wall synthesis, uptake and metabolism of inorganic substances, and defense against herbivores. Many of the genes in *Arabidopsis* can be found in rice, *Oryza sativa,* and poplar, *Populus trichocarpa* (see Figure 12.10 for overlap among plant genomes.)

Eukaryotes have gene families—sets of duplicate or closely related genes (such as the globin genes). Gene families provide an organism with a functional gene while allowing mutations in other members of the gene family. Some of these mutations may create new gene variants that are advantageous to the organism, while others may create pseudogenes. Pseudogenes are nonfunctional because they lack promoters and/or recognition sites for intron removal.

In humans, the globin gene family consists of three $\alpha$-globin genes and five $\beta$-globin genes (see Figure 12.11). Different globins are expressed at different times in development. $\lambda$-globin, which is expressed in the fetus, binds oxygen more tightly to ensure oxygen transfer from the mother across the placenta to the fetus.

Highly repetitive sequences are short (less than 100 bp) and are repeated thousands of times in tandem in the genome. They can be densely packed in heterochromatin regions or scattered around the chromosome (as seen for short tandem repeats, STRs). The number of repeats at a given location in the genome varies among individual organisms and is heritable, providing unique molecular markers that can be used to identify individuals.

Moderately repetitive DNA sequences include sequences that are repeated 10–1,000 times and include tRNA and rRNA genes. Multiple copies of the tRNA and rRNA genes are needed to provide the cell with high concentrations of components needed for protein translation.

Most moderately repetitive sequences are not stably integrated into the DNA but can move about in the genome via transposons. Transposons make up about 40 percent of the human genome. There are two main types of transposons in eukaryotes, retrotransposons and DNA transposons (see Table 12.3).

A retrotransposon makes an RNA copy of itself which is then re-copied into DNA before it is inserted into a different site in the genome. There are two kinds of retrotransposons, LTR and non-LTR. LTR retrotransposons have long terminal repeats (LTR) of DNA at each end. Non-LTR retrotransposons lack these repeats.

Non-LTR retrotransposons are further subdivided into SINEs and LINEs. SINEs are short *in*terspersed *e*lements of up to 500 bp; they are transcribed but not translated. They include the 300-bp *Alu I* element, of which there are a million copies accounting for 11 percent of the human genome. LINEs are *l*ong *in*terspersed *e*lements of up to 7,000 bp and some are transcribed and translated. LINEs make up about 17 percent of the human genome.

DNA transposons do not replicate when they move to a new site on the chromosome, nor do they use RNA as an intermediate. Like some prokaryote transposons, they may be excised and moved to a new location without being replicated.

**Question 6.** Compare and contrast gene families with repetitive sequences in eukaryote genomes.

**Question 7.** Which model organism would you use to study the functional genomics of neuron signaling and why?

### 12.4 The Human Genome Sequence Has Many Applications

The human genome sequence held some surprises

Human genomics has potential benefits in medicine

DNA fingerprinting uses short tandem repeats

Genome sequencing is at the leading edge of medicine

The human genome contains about 24,000 protein-coding genes. This is a small number compared with the number of proteins in humans, so posttranscriptional mechanisms such as alternative splicing must provide the additional proteins.

The average gene has 27,000 base pairs, although the size ranges from 1,000 to 2.4 million bp. Virtually all human genes have numerous introns, and at least 3.5 percent of the genome is functional but noncoding, having roles in gene regulation. Over 50 percent of the genome contains highly repetitive sequences.

Even though almost all genes (99 percent) are the same in all people, scientists have mapped over 3.3 million single nucleotide polymorphisms (SNPs) in humans. Single nucleotide polymorphisms are when two or more sequences differ by one nucleotide; for example, if one section of a noncoding DNA strand is ATTGCGC in one person and ATAGCGC in another person, that locus is polymorphic in the population.

Comparisons of genes from different organisms have revealed evolutionary relationships (see Figure 12.12). Ninety-five percent of the human genome is shared with the chimpanzee.

Because complex phenotypes, such as susceptibility to disease, are determined by multiple genes interacting with the environment, human genomics has potential benefits in medicine. One approach taken by medical genetics researchers is haplotype mapping. This method uses SNPs that are linked (inherited together because they are located close together on individual chromosomes). The piece of chromosome with the SNPs is called a haplotype.

Over 500,000 SNPs from different individuals can be placed on a chip and used to analyze disease states. Statistical measures of association of SNP data can be used to determine increased risk for particular diseases (see Figure 12.13). SNP testing will eventually be replaced with DNA sequencing.

Pharmagenomics is the study of how an individual's genome can affect his or her response to drugs or other outside agents (see Figure 12.14). The enzyme variants produced by different individuals may affect the activity of drugs that are processed by the enzyme, so physicians can use this information to tailor drug dose to the individual.

Because of alternative splicing and posttranslational modifications, the sum total of proteins produced (the proteome) is more complex than the genome. Comparative proteome analysis has revealed a common set of proteins that provide the basic metabolic functions of a eukaryotic cell in humans, worms, flies, and yeast. The unique proteins in each organism may result from a reshuffling of the same domains that exist in proteins from other organisms.

**Question 8.** Analysis of the human genome has revealed many genes that cause disease when mutated. Nevertheless, this knowledge has not enabled us to eliminate these diseases. Why?

**Question 9.** Why do different organisms have more similarities in their proteomes than in their genomes?

**Question 10.** Both dogs shown in the figure below are whippets. Propose a mechanism by which the muscle-bound dog's phenotype was achieved.

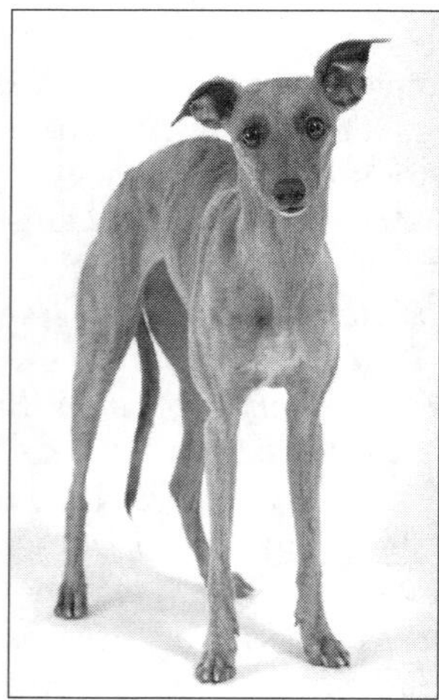

## Test Yourself

1. Functional genomics
   a. assigns functions to DNA sequences.
   b. assigns functions to regulatory sequences.
   c. compares genes in different organisms to see how those organisms are related physiologically.
   d. assigns functions to protein sequences.

e. compares proteins in the same organisms.

***Textbook Reference:*** *Concept 12.2 Prokaryotic Genomes Are Small, Compact, and Diverse; Prokaryotic genomes are compact*

2. Comparative genomics
   a. assigns functions to DNA sequences.
   b. assigns functions to regulatory sequences.
   c. compares genes in different organisms to see how those organisms are related physiologically.
   d. assigns functions to protein sequences.
   e. compares proteins in the same organisms.

   ***Textbook Reference:*** *Concept 12.2 Prokaryotic Genomes Are Small, Compact, and Diverse; Prokaryotic genomes are compact*

3. Which of the following is *not* a way in which proteomic studies identify proteins that are coded by specific genes?
   a. Reading the genetic code that specifies the amino acid sequence
   b. Using PCR to amplify the protein
   c. Separating the expressed protein by gel electrophoresis and then analyzing and sequencing the protein
   d. Knockout studies that prevent gene expression
   e. Mass spectrometry characterization of the expressed protein and comparisons with known proteins

   ***Textbook Reference:*** *Concept 12.1 There Are Powerful Methods for Sequencing Genomes and Analyzing Gene Products; Phenotypes can be analyzed using proteomics and metabolomics*

4. In a certain patient, a certain medication has an extended effect. Based on this effect, what can you conclude about the gene for an enzyme that metabolizes the medication?
   a. The gene is mutated, resulting in an inactive enzyme, thus the medication is not metabolized.
   b. The gene is mutated, resulting in a highly active enzyme, thus the medication is metabolized quickly.
   c. The gene is mutated, resulting in a less active enzyme, thus the medication is metabolized slowly.
   d. The gene is not mutated, but the medication is still slowly metabolized.
   e. Mutation of a gene and changes in the resulting protein do not affect metabolism of drugs.

   ***Textbook Reference:*** *Concept 12.4 The Human Genome Sequence Has Many Applications; The human genome sequence held some surprises*

5. Haplotype mapping of the human genome has allowed scientists to discover
   a. single gene associated diseases.
   b. a single gene that causes a disease.
   c. evolutionary relationships by comparing genes within a single patient.
   d. individual genes and their gene product.
   e. multiple gene interactions associated with disorders such as diabetes.

   ***Textbook Reference:*** *Concept 12.4 The Human Genome Sequence Has Many Applications; Human genomics has potential benefits in medicine*

6. Proteomics has been used to compare
   a. DNA sequences between closely related species.
   b. gene expression during embryonic development.
   c. protein sequences between closely related species.
   d. shotgun cloned sequences.
   e. prokaryotic genomes.

   **Textbook Reference:** *Concept 12.1 There Are Powerful Methods for Sequencing Genomes and Analyzing Gene Products; Phenotypes can be analyzed using proteomics and metabolomics*

7. Which of the following is *not* a component of high-throughput sequencing?
   a. Cutting of DNA into 100-bp fragments
   b. cDNA cloning
   c. Computer alignment of overlapping pieces of chromosomes
   d. Sequencing of DNA with dideoxy nucleotides
   e. Bioinformatics

   ***Textbook Reference:*** *Concept 12.1 There Are Powerful Methods for Sequencing Genomes and Analyzing Gene Products; Methods have been developed to rapidly sequence DNA*

8. Which of the following about transposable elements is *false*?
   a. They can inactivate genes into which they are inserted.
   b. They can contain gene sequences.
   c. They are mobile genetic elements that move from RNA molecule to RNA molecule.
   d. They may be spliced out of one region of the genome and inserted into another.
   e. They replicate themselves before moving to another site on the genome.

   ***Textbook Reference:*** *Concept 12.2 Prokaryotic Genomes Are Small, Compact, and Diverse; Some sequences of DNA can move about the genome*

9. Gene mutation studies have allowed us to do all of the following *except*
   a. determine the minimal number of genes humans need to survive.
   b. determine the causative genes for diseases with a complex phenotype.
   c. investigate the minimal number of genes needed to sustain life.
   d. create artificial life in a test tube.
   e. demonstrate that *C. elegans* needs most of its genes.

   ***Textbook Reference:*** *Concept 12.2 Prokaryotic Genomes Are Small, Compact, and Diverse; Will defining the genes required for cellular life lead to artificial life?*

10. Comparisons of yeast and bacterial cell genomes have revealed that
   a. yeast cells have more genes devoted to the basic functions of survival than bacteria do.
   b. eukaryotic cells are structurally similar to bacterial cells in terms of complexity.
   c. there are more genes for targeting proteins to organelles in yeast than in bacteria.
   d. the histones of bacteria are very similar to those of yeast.
   e. bacteria and yeast are both haploid.

   ***Textbook Reference:*** *Concept 12.3 Eukaryotic Genomes Are Large and Complex; Model organisms reveal many characteristics of eukaryotic genomes*

11. Which of the following statements does *not* represent information that we have gained from genome sequencing?
   a. The nematode genome is eight times larger than the yeast genome.
   b. There is extensive genetic exchange between different kinds of bacteria.
   c. Genomes can be sequenced even when organisms cannot be cultured.
   d. Much of the eukaryotic genome contains coding sequences.
   e. The *Drosophila* genome contains many genes encoding transcription factors.

   ***Textbook Reference:*** *Concept 12.3 Eukaryotic Genomes Are Large and Complex; Model organisms reveal many characteristics of eukaryotic genomes*

12. Which of the following about the discoveries that resulted from the genome sequencing of plants is *false?*
   a. There are more protein-coding genes in animals than in plants.
   b. Many *Arabidopsis* genes are duplicated due to chromosomal rearrangements.
   c. There are more genes in plants that are similar to each other than there are genes that are unique.
   d. Plants have many genes whose products are used for defense against microbes and herbivores.
   e. Plants have over 27,000 protein-coding genes, but only about 15,000 are unique.

   ***Textbook Reference:*** *Concept 12.3 Eukaryotic Genomes Are Large and Complex; Model organisms reveal many characteristics of eukaryotic genomes*

13. Which of the following statements about eukaryote retrotransposons is *false?*
   a. They are translated before they are transcribed.
   b. They use an RNA intermediate to move from one region of the genome to another.
   c. They are very rare, making up less than 1 percent of the total DNA content of humans.
   d. They can encode gene products that are required for their own transposition.
   e. They are highly repetitive sequences found throughout the genome.

   ***Textbook Reference:*** *Concept 12.3 Eukaryotic Genomes Are Large and Complex; Eukaryotic genomes contain many repetitive sequences*

14. Which of the following statements about the human genome is *false?*
   a. Over 50 percent of the genome contains transposons.
   b. Almost every gene has introns.
   c. About 97 percent of the genome is shared among human individuals, while about 90 percent is shared with chimpanzees.
   d. About 2 percent of the genome codes for genes.
   e. Humans have about the same number of genes as fruit flies have.

   ***Textbook Reference:*** *Concept 12.4 The Human Genome Sequence Has Many Applications*

15. Which of the following statements about single nucleotide polymorphisms (SNPs) is *false?*
   a. They can be used to map unlinked genes in order to follow the inheritance of disease traits.
   b. They can be used to predict if a patient is at risk for a particular disease.
   c. They can be linked in order to generate gene sequences.
   d. They have limited sequence variations.
   e. They rarely occur in the genome and can thus be used for genotype mapping.

   ***Textbook Reference:*** *Concept 12.4 The Human Genome Sequence Has Many Applications; The human genome sequence held some surprises*

16. The study of proteomes allows scientists to compare
   a. the proteome with the genome to see if the gene sequences are correct.
   b. proteome sequences between species to see if similar proteins are expressed in all species.
   c. transcriptional patterns in different organisms.
   d. highly repetitive DNA sequences in different organisms.
   e. how noncoding regions of the genome differ in different organisms.

   ***Textbook Reference:*** *Concept 12.1 There Are Powerful Methods for Sequencing Genomes and Analyzing Gene Products; Phenotypes can be analyzed using proteomics and metabolomics*

17. Which of the following is *not* a characteristic you would look for when choosing the next "model" plant genome to study agriculturally important traits?
   a. A small genome
   b. A genome with little redundancy
   c. Ease of laboratory culture
   d. Rapid life cycle

e. Large numbers of transposons

***Textbook Reference:*** *Concept 12.3 Eukaryotic Genomes Are Large and Complex; Model organisms reveal many characteristics of eukaryotic genomes*

18. Which of the following is *not* something you would expect about susceptibility to a disease revealed by haplotype mapping?
    a. The SNP patterns would have a good chance of being altered by transposable elements.
    b. They might involve interactions with environmental cues.
    c. The susceptibility would be shared by genetically related individuals.
    d. Some of the SNP patterns would impact posttranscription events.
    e. The susceptibility might be greater if the individual was homozygous for the haplotype.

    ***Textbook Reference:*** *Concept 12.4 The Human Genome Sequence Has Many Applications; Human genomics has potential benefits in medicine*

## Answers

### *Key Concept Review*

1. The diagram should include the following steps:
    1. Cut the DNA into 100-bp fragments physically or using enzymes to hydrolyze phosphodiester bonds.
    2. Heat the DNA to break the hydrogen bonds holding the two strands together.
    3. Attach each end to a short adapter sequence that is anchored to a bead or flat surface.
    4. Amplify using PCR.
    5. Heat the fragments to denature them.
    6. Add a universal primer complementary to one of the adaptor sequences, DNA polymerase, and the four nucleotides each with an identifying fluorescent dye.
    7. DNA synthesis stops when a nucleotide is incorporated into the replicating strand.
    8. All unincorporated nucleotides are removed.
    9. Photograph the color of the added nucleotide.
    10. The dye is removed from the added nucleotide and the process is repeated from step 5.
    11. The sequence of nucleotides is recorded by the color of the nucleotides in the series of photographs taken.
    12. Use a computer program to analyze the photographs to determine how the sequences of the fragments overlap and to reconstruct the genome sequence.
2. Examine the genome sequence for start and stop codons for translation, which indicate the location of genes. Check each gene sequence for correct codons to produce the protein's sequence of amino acids.
3. The tuberculosis bacterium must use lipids as a source of energy-rich compounds; a drug that inhibits lipid uptake or metabolism in this bacterium may inhibit its growth.
4. Some transposons become duplicated, with two copies flanking one or more genes. The duplicated transposons together with the genes form a single transposable element. If the flanked genes contain antibiotic resistance factors, they could be transported along with the new large transposable element to different parts of the genome or onto plasmids.
5. 

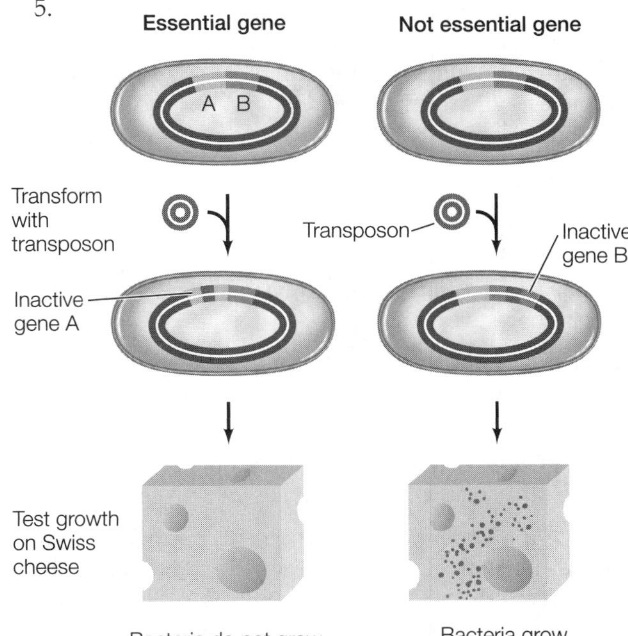


    1. Transform the bacteria with a transposon that can insert into random sites in the genome.
    2. Test subsamples of each transposon-mutated bacteria culture for growth and survival on Swiss cheese.
    3. Isolate DNA from bacteria cultures that show defective growth and survival to determine which gene was inactivated by the transposon. Such genes are likely to be important for growth and survival on Swiss cheese.
6. Gene families are duplicated genes that code for proteins and can include a few to hundreds of copies of the genes in a single genome. Usually different copies contain slight mutations that code for a functional variant of the gene product. Pseudogenes are nonfunctional duplicate genes. Repetitive sequences may be short, nontranscribed sequences that exist in thousands of tandem copies or they may be moderately repetitive (10–1,000 repeats) that code for tRNAs and rRNAs.

7. The simplest model organism that includes genes for a nervous system and intercellular communication is the nematode *Caenorhabditis elegans,* so this would be a good choice.
8. We have not been able to eliminate genetic diseases because, except in a few rare cases, we are not able to replace the mutated genes in the diseased individual.
9. More sequence similarities are revealed when proteomes are compared than when genomes are compared because the genetic code is redundant. There can be more than one codon for a particular amino acid, so the DNA sequences can vary and still generate the same amino acid sequences.
10. Either a mutation in the gene that inhibits the overdevelopment of muscles or a mutation in a gene that causes constitutive activation of muscle-promoting genes could result in the phenotype of the muscle-bound dog.

### *Test Yourself*

1. **a.** Functional genomics assigns functions to DNA sequencing, not to the regulatory sequences. Comparing the genes of different organisms is the work of comparative genomics.
2. **c.** Comparative genomics compares genes in different organisms to see how these organisms are related physiologically.
3. **b.** PCR is used to amplify DNA, not proteins. All of the other methods are used to study proteins.
4. **c.** A mutation in the gene that encodes this enzyme may make the enzyme less active and reduce the rate at which the active drug is modified to a less active form. For a given dose of the drug, a person with the mutation would have more active drug in his or her bloodstream than a person without the mutation. So the effective dose of the drug would be lower for this person.
5. **e.** Haplotype mapping can be used to identify multiple genes involved in diseases with a complex phenotype, such as diabetes.
6. **c.** Proteomics is used to compare protein sequences between different organisms. Answers a, b, and d all refer to techniques used in genomics.
7. **b.** cDNA cloning is used to determine which mRNAs are being expressed in cells. All of the other techniques were utilized to sequence the human genome and order those sequences.
8. **e.** Some transposons do contain gene sequences. They can be spliced out of one region and inserted into another and also replicate themselves and then move to another site. They do not move from one RNA molecule to the next.
9. **b.** The reason is that genes associated with complex phenotypes are identified using haplotype mapping.
10. **c.** A comparison of yeast and bacterial cell genomes revealed that the number of genes needed for survival in these two organisms was roughly the same but that yeast cells had many genes for targeting proteins to organelles. Yeast cells are structurally more complex than bacterial cells. Bacterial cells do not have histones. Yeast cells can be haploid or diploid; bacteria are haploid (which had been known long before comparative genomics was developed).
11. **d.** Only 2 percent of the human genome codes for protein-coding genes.
12. **a.** The reverse is true: There are more protein-coding genes in plant cells than in animal cells.
13. **c.** LTR retrotransposons constitute about 8 percent and non-LTR retrotransposons constitute about 15 percent of the DNA of humans.
14. **c.** Humans and chimpanzees share 95 percent of their genomes, while individuals within humans share 97 percent.
15. **e.** SNPs are not rare and there are actually currently at least 3.3 million known SNPs.
16. **b.** Proteomics studies can be used to compare proteins in different organisms. The gene sequence will determine the protein sequence. Proteomic studies cannot be used to study transcriptional patterns, which are cellular processes involving RNA. Noncoding regions of the genome do not specify proteins and highly repetitive sequences do not code for protein.
17. **e.** A plant with large numbers of transposons will be subject to gene interruptions and movement, which could complicate studies. A small genome is more manageable and means less redundancy and noncoding DNA, which simplifies the search for genes of interest. The ability to be cultivate many generations of a research organism easily minimizes the time and effort needed to observe expressed phenotypes.
18. **a.** Because the haplotype is a set of closely linked SNPs on a chromosome, they would not be likely to be interrupted by a transposable element.

# 13 Biotechnology

## The Big Picture

- The ability to isolate DNA from any organism, ligate it to vector DNA, introduce that DNA into host cells, and propagate the host cells has had an enormous impact on our understanding of genetics, molecular biology, and cell function and development. These techniques are being used to elucidate evolutionary relationships among different organisms and to understand gene regulation and function in greater depth.
- Biotechnology, the application of technologies that use "biological systems, living organisms, or derivatives thereof to make or modify products or processes," has wide applications that include the development and production of new medicines and diagnostics, agricultural products, and bioremediation.

## Study Strategies

- There are a variety of ways to clone DNA fragments, and trying to remember all the different cloning methods can be challenging. Consider the following as you review the different cloning procedures: the size of the cloned DNA, the host cell in which the DNA can be cloned, and the expression of the cloned DNA in the host cell. Different vectors can be used in different host cells to address each of these considerations.
- Many different experimental questions can be answered using cloning. Ask yourself what cloning strategies could be used to answer the following questions, and review the textbook for answers: What is the sequence of a gene? What sequences are important for regulation of the gene? What sequences are important for targeting the gene to a particular site in a eukaryotic cell? What is the difference in function between a mutant gene product and a wild-type gene? What kinds of genes are expressed during the development of an organism? How can cloned genes be expressed in the seeds of plants or in the milk of mammals?
- Outline the specific steps needed to clone a gene in a bacterial cell. Include how the gene is initially isolated, what kind of vectors can be used, how to introduce the recombinant DNA into the cell, and how to confirm that the recombinant molecule is in the host cell. Then outline the steps needed to clone a gene in a eukaryotic cell.
- Complementary base pairing is important for many aspects of biotechnology. Describe each technique in gene cloning that uses complementary base pairing, detailing specifically how base pairing is involved.
- Go to **LaunchPad** (or use the Web addresses listed) to review the following additional resources:

  Animated Tutorial 13.1 Separating Fragments of DNA by Gel Electrophoresis (PoL2e.com/at13.1)

  Animated Tutorial 13.2 DNA Chip Technology (PoL2e.com/at13.2)

  Activity 13.1 Expression Vectors (PoL2e.com/ac13.1)

  Analyze the Data in textbook Figure 13.4

  Apply the Concept in textbook Sections 13.1 and 13.2

## Key Concept Review

### 13.1 Recombinant DNA Can Be Made in the Laboratory

Restriction enzymes cleave DNA at specific sequences

Gel electrophoresis separates DNA fragments

Recombinant DNA can be made from DNA fragments

Recombinant DNA—single-molecule DNA sequences made from the DNA of at least two different organisms—is used for the genetic modification of organisms. Three widely used tools used, in sequence, in the construction of recombinant DNA are: restriction enzymes for cutting DNA into fragments (see Figure 13.1); gel electrophoresis for the analysis and purification of fragments (see Figure 13.2); and DNA ligase for joining DNA fragments to create novel combinations.

Restriction enzymes function as defense molecules of bacteria. They target invader bacteriophage double-stranded DNA and inactivate it by cutting it at particular recognition sequences or restriction sites. Each restriction enzyme recognizes a specific palindromic sequence in DNA and cuts at or near those sites. Some restriction enzymes cut between the same bases on both DNA strands, producing blunt ends. When restriction enzymes cut DNA, they often leave ends that have 5′ or 3′ overhangs of single-stranded DNA. These ends are called "sticky ends," and they can form complementary base pairs with other DNA molecules that have the same sticky ends (see Figure 13.3).

The bacterial cell protects itself from its restriction enzymes by methylating its own DNA restriction sites.

DNA fragments cut with the same restriction enzyme can be joined together by ligase, even if they are from different species. DNA ligase (the enzyme that covalently joins the Okazaki fragments during DNA replication and mends broken DNA; see Chapter 9) is used to form a covalent bond on each DNA strand of the recombinant molecule (see Figure 13.3).

The recombinant DNA molecule can be incorporated, using these tools, into the DNA of a vector (such as a plasmid used for bacterial gene insertions). The genes inserted by the vector into a host genome may be expressed by the host organism (see Figure 13.4).

**Question 1.** Sometimes laboratory technicians may try several restriction enzymes before they find a satisfactory one for the genome of interest. What are two reasons that a specific restriction enzyme might not cleave a given genome?

**Question 2.** You want to produce large quantities of an enzyme used in drug therapy. You have a DNA fragment that contains the gene that encodes the enzyme, but the DNA also contains multiple other genes. Describe how you would isolate the gene and create a recombinant plasmid DNA that expresses the enzyme for insertion into a bacterial host cell.

## 13.2 DNA Can Genetically Transform Cells and Organisms

Genes can be inserted into prokaryotic or eukaryotic cells

A variety of methods are used to insert recombinant DNA into host cells

Reporter genes help select or identify host cells containing recombinant DNA

Cloned genes are useful for sequence analysis, to produce quantities of their protein products for medicinal or agricultural use, or to create new transgenic organisms.

Recombinant DNA is inserted into a host cell by transformation (or transfection, if the host cell is an animal cell), creating a transgenic cell or organism. Because only a few of the cells exposed to the recombinant DNA are transformed, selectable markers, such as genes that confer resistance to antibiotics, are often included on the recombinant DNA molecule.

Theoretically, genes can be cloned into any cell or organism. Most research has centered on model organisms. Prokaryotes, especially *E. coli*, have been used to clone many genes and have plasmids that can easily insert genes into the host organism genome. Because of differences in gene expression between prokaryotes and eukaryotes, bacteria might not be suitable as hosts to express eukaryote genes. Eukaryote gene expression studies are often carried out in yeast cells, such as *Saccharomyces.* Yeasts have a rapid cell division cycle (2–8 hours), are easy to grow, have a small genome size (12 million base pairs), and have been used to clone many eukaryotic genes.

Many plant cells are totipotent, so unspecialized stem cells can be cultured from mature cells. They can be grown in culture, transformed with recombinant DNA, and manipulated to form an entire new transgenic plant containing the recombinant DNA molecule. Recombinant DNA that is incorporated into germ cells will be passed on to the next generation in the seeds. Recombinant DNA can also be incorporated into cultured animal cells and whole transgenic animals.

Methods for inserting recombinant DNA into host cells vary and can be challenging. The DNA must not only be incorporated into the cell, but also into a replicon, a replication unit containing an origin of replication. This can be accomplished either by inserting the recombinant DNA into the host chromosome, where it is replicated when the chromosome is replicated, or recombinant DNA molecules can enter the host cell as part of a vector that already has an origin of replication. Vectors can be plasmids or viruses. Plasmids are useful because they can replicate independently, have one or more restriction sites where DNA can be cut to insert genes, often contain selectable markers such as bacterial resistance, and are small in size.

Plasmids are capable of making many copies of plasmid DNA per cell. Scientists have produced desirable combinations of unique restriction enzyme sites, origins of replication for specific host organisms, and a variety of reporter and selectable marker genes in strains used in the laboratory.

*Agrobacterium tumefaciens*, a bacterium that causes crown gall disease in plants, harbors a Ti (tumor inducing) plasmid. This plasmid contains a segment, T DNA, which inserts itself into the host plant cell's DNA when the bacteria infect the plant. This plasmid has been modified to be nonpathogenic and is widely used to insert desirable genes into plant genomes.

Plasmid replication limits size of DNA insertions to about 10,000 base pairs. Most eukaryotic genes are larger than this. For larger DNA sequences (up to 20,000 bp of inserted DNA), virus vectors are used. Viruses infect cells naturally, allowing easy entry of cloned sequences into the cytoplasm of the cell.

Reporter genes identify host cells that contain recombinant DNA. These are important because not all vector copies contain the recombinant DNA and only a small proportion of potential host cells actually take up the vector. Selectable markers, such as antibiotic resistance genes, can be used to determine if the host cell contains the recombinant DNA molecule (see Figure 13.5). Other types of reporter genes include those for β-galactosidase and green fluorescent protein (GFP), which can be used as visual markers to detect uptake of recombinant DNA molecules in host cells. Reporter genes can also be attached to promoters of gene coding regions to visualize transcription or protein localization in cells.

**Question 3.** What are the critical elements needed for a useful plasmid vector for recombinant DNA?

**Question 4.** What are the constraints on the use of plasmids as vectors for inserting recombinant DNA into cells and organisms?

### 13.3 Genes Come from Various Sources and Can Be Manipulated

DNA fragments for cloning can come from several sources

Synthetic DNA can be made in the laboratory

DNA sequences can be manipulated to study cause-and-effect relationships

Genes can be inactivated by homologous recombination

Complementary RNA can prevent the expression of specific genes

DNA microarrays reveal RNA expression patterns

One of the reasons for cloning DNA is to study its function, including the proteins it codes and their regulatory sequences. Sources of DNA for cloning include genomic libraries, cDNA libraries, and synthetic DNA.

Genomic libraries are a collection of DNA fragments from the entire genome of an organism. After the DNA is cut or broken into fragments, each fragment is inserted into a vector, which is used to insert the gene into bacteria. DNA hybridization with colonies of transformed bacteria can identify fragments that contain particular coding sequences.

cDNA libraries consist of all the genes transcribed in a particular tissue. To make cDNA (complementary DNA), the messenger RNA must first be isolated from the cell. A DNA molecule complementary to the RNA is synthesized using reverse transcriptase, and the new molecule can then be cloned (see Figure 13.7). cDNA clones are used to compare gene expression in different tissues at different stages of development.

Synthetic DNA can be made using PCR to amplify a specific sequence. Artificial genes can be made if the amino acid sequence of the gene is known. Sequences for transcriptional and translational initiation and termination can be added to the gene sequence.

Recombinant technology permits generation of mutant genes in the lab. Mutants can be synthesized and compared to wild-type genes to analyze gene function.

Another method of studying gene function is to eliminate gene expression by creating knockout genes. This can be done in a variety of ways, one of which is called homologous recombination. In this case (used in the mouse model) a normal gene is replaced by an inactivated form of the gene.

In constructing transgenic mice to test the function of a gene, the gene of interest is disrupted by the insertion of a reporter or marker gene into the middle of the native gene. A plasmid containing the inactivated gene with the marker is transfected into a mouse stem cell (see Figure 13.8). If recombination occurs, the marker gene will be expressed. The transfected stem cell is then transplanted into an early mouse embryo, and the resulting phenotype is analyzed.

Antisense messenger RNA and RNAi can prevent the translation of specific genes. Antisense RNA will base pair with mRNA in the cytoplasm and form a double-stranded RNA molecule, which cannot be translated and will be degraded by the cell. Interference RNAs, known as small interfering RNAs (siRNAs), are short (20–25 bp) double-stranded RNA molecules that bind specific mRNAs and target them for degradation. Because siRNAs are more stable than antisense RNA, they are the preferred method in medicine and in the laboratory for inhibiting translation (see Figure 13.9).

DNA microarrays reveal RNA expression patterns using hybridization in a large number of sequences simultaneously. It is possible to examine patterns of expression in different tissues, under different conditions, and in individuals with known mutations. Microarrays are small glass chips containing thousands of copies of each DNA sequence per chip. These sequences (greater than 20 bp) are attached to the chip in spots in a precise order. Each spot on the chip contains a unique sequence. Chip technology has been used to look at gene expression in various forms of breast cancer to predict the prognosis for patients and determine treatments (see Figure 13.10).

**Question 5.** What techniques can be used to study gene expression during development?

**Question 6.** Hybridization is a useful technique in biotechnology. Describe three ways in which hybridization is used experimentally in DNA recombinant technology.

**Question 7.** You are studying the role of a protein in human fibroblasts. What are two techniques you could use to inhibit the expression of this protein in human fibroblasts cultured in your laboratory?

### 13.4 Biotechnology Has Wide Applications

Expression vectors can turn cells into protein factories

Medically useful proteins can be made by biotechnology

DNA manipulation is changing agriculture

There is public concern about biotechnology

Biotechnology is the use of living cells to produce or modify useful materials for people, including food, medicines, and chemicals. This includes turning organisms into "factories" by manipulating them to express genes at high levels.

For cells to produce a cloned gene product, the vector must have appropriate DNA sequences that allow the cloned gene to be expressed in the host organism. Prokaryotic expression vectors require a promoter, a termination site for transcription, and a ribosome-binding site (see Figure 13.11). Eukaryotic expression vectors require a poly A addition site, transcription factor binding sites, and enhancers. Modifications of expression vectors include the addition of inducible promoters (which respond to a specific signal), tissue-specific promoters to localize the expression to a specific tissue or time in development, and signal sequences that direct the product to the appropriate destination.

Medically useful products that have been cloned in expression vectors include tissue plasminogen activator, human insulin (see Figure 13.12), and vaccine proteins (see Table 13.1). Bacteria or other organisms can be transformed using recombinant DNA to produce the desired substance or its subunits that can then be extracted for human use.

Recombinant DNA offers breeders the opportunity to choose specific genes that will be incorporated into an organism, to introduce any gene into a plant or animal species, and to generate new organisms quickly.

Transgenic plants have been created that express toxins for insect larva (using the gene for the toxin naturally produced by the bacterium *Bacillus thuringiensis*), that produce extra nutrients (adding genes for β-carotene to rice to reduce vitamin A deficiencies in human populations), or that can tolerate drought or high-salt conditions (adding salt-tolerance genes to tomatoes).

The creation of transgenic plants has raised some concerns that these crops could be unsafe for human consumption, that it is unnatural to interfere with nature, and that transgenes could escape into other noxious plants. Most scientists agree that transgenic plants should be extensively field tested and that the technology should proceed cautiously.

Recombinant bacteria have been used to clean up the environment in composting programs, wastewater treatments, oil spills, and other such efforts.

**Question 8.** Suppose that your lab assistant has cloned gene *X* into yeast and confirmed that the recombinant DNA molecule is present in the yeast cells. However, the yeast cell is unable to synthesize protein X. Further investigation reveals that your lab assistant did not use a specialized yeast expression vector. Suggest why the yeast cannot synthesize protein X and what modifications are needed in the cloning procedure.

**Question 9.** Describe three useful products that have been produced using biotechnology. Outline two specific dangers that could result from producing organisms that contain foreign genes.

**Question 10.** You are working at a biotech company and your project is to clone a eukaryotic gene (*X*) so that you can isolate large amounts of protein X. First, describe the steps you would take to successfully complete this project using genomic DNA, a plasmid, and *E. coli* as the host organism. Next, assume that you have your clone but you realize that protein X, which is being made from this clone, is nonfunctional. What do you have to change in your procedure to clone gene *X* so that the expressed protein will be functional?

**Question 11.** Diagram the cloning steps you would take to create a transgenic *E. coli* cell using the isolated gene *X* fragment from Question 9.

## Test Yourself

1. Bacteria protect themselves from invading viruses with
   a. restriction enzymes.
   b. bacteriophage λ.
   c. RNA polymerases.
   d. transformation.
   e. plasmids.
   ***Textbook Reference:*** *Concept 13.1 Recombinant DNA Can Be Made in the Laboratory; Restriction enzymes cleave DNA at specific sequences*

2. Which of the following makes use of complementary base pairing?
   a. Ligation reactions with blunt-end DNA molecules
   b. Hybridization between DNA and transcription factors
   c. The cutting of cell walls with restriction enzymes
   d. Synthesis of cDNA molecules from mRNA templates
   e. Transcriptional activation of expression vectors
   ***Textbook Reference:*** *Concept 13.3 Genes Come from Various Sources and Can Be Manipulated; DNA fragments for cloning can come from several sources*

3. For a prokaryotic vector to be propagated in a host bacterial cell, the vector needs
   a. an origin of replication.
   b. telomeres.
   c. centromeres.
   d. drug-resistance genes.
   e. reporter genes.
   ***Textbook Reference:*** *Concept 13.2 DNA Can Genetically Transform Cells and Organisms; A variety of methods are used to insert recombinant DNA into host cells*

4. For a Ti plasmid to be propagated in a host plant cell, the vector needs
   a. telomeres.
   b. centromeres.
   c. an origin of replication.
   d. a reporter gene.
   e. drug-resistance genes.
   ***Textbook Reference:*** *Concept 13.2 DNA Can Genetically Transform Cells and Organisms; A variety of methods are used to insert recombinant DNA into host cells*

5. Reporter genes include genes for
   a. DNA polymerases.
   b. bioluminescence.
   c. DNA origins.
   d. restriction enzymes.
   e. RNA polymerases.
   ***Textbook Reference:*** *Concept 13.2 DNA Can Genetically Transform Cells and Organisms; Reporter genes help select or identify host cells containing recombinant DNA*

6. Genes can be inserted into plants using _______, which is isolated from the bacterium *A. tumefaciens.*
   a. a Ti plasmid
   b. a virus
   c. a pBR322 plasmid
   d. bacteriophage λ
   e. a chromosome
   ***Textbook Reference:*** *Concept 13.2 DNA Can Genetically Transform Cells and Organisms; A variety of methods are used to insert recombinant DNA into host cells*

7. A cDNA clone is
   a. mostly cytosine.
   b. a copy of the DNA identical to the nuclear gene.
   c. a copy of noncoding DNA.
   d. a DNA molecule complementary to an mRNA molecule.
   e. a fragment of DNA inserted into the host chromosome.
   ***Textbook Reference:*** *Concept 13.3 Genes Come from Various Sources and Can Be Manipulated; DNA fragments for cloning can come from several sources*

8. Proteins can be overexpressed in cells using
   a. antisense RNA.
   b. knockout genes.
   c. DNA microarrays.
   d. microRNA.
   e. expression vectors.
   ***Textbook Reference:*** *Concept 13.4 Biotechnology Has Wide Applications; Expression vectors can turn cells into protein factories*

9. Expression vectors are different from other vectors in that they contain
   a. drug-resistance markers.
   b. telomeres.
   c. regulatory regions that permit the cloned DNA to produce a gene product.
   d. DNA origins.
   e. reporter genes that are expressed in the host.
   ***Textbook Reference:*** *Concept 13.4 Biotechnology Has Wide Applications; Expression vectors can turn cells into protein factories*

10. Antisense RNAs
    a. inhibit the transcription of target genes.
    b. are produced only by prokaryotes.
    c. are produced only by viruses.
    d. inhibit the translation of target mRNA sequences.
    e. are only produced synthetically, by scientists.
    ***Textbook Reference:*** *Concept 13.3 Genes Come from Various Sources and Can Be Manipulated; Complementary RNA can prevent the expression of specific genes*

11. DNA microarray technology can be used to
    a. overexpress genes.
    b. show transcriptional patterns in an organism during different times of development.
    c. clone DNA.
    d. make transgenic plants.
    e. inhibit transcription of disease genes.
    ***Textbook Reference:*** *Concept 13.3 Genes Come from Various Sources and Can Be Manipulated; DNA microarrays reveal RNA expression patterns*

12. Which of the following is *not* an application of recombinant DNA technology?
    a. Generating large amounts of tissue plasmonigen factor to help dissolve blood clots
    b. Making plants more resistant to insect larva
    c. Creating vaccines for pathogens
    d. Reducing the salt in environmental soil so that plants can grow
    e. Creating bacteria that can accelerate the breakdown of wood chips and paper
    ***Textbook Reference:*** *Concept 13.4 Biotechnology Has Wide Applications; DNA manipulation is changing agriculture*

13. Expression vectors
    a. are useful for analyzing RNA transcription patterns.
    b. are useful for isolating large amounts of DNA.
    c. can be used to make large amounts of protein in *E. coli* cells.
    d. are useful only in prokaryotic cells and cannot be expressed in eukaryotic cells.
    e. are used to create oligonucleotides for DNA microarrays.
    ***Textbook Reference:*** *Concept 13.4 Biotechnology Has Wide Applications; Expression vectors can turn cells into protein factories*

14. Which of the following is required for a transgene to be expressed in a eukaryotic host?
    a. A tissue-specific promoter
    b. A reporter gene
    c. Transcription factor binding sites, enhancers, and a poly A recognition sequence
    d. An inducible promoter and repressor proteins
    e. Signal sequences
    ***Textbook Reference:*** *Concept 13.4 Biotechnology Has Wide Applications; Expression vectors can turn cells into protein factories*

15. Gel electrophoresis
    a. causes DNA to be pulled through the gel toward the negative end of the field.
    b. causes larger DNA fragments to move more quickly through the gel than smaller DNA fragments.
    c. is required for PCR reactions.
    d. is used to identify and isolate DNA fragments.

e. is used in allele-specific oligonucleotide hybridization.

***Textbook Reference:*** *Concept 13.1 Recombinant DNA Can Be Made in the Laboratory; Gel electrophoresis separates DNA fragments*

16. Which of the following is *not* a step in the process of creating a transgenic mouse with a nonfunctional gene for a protein of interest?
    a. Incorporation of an inactivated copy of the gene into a vector
    b. Addition of a response element to the gene sequence
    c. Homologous recombination
    d. Inactivation of the gene by insertion of a reporter gene
    e. Insertion of the vector into a mouse stem cell

    ***Textbook Reference:*** *Concept 13.3 Genes and Gene Expression Can Be Manipulated; Genes can be inactivated by homologous recombination*

17. Which statement about typical useful recombinant gene vectors is *false*?
    a. They are about 60,000 bp long.
    b. They have several restriction sites.
    c. They can reproduce independently.
    d. Some potentially harmful genes must be removed before they can be useful.
    e. They contain reporter genes.

    ***Textbook Reference:*** *Concept 13.2 DNA Can Genetically Transform Cells and Organisms; A variety of methods are used to insert recombinant DNA into host cells*

18. Bacterial restriction enzymes
    a. are used by bacteria to organize their DNA.
    b. cut methylated restriction sites.
    c. recognize palindromic sequences.
    d. are typically named after the phage they cut.
    e. cut only single-stranded DNA.

    ***Textbook Reference:*** *Concept 13.1 Recombinant DNA Can Be Made in the Laboratory; Restriction enzymes cleave DNA at specific sequences*

19. A researcher wants to identify genes in skin cells whose RNA expression is upregulated in response to exposure to ultra violet (UV) radiation. Which of the following techniques would be appropriate?
    a. Use of antisense RNA
    b. Creation of knockout genes
    c. Cloning
    d. Use of microRNA
    e. DNA microarray

    ***Textbook Reference:*** *Concept 13.3 Genes Come from Various Sources and Can be Manipulated; DNA microarrays reveal RNA expression patterns*

## Answers

### *Key Concept Review*

1. A given restriction enzyme may not be able to cleave the genome if the target sequences are methylated. Another possible reason is that the recognition sequence may not appear at all in a very small genome.
2. First, a DNA fragment containing only the gene that encodes the enzyme must be identified. A first step is to use a restriction enzyme to digest the human DNA. The resulting fragments are then separated by size and abundance using gel electrophoresis. Next, the fragments can be purified from the gel and their sequences analyzed to determine which fragment contains the gene. (Note that many different restriction enzymes may have to be tested to isolate a fragment that contains only the gene of interest). Once the appropriate sequence is identified, the fragment is inserted into a plasmid vector. This is accomplished by first cleaving the plasmid with the same restriction enzyme that had been used to isolate the DNA fragment containing the enzyme. The enzyme fragment and cut vector are mixed together and then the vector and DNA fragments are joined together using DNA ligase.
3. A successful plasmid vector must include 20 or more unique restriction sites, origins of replication (*ori*) sites, and reporter genes and selectable marker genes.
4. Constraints on plasmid replication limit the size of the new DNA that can be inserted to 10,000 bp, while most eukaryotic genes together with their introns and flanking sequences are bigger. In addition, plasmids do not naturally infect cells, and sometimes it is difficult to get them inside the desired host cells.
5. cDNA libraries can be made from different developing tissues in order to see which genes are being expressed. DNA microarrays can also be used to analyze transcriptional patterns during development.
6. Hybridization of complementary base pairs allows ligase to seal DNA fragments with complementary sticky ends to create a recombinant molecule. Genomic and cDNA libraries are screened for sequences of interest by the hybridization of DNA probes. Hybridization is also used in DNA microarray technology.
7. Antisense RNA can be used to inhibit the protein's expression by preventing translation of its mRNA. Both small interfering RNAs (siRNAs) and microRNAs (miRNAs) could be tested to determine which technique better decreases the protein levels in the fibroblast cells.
8. Gene *X* can be cloned into a yeast cell, but unless the vector has the appropriate regulatory signals (promoters, poly A addition sites, translational initiation, and

termination signals), no expression of gene *X* will occur. Recloning gene *X* in a specialized yeast expression vector will result in the expression of gene *X*.

9. Useful products include rice grains that produce β-carotene, plants that are resistant to herbicides and insect larvae, the production of human growth hormone in cow's milk, and others (see Tables 13.1 and 13.2). Dangers include the creation of genetically engineered foods that could adversely affect human nutrition, the transfer of herbicide- and insect-resistant genes from crop plants to noxious weeds, and the introduction into the wild of new organisms that might have unforeseen ecological consequences.

10. The steps would be the following: (1) isolate DNA containing gene *X* from your cell sample; (2) create a genomic library with DNA; (3) using a probe for gene *X*, select a colony containing gene *X* from the genomic library; (4) isolate DNA from the selected clone and cut it with a restriction enzyme; (5) purify the fragments and identify the correct one using gel electrophoresis and confirm the sequence by sequencing; (6) cut an expression vector with the same restriction enzyme as gene *X*; (7) ligate gene *X* into the expression vector; (8) identify recombinant vectors using a selectable marker; (9) transform *E. coli* with the recombinant vector; and (10) isolate protein X from *E. coli* cells.

    However, the reason that protein X is nonfunctional is that gene *X* was cloned from genomic DNA, which still had introns in the gene sequence. Bacterial DNA does not have introns, so bacteria do not have the machinery to properly edit the genomic DNA. A cDNA clone (which is made from the mRNA from gene *X* and thus lacks introns) should have been used as a source for gene *X*.

    Therefore, a successful procedure would have the following steps: (1) isolate mRNA from your cell sample and use RT-PCR to create cDNA; (2) utilizing the PCR technique and primers flanking gene *X*, amplify gene *X* from the cDNA; (3) purify the PCR fragment and cut with restriction enzyme to produce a ligatable fragment containing gene *X*; (5) cut an expression vector with same restriction enzyme as gene *X*; (6) ligate gene *X* into the expression vector; identify recombinant vectors using a selectable marker; (8) transform *E. coli* with the recombinant vector; and (9) isolate protein X from *E. coli* cells.

11. You would use an expression vector that has a promoter, a transcriptional termination site, and a ribosome binding site that will be recognized by the host cell (*E. coli*), so that the protein will be expressed in the host cell.

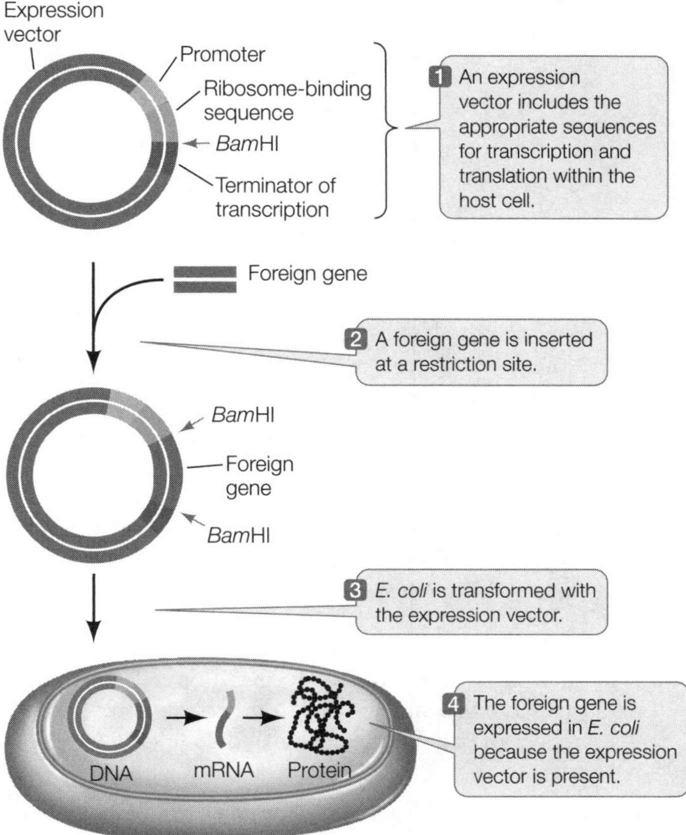


### *Test Yourself*

1. **a.** Bacteria use restriction enzymes to cleave invading viruses into smaller, noninfectious fragments

2. **d.** Reverse transcriptase synthesizes DNA by the complementary base pairing of DNA nucleotides with the RNA nucleotides of the mRNA template. No complementary base pairing can occur between blunt-end cut DNA molecules. Transcription factors are proteins that bind DNA through interactions with their side chains (which are amino acids) and the nucleotides of the DNA. Restriction enzymes cut double-stranded DNA, not cell walls. The activation of vectors does not require complementary base pairing.

3. **a.** A prokaryotic vector needs an origin of replication to be propagated in a prokaryotic cell. Prokaryotic vectors do not have telomeres or centromeres. Drug-resistance genes and reporter genes are helpful for cloning, but they are not necessary for a prokaryotic vector to be propagated in a prokaryotic cell.

4. **c.** The Ti plasmid requires an origin of replication but not reporter genes or drug-resistance genes to propagate itself inside the plant cell. Ti plasmids do not have telomeres or centromeres.
5. **b.** Reporter genes include genes that encode bioluminescence, such as green fluorescent protein (GFP).
6. **a.** A Ti plasmid is the vector isolated from the bacterium *Agrobacterium tumefaciens* that can be used for carrying DNA into plants.
7. **d.** cDNA clones are not clones that contain mostly cytosine, nor are they copies of noncoding genes. A cDNA clone is generated by making a DNA copy of a particular messenger RNA using reverse transcriptase. The cDNA clone is not identical to the nuclear gene because in the messenger RNA (which served as a template for the cDNA) the introns have been removed, leaving coding sequence and 5′ and 3′ flanking sequences.
8. **e.** Protein expression can be increased above normal expression levels by using an expression vector that contains the gene encoding the protein. Antisense RNA, which complementarily base pairs with the target mRNA, inhibits protein expression by making the mRNA inaccessible to the translation machinery in the cell. MicroRNAs bind mRNAs and target them for degradation. Knockout genes are genes that have been inactivated by the insertion of DNA into their coding sequences. DNA microarrays are used in hybridization experiments and do not affect gene expression.
9. **c.** Expression vectors may contain drug-resistance markers and reporter genes, and they must contain DNA origins, but none of these features distinguishes them from other vectors. Expression vectors are unique because they contain regulatory sequences that allow the cloned gene to be expressed in the host cell.
10. **d.** Antisense RNAs are expressed by complex eukaryotes and inhibit translation (not transcription). They can also be produced synthetically.
11. **b.** DNA chips can be used to analyze gene expression at different times in development. They are not used to make transgenic plants, to clone DNA, to overexpress genes, or to inhibit transcription of disease genes.
12. **d.** Soils have not been made less salty using biotechnology; plants have been made more salt-tolerant by means of recombinant DNA techniques.
13. **c.** Expression vectors have the appropriate control regions that allow them to be expressed (make protein) in both prokaryotic and eukaryotic host cells. They are not used to create DNA oligonucleotides, nor are they used to isolate DNA or study transcription.
14. **c.** Transcription factor binding sites, enhancers, and a poly A sequence are required for an expression vector to produce its cloned gene product in a host eukaryotic cell. Inducible promoters and repressors will regulate gene expression but are not required for expression. A tissue-specific promoter will cause the gene to be expressed in particular cells (and would be required for the expression vector to work in a particular tissue), and signal sequences will target the protein to particular compartments in the cell, but they are not part of the minimal requirements for expression. A reporter gene can be used to identify cells that carry the transgene, but is not required for propagation.
15. **d.** DNA fragments migrate toward the positive end of the electric field, with the smallest fragments migrating the fastest. PCR reactions do not require gel electrophoresis, although their products are analyzed using it. Allele-specific oligonucleotide hybridization does not require gel electrophoresis.
16. **b.** A response element is a short sequence of DNA that binds a specific transcription factor and induces the expression of genes, so it would not be an important step in a process that creates a nonfunctional gene.
17. **a.** Most useful vectors are small (<6,000 bp). This makes them too small to carry eukaryotic genes, but even virus vectors that can carry eukaryotic genes are about 45,000 bp. Some vectors, like Ti plasmid and viruses, have their genomes altered to eliminate the genes that cause tumors (in the case of Ti) and the genes that cause host cells to die and lyse, but this is not typical of a useful vector.
18. **c.** With two identical active sites on two subunits, bacterial restriction enzymes recognize and cut palindromic sequences of viral double-stranded DNA into smaller, noninfectious fragments.
19. **e.** To identify RNAs that are upregulated by fibroblasts in response to UV radiation, a researcher would to use a DNA microarray. The researcher would first harvest mRNA from the fibroblasts (with and without UV exposure), synthesize cDNA , and then hybridize the cDNA to a DNA microarray containing human DNA sequences.

# 14 Genes, Development, and Evolution

## The Big Picture

- The development of a mature organism from a fertilized egg involves the patterned activation of genes in response to environmental signals. This sort of gene activation determines the fate of the cell. The zygote and early embryonic cells are totipotent; they can develop into any structure in the adult organism. As development proceeds, the developmental potential of embryonic cells narrows. Adult stem cells are multipotent, meaning they are restricted to a subset of all possible cells in an organism. Positional information, often coming from inducers called morphogens, leads to changes in the expression of key developmental genes, which in turn control morphogenesis, the creation of body form.
- Evolutionary developmental biology is the study of how organisms evolved by examination of the differences in developmental gene expression and regulation between species. Animals have highly conserved genes, such as homeobox genes, that control development. Genetic switches determine where and when genes will be expressed. Changes in these switches can result in the evolution of species differences. Heterochrony is the phenomenon whereby phenotypic differences in organisms have arisen because of differences in the timing of developmental processes.

## Study Strategies

- It may be difficult to imagine how a single cell can divide and grow to produce a mature functional organism with highly differentiated tissues. At some point very early in development (either before fertilization or in one of the first sets of divisions), different genes begin to be expressed in different cells. This gene expression can be in response to cytoplasmic factors or signals from other cells. Early gene expression sets up positional determinants that activate another wave of genes that further divides regions of the embryo into different developmental areas. As cell division proceeds in the embryo, these developmental genes continue to be differentially expressed, resulting in differentiation and morphogenesis in the organism. Review the four key processes of development: (1) determination; (2) differentiation; (3) morphogenesis; and (4) growth.
- Make sure that you understand the different potentials of totipotent, pluripotent, and multipotent stem cells. Review these in the context of human development, focusing on when and where each type of stem cell occurs.
- Review some of the important genes whose products direct development in the model organisms discussed in the chapter. Identify orthologous genes from different organisms.
- The concept of heterochrony can be difficult to grasp. Try to remember that when we discuss heterochrony, we are talking about differences among species, not within species.
- Go to **LaunchPad** (or use the Web addresses listed) to review the following additional resources:

  Animated Tutorial 14.1 Cell Fates (PoL2e.com/at14.1)

  Animated Tutorial 14.2 Embryonic Stem Cells (Pol2e.com/at14.2)

  Animated Tutorial 14.3 Early Asymmetry in the Embryo (Pol2e.com/at14.3)

  Animated Tutorial 14.4 Pattern Formation in the *Drosophila* Embryo (Pol2e.com/at14.4)

  Animated Tutorial 14.5 Modularity (Pol2e.com/at14.5)

  Activity 14.1 Stages of Development (PoL2e.com/ac14.1)

  Activity 14.2 Plant and Animal Development (PoL2e.com/ac14.2)

  Analyze the Data in textbook Figure 14.4

  Apply the Concept in textbook Sections 14.3 and 14.4

## Key Concept Review

### 14.1 Development Involves Distinct but Overlapping Processes

Four key processes underlie development

Cell fates become progressively more restricted during development

### Cell differentiation is sometimes reversible

### Stem cells differentiate in response to environmental signals

During development, an organism progresses through successive forms as it moves through its life cycle (see Figure 14.1). The fertilized egg is called a zygote, and from the zygote an embryo develops. Development includes four processes: determination, differentiation, morphogenesis, and growth. Development ends with death.

The developmental fate of a cell is set during determination and is influenced by internal and external conditions. More specifically, differential gene expression and the extracellular environment set the fate of a cell. During differentiation, cells become specialized to contain specific structures and to perform particular functions. Morphogenesis is the creation of form as seen in body shape and organs. It can occur by cell division (an increase in the number of cells), cell expansion (an increase in the size of existing cells), cell movements, and programmed cell death (apoptosis). Growth is an increase in size due to cell division or cell expansion. Cell division is important in the development of plants and animals; cell expansion is particularly important in the development of plants.

Transplantation experiments have shown that the environment in which early embryonic cells exist can redirect them along different developmental paths (see Figure 14.2). Nevertheless, embryonic cells eventually become committed to a particular developmental fate, even though they are not yet differentiated. This is because their fate has been determined. When these cells from older embryos are transplanted, they continue to develop into the original differentiated tissue, regardless of their environment. Cell fate becomes apparent as cells differentiate.

Some cells, such as a zygote, are totipotent, meaning they can develop into any of the different kinds of cells in the mature organism. Developmental possibilities narrow as cell determination and differentiation occur.

In plants, differentiation is reversible in some cells (see Figure 14.3). Under certain conditions, plant cells in culture can give rise to a new, genetically identical plant (a clone). Plant cells first dedifferentiate, and then produce a callus (mass of cells), which develops into a plant embryo when placed in the appropriate medium. Thus, the original differentiated cells contained all the genetic information needed to express genes in the correct sequence to generate a whole plant; this is evidence of genomic equivalence.

Mammals can be cloned by fusing somatic cells with enucleated eggs and implanting the resulting embryos in surrogate mothers (see Figure 14.4). Several mammals, including sheep, mice, and cattle, have been cloned by the technique of nuclear transfer. Practical uses of cloning include increasing the number of valuable animals (for example, genetically engineered animals) and preservation of endangered species and pets.

Development occurs in adults as well as in embryos. In adult plants, the areas of undifferentiated cells in the growing tips of roots and stems are known as meristems. These cells can become any cell type found in the plant.

Undifferentiated dividing cells in mammals are known as stem cells. Stem cells in adult mammals are specific for the kinds of tissue they replace, typically skin, the lining of the intestines, and blood cells. These stem cells are multipotent, meaning that they can differentiate into a limited number of cell types. Signals from adjacent cells or from circulation influence the differentiation of stem cells.

Bone marrow contains two types of multipotent stem cells: hematopoietic and mesenchymal. Hematopoietic stem cells produce red and white blood cells, whereas mesenchymal stem cells produce bone and muscle cells. In hematopoietic stem cell transplantation, stem cells from a patient about to undergo high doses of cancer treatment are removed from the blood, stored, and encouraged to increase in number, and then they are returned to the depleted bone marrow once the treatment has been completed.

In embryonic mammals, cells from a part of the blastocyst (the inner cell mass) are pluripotent, which means they are capable of forming nearly every type of cell. These pluripotent embryonic stem cells (ESCs) can be taken directly from human embryos made available through in vitro fertilization or by making induced pluripotent stem cells (iPS cells) from skin cells. The latter technique does not destroy human embryos or provoke an immune response in recipients.

**Question 1.** Create a flow chart detailing the three different types of stem cells present in a developing human. Name and define each type of stem cell, and include the developmental stage (and location, when possible) at which each type of stem cell is present.

**Question 2.** Describe one application for human health of multipotent stem cells and one application for human health of cloned mammals.

**Question 3.** When mammals are cloned by the fusing of whole cells with enucleated eggs, why is the original nucleus removed from the egg?

## 14.2 Changes in Gene Expression Underlie Cell Fate Determination and Differentiation

### Cell fates can be determined by cytoplasmic polarity

### Inducers passing from one cell to another can determine cell fates

### Differential gene transcription is a hallmark of cell differentiation

Cells can be made to transcribe different sets of genes in two ways: (1) by asymmetrical distribution of cytoplasmic factors; and (2) by differential exposure to an external inducer. In the first way, cytoplasmic segregation of factors in the egg results in an unequal distribution of maternal elements in each of the cells of the embryo (see Figure 14.7). Each cell's fate and pattern of gene expression is determined by the amount of cytoplasmic determinants received. Unequal distribution of cytoplasmic determinants directs embryonic development and controls the polarity of the organism. In the second way, cells can induce other cells to differentiate by secreting chemical signals called inducers.

The nematode *Caenorhabditis elegans* is a model organism used in developmental studies, including those that demonstrate induction. The egg of *C. elegans* develops into a larva in 8 hours and an adult in 3.5 days. The animal has a transparent body, making it easy to follow the developmental fate of its cells. The adult is hermaphroditic, having female and male reproductive organs. Eggs are laid through a pore called the vulva, which is induced to form from a single cell called an anchor cell. If this cell is destroyed, no vulva develops. The anchor cell determines the fates of six cells on the ventral surface of *C. elegans* by producing a primary inducer (LIN-3 protein) that diffuses toward adjacent cells, establishing a concentration gradient. The closest cell is exposed to the highest concentration of LIN-3 and becomes the primary precursor cell. It produces a secondary inducer that causes the next closest cells to become secondary precursors. Descendants of primary and secondary precursor cells form the vulva. The final three cells (which are farthest from the anchor cell) become epidermal cells (see Figure 14.8). When LIN-3 binds to a receptor on the surface of the closest cell, it causes a signal transduction cascade, the end result of which is the differentiation of vulval cells in the nematode.

Differentiated cells retain all their original genetic content, even though they express a very small subset of their genes. Mechanisms that control gene expression leading to cell differentiation typically work at the level of transcription.

Transcription factors are proteins that bind DNA and regulate the expression of genes. In response to the transcription factor MyoD (*myo*blast-*d*etermining gene), undifferentiated muscle precursor cells stop dividing; this step is necessary for the eventual differentiation of these cells into mature muscle cells (see Figure 14.10). Sometimes a single transcription factor causes a cell to differentiate. In other cases, several different transcription factors promote differentiation.

**Question 4.** If you divided all eight cells in the 8-cell stage of sea urchin development, each cell would have a separate fate. Describe the resulting sea urchin embryos for each of the 8 cells.

**Question 5.** Activated MyoD has been found in muscle stem cells of adult vertebrates. What might its role be in this situation?

### 14.3 Spatial Differences in Gene Expression Lead to Morphogenesis

- Morphogen gradients provide positional information during development
- Multiple genes interact to determine developmental programmed cell death
- Expression of transcription factor genes determines organ placement in plants
- A cascade of transcription factors establishes body segmentation in the fruit fly

Pattern formation results in the spatial organization of a tissue or organism and morphogenesis is the creation of body form. These two closely linked processes arise from spatial differences in gene expression. For spatial differences in gene expression to occur, cells must know their relative locations in the body and must activate the pattern of gene expression appropriate for their location.

Positional information allows cells to determine where they are in the developing organism. Morphogens are signals that establish positional information. They act directly on the target cell, and different concentrations within the embryo cause different effects. Concentration gradients of the morphogen Sonic hedgehog, secreted by the zone of polarizing activity in the limb bud, determine the anterior–posterior axis of the developing vertebrate limb. According to the "French flag" model, cells in the limb bud form different digits depending on the concentration of the morphogen.

During morphogenesis, some cells are programmed to die through apoptosis. In human embryos, apoptosis occurs in the webs of skin that initially form between fingers and toes. Apoptosis also occurs in *C. elegans*. As the nematode develops from a fertilized egg into an adult, 1,090 cells are produced. However, 131 of these cells are programmed to die due to the sequential expression of *ced-4* and *ced-3* genes. The genes controlling apoptosis are crucial for proper development in many organisms. These genes have been conserved in organisms separated by 600 million years of evolution (nematodes and humans).

During plant development, organs such as leaves, roots, and flowers are produced. Flowers have four types of organs: sepals, petals, stamens, and carpels. Stamens are male reproductive organs and carpels are female reproductive organs. Flower organs occur in whorls organized around a central axis and are derived from meristematic tissue on the plant (see Figure 14.12).

Plant geneticists have studied the development of flower organs in *Arabidopsis*. There are four whorls of organs, and three classes of genes expressed in the whorls act as organ identity genes to guide differentiation in each whorl. Gene A is expressed in whorls 1 and 2, which form the sepals and petals, respectively; gene B is expressed in whorls 2 and 3, which form petals and stamens, respectively; and gene C is expressed in whorls 3 and 4, which form stamens and carpels, respectively. The three genes—A, B, and C—all encode transcription factors that are active as dimers (proteins with two polypeptide subunits). Gene regulation is combinatorial, and the combination of different dimers determines which genes are activated. A dimer of the transcription factor encoded by gene A will activate genes that make sepals. A dimer of transcription factor A with transcription factor B will result in petals. *LEAFY* is a transcription factor that regulates the transcription of the genes A, B, and C.

A cascade of transcription factors controls development in the fruit fly (*Drosophila melanogaster*). The body of *Drosophila* is segmented, consisting of a head (which is formed from fused segments), three thoracic segments, and eight abdominal segments. Three classes of genes, expressed in sequence, define the body segments: (1) maternal effect genes; (2) segmentation genes; and (3) Hox genes.

The first step in body segment definition in *Drosophila* is the establishment of anterior–posterior and dorsal–ventral polarity. Polarity is based on the cytoplasmic distribution of mRNA and proteins produced by maternal effect genes. Mutations in maternal effect genes (*bicoid* and *nanos*) create larvae lacking anterior structures (*bicoid*) or abdominal segments (*nanos*). *Bicoid* encodes a transcription factor that positively affects some genes (*hunchback*) and negatively affects others. Nanos protein inhibits the translation of *hunchback*.

Segmentation genes are expressed when the embryo is at the 6,000-nuclei stage and determine the number, boundaries, and polarity of the segments. Three types of segmentation genes regulate segmental development: gap genes, pair rule genes, and segment polarity genes. Gap genes organize large areas along the anterior–posterior axis of the developing embryo. Gap mutants produce larvae that are missing consecutive larval segments. Pair rule genes divide the larva into units of two segments each. Mutations in the pair rule genes produce larvae that are missing every other segment. Segment polarity genes determine the boundary and the anterior–posterior organization of each segment. Mutations in segment polarity genes produce larvae in which the posterior structures in the segments have been replaced by reversed anterior structures.

Differences between segments are encoded by Hox genes, which give each segment an identity. For instance, in response to Hox gene expression, cells in a thoracic segment produce legs and cells in the head segment produce eyes. Homeotic mutants include *Antennapedia* (producing legs in place of antennae; see Figure 14.14) and *bithorax* (producing a fly with an extra set of wings). A DNA sequence, the homeobox, encodes a 60-amino acid sequence, the homeodomain. The homeodomain includes a DNA-binding motif and acts as a transcription factor.

**Question 6.** The textbook describes development in several model organisms, including thale cress (*Arabidopsis thaliana*), fruit fly (*Drosophila melanogaster*), and nematode (*Caenorhabditis elegans*). How would you define "model organism"? What traits characterize model organisms?

**Question 7.** Develop a flow chart showing the gene cascade that controls body segmentation in the fruit fly. For each class of genes, include a brief description of general function.

## 14.4 Changes in Gene Expression Pathways Underlie the Evolution of Development

Developmental genes in distantly related organisms are similar

Genetic switches govern how the genetic toolkit is used

Modularity allows for differences in the pattern of gene expression among organisms

Evolutionary developmental biology (often shortened to evo-devo) is the contemporary study of evolution and development. The major findings of this field include: (1) the molecular mechanisms for morphogenesis and pattern formation are shared across diverse organisms; (2) the molecular pathways for different developmental processes operate independently from one another in modules; (3) new structures can evolve through changes in the spatial location and timing of expression of certain genes; and (4) structures change over time, largely as the result of modifications to existing developmental genes and pathways.

Many of the genes controlling development are conserved across different organisms. Thus, these genes can be very similar in organisms that have very different appearances. As an example, the development of an anterior–posterior axis in insects and mammals is due to the same types of homeobox-containing genes that are expressed at either the anterior or posterior end of the developing embryo (see Figure 14.15). The major differences in body form arise because genes are turned on and off at different times and locations during development. The organism's genes and transcription factors, along with the way they interact with extracellular signals, can be thought of as a genetic toolkit used to assemble the organism.

Genetic switches, which consist of promoters and the transcription factors that bind them, control the way that a gene is used during development. Each gene is controlled by multiple switches that influence where and when it is expressed. For example, genetic switches control spatial development of the embryo. The pattern of development depends on the Hox genes that are expressed within the module. This can be seen in the development of wings in *Drosophila*, in which Hox genes are differentially expressed in the three thoracic segments of the embryo (see Figure 14.16).

The many parts of the developing embryo can change independently of one another because the organism consists of developmental modules. On the time scale of evolution, modularity means that changes in the timing or spatial position of a particular developmental process can occur without disrupting the entire organism.

Heterochrony occurs when the relative timing of developmental processes independently shifts in different species. Like almost all mammals, the giraffe has seven cervical vertebrae in its neck. Thus, its long neck cannot be explained by the presence of additional vertebrae. Instead, it results from a delay in the signaling process that stops bone growth, which allows the vertebrae to grow longer (see Figure 14.17). In other words, the giraffe's long neck results from changes in the timing of expression of genes that control bone formation.

Evolutionary change can also occur as a result of changes in the spatial expression of developmental genes. Bird embryos have webbed feet. The webbing is retained in adult ducks but not in chickens. The signaling protein bone morphogenetic protein 4 (BMP4) causes the cells of the webbing to undergo apoptosis. *Gremlin* encodes a protein that inhibits BMP4. Although BMP4 is present in the webbing of embryonic ducks and chickens, it is only expressed in ducks.

**Question 8.** There are morphological differences between adult chickens and ducks that relate to the duck's aquatic lifestyle. Why were these adaptations able to occur without influencing the development of the rest of the duck's morphology?

**Question 9.** Evaluate the following statement: "Most changes in morphology over evolutionary time occur as a result of the introduction of radically new developmental mechanisms."

### 14.5 Developmental Genes Contribute to Species Evolution but Also Pose Constraints

Mutations in developmental genes can cause major morphological changes

Evolution proceeds by changing what's already there

Conserved developmental genes can lead to parallel evolution

Major evolutionary changes can occur as a result of mutations in developmental genes. The insect *Ultrabithorax (Ubx)* homeotic gene has a mutation that results in the repression of the gene involved in limb formation, the *Distal-less* gene. The *Ubx* gene is expressed in the abdomen during development, leading to the absence of limbs on the abdomen in insects. Other arthropods, such as centipedes, do not have this mutation and have abdominal legs. Thus, a mutation in a Hox gene changed the number of legs in insects relative to other arthropods (see Figure 14.19).

Evolution works on existing genes and their expression and can occur through changes in Hox gene expression in different modules. Thus, most evolutionary innovations are modifications of previously existing structures. For example, the wings of birds and bats are not new structures; rather, they are modified limbs (see Figure 14.20). Evolutionary losses are also controlled by Hox genes, such as the loss of forelimbs in the ancestors of present-day snakes.

Similar traits evolve repeatedly because of the highly conserved nature of the genetic code. This can lead to parallel phenotypic evolution of a trait, such as the evolutionary loss of body armor in numerous populations of freshwater stickleback fish. Freshwater populations of sticklebacks are descended from marine populations. In marine sticklebacks, the gene *Pitx1* codes for a transcription factor that stimulates the production of protective plates and spines. The *Pitx1* gene has changed in various freshwater populations of sticklebacks, leading to the loss of body armor in these populations (See Figure 14.21).

**Question 10.** Why do insects have legs only on thoracic segments, whereas centipedes have legs on thoracic and abdominal segments?

**Question 11.** Amphipods are small aquatic crustaceans. Researchers have discovered the independent reduction of eyes in different populations of a species of cave-dwelling amphipod. What phenomenon does this represent?

**Question 12.** The diagram below illustrates finger development in humans. How does each digit develop with regard to Sonic hedgehog (Shh)?

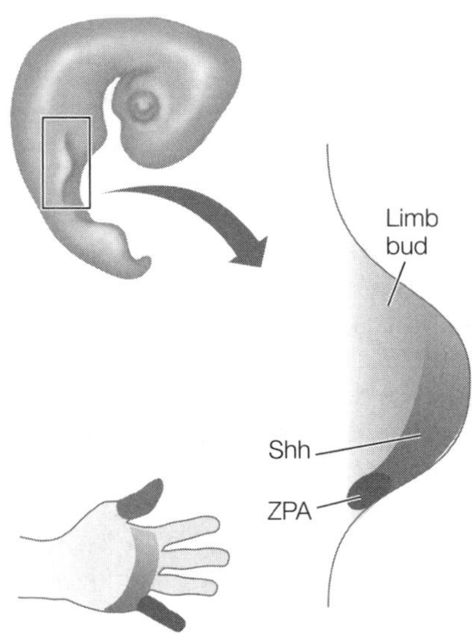


## Test Yourself

1. Which of the following statements about wings is true?
   a. Wings have evolved independently four times in vertebrates.
   b. Wings arose as a result of new "wing genes."
   c. Wings arose as modifications to already existing structures.
   d. Wings have different skeletal components in birds and bats.
   e. Like birds, pterosaurs have a flight surface made of feathers.

   ***Textbook Reference:*** *Concept 14.5 Developmental Genes Contribute to Species Evolution but Also Pose Constraints; Evolution proceeds by changing what's already there*

2. A totipotent cell is a cell
   a. whose developmental fate has been decided.
   b. that has differentiated into a specialized tissue.
   c. that is fated to form a particular structure.
   d. whose developmental potential is extremely broad.
   e. whose developmental potential is narrower than that of a pluripotent cell.

   ***Textbook Reference:*** *Concept 14.1 Development Involves Distinct but Overlapping Processes; Cell differentiation is sometimes reversible*

3. All of the following about heterochrony are true *except*
   a. it can explain the long neck of the giraffe.
   b. it involves shifts in the relative timing of developmental processes.
   c. it can lead to major morphological changes.

d. it explains the reduction in body armor of freshwater sticklebacks.
e. it explains the differential development stages or duration among species.
***Textbook Reference:*** *Concept 14.4 Changes in Gene Expression Pathways Underlie the Evolution of Development; Modularity allows for differences in the pattern of gene expression among organisms*

4. The fate of a cell
a. refers to its original type.
b. refers to its genetic makeup.
c. refers to the type of cell into which it will differentiate.
d. describes its death.
e. cannot by modified by its cytoplasmic environment.
***Textbook Reference:*** *Concept 14.1 Development Involves Distinct but Overlapping Processes; Cell fates become progressively more restricted during development*

5. Genes that regulate development are highly conserved. This means that
a. large differences have evolved among multicellular organisms.
b. they have changed very little over the course of evolution.
c. they are always turned on.
d. they have undergone mutation.
e. they have been lost in some lineages.
***Textbook Reference:*** *Concept 14.5 Developmental Genes Contribute to Species Evolution but Also Pose Constraints*

6. The protein MyoD
a. is a transcription factor that controls the expression of genes involved in the differentiation of muscle cells.
b. controls segment identity in mice.
c. is a transcription factor that activates a gene causing the cell cycle to start in muscle precursor cells.
d. causes the induction of vulval cells in nematodes.
e. is the only transcription factor involved in the differentiation of mesoderm cells into mature muscle cells.
***Textbook Reference:*** *Concept 14.2 Changes in Gene Expression Underlie Cell Differentiation in Development; Differential gene transcription is a hallmark of cell differentiation*

7. Induction occurs when
a. nuclear genes are lost in some tissues.
b. one cell contacts another and fails to alter its developmental fate.
c. a cell or tissue sends a chemical signal to another cell or tissue, causing differentiation of that cell or tissue.
d. two tissues come into contact, causing the release of transcription factors.
e. factors are unequally distributed in the cytoplasm of an egg.
***Textbook Reference:*** *Concept 14.2 Changes in Gene Expression Underlie Cell Differentiation in Development; Inducers passing from one cell to another can determine cell fates*

8. Which statement about plant development is *false*?
a. Both cell division and cell expansion contribute to growth in early plant embryos.
b. Differentiation in plant cells is irreversible.
c. Meristems at the tips of roots and stems consist of undifferentiated cells.
d. Flower development in a plant involves organ identity genes.
e. In plants, the fertilized egg is called a zygote.
***Textbook Reference:*** *Concept 14.3 Spatial Differences in Gene Expression Lead to Morphogenesis*

9. It may be possible to genetically engineer plants to produce more seeds and fruits by
a. providing more fertilizer in the spring.
b. manipulating their gap genes.
c. altering the homeotic genes that control carpel development.
d. inducing a mutation that eliminates *LEAFY* gene function.
e. altering the homeotic genes that control stamen development.
***Textbook Reference:*** *Concept 14.3 Spatial Differences in Gene Expression Lead to Morphogenesis; Expression of transcription factor genes determines organ placement in plants*

10. All of the following are true of morphogens *except*
a. they diffuse within the embryo to set up a concentration gradient.
b. they are expressed as positional signals in developing embryos.
c. they direct differentiated cells to form organs.
d. they must specifically affect target cells.
e. they can trigger apoptosis.
***Textbook Reference:*** *Concept 14.3 Spatial Differences in Gene Expression Lead to Morphogenesis; A cascade of transcription factors establishes body segmentation in the fruit fly*

11. Maternal effect genes
a. begin setting up positional axes in the egg prior to fertilization.
b. are the same as segmentation genes.
c. are masked by paternal genes.
d. operate late in the gene cascade regulating development of *Drosophila* embryos.
e. determine cell fate within each segment.
***Textbook Reference:*** *Concept 14.3 Spatial Differences in Gene Expression Lead to Morphogenesis; A cascade of transcription factors establishes body segmentation in the fruit fly*

12. All of the following are true of hox genes *except*
    a. they encode protein domains that are important in development and have been highly conserved over evolutionary time.
    b. they are found in diverse organisms.
    c. they can produce the wrong structure in the wrong place when mutated.
    d. they set up the major axes of the egg.
    e. they help determine cell fate within each segment of a developing *Drosophila* embryo.
    ***Textbook Reference:*** *Concept 14.3 Spatial Differences in Gene Expression Lead to Morphogenesis; A cascade of transcription factors establishes body segmentation in the fruit fly*

13. What principle of development states that the nuclei of cells do not lose any genetic information during the stages of development?
    a. Genomic equivalence
    b. Apoptosis
    c. Totipotency
    d. Transcription
    e. Fate mapping
    ***Textbook Reference:*** *Concept 14.1 Development Involves Distinct but Overlapping Processes; Cell differentiation is sometimes reversible*

14. What is the role of cytoplasmic segregation in determining the fate of a cell?
    a. It keeps cells totipotent.
    b. It determines polarity within the organism.
    c. It ensures that gradients do not develop in the developing organism.
    d. It stops cell division.
    e. It has no role in determining cell fate.
    ***Textbook Reference:*** *Concept 14.2 Changes in Gene Expression Underlie Cell Differentiation in Development; Cell fates can be determined by cytoplasmic polarity*

15. Module development has allowed evolutionary changes to occur while still resulting in a viable organism because
    a. all modules must change together.
    b. modules can change independently of one another.
    c. modules are unimportant for development.
    d. modules prevent heterochrony.
    e. developmental genes always exert their effects on more than one module.
    ***Textbook Reference:*** *Concept 14.4 Changes in Gene Expression Pathways Underlie the Evolution of Development; Genetic switches govern how the genetic toolkit is used*

16. Which of the following statements about sticklebacks is *false?*
    a. Body armor is absent from sticklebacks living in freshwater environments due to a genetic mutation.
    b. The *Pitx1* gene promotes the production of protective plates and spines in marine populations of sticklebacks.
    c. Similar traits are likely to evolve repeatedly in sticklebacks because of the highly conserved nature of developmental genes.
    d. The *Pitx1* gene is inactive in freshwater stickleback populations.
    e. Environmental conditions induce the loss of body armor in freshwater sticklebacks.
    ***Textbook Reference:*** *Concept 14.5 Developmental Genes Contribute to Species Evolution but Also Pose Constraints; Evolution proceeds by changing what's already there*

17. Cell differentiation
    a. results from the loss of particular genes from the nucleus of the differentiated cell.
    b. results from the differential expression of genes that are responsive to environmental signals.
    c. involves the persisting totipotency of early embryonic cells in the mature organism.
    d. results from mutations in genes that control the synthesis of DNA.
    e. precedes cell determination.
    ***Textbook Reference:*** *Concept 14.1 Development Involves Distinct but Overlapping Processes; Cell differentiation is sometimes reversible*

## Answers

### *Key Concept Review*

1.

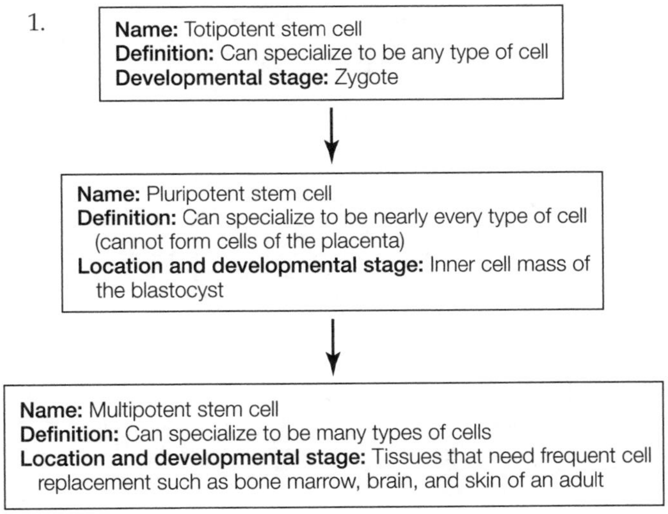

**Name:** Totipotent stem cell
**Definition:** Can specialize to be any type of cell
**Developmental stage:** Zygote

↓

**Name:** Pluripotent stem cell
**Definition:** Can specialize to be nearly every type of cell (cannot form cells of the placenta)
**Location and developmental stage:** Inner cell mass of the blastocyst

↓

**Name:** Multipotent stem cell
**Definition:** Can specialize to be many types of cells
**Location and developmental stage:** Tissues that need frequent cell replacement such as bone marrow, brain, and skin of an adult

2. Hematopoietic stem cells are multipotent stem cells that produce red and white blood cells. If left in the body, these dividing cells are killed during cancer treatment along with cancer cells. In hematopoietic stem cell transplantation, these stem cells are removed prior to cancer treatment, stored in a laboratory

where they are provided with growth factors, and then replaced in greater numbers once treatment has been completed. Through cloning we can increase the number of particularly valuable animals, such as transgenic cows that produce human growth hormone, and thereby increase the supply of growth hormone needed for treating hormone deficiencies.

3. The nucleus of the egg contains a haploid set of chromosomes. Fusing a diploid somatic cell with an egg cell that still contained a nucleus would result in a cell with three copies of each chromosome and little chance of developing normally.
4. If all 8 cells in an 8-cell stage sea urchin embryo were separated, the upper 4 cells would not develop at all, while the lower 4 would each develop into a very small sea urchin.
5. The discovery of activated MyoD in muscle stem cells of adult vertebrates suggests that this transcription factor might also be involved in muscle repair. In other words, its role is not restricted to stimulating muscle cell differentiation from mesoderm in vertebrate embryos.
6. Model organisms are research organisms that are expected to yield insights into how other organisms work. Model organisms generally share genetic, cellular, and/or molecular characteristics with these other organisms. They also have short generation times, are easy to obtain and maintain, are amenable to experimental manipulation, and are not of concern from a conservation standpoint.
7. 

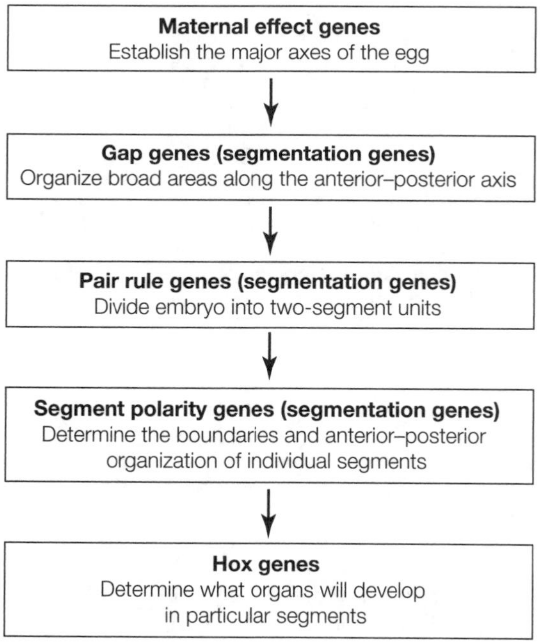

**Maternal effect genes**
Establish the major axes of the egg

↓

**Gap genes (segmentation genes)**
Organize broad areas along the anterior–posterior axis

↓

**Pair rule genes (segmentation genes)**
Divide embryo into two-segment units

↓

**Segment polarity genes (segmentation genes)**
Determine the boundaries and anterior–posterior organization of individual segments

↓

**Hox genes**
Determine what organs will develop in particular segments

8. Organisms are made up of modules. Change can occur in a module independent of other modules. The webbing found in duck feet is due to expression of the *Gremlin* gene in the foot module. Expression of *Gremlin* in the webbing cells of the feet results in a protein that inhibits the BMP4 protein responsible for prompting apoptosis.
9. The statement is incorrect. In fact, most evolutionary changes in morphology occur by modification of existing developmental genes and developmental pathways.
10. Leg development in arthropods involves the homeotic gene *Ultrabithorax* (*Ubx*) and the gene *Distal-less* (*Dll*). Expression of *Dll* in a segment results in the development of a pair of legs in that segment. In insects, the *Ubx* gene has a mutation that suppresses the expression of *Dll* in the abdomen. As a result, legs will not develop in the abdomen. Centipedes do not have this mutation. Therefore, legs develop in thoracic and abdominal segments.
11. The independent reduction of eyes in several populations of a cave-dwelling amphipod is an example of parallel phenotypic evolution, the phenomenon whereby similar traits evolve repeatedly across multiple populations of a species.
12. Digit development is a response to the concentration of Shh, a morphogen. At the little finger there is a zone of polarizing activity (ZPA) that is secreting Shh. A high concentration of Shh results in little finger development. As Shh diffuses away from the ZPA, its concentration decreases. The changing concentration along a diffusion gradient is responsible for the different digits on the human hand. A thumb (i.e., the digit farthest from the ZPA) forms when Shh is lacking.

### *Test Yourself*

1. **c.** Wings arose as modifications of preexisting structures. In vertebrates, wings are modified forelimbs.
2. **d.** A totipotent cell is a cell that can develop in any number of ways. Its fate has not been determined, it has not differentiated, and it can develop into more cell types than a pluripotent cell can.
3. **d.** Heterochrony is the process by which the timing of gene expression differs between two or more species. Heterochrony can explain the long neck of the giraffe (delays occur in the signaling process that stops bone growth, so the neck vertebrae grow longer) but not the reduction in body armor in freshwater populations of sticklebacks.
4. **c.** The fate of a cell is the type of cell it will eventually become once it has differentiated.
5. **b.** Conserved genes are genes that are found in many organisms and have undergone very little change over time.
6. **a.** MyoD is a transcription factor that controls the expression of genes involved in the differentiation of mesoderm cells into mature muscle cells. It activates a gene that causes the cell cycle to stop, allowing

differentiation to begin. It is one of several involved in the differentiation of muscle cells.

7. **c.** Induction occurs when one cell or tissue secretes factors that influence the developmental fate of other cells or tissues. Nuclear genes are not lost in developing cells (red blood cells are an exception). Transcription factors are found in the nucleus and are not released to adjacent tissues.

8. **b.** In general, it is easier to reverse differentiation in plant cells than in animal cells, as demonstrated by the cloning of a carrot from a differentiated storage cell in its root.

9. **c.** Fertilization is not genetic engineering. Gap genes are found in *Drosophila*, not plants. The LEAFY protein transcriptionally activates the homeotic genes that control organ development (and subsequent seed and fruit formation). In its absence, no fruit or seeds would develop. Altering the homeotic genes that control carpel (not stamen) development could lead to more fruit and seeds.

10. **e.** Morphogens diffuse in the embryo to form concentration gradients that act as positional signals to direct differentiated tissues to form organs. They must specifically affect target cells. They do not trigger apoptosis.

11. **a.** Beginning before fertilization, maternal effect genes determine the anterior–posterior axis. These genes, which operate early in the gene cascade, also induce segmentation genes. The paternal genotype does not affect the expression of maternal gene products in the egg. Hox genes, and not maternal effect genes, determine cell fate within each segment.

12. **d.** Hox genes encode proteins that are highly conserved and important for directing development in many different organisms. Mutations in these genes can produce an abnormal developmental event in the organism. In *Drosophila* embryos, Hox genes determine segment identity. Maternal effect genes set up the major axes of the egg.

13. **a.** Genomic equivalence is a fundamental principle of developmental biology. It states that none of the genetic information contained in the original zygote is lost during subsequent cell division and growth.

14. **c.** Cytoplasmic segregation refers to the unequal distribution of factors within the developing embryo. This asymmetry helps set up, for example, the polarity (distinct top and bottom) of the developing organism.

15. **b.** Embryos are composed of self-contained units called modules that can change independently of one another. This allows for one module to evolve without disruption of the other modules.

16. **e.** Absence of body armor in freshwater sticklebacks results from a mutation in the *Pitx1* gene; the absence of armor is not induced by environmental conditions.

17. **b.** Genes are not lost from differentiated cells. There are no cells left from the embryo in a mature organism. Mutations in cells that affect DNA replication would result in cells that were unable to replicate their DNA. Cell differentiation occurs after cell determination.

# 15 Processes of Evolution

## The Big Picture

- Evolution through natural selection remains the most important unifying concept in biology.
- Mutations in the genome produce the genetic variation upon which natural selection acts. The allele frequency of a population can be altered by gene flow, genetic drift, the founder effect, and sexual selection. The Hardy–Weinberg equilibrium principle can be used to assess if changes in allele frequencies, and thus evolution, are occurring in a population.
- Selection can be stabilizing, directional, or disruptive. These types of selection alter the distribution of phenotypes in a population. The neutral theory states that the majority of variants in a population are selectively neutral. Heterozygotes can be at a selective advantage over homozygotes.
- New features in a population can arise from sexual recombination, lateral gene transfer from other organisms, and gene duplication.
- Multiple applications exist for evolutionary theory, including the study of protein function, in vitro evolution, and medical sciences.

## Study Strategies

- A common mistake when attempting to solve Hardy–Weinberg problems is the use of the wrong equation. For example, if you are given phenotypic frequencies for a trait that is at genetic equilibrium and shows dominance, remember that the frequency of the recessive phenotype (and genotype) is equal to $q^2$, not $q$. By taking the square root of $q^2$, you can obtain the frequency of the recessive allele and then determine the frequency of the dominant allele by subtraction ($p = 1 - q$).
- It is important to bear in mind that for any gene locus with two alleles, the allele frequencies must total 1 (i.e., it is always the case that $p + q = 1$). The equation that specifies genotype frequencies ($p^2 + 2pq + q^2 = 1$) is valid only for a population at genetic equilibrium.
- The different means by which evolution can occur can become overwhelming. To help organize them, construct a table listing the different means by which a population can evolve.
- Go to **LaunchPad** (or use the Web addresses listed) to review the following additional resources:

  Animated Tutorial 15.1 Natural Selection (PoL2e.com/at15.1)

  Animated Tutorial 15.2 Genetic Drift Simulation (PoL2e.com/at15.2)

  Animated Tutorial 15.3 Hardy–Weinberg Equilibrium (PoL2e.com/at15.3)

  Activity 15.1 Darwin's Voyage (PoL2e.com/ac15.1)

  Activity 15.2 Gene Tree Construction (PoL2e.com/ac15.2)

  Analyze the Data in textbook Figures 15.10 and 15.20

  Apply the Concept in textbook Sections 15.3 and 15.5

## Key Concept Review

### 15.1 Evolution Is Both Factual and the Basis of Broader Theory

Darwin and Wallace introduced the idea of evolution by natural selection

Evolutionary theory has continued to develop over the past century

Evolutionary theory encompasses our understanding of the mechanisms that lead to biological changes in populations over time. It provides biologists with a means to understand how life diversified. Geological, morphological, and molecular data provide strong support for evolution.

Charles Darwin studied natural history during his five-year voyage on the HMS *Beagle*. Darwin's main contribution to biology was the theory of evolution by natural selection. This is based on the evidence that species change over time, that divergent species share a common ancestor, and that differential survival and reproduction in the population are based on variation of traits. In short, natural selection is the differential contribution of offspring to the next generation by various genetic types belonging to the same population. Russel Wallace independently came up with the concept of natural selection at roughly the same time as Darwin. The work of Gregor Mendel, Watson and Crick, and E. O. Wilson has led to continued development of evolutionary theory.

**Question 1.** We call the mechanisms of evolution the "theory" of evolution. Strictly speaking, however, the term "theory" is incorrect. Why?

**Question 2.** What contributions did the work of Mendel, T. H. Morgan, Watson and Crick, and E. O. Wilson make to our understanding of evolutionary theory?

## 15.2 Mutation, Selection, Gene Flow, Genetic Drift, and Nonrandom Mating Result in Evolution

Mutation generates genetic variation

Selection on genetic variation leads to new phenotypes

Natural selection increases the frequency of beneficial mutations in populations

Gene flow may change allele frequencies

Genetic drift may cause large changes in small populations

Nonrandom mating can change genotype or allele frequencies

Biologically, evolution results in a change in the genetics of a population over time. A population is a group of individuals of a single species living and interbreeding in a particular area at the same time. Populations evolve, whereas individuals do not.

The ultimate source of genetic variation in a population is mutation. Because mutations are random changes in genetic material, most are harmful or neutral, but the environment determines whether a particular mutation is disadvantageous or adaptive. Rates of mutation can be high, as in a virus, or they can be low. However, even low rates of mutation can lead to genetic variation in a population. Mutations lead to the production of different forms a gene or alleles. The allele frequency is the proportion of each allele in a population, while the genotype frequency is the proportion of each genotype.

New phenotypes can arise as a result of selection on genetic variation. When selection of individuals with desirable traits is carried out by humans, such as plant and animal breeders, it is referred to as artificial selection. Slight differences in characteristics among individuals may favor survival and reproduction over others. An adaptation is a characteristic that helps its bearer survive and reproduce; the term also refers to the evolutionary process that produces such characteristics.

A number of other mechanisms can also result in evolutionary change within populations. Gene flow occurs when individuals migrate from one population to another and breed in their new location. Gene flow can add new alleles to a population's gene pool, change the frequencies of alleles already present, or both. Genetic drift is caused by chance events that alter allele frequencies in a population. It has its greatest impact on small populations, in which it may even cause harmful alleles to increase in frequency. During a population bottleneck, when a large population is severely reduced in size, allele frequencies may shift drastically, and genetic variation may be reduced as a result of genetic drift. The change in genetic variation that occurs when a few individuals originate a new population is called the founder effect. As in the case of a population bottleneck, some alleles found in the source population will be missing from the founding population, and others will occur with altered frequencies. The preferential mating of individuals with others, either of the same genotype or of a different genotype, is called nonrandom mating. The effect of self-fertilization, another form of nonrandom mating, is a reduction in the frequency of heterozygotes. Sexual selection favors traits that benefit their bearers (generally males) in the competition for access to members of the other sex, or make their bearers more attractive to members of the other sex. Sexual selection often results in sexually dimorphic species, in which males and females differ in appearance.

**Question 3.** Genetic drift, population bottlenecks, and the founder effect are all means by which allele frequencies can change. Are these processes random or nonrandom? Explain.

**Question 4.** Describe what is meant by sexual selection and discuss what behavioral ecologists discovered about sexual selection of tail length from studies of male widowbirds.

## 15.3 Evolution Can Be Measured by Changes in Allele Frequencies

Evolution will occur unless certain restrictive conditions exist

Deviations from Hardy–Weinberg equilibrium show that evolution is occurring

Evolution can be measured as the changes in allele frequencies in a population. The sum of all allele frequencies at a locus is equal to 1, as is the sum of all genotype frequencies. For a locus with two alleles, the frequencies of the dominant and recessive alleles typically are represented by $p$ and $q$, respectively; thus $p + q = 1$.

Biologists estimate allele frequencies by measuring numbers of alleles in a sample of individuals from the population. A given allele frequency is then calculated by taking the number of copies of the allele in the population and dividing it by the total number of copies of that allele in the population. If there is only one allele at a locus, the population is *monomorphic* and the allele is *fixed*. If two or more alleles exist, the population is *polymorphic* at that locus. The genetic structure of a population is described by the frequencies of different alleles at each locus and the frequencies of different genotypes.

A population at Hardy–Weinberg equilibrium is not changing genetically and hence is not evolving. To be at Hardy–Weinberg equilibrium, a population must meet five conditions: (1) no mutation; (2) no differential selection among genotypes; (3) no gene flow; (4) population size must be infinite; and (5) random mating. If these conditions are met, allele frequencies at a locus remain the same from generation to generation. Moreover, after one generation of random mating, the genotype frequencies will remain in the proportions $p^2 + 2pq + q^2 = 1$, where $p^2$, $2pq$, and $q^2$ represent the frequencies of the homozygous dominant, heterozygous, and homozygous recessive genotypes, respectively. Though populations in nature can never fully meet the conditions of Hardy–Weinberg equilibrium, this equation is often useful for predicting the approximate genotype frequencies in a population. It is also important because deviations from it may show that evolution is occurring. Moreover, the pattern of deviations is useful in identifying the agents of evolutionary change operating on the population.

**Question 5.** Discuss the main application of the Hardy–Weinberg rule in evolutionary biology.

**Question 6.** In a population with 600 members, the numbers of individuals of three different genotypes are $AA$ = 350, $Aa$ = 100, $aa$ = 150. Answer the following questions about this population:

(1) What are the genotype frequencies of *AA*, *Aa*, and *aa*?
(2) What are the frequencies of the *A* and *a* alleles?
(3) What would be the expected genotype frequencies of *AA*, *Aa*, and *aa* if this population were in genetic equilibrium?
(4) Is this population in genetic equilibrium? Explain.

### 15.4 Selection Can Be Stabilizing, Directional, or Disruptive

Stabilizing selection reduces variation in populations
Directional selection favors one extreme
Disruptive selection favors extremes over the mean

Natural selection, unlike other agents of evolution, adapts organisms to their environment. In cases in which the distribution of a phenotype approximates a bell-shaped curve because it is controlled by many gene loci, selection can produce any one of three results.

Stabilizing selection reduces variation in the population by favoring average individuals. This can result in the selection against deleterious mutations resulting in purifying selection.

Directional selection changes the mean value for a character by favoring individuals that vary in one direction. This can result in positive selection of a genetic variant at a single locus. Many generations of directional selection will result in an evolutionary trend.

Disruptive selection, by favoring both extremes, leads to a population with two peaks in the distribution of the character. This bimodal distribution will be maintained within the population by disruptive selection.

**Question 7.** Construct a concept map with the theme of "evolutionary agents." Include in your map the following terms: evolutionary agents, allele frequencies, directional, disruptive, founder effects, gene flow, genetic drift, genetic structure of a population, genotype frequencies, mutation, natural selection, nonrandom mating, phenotypic variation, population bottlenecks, and stabilizing. Connect these terms with verbs or short phrases to indicate the relationships among them.

**Question 8.** Explain how stabilizing selection in human birth weights could be considered purifying selection.

### 15.5 Genomes Reveal Both Neutral and Selective Processes of Evolution

Much of molecular evolution is neutral
Positive and purifying selection can be detected in the genome
Heterozygote advantage maintains polymorphic loci
Genome size and organization also evolve

Molecular evolution occurs by diverse mechanisms. A point mutation of a single nucleotide is known as a nucleotide substitution. A synonymous (silent) mutation replaces a nucleotide base in a codon but does not change the amino acid specified by the codon. Because synonymous mutations do not affect the functioning of a protein, they are unlikely to be affected by natural selection. A nonsynonymous mutation changes the amino acid specified by the codon. Though such mutations are likely to be harmful, they are sometimes selectively neutral, or nearly so, and are occasionally advantageous.

Within functional genes, nucleotide substitution rates are highest at nucleotide positions that do not change the amino acid being expressed. The rate of substitution is higher in pseudogenes—duplicate copies of genes that are never expressed—than in functional genes. The neutral theory of molecular evolution postulates that most evolutionary change in macromolecules, as well as much of the genetic variation within species, is the result of random genetic drift, rather than natural selection. The rate of fixation of neutral mutations is theoretically constant and equal to the mutation rate. The rate of evolution of genes or proteins can thus be used as a molecular clock of evolution.

According to the neutral theory of molecular evolution, it is possible to distinguish among evolutionary processes by comparing the rates of synonymous and nonsynonymous substitutions in a protein-coding gene. If an amino acid substitution is neutral in its effect on fitness, then the rates of synonymous and nonsynonymous substitutions in the corresponding DNA sequences are expected to be very similar. If an amino acid position is under strong selection for change, then the rate of nonsynonymous substitutions in the corresponding DNA is expected to exceed the rate of synonymous substitutions. If an amino acid position is under purifying selection, then the rate of synonymous substitutions in the corresponding DNA is expected to be much higher than the rate of nonsynonymous substitutions.

The enzyme lysozyme, while serving in almost all animals as an important first line of defense against invading bacteria, also has evolved to take on an essential role in the digestive process of several groups of foregut fermenters. By comparing the lysozyme-coding sequences in foregut fermenters with several of their nonfermenting relatives, molecular evolutionists have discovered that neutral evolution, purifying selection, and selection for change have all occurred as lysozyme evolved to take on its new function. The independent evolution in several groups of foregut fermenters of a type of lysozyme adapted to its new environment and function shows that convergent evolution occurs at the molecular level.

In some cases, heterozygous individuals have an advantage over homozygous individuals. Heterozygous individuals may be polymorphic for one or more genes. This polymorphism will allow the heterozygous individual to produce two different forms of the same protein, thereby allowing that individual to cope with a wider range of environmental conditions.

Genome size and organization also evolve. The size of the coding portion of the genome is larger in more complex organisms. Thus, eukaryotes have many times more genes than prokaryotes, and multicellular eukaryotes with tissue organization have more genes than single-celled eukaryotes. Most of the variation in genome size of various organisms is due not to differences in the number of functional genes, but in the amount of noncoding DNA. Although much of the noncoding DNA appears to be nonfunctional, it may alter the expression of surrounding genes. Important categories of noncoding DNA include pseudogenes and parasitic transposable elements.

**Question 9.** Explain why the majority of mutations are selectively neutral.

**Question 10.** Using the example of *Colias* butterflies, provide an example of heterozygote advantage in a population. What are the benefits that a heterozygous butterfly has over a homozygous butterfly?

### 15.6 Recombination, Lateral Gene Transfer, and Gene Duplication Can Result in New Features

Sexual recombination amplifies the number of possible genotypes

Lateral gene transfer can result in the gain of new functions

Many new functions arise following gene duplication

Sexual recombination generates variety in the genotype combinations in a population. Sexual reproduction has a number of disadvantages as compared to asexual reproduction. Adaptive combinations of genes can be broken up, the rate of gene flow to the next generation is slower, and reproduction rate is lower. However, sexual reproduction has a number of benefits as well. It may have evolved to facilitate DNA repair, promote the elimination of deleterious mutations, or to increase the variety of genetic combinations. Sexual reproduction generates new combinations of alleles upon which natural selection can act.

Lateral gene transfer occurs when a species picks up fragments of foreign DNA directly from the environment, or via a virus, or through hybridization with another species. This process increases the genetic variability of the species and thus provides additional raw material on which natural selection can act. It can also result in the spread of genetic functions between distantly related species.

Gene duplication can produce new functions. When a gene is duplicated, four evolutionary outcomes are possible: (1) both copies can retain the gene's original function; (2) both copies can retain the ability to produce the original gene product, but the expression of the genes may diverge in different tissues or at different times of development; (3) one copy can become a functionless pseudogene; or (4) one copy can retain its original function, while the other mutates so extensively that it can perform a different function. When an entire genome is duplicated (as in polyploid organisms), there are major opportunities for new gene functions to evolve.

Genome duplication events that occurred in the ancestor of the jawed vertebrates have permitted many individual vertebrate genes to become highly tissue-specific in their expression. Successive rounds of gene duplication and mutation can result in a gene family—a group of homologous genes with related functions. The globin gene family shows that gene diversification can produce molecules with different functions (hemoglobin and myoglobin) as well as functionless pseudogenes.

**Question 11.** Describe the advantages and disadvantages of sexual and asexual reproduction.

**Question 12.** How could lateral gene transfer occur in eukaryotes?

### 15.7 Evolutionary Theory Has Practical Applications

Knowledge of gene evolution is used to study protein function

In vitro evolution produces new molecules

Evolutionary theory provides multiple benefits to agriculture

Knowledge of molecular evolution is used to combat diseases

The principles of molecular evolution help us understand function and diversification of function in many proteins. For example, detection of strong selection for change in a nucleotide sequence can help us identify molecular changes that have resulted in functional changes. Molecular evolutionary principles underlie the field of in vitro evolution, in which new molecules are produced in the laboratory to perform particular desired functions. The basis of in vitro evolution is the creation of random molecular variation followed by selection by the experimenter. Agriculture has used evolutionary theory to combat pesticide resistance. Biomedical scientists are using principles of molecular evolution to identify and combat human diseases.

**Question 13.** Compare in vitro evolution with molecular evolution in organisms.

**Question 14.** How is the study of molecular evolution important in efforts to combat HIV and other viral pathogens that have recently emerged?

## Test Yourself

1. When developing his theory of evolutionary change, Darwin concluded all of the following *except*
   a. there was increased survival of some individuals compared with others, based on differences in their traits.
   b. in most species, many more individuals are born than survive to reproduce.
   c. species change over time.
   d. divergent species share a common ancestor and diverged from one another gradually over time.
   e. offspring tend to be identical to their parents.
   ***Textbook Reference:*** *Concept 15.1 Evolution is Both Factual and the Basis of Broader Theory; Darwin and Wallace introduced the idea of evolution by natural selection*

2. Evolution occurs at the level of
   a. the individual genotype.
   b. the individual phenotype.
   c. environmentally based phenotypic variation.
   d. the population.
   e. the species.
   ***Textbook Reference:*** *Concept 15.2 Mutation, Selection, Gene Flow, Genetic Drift, and Nonrandom Mating Result in Evolution*

3. Natural selection acts on
   a. the gene pool of the species.
   b. the genotype.
   c. the phenotype.
   d. multiple gene inheritance systems.
   e. the environment.
   ***Textbook Reference:*** *Concept 15.2 Mutation, Selection, Gene Flow, Genetic Drift, and Nonrandom Mating Result in Evolution; Selection on genetic variation leads to new phenotypes*

4. Which of the following ranks the organisms correctly in terms of the expected total amount of coding DNA in their genomes (from least coding DNA to most coding DNA)?
   a. Bacterium, single-celled eukaryote, *Drosophila*, bird
   b. Bacterium, *Drosophila*, bird, single-celled eukaryote
   c. Single-celled eukaryote, bacterium, *Drosophila*, bird
   d. *Drosophila*, single-celled eukaryote, bird, bacterium
   e. *Drosophila*, bacterium, single-celled eukaryote, bird
   ***Textbook Reference:*** *Concept 15.5 Genomes Reveal Both Neutral and Selective Processes of Evolution; Heterozygote advantage maintains polymorphic loci*

5. The ability to taste the chemical PTC (phenylthiocarbamide) is determined in humans by a dominant allele *T*, with tasters having the genotypes *Tt* or *TT* and nontasters having *tt*. If 36 percent of the members of a population cannot taste PTC, then according to the Hardy–Weinberg rule, the frequency of the *T* allele should be
   a. 0.36.
   b. 0.4.
   c. 0.6.
   d. 0.64.
   e. 0.8.
   ***Textbook Reference:*** *Concept 15.3 Evolution Can Be Measured by Changes in Allele Frequencies; Evolution will occur unless certain restrictive conditions exist*

6. A gene in humans has two alleles, *M* and *N*, that code for different surface proteins on red blood cells. If you know that the frequency of allele *M* is 0.2, according to the Hardy–Weinberg rule, the frequency of the genotype *MN* in the population should be
   a. 0.16.
   b. 0.2.
   c. 0.32.
   d. 0.64.
   e. 0.8.
   ***Textbook Reference:*** *Concept 15.3 Evolution Can Be Measured by Changes in Allele Frequencies; Evolution will occur unless certain restrictive conditions exist*

7. Random genetic drift would probably have its greatest effect on a
   a. small, isolated population.
   b. large population in which mating is nonrandom.
   c. large population in which mating is random.
   d. large population with regular immigration from a neighboring population.
   e. large population with a high mutation rate.
   ***Textbook Reference:*** *Concept 15.2 Mutation, Selection, Gene Flow, Genetic Drift, and Nonrandom Mating Result in Evolution; Gene flow may change allele frequencies*

8. Allele frequencies for a gene locus are *least* likely to be significantly changed by
   a. mutation.
   b. the founder effect.
   c. self-fertilization.
   d. gene flow.
   e. natural selection.
   ***Textbook Reference:*** *Concept 15.2 Mutation, Selection, Gene Flow, Genetic Drift, and Nonrandom Mating Result in Evolution; Selection on genetic variation leads to new phenotypes*

9. Which of the following evolutionary agents would produce nonrandom changes in the genetic structure of a population?
   a. Self-fertilization
   b. Population bottlenecks
   c. Mutation
   d. Gene flow
   e. Founder effect
   ***Textbook Reference:*** *Concept 15.2 Mutation, Selection, Gene Flow, Genetic Drift, and Nonrandom Mating Result in Evolution; Genetic drift may cause large changes in small populations*

10. Genetic variation within a population may be maintained by all of the following *except*
   a. frequency-dependent selection.
   b. the accumulation of neutral alleles.
   c. sexual reproduction.
   d. homozygote advantage.
   e. hybrid sterility.
   ***Textbook Reference:*** *Concept 15.6 Recombination, Lateral Gene Transfer, and Gene Duplication Can Result in New Features; Sexual recombination amplifies the number of possible genotypes*

11. The graph below shows the range of variation among population members for a trait determined by multiple genes.

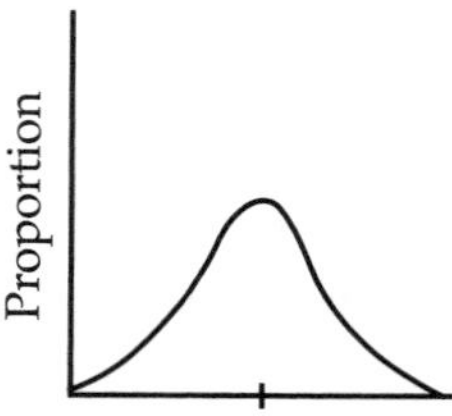


If this population is subject to stabilizing selection for several generations, which distribution(s) would most likely result?

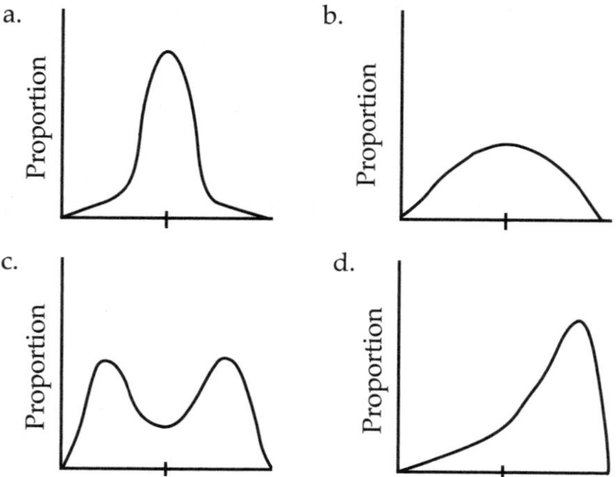


   e. Both a and b
   ***Textbook Reference:*** *Concept 15.4 Selection Can Be Stabilizing, Directional, or Disruptive; Disruptive selection favors extremes over the mean*

12. All of the following are disadvantages of sexual reproduction *except*
   a. when it involves separate genders, it increases the overall reproductive rate.
   b. it breaks up adaptive combinations of genes.
   c. it increases the rate at which females pass genes on to their offspring.
   d. it increases the difficulty of eliminating harmful mutations from the population.
   e. it leads to a uniform gene pool.
   ***Textbook Reference:*** *Concept 15.6 Recombination, Lateral Gene Transfer, and Gene Duplication Can Result in New Features; Sexual recombination amplifies the number of possible genotypes*

13. In a eukaryote, one would expect to find the lowest rate of nonsynonymous nucleotide substitutions in an
   a. intron of a protein-coding gene.
   b. exon of a protein-coding gene.
   c. intron of a pseudogene.
   d. exon of a pseudogene.
   e. intron or exon of a protein-coding gene.
   ***Textbook Reference:*** *Concept 15.5 Genomes Reveal Both Neutral and Selective Processes of Evolution*

14. Which of the following statements about mutations is *false?*
   a. A synonymous mutation results in no change in the amino acid sequence of a protein.
   b. According to the neutral theory of molecular evolution, most substitution mutations are selectively neutral and accumulate through genetic drift.
   c. A base substitution mutation in the third codon position is more likely to be neutral than a substitution at the first or second codon position.
   d. Nonsynonymous mutations are virtually always deleterious to the organism.
   e. The rate of fixation of neutral mutations is equal to the mutation rate and is independent of population size.
   ***Textbook Reference:*** *Concept 15.5 Genomes Reveal Both Neutral and Selective Processes of Evolution; Much of molecular evolution is neutral*

15. Which of the following statements about the enzyme lysozyme is true?
   a. A small group of closely related mammals has evolved a special form of lysozyme that functions in digestion.
   b. The lysozymes found in the foregut fermenters resulted from convergent evolution.
   c. Lysozyme could not have evolved a secondary function if it had been an enzyme with a vital primary function.
   d. A higher mutation rate in the foregut fermenters allows their lysozymes to evolve rapidly.
   e. Lysozyme first evolved as a defense against bacteria in the common ancestor of mammals.
   ***Textbook Reference:*** *Concept 15.5 Genomes Reveal Both Neutral and Selective Processes of Evolution; Positive and purifying selection can be detected in the genome*

16. In comparing several populations of the same species, the population with the greatest genetic variation will have the
    a. greatest number of genes.
    b. greatest number of alleles per gene.
    c. greatest number of population members.
    d. largest gene pool.
    e. smallest gene pool.
    ***Textbook Reference:*** *Concept 15.2 Mutation, Selection, Gene Flow, Genetic Drift, and Nonrandom Mating Result in Evolution; Selection on genetic variation leads to new phenotypes*

17. Which of the following would be the *least* likely result of gene duplication?
    a. Production of less of the gene product than was produced before duplication
    b. Expression of the two copies of the gene at different stages of an organism's development
    c. One copy retaining its original function and the other copy acquiring a different function
    d. One copy remaining functional and the other copy evolving into a functionless pseudogene
    e. Both copies of the gene retain their original functions.
    ***Textbook Reference:*** *Concept 15.6 Recombination, Lateral Gene Transfer, and Gene Duplication Can Result in New Features; Many new functions arise following gene duplication*

18. Which of the following about in vitro evolution is *false?*
    a. It can produce both nucleic acid and protein molecules known in living organisms.
    b. It requires many rounds of production of variant molecules and the selection of those that appear promising in terms of the desired properties.
    c. It often involves techniques and molecules employed in recombinant DNA technology, such as PCR and cDNA.
    d. It has potential pharmaceutical, agricultural, and industrial uses.
    e. It is based on principles of molecular evolution.
    ***Textbook Reference:*** *Concept 15.7 Evolutionary Theory Has Practical Applications; In vitro evolution produces new molecules*

19. Evolution *cannot* be measured by
    a. examining rates at which new mutations arise.
    b. observing long term morphological changes in the fossil record reflecting underlying genetic change.
    c. observing the spread of new genetic variants through a population.
    d. observing changes in a single individual over its lifetime.
    e. observing the effects of genetic change on the form and function of organisms.
    ***Textbook Reference:*** *Concept 15.1 Evolution Is Both Factual and the Basis of Broader Theory; Darwin and Wallace introduced the idea of evolution by natural selection*

## Answers

### *Key Concept Review*

1. In science, a theory typically refers to an untested hypothesis in science. The concept of evolutionary theory is not limited to a single hypothesis, but refers to many different concepts. Additionally, there are vast amounts of data that support evolutionary theory.
2. Each of these scientists further contributed to our understanding of evolutionary theory. The rediscovery of Mendel's work led to an understanding of the basic process of genetic inheritance. Morgan's studies on fruit flies led to the discovery of the role of chromosomes in inheritance. Watson and Crick's discovery of the structure of DNA led to a detailed understanding of molecular evolutionary processes. E. O. Wilson's work spurred research in the evolution of behavior.
3. These processes are random and typically occur in small populations. Genetic drift refers to random changes in allele frequencies. A population bottleneck occurs after an environmental event in which a large percentage of a population has died. In this case, the survival of an individual is due solely to chance, as are the resulting allele frequencies among these survivors. In the founder effect, the small number of individuals that found a population is unlikely to have the same allele frequency as the source population. As with the other two mechanisms, the alleles present in the founding population is going to be random.
4. Sexual selection is the spread of a trait that improves the reproductive success of an individual. The trait may improve the ability of its bearer to compete with other members of its sex for access to mates, or it may make its bearer more attractive to members of the opposite sex. By artificially lengthening or shortening the tails of male widowbirds, behavioral ecologists were able to show that increased tail length in males made them more attractive to females but did not give the long-tailed males an advantage in their interactions with other males.
5. No population meets the Hardy–Weinberg conditions for genetic equilibrium. However, by determining how a population deviates from the expectations of Hardy–Weinberg, evolutionary biologists can identify which evolutionary agents are affecting the population.

6.

(1) The genotype frequencies in this population are the following:

$AA = 350/600 = 0.58$

$Aa = 100/600 = 0.17$

$aa = 150/600 = 0.25$

(2) The allele frequencies in this population are the following:

$A = (700 + 100)/1200 = 0.67$

$a = (300 + 100)/1200 = 0.33$

(3) The expected genotype frequencies of this population in genetic equilibrium would be the following:

$AA = p^2 = (0.67)^2 = 0.45$

$Aa = 2pq = 2 \times 0.67 \times 0.33 = 0.44$

$aa = q^2 = (0.33)^2 = 0.11$

(4) The population is not in genetic equilibrium. Using the observed allele frequencies for $p$ and $q$, the observed genotypic frequencies differ from those predicted by Hardy–Weinberg.

7.

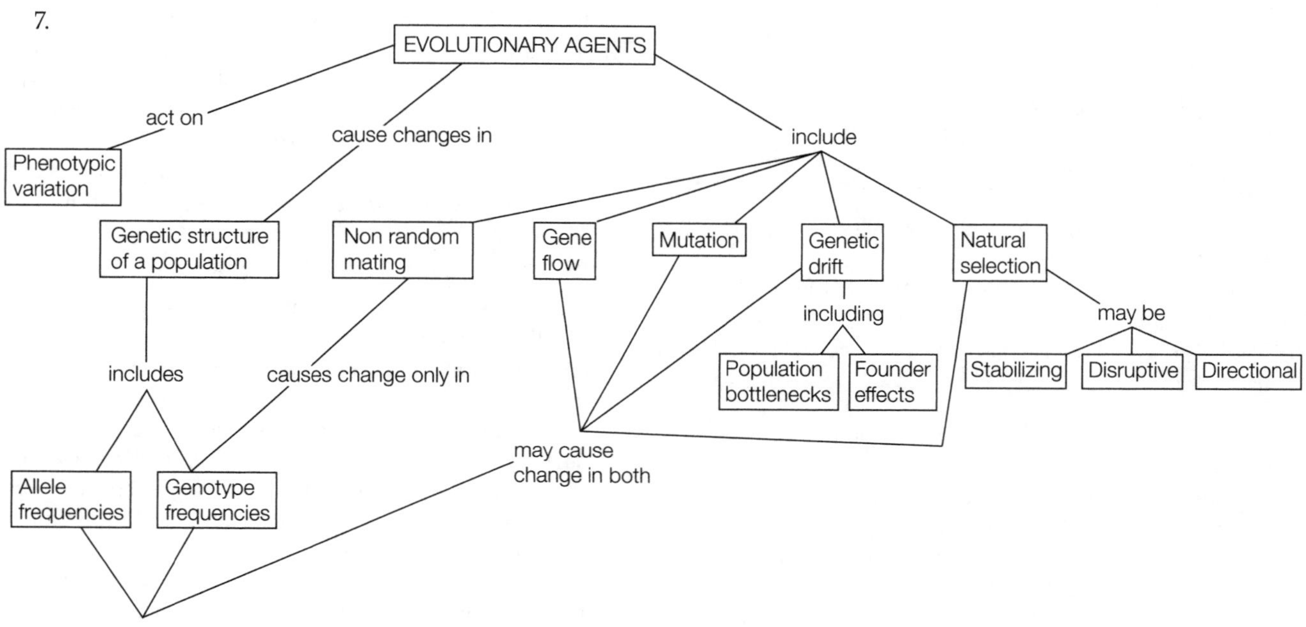


8. Stabilizing selection reduces variation in a population, but does not change the mean. Since human babies that are born lighter or heavier than normal die at higher rates than babies whose weights are closer to the mean birth weight, there is stabilizing selection for birth weight. If high or low birth weight were linked to a deleterious mutation in a gene, then this gene would be eliminated from the population, resulting in purifying selection.

9. When a mutation occurs it can be either synonymous (silent), with no change in the encoded amino acid, or nonsynonymous, with a change in the encoded amino acid. One type of nonsynonymous substitution results in a protein with an altered or different function. Some nonsynonymous substitutions of amino acids are neutral because the resulting protein function has not been altered. When a synonymous substitution or neutral nonsynonymous substitution occurs, there is neither an advantage nor a disadvantage to the change. Therefore, there is no selection against these mutations and they tend to accumulate in the population. In the case of deleterious mutations, they are rapidly selected against and removed from the population.

10. Butterflies of the genus *Colias* are polymorphic for a gene that encodes phosphoglucose isomerase (PGI), which affects flight at different temperatures. Heterozygous individuals can fly over a greater range of temperatures, which gives them an advantage in both finding mates and foraging.

11. In asexual reproduction, an organism's offspring is genetically identical to the parent. One advantage is the rapid rate of reproduction possible in asexual species. The major disadvantage is that deleterious mutations cannot be easily eliminated from the population. Sexual reproduction has the disadvantage of being slower and potentially breaking up advantageous combinations of genes. It has the advantage

of creating greater genetic variation and a means to eliminate deleterious alleles from the population.

12. Lateral gene transfer is most common in bacteria and viruses, but it may occur in eukaryotes. The major endosymbioses that gave rise to mitochondria and chloroplasts involved lateral gene transfer of entire bacterial genomes to eukaryotic cells. Because of the high levels of hybridization among closely-related plant species, lateral transfer through the exchange of many genes among recently separated plant lineages could also occur.
13. Both in vitro evolution and molecular evolution involve variation and selection in organisms. With in-vitro evolution, a huge number of variant molecules are created in the laboratory, and the experimenters select those that show any sign of having the desired property. Many repetitions of these two steps—production of variants and selection—eventually produce the targeted molecule. In molecular evolution in organisms, naturally occurring mutations in previously existing DNA sequences are the ultimate source of variation, and the environment determines through natural selection, which mutations are the favorable variants. However, according to the neutral theory of molecular evolution, much evolutionary change in DNA (and in the encoded proteins) is not adaptive but rather the result of genetic drift.
14. The principles of molecular evolution are critical for understanding the origin and evolutionary development of such pathogenic viruses as hantaviruses, the SARS virus, and HIV. By studying the evolutionary changes that occur in such viruses, medical researchers gain valuable insights into such questions as how some viruses are able to switch from an animal to a human host and how vaccines can remain effective as viruses evolve. In the future, as more extensive genomic databases and evolutionary trees for viruses are developed, it will be possible to identify and treat a much wider array of human diseases.

### *Test Yourself*

1. **e.** Although offspring resemble their parents, they are not identical to one another or to either parent.
2. **d.** Evolution is defined as change in the genetic structure of a population over time, so evolution occurs at the level of the population.
3. **c.** Natural selection acts on phenotypes, not genotypes. For example, a harmful recessive allele is "invisible" to natural selection when it occurs in a heterozygote, where its harmful effect is masked by the dominant allele.
4. **a.** There is a rough relationship between the amount of coding DNA and organismal complexity.
5. **b.** The frequency of the *tt* genotype is $q^2 = 0.36$, so the frequency of the *t* allele is $q$ (0.6, the square root of 0.36). If there are only two alleles for this trait, then $T + t = 1$. The frequency of the *T* allele is therefore $1 - t = 1 - 0.6 = 0.4$.
6. **c.** Because $p = 0.2$, $q = 1 - 0.2 = 0.8$, the frequency of the *MN* genotype is $2pq$, or $2 \times 0.2 \times 0.8 = 0.32$.
7. **a.** Genetic drift is most significant in small populations.
8. **c.** Unlike the founder effect, mutation, gene flow, and natural selection, all of which may change allele frequencies in a population, self-fertilization (like other types of nonrandom mating, with the exception of sexual selection) only causes a deviation from the frequency of heterozygotes predicted by Hardy–Weinberg equilibrium.
9. **a.** Individuals in populations do not mate randomly, so self-fertilization leads to increased numbers of homozygous individuals, changing the genetic structure of a population.
10. **e.** Frequency-dependent selection, the accumulation of neutral alleles, sexual recombination, and heterozygote advantage are the four major forces that maintain genetic variation in a population; hybrid sterility does not.
11. **a.** Stabilizing selection results when individuals that are intermediate in phenotype make a larger contribution to future generations than individuals of a more extreme phenotype. This leads to reduced variation for the trait and causes the curve to be higher and narrower. Curve b shows greater variation, curve c would result from disruptive selection, and curve d would result from directional selection.
12. **d.** Sexual recombination produces some individuals in a population that are less fit than others because they carry a greater than average number of deleterious mutations. Because these individuals are selected against, sexual reproduction is able to reduce the number of deleterious mutations in the population over time.
13. **b.** Eukaryotic genes usually contain both protein-coding regions (exons) and noncoding regions (introns). A substantial proportion of nonsynonymous substitutions occurring in an exon of a gene would most likely be deleterious and therefore be eliminated by selection. Because introns and pseudogenes are not expressed, nonsynonymous substitutions occurring in them cannot be selected against.
14. **d.** A nonsynonymous substitution mutation can be selectively neutral if, as sometimes happens, it results in an amino acid change that has no significant effect on the shape (and hence the functional properties) of the protein.
15. **b.** Because the animals in which a similar lysozyme has evolved do not share a recent common ancestor, the mechanism involved is convergent evolution. All the other statements are false.

16. **b.** Genetic variation is related to the number of different alleles per gene. Answer a is incorrect because all members of a species have the same number of genes. Population size per se has little to do with genetic variation, so choices c and d are also incorrect.

17. **a.** If there are two copies of the gene, it is more likely to produce more of its product than less of it.

18. **a.** In vitro evolution can create novel molecules unknown to occur in living organisms.

19. **d.** Since evolution is a change in the genetic composition of populations over time, it is not possible to measure it at the level of the individual.

# 16 Reconstructing and Using Phylogenies

## The Big Picture

- Advances in the field of systematics include cladistics, new methods of sequencing DNA and RNA, and application of sophisticated mathematical techniques. As a result, phylogenetics is now contributing to a wide array of biological studies, including the origins and types of human immunodeficiency virus (HIV) and other infectious organisms.
- The parsimony principle (that the simplest hypothesis capable of explaining the known facts is the preferred explanation) is fundamental not only to the reconstruction of phylogenies but also to every other field of scientific research.
- The concept of evolution has revolutionized taxonomy. Linnaeus and others developed "natural" systems of classification based on similarity of morphology. Modern biologists view similarities of organisms as the result of descent from a common ancestor.
- Knowing that organisms are evolutionarily related enables biologists to make predictions about their characteristics, providing important hints in the search for organisms with valuable properties, such as the ability to produce medically useful drugs.

## Study Strategies

- The concept of homology can be confusing. Figure 16.2 provides a helpful image of the difference between homologous and homoplastic traits.
- Distinguishing monophyletic, paraphyletic, and polyphyletic groups is difficult for many students. Bear in mind that a monophyletic group is analogous to a branch (or twig) of a tree: a single "cut" can remove it from a phylogenetic tree. This analogy should help you work out which kinds of "cuts" would result in paraphyletic and polyphyletic groups.
- Here is a mnemonic for the hierarchy of taxa (kingdom, phylum, class, order, family, genus, species) in the Linnaean system of classification: "Kindly Professors Cannot Often Fail Good Students." Note that the plural of genus is genera, and the words *general* and *generic* come from the same root; this can help you remember that the genus is the more general taxon in the genus/species binomial. *Species* and *specific* also come from the same root, and the species is, of course, the most specific taxon.
- Go to **LaunchPad** (or use the Web addresses listed) to review the following additional resources:

  Animated Tutorial 16.1 Phylogeny and Molecular Evolution (PoL2e.com/at16.1)

  Animated Tutorial 16.2 Using Phylogenetic Analysis to Reconstruct Evolutionary History (PoL2e.com/at16.2)

  Activity 16.1 Constructing a Phylogenetic Tree (PoL2e.com/ac16.1)

  Activity 16.2 Types of Taxa (PoL2e.com/ac16.2)

  Analyze the Data in textbook Figure 16.5

  Apply the Concept in textbook Sections 16.2 and 16.4

## Key Concept Review

### 16.1 All of Life Is Connected through Its Evolutionary History

Phylogenetic trees are the basis of comparative biology

Derived traits provide evidence of evolutionary relationships

A phylogeny is a description of the evolutionary history of relationships among organisms or their genes. Phylogenetic trees display the order in which lineages are hypothesized to have split. Each split (or node) in a phylogenetic tree represents a point at which lineages diverged in the past. The common ancestor of all the organisms in the tree forms the root of the tree. The timing of separations between lineages of organisms is shown by the positions of nodes on a time or divergence axis. A taxon is any named group of species. A taxon consisting of all the evolutionary descendants of a common ancestor is called a clade. Just as species that are each other's closest relatives are called sister species, so clades that are each other's closest relatives are called sister clades. Systematics is the study and classification of the diversity of life.

All of life is connected through its evolutionary history, known as the tree of life. The evolutionary relationships among species, as shown by the tree of life, form the basis for biological classification. Knowledge about these relationships is important when making comparisons of species, populations, or genes. Homologous traits are features that are shared by members of a lineage because of their descent from a common ancestral trait. As such, these traits are key to reconstructing phylogenetic trees. A trait that differs from its ancestral form is called a derived trait. Conversely, a trait that was present in the ancestor of a group is known as an ancestral trait for that group. Synapomorphies are derived traits that are shared among a group of organisms and are viewed as evidence of the common ancestry of the group. Homoplasies (homoplastic traits) create confusion in reconstructing the evolutionary history of a lineage because they are features that are similar for some reason other than descent from a common ancestral trait. Two processes generate homoplasies: convergent evolution and evolutionary reversals. In convergent evolution, features that evolved independently become superficially similar. In an evolutionary reversal, a character reverts from a derived state to an ancestral one.

**Question 1.** All but one of the trees shown in the diagram below portray the same phylogenetic relationships among taxa A, B, C, D, E, F, and G. Which one depicts a different phylogeny?

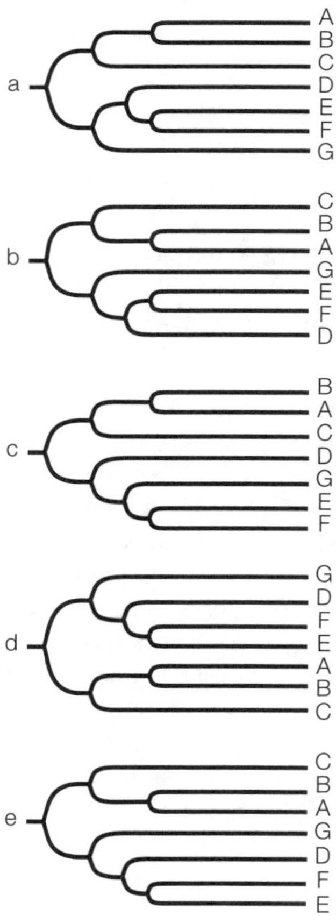


**Question 2.** In the phylogeny shown below, which group of taxa would *not* constitute a clade?

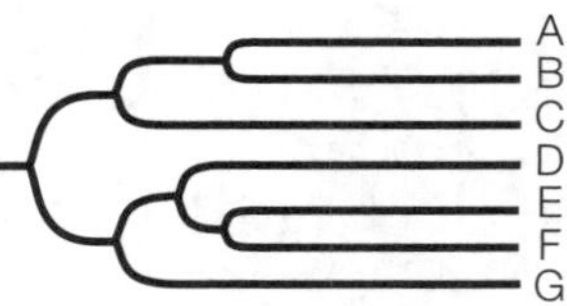


a. A, B, C, D, E, F, and G and their common ancestor
b. E, F, and G and their common ancestor
c. A, B, and C and their common ancestor
d. C, D, E, and F
e. C, D, E, F, and G

**Question 3.** You are constructing a phylogenetic tree for several animals, including two flying species. You determine that the presence of wings in the two flying species can be explained by convergent evolution. Can this trait help you create the phylogenetic tree for the species?

## 16.2 Phylogeny Can Be Reconstructed from Traits of Organisms

Parsimony provides the simplest explanation for phylogenetic data

Phylogenies are reconstructed from many sources of data

Mathematical models expand the power of phylogenetic reconstruction

The accuracy of phylogenetic methods can be tested

Reconstructing accurate phylogenies involves several steps. The first step is the choice of the ingroup and the appropriate outgroup. The ingroup is an assemblage of organisms whose phylogeny is to be determined. The outgroup can be any species or group of species outside the group of interest. One method of distinguishing ancestral and derived traits is to compare the ingroup to an outgroup. Ancestral traits should be present in both, whereas derived traits should occur only in the ingroup. The root of the tree is determined by the relationship of the ingroup to the outgroup.

In reconstructing phylogenies, systematists are guided by the parsimony principle, which states that the preferred explanation of the observed data is the simplest explanation. In practice, this means that the best phylogenetic reconstruction is the one that minimizes the number of evolutionary changes that need to be assumed over all characters in all groups in the tree. In other words, the best hypothesis is the one that requires the fewest homoplasies.

Phylogenies are constructed from many sources of data. Morphology—the presence, size, shape, and other attributes of body parts—is an important source of traits for phylogenetic analysis. The morphology of fossils is particularly useful in helping to distinguish ancestral and derived traits. The fossil record also reveals when lineages diverged. In groups with few living representatives, information on extinct species may be critical to an understanding of the large divergences among the surviving species.

Similarities in developmental pattern also may reveal evolutionary relationships, since structures in early developmental stages may reveal relationships that are not evident in adults. Behavior, if genetically determined, is another useful source of information.

The complete genome of an organism contains an enormous set of traits (the individual nucleotide bases of DNA) that can be used to analyze phylogenies. Like morphological characters, molecules are heritable characteristics that may diverge among lineages. Both nuclear and organelle DNA sequences are used in phylogenetic studies, as are sequences in gene products (such as the amino acid sequences of proteins). Because the chloroplast genome has changed slowly over evolutionary time, it is often used for the study of relatively ancient phylogenetic relationships among plants. Animal mitochondrial DNA has changed more rapidly, making it useful for studies of evolutionary relationships among closely related animal species.

Biologists have conducted experiments both in living organisms and with computer simulations that have demonstrated the effectiveness and accuracy of phylogenetic methods. The maximum likelihood method uses computer analysis for the reconstruction of phylogenies. A likelihood score of a tree is based on the probability that the observed data evolved on the specified tree, given an explicit mathematical model of evolution for the characters. An advantage of maximum likelihood analyses is that they incorporate more information about evolutionary change than parsimony methods do.

**Question 4.** The phylogenetic tree below shows the evolutionary relationships of five species (A–E) relative to five traits (1–5). Based on the tree, fill in the table below it, using 1 to indicate the presence of a derived trait and 0 to indicate the presence of an ancestral trait.

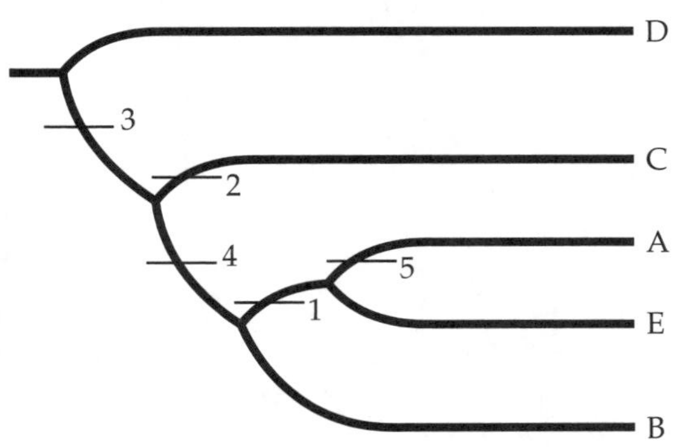


| SPECIES | TRAIT 1 | 2 | 3 | 4 | 5 |
|---|---|---|---|---|---|
| A | | | | | |
| B | | | | | |
| C | | | | | |
| D | | | | | |
| E | | | | | |

**Question 5.** The following table shows the ancestral and derived traits of five species (A–E). Based on the table, and following conventions presented in the textbook, construct a phylogenetic tree that represents the evolutionary relationships of this group. In this table, the ancestral state of each trait is indicated by 0 and the derived state is indicated by 1.

| SPECIES | TRAIT 1 | 2 | 3 | 4 | 5 |
|---|---|---|---|---|---|
| A | 1 | 1 | 1 | 0 | 0 |
| B | 0 | 0 | 0 | 0 | 0 |
| C | 1 | 1 | 0 | 1 | 0 |
| D | 0 | 1 | 0 | 0 | 0 |
| E | 1 | 1 | 0 | 1 | 1 |

**Question 6.** Discuss the application of the parsimony principle in the construction of phylogenetic trees.

## 16.3 Phylogeny Makes Biology Comparative and Predictive

- Phylogenies are important for reconstructing past events
- Phylogenies allow us to understand the evolution of complex traits
- Phylogenies can reveal convergent evolution
- Ancestral states can be reconstructed
- Molecular clocks help date evolutionary events

Phylogenetic trees provide information that helps biologists answer many kinds of questions, such as how many times a particular trait has evolved within a lineage. Using the phylogeny of angiosperm fertilization, scientists have determined that self-incompatible mating is the ancestral state and that self-compatibility evolved three times. Phylogenetic trees can also be used to determine when, where, and how zoonotic diseases (diseases caused by infectious organisms that have been transferred to humans from another animal host) first entered human populations. Studies of the transmission of HIV have benefited from phylogenetic analysis.

Phylogenetic analysis is used by biologists to help them make comparisons among genes, populations, and species, and to provide an understanding of how complex traits evolve. Phylogenetic methods are used not only to discover the evolutionary relationships among lineages of living organisms, but also to reconstruct morphological, behavioral, and molecular characteristics of ancestral species. In a molecular clock analysis, biologists assume that particular DNA sequences evolve at a reasonably constant rate and can be used as a metric to gauge the time of divergence for a particular split in a phylogeny. Molecular clocks must be calibrated with independent data such as the fossil record, known times of divergence, or biogeographic dates.

**Question 7.** How does knowledge of the phylogeny of a group of organisms contribute to one's biological understanding of the organisms?

**Question 8.** How can molecular clocks be used to construct a phylogenetic tree?

## 16.4 Phylogeny Is the Basis of Biological Classification

Evolutionary history is the basis for modern biological classification

Several codes of biological nomenclature govern the use of scientific names

Biological classification systems are designed to express evolutionary relationships among organisms. The system of binomial nomenclature developed by Linnaeus in 1758 assigns two names to each species, one identifying the species itself and the other the genus to which it belongs. The generic name (always capitalized) is followed by the species name (lower case), and both are italicized. In the Linnaean system of classification, species are grouped into higher-order taxa. The hierarchy of taxa, ranked from most to least inclusive, is kingdom, phylum, class, order, family, genus, species. Thus, a genus includes one or more species, a family includes one or more genera, and so forth.

Biologists today recognize the tree of life as the basis for classification and often name clades without placing them into any Linnaean rank. Taxa in biological classifications are expected to be monophyletic. A monophyletic taxonomic group (a clade) contains an ancestor, all descendants of that ancestor, and no other organisms. A polyphyletic group does not include its common ancestor, whereas a paraphyletic group includes some, but not all, descendants of a particular ancestor.

Polyphyletic and paraphyletic groups are inappropriate as taxonomic units. Several sets of rules govern the use of scientific names, with the goal of providing unique and universal names for biological taxa. One such rule is that if a species is named more than once, the valid name is the first name proposed. In the past, different sets of taxonomic rules were developed by zoologists, botanists, and microbiologists, resulting in many duplicated names. Today, taxonomists are developing rules to ensure that every taxon has a unique name.

**Question 9.** Why are polyphyletic and paraphyletic groups inappropriate as taxonomic units?

**Question 10.** Why do scientists use explicit rules when assigning scientific names to organisms?

## Test Yourself

1. The most important attribute of a biological classification scheme is that it
   a. avoids the ambiguity created by the use of common names.
   b. reflects the evolutionary relationships among organisms.
   c. helps us organize our knowledge about organisms and their traits.
   d. improves our ability to make predictions about the morphology and behavior of organisms.
   e. groups together organisms with similar traits.
   ***Textbook Reference:*** *Concept 16.4 Phylogeny Is the Basis of Biological Classification; Several codes of biological nomenclature govern the use of scientific names*

2. Suppose you are writing a scientific paper about a unicellular green alga called *Chlamydomonas reinhardtii.* What is the proper way to refer to this species after the full binomial has been used once?
   a. *Chlamydomonas reinhardtii*
   b. *Chlamydomonas* spp.
   c. *Chlamydomonas* sp.
   d. *C. reinhardtii*
   e. *Chlamydomonas r.*
   ***Textbook Reference:*** *Concept 16.4 Phylogeny Is the Basis of Biological Classification*

3. Members of genus *X*, a hypothetical taxon of invertebrates, have antennae with a variable number of segments. Species A and B have 10 segments; species C and D have 9 segments; and species E has 8 segments. In all other genera in this family (including genus *Y*), all species have antennae with 10 segments. Which of the following character states is a synapomorphy that would be useful for determining evolutionary relationships within genus *X*?
   a. 10-segment antennae in species A and B
   b. 10-segment antennae in genus *Y* and in two species of genus *X*
   c. Antennae with fewer than 10 segments in species C, D, and E
   d. 8-segment antennae in species E
   e. Species A, C, and D with 9 segments
   ***Textbook Reference:*** *Concept 16.1 All of Life Is Connected through Its Evolutionary History; Derived traits provide evidence of evolutionary relationships*

4. Evolutionary relationships can be revealed through study of all the following areas *except* for
   a. behavior.
   b. molecular data.
   c. morphology.
   d. maximum likelihood.
   e. paleontology.
   ***Textbook Reference:*** *Concept 16.2 Phylogeny Can Be Reconstructed from Traits of Organisms; Phylogenies are reconstructed from many sources of data*

5. Which of the following lists of taxonomic categories represents the correct ordering, from most inclusive to least inclusive?
   a. Phylum, order, family, genus
   b. Class, phylum, order, species
   c. Order, class, family, genus
   d. Family, order, class, kingdom
   e. Kingdom, class, species, genus
   ***Textbook Reference:*** *Concept 16.4 Phylogeny Is the Basis of Biological Classification*

6. Which of the following statements about reconstructing phylogenies is *false*?
   a. Traits found in the outgroup as well as in the ingroup are likely to be ancestral traits.
   b. Shared traits are generally assumed to be homoplastic until they can be proven to be homologous.
   c. Phylogenetic trees do not always provide an explicit time scale by which to date the splits between lineages.
   d. In a phylogenetic tree, branches can be rotated around any node without changing the meaning of the tree.
   e. A particular trait may be either ancestral or derived depending on the point of reference of the phylogeny.

   ***Textbook Reference:*** *Concept 16.2 Phylogeny Can Be Reconstructed from Traits of Organisms; The accuracy of phylogenetic methods can be tested*

7. Which of the following is the most significant limitation of fossils as a source of information about evolutionary history?
   a. In some cases it is impossible to determine when a fossil organism lived.
   b. The fossil record for many groups is fragmentary or even nonexistent.
   c. Most fossils contain no nucleic acids or proteins and therefore are cannot be used for studies of molecular evolution.
   d. It is impossible to determine if morphologically similar fossils belong to the same species, because one cannot know if the fossil species interbred.
   e. Most fossils provide no information about the morphology of soft anatomical structures or about external characteristics, such as color.

   ***Textbook Reference:*** *Concept 16.2 Phylogeny Can Be Reconstructed from Traits of Organisms; Phylogenies are reconstructed from many sources of data*

8. Which of the following sources of molecular data would be most helpful in a study of the evolutionary relationships of closely related animal species?
   a. Chloroplast DNA
   b. Mitochondrial DNA
   c. The amino acid sequences of a protein found in all animals, such as cytochrome *c*
   d. Ribosomal RNA sequences
   e. Messenger RNA sequences

   ***Textbook Reference:*** *Concept 16.2 Phylogeny Can Be Reconstructed from Traits of Organisms; Phylogenies are reconstructed from many sources of data*

9. Which of the following statements about the role of molecular clocks in phylogenetic analyses is true?
   a. A given gene usually evolves at the same rate in two different species regardless of differences in generation time of the species.
   b. Because changes in DNA sequences occur very slowly, molecular clocks can be used only to date evolutionary divergences that occurred millions of years ago.
   c. Molecular clocks must be calibrated with independent data, such as the fossil record.
   d. Even in a group of closely related species, different genes have been found to evolve at different rates.
   e. Molecular clocks can only be used with DNA samples.

   ***Textbook Reference:*** *Concept 16.3 Phylogeny Makes Biology Comparative and Predictive; Phylogenies can reveal convergent evolution*

10. Which of the following statements does *not* describe a purpose for which biologists use phylogenetic trees?
    a. They are helpful in determining when and where an infectious organism once found in other animals first entered human populations.
    b. They are useful for determining how many times a particular trait may have evolved independently within a lineage.
    c. They can be used to reconstruct ancestral traits.
    d. They can be used in conjunction with molecular clocks to estimate the timing of evolutionary events.
    e. They can be recreated from any trait an organism possesses.

    ***Textbook Reference:*** *Concept 16.3 Phylogeny Makes Biology Comparative and Predictive; Molecular clocks help date evolutionary events*

11. A group that consists of all the evolutionary descendants of a common ancestor is called a(n)
    a. grade.
    b. taxon.
    c. homology.
    d. ingroup.
    e. clade.

    ***Textbook Reference:*** *Concept 16.1 All of Life Is Connected through Its Evolutionary History*

12. A synapomorphy is
    a. the product of convergent evolution.
    b. the result of an evolutionary reversal.
    c. a shared derived characteristic.
    d. a trait that was present in the ancestor of a group.
    e. a phylogenetic tree.

    ***Textbook Reference:*** *Concept 16.1 All of Life Is Connected through Its Evolutionary History; Derived traits provide evidence of evolutionary relationships*

13. The organisms that make up a class are _______ diverse and _______ numerous than those in a family within that class. The organisms that make up a phylum all diverged from a common ancestor _______ recently than did the organisms in an order within that phylum.
    a. more; more; less
    b. more; more; more
    c. more; less; less

d. less; less; more
e. less; less; less
***Textbook Reference:*** *Concept 16.4 Phylogeny Is the Basis of Biological Classification*

14. A derived trait is one that
a. differs from its ancestral form.
b. is homologous with another trait found in a related species.
c. is the product of an evolutionary reversal.
d. has the same function, but not the same evolutionary origin, as a trait found in another species.
e. is found only in members of the outgroup.
***Textbook Reference:*** *Concept 16.1 All of Life Is Connected through Its Evolutionary History; Derived traits provide evidence of evolutionary relationships*

15. The ratites are a group of flightless birds comprising the ostrich, emu, cassowaries, rheas, and kiwis. All share certain morphological similarities (such as a breastbone without a keel) not found in other birds, but they live on different continents. In the past, some ornithologists regarded their similarities as homoplasies, but they are now thought to be synapomorphies. Based on this information, you would conclude that the ratites were once regarded as a _______ group but are now believed to be _______.
a. polyphyletic; paraphyletic
b. paraphyletic; monophyletic
c. polyphyletic; monophyletic
d. monophyletic; polyphyletic
e. monophyletic; paraphyletic
***Textbook Reference:*** *Concept 16.4 Phylogeny Is the Basis of Biological Classification; Evolutionary history is the basis for modern biological classification*

## Answers

### *Key Concept Review*

1. **c.** Recall that branches of a phylogenetic tree can be rotated around any node without changing the meaning of the tree. Tree "c" is different from all the others because it shows G (rather than D) as the sister taxon to taxa E and F.

2. **e.** E, F, and G and their common ancestor constitute a paraphyletic group. To be a clade, D would have to be included in the group. C, D, E, and F represent a polyphyletic group because the common ancestor of these taxa is not included in the group. (If the common ancestor of these four taxa were included, they would make up a paraphyletic group.)

3. When two animals have traits that can be explained by convergent evolution, the traits cannot be used to construct a phylogenetic tree. This is because they have evolved independently and were not shared by a common ancestor of the two species.

4.

| | TRAIT | | | | |
|---|---|---|---|---|---|
| SPECIES | 1 | 2 | 3 | 4 | 5 |
| A | 1 | 0 | 1 | 1 | 1 |
| B | 0 | 0 | 1 | 1 | 0 |
| C | 0 | 1 | 1 | 0 | 0 |
| D | 0 | 0 | 0 | 0 | 0 |
| E | 1 | 0 | 1 | 1 | 0 |

5.

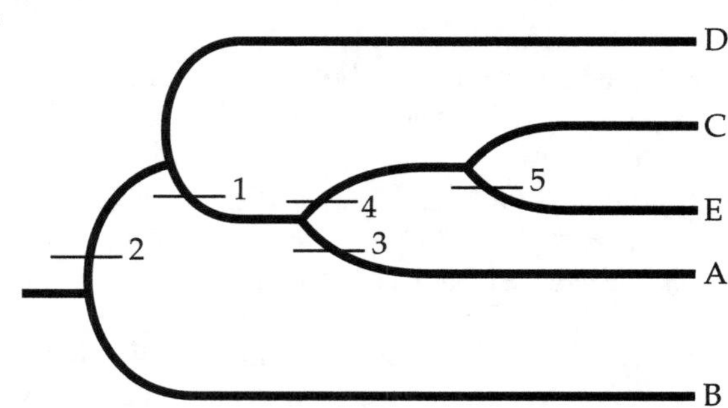


6. In the construction of a phylogenetic tree, the initial assumption is that derived traits appear only once and never disappear. Given a set of traits for a group of species, these restrictions sometimes must be relaxed to produce a phylogenetic tree for the group. Parsimony involves arranging the species so that the number of required reversals and multiple origins is minimized. Generally, the simplest explanation is likely to be the most accurate.

7. Knowledge of the phylogeny of a group of organisms can provide information about the origin and evolution of different traits. In medicine, the history of infectious pathogens can be determined using a phylogeny. Phylogenies can be used to determine ancestral traits and the evolution of complex adaptations.

8. Molecular clocks are the average rate at which a gene or protein accumulates change. By knowing the rate of change, scientists can gauge the time of divergence of closely related species and estimate the time of a particular split in a phylogeny.

9. Monophyletic groups are taxa that contain all the ancestors and all the descendants of that ancestor. It consists of all of the related organisms on a tree. A polyphyletic group does not contain the common ancestor, and a paraphyletic group does not include all of the descendants of a common ancestor—thus they provide an incomplete picture of the phylogenetic tree.

10. Scientists use specific rules in assigning scientific names to organisms to facilitate communication and dialogue. Typically there are many common names, in different languages, for an organism. The use of binomial nomenclature to assign a single name to a species eliminates confusion.

### *Test Yourself*

1. **b.** Although all of the statements listed are important attributes of biological classification schemes, the most important attribute is that biological classification reflects evolutionary relationships.
2. **d.** After a scientific name is referenced once in a text, the genus typically is abbreviated but the species name is given in full.
3. **c.** Because antennae with 10 segments are found in all genera in this family except for genus X, the trait of 10-segment antennae is best regarded as ancestral; hence, having fewer than 10 segments in the antennae is a derived trait. The presence of 8-segment antennae in species E is not a synapomorphy because it is a trait found only in that species.
4. **d.** Maximum likelihood is a mathematical method used to predict how a phylogenetic tree might evolve given observed data.
5. **a.** The complete hierarchy of taxonomic categories, from most to least inclusive, is: kingdom, phylum, class, order, family, genus, species.
6. **b.** Most shared traits, especially in species with a recent common ancestor, are likely to be homologous, not homoplastic. Therefore, the assumption that traits are homologous until proven homoplastic is more consistent with the parsimony principle than the reverse assumption is.
7. **b.** That the fossil record is incomplete is by far the greatest limitation of its usefulness in determining phylogenies.
8. **b.** Mitochondrial DNA in animals changes rapidly over evolutionary time and hence would be most useful in determining evolutionary relationships among species that have diverged from one another only recently.
9. **e.** It is true that molecular clocks must be calibrated with independent data, and it is true that different genes and other DNA sequences evolve at different rates. However, it is not true that the rate of gene evolution is independent of generation time or several other biological factors, or that molecular clocks cannot be used to date comparatively recent events.
10. **e.** Only traits that are genetically determined, and thus heritable, can be used in phylogenetic analysis.
11. **e.** A clade can be thought of as a complete branch on the tree of life. It includes the ancestor of a group, all the ancestor's descendants, and no other organisms.
12. **c.** Synapomorphies are traits that are not found in the ancestor of a group (hence they are derived) and that are found in more than one member of a group (hence they are shared).
13. **a.** Organisms in a higher taxon are *more* diverse, diverged from a common ancestor *less* recently, and include *more* species than organisms in a lower included taxon.
14. **a.** Derived traits are those that have undergone a change during evolution from the ancestral (original) character state.
15. **c.** If the shared characteristics of the ratites are homoplasies (meaning that they evolved independently by convergent evolution), then the group did not have a single common ancestor and is polyphyletic. If (as is now accepted) the shared traits are synapomorphies shared among all ratites and not found in other birds, then the group is monophyletic.

# 17 Speciation

## The Big Picture

- Speciation is the process that has produced the millions of life forms—each adapted to a particular environment and way of life—that constitute life on Earth. A number of concepts have been proposed to define a species, including the morphological species concept, the biological species concept, and the lineage species concept. Speciation occurs when there is reproductive isolation of one population into new lineages that cannot interbreed successfully.
- Allopatric speciation involves separation of populations due to geographical barriers and isolation of populations. Sympatric speciation occurs without any physical isolation and can involve polyploidy. When divergent populations come into contact, reproductive isolation is reinforced by prezygotic and postzygotic isolating mechanisms.

## Study Strategies

- You may find have difficulty understanding why hybrids between species that have different numbers of chromosomes are inevitably sterile unless the hybrid is an allopolyploid. Recall that in meiosis I, homologous chromosomes undergo synapsis. This process cannot occur properly if the haploid sets of chromosomes inherited from the parents contain different numbers of chromosomes, because it is then impossible for every chromosome to have a homolog. As a consequence, meiosis does not proceed normally and few if any normal gametes will be produced. Because allopolyploids possess four sets of chromosomes (two from each parent), their chromosomes can synapse normally and they can produce viable gametes.
- The heart of this chapter is the discussions of allopatric (geographic) speciation and reproductive isolating mechanisms, so be sure to focus on these topics.
- Go to **LaunchPad** (or use the Web addresses listed) to review the following additional resources:

  Animated Tutorial 17.1 Speciation Simulation (PoL2e.com/at17.1)

  Animated Tutorial 17.2 Founder Events and Allopatric Speciation (PoL2e.com/at17.2)

  Animated Tutorial 17.3 Speciation Mechanisms (PoL2e.com/at17.3)

  Activity 17.1 Concept Review (PoL2e.com/ac17.1)

  Analyze the Data in textbook Figure 17.12

  Apply the Concept in textbook Sections 17.3 and 17.4

## Key Concept Review

### 17.1 Species Are Reproductively Isolated Lineages on the Tree of Life

We can recognize many species by their appearance

Reproductive isolation is key

The lineage approach takes a long-term view

The different species concepts are not mutually exclusive

Speciation is the process by which one species splits into two species. Determining whether two populations constitute different species may be difficult because speciation is frequently a gradual process. The morphological concept of species, used by early biologists, grouped organisms into species on the basis of their appearance. This concept has limitations, because in some instances not all members of a species look alike and in other instances two or more cryptic species are morphologically indistinguishable but do not interbreed. The biological species concept defines a species as a group of actually or potentially interbreeding natural populations that are reproductively isolated from other such groups. It emphasizes the significance of reproductive isolation in keeping sexual lineages separated from one another. This definition cannot be applied to organisms that reproduce asexually, and it is limited to a single point in evolutionary time. The lineage species concept regards species as the smallest branches on the tree of life. The lineage splitting may be sudden or gradual, but in either case the lineages are thereafter independent of each other, allowing biologists to consider species over evolutionary time. These concepts of species are not mutually exclusive.

**Question 1.** There are a number of competing concepts that describe a species. What are the limitations of each concept?

**Question 2.** In the eastern United States, populations of the white-footed mouse, *Peromyscus leucopus*, may be found almost continuously from Maine to Georgia. Based on the biological species concept, why would a population of this species from Maine *not* be considered a separate species from a population in Georgia?

## 17.2 Speciation Is a Natural Consequence of Population Subdivision

### Incompatibilities between genes can produce reproductive isolation

### Reproductive isolation develops with increasing genetic divergence

Evolutionary change can occur without speciation. A single lineage may change through time without diverging into two species. Speciation requires that the gene pool of the original species divide into two isolated gene pools. According to the Dobzhansky–Muller model, after speciation the isolated populations will accumulate allelic differences at gene loci (or chromosomal differences) that eventually will make it impossible for members of the two populations to successfully interbreed if they come together again. The rate of reproductive isolation can be gradual, over millions of years, or as short as a few generations.

**Question 3.** Describe how centric fusion can play a role in speciation.

**Question 4.** Explain how incompatibilities between genes can produce reproductive isolation in a population.

## 17.3 Speciation May Occur through Geographic Isolation or in Sympatry

### Physical barriers give rise to allopatric speciation

### Sympatric speciation occurs without physical barriers

In allopatric speciation, the population is initially divided by a geographic barrier. Evidence suggests that allopatric speciation is the most common mechanism of speciation in most groups of organisms. The barrier can result from a geological or climatic change, and it results in two isolated populations that often are large and genetically similar. These populations diverged not only because of genetic drift, but also because the environments in which they live are, or became, different. Alternatively, separation may occur when some members of a population cross a barrier and found a new, isolated population. The speciation of the 14 species of Darwin's finches in the Galápagos are an example of speciation through geographic isolation.

Sympatric speciation occurs without geographic subdivision of the gene pool of the original species. Disruptive selection, in which different genotypes have high fitness with one of two different food resources, may be a widespread mechanism of sympatric speciation among insects. Sympatric speciation by polyploidy, the production of duplicate sets of chromosomes within an individual, is common in plants. Polyploidy produces new species because the polyploid organisms cannot interbreed with members of the parent species. Polyploid species that have a single ancestor are called autopolyploids, whereas those that have resulted from the hybridization of two species are referred to as allopolyploids. New species are much more likely to arise by polyploidy among plants than among animals because plants of many species can reproduce by self-fertilization.

**Question 5.** Discuss the conditions on the Galápagos Islands that led to the evolution of the birds known as Darwin's finches.

**Question 6.** In autopolyploidy, a new species of plant can arise by the doubling of chromosome numbers in a single individual of one species (provided that the individual is capable of self-fertilization). Why is it virtually impossible for such a tetraploid plant to interbreed successfully with diploid individuals of the "same" species?

## 17.4 Reproductive Isolation Is Reinforced When Diverging Species Come into Contact

### Prezygotic isolating mechanisms prevent hybridization between species

### Postzygotic isolating mechanisms result in selection against hybridization

### Hybrid zones may form if reproductive isolation is incomplete

If two populations reestablish contact before reproductive isolation is complete, several outcomes are possible. If hybrid offspring are not at a selective disadvantage, they may spread through both populations with the result that the gene pools of the populations combine. Thus, no new species would result from the period of isolation. If hybrid offspring are less successful, reinforcement may strengthen prezygotic reproductive barriers.

Prezygotic isolating mechanisms prevent members of different species from mating. Differences in reproductive organs may prevent interbreeding (mechanical isolation). Species may not be able to interbreed because they are fertile at different times (temporal isolation). The two species may not recognize or respond to each other's mating behaviors, or in flowering plants, the behavioral preferences of the pollinating animals may prevent interbreeding (behavioral isolation). Species may simply mate in different areas or different parts of a habitat (habitat isolation). The sperm and egg may be chemically incompatible (gametic isolation). Postzygotic barriers can prevent effective gene flow between species, even if mating occurs. Hybrid zygotes may not mature normally (low hybrid zygote viability). Hybrids may survive less well than either parent species (low hybrid adult viability). Hybrids may be infertile (hybrid infertility).

The evolution of more effective prezygotic reproductive barriers is known as reinforcement. It may occur if the hybrid offspring of two species survive poorly. If hybrid offspring are at a disadvantage but reinforcement fails to occur, a stable, narrow hybrid zone may form.

**Question 7.** Suppose that members of two populations are separated by a geographic barrier and begin to diverge genetically. Many generations later, when the barrier is removed, the two populations can interbreed, but the hybrid offspring do not survive and reproduce well. Explain how natural selection might lead to the evolution of more effective prezygotic barriers in these species.

**Question 8.** The yellow-rumped warbler was formerly split into two species (Myrtle warbler and Audubon's warbler), but in 1973 they were reclassified as a single species. Myrtle warblers and Audubon's warblers have largely allopatric ranges but hybridize where they are sympatric in the Canadian Rockies. They are similar in appearance but are readily distinguished by experienced birders. What further data about these two forms should ornithologists collect and analyze in order to decide whether they should continue to be classified as a single species?

**Question 9.** Construct a concept map whose theme is "species." Include in your map the following terms: species, allopatric, allopolyploidy, autopolyploidy, biological, concepts, founder events, independent evolution, interruption of gene flow, lineage, morphological, physical similarity, postzygotic barriers, prezygotic barriers, reproductive isolation, speciation, and sympatric. Connect these terms by verbs or short phrases to indicate the relationships among them.

## Test Yourself

1. It is difficult to apply the biological species concept to groups of organisms that
   a. reproduce sexually.
   b. produce hybrids only in captivity.
   c. show little morphological diversity.
   d. exist only in the fossil record.
   e. occur in wide geographic regions.
   ***Textbook Reference:*** *Concept 17.1 Species Are Reproductively Isolated Lineages on the Tree of Life; Reproductive isolation is key*

2. Which of the following statements about allopatric speciation is *false*?
   a. It can sometimes involve small populations.
   b. It occurs only in species that are widely distributed.
   c. It always involves a physical barrier that interrupts gene flow.
   d. It sometimes can involve chance events.
   e. It is the dominant mode of speciation in most groups of organisms.
   ***Textbook Reference:*** *Concept 17.3 Speciation May Occur through Geographic Isolation or in Sympatry; Physical barriers give rise to allopatric speciation*

3. A long, narrow hybrid zone exists in Europe between the ranges of the fire-bellied toad and the yellow-bellied toad. The persistence of this zone can be attributed to which of the following factors?
   a. Reinforcement strengthens the prezygotic barriers between the two species.
   b. Hybrid offspring have the same fitness as nonhybrid offspring.
   c. Both species travel long distances over the course of their lives.
   d. Individuals from outside the hybrid zone regularly move into the hybrid zone.
   e. Mutations regularly occur in each population.
   ***Textbook Reference:*** *Concept 17.4 Reproductive Isolation Is Reinforced When Diverging Species Come into Contact; Prezygotic isolating mechanisms prevent hybridization between species*

4. Which type of speciation is most common among flowering plants?
   a. Geographic
   b. Sympatric
   c. Allopatric
   d. Disruptive
   e. Parapatric
   ***Textbook Reference:*** *Concept 17.3 Speciation May Occur through Geographic Isolation or in Sympatry; Sympatric speciation occurs without physical barriers*

5. Which of the following would *not* be considered an example of a prezygotic reproductive isolating mechanism?
   a. One bird species forages in the tops of trees for flying insects, whereas another forages on the ground for worms and grubs.
   b. The males of one species of moth cannot detect and respond to the sex attractant chemicals produced by the females of another species.
   c. Sperm of one species of sea urchin are unable to penetrate the egg plasma membrane of another species.
   d. Mosquitoes of one species are active in foraging and searching for mates at dusk, whereas those of another species are active at dawn.
   e. Flowers of one orchid species mimic female bees of species A, whereas flowers of another orchid species mimic female bees of species B.
   ***Textbook Reference:*** *Concept 17.4 Reproductive Isolation Is Reinforced When Diverging Species Come into Contact; Prezygotic isolating mechanisms prevent hybridization between species*

6. The species definition that includes similarity of appearance among individual members is the basis for
   a. the lineage species concept.
   b. reproductive isolation.
   c. the morphological species concept.
   d. centric fusion.
   e. the biological species concept.
   ***Textbook Reference:*** *Concept 17.1 Species Are Reproductively Isolated Lineages on the Tree of Life; We can recognize many species by their appearance*

7. The Dobzhansky–Muller model describes how
   a. two independent lineages come together to form an ancestral population.
   b. an ancestral population separates to become two independent lineages.
   c. geographical barriers lead to speciation.
   d. species should be defined.
   e. allopatric speciation occurs.

   ***Textbook Reference:*** *Concept 17.2 Speciation Is a Natural Consequence of Population Subdivision; Incompatibilities between genes can produce reproductive isolation*

8. More than 800 species of *Drosophila* occur in the Hawaiian Islands, representing 30 to 40 percent of all the species in this genus. The occurrence of so many *Drosophila* species in this island chain is
   a. the result of a single founder event followed by genetic divergence.
   b. an example of an evolutionary radiation.
   c. largely the result of sympatric speciation.
   d. evidence that the genus *Drosophila* first evolved in the Hawaiian Islands.
   e. the result of allopatric speciation.

   ***Textbook Reference:*** *Concept 17.3 Speciation May Occur through Geographic Isolation or in Sympatry; Sympatric speciation occurs without physical barriers*

9. Which of the following statements about speciation is *false*?
   a. A small founding population can be involved in speciation.
   b. Speciation always involves interruption of gene flow between different groups of organisms.
   c. The rate of speciation can vary for different groups of organisms.
   d. Speciation always requires many generations.
   e. Speciation may occur because certain genotypes within a population prefer distinct microhabitats where mating takes place.

   ***Textbook Reference:*** *Concept 17.3 Speciation May Occur through Geographic Isolation or in Sympatry; Physical barriers give rise to allopatric speciation*

10. Which of the following observations would constitute conclusive evidence that two overlapping populations that have been geographically separated have *not* diverged into distinct species?
    a. Matings between members of the two populations produce viable hybrids.
    b. A hybrid zone exists where their ranges overlap.
    c. Interbreeding is common between members of the two populations.
    d. There are distinct morphological differences between the two populations.
    e. Matings between members of the two populations produce fertile offspring.

    ***Textbook Reference:*** *Concept 17.4 Reproductive Isolation Is Reinforced When Diverging Species Come into Contact; Hybrid zones may form if reproductive isolation is incomplete*

11. Which of the following processes is likely to be the *least* important in allopatric speciation?
    a. A founder event
    b. Allopolyploidy
    c. Behavioral isolation
    d. Genetic drift
    e. Habitat isolation

    ***Textbook Reference:*** *Concept 17.3 Speciation May Occur through Geographic Isolation or in Sympatry*

12. A field contains two related species of flowering plants. Species A has a diploid chromosome number of 16, and species B has a diploid number of 18. If a third species arises as a result of hybridization between A and B, how many chromosomes will it have?
    a. 17
    b. 32
    c. 34
    d. 36
    e. 68

    ***Textbook Reference:*** *Concept 17.3 Speciation May Occur through Geographic Isolation or in Sympatry; Sympatric speciation occurs without physical barriers*

13. Speciation by polyploidy occurs far more often in plants than in animals because
    a. plants are more likely to be capable of self-fertilization than animals are.
    b. plant cells can tolerate extra sets of chromosomes, whereas animal cells cannot.
    c. plants as a rule have higher reproductive rates than animals do.
    d. many plants are specialized with respect to their pollinating agent.
    e. fertilization in plants involves multiple gametes.

    ***Textbook Reference:*** *Concept 17.3 Speciation May Occur through Geographic Isolation or in Sympatry; Sympatric speciation occurs without physical barriers*

14. Two species of narrowmouth frogs in the United States have mating calls that differ more in their region of sympatry than in those parts of their ranges that do not overlap. If this difference in their vocalizations has the function of preventing hybridization between the two species, it is an example of
    a. a hybrid zone.
    b. reinforcement.
    c. sympatric speciation.

d. a postzygotic reproductive barrier.
e. allopatric speciation.
***Textbook Reference:*** *Concept 17.4 Reproductive Isolation Is Reinforced When Diverging Species Come into Contact; Prezygotic isolating mechanisms prevent hybridization between species*

15. Centric fusion of acrocentric (one-armed) chromosomes results in
a. mechanical isolation.
b. behavioral isolation.
c. temporal isolation.
d. reproductive isolation.
e. allopatric speciation.
***Textbook Reference:*** *Concept 17.2 Speciation Is a Natural Consequence of Population Subdivision; Incompatibilities between genes can produce reproductive isolation*

## Answers

### *Key Concept Review*

1. Limitations of the morphological species concept include the fact that the different sexes of the same species may not look alike or two species may look very much alike but are unable to interbreed. The limitation of the biological species concept is that it does not apply to asexual species. The lineage species concept is limited because of lack of knowledge as to when a lineage actually split or whether the incipient species will continue to diverge or merge at some point in the future.
2. According to the biological species concept, species are groups of "actually or potentially" interbreeding individuals. While individual mice from populations in Maine are not *actually* breeding with individuals from populations in Georgia, there is the *potential* for individuals from both populations, if brought together, to interbreed and produce viable offspring.
3. Centric fusion is the fusion of two acrocentric (one-armed) chromosomes to form a metacentric chromosome. If centric fusion becomes fixed at one chromosome in one population of organisms but at a different chromosome in another population, individuals from the two populations will be unable to produce viable offspring. This is because the offspring will not be able to produce normal gametes in meiosis.
4. If an initial population becomes subdivided into two populations by some type of barrier to gene flow, new alleles at different loci can arise and become fixed in the separated populations. Neither new allele may cause reproductive incompatibility. If the two new alleles from the isolated subpopulations were to come together, they might not be compatible, leading to functionally inferior offspring or lethality.
5. The relatively great distance between the Galápagos Islands and the South American mainland and between each of the islands in the archipelago ensured that once immigrants had arrived on an island, they would be genetically isolated for a substantial period of time. Also, because the islands differ greatly in climate and vegetation, the resident birds were subject to different selection pressures. This, in combination with reduced gene flow between the islands, led to a rapid evolutionary radiation of finches.
6. Recall from Chapter 7 that pairs of homologous chromosomes synapse during prophase and metaphase of the first division of meiosis. The homologs then separate, so that each cell resulting from meiosis I is haploid, as are the products of meiosis II. In tetraploids, as in diploids, meiosis is normal because pairing of homologs can occur. Any offspring of a cross between tetraploid and diploid individuals will be triploid, and therefore sterile however, because correct synapsis of homologs cannot occur. Because the tetraploid product of autopolyploidy is reproductively isolated from its diploid relatives, it is a new species.
7. Recall that natural selection tends to remove traits that reduce survival or reproductive success from a population. Individuals that interbreed between populations will have lower fitness (they will contribute fewer offspring to future generations) than those who breed within their own population. If the tendency to avoid interbreeding is heritable (and not just the result of chance), the frequency of alleles that prevent interbreeding will increase in each population. How might such a trait be heritable? Any of the prezygotic barriers to interbreeding might be heritable traits. For example, if the species in question are frogs, and if their mating calls started to diverge while the populations were separated, the following traits might be heritable: a tendency to make a call that is more distinct from that of the other population, or the ability to distinguish between the existing calls of the two populations (coupled with a preference for the call of one's own population). As alleles for these traits increased in frequency, they would contribute to the behavioral isolation of the two populations and perhaps eventually to more complete speciation.
8. During allopatric speciation, divergence of two populations often occurs gradually. In such cases it is inevitable that intermediate stages of speciation occur, and it may be a matter of opinion whether two forms have diverged sufficiently to be considered separate species. With respect to these two warbler populations, biologists would seek answers to these questions: Are hybrid offspring as fit as those resulting from mating of individuals of the same population? Is there evidence that the zone of hybridization is expanding, indicating that the gene pools of the populations are combining? Is there evidence of reinforcement of prezygotic

barriers to interbreeding (e.g., a greater difference in the songs of the two forms in the area of hybridization than in allopatric parts of their ranges)? Lesser fitness of hybrid offspring, a stable, narrow zone of hybridization, and evidence of reinforcement would favor the conclusion that the populations are best regarded as separate species.

9.

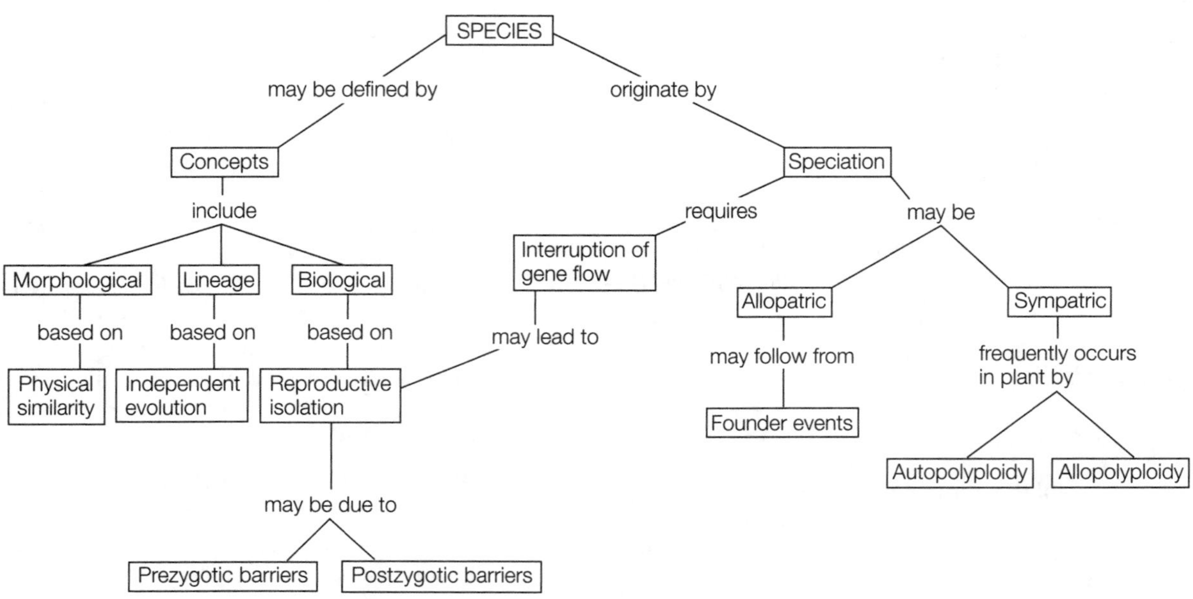


### *Test Yourself*

1. **d.** The key criterion of a biological species is that its members are reproductively isolated from other such groups. This criterion is impossible to evaluate in fossil species.
2. **b.** A wide distribution is not a requisite for allopatric speciation.
3. **d.** These toads are an example of related species in which reinforcement does not strengthen prezygotic barriers, even though hybrid offspring are only half as fit as nonhybrid offspring. The reason for this is that toads from outside the hybrid zone (not subject to the selective pressure against hybridizing) regularly move into the hybrid zone and mate with members of the other species. The hybrid zone remains narrow because toads do not travel long distances; natural selection removes hybrids from the population before they can disperse very far.
4. **b.** Sympatric speciation is most common in flowering plants. It has been estimated that about 70 percent of all flowering plant species are polyploid.
5. **a.** Provided that the two species are active in the same locality at the same time, a difference in the habitat in which they forage would not in itself be a barrier to interbreeding (though seeking mates in different habitats might well be a barrier to interbreeding). All the other choices describe reproductive barriers that would act prior to fertilization.
6. **c.** The morphological species concept assumes that individuals that look alike can be considered as belonging to the same species.
7. **b.** The Dobzhansky–Muller model describes how changes in different alleles can produce two independent lineages from a single ancestral population.
8. **b.** Island groups are frequently sites of evolutionary radiations through repeated allopatric speciation events that are initiated by individuals (or groups) dispersing from one island to another. The numerous species of *Drosophila* in the Hawaiian Islands are believed to have originated in this way.
9. **d.** New species formed by polyploidy can arise in only two generations.
10. **e.** Interbreeding, production of viable hybrids, establishment of a hybrid zone, and morphological differences do not necessarily mean that speciation is incomplete. If, however, the hybrids were fertile, their gene pools would merge and you would conclude that speciation had not taken place.
11. **b.** Genetic drift, founder events, and behavioral isolation may all play a role in allopatric speciation. Allopolyploidy, however, can occur only as the result of

hybridization between individuals of different species; hence the two parent species cannot be allopatric.

12. **c.** If haploid gametes of species A and B joined, the result would be a zygote with 17 chromosomes. A mature plant with 17 chromosomes would be sterile, because chromosomes would be unable to pair properly during prophase and metaphase of meiosis I. A fertile allopolyploid would therefore have to have 34 chromosomes so that each chromosome would have a homolog with which to pair. Autopolyploids of species A and B would have 32 and 36 chromosomes, respectively.

13. **a.** If a polyploid plant or animal is capable of self-fertilization, then a new species can arise from a single individual. The ability to self-fertilize is far more common among plants than animals.

14. **b.** Reinforcement is defined as the evolutionary strengthening of prezygotic barriers to interbreeding within the zone of sympatry of two closely related species.

15. **d.** When two chromosomes fuse in one lineage of a population, this lineage is unable to successfully produce viable offspring with the other lineage.

# 18 The History of Life on Earth

## The Big Picture

- Radiometric dating provides a means of dating rocks, allowing scientists to construct a geological time scale and determine the age of fossils.
- Throughout Earth's history, its physical environment has changed, resulting in a number of mass extinctions. These changes include the movement of major land masses as a result of continental drift, variation in climate, volcanic eruptions, and extraterrestrial events. Initially, there was no oxygen in Earth's atmosphere. With the evolution of photosynthetic bacteria, oxygen levels rose, allowing for the evolution of multicellular organisms.
- Life first appeared on Earth about 3.8 bya. In the Precambrian, life was found only in the oceans. The Cambrian explosion resulted in rapid diversification of life. Multiple mass extinctions occurred in the Ordovician, Devonian, and Permian. A single supercontinent, Pangea, was formed during the Permian period, breaking into two large continents during the Triassic and Jurassic periods. Modern organisms evolved during the Cenozoic era.

## Study Strategies

- The concept of the half-life of radioactive isotopes may seem confusing. Bear in mind that if half of a radioisotope decays in a given time period (its half-life), then at the end of the first half-life only one-half of the original quantity of the isotope is still present, one-half of which will decay in the second half-life. Thus after two half-lives, one-quarter of the original quantity of radioisotope is still present. The same line of reasoning applies to all further half-lives. Study Figure 18.1, which explains this concept.
- This chapter contains many names of geological eras and periods and many dates. Instructors vary with respect to the importance they place on memorization of this information. In general, it is best to focus on broad evolutionary patterns and trends. Most instructors will also place more importance on the phylogenetic sequences (e.g., amphibians gave rise to amniotes, which gave rise to reptiles and to mammals) than on the dates at which these events occurred.
- Making a table, chart, or timeline to summarize and organize the events that occurred in each geological era and period is an excellent study strategy.
- Go to **LaunchPad** (or use the Web addresses listed) to review the following additional resources:

  Animated Tutorial 18.1 Evolution of the Continents (PoL2e.com/at18.1)

  Activity 18.1 Concept Matching (PoL2e.com/ac18.1)

  Analyze the Data in textbook Figure 18.8

  Apply the Concept in textbook Sections 18.1 and 18.2

## Key Concept Review

### 18.1 Events in Earth's History Can Be Dated

Radioisotopes provide a way to date rocks

Radiometric dating methods have been expanded and refined

Scientists have used several methods to construct a geological time scale

Evolutionary changes may occur over both short and long time frames. Short-term evolutionary changes occur rapidly enough to be studied directly. Long-term evolutionary changes involve the appearance of new species and evolutionary lineages. The fossil record provides evidence of such changes. Several types of evidence have been used to estimate the age of rocks, of fossils, and of Earth itself. Sedimentary rocks (formed by the accumulation of sediments at the bottom of bodies of water) are deposited in strata (layers). The oldest layers are found at the bottom, and successively higher strata are progressively younger. Fossils, which are the preserved remains of ancient organisms, are useful for establishing the relative ages of the sedimentary rocks in which they occur. The fossil record shows that an organism of any specific type can predictably be found in rocks of a particular age. Living organisms resemble recent fossils more closely than they resemble fossils from more ancient periods.

The regular pattern of decay of radioactive isotopes provides a means of estimating the absolute ages of fossils and rocks. The half-life of a radioisotope is the time period during which half of the remaining radioactive material decays to become a different, stable isotope (see Figure 18.1). The ratio of radioactive carbon-14 to its stable isotope, carbon-12, is used to date fossils less than 60,000 years old. The decay of potassium-40 to argon-40 is widely used to date ancient evolutionary events. Paleomagnetic dating is based on the fact that the age of sedimentary and igneous rocks can be determined because they preserve a record of Earth's magnetic field at the time they were formed.

Earth's geological history is divided into eras, which are subdivided into periods. The boundaries between these divisions are marked by changes in the types of fossils found in successive layers of sedimentary rock. The first forms of life evolved early in the Precambrian era, which lasted for more than 3 billion years and was marked by enormous physical changes on Earth.

**Question 1.** For each geological period, make a chart summarizing its major geological and evolutionary events.

**Question 2.** $^{14}C$ decays to $^{14}N$ with a half-life of about 5,700 years. Suppose you find a fossil in which the amount of $^{14}C$ is only 1/16 of what you would find in a living organism with the same carbon content. Approximately how old is the fossil?

**Question 3.** In 2006, paleontologists announced that fossils of *Tiktaalik roseae*, an important link between fish and tetrapods, had been discovered in freshwater sediments in the Canadian Arctic that were 375 million years old. How were geologists able to determine the age of these sediments?

## 18.2 Changes in Earth's Physical Environment Have Affected the Evolution of Life

The continents have not always been where they are today

Earth's climate has shifted between hot and cold conditions

Volcanoes have occasionally changed the history of life

Extraterrestrial events have triggered changes on Earth

Oxygen concentrations in Earth's atmosphere have changed over time

The theory of plate tectonics proposes that Earth's crust consists of solid lithospheric plates floating on a fluid layer of magma (see Figure 18.2). Convection currents in the magma result from heat emanating from Earth's core. The currents cause continental drift, which is the gradual shift in the position of the plates and the continents they contain. The drifting of continents has had profound effects on climate, sea level, oceanic circulation, and the distributions of organisms.

Changes in the physical parameters on Earth have resulted in numerous mass extinctions. The climate of Earth has alternated between hot/humid and cold/dry conditions. Though most major climatic changes have been gradual, some have occurred within periods of 5,000 to 10,000 years or less. The rapid climate change occurring today is thought to be caused by a buildup of atmospheric $CO_2$, primarily from the burning of fossil fuels. The climatic shifts caused by massive volcanic eruptions associated with continental drift are implicated in several mass extinctions. The late Permian mass extinction was in part caused by volcanism triggered by the collision of continents that formed the supercontinent Pangaea. Several mass extinctions have likely been caused by collisions of Earth with meteorites or comets, such as the event 65 mya that is thought to have caused the extinction of dinosaurs.

No free oxygen was present in the atmosphere until certain bacteria evolved the ability to use water as a source of hydrogen ions for photosynthesis (see Figure 18.6). The gradual increase in oxygen concentration resulted in the dominance of organisms using aerobic metabolism and in the evolution of larger eukaryotic cells and of multicellular organisms. The exceptionally high oxygen concentrations that occurred during the Carboniferous and Permian periods are associated with the evolution of giant amphibians and flying insects.

**Question 4.** Scientists believe that if there are no controls on the emission of $CO_2$ from the burning of fossil fuels, the concentration of this gas could double by the end of the current century, leading to a significant rise in the average temperature of Earth. What would be some of the likely evolutionary effects of this climatic change?

**Question 5.** In what way was the evolution of eukaryotic cells linked to the increase in the oxygen concentration in the atmosphere that occurred during the Precambrian?

## 18.3 Major Events in the Evolution of Life Can Be Read in the Fossil Record

Several processes contribute to the paucity of fossils

Precambrian life was small and aquatic

Life expanded rapidly during the Cambrian period

Many groups of organisms that arose during the Cambrian later diversified

Geographic differentiation increased during the Mesozoic era

Modern biotas evolved during the Cenozoic era

The tree of life is used to reconstruct evolutionary events

The assemblage of organisms found in a particular place or time constitutes a biota. The plant component of a biota is its flora, and the animal component is its fauna. The 300,000 known fossil species represent only a tiny fraction of the species that have ever lived. Some groups, such as hard-shelled marine animals, are much better represented in the fossil record than others. Because an oxygen-rich environment favors rapid decomposition, organisms that become fossils are likely either to have lived in a poorly oxygenated environment or to have been transported to such a site soon after death.

The earliest life on Earth appeared about 3.8 bya, but the fossil record of organisms that lived in the Precambrian is fragmentary. The first organisms were unicellular prokaryotes. For most of the Precambrian, which lasted for over 3 billion years, life consisted of microscopic prokaryotes. The first unicellular eukaryotes evolved about 1.5 billion years ago. The best Precambrian fossil deposits, dating from about 600 mya, contain diverse soft-bodied invertebrates, some of which may represent lineages with no living descendents.

During the Cambrian period, the oxygen concentration in the atmosphere approached its current level, and several large continents formed. The rapid increase in the diversity of multicellular life forms that occurred at this time is known as the "Cambrian explosion." (see Figure 18.11)

An evolutionary radiation occurs when there is a rapid diversification of organisms. The Ordovician period was marked by a proliferation of marine filter feeders living on the sea floor. At the end of this period, massive glaciers formed over the southern continents, the sea level and ocean temperatures dropped, and the majority of animal species became extinct. The Silurian period witnessed the evolution of swimming marine animals and the diversification of jawless fish. It also marked the appearance of the first terrestrial arthropods and vascular plants. During the Devonian period, all major groups of fishes evolved. On land, the first insects and amphibians evolved, and forests of club mosses, horsetails, and tree ferns appeared. A mass extinction of approximately three-quarters of all marine species occurred at the end of this period, possibly caused by the collision of two large meteorites with Earth.

In the Carboniferous period, swamp forests consisting largely of giant tree ferns and horsetails became widespread. The fossilized remains of these plants formed coal. The first winged insects evolved during this period, while amphibians became better adapted to life on land and gave rise to the lineage leading to the amniotes (whose eggs can be laid in dry places). The Permian period was marked by the formation of a single supercontinent called Pangaea. As the climate cooled drastically, amniotes split into two lineages—the reptiles and one leading to the mammals. The occurrence of the most extensive of all mass extinctions brought the Permian period to a close.

During the Mesozoic era (251–65 mya), the continents that formed Pangaea slowly separated to form Laurasia and Gondwana, and distinct assemblages of plants and animals evolved on each continent. During the Triassic period, seed ferns and conifers were the dominant forms of terrestrial vegetation, and reptiles were the dominant vertebrates. A mass extinction occurred at the end of the Triassic. The Jurassic period was marked by the complete separation of Pangaea to form Laurasia in the north and Gondwana in the south. It also witnessed the radiation of the dinosaurs, the evolution of flying reptiles, and the first appearance of mammals. The earliest fossils of flowering plants date from late in this period. In the sea, ray-finned fishes also began a great radiation. By the early Cretaceous period, Laurasia and Gondwana had begun to break apart into the present-day continents. The flowering plants began the radiation that led to their current dominance. The dinosaurs continued as the dominant land vertebrates, despite the presence of many groups of mammals. At the end of this period, the collision of a large meteorite with Earth caused the extinction of the dinosaurs and of many other animal and plant lineages.

Modern groups of plants and animals evolved during the Cenozoic era (65 mya–present). During the Tertiary period, the continents drifted toward their present positions. As the climate became cooler and drier, extensive grasslands appeared. Many groups of land vertebrates—especially frogs, snakes, lizards, birds, and mammals—radiated extensively. The current geological period, the Quaternary, is divided into the Pleistocene and Holocene (Recent) epochs. Modern humans evolved during the Pleistocene, which was a time of severe climatic fluctuations, including four major periods of extensive glaciation. As humans spread geographically, many species of large birds and mammals became extinct.

Phylogenetic trees help reconstruct the timing of evolutionary events and clarify relationships among modern species.

**Question 6.** Below is a chronologically scrambled list of important events in the history of life on Earth. Draw four lines on a sheet of paper to represent the time lines of events occurring in the Precambrian, the Paleozoic era, the Mesozoic era, and the Cenozoic era. Place the listed events on the correct line and in proper sequence.

Conifers become dominant
Cambrian "explosion"
Evolution of *Homo*
Fifth mass extinction
First eukaryotes
First flowering plant fossils
First forests ("fern" forests); first jawed fish
First fossils of multicellular animals
First mammals, dinosaurs diversify
First mass extinction
First photosynthetic eukaryotes
First vascular plants and terrestrial arthropods
Fourth mass extinction
Grasslands spread
Origin of amniotes
Origin of life
Origin of photosynthesis
Second mass extinction
Third mass extinction
Rapid radiation of mammals

**Question 7.** How did the mass extinctions of the Ordovician, Devonian, and Permian periods impact life on Earth?

## Test Yourself

1. The half-life of an isotope is the
   a. time it takes a fixed fraction of isotope material to change from one form to another.
   b. time interval during which the isotope is useful for dating rocks.
   c. ratio of one isotope species to another in a sample of organic matter.
   d. frequency at which it is found in multiple samples.
   e. time it takes for the sample to bind with oxygen.
   ***Textbook Reference:*** *Concept 18.1 Events in Earth's History Can Be Dated; Radioisotopes provide a way to date rocks*

2. Mountain ranges are fundamentally the result of
   a. plates in Earth's crust moving against one another on top of a fluid layer of molten rock.
   b. climate changes and the movement of glacial ice sheets.
   c. leftover debris from ancient collisions with an asteroid or meteor.
   d. the breakup of Laurasia and Gondwana.
   e. erosion due to rainfall and wave action.
   ***Textbook Reference:*** *Concept 18.2 Changes in Earth's Physical Environment Have Affected the Evolution of Life; The continents have not always been where they are today*

3. Fossilization is most likely to occur in which of the following environments?
   a. The surf zone along a sandy beach
   b. A shallow, cool swamp with high deposition rates of mud sediments
   c. The bottom of a hot, dry cave with no running water
   d. A fast-running mountain stream
   e. A high mountain top with seasonally deep snow
   ***Textbook Reference:*** *Concept 18.3 Major Events in the Evolution of Life Can Be Read in the Fossil Record; Several processes contribute to the paucity of fossils*

4. Despite being incomplete as a whole, the fossil record is rather detailed for
   a. soft-bodied insects.
   b. cnidarians and sponges.
   c. most terrestrial animals.
   d. hard-shelled mollusks.
   e. vascular plants.
   ***Textbook Reference:*** *Concept 18.3 Major Events in the Evolution of Life Can Be Read in the Fossil Record; Several processes contribute to the paucity of fossils*

5. During the _______ geological period, Pangaea became fully divided and distinctive assemblages of plants and animals begin to arise on different continents.
   a. Cambrian
   b. Tertiary
   c. Devonian
   d. Permian
   e. Jurassic
   ***Textbook Reference:*** *Concept 18.3 Major Events in the Evolution of Life Can Be Read in the Fossil Record; Geographic differentiation increased during the Mesozoic era*

6. Which of the following statements about the Ordovician period is *false?*
   a. Marine filter feeders flourished.
   b. The number of classes and orders increased.
   c. Modern mammals appeared.
   d. Many groups became extinct at the end of the period.
   e. The continents were located primarily in the Southern Hemisphere.
   ***Textbook Reference:*** *Concept 18.3 Major Events in the Evolution of Life Can Be Read in the Fossil Record; Many groups of organisms that arose during the Cambrian later diversified*

7. Most of human evolution occurred during the
   a. Paleozoic era.
   b. Devonian period.
   c. Quaternary period.
   d. Carboniferous period.
   e. Cretaceous period.
   ***Textbook Reference:*** *Concept 18.3 Major Events in the Evolution of Life Can Be Read in the Fossil Record; Modern biotas evolved during the Cenozoic era*

8. For terrestrial animals and plants, the most recent mass extinction event that occurred prior to the evolution of humans took place approximately _______ mya.
   a. 10
   b. 65
   c. 200
   d. 250
   e. 400
   ***Textbook Reference:*** *Concept 18.3 Major Events in the Evolution of Life Can Be Read in the Fossil Record; Geographic differentiation increased during the Mesozoic era*

9. One of the main factors that distinguishes the Cambrian explosion from all others is that
   a. evolutionarily, it was the most recent.
   b. many new major groups of animals appeared at this time.
   c. it was when the dinosaurs became extinct.
   d. it saw a dramatic drop in species diversity, especially among marine organisms.
   e. it was a time of major volcanic eruptions.
   ***Textbook Reference:*** *Concept 18.3 Major Events in the Evolution of Life Can Be Read in the Fossil Record; Life expanded rapidly during the Cambrian period*

10. During which of the following geological times did the most new kinds of body plans appear?
    a. Carboniferous
    b. Triassic
    c. Jurassic
    d. Devonian
    e. Cambrian
    ***Textbook Reference:*** *Concept 18.3 Major Events in the Evolution of Life Can Be Read in the Fossil Record; Life expanded rapidly during the Cambrian period*

11. The sudden disappearance of the dinosaurs some 65 mya may have been the result of
    a. Earth's collision with a large meteorite.
    b. slow climate changes due to planetary cooling.
    c. competition from better-adapted organisms.
    d. the rise of birds and mammals.
    e. the formation of Pangaea.
    ***Textbook Reference:*** *Concept 18.3 Major Events in the Evolution of Life Can Be Read in the Fossil Record; Geographic differentiation increased during the Mesozoic Era*

12. Fossil insects many times larger than any insects alive today have been dated to the Carboniferous and Permian periods. These fossils are evidence that during these periods
    a. the climate was much warmer than it is today.
    b. there were fewer predators of insects than there are today.
    c. the carbon dioxide concentration of the atmosphere was lower than it is today.
    d. insects had less competition for food than modern insects have.
    e. the oxygen concentration of the atmosphere was significantly higher than it is today.
    ***Textbook Reference:*** *Concept 18.2 Changes in Earth's Physical Environment Have Affected the Evolution of Life; Oxygen concentrations in Earth's atmosphere have changed over time*

13. Which of the following statements about patterns or processes in the evolution of life is *false*?
    a. $^{14}C$ can be used to date the age of dinosaur bones.
    b. The supercontinent Pangaea formed during the Permian period.
    c. Mass extinctions of marine organisms have coincided with periods of low sea levels.
    d. The ends of five geological periods have been marked by mass extinctions.
    e. The size of many organisms can be correlated with atmospheric oxygen levels.
    ***Textbook Reference:*** *Concept 18.3. Major Events in the Evolution of Life Can Be Read in the Fossil Record; Geographic differentiation increased during the Mesozoic era*

14. Which of the following paired organisms were *not* present on Earth in their living forms at the same time?
    a. Tree ferns and ray-finned fish
    b. Amphibians and birds
    c. Gymnosperms and insects
    d. Humans and dinosaurs
    e. Jawless fish and vascular plants
    ***Textbook Reference:*** *Concept 18.3 Major Events in the Evolution of Life Can Be Read in the Fossil Record; Modern biotas evolved during the Cenozoic era*

15. The most severe mass extinction event, linked to the formation of Pangaea and massive volcanic eruptions, occurred at the end of the _______ period.
    a. Ordovician
    b. Devonian
    c. Permian
    d. Triassic
    e. Cretaceous
    ***Textbook Reference:*** *Concept 18.3 Major Events in the Evolution of Life Can Be Read in the Fossil Record; Many groups of organisms that arose during the Cambrian later diversified*

## Answers

### *Key Concept Review*

1. See Table 18.1 in the textbook.
2. If only $\frac{1}{16}$ of the $^{14}C$ remains, then four $^{14}C$ half-lives have passed since the fossil was formed ($\frac{1}{2} \times \frac{1}{2} \times \frac{1}{2} \times \frac{1}{2} = \frac{1}{16}$). Because the half-life of $^{14}C$ is roughly 5,700 years, the fossil must be about 22,800 years old ($4 \times 5{,}700 = 22{,}800$).
3. Recall that radioisotope dating is used to determine the absolute age of rocks. Igneous rocks, formed from volcanic ash or lava flows, are required for this purpose. Such rocks may have been found near the discovery site in Canada, or they may have been found at one or more sites that have sediments of the same relative age. The relative age of the sediments can be determined by the types of fossils found in them.
4. The fossil record shows that major environmental changes occurring over a short time interval sometimes lead to large-scale rapid extinctions that appear "instantaneous" in the fossil record. The possible effects of global warming on biodiversity are discussed in more detail in Chapter 45.
5. Small prokaryotic cells can obtain enough oxygen by diffusion, even when oxygen concentrations are very low. Since eukaryotic cells are larger, they have a lower surface area-to-volume ratio and hence need a higher concentration of oxygen in their environment for diffusion to meet their requirement for this gas.

6.

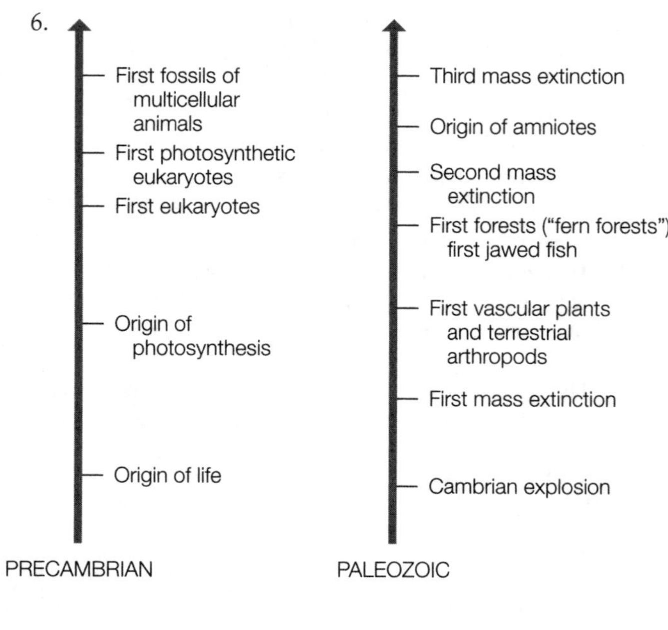


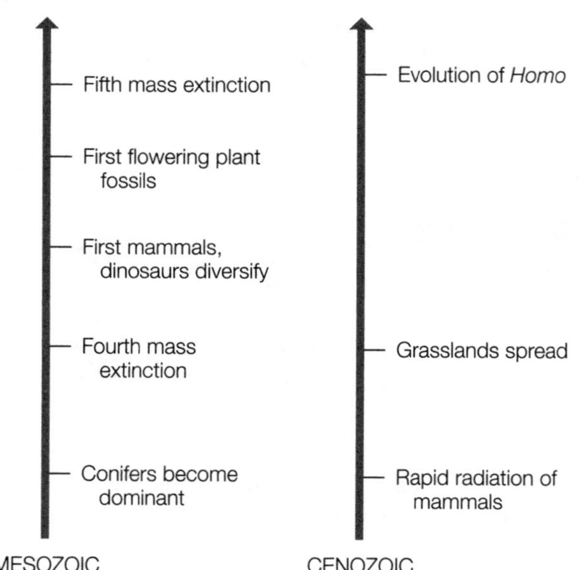


Also see Figures 18.10 and 18.12 in the textbook.

7. The mass extinction associated with the Ordovician is thought to be the result of a the drop in sea level and cooling of the oceans, both associated with glaciation. The cause of the extinction during the Devonian, while uncertain, is thought to be the result of meteorite collisions with Earth. At the end of the Permian, major changes, including volcanic eruptions that blocked sunlight and cooled climate, death of forests that depleted atmospheric oxygen, and loss of photosynthetic organisms to replace oxygen, resulted in the disappearance of about 96 percent of multicellular species. At the times of these extinctions, the diversity of life was greatly reduced. As environments changed, life rebounded.

### *Test Yourself*

1. **a.** Radioactive decay is measured as the time it takes for one-half the amount of a substance to spontaneously convert into another substance.
2. **a.** As the crustal plates are pushed together, one may move underneath the other, pushing up mountain ranges.
3. **b.** Fossilization is most likely to occur in areas of low oxygen concentration and rapid sedimentation, where scavengers cannot destroy the specimen.
4. **d.** Hard-shelled animals, such as mollusks, are good candidates for fossilization because their shells can withstand decay long enough for them to be buried. Many also tend to live in quiet, shallow waters.
5. **e.** The division of Pangaea began during the Triassic period, but Laurasia and Gondwana did not become completely separated until the Jurassic.
6. **c.** The modern mammals did not appear until the Tertiary, more than 350 million years after the Ordovician.
7. **c.** The Pleistocene epoch, which is part of the Quaternary, was the time of most of hominid evolution.
8. **b.** The mass extinction at the end of the Cretaceous, 65 mya, was the most recent mass extinction affecting terrestrial life before the evolution of humans.
9. **b.** The Cambrian explosion produced many novel groups of animals characterized by distinctive body plans. Later explosions caused an increase in diversity only within already-existing major lineages.
10. **e.** The Cambrian explosion produced many new types of body plans.
11. **a.** Paleontologists believe that a large meteorite collided with Earth and so altered conditions that the dinosaurs rapidly succumbed. This was during what we know as the great Cretaceous mass extinction.
12. **e.** There is experimental evidence that insects can grow larger when raised in hyperbaric conditions (see Figure 18.8). The current level of atmospheric $O_2$ appears to limit the evolution in body size of flying insects.
13. **a.** Carbon dating is generally reliable only for fossils less than 60,000 years old.
14. **d.** Dinosaurs disappeared at the end of the Cretaceous, over 60 million years before humans evolved.
15. **c.** Though mass extinctions occurred at the end of all the periods listed, the highest percentage of species became extinct during the Permian period.

# 19 Bacteria, Archaea, and Viruses

## The Big Picture

- All organisms fall within one of three domains: the Archaea, the Bacteria, and the Eukarya. This chapter focuses on the two prokaryotic domains (Archaea and Bacteria) and also on the nonliving viruses. Archaea and Bacteria retain many features of the earliest forms of life on Earth. They have circular chromosomes and cell walls made of peptidoglycans or proteins, and they lack membrane-enclosed organelles. Their structures are simple, and they exist as single cells. Their metabolic processes are highly varied, as are their modes for procuring energy. Many prokaryotic organisms make valuable contributions to the environment through nitrogen fixation, photosynthesis, and other metabolic processes. Most are free-living, but some live in mutualistic relationships and others are pathogenic to other organisms.
- The evolutionary relationships among prokaryotes are still highly disputed and best understood at the genetic level. The most ubiquitous extant prokaryotes are the bacteria. They are grouped according to physical and biochemical characteristics, but these groupings do not necessarily reflect their evolutionary relationships. The archaea are more closely related to the eukarya, and exist in some of the harshest known environments on Earth.
- While there is no single definition of "life" that all scientists agree upon, many do not consider viruses living organisms, even though they arose from cells and require a host cell to carry out their basic functions. Viral genomes are varied in form and structure. They can be made of either RNA or DNA; RNA viruses can be positive- or negative-sense single-stranded, single-stranded retroviruses, or double-stranded. Currently, the genome structure is the basic classification system used, since the evolutionary history of viruses is unknown and unlikely to be determined.

## Study Strategies

- Because this chapter covers a large number of diverse microscopic organisms included in two of the three domains of life, the details and exceptions can be overwhelming. Many of the names of these groups are long, complex, and unfamiliar. Many of the distinguishing features of the different groups of bacteria and archaea are biochemical, so it may be helpful to review information from Part I of the textbook. Use flash cards to practice identifying the groups by their defining traits.
- It is tempting to focus on pathogenic organisms, but remember that only a very small percentage of prokaryotes are pathogenic. Pay attention to the ecological services performed by members of different groups as well as the vast array of different conditions in which they thrive. Again, use flash cards to sort different organisms into groups according to where they live or what they do.
- Go to **LaunchPad** (or use the Web addresses listed) to review the following additional resources:

  Animated Tutorial 19.1 The Evolution of the Three Domains (PoL2e.com/at19.1)

  Activity 19.1 Gram Stain and Bacteria (PoL2e.com/ac19.1)

  Analyze the Data in textbook Figure 19.14

  Apply the Concept in textbook Sections 19.3 and 19.4

## Key Concept Review

### 19.1 Life Consists of Three Domains That Share a Common Ancestor

The two prokaryotic domains differ in significant ways

The small size of prokaryotes has hindered our study of their evolutionary relationships

The nucleotide sequences of prokaryotes reveal their evolutionary relationships

Lateral gene transfer can lead to discordant gene trees

The great majority of prokaryote species have never been studied

Prokaryotes include members of the domains Archaea and Bacteria. Both groups of organisms have single chromosomes that are often circular and replicate by binary fission rather than mitosis. Additional DNA plasmids may be present in some. Because they lack membrane-bound compartments, they rely on infoldings of the membrane to carry out many metabolic processes.

These similarities lead many students to treat Archaea and Bacteria as a single group, but there are some significant differences between them (see Table 19.1). The cell wall in bacteria is made of peptidoglycan, which is not found in archaea. In addition, the cellular machinery in archaea is more like that of eukaryotes than of bacteria. The last common ancestor of all three domains (Archaea, Bacteria, and Eukarya) is estimated to have lived about three billion years ago.

The classification of prokaryotes was originally based on shape, color, motility, nutritional requirements, and the amount of peptidoglycan found in the bacterial cell wall. For instance, the Gram stain is a classic technique used to identify bacteria. Gram-positive bacteria appear purple because the thick layer of peptidoglycan in the cell wall retains a purple dye. Gram-negative bacteria have an extra membrane exterior to the cell wall and stain pink or red. Modern techniques allow for the study of genomes, which is aiding our understanding of evolutionary relationships. Sequences of rRNA genes have been particularly useful in phylogenetic studies. Many archaea have been identified only though analysis of DNA samples collected from the environment. We now know that some genes have been transferred between species via lateral gene transfer and that most genomes have a stable core of genes that lets us trace the phylogeny despite this transfer. A large proportion of prokaryotic species have yet to be isolated or identified.

**Question 1.** Runoff into a stream from a wastewater treatment plant has been killing the fish in the stream. The organism responsible has been isolated, and consulting scientists are examining it under a microscope. How can they tell if the organism is a prokaryote or a eukaryote?

**Question 2.** Describe how Gram staining depends on bacterial cell wall structure. Would Gram staining work for identifying archaea?

**Question 3.** Label the three basic shapes of bacteria shown in the micrograph below.

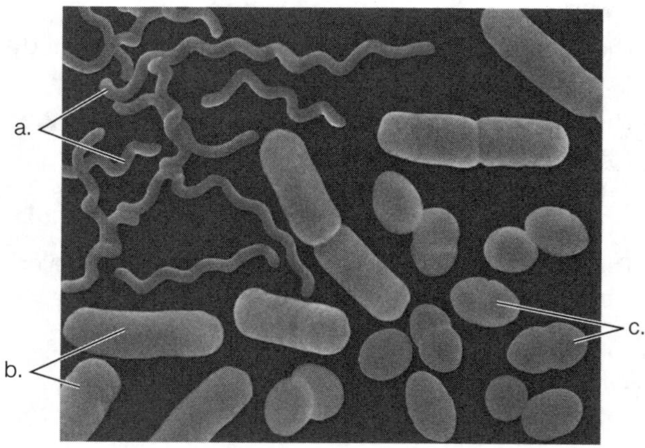


**Question 4.** In the diagrams below, label the bacteria as Gram-positive or Gram-negative and identify the structures in each.

This is a Gram-_______ bacteria.

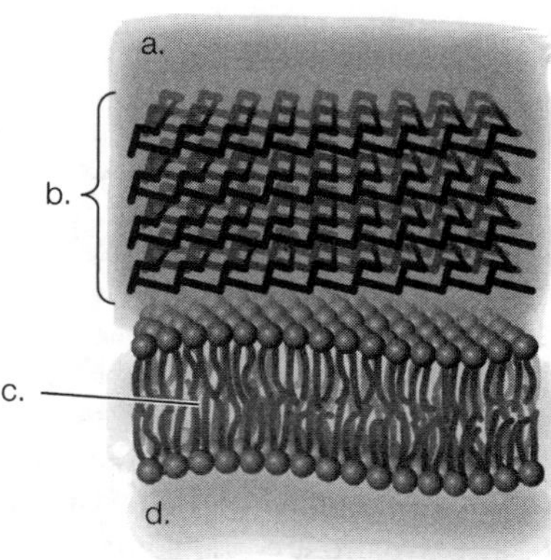


This is a Gram-_______ bacteria.

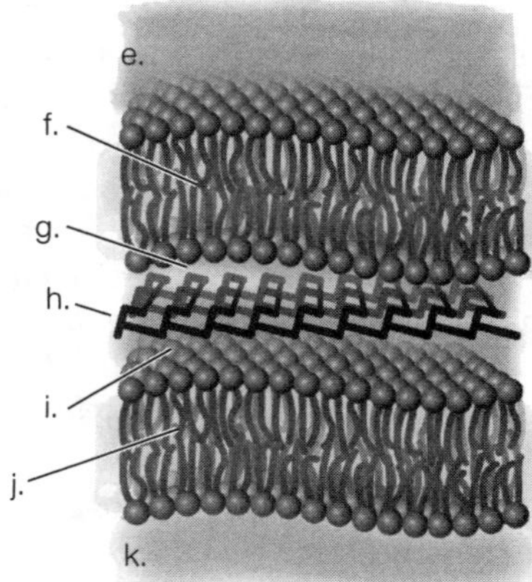


### 19.2 Prokaryote Diversity Reflects the Ancient Origins of Life

- The low-GC Gram-positives include the smallest cellular organisms
- Some high-GC Gram-positives are valuable sources of antibiotics
- Hyperthermophilic bacteria live at very high temperatures
- Hadobacteria live in extreme environments
- Cyanobacteria were the first photosynthesizers
- Spirochetes move by means of axial filaments
- Chlamydias are extremely small parasites
- The proteobacteria are a large and diverse group
- Gene sequencing enabled biologists to differentiate the domain Archaea
- Most crenarchaeotes live in hot or acidic places
- Euryarchaeotes are found in surprising places
- Several lineages of Archaea are poorly known

On an individual basis, there far more bacteria and archaea than eukaryotes. Knowledge about prokaryotes is very incomplete, and there are many different hypotheses about the relationships among them. This book uses one classification system that is based on sequence data. The eight groups of bacteria are: (1) low-GC Gram-positives, (2) high-GC Gram-positives, (3) hyperthermophilic bacteria, (4) hadobacteria, (5) cyanobacteria, (6) spirochetes, (7) chlamydias, and (8) proteobacteria. The two groups of archaea that are most studied are Crenarchaeota and Euryarchaeota.

Low-GC Gram-positives, or firmicutes, include members that can form endospores, staphylococci that live on human skin, and mycoplasmas that lack cell walls. High-GC Gram-positives are known as actinobacteria. These species develop branched filaments that resemble fungi. Some members include several species that are sources of antibiotics and the human pathogen *Mycobacterium tuberculosis.* Hyperthermophilic bacteria, hadobacteria, and some archaea are extremophiles; they thrive in extreme environments where the temperatures, radiation levels, or toxin levels are fatal to other organisms. Cyanobacteria are believed to be the earliest photosynthetic organisms; they have a highly organized system of internal membranes called photosynthetic lamellae. Cyanobacteria can live as single cells or in multicellular colonies. Some filamentous colonies have three differentiated cell types: vegetative cells that perform photosynthesis, spores that can survive harsh conditions, and heterocysts that perform nitrogen fixation. Spirochetes are characterized by axial filaments that enable a corkscrew-like movement. Some species are the cause of syphilis and Lyme disease. Chlamydias are extremely small obligate parasites known to cause some human STDs. Their life cycle is unique among prokaryotes and has two stages: elementary bodies and reticulate bodies. Proteobacteria are the largest group based on numbers of identified species. The members of this clade are highly diverse and include, among many others, the nitrogen-fixing *Rhizobium, Escherichia coli,* and many species that infect plants.

Archaea phylogeny was originally based on rRNA sequences and is still being developed. The current classification scheme divides the archaea into two groups: Crenarchaeota and Euryarchaeota. More recently discovered groups are the Korarchaeota, Nanoarchaeota, and Thaumarchaeota. All archaea share the absence of proteto-glycan in the cell wall and the presence of unique lipids in their cell membranes. While most bacterial and eukaryotic lipids have ester linkages joining the fatty acids to glycerol, some archaeal lipids have ether linkages. The long-chain hydrocarbons in archaeal lipids are branched; one class has glycerol at both ends. These lipids form a lipid monolayer that is compatible with typical phospholipids because of the total length. Crenarchaeotes are generally thermophilic and/or acidophilic. Euryarchaeotes include methanogens that produce methane, extreme halophiles that tolerate extremely salty environments, and other varieties of extremophile. The other groups of archaea are poorly understood, and are relatively small groups in terms of species discovered to date.

**Question 5.** Complete the table below.

| Domain | Group Name | Key Features | Example Genera or Associated Diseases |
|---|---|---|---|
| | Chlamydias | | |
| | Crenarchaeota | | |
| | Cyanobacteria | | |
| | Euyarchaeota | | |
| | High-GC Gram-positives | | |
| | Korarchaeota | | |
| | Low-GC Gram-positives | | |
| | Nanoarchaeota | | |
| | Proteobacteria | | |
| | Spirochetes | | |

**Question 6.** You discover a new prokaryotic organism. What sorts of traits will you look for in order to determine which bacterial or archaeal group this organism belongs to? (Assume you cannot sequence the genome but must use other criteria for your initial classification.)

### 19.3 Ecological Communities Depend on Prokaryotes

Many prokaryotes form complex communities

Microbiomes are critical to the health of many eukaryotes

A small minority of bacteria are pathogens

Prokaryotes have amazingly diverse metabolic pathways

Prokaryotes play important roles in element cycling

Prokaryotes as a group perform many different roles in the larger ecosystem. Some are able to form biofilms, aggregations of cells that bind to a solid surface and secrete a sticky matrix that protects them from antibiotics and other agents that might harm them. Biofilms are the subject of much current research since many biofilm-forming bacteria are human pathogens or cause corrosion in industrial equipment. Fossil stromatolites are the remnants of biofilms. Another aspect of bacterial biology currently under study is quorum sensing, communication among large numbers of bacteria using chemical signals.

Current research is uncovering that the microbiome—the communities of bacteria and archaea that live on and in the body—is of great importance in human health. Recent counts estimate more than 10,000 species living on humans. These bacteria play important roles in digestion, the immune system, and general regulation of gene expression. Breastfeeding is important in the establishment of a healthy microbiome in an infant. Other animals have their own microbiomes; for instance, cattle have bacteria in their rumen that allow them to digest cellulose from their plant diet.

A limited number of prokaryotes are pathogens and thus are important in human health, agriculture, and animal husbandry. The work of Robert Koch was key in developing a method of determining whether or not a prokaryotic organism is responsible for a particular disease. Even in modern times, when it is well known that microbes can cause disease, Koch's postulates help researchers determine if a specific organism is the cause of a particular disease. For example, Koch's postulates were used to demonstrate that *Helicobacter pylori* causes stomach ulcers. The consequences of infection are based on several aspects of a bacterial species' biology, including its invasiveness and toxigenicity. The soluble protein products secreted by replicating cells, called exotoxins, are often fatal to the host. Endotoxins, which are lipopolysaccharides that are released when cells grow or lyse, are less dangerous to humans.

The long evolutionary history of bacteria and archaea has led to the wide diversity of their metabolic "lifestyles"—their use or nonuse of oxygen, their energy sources, their sources of carbon atoms, and the materials they release as waste products. Obligate anaerobes cannot live in an environment with oxygen gas. Facultative anaerobes can alternate between aerobic and anaerobic metabolism as conditions require. Aerotolerant anaerobes do not perform aerobic respiration, but are not poisoned by the presence of oxygen. Other bacteria are obligate aerobes, which require oxygen for cellular respiration. There are four broad nutritional categories, and there are bacterial or archaeal examples of all four. Photoautotrophs perform photosynthesis, and use light for their energy source and carbon dioxide for their carbon source. Photoheterotrophs use light for energy but obtain their carbon from organic compounds (e.g., fatty acids or alcohols). Chemoautotrophs obtain energy from inorganic compounds; they can use many different types of carbon sources, including carbon dioxide. Finally, chemoheterotrophs obtain energy and carbon from organic sources; animals are also chemoheterotrophs. Organisms that obtain energy from oxidizing inorganic molecules (as opposed to organic molecules from other organisms) are known as lithotrophs.

Some prokaryotes serve vital roles in the cycling of elements in the biosphere. Denitrifiers release nitrogen gas into the air and keep the nitrogen cycling through the environment. Nitrogen fixers convert nitrogen gas into ammonia, and prokaryotic nitrifiers oxidize this ammonia to nitrites and then nitrates. The prokaryotes performing these reactions can convert energy into ATP by performing these reactions.

**Question 7.** Differentiate between the following terms: obligate anaerobe versus obligate aerobe; photoautotroph versus photoheterotroph; chemolithotroph versus chemoheterotroph.

**Question 8.** A friend with young children tells you that she intends to protect their health by eradicating all bacteria from her home and their immediate environment. What do you tell her?

**Question 9.** Biologists from the Centers for Disease Control in Atlanta have been called to a remote area of Uganda to study a mysterious disease that is causing respiratory ailments in a small village. They isolate a bacterium from several patients that seems to be the likely cause of the illness. How can they determine if this is indeed the pathogen?

**Question 10.** A U.S. Department of Agriculture field representative counsels an inexperienced farmer to plant alfalfa in fields with soils that have low nitrogen levels because alfalfa roots are hosts for nitrogen-fixing bacteria. How will the alfalfa and its associated bacteria help "fertilize" this farmer's soil?

**Question 11.** A patient with a bacterial infection has caused a great deal of concern to his physician because the bacteria responsible for the infection produce an exotoxin. Why are exotoxins often more dangerous than endotoxins?

### 19.4 Viruses Have Evolved Many Times

Many RNA viruses probably represent escaped genomic components

Some DNA viruses may have evolved from reduced cellular organisms

Viruses can be used to fight bacterial infections

Viruses lack a cellular structure and are not considered living organisms by many scientists. (Note that they are not even named according the binomial nomenclature system used for the three domains of life.) Even so, they have some traits of living things, including a genetic code and the ability to reproduce. Viruses are classified on the basis of their genomes, since their evolutionary history is complex, and viruses most likely arose multiple times from multiple branches of the tree of life. RNA viruses can be positive- or negative-sense single-stranded, or double-stranded. The single-stranded retroviruses contain an enzyme that allows them to make DNA copies of the RNA genome and insert them into the host genome. DNA viruses are double-stranded.

**Question 12.** Why would a physician refuse to prescribe antibiotics for a cold?

**Question 13.** Why aren't the groups of viruses based on true phylogenetic classification?

## Test Yourself

1. Which of the following is a characteristic that is unique to prokaryotes?
   a. Lack of membrane-enclosed organelles
   b. Presence of cell walls
   c. Presence of cell membranes
   d. Presence of a cytoskeleton
   e. Absence of ribosomes
   ***Textbook Reference:*** *Concept 19.1 Life Consists of Three Domains That Share a Common Ancestor*

2. One of the major diagnostic characteristics that distinguishes archaea from bacteria is the absence of
   a. peptidoglycan cell walls in archaea.
   b. peptidoglycan cell walls in bacteria.
   c. ribosomes in archaea.
   d. chemoautotrophy in bacteria.
   e. heterotrophy in archaea.
   ***Textbook Reference:*** *Concept 19.2 Prokaryote Diversity Reflects the Ancient Origins of Life; Gene sequencing enabled biologists to differentiate the domain Archaea*

3. Which statement concerning prokaryotes is true?
   a. Because prokaryotes do not contain organelles, they cannot photosynthesize or carry out cellular respiration.
   b. Prokaryotes have no chromosomes and therefore lack DNA.
   c. No prokaryote performs glycolysis.
   d. Prokaryotes undergo mitosis.
   e. Two of the three domains of life are prokaryotic.
   ***Textbook Reference:*** *Concept 19.1 Life Consists of Three Domains That Share a Common Ancestor*

4. Gram-negative bacteria stain pink because
   a. they have specialized lipids in their cell walls.
   b. their peptidoglycan layer is thin.
   c. their peptidoglycan layer is thick.
   d. they are receptive to antibiotics.
   e. their cell walls are composed largely of proteins.
   ***Textbook Reference:*** *Concept 19.1 Life Consists of Three Domains That Share a Common Ancestor; The small size of prokaryotes has hindered our study of their evolutionary relationships*

5. Archaea are more closely related to _______ than to any other monophyletic group.
   a. bacteria
   b. eukaryotes
   c. bacteria and eukaryotes
   d. fungi
   e. protists
   ***Textbook Reference:*** *Concept 19.1 Life Consists of Three Domains That Share a Common Ancestor; The two prokaryotic domains differ in significant ways*

6. The dense films laid down by many prokaryotes are
   a. endotoxins.
   b. denitrifiers.
   c. biofilms.
   d. pathogens.
   e. endospores.
   ***Textbook Reference:*** *Concept 19.3 Ecological Communities Depend on Prokaryotes; Many prokaryotes form complex communities*

7. Which group of bacteria most likely has the highest proportion of pathogens?
   a. Proteobacteria
   b. Cyanobacteria
   c. Chlamydias
   d. High-GC Gram-positives
   e. Low-GC Gram-positives
   ***Textbook Reference:*** *Concept 19.2 Prokaryote Diversity Reflects the Ancient Origins of Life; Chlamydias are extremely small parasites*

8. The mitochondria of eukaryotes were derived by endosymbiosis from
   a. proteobacteria.
   b. chemoheterotrophs.
   c. eukaryotes.
   d. archaea.
   e. viruses.
   ***Textbook Reference:*** *Concept 19.2 Prokaryote Diversity Reflects the Ancient Origins of Life; The proteobacteria are a large and diverse group*

9. In many states, autoclaves, which sterilize medical and laboratory equipment by means of pressurized heat, must pass a "spore test" to demonstrate that they work correctly. The spore-producing bacteria used for this test are most likely taken from which group?
   a. Low-GC Gram-positive bacteria
   b. Proteobacteria
   c. Cyanobacteria

d. Chlamydias
e. High-GC Gram-positive bacteria
***Textbook Reference:*** *Concept 19.2 Prokaryote Diversity Reflects the Ancient Origins of Life; The low-GC Gram-positives include the smallest cellular organisms*

10. Archaea that live in extremely salty conditions are referred to as
a. thermophiles.
b. halophiles.
c. salinophiles.
d. brinophiles.
e. sodiophiles.
***Textbook Reference:*** *Concept 19.2 Prokaryote Diversity Reflects the Ancient Origins of Life; Euryarchaeotes are found in surprising places*

11. Methane gas contributes to the greenhouse effect that is raising atmospheric temperatures. A large portion of all methane emission is from grazing cattle because cows harbor methane-producing archaea from the _______ group.
a. Crenarchaeota
b. Euryarchaeota
c. Anarchaeota
d. Proteobacteria
e. Nanoarchaeota
***Textbook Reference:*** *Concept 19.2 Prokaryote Diversity Reflects the Ancient Origins of Life; Euryarchaeotes are found in surprising places*

12. Rod-shaped bacteria are referred to as
a. bacilli.
b. cocci.
c. spiral.
d. helici.
e. roddi.
***Textbook Reference:*** *Concept 19.1 Life Consists of Three Domains That Share a Common Ancestor; The small size of prokaryotes has hindered our study of their evolutionary relationships*

13. A bacterium that requires a carbon source other than carbon dioxide, yet can convert light energy to chemical energy, is called a
a. photoautotroph.
b. photoheterotroph.
c. chemoautotroph.
d. chemoheterotroph.
e. chemolithotroph.
***Textbook Reference:*** *Concept 19.3 Ecological Communities Depend on Prokaryotes; Prokaryotes have amazingly diverse metabolic pathways*

14. A bacterium that cannot live in the presence of oxygen is called a(n)
a. obligate aerobe.
b. facultative aerobe.
c. obligate anaerobe.
d. facultative anaerobe.
e. aerotolerant anaerobe.
***Textbook Reference:*** *Concept 19.3 Ecological Communities Depend on Prokaryotes; Prokaryotes have amazingly diverse metabolic pathways*

15. Which virus type inserts a double-stranded DNA copy of its genome into the host cell's genome?
a. Double-stranded RNA viruses
b. Double-stranded DNA viruses
c. Retroviruses
d. Negative-sense single-stranded RNA viruses
e. Positive-sense single-stranded RNA viruses
***Textbook Reference:*** *Concept 19.4 Viruses Have Evolved Many Times; Many RNA viruses probably represent escaped genomic components*

16. Which of the following statements is *not* one of Koch's postulates?
a. The introduction of the disease-causing microorganism to a new, healthy host causes the same disease that existed in the original host.
b. The disease-causing microorganism is always found in the person with the disease.
c. After the disease-causing microorganism has been introduced into a healthy host who then gets the disease, the same microorganism can be isolated from this individual.
d. The disease-causing microorganism is susceptible to antibiotic treatment.
e. The disease-causing microorganism can be isolated from an infected individual and grown in culture.
***Textbook Reference:*** *Concept 19.3 Ecological Communities Depend on Prokaryotes; A small minority of bacteria are pathogens*

## Answers

### *Key Concept Review*

1. The most immediately apparent difference between a prokaryote and a eukaryote is the much smaller size and absence of membrane-enclosed organelles in the prokaryote. With proper staining techniques, the nuclei of any eukaryote would likely be visible with the light microscope.
2. Gram staining results depend on the arrangement and amounts of peptidoglycans in the bacterial cell wall. This technique does not work with archaea because they do not have peptidoglycan in their cell walls.
3. 
a. Spirilla
b. Bacilli
c. Cocci

4. positive
   a. Outside of cell
   b. Cell wall (peptidoglycan)
   c. Cell membrane
   d. Inside of cell

   negative
   e. Outside of cell
   f. Outer membrane of cell envelope
   g. Periplasmic space
   h. Peptidoglycan layer
   i. Periplasmic space
   j. Cell membrane
   k. Inside of cell

5.

| Domain | Group Name | Key Features | Example Genera or Associated Diseases |
|---|---|---|---|
| Bacteria | Chlamydias | Gram-negative cocci. Very small, obligate parasites. Life cycle involves two forms of cells called elementary bodies and reticulate bodies. | Cause human eye disease, the STD chlamydia, and some types of pneumonia |
| Archaea | Crenarchaeota | Most live in hot and/or acidic environments such as sulfur hot springs. Internal cellular environment is close to pH 7. | *Sulfolobus* |
| Bacteria | Cyanobacteria | Photoautotrophs. Use chlorophyll *a* for photosynthesis. May form colonies with differentiated cells (see Figure 19.9) . Some also fix nitrogen. Thought to be the source of eukaryotic chloroplasts. | *Anabaena* (see Figure 19.9A) |
| Archaea | Euyarchaeota | Many are methanogens living in the guts of ruminant herbivores or termites and cockroaches. Others are extreme halophiles. *Thermoplasma* lacks a cell wall and lives in coal deposits. | *Methanopyrus thermoplasma* |
| Bacteria | High-GC Gram-positives | Gram-positive, GC-rich genomes, may form spores at tips of filaments. Source of most of our antibiotics. | *Mycobacterium tuberculosis*, the cause of tuberculosis and *Streptomyces*, the source of streptomycin. |
| Archaea | Korarchaeota | Poorly characterized, only DNA has been isolated directly from hot springs. | None available |
| Bacteria | Low-GC Gram-positives | Gram-positive, although some are Gram negative and lack a cell wall. Genome is not GC-rich. May produce endospores capable of surviving extreme conditions. Mycoplasmas are the smallest cellular organisms known. | *Bacillus anthracis:* causes anthrax, *Clostridium:* a cause of food poisoning, *Staphylococcus aureus*: can colonize healthy individuals or cause infection. |
| Archaea | Nanoarchaeota | Poorly characterized. Lives attached to cells of *Ignicoccus,* a crenarchaeote. Discovered at a deep-sea thermal vent near Iceland's coast. | *Nanoarchaeum equitans* |
| Bacteria | Proteobacteria | Largest group of identified species. Very diverse group classified into five groups based on their metabolic pathways. | *Rhizobium:* important for nitrogen fixation, *Yersinia pestis:* cause of bubonic plague, *Vibrio cholerae:* cause of cholera, *Salmonella typhinurium*: cause of gastrointestinal disease, and *Escherichia coli:* well characterized. |
| Bacteria | Spirochetes | Gram-negative, helical structure (See Figure 19.3) hemoheterotrophic. Move via axial filaments. | Some species cause syphilis and Lyme disease. |

6. A Gram stain would allow you to determine if the prokaryote is a Gram-positive bacteria or archaea or Gram-negative bacteria. You could also use information about the environment where the new organism was found. Some bacteria and archaea are known to live in extreme environments, while other bacteria can form endospores, fix nitrogen, or have unique structures, like axial filaments. Using the environmental clues, information from Gram staining, and the shape and motility of the cells, you would probably be able to predict to which group this new organism belongs.
7. Obligate anaerobes die in the presence of oxygen gas, whereas obligate aerobes die without oxygen gas for

cellular respiration. Photoautotrophs convert light energy to chemical energy using carbon dioxide as their carbon source; photoheterotrophs rely on an outside carbon source other than carbon dioxide but still can convert light energy to chemical energy. Chemolithotrophs can produce what they need from inorganic molecules, whereas chemoheterotrophs depend on carbon-containing molecules from other organisms.

8. This effort to eradicate all bacteria will not be successful, since humans are hosts to innumerable bacteria both on and in our bodies. Furthermore, most bacteria are not harmful. In fact, many are beneficial. For example, bacteria are essential for nutrient cycling in the ecosystem. In some cases, the attempt to eradicate bacteria has created dangers of its own. The abuse and overuse of antibiotics, for instance, creates selective pressures that lead to an increase in antibiotic-resistant pathogenic bacteria, rendering previously effective treatments useless. Your friend would be better served by declaring a truce with the bacteria and recognizing them as part of the world we inhabit.
9. Koch's postulates stipulate that in order for an organism to be identified as a disease-causing agent: (1) it must always be found in individuals with the disease; (2) it must be taken from the host and grown in pure culture; (3) a sample of the culture must produce disease if injected into a healthy individual; and (4) the newly infected individual must yield a new pure culture of the same organism. That said, it is unethical to knowingly infect a person with a pathogen just for the purposes of identification! Therefore, a diagnosis can be made based on the first two postulates but not on the second two.
10. Nitrogen-fixing bacteria can convert atmospheric nitrogen into ammonia. Other bacteria can then convert the ammonia into nitrates that plants can use. This topic is covered in greater detail in Chapter 25.
11. Exotoxins are produced and released continuously by the infecting bacteria, whereas endotoxins are released only at cell death. Only a limited number of endotoxin molecules are released, but exotoxins can be released until the host dies.
12. Antibiotics such as penicillin and ampicillin interfere with the synthesis of the peptidoglycan-based cell walls of bacteria. Viruses do not have a cellular structure, and they use the host cell's machinery to replicate and spread. Since the cold virus is replicating in a person's cells, an antibiotic would have no effect on a cold and would indeed kill many of the "good" bacteria living in the body.
13. Viruses are thought to have evolved several times within each group of living organisms. Additionally, the small sizes of viral genomes make sequence-based phylogenetic comparisons difficult. Viral groups are instead based on functional classification of the genome structure.

### *Test Yourself*

1. **a.** Prokaryotes are differentiated from eukaryotes by their lack of membrane-enclosed organelles. They also have a single circular piece of genomic DNA and lack a cytoskeleton.
2. **a.** Archaea lack peptidoglycans. Instead they have a unique lipid in their cell walls.
3. **e.** The domains Bacteria and Archaea both contain prokaryote members. Bacteria do carry out both photosynthesis and multiple forms of cellular respiration. All life forms have DNA. There are bacteria that perform glycolysis. Bacteria replicate by binary fission.
4. **b.** Gram-negative bacteria have thin peptidoglycan layers that do not retain crystal violet stain.
5. **b.** The Archaea and the Eukarya share a more recent common ancestor with each other than either does with the Bacteria. One piece of evidence for this is that a signature sequence from rRNA has been found in all archaea and eukaryotes tested so far, but in none of the bacteria.
6. **c.** Biofilms are gel-like polysaccharide matrices that are laid down by prokaryotes and trap other bacteria.
7. **c.** Chlamydias are all parasitic and cause several human diseases.
8. **a.** Endosymbiosis of proteobacteria gave rise to the mitochondria of eukaryotes.
9. **a.** Among the groups listed, only low-GC Gram-positive bacteria, high-GC Gram-positive bacteria, and cyanobacteria produce spores. Low-GC Gram-positive spores can withstand extreme environmental condition, so they would be useful for testing an autoclave's ability to sterilize equipment.
10. **b.** The term "halophile" means "salt loving."
11. **b.** Methanogenic bacteria that reside in the guts of cows belong to the Euryarchaeota.
12. **a.** A rod-shaped bacterium is known as a bacillus (plural, bacilli). Cocci are roughly spherical, and the spirilla are corkscrew shaped.
13. **b.** Photoheterotrophs require a carbon source other than carbon dioxide, yet are able to harvest light energy.
14. **c.** Oxygen gas is toxic to obligate anaerobes.
15. **c.** Retroviruses, such as HIV, insert a form of their genome into the host's DNA. This provirus is replicated every time the cell divides by mitosis.
16. **d.** Koch's postulates were developed in the 1880s, a time when even the role of microorganisms in disease was not understood. Antibiotic therapies to target these microorganisms came later.

# 20 The Origin and Diversification of Eukaryotes

## The Big Picture

- The many lineages of protists, as well as all fungi, animals, and plants, are eukaryotic. Eukaryotic cells are characterized by membrane-enclosed organelles, the presence of a cytoskeleton, and a nuclear envelope. The evolution of the eukaryotic cell, which acquired features of both archaea and bacteria, was a major evolutionary milestone that occurred in the Precambrian. The eukaryotic cell allows for compartmentalization of processes and increased adaptability. Though the evolutionary lineage is still debated, it is thought that eukaryotes evolved via multiple symbiotic events.
- Protists as a group are not monophyletic. The most recent common ancestor of protists also gave rise to the plants, animals, and fungi, groups that are not included in the protists. Protists are diverse, ranging from microscopic single-celled organisms to complex multicellular organisms. Several subgroups of protists are covered in this chapter, some of which are monophyletic and some that are not. The five major eukaryotic clades designated protists are: alveolates, stramenophiles, rhizarians, excavates, and amoebozoans.
- Certain body plans and modes of nutritional acquisition are found repeatedly throughout the microbial eukaryote groups and are good examples of form following function. Protists as a group play a vast role in the environment and medicine, are economically valuable, and include some of the most challenging disease organisms.

## Study Strategies

- It is easy to become overwhelmed by organism names and characteristics. Focus on the trends outlined in the chapter and use lab opportunities to understand the organism groupings.
- This chapter contains a great deal of information. It is best not to try to learn it all at once, but to break it up into smaller pieces and study it during several sessions.
- To learn the organism groups, create a chart with key characteristics. This will help you compare and contrast the groups.
- On one side of a 4 × 6 index card, write down a particular clade or subgroup; on the other side, note the defining characteristics. Do the same with sample organisms, with the genus specified on one side of the card and the clade and subgroup on the other side. Then shuffle the deck and organize them into different categories. For example, you can sort all of the subgroups into their proper clades. You can match sample organisms with their subgroups or clades. Or you can sort the groups according to the presence or absence of morphological characteristics, life cycle strategies, or other criteria of your choosing.
- Go to **LaunchPad** (or use the Web addresses listed) to review the following additional resources:

  Animated Tutorial 20.1 Family Tree of Chloroplasts (PoL2e.com/at20.1)

  Animated Tutorial 20.2 Digestive Vacuoles (PoL2e.com/at20.2)

  Animated Tutorial 20.3 Life Cycle of the Malarial Parasite (PoL2e.com/at20.3)

  Activity 20.1 Anatomy of a *Paramecium* (PoL2e.com/ac20.1)

  Analyze the Data in textbook Figure 20.19

  Apply the Concept in textbook Sections 20.1 and 20.4

## Key Concept Review

### 20.1 Eukaryotes Acquired Features from Both Archaea and Bacteria

#### The modern eukaryotic cell arose in several steps

#### Chloroplasts have been transferred among eukaryotes several times

Protists are a paraphyletic group of very diverse organisms. The term "protist" lacks any real taxonomic meaning; it is shorthand for all eukaryotes that are not land plants, fungi, or animals. Because of the "catch-all" nature of this grouping, it is extremely diverse, containing many different organisms that are not necessarily closely related to one another.

The origin of the eukaryotic cell is still being debated, but evidence suggests that eukaryotes are monophyletic. The evolution of the eukaryotic cell occurred at a time of great environmental change. Important events include: (1) the origin of a flexible cell surface; (2) the origin of a cytoskeleton; (3) the origin of a nuclear envelope; (4) the appearance of digestive vesicles; and (5) endosymbioses that led to the development of organelles.

The first step in the path toward a eukaryotic cell may have involved loss of the rigid cell wall, which allowed cells to become larger and also allowed infoldings of the cell surfaces. Such infoldings increased the surface area-to-volume ratio and also may have been the origin of the nuclear envelope, due to folding around DNA that is anchored to the cell membrane. This also permitted vesicles to form from pinched-off bits of infolded membrane.

New structures then evolved; ribosomes became associated with internal membranes to form the endoplasmic reticulum, a complex cytoskeleton of actin and microtubules formed, and digestive vesicles developed. Although it was previously thought that only eukaryotes have a cytoskeleton, research has now revealed the presence of homologous proteins in prokaryotes. The microtubules of the cytoskeleton also formed a flagellum. The next step was probably the ability to perform phagocytosis; subsequent endosymbiotic events may have led to the evolution of mitochondria and chloroplasts (see Figure 20.1).

Some chloroplasts are enclosed in double membranes, and others are enclosed in triple membranes. This phenomenon can be explained by endosymbiosis. All chloroplasts can be traced to the engulfment of an ancestral cyanobacterium, called primary endosymbiosis. The cyanobacterium was Gram-negative and thus had two membranes, leading to the presence of two membranes in the original chloroplasts. Primary endosymbiosis gave rise to the green and red algae, and to land plants. Photosynthetic euglenoids, with their triple membranes, arose from secondary endosymbiosis (see Figure 20.2). In this case, a unicellular green alga was ingested with its housed chloroplast. Ultimately, all of the constituents of the alga except the chloroplast were lost, resulting in three membranes. Tertiary endosymbiosis occurred when a dinoflagellate lost its chloroplast and took up a protist that had acquired its chloroplast through secondary endosymbiosis.

**Question 1.** Explain one line of thought regarding the origin of the eukaryotic cell.

**Question 2.** Describe the origin of double-membrane-enclosed and triple-membrane-enclosed chloroplasts.

## 20.2 Major Lineages of Eukaryotes Diversified in the Precambrian

Alveolates have sacs under their cell membranes

Stramenopiles typically have two unequal flagella, one with hairs

Rhizarians typically have long, thin pseudopods

Excavates began to diversify about 1.5 billion years ago

Amoebozoans use lobe-shaped pseudopods for locomotion

Most eukaryotes can be classified in one of eight major clades, and five of these clades are collectively called "protists." The five protist clades are: alveolates, stramenopiles, rhizarians, excavates, and amoebozoans. These clades consist of extremely diverse organisms in terms of mobility, nutritional lifestyle, and number of cells. Most protists are microscopic but a few are very large (such as giant kelps). Unicellular protists are called microbial eukaryotes. Multicellularity has arisen many times during eukaryotic evolutionary history. Many familiar eukaryotes are multicellular: plants, animals, fungi, and brown algae. Other eukaryotic species retain individual cell identity but associate in multicellular colonies. There is, however, a continuum between true unicellularity and multicellularity. Newer experimental techniques, such as DNA sequencing, have allowed biologists to find new patterns of evolutionary relatedness among protists. Many areas of uncertainty in the phylogeny of eukaryotes still remain. Lateral gene transfer among species complicates these relationships.

The synapomorphy that defines the alveolates is the possession of cavities called alveoli just below their cell surfaces. All alveolates are unicellular and most are photosynthetic. The three groups considered here are the dinoflagellates, apicomplexans, and ciliates. The majority of dinoflagellates are photosynthetic marine organisms with two flagella. They are golden brown in color due to photosynthetic and accessory pigments and are important primary producers in marine environments. Some cause the harmful red tides that can poison fish and other marine species. Many live in symbiosis with other organisms—for example, with corals. Some dinoflagellates are parasitic. Dinoflagellates have a distinct appearance due to the special arrangement of their two flagella: one in an equatorial groove around the cell and another that passes through a longitudinal groove before extending beyond the cell into the surroundings. Some dinoflagellates can change form, including amoeboids, depending on environmental conditions. *Pfisteria piscicida* has been said to exist in at least two dozen distinct forms; this dinoflagellate species is harmful to fish. Apicomplexans are parasitic protists. Their name derives from the cluster of organelles at the apex of their cells, which assists with invasion of their host organism. The life cycles of these parasites are complex and often involve infection of multiple hosts to complete the life cycle. Apicomplexans in the genus *Plasmodium* cause malaria. Ciliates move via cilia and are distinguished by their possession of two types of nuclei. Almost all ciliates are heterotrophic. *Paramecium* is a common and well-studied ciliate genus. *Paramecium* moves by means of cilia that are well coordinated to provide precise control of movement. Its membranes are protected by a pellicle, a structure composed of an outer membrane and an inner layer of closely packed membrane-enclosed sacs. It also protects itself with trichocysts present in the pellicle, which act as sharp darts (see Figure 20.6). A specialized organelle, the contractile vacuole, excretes excess water that flows in due to its hypotonic environment. *Paramecium* and many other protists engulf their food via endocytosis, forming a digestive vacuole to break down the food particles. Smaller vesicles pinch away and provide a large surface area to allow the digestive contents to be absorbed.

Stramenopiles are characterized by their rows of tubular hairs on the longer of their two flagella. Those that are not flagellated have lost their flagella over the course of evolution. The stramenopile groups are the photosynthetic diatoms and brown algae, and the nonphotosynthetic oomycetes. Diatoms are yellowish brown and store carbohydrates and oils as photosynthetic products. They are most noted for their silica-containing cell walls, which have two halves that fit together like a petri plate (see Figure 20.7). All diatoms are symmetrical and unicellular. They reproduce both asexually and sexually, but since each asexual reproduction event results in size reduction, occasional sexual reproduction is required. Diatoms are major photosynthetic producers. Brown algae are multicellular and some are extremely large, such as giant kelps. Brown algae are almost exclusively marine. They produce branched filaments (Figure 20.8A) or leaflike growths (Figure 20.8B). Brown algae are brown due to the presence of chlorophylls *a* and *c* and the carotenoid fucoxanthin. They have specialized regions called holdfasts, which anchor them to a substrate. Brown algae are commercially important as a source of alginic acid, which is used as a binder in many food and cosmetic products. Oomycetes are a nonphotosynthetic group of stramenopiles. These are commonly known as water molds and downy mildews, but their resemblance to fungi is only superficial. Unlike the cell walls of fungi, which contain chitin, the cell walls of oomycetes typically contain cellulose. Water molds are absorptive heterotrophs; they secrete enzymes that digest food molecules which are then absorbed. All are aquatic and saprobic (feed on dead organic matter). Terrestrial oomycetes include the downy mildews, some of which are agriculturally significant plant parasites.

The rhizarians are unicellular and mostly aquatic; the three primary groups are the cercozoans, foraminiferans, and radiolarians. Cercozoans are very diverse, with many forms, and occur in aquatic or soil habitats; one group contains chloroplasts derived from secondary endosymbiosis. Limestone deposits are often the result of discarded foraminiferan shells made of calcium carbonate. Some foraminiferans live as plankton, while others dwell on the sea floor and have been found at the deepest parts of the oceans. Their threadlike pseudopods are used to trap food. Radiolarians have stiff, microtubule-reinforced pseudopods that help the cells float in their marine environments and provide additional surface area. Their glassy endoskeletons sometimes have elaborate geometric designs and are as varied as snowflakes. They include some of the largest unicellular eukaryotes, up to several millimeters across.

The excavates include diverse groups that split from one another soon after the eukaryotes originated. Several of these groups lack mitochondria, although there is evidence that this condition evolved subsequent to diversification. The continuing existence of excavates demonstrates that eukaryotic life is possible without mitochondria. Five major subgroups are aggregated into three main groupings: the diplomonads and parabasalids; the heteroloboseans; and the euglenids and kinetoplastids. Diplomonads and parabasalids lack mitochondria. *Giardia lamblia* is a well-known diplomonad with two nuclei that causes the human intestinal disorder giardiasis. *Trichomonas vaginalis* is a parabasalid responsible for a sexually transmitted disease in humans. Both groups have multiple flagella, and parabasalids use an undulating membrane in locomotion. Heteroloboseans have a life cycle that alternates between an amoeboid stage and a flagellated stage. One species of *Naegleria* can cause a fatal disease of the nervous system in humans. Euglenids and kinetoplastids together constitute a clade of unicellular excavates with flagella. They reproduce asexually through binary fission. Their mitochondria contain disc-shaped cristae and their flagellae contain a unique crystalline rod. The two euglenid flagella are found at the anterior end and may be used for propulsion and as an anchor. Some euglenids are purely heterotrophic, and some are flexible in nutritional requirements and may switch between autotrophism and heterotrophism. Euglenids have spiraling strips of proteins under the cell surface that control cell shape. Kinetoplastids are parasitic and are characterized by their single large mitochondrion. The mitochondrion is unique in having a kinetoplast housing multiple circular DNA molecules and associated proteins. Many tropical human diseases are caused by kinetoplastids, namely by trypanosomes (see Table 20.1). One of the ways these organisms evade control efforts is by frequently altering their cell surface recognition proteins.

Amoebozoans include the loboseans, the plasmodial slime molds, and the cellular slime molds. Their relationships to other groups of eukaryotes are unclear. Loboseans live as independent single cells and feed by phagocytosis. Most exist as predators, parasites, or scavengers. A few produce shells by gluing sand grains together or by secretion (see Figure 20.15). Plasmodial slime molds form multinucleate masses (see Figure 20.16A). During the vegetative stage, a plasmodial slime mold is a single wall-less mass of cytoplasm with multiple diploid nuclei, and is thus a coenocyte. It oozes as a network of strands called a plasmodium. This phenomenon is called cytoplasmic streaming and is used to move and engulf food. If exposed to adverse environmental conditions, the slime mold will form a dormant hardened mass from which it can turn back into a plasmodium when environmental conditions again become favorable. It may also transform into a fruiting structure and shed spores (see Figure 20.16B). The spores germinate into haploid swarm cells that can divide to form more swarm cells or function as gametes and fuse to create a diploid zygote. Cellular slime molds are made of large numbers of cells called myxamoebas with single haploid nuclei. They reproduce by mitosis and fission, and this life cycle can continue as long as conditions are favorable. Once conditions become unfavorable, myxamoebas aggregate and form a motile slug, or pseudoplasmodium, that eventually produces fruiting bodies. Spores from the fruiting bodies germinate to form more myxamoebas. Sexual reproduction can occur by the fusion of two myxamoebas, which then undergo meiosis to release haploid myxamoebas (see Figure 20.17).

**Question 3.** Why are scientists unsure of the evolutionary relationships among the protists?

**Question 4.** What would be the criteria for placing a newly discovered organism within the protists?

**Question 5.** Many protists are motile. Describe three types of mobility and the function that each type enhances (e.g., food acquisition, mate acquisition, etc.).

**Question 6.** Many excavates lack mitochondria and yet are very successful. How might you explain this for at least some excavates?

**Question 7.** You have found a new organism that has chloroplasts but lives equally well with or without sunlight and appears to perform phagocytosis. It has two flagella and spiraling strips of protein under the cell membrane. Into which group of protists would you place this organism?

## 20.3 Protists Reproduce Sexually and Asexually

### Some protists have reproduction without sex and sex without reproduction

### Some protist life cycles feature alternation of generations

Most protists reproduce both sexually and asexually, although sexual reproduction has not been confirmed for some protists. Modes of asexual reproduction in protists include binary fission, multiple fission, budding, and sporulation. All of these asexual modes of reproduction lead to lines of genetically identical individuals (except for rare mutants that may arise) called clonal lineages. Sexual reproduction in protists is variable: gametes may be the only haploid cell in the life cycle; the zygote may be the only diploid cell in the life cycle; and multicellular haploid and/or diploid generations may occur between sexual reproduction events.

Conjugation is an elaborate and unusual way of exchanging and rearranging genetic material (see Figure 20.18), but does not result in more cells and so is not reproduction. Many multicellular protists (such as brown algae), some fungi, and all land plants undergo alternation of generations, in which a multicellular, diploid organism gives rise to a multicellular, haploid, gamete-producing organism. Two gametes then fuse to form a new diploid organism. The haploid stage, diploid stage, or both can also perform asexual reproduction. Heteromorphic alternation of generations occurs when the two generations differ morphologically, and isomorphic alternation of generations occurs when the two generations are morphologically similar. The gamete-producing generation is haploid and produces gametes by mitosis. The spore-producing generation is diploid and its specialized cells (called sporocytes) each produce four haploid spores by meiotic division. In order to produce a new organism two gametes must fuse; spores can divide to create a multicellular organism.

**Question 8.** Explain how the process of sexual reproduction in *Paramecium*, as diagrammed in the figure below, results in genetic change. At which stages are there opportunities for genetic rearrangement? Are the resulting daughter cells genetically identical? Why or why not?

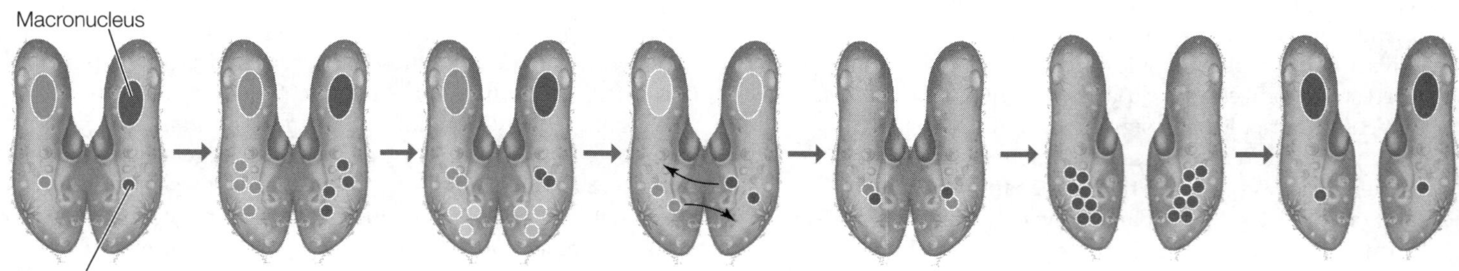


**Question 9.** Differentiate between asexual and sexual reproduction. Describe one mode of microbial eukaryotic sexual reproduction.

**Question 10.** Explain alternation of generations. Why are the protists the first group in which this process could be observed?

### 20.4 Protists Are Critical Components of Many Ecosystems

Phytoplankton are primary producers

Some microbial eukaryotes are deadly

Some microbial eukaryotes are endosymbionts

We rely on the remains of ancient marine protists

Diatoms produce about one-fifth of all fixed carbon on Earth, and other protist members of the phytoplankton are also important producers. These primary producers are the gateway for the sun's energy into living organisms. Many important plant and animal pathogens are also microbial eukaryotes. These include *Plasmodium,* the protist that causes malaria, dinoflagellates that produce neurotoxins that can affect fish and humans, and trypanosomes (see Table 20.1). The *Plasmodium* life cycle includes stages in both mosquitoes and humans.

Protists are also important symbionts with many organisms. Many are endosymbionts with other protists and animals, including corals.

Ancient deposits of protists provide oil, natural gas, diatomaceous earth, and limestone. They also provide important clues to Earth's evolutionary history and past climate.

**Question 11.** Many protists are significant human pathogens. Describe the pathogenic protist that causes malaria, describe its life cycle, and identify the microbial eukaryote group to which it belongs.

**Question 12.** Apicomplexans, the group containing *Plasmodium,* the disease organism responsible for malaria, contain a much reduced, nonphotosynthetic chloroplast. The chapter notes that researchers are targeting this organelle in efforts to develop anti-malarial drugs. How would you explain the presence of chloroplasts in *Plasmodium*?

## Test Yourself

1. *Ciliates,* as represented by *Paramecium,* have defensive organelles in their pellicles called
   a. trichonympha.
   b. tridents.
   c. trichomes.
   d. trichocysts.
   e. trochlea.
   ***Textbook Reference:*** *Concept 20.2 Major Lineages of Eukaryotes Diversified in the Precambrian; Alveolates have sacs under their cell membranes*

2. Which statement about protists is *false*?
   a. Apicomplexans are the only microbial eukaryote group without parasitic representatives.
   b. Foraminiferans and radiolarians are shelled protists.
   c. Ciliates have great control over the direction of their beating cilia.
   d. Although amoebas appear structurally simple, they are not primitive organisms.
   e. All diatoms are unicellular, although a few individual species associate in filaments.
   ***Textbook Reference:*** *Concept 20.2 Major Lineages of Eukaryotes Diversified in the Precambrian; Amoebozoans use lobe-shaped pseudopods for locomotion*

3. Which of the following is *not* one of the adaptations that occurred during the evolution of eukaryotes from prokaryotes?
   a. Infolding of the flexible cell membrane
   b. Loss of the cell wall
   c. A switch from aerobic to anaerobic metabolism
   d. Endosymbiosis of once free-living prokaryotes
   e. Development of a cytoskeleton
   ***Textbook Reference:*** *Concept 20.1 Eukaryotes Acquired Features from Both Archaea and Bacteria; The modern eukaryotic cell arose in several steps*

4. A major difference between the vegetative states of plasmodial and cellular slime molds is that plasmodial slime molds _______, whereas cellular slime molds _______.
   a. have haploid nuclei; have diploid nuclei
   b. produce fruiting bodies; do not produce fruiting bodies
   c. undergo aggregation under adverse conditions; do not undergo aggregation under adverse conditions
   d. exist as a coenocytic mass; exist as individual myxamoebas
   e. grow almost indefinitely in favorable conditions; must aggregate into myxamoebas every seven generations
   ***Textbook Reference:*** *Concept 20.2 Major Lineages of Eukaryotes Diversified in the Precambrian; Amoebozoans use lobe-shaped pseudopods for locomotion*

5. Why is sexual reproduction in diatoms required periodically?
   a. This is the only way they can evade parasitic excavates.
   b. Half of the asexually produced offspring sink to the sea floor to become diatomaceous earth.
   c. During asexual reproduction, the cells decrease in size each generation.
   d. This allows them to move, since only the male gametes have flagella.
   e. Diatoms lose their symmetry after a certain number of generations of asexual reproduction.
   ***Textbook Reference:*** *Concept 20.2 Major Lineages of Eukaryotes Diversified in the Precambrian; Stramenopiles typically have two unequal flagella, one with hairs*

6. Red tides often cause massive fish kills and human illness in those eating shellfish. Which group of protists is responsible for red tides?
   a. Parabasalids
   b. Red algae
   c. Euglenozoans
   d. Dinoflagellates

e. Radiolarians
***Textbook Reference:*** *Concept 20.4 Protists Are Critical Components of Many Ecosystems; Some microbial eukaryotes are deadly*

7. Holdfasts and alginic acid are characteristic of which group of protists?
a. Parabasalids
b. Red algae
c. Brown algae
d. Stramenopiles
e. Unikonts
***Textbook Reference:*** *Concept 20.2 Major Lineages of Eukaryotes Diversified in the Precambrian; Stramenopiles typically have two unequal flagella, one with hairs*

8. Three kinetoplastid trypanosomes cause diseases that result in more than 150,000 deaths annually. Which characteristic of these excavates has made it difficult to kill them and therefore to eradicate these diseases?
a. Their lack of mitochondria
b. Their ability to frequently change cell-surface recognition molecules
c. Their asexual reproduction by binary fission
d. Their nuclear genes, which replace the function of mitochondria
e. Their "guide proteins," which edit mRNA in the mitochondria
***Textbook Reference:*** *Concept 20.2 Major Lineages of Eukaryotes Diversified in the Precambrian; Excavates began to diversify about 1.5 billion years ago*

9. Which of the following statements about coral bleaching is *false*?
a. It results from the loss of a dinoflagellate.
b. It results from the loss of an endosymbiont.
c. It results from food depletion.
d. It results from environmental conditions such as rising water temperatures or increased water turbidity.
e. It results from a parasitic invasion.
***Textbook Reference:*** *Concept 20.4 Protists Are Critical Components of Many Ecosystems; Some microbial eukaryotes are endosymbionts*

10. Which statement regarding protists is true?
a. They are always parasitic.
b. They are all single-celled.
c. They are all heterotrophic.
d. They are always photosynthetic.
e. They are always aquatic for at least some part of their life cycle.
***Textbook Reference:*** *Concept 20.2 Major Lineages of Eukaryotes Diversified in the Precambrian*

11. Which statement about the evolution of eukaryotes is true?
a. The alveolates are more closely related to the stramenopiles than to any other group.
b. The most recent common ancestor of the parabasilids and euglenids also gave rise to the animals.
c. The opisthokonts are polyphyletic.
d. Euglenids and Cercozoans are subgroups of the clade Rhizaria.
e. All brown algae are unicellular.
***Textbook Reference:*** *Concept 20.2 Major Lineages of Eukaryotes Diversified in the Precambrian*

12. Which statement about oomycetes is *false*?
a. They are more distantly related to fungi than are humans.
b. Their cell walls are typically made of chitin.
c. They are absorptive heterotrophs.
d. They can be aquatic or terrestrial.
e. Some are plant parasites.
***Textbook Reference:*** *Concept 20.2 Major Lineages of Eukaryotes Diversified in the Precambrian; Stramenopiles typically have two unequal flagella, one with hairs*

13. Which statement about protists is *false*?
a. They are photosynthetic and are one of the foundations of the marine food web.
b. They are synthesizers of materials that form many sandy beaches
c. They neutralize the sulfuric acid in deep sea vents.
d. They are used as nutritional supplements for humans.
e. They aid in the formation of what we call crude oil.
***Textbook Reference:*** *Concept 20.4 Protists Are Critical Components of Many Ecosystems; We rely on the remains of ancient marine protists*

14. Which statement about the protist cytoskeleton is *false*?
a. It allows for the formation of pseudopods.
b. It is the main structural component of cilia.
c. It is the primary component of the outer shell of diatoms.
d. It is the main structural component of flagella.
e. It helps in removing excess water via the contractile vacuoles.
***Textbook Reference:*** *Concept 20.2 Major Lineages of Eukaryotes Diversified in the Precambrian; Stramenopiles typically have two unequal flagella, one with hairs*

15. Which statement regarding conjugation in *Paramecium* is *false*?
a. It results in genetic recombination.
b. It does not result in clones.
c. It results in offspring.
d. It is a sexual process.
e. It does not result in the production of new cells.
***Textbook Reference:*** *Concept 20.3 Protists Reproduce Sexually and Asexually; Some protists have reproduction without sex and sex without reproduction*

16. Slime molds were once classified with the fungi. Which characteristics favor their placement instead among the amoebozoans?

a. Cytoplasmic streaming, endocytosis, and alternation of generations
b. Cytoplasmic streaming, phagocytosis, and pseudopods
c. Cell walls, sporangia, and pseudopods
d. Creeping locomotion and engulfing of food particles
e. Production of thick-walled spore-bearing structures

***Textbook Reference:*** *Concept 20.2 Major Lineages of Eukaryotes Diversified in the Precambrian; Amoebozoans use lobe-shaped pseudopods for locomotion*

## Answers

### *Key Concept Review*

1. One hypothesis is that Eukarya split from Archaea and that after the split, endosymbioses with bacterial lineages resulted in mitochondria and chloroplasts. Another hypothesis proposes that eukaryotes resulted from the fusion of lineages from Archaea and Bacteria. Regardless of these early relationships, we can infer that the evolution of eukaryotic cells resulted from a number of events. These include the origin of a flexible cell surface, a cytoskeleton, and a nuclear envelope enclosing a chromosomal genome; the development of digestive vacuoles; and the endosymbiotic development of organelles, notably mitochondria and chloroplasts.
2. Double-membrane-enclosed chloroplasts most likely arose from the endosymbiosis of a cyanobacteria in a process called primary endosymbiosis; triple-membrane-enclosed chloroplasts most likely arose from the endosymbiosis of a unicellular alga.
3. The evolutionary relationships are difficult to understand due to limited fossilization and the contributions of symbiotic prokaryotes. Sometimes genes from previously endosymbiotic microbes are found in the genome of the former host organism. Lateral gene transfer complicates the phylogenetic analysis of protist genomes.
4. The organism must be a eukaryote and not fit the criteria for land plants, animals, or fungi.
5. Protists move by flagella, cilia, undulating membranes, pseudopodia, and cytoplasmic streaming.

   Flagella: Many protists possess one or more flagella in at least at one stage of the life cycle. Flagella propel them through the water to locate food and mates or escape enemies. In *Euglena*, flagella may also serve as an anchor to hold the organism in place.

   Cilia: Cilia provide *Paramecium* with a form of locomotion that is generally more precise than locomotion by flagella or pseudopods. A *Paramecium* can coordinate the beating of its cilia to propel itself either forward or backward in a spiraling manner. This movement can help *Paramecium* align for conjugation.

   Undulating membranes: In addition to possessing flagella and a cytoskeleton, the parabasalids have undulating membranes that contribute to the cell's locomotion.

   Pseudopodia: In foraminiferans, long, threadlike, branched pseudopods extend through microscopic apertures in the shell and interconnect to create a sticky net that is used to obtain food. The pseudopods can also provide locomotion. Loboseans use their pseudopods to engulf small organisms and food particles by phagocytosis.

   Cytoplasmic streaming: In *Plasmodium*, the outer cytoplasmic region of a plasmodium becomes more fluid in places, and cytoplasm rushes into those areas, stretching the plasmodium. Microfilaments and a contractile protein interact to produce the streaming movement. As it moves, the plasmodium engulfs food particles by endocytosis.
6. Many excavates, including the trypanosomes and the kinetoplastids, are parasitic and may be able to obtain energy without mitochondria. In addition, mitochondrial genes have been discovered in the nuclear genome, and so the lack of mitochondria may not reflect the lack of energy-generating reactions in the cell.
7. This organism is most likely a euglenid, due to the ability to be autotrophic or heterotrophic, the presence of two flagella, and the presence of spiraling protein strips under the cell surface.
8. Ciliates in the genus *Paramecium* reproduce sexually in a process called conjugation, during which two single-celled organisms exchange genetic material. Although each *Paramecium* has both a macronucleus and many micronuclei, only the micronuclei are exchanged. The macronuclei then incorporate the new genetic material from the micronuclei. Genetic change is likely to occur during crossing over during meiotic divisions of a single micronucleus within each individual and also when the haploid micronuclei from each individual are exchanged and fuse into a novel diploid cell. The resulting daughter cells are not identical because their diploid micronuclei are the result of fusions of different haploid micronuclei.
9. Asexual reproduction results in clones of the original organism, and there is no genetic recombination or variation associated with the creation of offspring. Sexual reproduction allows for new genetic combinations. Protists undergo varied types of sexual reproduction, from fusing haploid myxamoebas in cellular slime molds to alternation of generations in brown algae. Conjugation in *Paramecium* is an example of sexual recombination without reproduction.
10. An organism that exhibits alternation of generations exists in a haploid, gamete-producing form and a diploid, spore-producing form. Prokaryotes have a single chromosome, so only in eukaryotes, which have multiple copies of chromosomes, is a diploid stage of a

life cycle possible. Protists were the first group of eukaryotes to evolve, and so this is where we would first expect to see alternation of generations.

11. The protist that causes malaria is a member of the genus *Plasmodium*. *Anopheles* mosquitoes are the vector for *Plasmodium* and transfer it to the human circulatory system through a bite. The parasites then move to the liver and lymph system and multiply. They then reenter the bloodstream and infect red blood cells, where they multiply again; the red blood cells burst, releasing more parasites. When another *Anopheles* mosquito bites the infected human, it takes in *Plasmodium* cells, which develop into gametes that produce a zygote. The zygotes move into the mosquito salivary glands, ready to be passed to another human host. *Plasmodium* is an apicomplexan protist, a member of the alveolates.
12. The apicomplexans (which are exclusively parasitic) are closely related to the dinoflagellates (which are photosynthetic), so it is not surprising that apicomplexans would have once possessed chloroplasts for photosynthesis. The question is why they would retain the chloroplast after adopting a parasitic lifestyle. One possibility is that the original photosynthetic endosymbiont provided other functions, such as synthesis of fatty acids, lipids, or other cellular molecules.

### *Test Yourself*

1. **d.** Trichocysts are defensive barbs ejected from ciliates when they are disturbed.
2. **a.** Malaria is caused by *Plasmodium*, which is an apicomplexan. In fact, all apicomplexans are parasitic.
3. **c.** At the time the first eukaryotes evolved, the environment was becoming oxygen-rich. There was a switch from anaerobic to aerobic metabolism, not vice versa.
4. **d.** The vegetative (feeding) state of an acellular slime mold is called a plasmodium; it consists of multiple diploid nuclei enclosed in a single membrane. The vegetative state of a cellular slime mold is a myxamoeba with a single haploid nucleus. Although the myxamoebas of cellular slime molds do aggregate to form fruiting structures, individual myxamoebas never fuse into multinucleated structures.
5. **c.** Diatoms have a stiff cell wall and are composed of two halves that fit together like petri dishes. Neither half can grow, and the new cells must fit inside each parent cell half. Thus, the cells reduce in size each generation until sexual reproduction occurs, at which point the cell walls are shed during gamete production and the zygote is able to grow before a new cell wall is laid down.
6. **d.** Dinoflagellates are the cause of red tides.
7. **c.** The brown algae are multicellular protists notable for their organ and tissue differentiation and the presence of alginic acid in their cell walls.
8. **b.** Sleeping sickness, Chagas' disease, and Leishmaniasis are all caused by kinetoplastids, many of which are able to frequently change their cell-surface recognition molecules.
9. **e.** Coral rely on the photosynthetic products of their dinoflagellate endosymbionts. Coral bleaching occurs when the dinoflagellates die or are expelled by the coral cells due to a change in conditions, such as warm temperatures or increased water turbidity. When the corals are bleached, they lose part of their food supply.
10. **e.** Protists do not have a unifying characteristic other than being eukaryotic organisms that do not fit into the kingdoms Plantae, Animalia, or Fungi. If an aquatic environment is considered to be a single droplet of water, however, protists would meet this criterion. For a single-celled organism, a droplet of water is a very large environment.
11. **a.** Alveolates are more closely related to the stramenopiles than to any other group.
12. **b.** Oomycetes, whose cell walls contain cellulose, are distantly related to fungi, whose cell walls contain chitin.
13. **c.** Protists are a very diverse group, but they do not release bases or neutralize environmental acids. Members of Archaea are likely to be found in an acidic environment.
14. **c.** The cytoskeleton, consisting of actin filaments, microtubules, and intermediate filaments, is a feature of eukaryotic cells. In protists, the cytoskeleton is important in anchoring organelles, movement of materials in the cell, movement of the cell in the environment, and controlling the shape of the cell. The outer shell of diatoms is cell wall impregnated with silica.
15. **c.** Conjugation does result in genetic recombination but does not result in the production of clones or offspring.
16. **d.** These are the two characteristics that are common among the amoebozoans, but are not characteristic of fungi.

# 21 The Evolution of Plants

## The Big Picture

- Land plants are photosynthetic eukaryotes that utilize chlorophylls *a* and *b*, undergo alternation of generations, and develop from multicellular embryos that are protected by the parent plant. The ten extant (surviving) groups can be classified as nonvascular plants (those without highly developed vascular tissue) and vascular plants (those with highly developed vascular tissue). This chapter focuses on the nonvascular plants and the nonseed vascular plants.
- In order to colonize land, plants had to evolve strategies for coping with desiccation and gravity. This involved mechanisms for extracting water from soil, means of transporting water throughout the plant, methods of ensuring fertilization, and modes of protecting developing embryos.
- This chapter introduces and describes the liverworts, hornworts, and mosses, all of which are considered nonvascular plants, and the lycophytes, horsetails, and ferns, which are the nonseed vascular plants.

## Study Strategies

- Rather than memorizing land plant types and characteristics, focus on the trends and relationships among them, particularly the evolutionary relationships.
- Plant terminology is probably less familiar than the vocabulary that refers to animals. Focus on the meaning of the word parts. For example, in the word "glaucophyte," the root of the word ("phyte") means *plant* and the prefix ("glauca") means *bluish* white or pale green, so unicellular glaucophytes are so named because of the distinguishing color of the peptidoglycan found between the inner and outer cell membranes. A gametophyte is the gamete-producing plant stage, and a sporophyte is the spore-producing plant stage. Understanding how the words are constructed will help you understand and retain their meanings.
- Focus on relationships among the land plants and how to differentiate one group from another. Many of the important distinctions between groups are related to adaptations for terrestrial life.
- When studying life cycles, be sure to note whether the sporophyte or the gametophyte is dominant and make note of the point at which mitosis and meiosis take place. Make diagrams of life cycles indicating haploid and diploid stages. Keep track of spore formation and the gametophyte in the seed plants.
- Go to **LaunchPad** (or use the Web addresses listed) to review the following additional resources:

  Animated Tutorial 21.1 Life Cycle of a Moss (PoL2e.com/at21.1)

  Animated Tutorial 21.2 Life Cycle of a Conifer (PoL2e.com/at21.2)

  Animated Tutorial 21.3 Life Cycle of an Angiosperm (PoL2e.com/at21.3)

  Activity 21.1 The Fern Life Cycle (PoL2e.com/ac21.1)

  Activity 21.2 Homospory (PoL2e.com/ac21.2)

  Activity 21.3 Heterospory (PoL2e.com/ac21.3)

  Activity 21.4 Life Cycle of a Conifer (PoL2e.com/ac21.4)

  Activity 21.5 Flower Morphology (PoL2e.com/ac21.5)

  Analyze the Data in textbook Figure 21.1

  Apply the Concept in textbook Sections 21.4 and 21.5

## Key Concept Review

### 21.1 Primary Endosymbiosis Produced the First Photosynthetic Eukaryotes

Several distinct clades of algae were among the first photosynthetic eukaryotes

There are ten major groups of land plants

The development of photosynthetic eukaryotes contributed greatly to the development of terrestrial life. Primary endosymbiosis is a shared derived trait of all members of the group Plantae, which includes many aquatic glaucophytes, red algae, and green algae in addition to the terrestrial organisms we commonly refer to as plants. The unicellular glaucophytes are distinguished by the small amount of peptidoglycan found between the inner and outer membranes of their chloroplasts—the same arrangement found in cyanobacteria. In contrast to the glaucophytes, red algae are mostly multicellular and are characterized by the photosynthetic accessory pigment *phycoerythrin*. Red algae also contain chlorophyll *a* and several other accessory pigments. The ratio of phycoerythrin to chlorophyll *a* determines the color of red algae, and depends on light exposure.

Green algae are the other algal members of Plantae. They are distinguished by the possession of chlorophylls *a* and *b* and their storage of photosynthetic products in their chloroplasts as starches. Three important clades of green algae are the chlorophytes, coleochaetophytes, and stoneworts. All the green plants other than chlorophytes are collectively called streptophytes. The stoneworts are thought to be the closest relatives to land plants and exhibit branched apical growth, which is typical in land plants.

All land plants develop from an embryo that is protected by tissues of the parent plant. This is a key shared trait, or synapomorphy, of land plants. For this reason some people refer to land plants as embryophytes. The textbook distinguishes green plants (including green algae) from land plants (embryophytes).

There are ten extant groups of land plants (see Table 21.1). Seven of these clades are the vascular plants, or tracheophtyes; they contain fluid-conducting tracheids and form a single clade. The three remaining clades (liverworts, hornworts, and mosses) do not form a single clade.

**Question 1.** Chloroplast-containing organisms are often referred to as plants, and green plants, streptophytes, and land plants are sometimes referred to collectively as the "plant kingdom." What are the differences among them?

**Question 2.** Differentiate between glaucophytes and red algae.

## 21.2. Key Adaptations Permitted Plants to Colonize Land

Adaptations to life on land distinguish land plants from green algae

Life cycles of land plants feature alternation of generations

Nonvascular land plants live where water is readily available

The sporophytes of nonvascular land plants are dependent on the gametophytes

To survive on land, aquatic organisms had to reduce their dependence on water and develop mechanisms to avoid lethal desiccation (drying). Terrestrial plants also needed support to resist gravity in order to grow upward, and they required some mechanism other than swimming gametes (through water) to move the gametes from one plant to another. The first land plants that met at least some of these challenges were the nonvascular land plants.

Many characteristics that distinguish land plants from green algae are adaptations to a terrestrial environment. The earliest land plants developed the following modifications for terrestrial life: a waxy cuticle, the presence of stomata, gametangia, embryos, protective pigments, thick spore walls, and mutualistic associations with fungi. Waxy cuticles (waxy lipid coatings of leaves and stems) prevent water loss from tissues exposed to dry air. The development of a cuticle was likely one of the earliest and most important adaptations to land.

Land plants undergo alternation of generations (see Figure 21.4). Haploid gametophytes, which grow from haploid spores, produce haploid gametes through mitosis; diploid sporophytes, which arise from the fusion of gametes, produce haploid spores through meiosis. The sporophyte and gametophyte generations differ genetically: the sporophyte has diploid cells while the gametophyte has haploid cells. In plant evolution, the trend is toward reduction of the gametophyte generation. The life cycle of nonvascular land plants is dominated by the gametophyte generation. The sporophyte is very tiny and is completely dependent on the gametophyte. Figure 21.6 illustrates the life cycle of a moss as an example of the nonvascular plant life cycle.

Nonvascular land plants lack true leaves, stems, and roots. Land plants without vascular tissue depend on capillary action to move water through the plant. They lack the support of lignin and hug the ground closely. The cuticle is either lacking or thin and ineffectual in reducing water loss. Mutualisms with fungi probably facilitate water and mineral absorption.

Liverworts have green, leaflike gametophytes which lie flat (see Figure 21.5 A). Liverwort sporophytes remain attached to the larger gametophyte and rarely exceed a few millimeters in length. Liverworts can reproduce asexually as well as sexually.

Mosses are found in almost all types of terrestrial environments and are the most familiar and widespread nonvascular plants. The specialized cells called hydroids are functionally similar to the tracheids seen in vascular plants, but they lack lignin and the cell wall structure of tracheids. Like a tracheid, the hydroid cell dies and leaves a tiny channel that can transport water through the plant. Mosses (see Figure 21.5B) are the sister lineage to vascular plants plus the hornworts and share with those groups the presence of stomata that liverworts lack.

Hornworts are distinguished from other nonvascular plants by two characteristics: the presence of a single platelike chloroplast in each cell, and sporophytes that are capable of indeterminate growth. The sporophyte produces new spore-bearing tissue from a basal region of cell division. In very moist environments, the sporophyte can be as tall as 20 centimeters; growth is limited only by the lack of a true vascular system. The hornwort sporophyte resembles a long, slender horn (see Figure 21.5C). Hornworts often have symbiotic relationships with cyanobacteria (contained in mucilage-filled internal cavities) that are able to fix atmospheric nitrogen and make it available to the hornwort.

In nonvascular land plants, the dominant photosynthetic and nutritionally independent generation is the haploid gametophyte. The diploid sporophyte may be photosynthetic, but it is always nutritionally dependent on the gametophyte and remains attached to it. The sporophyte produces unicellular haploid spores as products of meiotic division that occur within a sporangium. The germinating spore gives rise to a multicellular haploid gametophyte whose cells contain chloroplasts. The gametophyte forms specialized sex organs called gametangia, within which gametes form via mitotic cell division. The archegonium, a multicellular flask-shaped organ with a long neck and a swollen base, produces a single egg. The male sex organ, the antheridium, produces sperm in large numbers. Each sperm bears two flagella and swims to the archegonium and down the canal to fertilize the egg. The eggs release chemical attractants to guide them. Water is critical to all these events. Each gametophyte individual produces both archegonia and antheridia, but the density of gametophytes helps ensure some cross fertilization that maintains genetic variation. Once the sperm nucleus fuses with the egg nucleus, the diploid zygote produces a multicellular embryo that matures into a sporophyte that then repeats the cycle.

**Question 3.** Nonvascular plants faced many problems as they colonized land. Describe three problems and the mechanisms that land plants evolved to surmount those problems.

**Question 4.** Explain why the largest mosses are less than a meter tall.

**Question 5.** Label the structures in the moss life cycle in the diagram below. Indicate the sporophyte generation, gametophyte generation, and diploid or haploid status. Also indicate, in the open boxes, if the process occurring is meiosis, mitosis, or fertilization.

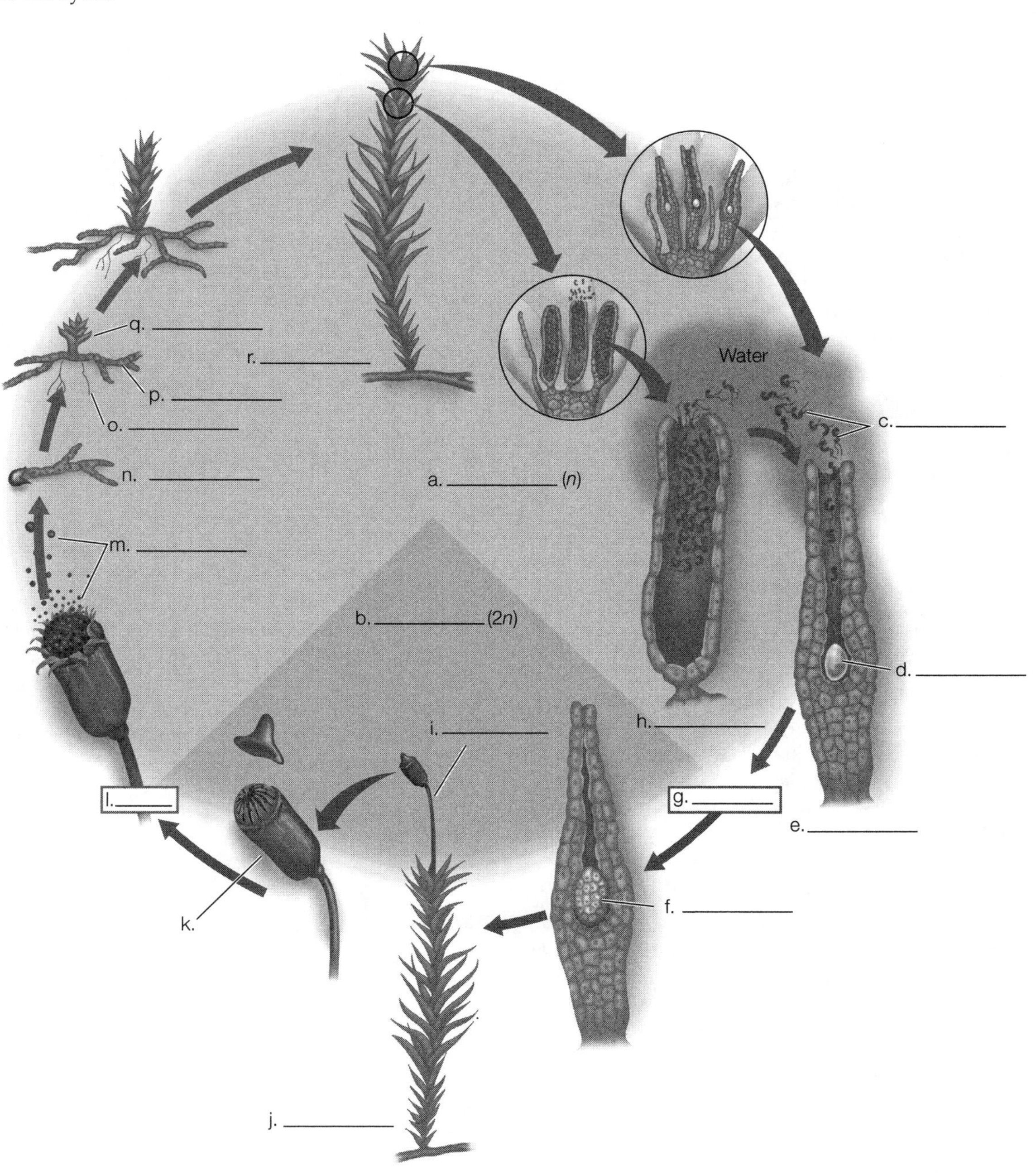

## 21.3 Vascular Tissues Led to Rapid Diversification of Land Plants

Vascular tissues transport water and dissolved materials

The diversification of vascular plants made land more suitable for animals

The earliest vascular plants lacked roots

The lycophytes are sister to the other vascular plants

Horsetails and ferns constitute a clade

The vascular plants branched out

Heterospory appeared among the vascular plants

The development of vascular tissue was key to the further adaptation to a terrestrial existence. Vascular tissue provides for the transport of water and food through a plant. Xylem is a vascular tissue responsible for the transport of water and minerals from the soil to the aerial parts of the plant. Xylem that contains lignin allows for structural support by the vascular system. Vascular plants also contain phloem, which conducts the products of photosynthesis from the location where they are produced to where they are stored.

Tracheids, which evolved about 430 mya, are strawlike cells that are the principal water-conducting elements of the xylem in all vascular plants except the angiosperms and a small group of gymnosperms, although they do persist in those groups in a more specialized form. The cell walls of tracheids provide rigid structural support, which allows vascular plants to grow upward and compete for sunlight. Increased height also increases the distance over which spores can disperse. Vascular plants are also characterized by branching, independent sporophytes that can produce more spores than an unbranched body and develop more complex architecture.

The presence of terrestrial plants modified the environment to make it more hospitable to animals. By the Devonian period, trees appeared, including the giant lycophytes and horsetails that later dominated the landscape and would become the source of coal. During the Permian period the climate dried and cooled and gymnosperms replaced the lycophyte fern forests.

Significant new features that arose in early vascular plants included roots and true leaves. Early vascular plants, like rhyniophytes, did not have roots and were anchored by horizontal portions of stem called rhizomes. Rhizomes contain water-absorbing unicellular filaments called rhizoids. Their branching pattern was dichotomous. Lycophytes (club mosses and their relatives, spike mosses, and quillworts) have dichotomously branching true roots, simple vascular tissue, and simple leaflike structures called microphylls. Growth and branching occurs by apical cell division. Sporangia of many club mosses aggregate in conelike structures called strobili (see Figure 21.8A).

The horsetails and ferns were once thought to be only distantly related but are now placed in their own clade, the monilophytes. As with seed plants, there is differentiation between the main stem and side branches. Horsetails are represented by few extant species. They have true roots, their sporophytes are large and independent, and their gametophytes are highly reduced but also independent. The leaves of horsetails are simple, forming whorls around the stem (see Figure 21.8B).

Ferns are characterized by relatively large, complex leaves with branching vascular strands. Like all other nonseed vascular plants and nonvascular plants, ferns continue to be dependent on water to carry motile sperm. They have advanced vascular structures and can reach great heights, but they do not produce true wood, and their root systems are poorly developed. The ferns consist of more than 12,000 species today. The fern life cycle is dominated by the sporophyte, but the gametophyte is an independent photosynthetic structure (see Figure 21.9). In most species of ferns, the sporangia are found in clusters called sori. The euphyllophyte clade includes the monilophytes and seed plants. One important synapomorphy of euphyllophytes is overtopping—when one branch differentiates and grows higher than the others, giving it an advantage in seeking sunlight. This led to the evolution of a larger, more complex leaf called a megaphyll.

Most early vascular plants exhibit homospory, in which spores (and the gametophytes that grow from them) are all of the same type; gametophytes produce both archegonia (female organs) and antheridia (male organs). Heterospory appears to have evolved several times in the vascular plants (see Figure 21.11). In heterospory, one spore type called a megaspore gives rise to the female, egg-producing megagametophyte; another spore type called a microspore develops into a male, sperm-producing microgametophyte. The sporophyte produces megaspores in megasporangia and microspores in microsporangia. Land plants typically produce many more microspores than megaspores. Heterospory offers more selective advantages to a plant than homospory does.

**Question 6.** Describe the major consequences of the evolution of a branching sporophyte.

**Question 7.** Label the structures in the homosporous fern life cycle in the diagram below. Indicate diploid or haploid status. Also indicate, in the open boxes, if the process occurring is meiosis, mitosis, or fertilization.

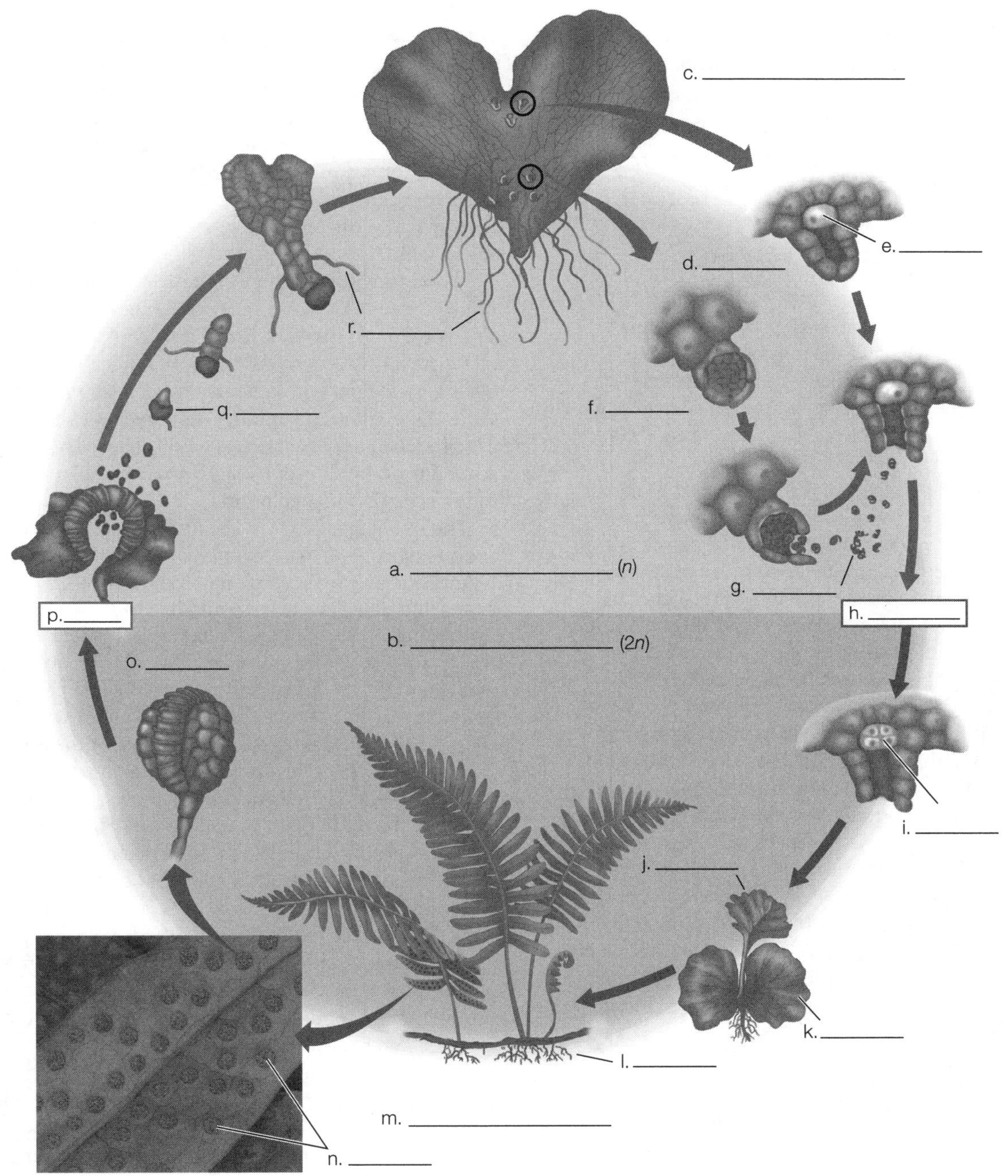

**Question 8.** Compare and contrast homospory and heterospory. Which reproductive structures result from meiosis in each type of life cycle, and which structures result from mitosis?

## 21.4 Seeds Protect Plant Embryos

Features of the seed plant life cycle protect gametes and embryos

The seed is a complex, well-protected package

A change in stem anatomy enabled seed plants to grow to great heights

Gymnosperms have naked seeds

Conifers have cones and lack swimming sperm

Seed plants have characteristics that set them apart from nonseed plants. They consist of two groups of vascular plants: the gymnosperms and the angiosperms.

Seed plants have highly reduced gametophyte generations that are nutritionally dependent on the sporophyte for survival. This nutritional dependence is what sets the seed plants apart from the seedless vascular plants (see Figure 21.12). Very few seed plants (e.g., the cycads and ginkgos) have retained swimming sperm; most have evolved other means for dispersing male gametes. All seed plants are heterosporous in that they produce two types of spores. One becomes the female gametophyte and the other becomes the male gametophyte. Microspores and megaspores develop in specialized cones or in flowers. Microspores develop into pollen grains, the male gametophytes. Pollen grains consist of sperm and supporting cells; they are dispersed by wind or animals. The wall of the pollen grain contains sporopollenin. Sporopollenin is one of the most chemically resistant biological compounds known, and represents a major advantage for land colonization by plants. In the megasporangium, meiosis produces four megaspores, but in most seed plants only one of the four megaspores is retained. This megaspore divides by mitosis to form the multicellular (yet tiny) megagametophyte (female gametophyte), which produces the eggs. When the eggs are ready to be fertilized, they are surrounded by megagametophyte cells that are still within the megasporangium. The megasporangium is surrounded by sterile sporophyte tissues (the integument). Together, the megasporangium and the integument constitute the ovule.

Because the female gametophyte is retained within sporophyte tissue, pollen grains do not have direct access to gametophytes and their eggs. The sporophyte housing a gametophyte creates tissue for receiving pollen grains; upon reaching this tissue, pollen produces pollen tubes that deliver the sperm through the sporophyte tissue to the eggs for fertilization. This process is known as pollination (see Figure 21.14). The embryo resulting from fertilization grows to a certain size and then becomes dormant within the surrounding tissues. This dormant, protected embryo together with its surrounding tissues constitutes the seed.

Seeds are complex structures that may contain tissue from three plant generations. Recall that alternation of generations involves a multicellular sporophyte generation that alternates with a multicellular gametophyte generation. The embryo within a seed represents the beginning of a new sporophyte generation. It is still surrounded by the female gametophyte tissue, which will provide the embryo with nutrients to begin growth (particularly if it is a gymnosperm). The tough coat, or seed coat, that surrounds the seed consists of tissue provided by the embryo's sporophyte parent. This tissue is derived from the integument of the diploid parent. Seeds protect the embryo until conditions are right for germination. Seeds can remain viable or dormant for many years, waiting for favorable conditions to germinate. Many seeds have adaptations that allow dispersal by wind or another vector.

Wood provides support to many seed plants, allowing them to grow taller to capture more light for photosynthesis. The younger portions of wood consist of vascular tissue, allowing for water transport. As wood becomes older, it becomes clogged with resins or other materials and provides the plant with support. New layers of vascular tissue are produced underneath the wood. The existence of this secondary growth is one reason that seed plants became today's dominant vegetation.

The two major types of seed plants are the gymnosperms (such a pines) and angiosperms (flowering plants). Gymnosperms are seed plants that do not form flowers or true fruits. They are diverse and live in a variety of habitats, from sparse deserts to vast forests. There are four groups within the gymnosperms (see Figure 21.16): cycads, ginkgos, gnetophytes, and conifers, with the last being the most abundant. With the exception of the gnetophytes, their vascular tissue is less complex than that of the angiosperms. Tracheids are the only support and water-conducting cells within the xylem. The gymnosperm life cycle can be illustrated by that of the pine, a conifer (see Figure 21.18). Conifers differ from other gymnosperms in that they produce woody cones (megastrobili) and smaller, pollen-bearing cones (microstrobili), which are specialized structures for reproduction (see Figure 21.17). Woody cones house the megasporangia and produce megaspores, megagametophytes, and eggs; pollen-bearing cones house the microsporangia and produce microspores, microgametophytes (pollen), and sperm.

Pines do not have swimming sperm, and their pollen is modified for wind dispersal. Pollen lands on the micropyle of the female cone and the pollen tube then grows through the maternal sporophyte tissue to the female gametophyte, where it releases two sperm. Only one of the two sperm will fertilize an egg. The resulting zygote develops into an embryo that remains encased in the tissues of the megasporangium and gametophyte. The seed is also protected by the scale of the cone (sporophyte tissue) until it is mature and ready for dispersal. Some conifer species have soft, fleshy modifications that surround the seed. Examples include the "berries" found on yew and juniper plants. These are not true fruits, however. The fruits of angiosperms are ripened ovaries, and gymnosperms do not have ovaries.

**Question 9.** Seed plants are heterosporous. Compare the protection provided to the microspores with the protection provided to the megaspores in seed plants.

**Question 10.** The sperm of seed plants is transported in the pollen grain. Describe, in detail, the two parts of the journey from the paternal parent to the egg in the maternal parent plant.

### 21.5 Flowers and Fruits Increase the Reproductive Success of Angiosperms

- Angiosperms have many shared derived traits
- The sexual structures of angiosperms are flowers
- Flower structure has evolved over time
- Angiosperms have coevolved with animals
- The angiosperm life cycle produces diploid zygotes nourished by triploid endosperms
- Fruits aid angiosperm seed dispersal
- Recent analyses have revealed the phylogenetic relationships of angiosperms

The production of flowers and fruits is the most obvious feature that distinguishes the angiosperms, which currently are the dominant plant form on Earth. Angiosperms have the most reduced gametophyte generation, with the female gametophyte consisting of just seven cells. Angiosperms differ from all other plants in that they produce flowers and fruits, have their ovules and seeds enclosed in a carpel, have highly reduced gametophytes, germinate pollen on a stigma, have double fertilization, produce triploid endosperm (the nutritive tissue), and have xylem and phloem with multiple modified cell types, including vessel elements, fibers, and companion phloem cells.

Flowers may occur singly on a plant, or be clustered together in an inflorescence (see Figure 21.19). All flower parts are modified leaf structures. Although flowers are very diverse, they all consist of the same small set of structures that have evolved over time. The male structures in a flower are stamens. Each stamen consists of a filament and an anther; the anther contains microsporangia, which produce pollen. Flowers often have several stamens.

The female structure in a flower is the carpel, which consists of a stigma, a style, and an ovary. The stigma is a surface modified to receive pollen. The style is a stalk that separates the stigma from the ovary. The ovary contains one or more ovules, each of which houses a megasporangium. The pistil is another name for a single carpel or fused carpels. Nonreproductive floral leaves include petals (collectively, the corolla) and the sepals (collectively, the calyx). Petals and sepals are often modified to attract animal pollinators. Sepals also serve to protect the developing flower bud. The sepals, petals, stamens, and carpels are arranged in circular whorls attached to a central stalk. Flowers that have both megasporangia and microsporangia are called "perfect" or hermaphroditic. If either structure is missing, the flower is called "imperfect." Species that have both megasporangiate flowers and microsporangiate flowers on the same plant are called monoecious. If the megasporangia and microsporangia are on separate plants, the species is dioecious.

Early flowers had greatly varying numbers of organs. Evolution of the flower has reduced the number of floral organs to a fixed number, differentiated petals from sepals, and altered symmetry from radial to bilateral in some cases. The evolution of the flower structure is still under debate. One hypothesis states that carpels formed when leaves with sporangia at their margins folded onto themselves and then fused with one another to form the ovary (see Figure 21.21). Some flowers have floral organs attached to the top, rather than the bottom, of the ovary. A perfect flower must strike a balance between the male and female functions, as well as minimize self-pollination (in order to increase diversity). Some species have evolved ways to ensure that pollinators both bring pollen from other individuals to the stigma and take pollen to other individuals.

Many plants and animals have coevolved in such a way that the nutrition of the animal and pollination of the plant are interdependent. Some plants have evolved methods of limiting pollination to a single species of insect, but most can be pollinated by a range of species. Many flowers entice animals by providing food rewards, such as nectar. Pollen grains themselves can be a food reward, and this pollen can be carried from one plant to another. Bee-pollinated flowers often have nectar guides that can be seen only in the ultraviolet spectrum visible to bees.

Angiosperms are heterosporous, with ovules contained within carpels. Pollination occurs when a pollen grain (microgametophyte) arrives on the stigma. Pollination is followed by the growth of a pollen tube to reach the megagametophyte, then by fertilization to create the seed. The angiosperm life cycle differs from that of all other plants in that double fertilization occurs (see Figure 21.25). In double fertilization, one sperm unites with the egg to produce a diploid zygote; the other sperm unites with two other haploid cells of the female gametophyte to produce a triploid cell. This triploid cell divides mitotically to create the (triploid) endosperm tissue. The endosperm provides nutrition for the developing embryo. Angiosperm embryos have one or two seed leaves called cotyledons, which have different fates depending on the species.

Fruits develop from the ovary and its supporting tissues. Fruit types (see Figure 21.26) depend on the number of carpels associated with the fruit and the extent of support tissue incorporated into the fruit structure. A simple fruit, such as a plum or peach, develops from a single carpel or set of fused carpels. An aggregate fruit, such as a raspberry, develops from several carpels of a single flower. A fruit formed from an inflorescence, like a pineapple, is a multiple fruit. Accessory fruits, such as apples, pears, and strawberries, are derived from the carpel and additional parts of the flower.

Different phylogenetic approaches have led to different conclusions regarding angiosperm phylogeny. Investigators are still working on the question of how angiosperms first arose. Some angiosperms are neither monocots nor eudicots. The evolutionary relationships among the angiosperm clades are shown in Figure 21.27. Molecular and morphological evidence now points to the woody shrub *Amborella* as the living species most similar to the first angiosperms. Among the earliest branching angiosperm groups are the magnoliids. Most angiosperm species belong to the clades of the monocots and the eudicots. Monocots have one cotyledon; eudicots have two.

**Question 11.** Describe the differences between the seeds of gymnosperms and angiosperms. For each difference, identify the characteristic of the angiosperm seed that might contribute to the success of this group.

**Question 12.** In double fertilization, triploid endosperm results. Do the seed parent and the pollen parent contribute equally to the endosperm, and if not, how might this unequal contribution affect the offspring?

## Test Yourself

1. Species in which both megasporangiate and microsporangiate flowers occur on the same plant are called
   a. perfect.
   b. dioecious.
   c. monoecious.
   d. imperfect.
   e. sterile.
   ***Textbook Reference:*** *Concept 21.5 Flowers and Fruits Increase the Reproductive Success of Angiosperms; The sexual structures of angiosperms are flowers*

2. Which characteristic was *not* necessary in order for plants to colonize land?
   a. Vascular tissue for moving water throughout the plant
   b. A waxy cuticle to reduce water loss
   c. The ability to screen ultraviolet radiation
   d. The development of thick spore walls to protect the spores from dehydration
   e. Development of embryos protected inside other tissues
   ***Textbook Reference:*** *Concept 21.2 Key Adaptations Permitted Plants to Colonize Land; Adaptations to life on land distinguish land plants from green algae*

3. The main difference between nonvascular plants and vascular plants is that vascular plants
   a. lack gametophytes.
   b. produce spores.
   c. have tracheids.
   d. reproduce sexually.
   e. possess stomata.
   ***Textbook Reference:*** *Concept 21.3 Vascular Tissues Led to Rapid Diversification of Land Plants; Vascular tissues transport water and dissolved materials*

4. Land plant life cycles feature alternation of generations. Which statement about alternation of generations is *false*?
   a. The life cycle includes both a diploid and haploid multicellular stage.
   b. Gametes are not produced by mitosis.
   c. Gametes fuse to form a zygote.
   d. The life cycle includes a multicellular spore-producing generation.
   e. Sporangia undergo meiosis to produce haploid unicellular spores.
   ***Textbook Reference:*** *Concept 21.2 Key Adaptations Permitted Plants to Colonize Land; Life cycles of land plants feature alternation of generations*

5. Two characteristics that help distinguish the mosses from the liverworts are _______ and _______.
   a. the presence of hydroids; gametophytic dominance
   b. gametophyte dominance; the presence of stomata
   c. sporophyte dominance; the presence of stomata
   d. the presence of stomata; hydroids
   e. sporophytic dominance; the presence of hydroids
   ***Textbook Reference:*** *Concept 21.2 Key Adaptations Permitted Plants to Colonize Land; Nonvascular land plants live where water is readily available*

6. During a plant's life cycle, meiosis takes place in the _______ and produces _______.
   a. gametophyte; haploid gametes
   b. sporophyte; haploid gametes
   c. sporophyte; haploid spores
   d. gametophyte; diploid spores
   e. gametophyte; haploid spores
   ***Textbook Reference:*** *Concept 21.2 Key Adaptations Permitted Plants to Colonize Land; The sporophytes of nonvascular land plants are dependent on the gametophytes*

7. Which set of characteristics best represents the rationale for placing the stoneworts as the sister group to the land plants?
   a. Retention of the egg within the parental organism and DNA evidence
   b. Apical branching growth, plasmodesmata, structural similarities, and DNA evidence
   c. Flattened growth form, plasmodesmata, and DNA evidence
   d. Presence of chlorophyll *b*, growth form, retention of the egg within the parental organism, and DNA evidence
   e. Structural similarities and DNA evidence
   ***Textbook Reference:*** *Concept 21.1 Primary Endosymbiosis Produced the First Photosynthetic Eukaryotes; Several distinct clades of algae were among the first photosynthetic eukaryotes*

8. Which characteristic enhances the ability of the hornworts to produce elongated sporophytes?
   a. A basal region where cells divide indefinitely
   b. Waxy cuticle

c. Developmental genes that increase growth of the sporophyte
d. Branched apical growth
e. Plasmodesmata
***Textbook Reference:*** *Concept 21.2 Key Adaptations Permitted Plants to Colonize Land; Nonvascular plants live where water is readily available*

9. You are walking along a roadside and find a plant with the following characteristics: stomata; reduced, simple leaves in whorls around a central stem; independent sporophytes and gametophytes; and silica in the cell walls. This plant is most likely a member of which of the following groups?
a. Mosses
b. Monilophytes
c. Hornworts
d. Lycophytes
e. Cycads
***Textbook Reference:*** *Concept 21.3 Vascular Tissues Led to Rapid Diversification of Land Plants; Horsetails and ferns constitute a clade*

10. Which structure is common to all land plants *except* the seed plants?
a. Sporangium
b. Antheridium
c. Embryo
d. Sperm
e. Archegonium
***Textbook Reference:*** *Concept 21.5 Flowers and Fruits Increase the Reproductive Success of Angiosperms; The angiosperm life cycle produces diploid zygotes nourished by triploid endosperms*

11. Which change contributed most to the overall success of seed plants?
a. Development of secondary growth
b. Development of double fertilization
c. Loss of the antheridium in the microsporangia
d. Reduction of gametophyte generation
e. Development of heterospory
***Textbook Reference:*** *Concept 21.4 Seeds Protect Plant Embryos; A change in stem anatomy enabled seed plants to grow to great heights*

12. The female cone in gymnosperms is known as the _______ and the male cone is known as the _______
a. megastrobilus; microstrobilus
b. microstrobilus; megastrobilus
c. megasporangium; microsporangium
d. microstrobilus; integumen
e. integument; micropyle
***Textbook Reference:*** *Concept 21.4 Seeds Protect Plant Embryos; Conifers have cones and lack swimming sperm*

13. A land plant may be most reliably distinguished from green algae by which characteristic?
a. Chlorophyll type
b. The presence of an embryo protected by parent tissue
c. The presence of roots
d. Branched apical growth
e. Storage of photosynthetic products as starch in chloroplasts
***Textbook Reference:*** *Concept 21.1 Primary Endosymbiosis Produced the First Photosynthetic Eukaryotes; There are ten major groups of land plants*

14. In which of the following groups are sporangia arranged in strobili?
a. Ginkgos
b. Whisk ferns
c. Hornworts
d. Club mosses
e. Ferns
***Textbook Reference:*** *Concept 21.3 Vascular Tissues Led to Rapid Diversification of Land Plants; The lycophytes are sister to the other vascular plants*

15. Which group has large leaves with branching vascular strands?
a. Mosses
b. Lycophytes
c. Hornworts
d. Club mosses
e. Ferns
***Textbook Reference:*** *Concept 21.3 Vascular Tissues Led to Rapid Diversification of Land Plants; Horsetails and ferns constitute a clade*

16. Which piece of evidence would be most helpful in determining the relationship between a newly discovered green aquatic organism and land plants?
a. Chlorophyll type
b. Presence or absence of chloroplasts
c. Primary endosymbiosis
d. Retention of the egg in the parent organism
e. Presence of starch granules in chloroplasts
***Textbook Reference:*** *Concept 21.1 Primary Endosymbiosis Produced the First Photosynthetic Eukaryotes; Several distinct clades of algae were among the first photosynthetic eukaryotes*

## Answers

### *Key Concept Review*

1. The presence of chloroplasts as a result of primary endosymbiosis is common to all photosynthetic eukaryotes (including protists), and the groups that evolved chlorophyll *b* with starch stored in chloroplasts are known as the green plants. The streptophytes include the nonchlorophyte green algae and all other green plants. Streptophytes that exhibit a number of additional synapomorphies, including a protected embryo,

a cuticle, and a multicellular sporophyte, are called land plants.

2. Glaucophytes are unicellular, microscopic, freshwater algae whose characteristic whitish-blue color is due to their retention of a small amount of peptidoglycan between the inner and outer membranes of their chloroplasts. Red algae are almost exclusively multicellular organisms that inhabit saltwater environments, lack peptidoglycan in their chloroplasts, and are usually red in color due to their characteristic pigment, phycoerythrin.

3. Land plants required protection against water loss, support against gravity, the ability to disperse in the absence of water, protection from ultraviolet radiation and desiccation, and the ability to acquire water and nutrients from a dry environment. Some of the characteristics that plants developed that address these problems include: the cuticle (a waxy coating) to protect against water loss; stomata (openings to regulate gas exchange and water loss); multicellular enclosures to protect gametes and embryos from drought; pigments for protection against ultraviolet radiation; thick-walled spores to protect against desiccation and decay; and mutualisms with fungi to aid in the acquisition of resources.

4. Mosses have very rudimentary water transport cells called hydroids. Hydroids lack the waterproofing and support molecule lignin. Because of this, they can carry water only short distances and cannot support tall growth.

5.
   a. Haploid; Gametophyte generation
   b. Diploid; Sporophyte generation
   c. Sperm (*n*)
   d. Egg (*n*)
   e. Archegonium (*n*)
   f. Embryo (2*n*)
   g. Fertilization
   h. Antheridium (*n*)
   i. Sporophyte (2*n*)
   j. Gametophyte (*n*)
   k. Sporangium
   l. Meiosis
   m. Ungerminated spores
   n. Germinating spore
   o. Rhizoid
   p. Protonema
   q. Bud
   r. Gametophytes (*n*)

6. A branching sporophyte offers the opportunity for increasing structural complexity. Branching rhizomes or roots can provide increased anchoring for an independent sporophyte. A branching aerial portion of the sporophyte can support the production of an increased number of spores and the development of megaphylls, which can increase photosynthetic potential.

7.
   a. Haploid
   b. Diploid
   c. Mature gametophyte
   d. Archegonium
   e. Egg
   f. Antheridium
   g. Sperm
   h. Fertilization
   i. Embryo
   j. Sporophyte
   k. Gametophyte
   l. Roots
   m. Mature sporophyte
   n. Sori
   o. Sporangium
   p. Meiosis
   q. Germinating spore
   r. Rhizoids

8. In homospory, meiosis results in a single type of spore, but in heterospory, meiosis results in megaspores and microspores. Mitosis occurs throughout both types of life cycle as cells divide and plants grow, but the reproductive cells that result from mitosis in both homospory and heterospory are sperm and eggs. In homospory, the antheridium produces sperm by mitosis, and the archegonium produces eggs; both structures are found on the same gametophyte. In heterospory, the microgametophyte produces sperm and the megagametophyte produces eggs; these are distinct gametophyte forms that arise from either microsporangia or megasporangia.

9. In seed plants, the microspores are protected by the spore wall which contains sporopollenin, the most chemically resistant biological compound known. The megaspore is enclosed in the megasporangium, which is enclosed by a layer of sporophytic tissue called the integument. Together, the megasporangium and the integument constitute the ovule, which after fertilization and maturation becomes a seed.

10. Most gymnosperms rely on wind or animals for transfer of the male gametes to the female parent plant. In the first part of the journey, the male gametes are transported from the microsporangia in the microstrobilus in the form of pollen grains. When the pollen grains reach the megastrobilus, they germinate, and the second part of the journey is accomplished via the growth of the pollen tube down through the maternal sporophytic tissue. When the pollen tube reaches the female gametophyte, it releases two sperm, one of which unites with the egg; the other sperm disintegrates.

11. Pollination in angiosperms involves double fertilization, which provides additional nutrition supplied to the offspring. Carpels in the angiosperm that enclose ovules and seeds provide protection as well as an opportunity for the maternal parent to control pollen access to the eggs, including selecting against self-pollination.

    In angiosperms, pollen germination on a stigma means that the male gametophyte journey is simplified. Pollen tube growth through the stigma provides an opportunity to screen out inappropriate matches (e.g., self-pollination or distantly related pollen). Pollen tube growth through the stigma is much faster than growth in gymnosperms, which requires traversing the megasporangium, the female gametophyte, and the archegonium before reaching the egg.

Angiosperm flowers incorporate potential animal attractants such as color, shape, scent, and nutritional rewards that offer many options for animal-assisted pollination. The arrangement of flowers and their stamens and carpels offers opportunities to facilitate appropriate pollination.

Fruits provide additional protection for the embryo and increase the likelihood of water, wind, or animal dispersal of seeds by providing morphological transport enhancement.

In angiosperms, the reduction of the microgametophyte to 2–3 cells and the megagametophyte to 7–8 cells that are completely dependent on the sporophyte means that the haploid stage of the angiosperm life history requires few resources and can be completed very quickly, reducing the generation time and decreasing the importance of environmental conditions for the gametophytes.

12. The seed parent contributes two copies of its DNA($n$), while the pollen parent contributes one copy of its haploid DNA. This means that the endosperm resembles the seed parent more closely than the pollen parent, so the seed parent characteristics may have more impact on the fitness of the offspring. The benefits could include (among others) longevity of the nutrients available to the embryo.

### *Test Yourself*

1. **c.** "Monoecious" means "one house" and refers to species that possess separate male and female flowers on the same plant.
2. **a.** Several successful groups of terrestrial plants lack vascular tissue.
3. **c.** Tracheids are found only in the vascular plants. All the other characteristics are shared by at least some nonvascular and vascular plants.
4. **b.** In alternation of generations gametes are produced by mitosis, not by meiosis.
5. **e.** Mosses possess hydroids and stomata, both of which are lacking in liverworts.
6. **c.** The outcome of meiosis is four cells, each of which has half the genetic material of the parent cell. A haploid cell already has only half the normal number of chromosomes of most eukaryotes, so a cell must be diploid (or have higher ploidy) to undergo meiosis. In all plant life cycles, the sporophyte is diploid and the gametophyte is haploid; therefore, only the sporophyte can undergo meiosis. Spores are the products of sporophyte meiosis.
7. **b.** Both coleochaetophytes and stoneworts as well as land plants retain the egg on the parental plant. Coleochaetophytes have flattened growth patterns similar to liverworts but not most land plants. All green plants possess chlorophyll *b*. Apical, branching growth, plasmodesmata, and DNA evidence all support stoneworts as the closest algal group to land plants.
8. **a.** The indefinitely dividing group of sporophyte-generating cells contained in the gametophyte contributes to the hornwort's ability to produce elongated sporophytes.
9. **b.** The plant is a horsetail, which is very common along roadsides in damp ditches, particularly in the Midwest.
10. **b.** In seed plants, the microgametophyte is reduced to 2–4 cells and no longer includes the sterile jacket of cells that protect the gametes in nonseed plants. The megagametophyte does produce an archegonium that protects the egg. In angiosperms, there is no longer an archegonium.
11. **a.** Development of secondary growth enabled seed plants to increase their diameter and reach greater heights, allowing them to become better competitors for light. The seed coat protects the embryo from desiccation and herbivory, and also allows greater control over germination conditions. Double fertilization is very rare in gymnosperms.
12. **a.** The female cone is called the megastrobilus because it contains the megasporangia enclosed in the integument (together called the ovule). The smaller male cone is called the microstrobilus because it contains the microsporangia that bear pollen.
13. **b.** All land plants produce embryos that are protected by tissue of the parent plant. Green algae and land plants make use of the same types of chlorophyll. Not all land plants have roots, so a plantlike organism lacking roots does not necessarily belong to the green algae. Stoneworts, one group of the green algae, have branched apical growth like many land plants. Both land plants and green algae store starch in their chloroplasts.
14. **d.** The sporangia of many club mosses are aggregated in conelike structures called strobili.
15. **e.** Some ferns uncurl as their leaves grow.
16. **d.** Land plants retain the egg in the parent organism, so this characteristic would provide insight as to whether the new plant belonged in that group. All photosynthetic eukaryotes have chloroplasts and the chlorophyll type may not be helpful. Both green algae and land plants store starch in their chloroplasts. All members of Plantae are the result of primary endosymbiosis.

# The Evolution and Diversity of Fungi

## The Big Picture

- Fungi are heterotrophic organisms that absorb nutrients from the environment. Fungi can be unicellular or multicellular. The mycelium is the body of a multicellular fungus and is composed of many tubular filaments known as hyphae. Many fungi form symbiotic relationships with photosynthetic organisms, creating mycorrhizae and lichens.
- Reproduction in the fungi occurs both sexually and asexually. Asexual reproduction involves the production of spores, breakage, fission, or budding. Sexual reproduction in multicellular fungi requires the union of hyphae with two different mating types.

## Study Strategies

- The numerous ways that fungi reproduce can seem confusing. It is helpful to create flow charts showing the ploidy level at each stage of the life cycle in the different phyla of fungi. Make sure to include both the sexual and asexual stages in your charts. Comparisons of your flow charts for the different phyla will help you learn the differences among the groups.
- Organisms are placed in particular systematic groupings because they share unique characteristics. Look for patterns that distinguish the six major fungal groups from one another. Most fungi are grouped according to the reproductive structures that they produce.
- Go to **LaunchPad** (or use the Web addresses listed) to review the following additional resources:

  Animated Tutorial 22.1 Life Cycle of a Zygospore Fungus (PoL2e.com/at22.1)

  Activity 22.1 Fungal Phylogeny (PoL2e.com/ac22.1)

  Activity 22.2 Life Cycle of a Dikaryotic Fungus (PoL2e.com/ac22.2)

  Apply the Concept in textbook Section 22.4

## Key Concept Review

### 22.1 Fungi Live by Absorptive Heterotrophy

Unicellular yeasts absorb nutrients directly

Multicellular fungi use hyphae to absorb nutrients

Fungi are in intimate contact with their environment

Fungi are grouped with animals and choanoflagellates in the opisthokonts and probably share a unicellular protist common ancestor with these two groups. The synapomorphies that unite fungi are absorptive heterotrophy and the presence of chitin in their cell walls. Many fungi are saprobes, absorbing nutrients from dead organic matter.

Although most fungi are multicellular, unicellular fungi are found in almost all fungal groups and are commonly referred to as yeasts. Fungi can have yeast life stages in addition to multicellular life stages. Yeasts are model eukaryotic organisms because they are as easy to grow as bacteria yet have eukaryotic cellular traits.

The body of a multicellular fungus is called a mycelium. A mycelium is made up of individual tubular filaments called hyphae. Hyphae grow rapidly into a substrate; the hyphae in a single mycelium may collectively grow as much as 1 kilometer per day. Hyphal cell walls are strengthened by the polysaccharide chitin. Hyphae may be septate (divided into compartments by chitinous walls called septa) or coenocytic (continuous and multinucleate) (see Figure 22.3B). Rhizoids are modified hyphae that anchor a fungus to a substrate; they are not like the rhizoids of plants and cannot absorb nutrients and water. Hyphae provide a large surface area-to-volume ratio, and thus a large surface area for absorption of nutrients from the substrate. This also means that they can lose water very rapidly, and so grow best in a moist environment. They are also relatively tolerant of hypertonic environments and temperature extremes, unlike bacteria. Fungi will grow on food in the refrigerator more easily than most bacterial species.

**Question 1.** Early taxonomists considered fungi to be members of the plant kingdom. What evidence indicates that they are in fact more closely related to animals?

**Question 2**. Label the diagram below using the following terms: hypha, nuclei, cell wall, septa.

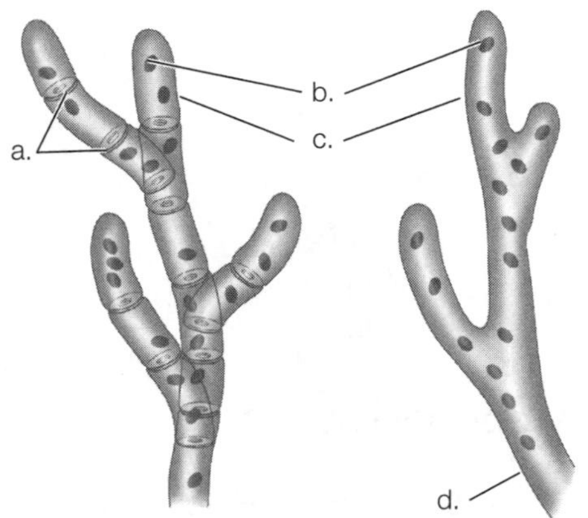


## 22.2 Fungi Can Be Saprobic, Parasitic, Predatory, or Mutualistic

Saprobic fungi are critical to the planetary carbon cycle

Some fungi engage in parasitic or predatory interactions

Mutualistic fungi engage in relationships beneficial to both partners

Endophytic fungi protect some plants from pathogens, herbivores, and stress

Fungi are important recyclers of elements in the ecosystem. They absorb the nutrition needed for their survival from dead matter (saprobes), from living hosts (parasites), and from mutually beneficial symbiotic relationships with other organisms (mutualists).

Many fungi are beneficial to other organisms. Saprobic fungi secrete enzymes into the environment that help in the absorption of dead matter and the decomposition and recycling of elements—especially carbon—used by living organisms. As decomposers, saprobic fungi are essential to life on Earth. During the Carboniferous period, acidification of swamps led to the formation of large peat deposits because of the lack of saprobic fungal activity to decompose the organic matter. Fungi increase spore production when food supply dwindles in order to overcome the poor growth conditions. The spores are easily spread by wind and water, ensuring wide dispersal.

Parasitic fungi are either facultative (possessing the ability to grow independently) or obligate (dependent on a living host for growth). Adaptations of hyphae allow fungi to exploit different sources of nutrition. Some plant parasites, for example, have hyphae that germinate from spores on the leaf surface and enter the leaf thorough stomata. The fungus forms a mycelium within the leaf (see Figure 22.5). Some hyphae produce branching projections nestled into the plant cell membranes called haustoria. Some parasitic fungi, called pathogens, not only derive nutrition from their hosts, but also sicken or kill the host. Other fungi are active predators that trap microscopic protists or animals in a sticky substance; some form a constricting ring around their prey (see Figure 22.7).

Fungi also form two crucial types of symbiotic (in close contact) and mutualistic (mutually beneficial) relationships: lichens and mycorrhizae. Lichens are associations of a fungus with a unicellular photosynthetic alga or cyanobacteria. The fungus provides minerals and water to its photosynthetic partner, and the photosynthesizing organism provides the fungus with organic compounds. Lichens are among Earth's hardiest organisms, and can thrive in barren and extreme environments, such as Antarctica. Lichens are characterized by their appearance as crustose (crusty), foliose (leafy), or fruticose (shrubby) (see Figure 22.8). Lichens can reproduce by fragmentation of the thallus (leaflike blade) or by production of soredia (one or more photosynthetic cells surrounded by fungal hyphae). Mycorrhizae are associations between fungi and the roots of plants in which a fungus obtains the products of photosynthesis from the plant and provides minerals and water to the plant. The mycorrhizal symbiosis is essential to the survival of most plants, and the evolution of this relationship may have been the most important step in allowing plants to colonize land. Ectomycorrhizal fungi wrap their hyphae around a plant root. The hyphae of arbuscular mycorrhizal fungi penetrate the root cell walls and form treelike (arbuscular) structures that provide the plant with nutrients (see Figure 22.10).

Endophytic fungi are symbionts living within the plant parts that are aboveground. They help certain plants (especially grasses) resist pathogens, herbivores, and stresses such as drought and salty soil; their role in other plants is not well understood.

**Question 3.** Explain how a haustorium, which does not enter the host cell, obtains nutrients from the host.

**Question 4.** How can you distinguish between a mutualistic and a pathogenic relationship between a fungus and its partner?

## 22.3 Major Groups of Fungi Differ in Their Life Cycles

Fungi reproduce both sexually and asexually

Microsporidia are highly reduced, parasitic fungi

Most chytrids are aquatic

Some fungal life cycles feature separate fusion of cytoplasms and nuclei

Arbuscular mycorrhizal fungi form symbioses with plants

The dikaryotic condition is a synapomorphy of sac fungi and club fungi

The sexual reproductive structure of sac fungi is the ascus

The sexual reproductive structure of club fungi is the basidium

Fungi are classified into six major groups on the basis of molecular evidence and type of life cycle: microsporidia, chytrids, Zygomycota, Glomeromycota, Ascomycota, and

Basidiomycota. Fungi have many different life cycles and can reproduce both sexually and asexually. Fungi have several means of asexual reproduction. They can produce spores enclosed within sporangia or naked spores at the tips of hyphae known as conidia. Unicellular fungi can reproduce by fission or budding. Just about any part of a mycelium is capable of living independently of the rest; therefore, simply the division of a mycelium into two or more parts is a method of reproduction.

Members of the different fungal groups can be distinguished from one another by their mechanisms of sexual reproduction, which involves two or more mating types (see Figure 22.12). Self-fertilization is prevented by the incapacity of individuals of the same mating type to mate with each other. Sexual reproduction is rare or unknown in some fungal groups. Sexual stages have not yet been identified in many fungi, so these species have been classified according to data from DNA sequence analysis.

Microsporidia are unicellular parasitic fungi and are among the smallest eukaryotes known. They lack true mitochondria but have reduced mitochondrial-like structures called mitosomes; they are obligate intracellular parasites of animals. They use a polar tube projection from their cell to invade the host and convey the contents of the microsporidia (sporoplasm) into the host cell. Inside the host cell the sporoplasm replicates and produces more spores. These species are thought to be mostly asexual, but some species have poorly understood sexual and asexual cycles.

Chytrids are aquatic fungi, and are the only fungi with flagellated gametes. Chytrids likely resemble the common ancestor of all fungi more closely than any other fungal group. There are actually three clades that constitute the chytrids, although some mycologists use the term "chytrid" to refer to only one of these clades, the Chytridiomycota. Chytrids have chitinous cell walls. The haploid zoospore is also flagellated. Chytrids reproduce both sexually and asexually and are diverse in form; they can be saprobic or parasitic.

The remaining four groups of fungi are generally terrestrial and do not have motile gametes; they reproduce sexually via fusion of the cytoplasms of opposite mating types (plasmogamy) followed by fusion of nuclei (karyogamy).

Sexual reproduction in zygospore fungi occurs between adjoining individuals with different mating types (see Figure 22.14B). Branches from opposite mating types are attracted to each other by chemical signals, and the individuals grow toward each other until they produce gametangia. The gametangia fuse (plasmogamy), and a thick-walled zygosporangium containing many haploid nuclei of each mating type forms at the fusion site. Individual haploid nuclei of different mating types fuse within the zygosporangium (karyogamy), and a unique multinucleate zygospore forms. The zygospore nuclei undergo meiosis, and a sporangium, containing haploid nuclei incorporated into new spores, sprouts from the stalked sporangiophore. The spores disperse and germinate to form a new generation of haploid hyphae. The Zygomycota include black bread mold (*Rhizopus stolonifer*) and more than 1,000 other saprobic, parasitic, and mutualistic species. The hyphae are coenocytic and usually do not form a fleshy fruiting body.

Glomeromycota consist of the arbuscular mycorrhizal fungi, which are terrestrial species that form symbiotic, mutualistic associations with the roots of 80 to 90 percent of all plants (see Figure 22.10B). The fewer than 200 species of Glomeromycota are coenocytic and are only known to reproduce asexually.

The sac fungi and club fungi share the synapomorphy of a dikaryotic hyphal stage. The cytoplasms of two individuals fuse, but the nuclei do not fuse immediately, and the resulting hyphae contain genetically different haploid nuclei. Two genetically distinct nuclei coexist and divide in each cell of the mycelium. This stage of the life cycle is called the dikaryon. This unique dikaryon condition is the first stage of sexual reproduction (see Figure 22.16). These nuclei eventually fuse in the "fruiting" structure of the fungus to form a diploid zygote that undergoes meiosis. The hyphae are neither haploid ($n$) nor diploid ($2n$), but dikaryotic ($n + n$). The dikaryon life history has no gamete cells, only gamete nuclei. The zygote is the only diploid cell. The dikaryotic mycelium can have characteristics not shared by the haploid or diploid stages.

Ascomycota (sac fungi) are found in marine, freshwater, and terrestrial habitats and include baker's yeast, truffles, and *Penicillium*. The Ascomycota are distinguished by their ascus, a sexual reproductive sac structure (see Figure 22.16A). Meiosis and spore cleavage occur after formation of the sac, producing haploid ascospores. Modern taxonomy has revised the traditional groupings, which were based on the presence of an ascoma (specialized fruiting structure) and its morphology. Some sac fungi are unicellular yeasts, such as the species of yeast used to make bread and alcoholic beverages, *Saccharomyces cerevisiae*. Reproduction in these yeasts is asexual and accomplished by budding. Sexual reproduction of these yeasts occurs when two haploid cells of opposite mating types fuse and then form an ascus with either four or eight ascospores.

Most sac fungi are filamentous. Their ascomata are cup-shaped and can be several centimeters in diameter, although most species are much smaller. Some ascomata are edible, such as morels and truffles. Sexual reproduction of filamentous sac fungi involves two different mating types that form a dikaryotic mycelium. This mycelium typically forms an ascoma, which then produces asci. The resulting ascospores germinate and grow into new haploid mycelia. Ascomycota include molds, some of which have a number of uses for humans. Mold from the genus *Penicillium* produces the antibiotic penicillin; other molds are important in making cheese. Some molds are plant parasites, such as chestnut blight and Dutch elm disease, and powdery mildews. Brown molds in the genus *Aspergillus* are used in brewing sake (a Japanese alcoholic beverage) and soy sauce. Asexual reproduction in filamentous sac fungi involves the production of conidia at the tips of hyphae (see Figure 22.18).

The fruiting structures of the club fungi (Basidiomycota), called basidiomata, are familiar as mushrooms. Included in the 30,000 species of Basidiomycota are puffballs and mushrooms, which produce the group's most spectacular fruiting structures. Bracket fungi are important in the decay of wood. Some club fungi are parasites and some are important to the survival of plants as ectomycorrhizae. The septate hyphae of the club fungi typically have small pores. As the hyphae grow, haploid nuclei meet and fuse, forming dikaryotic hyphae. When triggered by environmental conditions, the mycelium forms a basidioma. The dikaryon stage can last for years or centuries. A swollen cell at the tip of a specialized hypha, the basidium, is the reproductive structure. In mushrooms, these form on the gills. The basidium is similar to the ascus in sac fungi and the zygosporangium in zygospore fungi; it is the site of nuclear fusion and meiosis. After nuclear fusion, the diploid nucleus undergoes meiosis and the resulting nuclei become part of the basidiospores. The basidiospores form on the outside of the basidium and are forcibly discharged to disperse the spores, giving rise to new haploid hyphae.

**Question 5.** Examine the four life histories shown in Figures 22.14 and 22.16 and describe one feature in each that is unique among these groups of fungi.

**Question 6.** Contrast the sporangiophore produced in the Zygomycota with the conidia produced by Ascomycota.

### 22.4 Fungi Can Be Sensitive Indicators of Environmental Change

Lichen diversity and abundance indicate air quality

Fungi record and help remediate environmental pollution

Reforestation may depend on mycorrhizal fungi

Although lichens and fungi can survive in extreme environments, their absorptive feeding leaves them vulnerable to environmental pollutants. Lichens are sensitive to environmental pollutants and can serve as air quality indicators. Collections of fungi in museums contain a historical record of environmental pollutants. Fungi are also used in remediation efforts because of their ability to break down organic matter. Forest restoration projects may depend on providing appropriate mycorrhizal symbionts for tree establishment.

**Question 7.** What difficulties might you encounter when trying to reestablish trees and their mycorrhizal associate in a degraded environment?

**Question 8.** What characteristics make fungi a good source of information about patterns of pollution?

## Test Yourself

1. Which of the following would *not* be found in any of the typical fungal life cycles?
   a. Haploid nuclei
   b. Diploid nuclei
   c. Spores
   d. Chloroplasts
   e. A dikaryotic stage
   ***Textbook Reference:*** *Concept 22.3 Major Groups of Fungi Differ in Their Life Cycles; Fungi reproduce both sexually and asexually*

2. Fungi are absorptive heterotrophs. Which adaptation greatly aids this mode of nutrient procurement?
   a. Dikaryosis
   b. A large surface area-to-volume ratio
   c. Conjugation
   d. A complex life cycle
   e. A small surface area-to-volume ratio
   ***Textbook Reference:*** *Concept 22.1 Fungi Live by Absorptive Heterotrophy; Fungi are in intimate contact with their environment*

3. Assume that two normal hyphae of different fungal mating types meet. After a period of time, the cell walls between these hyphae will dissolve, producing a
   a. mycelium.
   b. fruiting body.
   c. zygote.
   d. spore.
   e. dikaryotic cell.
   ***Textbook Reference:*** *Concept 22.3 Major Groups of Fungi Differ in Their Life Cycles; The dikaryotic condition is a synapomorphy of sac fungi and club fungi*

4. Which notation is the best way to represent the ploidy of dikaryotic hyphae?
   a. ½ $n$
   b. $n/n$
   c. $n$
   d. $n + n$
   e. $2n$
   ***Textbook Reference:*** *Concept 22.3 Major Groups of Fungi Differ in Their Life Cycles; The dikaryotic condition is a synapomorphy of sac fungi and club fungi*

5. All of the following about sexual reproduction in fungi are true *except*
   a. motile gametes are present in all fungal species.
   b. an aquatic environment is not required for fertilization to occur in most fungi.
   c. there is no true diploid tissue in the life cycle of most sexually reproducing fungi.
   d. sexual reproduction often begins with contact between hyphae of different mating types.
   e. alternation of generations and sexual reproduction can occur in the same species.
   ***Textbook Reference:*** *Concept 22.3 Major Groups of Fungi Differ in Their Life Cycles; Some fungal life cycles feature separate fusion of cytoplasms and nuclei*

6. A mycorrhiza is
   a. a specialized type of lichen.
   b. the fruiting structure of a basidiomycota.

c. a symbiotic association between a fungus and cyanobacterium or green alga.
d. a reproductive stage of sac fungi.
e. a symbiotic association between a fungus and a plant.
***Textbook Reference:*** *Concept 22.2 Fungi Can Be Saprobic, Parasitic, Predatory, or Mutualistic; Mutualistic fungi engage in relationships beneficial to both partners*

7. Suppose a scientist investigating the classification of a fungus that has never been observed to reproduce sexually discovers that it has DNA sequences characteristic of basidiomycota. If this scientist could coax fungi of this species to reproduce sexually, which characteristic would most likely be observed?
a. Dikaryotic hyphae segmented by septa
b. Dikaryotic hyphae without septa
c. Asymmetrical cell division
d. A dikaryotic ascus
e. Flagellated gametes
***Textbook Reference:*** *Concept 22.3 Major Groups of Fungi Differ in Their Life Cycles; The sexual reproductive structure of club fungi is the basidium*

8. Which fungi have coenocytic hyphae and stalked sporangiophores?
a. Chytrids
b. Zygomycota
c. Ascomycetes
d. Basidiomycota
e. Glomeromycota
***Textbook Reference:*** *Concept 22.3 Major Groups of Fungi Differ in Their Life Cycles; Some fungal life cycles feature separate fusion of cytoplasms and nuclei*

9. Which fungi display alternation of generations?
a. Chytrids
b. Glomeromycota
c. Ascomycota
d. Basidiomycota
e. Zygomycota
***Textbook Reference:*** *Concept 22.3 Major Groups of Fungi Differ in Their Life Cycles; Most chytrids are aquatic*

10. A saprobe is an organism that
a. absorbs nutrients from the sap of a host plant.
b. reproduces in the sap of a plant.
c. undergoes asexual reproduction.
d. is mutualistic.
e. absorbs nutrients from dead organic matter.
***Textbook Reference:*** *Concept 22.1 Fungi Live by Absorptive Heterotrophy*

11. Which of the following is *not* characteristic of asexual reproduction in fungi?
a. Budding
b. Formation of haploid spores in sporangia
c. Formation of dikaryotic mycelia
d. Fission
e. Formation of haploid spores in conidia
***Textbook Reference:*** *Concept 22.3 Major Groups of Fungi Differ in Their Life Cycles; The dikaryotic condition is a synapomorphy of sac fungi and club fungi*

12. Which statement about fungi is *false* (i.e., a common misconception)?
a. Fungi grow only in warm, wet environments.
b. Fungi lose water rapidly in a dry environment.
c. Fungi can grow in environments too hypertonic to sustain bacteria.
d. Fungi are eukaryotes and have multiple mating types.
e. Male and female fungi have no distinctive morphology.
***Textbook Reference:*** *Concept 22.2 Fungi Can Be Saprobic, Parasitic, Predatory, or Mutualistic; Mutualistic fungi engage in relationships beneficial to both partners*

13. Which statement is the best explanation for why lichens can grow in extreme environments, but cannot tolerate poor air quality?
a. Algae cannot get enough $CO_2$ when the air quality is poor.
b. Lichens cannot excrete toxins.
c. Fungal spores cannot germinate at low pH.
d. The ascus of the fungal component is sensitive to pollution.
e. Air pollutants are the most extreme toxins on Earth.
***Textbook Reference:*** *Concept 22.4 Fungi Can Be Sensitive Indicators of Environmental Change; Lichen diversity and abundance indicate air quality*

14. Which of the following scientific groups is correctly matched with its correct common name?
a. Chytrids – microspore fungi
b. Zygomycota – sac fungi
c. Glomeromycota – mycorrhizal fungi
d Basidiomycota – sac fungi
e. Ascomycota – club fungi
***Textbook Reference:*** *Concept 22.3 Major Groups of Fungi Differ in Their Life Cycles*

15. Suppose scientists are contending with a new fungal plant disease that has appeared on rhododendrons in Asia and in the United States. How would they best determine whether this is a recent spread of a disease or one that has not been previously noted?
a. By attempting crosses among various similar strains
b. By observing whether the various strains are resistant to a particular bacterium
c. By testing the fungus on European rhododendron to see if it can infect them
d. By sequencing the DNA of this and other similar disease-causing strains and comparing them for similarities

e. By testing the temperature tolerance of the various similar strains

***Textbook Reference:*** *Concept 22.2 Fungi Can Be Saprobic, Parasitic, Predatory, or Mutualistic; Some fungi engage in parasitic or predatory interactions*

16. Damp weather increases the incidence of mildew fungal disease in plants because
a. chytrids are flagellated and need water for motility.
b. fungal life histories require water for karyogamy to occur.
c. fungal sporangiophores require water for support.
d. basidiomycetes produce mushrooms in damp weather.
e. in damp conditions the stomata are open, giving the fungus easy access to plant leaf cells.

***Textbook Reference:*** *Concept 22.2 Fungi Can Be Saprobic, Parasitic, Predatory, or Mutualistic; Some fungi engage in parasitic or predatory interactions*

## Answers

### Key Concept Review

1. Similar to animals and choanoflagellates, fungi are believed to have had a unicellular flagellated protist ancestor with a posterior flagellum. These three taxonomic groups constitute the opisthokonts.

2.
a. Septa
b. Nuclei
c. Cell wall
d. Hypha

3. Haustoria are branched hyphae that penetrate the target cell's wall, but not the cell's membrane. They invaginate into the membrane so that materials can cross the membranes of the cell and the hyphae.

4. Mutualistic fungi provide some benefit to the partner, while pathogenic ones lead to reduced growth and vigor of the host. In some cases this distinction is cloudy. For example, the photosynthetic partners in lichens sometimes grow faster on their own than they do when associated with the fungus in a lichen. However, the photosynthetic organism may not be able to survive a period of drought without the fungus. Endophytic fungi may or may not reduce growth in a plant while conferring resistance against herbivory and stress.

5. Chytrids have motile gametes and haploid zoospores. Zygospore fungi have a unique unicellular, multinucleate zygospore. Ascomycota have a unique microscopic spore-bearing sac called an ascus. Basidiomycota bear their spores on unique clublike structures called basidia.

6. The sporangiophore is produced as part of sexual reproduction in the Zygomycota. It sprouts from the zygospore and generates the sporangium on its tip. It contains numerous haploid spores that germinate to form haploid hyphae. The conidia produced by Ascomycota develop in chains at the tips of specialized hyphae. These contain spores that are produced through mitosis and are part of asexual reproduction.

7. It might be difficult to reestablish the mycorrhizae before there are trees, and vice-versa. Often bringing in some soil from a healthy forest can help. Until the mycorrhizae are established, the addition of nutrients and water can provide a temporary boost to trees lacking their ectomycorrhizal associates. While fewer than 200 species of arbuscular mycorrhizae have been identified, finding the appropriate associate might be difficult.

8. Fungi are unable to secrete any toxic substances they absorb from the environment, and so they are sensitive to pollution. Lichen samples can be analyzed for the toxins they contain, serving as a record of the types of pollution present during their growth. Scientists have collected fungal species for many decades, and thus have inadvertently supplied a diary of environmental conditions. This may be useful as we discover the importance of particular toxins; the library of fungal samples can be analyzed for the presence of compounds well after the samples were collected.

### Test Yourself

1. **d.** Fungi have haploid and diploid nuclei at different stages of their life cycle, and many have a dikaryotic stage. They also produce spores. The chloroplasts are not found in the fungi but in plants and algae.

2. **b.** The large surface area-to-volume ratio of the hyphae increases the ability of a fungus to absorb nutrients.

3. **e.** When two hyphae of different mating types fuse, they form dikaryotic hyphae.

4. **d.** Dikaryotic hyphae are neither truly diploid ($2n$) nor haploid ($n$). Because dikaryotic hyphae include genetic material from two haploid nuclei that remain separate, the best way to represent their ploidy is $n + n$.

5. **a.** Not all fungi have motile gametes; only the gametes of chytrids are motile.

6. **e.** Mycorrhizae are associations between fungi and the roots of plants. Lichens are symbiotic relationships between fungi and cyanobacteria or green algae.

7. **a.** If this fungus is indeed a basidiomycota, the fusing of hyphae of different mating types would most likely result in dikaryotic hyphae that are segmented by septa.

8. **b.** Coenocytic hyphae are characteristic of both the chytrids and the zygomycota, but only the zygomycota regularly produce sporangiophores.

9. **a.** Among the fungi, only the chytrids have a multicellular haploid stage and a true multicellular diploid stage.
10. **e.** Saprobes are organisms that absorb nutrients from dead matter. Some bacteria are saprobes.
11. **c.** The dikaryotic mycelium is a structure formed in the sexual reproduction life cycle.
12. **a.** Fungi are plentiful in warm, wet environments, but many also survive extreme temperatures and very dry conditions.
13. **b.** The fungal component of lichens surrounds the algal component. Fungi absorb material from their environment and are unable to excrete compounds that may be toxic.
14. **c.** The correct associations are as follows: Microsporidia/microspore fungi; Chytrids/chytrids; Zygomycota/zygospore fungi; Glomeromycota/mycorrhizae; Ascomycota/sac fungi; Basidiomycota/club fungi.
15. **d.** The more recently the fungi have spread, the more likely they are to share genetic similarities.
16. **e.** Mildew enters the plant via stomata. In dry weather, plants close their stomata for at least part of the day. In damp weather, stomata may stay open longer, allowing greater access for the fungus.

# Animal Origins and Diversity

## The Big Picture

- The animals are a monophyletic group, sharing a number of morphological and genetic traits. Animals are motile, multicellular organisms that must ingest nutrients. They are classified according to their early embryonic patterns and body plan characteristics, including symmetry, body cavity structure, segmentation, and type of appendages.
- The eumetazoans encompass all animals except the sponges. Ctenophores and cnidarians are diploblastic eumetazoans that have radial symmetry. Bilaterians are divided into protostomes (mouth develops from the blastopore) and deuterostomes (anus develops from the blastopore).
- Protostomes include lophophores with wormlike body forms and echdysozoans with a rigid cuticle covering their bodies. The ecdysozoan group of arthropods dominate life on Earth today, both in numbers of species and numbers of individuals.
- There are many fewer deuterostomes than protostomes, but many deuterostomes, including vertebrates, are large and ecologically important. Terrestrial vertebrate life is dominated by reptiles and mammals.

## Study Strategies

- Many of the same evolutionary "themes" can be found in widely divergent species. Venomous or venom-producing structures, for example, are found in many different animal groups, from the cnidarians to spiders to snakes. As the different animal groups are described, create a table of the various characteristics or mechanisms that consistently appear. Use this table to help you organize your thinking about evolution and diversity.
- The phylogenetic trees in the textbook are good frameworks to which more information can be added. They can provide a point of reference when studying the different groups. Remember that animals have a set of traits shared with and inherited from their ancestors, as well as distinctive derived traits.
- It is important to understand what features of an animal group may give it an advantage in its environment and the compromises (trade-offs) that are made in other areas. Don't forget that even though particular animals may be placed in a group or clade, not all species have all of the features that characterize the group. For example, think about the reasons that sponges are considered animals even though they lack many of the features of other animals.
- Sometimes learning the Latin root of a group can help you organize your thinking. Although the names are unfamiliar at first, they will make sense if their meanings are understood.
- Go to **LaunchPad** (or use the Web addresses listed) to review the following additional resources:

  Animated Tutorial 23.1 Life Cycle of a Cnidarian (PoL2e.com/at23.1)

  Animated Tutorial 23.2 An Overview of the Protostomes (PoL2e.com/at23.2)

  Animated Tutorial 23.3 An Overview of the Deuterostomes (PoL2e.com/at23.3)

  Animated Tutorial 23.4 Life Cycle of a Frog (PoL2e.com/at23.4)

  Activity 23.1 Animal Body Cavities (PoL2e.com/ac23.1)

  Activity 23.2 Sponge and Diploblast Classification (PoL2e.com/ac23.2)

  Activity 23.3 The Amniote Egg (PoL2e.com/ac23.3)

  Activity 23.4 The Major Groups of Organisms (Pol2e.com/ac23.4)

  Apply the Concept in textbook Sections 23.3 and 23.4

## Key Concept Review

### 23.1 Distinct Body Plans Evolved among the Animals

Animal monophyly is supported by gene sequences and cellular morphology

Basic developmental patterns and body plans differentiate major animal groups

Most animals are symmetrical

The structure of the body cavity influences movement

Segmentation improves control of movement

Appendages have many uses

Nervous systems coordinate movement and allow sensory processing

Although there are exceptions, the following characteristics are commonly associated with animals: multicellurity, heterotrophy, internal digestion, and the capacity either to move to their food or to bring it to them. However, some animals have stages in their life cycle in which movement does not take place.

Phylogenetic analyses of animal gene sequences support the conclusion that animals are monophyletic (see Figure 23.1). Animals generally share the following morphological and genetic synapomorphies: tight junctions, desmosomes, and gap junctions between their cells; extracellular matrix molecules, including collagen and proteoglycans; and Hox genes that specify body pattern and axis formation. These traits were probably possessed by the common ancestor of all animals but have been lost in some groups.

The common ancestor of modern animals may have been a colonial flagellated protist such as a choanoflagellate (see Figure 23.2). Differences in patterns of embryonic development, including cleavage patterns, gastrulation patterns, and the number of cell layers present, show the evolutionary relationships among animals. On these bases, animals can be grouped as diploblastic or triploblastic.

Diploblastic animals have two cell layers: the endoderm and the ectoderm. Triploblastic animals have three cell layers: the endoderm, mesoderm, and ectoderm. Triploblastic animals are divided further into protostomes ("mouth first"; the blastopore becomes the mouth and the anus forms later) and deuterostomes ("mouth second"; the blastopore becomes the anus and the mouth forms later). Sequencing data indicate that the protostomes and deuterostomes are two different animal clades. Together, they are known as the bilaterians (see Figure 23.1) and account for most of the animal species.

Basic developmental patterns and body plans differentiate major animal groups. The overall organization of an animal's body is known as its body plan. The features of an animal's body plan include symmetry or asymmetry, body cavity structure, support (skeletal) structure, segmentation, and the presence or absence of appendages.

Most animals are symmetrical. An animal that can be divided along at least one plane into similar halves is said to be symmetrical. Animals exhibiting bilateral symmetry can be divided into two mirror images by a single plane that passes through the midline of the body. Bilateral symmetry is common among animals that are able to move quickly. Bilaterally symmetrical animals often have sense organs and nervous tissue concentrated at the anterior end; this type of organization is known as cephalization (from the Greek word for "head").

Animals have one of three types of body cavities (see Figure 23.4). Acoelomates have no enclosed body cavity. Pseudocoelomates have a liquid-filled space known as the pseudocoel in which many of the internal organs are located. Coelomates have a true body cavity, a coelom, which develops within the mesoderm. In coelomates the internal organs are in pouches of the peritoneum. The structure of the body plan influences an animal's movement.

Fluid-filled body cavities act as hydrostatic skeletons for many animals. Other animals evolved rigid supportive skeletons that can be internal (bones or cartilage) or external (a shell or cuticle). Muscles attached to hard skeletons allow the animal to move.

Segmentation allows for specialization of the different body regions and can improve control of movement. In some animals segments are not apparent (e.g., vertebrae column). In other animals similar body segments are repeated many times, and in yet others the body segments differ (see Figure 23.5).

Appendages, especially jointed limbs, enhance locomotion. Jointed limbs in the arthropods and vertebrates are a major factor in their evolutionary success. Other appendages are specialized and can be used to sense the environment (e.g., antennae) or can be used to capture prey.

**Question 1.** Explain the fundamental difference between protostomes and deuterostomes. What evidence supports the division of organisms into these two clades?

**Question 2.** What are some of the limitations imposed by a hydrostatic skeleton? Are the limitations the same for terrestrial animals?

## 23.2 Some Animal Groups Fall Outside the Bilataria

Sponges are loosely organized animals

Ctenophores are radially symmetrical and diploblastic

Placozoans are abundant but rarely observed

Cnidarians are specialized carnivores

The simplest animals are the sponges, which have no body symmetry and no distinct cell layers. Placozoans have only four cell types and weakly differentiated layers of tissue. All animals other than the sponges and placozoans make up the eumetazoans. Eumetazoans have body symmetry, a defined gut, a nervous system, and distinct organs.

The Bilateria are a monophyletic group that includes all of the eumetazoans except the ctenophores and the cnidarians. Bilaterian synapomorphies include bilateral symmetry, three cell layers, and the presence of at least seven Hox genes. The bilaterians have two major subgroups: the protostomes and the deuterostomes.

Sponges and placozoans are weakly organized animals. Sponges can be classified in three different groups based on molecular evidence and the morphology of their spicules. They have differentiated cells but no true organs. Although the sponge body plan is relatively simple, the three sponge groups all have cells that are differentiated for specific functions. Most of the 8,000 species of sponges are marine filter feeders that remove small organisms and nutrient particles from seawater as it flows through pores in the walls of their inner cavity (see Figure 23.2). A few sponges are carnivores and can trap prey on hook-shaped spicules that are on the outside of the body surface. Sponges have a supporting skeleton composed of branching spines (spicules). The spicules of glass sponges and demosponges (the largest sponge group) are made of silicon. The calcareous sponges take their name from their calcium carbonate spicules and are the sponge group most closely related to the eumetazoans. In addition to spicules, sponges also have an extracellular matrix composed of collagen, adhesive glycoproteins, and other molecules that hold the cells together.

The huge variety of sponge body sizes and shapes is a response to the different movement patterns of water, specifically tides and currents. Sponges that live in environments that have strong wave action tend to be firmly attached to substratum, while those that live in slowly moving water are generally flat and oriented at right angles to the current flow. Sponges reproduce sexually by producing both egg and sperm, and asexually by budding and fragmentation.

Placozoans are structurally very simple, with a diploblastic body plan. Placozoans do not have a mouth, gut, or true nervous system, and they consist of only four distinct cell types. They have upper and lower epithelial cell layers with contractile fiber cells between the layers. Based on phylogenetic analysis, their structural simplicity may have been secondarily derived (i.e., some of their common features were probably lost sometime during evolution from a common ancestor). The life cycle of placozoans remains relatively unknown because their transparency makes it difficult to observe them in nature (see Figure 23.8B). It is known that they have a pelagic (free-swimming) stage in the ocean and that they are capable of sexual and asexual reproduction.

The ctenophores (comb jellies) are marine diploblastic animals that are radially symmetrical (see Figure 23.7). There are 150 known species, and they are found primarily in the open ocean, where they feed on planktonic organisms that are filtered by sticky filaments on the tentacles. Ctenophores have two cell layers (ectoderm and endoderm) separated by mesoglea. Although they are classified as eumetazoans, they lack most of the Hox genes. However, they have a complete gut (i.e., a gut with an entrance and exit, or mouth and anus). Ctenophores get their name from the eight rows of comblike plates of cilia, known as ctenes. The cilia are used to propel the animal through the water. Prey is caught on sticky filaments on the tentacles or body (see Figure 23.7). Ctenophores have a simple life cycle and reproduce sexually. In most species the externally fertilized egg hatches into a miniature ctenophore (i.e., direct development).

The cnidarian life cycle has two stages: the polyp and the medusa. All cnidarians are diploblastic and have radial symmetry. They include sea anemones, corals, and jellyfish. The cnidarian gastrovascular cavity is a blind sac, so they do not have a complete gut. There is only one opening, which serves as both mouth and anus. The gastrovascular cavity functions in food digestion, respiratory gas exchange, and circulation. It also lends support as a hydrostatic skeleton. As in the ctenophores, a large amount of mesoglea is found between the two cell layers of cnidarians. Cnidarian tentacles have specialized cells, called cnidocytes, which inject toxins into their prey with the help of stingers called nematocysts (see Figure 23.10).

The life cycle of most cnidarians includes a sessile polyp stage and a motile medusa stage (see Figure 23.9). The polyp stage usually reproduces asexually. The medusa stage reproduces sexually, with the fertilized egg becoming a planula larva that eventually develops into a polyp.

Cnidarians possess muscle fibers that enable them to move and simple nerve nets that integrate their activities. All but a few of the 11,000 or so species of cnidarians are marine. The smallest individuals are almost microscopic, while individual jellyfish can be quite large. Many cnidarians are colonial.

Anthozoans comprise about 6,000 species of sea anemones, sea pens, and corals.

Scyphozoans include the marine jellyfish. Hydrozoans include both freshwater and marine species.

**Question 3.** Discuss the two major strategies animals use to get food, and indicate which of these would most likely be used by sessile organisms.

**Question 4.** How does the body structure of sponges reflect their mode of food acquisition?

## 23.3 Protostomes Have an Anterior Brain and a Ventral Nervous System

### Cilia-bearing lophophores and trochophore larvae evolved among the lophotrochozoans

### Ecdysozoans must shed their cuticles

Protostomes are highly diverse, but most are bilaterally symmetrical, possessing two major derived traits: an anterior brain surrounding the entrance to the digestive tract and a ventral nervous system consisting of paired or fused longitudinal nerve cords. Many species in both groups have a wormlike appearance. The blastopore of the embryo develops into the mouth in almost all protostomes.

Most protostomes belong to one of two major groups: the lophotrochozoans and the ecdysozoans. The arrow worms are not placed in either group; they may be sister to the protostomes as a whole, or they may be more closely related to the lophotrochozoans. The 100 or so species are small marine predators.

The common ancestor of the protostomes had a coelom (a fluid-filled cavity within the mesoderm). However, the protostomes include some groups that are coelomate and some that are pseudocoelomate. One important group, the flatworms, is acoelomate (lacks a coelom). Two prominent protostome groups, the arthropods and the mollusks, have had secondary evolutionary modifications of the coelom. In the arthropods, the coelom has become a hemocoel; the mollusks have returned secondarily to a virtually open circulatory system.

The lophotrochozoans get their name from two structures: the lophophore (a circular or U-shaped ring of ciliated, hollow tentacles used for feeding and gas exchange found in a number of groups in this clade) (see Figure 23.11A) and the trochophore larvae (see In-text Art, p. 481). A number of lophotrochozoan groups, including the annelids (segmented worms) and mollusks, undergo spiral cleavage during early development.

The 4,500 species of bryozoans are colonial; strands of tissue connect individuals in each colony, and in some species individuals are specialized for feeding, reproduction, defense, and support (see Figure 23.11B). Individuals have a great deal of control in manipulating their lophophores to increase contact with prey. Colonies grow via asexual reproduction of the founding members. Sexual reproduction also occurs; eggs are brooded internally and larvae emerge to seek suitable sites to form new colonies. Bryozoans can cover large areas of coastal rock and can even form small reefs in shallow seas.

Flatworms lack respiratory organs and have only simple cells for waste removal. The flat shape of the animal helps in oxygen transport and waste removal (see Figure 23.12A). The digestive tract consists of a mouth that opens into a blind sac that has many branches and aids in nutrient absorption. Although there are some free-living flatworm species, most are parasites, including about 25,000 species of tapeworms or flukes. Parasitic flatworms feed on the nutrient-rich body tissues of their host animal and disperse their eggs in the host's feces.

Most of the 1,800 species of rotifers are very small (some smaller than single-celled protists), but they have specialized internal organs, including a complete gut and a pseudocoel that serves as a hydrostatic skeleton (see Figure 23.13). Cilia are used to propel the rotifers through the water. Most species live in freshwater habitats and feed using a ciliated organ called a corona. Some species have both males and females; some have only one sex and reproduce asexually. Rotifers are the only group of animals known to have existed for millions of years without the benefits of sexual reproduction.

Nemerteans, or ribbon worms, have a complete digestive tract with two openings. Small ribbon worms use their cilia for movement, and large ribbon worms move by means of muscle contractions. The feeding organ of the ribbon worms is a proboscis, which lies within a rhynchocoel, or fluid-filled cavity (see Figure 23.14). The proboscis has a sharp stylet and can be forcefully ejected from the body to catch prey. Most of the 1,000 or so species are marine, although a few species are found in fresh water or on land. Most are small, but some species can be up to 20 meters long.

The phoronids include 20 species of tiny sessile worms that live in chitinous tubes and extract food from the water with their lophophores (see Figure 23.16). While the lophophore is similar in function to that of the bryozoans, these groups are not closely related based on genetic analyses. This indicates that the lophophore structure has evolved more than once.

Brachiopods are solitary marine animals that live attached to the substratum. Their divided shell gives them a superficial resemblance to bivalve mollusks (e.g., clams), but the two halves are dorsal and ventral instead of lateral (see Figure 23.15). The lophophore is located in the shell, and cilia help draw water and food into the shell. More than 26,000 fossil brachiopod species have been described, but only about 335 species are known to exist today.

The annelids consist of approximately 16,500 species of segmented worms living in marine, freshwater, and moist terrestrial environments. Their thin body wall serves as a surface for gas exchange, and they are restricted to these environments because the thin covering causes them to lose moisture rapidly when exposed to air. A segmented body plan gives these worms extremely good control of their movement. A separate nerve center called a ganglion controls the movement of each segment, and in most cases each segment also contains an isolated coelom (see Figure 23.17). Many annelids have more than one pair of eyes and tentacles (see Figure 23.18A). Outgrowths called parapodia, used in gas exchange, extend laterally from segments over much of the body. Setae extending from the parapodia help attach the animal to the substrate and aid in movement.

Pogonophorans have secondarily lost their digestive tract and secrete tubes made of chitin and other substances that come from their surroundings (see Figure 23.18B). Found in the deep ocean near hydrothermal vents, they harbor a number of endosymbiont bacteria in a specialized organ known as a trophosome. The bacteria provide much of their nutrition.

The oligochaetes ("few hairs") include the most familiar annelids, the earthworms. Oligochaetes live mainly in freshwater and terrestrial environments and are hermaphroditic, with both male and female reproductive organs in the same individual. Sperm is exchanged between two individuals outside the worm's body in a cocoon secreted by the clitellum.

Leeches are also hermaphroditic species that live either in fresh water or on land (see Figure 23.18C). The coelom of these parasitic annelids is not segmented but is composed of undifferentiated tissue. Clusters of segments at their anterior and posterior ends are modified into suckers, which the leech attaches to the substratum for movement, or to a host mammal from which it sucks blood (its nutritional source). To keep the blood from clotting, the leech secretes an anticoagulant.

About 100,000 species are included in the four main groups of mollusks: chitons, gastropods, bivalves, and cephalopods. Mollusks have evolved into this morphologically diverse group based on a distinctive three-part body plan including a foot, a visceral mass, and a mantle (see Figure 23.19A). All species have a large muscular foot. In some groups, such as the clams, the foot is a burrowing organ, while in squids and octopuses it has been modified and consists of arms and tentacles; the tentacles bear complex sensory organs. Organs such as the heart, the digestive tract, and the reproductive system are concentrated centrally in a visceral mass. A tissue fold, known as the mantle, covers a visceral mass of internal organs. In many species, the mantle secretes a hard, calcareous shell. In most species the mantle is extended to create a mantle cavity holding the gills used in respiration. Mollusks have a secondarily reduced coelom (see Figure 23.19A). The blood vessels of the mollusks do not form a closed circulatory system. Instead, blood and other fluids empty into a hemocoel through which fluid moves around the animal to deliver oxygen to the internal organs. Mollusks have a heart that moves the blood back into the blood vessels.

Chitons are marine mollusks that feed on algae, bryozoans, and other organisms that they scrape off rocks using a razorlike body structure called the radula. A chiton's shell consists of eight overlapping plates that are surrounded by a girdle (see Figure 23.19B). They have simple internal organs, multiple gills, and bilateral symmetry.

Gastropods are the most species-rich and widely distributed of the mollusks, and are found in all environments. There are shelled and unshelled gastropod species, including the snails and slugs (the only terrestrial mollusks), as well as the marine nudibrachs (sea slugs), whelks, limpets, and abalones. Gastropods use their foot either to crawl or swim (see Figure 23.19C). Land snails and slugs are able to survive on land, as the mantle tissue is modified into a highly vascularized lung.

The 30,000 living species of bivalve mollusks include the familiar clams, oysters, scallops, and mussels. They are found in both salt and fresh water, but they are all aquatic. Bivalves use an opening called an incurrent siphon to bring water into their two-part hinged shells (see Figure 23.19D); they are filter feeders that extract foodstuffs from these water currents. Large gills inside the shell extract the food and also function as respiratory organs. Water and gametes exit from an excurrent siphon. Among the clams, the molluscan foot has been modified into a digging device that allows the animal to burrow into mud or sand.

Cephalopods include the octopuses, squids, and nautiluses. The excurrent siphon is modified to give cephalopods the ability to control water movement into the mantle, allowing them to use ejected water as a jet for propulsion (see Figure 23.19E). They capture prey with their tentacles (see Figure 23.20B), and their greatly enhanced mobility makes them dominant ocean predators. They are also able to control gas movement in the mantle, which helps in buoyancy control. As is typical of active, rapidly moving predators, cephalopods have a head with complex sensory organs, most notably the eyes, which are in many ways comparable to those of vertebrates.

Nautiluses are the surviving cephalopods with external chambered shells. Shells have been lost by species in several molluscan groups, including gastropods (slugs and nudibranchs) and cephalopods (octopus)—shell protection in these groups is replaced by protective strategies including toxins and evasion tactics.

The ecdysozoans include more species than all other lineages combined. They are characterized by a rigid external covering (cuticle or exoskeleton), which must be molted periodically in order to allow the animal to grow (*ecdysis* is the Greek word for "shedding"). The new exoskeleton is temporarily soft after molting, leaving the animal vulnerable (see Figure 23.21B). Increasing molecular evidence, including a set of Hox genes shared by all ecdysozoans, supports the monophyly of these animals and suggests that cuticle molting is a lifestyle that may have evolved only once.

Wormlike ecdysozoans, which include the priapulids and the kinorhynchs, are unsegmented. They have a thin cuticle that allows gas, mineral, and water exchange but affords little protection and support and restricts the animals to moist environments.

Arthropods have a hard exoskeleton of chitin that provides protection, waterproofing, and muscle attachment sites. Their bodies are segmented, with individual muscles attached to the exoskeleton that operate each segment. The jointed appendages that give the group its name (*arthros* = joint, *poda* = limb) allow for a great range of movement and the specialization of different appendages for different purposes. Chitin is waterproof and keeps the animal from dehydrating. Arthropods are Earth's dominant animals in both number of species and number of individuals.

Nematodes, or roundworms, range in size from microscopic to up to 9 meters in length. About 25,000 species of this diverse group have been described, and these may represent less than a quarter of extant nematode species. They live in soil, on the bottoms of lakes and streams, and in marine sediments. Nematodes use their gut for both gas exchange and nutrient uptake (see Figure 23.22). Many species prey upon protists and other microscopic organisms, but of most significance to humans is the large number of parasitic species; several of these, including *Trichinella spiralis*, are dangerous to humans. The free-living nematode *Caenorhabditis elegans* is a model organism widely used by geneticists and developmental biologists.

The 320 species of horsehair worms are long and very thin, as their name suggests. They live in fresh water or very damp soil near the edges of ponds and streams and feed in the larval stage as parasites of terrestrial and aquatic insects and crabs (see Figure 23.23). Adults have no mouth and reduced guts. In some species the adults may not feed. The adults of other species continue to grow, so it is also possible that adult worms may absorb nutrients from the environment during the time between the molting of the old cuticle and the hardening of the new one.

Some marine ecdysozoans have a thin exoskeleton, called a cuticle, that is molted periodically. The cuticle allows for gas exchange with the environment, but it offers little protection. The priapulids, kinorhynchs, and loriciferans are groups of tiny wormlike marine animals that live burrowed in ocean sediments (see Figure 23.24). Most priapulids have a larval stage; the kinorhynchs do not. The loriciferans were only recently discovered (in 1983), and they are still being described.

**Question 5.** The figure below shows a generalized molluscan body plan (A) and the body plans that are characteristic of gastropods (B) and cephalopods (C). Label each structure in the figure.

(A) Generalized molluscan body plan

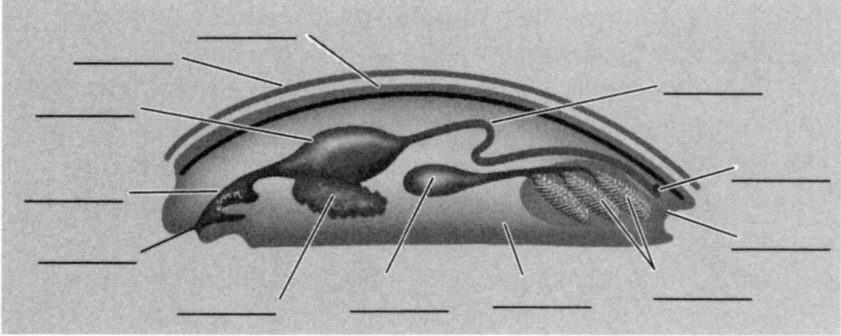

(B) Gastropods

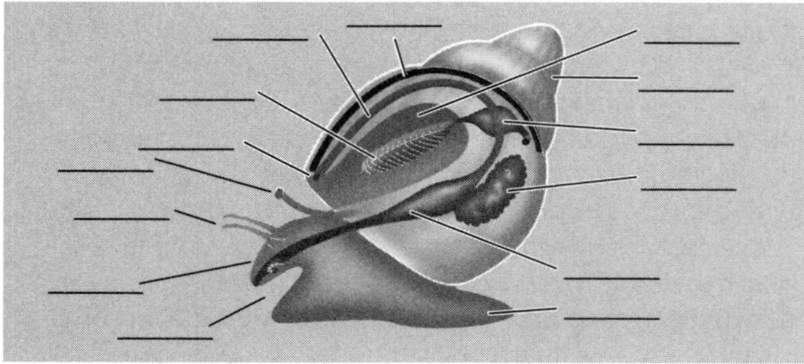

(C) Cephalopods

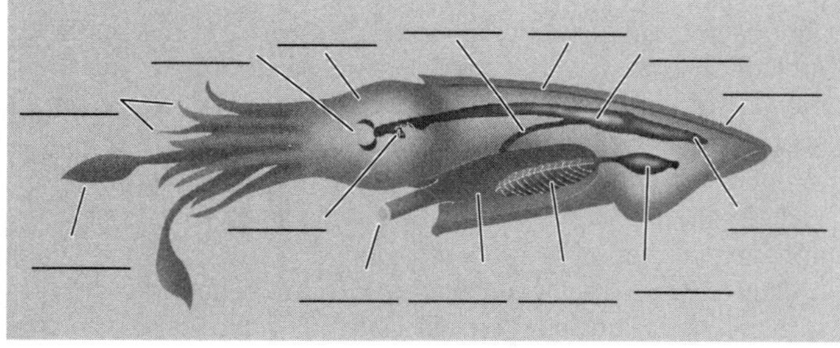

**Question 6.** What advantages did the incorporation of chitin into a cuticle have for ecdysozoans and what were some disadvantages?

## 23.4 Arthropods Are Diverse and Abundant Animals

Arthropod relatives have fleshy, unjointed appendages

Chelicerates are characterized by pointed, nonchewing mouthparts

Mandibles and antennae characterize the remaining arthropod groups

More than half of all described species are insects

The arthropods and their relatives are ecdysozoans with paired appendages.

Today four arthropod groups—crustaceans, hexapods, myriapods, and chelicerates—are all numerous and are found in all environments. Several key features have contributed to their success. Their bodies are segmented, and their muscles are attached to the inside of their rigid exoskeletons. Each segment has muscles that operate that segment and the jointed appendages attached to it. A rigid exoskeleton (composed of chitin) provides waterproofing and protection from predators. The relationships among the arthropod groups are currently being revised based on gene sequence data, and current data support the idea that arthropods may be monophyletic.

Arthropod relatives have fleshy, unjointed appendages. The onychophorans (velvet worms) were once thought to be more closely related to the annelids, but recent genetic analysis links them to the arthropods. They have unjointed legs and thin cuticles composed of chitin (see Figure 23.25A). Onychophorans have probably changed relatively little from their common ancestor with the arthropods. The tardigrades have fleshy, unjointed legs and use their fluid-filled body cavities as hydrostatic skeletons (see Figure 23.25B). Tardigrades lack a circulatory system and do not have gas exchange organs. When their environment dries out, they shrink in size drastically and can survive in a dormant state for over a decade.

Four major arthropod groups survive: the myriapods (centipedes and millipedes), the chelicerates (including the arachnids—spiders, mites, ticks, etc.), the crustaceans (crabs, lobsters, scallops, barnacles, etc.), and the hexapods (insects and their relatives).

Chelicerates are characterized by pointed, nonchewing mouthparts. The 98,000 described species are placed into three major clades: pycnogonids, horseshoe crabs, and arachnids. All chelicerates have two body parts, and most have eight legs. The pycnogonids, or sea spiders, are a group of about 1,000 exclusively marine species. Most are very small and are rarely seen (see Figure 23.26A). There are only four living species of horseshoe crabs. These animals have changed so little in morphology over evolutionary time that they are often referred to as "living fossils" (see Figure 23.26B). The arachnids are the most prominent chelicerates and include the spiders, scorpions, mites, and ticks (see Figure 23.27). They have a simple life cycle, with young that resemble small adults. Spiders, the most familiar arachnids, build webs of protein threads that they use to capture prey. Spider webs are strikingly varied, often species-specific, and increase the predatory ability of spiders in many different environments. Spiders use hollow chelicerae to inject venom into prey. Mites and ticks are vectors for a variety of organisms that cause diseases in plants and animals.

Mandibles and antennae characterize the remaining arthropod groups. The myriapods include 3,000 described species of centipedes and 11,000 described species of millipedes. Members of both groups have similar body plans, consisting of a well-formed head and a long, segmented trunk. Centipedes have one pair of legs on each trunk segment (see Figure 23.28A); millipedes have two pairs on each segment (see Figure 23.28B). Crustaceans include many familiar animals (decapods, which include shrimp, lobsters, crabs, barnacles; see Figure 23.29A) as well as the less-familiar isopods (which include sowbugs; see Figure 23.29B), amphipods, ostracods, copepods (see Figure 23.29C), and branchiopods (see Figure 23.29D). About 50,000 species have been described so far. Barnacles are an unusual crustacean that is sessile as an adult (see Figure 23.29E).

According to recent gene sequencing, crustaceans may be paraphyletic with respect to the hexapods. The crustacean body is divided into the head, thorax, and abdomen (see Figure 23.30A). In many species, a carapace extends dorsally from the head to protect and cover the body. The thorax and abdomen have one pair of appendages each; crustacean appendages are specialized for walking, swimming, feeding, sensation, and gas exchange. In some species complex branched appendages have evolved. Fertilized eggs remain attached to the body of the female during the early stages of development. Upon hatching, the young are released as larvae in some species. In other species, the juveniles are similar in form to the adults. In some species, the fertilized eggs are released into the water or attached to an object.

Insects are six-legged terrestrial hexapods. More than half of all described species are insects. They are prominent occupants of terrestrial and freshwater environments. More than a million species of insects have been described so far, and many biologists believe this is only a small fraction of the species that actually exist. (See Table 23.2 and Apply the Concept, p. 496.)

Three groups of wingless hexapods—the springtails, two-pronged bristletails, and proturans—are related to the insects and probably resemble the insect ancestral form most closely. Members of these three groups differ from insects in having internal, rather than external, mouthparts. Springtails, which are probably the most abundant hexapod, have a very simple lifestyle in which the juveniles resemble the adults.

Like the crustaceans, the insect body has three parts: the head with a pair of antennae, the thorax with pairs of legs, and the abdomen. Unlike other arthropods, the abdominal segments do not bear appendages (see Figure 23.30B). Gas exchange occurs in a system composed of a series of air sacs and tubular channels, called tracheae, that extend from external openings called spiracles. Other distinguishing characteristics of insects include paired antennae with a sensory receptor known as Johnston's organ, three pairs of legs on the thorax, and external mouthparts.

There are two classes of insects: the wingless apterygotes and the winged pterygotes (some species of which have secondarily become wingless). The apterygotes, including the jumping bristletails and silverfish, have simple life cycles. Pterygotes typically have two pairs of wings attached to the thorax, but in some groups (e.g., parasitic lice and fleas, some beetles, and worker ants), one or both pairs of wings have been secondarily lost.

Hatching pterygotes do not look like the adults, and they undergo changes at each molt. Each stage between molts is called an instar. If the changes are gradual, the insect is said to undergo incomplete metamorphosis. If a drastic change occurs between some instars, then the insect is said to undergo complete metamorphosis. The most dramatic example is the change that occurs when a caterpillar transforms into a pupa in which the adult form develops. In insects that undergo complete metamorphosis, each stage is often specialized with regard to the environment and food source.

Pterygote insects were the first animals to evolve the ability to fly. Homologous genes control the development of insect wings and crustacean appendages and there is evidence that insect wings evolved from a dorsal branch of a crustacean appendage that functioned in gas exchange.

Most flying insects have two pairs of stiff, membranous wings attached to the thorax. Those known collectively as the neopterans can fold their wings over their bodies, allowing them to fit into tight crevices. True flies have one pair of wings plus a pair of stabilizers called haltares. In beetles one pair of wings forms a hardened cover. Insects that cannot fold their wings over their bodies include dragonflies and mayflies. The aquatic larvae can be predatory or herbivorous. This is an ancestral form for pterygotes. Dragonflies are active predators, while adult mayflies do not have a functional digestive tract.

Neopteran insects that undergo incomplete metamorphosis include grasshoppers (see Figure 23.31B), termites, stone flies, earwigs, thrips, true bugs (see Figure 23.31C), aphids, cicadas, and many others. In these groups, the hatchlings resemble small, usually wingless adults; as they molt from one stage to the next they gradually acquire more adult characteristics.

More than 80 percent of neopterans undergo complete metamorphosis (see Figure 23.30D), including beetles (see Figure 23.31E), lacewings, caddisflies, butterflies, moths (see Figure 23.31F), sawflies and true flies (see Figure 23.31G), wasps, bees, and ants (see Figure 23.31H). In these groups the larvae and adult forms are substantially different; the young pass through at least two stages (larva and pupa) before becoming adults. They form a subgroup of the neopterans called the holometabolous insects.

**Question 7.** What are some advantages to segmentation?

**Question 8.** Distinguish the body plan of insects from that of crustaceans and other arthropods.

## 23.5 Deuterostomes Include Echinoderms, Hemichordates, and Chordates

- Echinoderms have unique structural features
- Hemichordates are wormlike marine deuterostomes
- Chordate characteristics are most evident in larvae
- Adults of most lancelets and tunicates are sessile
- The vertebrate body plan can support large, active animals
- There are two groups of living jawless fishes
- Jaws and teeth improved feeding efficiency
- Fins and swim bladders improved stability and control over locomotion

The deuterostomes include the echinoderms, hemichordates, and chordates.

There are many fewer species of deuterostomes than there are of protostomes, but the deuterostomes are of special interest because mammals—a group that includes the largest living animals as well as the human lineage—are deuterostomes. Deuterostomes are united based on early developmental patterns.

Deuterostomes fall into three major clades: the echinoderms, the hemichordates, and the chordates. All are triploblastic, coelomate animals (see Figure 23.4C). Skeletal support features, when present, are internal. A few species have segmented bodies, but the segments are not visible. Echinoderms and hemichordates (collectively known as ambulacrarians) have bilaterally symmetrical, ciliated larvae (see Figure 23.32A). Adult hemichordates are bilaterally symmetrical, while echinoderm adults have unique pentaradial symmetry. Other deuterostome groups retain bilateral symmetry.

While 23 major groups consisting of about 13,000 species of echinoderms have been described from the fossil record, only six groups containing 7,000 species live today. All known species occur in marine environments.

In addition to their pentaradial symmetry, echinoderms have two unique structural characters: a system of water-filled canals called a water vascular system that functions in feeding, gas exchange, and locomotion, and an internal skeleton made up of calcified plates (see figure 23.32B).

Echinoderms are divided into two major clades. The crinoids include about 80 species of stalked, sessile animals called sea lilies, and 600 species of feather stars, which have flexible appendages that allow for limited movement (see Figure 23.33A).

The remaining echinoderms are generally mobile and are divided into two main groups: the echinozoans (sea urchins and sea cucumbers; see Figures 23.33B, C) and the asterozoans (sea stars and brittle stars; see Figures 23.33D, E). Hemichordates are wormlike marine deuterostomes. The 100 species of hemichordates include the acorn worms and the pterobranchs. Both groups have a body plan composed of a trunk, a collar, and a sticky proboscis used for catching prey and digging (see Figure 23.34A). In the acorn worm, cilia move food from the proboscis to the mouth. Behind the mouth are pharynx and an intestine. The pharynx can open to the outside through pharyngeal slits that contain vascularized tissue for gas exchange. Acorn worms breathe by pumping water through the mouth and out the pharyngeal slits. They burrow in sand or mud and extract food items from the substrate.

Pterobranchs are sedentary marine animals that are either solitary or colonial (see Figure 23.34B). Behind the proboscis are one to nine pairs of arms that are used to capture prey and also function in gas exchange. Chordate characteristics are most evident in larvae.

The features that reveal the evolutionary relationship between the echinoderms and chordates, as well as among the chordates, are primarily observed in the larval stage or during the early states of development.

The three principle chordate clades are the cephalochordates, the urochordates, and the vertebrates. All three chordate clades share the following at some point during development (see Figure 23.35): (1) a hollow dorsal nerve cord; (2) a tail that extends beyond the anus; (3) a dorsal supporting rod called the notochord; and (4) pharyngeal slits.

The notochord is the main distinguishing characteristic of the chordates. In urochordates, the notochord is lost during the metamorphosis from larval to adult forms. In vertebrates, the notochord is replaced by skeletal structures called the vertebrae (spinal column).

Ancestral pharyngeal slits are generally lost in adults, but are retained in some urochordates and cephalochordates (see Figure 23.35). A pharynx develops around the slits and becomes large in some chordates, forming a pharyngeal basket.

Adults of most cephalochordates and urochordates are sessile. The 30 species of cephalochordates (lancelets) are small fishlike animals that retain the notochord for their entire life and use their pharyngeal baskets to catch prey (see Figure 23.35B). Fertilization of eggs takes place in the water.

The three major urochordate groups are the ascidians (sea squirts, also known as tunicates), thaliaceans, and larvaceans. All members of these three groups are marine, and more than 90 percent of urochordate species are ascidians. The body form of an adult ascidian is baglike in shape and surrounded by a tough tunic (hence the name "tunicate") (see Figure 23.36).

Ascidian larvae have pharyngeal slits, a nerve cord, and a notochord; it is the larva, not the adult form, that suggests their chordate ancestry (see Figure 23.35A). The adults lose the notochord and nerve cord but have an enlarged pharynx—the baglike pharyngeal basket—for catching food. Ascidians sometimes reproduce asexually by budding, thus forming colonies.

A dorsal supporting structure replaces the notochord in vertebrates. The vertebrates take their name from the unique jointed dorsal vertebral column that replaces the notochord during early development and provides support.

The vertebrates are divided into two groups: the jawless fishes and the gnathostomes. (See Figure 23.38 for the phylogenetic tree that shows the evolutionary relationships among the vertebrates.)

The elongate, eel-like hagfishes are the sister group of the remaining vertebrates (see Figure 23.39A). These jawless fishes look superficially like the lampreys, which are also jawless, but hagfishes lack true vertebrae and have only a partial cranium (skull). Lampreys have true vertebrae (see Figure 23.39B). The hagfishes also have a weak circulatory system with three small accessory hearts; they lack a stomach, and their skeleton is composed of cartilage. Although they lack a jaw, they have a tonguelike structure with toothlike rasps that they use to capture prey. There is some debate as to the accuracy of placing the hagfishes with other vertebrates in the same phylogenetic tree. The analyses of gene sequences suggest that the hagfishes and lampreys are related, and it is possible that during evolution the ancestor of the hagfishes lost some of the structural characteristics that define vertebrates.

Lampreys resemble hagfishes, but they differ biologically. They have a complete braincase and distinct vertebrae made of cartilage. Unlike hagfishes, which undergo direct development, lampreys undergo complete metamorphosis. The larvae, called ammocoetes, are filter feeders, while the adult forms are often parasitic. The mouth is a rasping, sucking organ used by the lampreys to attach themselves to prey. The adults of a few lamprey species do not feed as adults, surviving only long enough to breed.

The four key features of the vertebrate body plan are: (1) an anterior head with a large brain and protective skull; (2) a rigid internal skeleton supported by a vertebral column; (3) a large coelom in which the internal organs are suspended; and (4) a closed circulatory system. This body plan is capable of supporting large-sized animals.

After the Devonian period, jaws evolved from the skeletal arches that supported the gills. The jawed vertebrates fall into the category of gnathostomes, or "jaw mouths." The evolution of the jaw—and then of teeth—led to greatly enhanced feeding efficiency (see Figure 23.40), and jawed fishes were the major predators of the Devonian. Most aquatic gnathostomes have unjointed appendages, called fins, that help control movement through the water (see Figure 23.41).

The chondrichthyan fishes, including sharks, skates, rays, and chimaeras, have flexible and leathery skin and a skeleton composed of cartilage (see Figure 23.41A). Sharks move forward by laterally undulating their body and tail (caudal) fin; skates and rays move vertically by undulating their pectoral fins. Almost all chondrichthyans are marine, but a few live in estuarine waters or migrate to lakes and rivers.

Lunglike gas-filled sacs called swim bladders evolved in the ancestral fishes of the bony fishes. With the aid of the swim bladder and fins, many fish can control their position in the water column with minimal energy expenditure.

The ray-finned fishes and virtually all other vertebrates have a bony skeleton. The outer body is covered by thin, lightweight scales, which aid in movement and provide protection. A hard flap called the operculum, which covers the gills, allows for the movement of water over the gills and thus for gas exchange. The approximately 30,000 species of ray-fins encompass a remarkable variety of shapes, sizes, and lifestyles, and they exploit every kind of food source in the aquatic environment (see Figure 23.42).

**Question 9.** Ascidians, or sea squirts, are armless and legless organisms that spend their adult lives attached to a substrate under water. They feed by pulling water into one tube, filtering out planktonic organisms, and pushing the water out another tube. Grasshoppers have legs, move around freely on land, and feed by ingesting food through a mouth. What evidence has led biologists to believe that humans are more closely related to sea squirts than to grasshoppers?

**Question 10.** Suppose you find an undiscovered fossil of an organism that has radial symmetry and was in a marine environment (as shown by examining the surrounding rock and adjacent fossils in the rock layer). What other features would you look for in order to determine if this fossil is an echinoderm?

**Question 11.** The evolution of a hinged jaw resulted in an extensive radiation of the fishes into the many modern jawed forms. What was the main advantage and significance of hinged jaws in this radiation?

## 23.6 Life on Land Contributed to Vertebrate Diversification

Jointed fins enhanced support for fishes

Amphibians adapted to life on land

Amniotes colonized dry environments

Reptiles adapted to life in many habitats

Crocodilians and birds share their ancestry with the dinosaurs

The evolution of feathers allowed birds to fly

Mammals radiated as non-avian dinosaurs declined in diversity

Most mammals are viviparous

The evolution of lunglike swim bladders in some ray-finned fishes set the stage for the move to the land. Some fishes may have used these sacs to supplement their oxygen supply in low-oxygen freshwater environments, and the sacs may have allowed some fishes to survive out of water. The fins of these fishes were unjointed, so they could only flop around on land.

Two pairs of jointed fins evolved in the ancestors of the coelocanths and the lungfishes, which along with the tetrapods are known as sarcopterygians. The jointed fins of lungfishes are connected to the body by a single large bone, and they have lungs as well as gills. During dry periods when the water from ponds evaporates, they burrow deep in the mud and can survive in an inactive state for many months, breathing air.

Coelacanths were thought to have died out 65 mya, but two species were discovered in the twentieth century. One genus, *Latimeria*, is a predator and can weigh up to 80 kilograms (see Figure 23.43A). Unlike other fishes, *Latimeria* has a skeleton composed primarily of cartilage; this is thought to be a derived feature, since the ancestors to this group had bony skeletons. Only six species of lungfishes remain, all of them in the southern hemisphere (see Figure 23.43B).

The change in fin structure allowed these fishes to support themselves in shallow water and eventually to move onto land and exploit food sources there. These early land-dwelling fishes gave rise to the tetrapods—the four-legged vertebrates.

Amphibians have a wide variety of life histories, but most spend at least part of their life cycle in the water (see Figure 23.45). Many species return to water to lay their eggs, producing aquatic larvae. The skin surface is a common site of respiratory gas exchange in amphibians, but many also have lungs.

There are approximately 6,500 species of extant amphibians, falling into three groups. The caecilians are wormlike, legless, burrowing or aquatic amphibians found only in moist tropical regions (see Figure 23.46A). Anurans include the tail-less frogs and toads, and account for the vast majority of amphibian species (see Figure 23.46B). Anurans undergo metamorphosis from an aquatic to a terrestrial life form. The adults have a very short vertebral column and a pelvic region modified for hopping on the hind legs. Some species have adapted to life in very dry, even desert environments, while others have returned to an entirely aquatic lifestyle. Salamanders are tailed amphibians that exchange respiratory gases through both lungs and gills (see Figures 23.46C, D). One group relies on gas exchange through its skin and mouth as all amphibians do, but it does not have lungs. Paedomorphic evolution (retention of the juvenile form in the adult) has led to several entirely aquatic salamander species. Most species have internal fertilization. Sperm is transferred in a jellylike capsule called a spermatophore.

The social behaviors of many amphibians are quite complex; males make species-specific calls to females and can defend their own breeding territories. The amount of care given to the eggs varies. Some amphibians lay large numbers of eggs that are abandoned, while others guard a few fertilized eggs. A few species are viviparous, meaning that the adult gives birth to well-developed juveniles.

Amphibians are the subject of much attention today, because globally over a third of all species has gone extinct or declined to endangered levels in the last few decades (see Figure 23.46B). Scientists are investigating several hypotheses about their decline, including infection from a pathogenic chytrid fungus.

A key innovation that allowed colonization of dry environments was the amniote egg (see Figure 23.47). The amniote egg has a protective calcium-based shell that inhibits dehydration while allowing oxygen and carbon dioxide to pass through. Within the shell, extraembryonic membranes further protect the embryo, and large quantities of yolk provide nutrition. The amniote animals—reptiles (including birds) and mammals—were thus freed from reliance on water to reproduce.

In several groups of amniotes the egg became modified to allow the embryo to grow inside the mother. In mammals, the egg has lost the shell and the embryonic membranes have been retained and modified.

Other evolutionary adaptations to life on land appeared among the amniotes, including a tough impermeable skin covered with scales or modified scales (i.e., hair or feathers). The excretory systems of amniotes also evolved adaptations that allowed these animals to excrete nitrogenous wastes in the form of concentrated urine with a minimal loss of valuable water.

During the Carboniferous period, about 250 mya, the amniote animals diverged into two groups, the reptiles and the mammals. Birds, the only living members of the now-extinct dinosaurs, evolved from one of the reptilian groups (see Figure 23.48).

Reptiles adapted to life in many habitats. One relatively small reptilian group, the turtles and tortoises, has a unique body plan that has changed very little over the millennia (see Figure 23.49D). These animals are characterized by dorsal and ventral bony plates that evolved from the ribs to form a protective shell. The relationship of turtles and tortoises to other reptiles is not clear. For example, it is not known how, unlike in other vertebrates, the pectoral girdles evolved to be inside the ribs.

The lepidosaurs include the squamates (lizards, snakes, and amphisbaenians—another group of legless, wormlike burrowers), and the tuataras (see Figure 23.49A). The body is covered in scales that greatly reduce the loss of water but make their skin surface unavailable for gas exchange. Gases are exchanged through the lung, which is larger in surface area compared to that of the amphibians. The lepidosaurs have a three-chambered heart that partially separates oxygenated from deoxygenated blood. The limbless condition of snakes is a product of secondary evolution. Most lizards and all snakes are carnivores, and adaptations to the jaws of snakes allow them to swallow prey much larger than themselves. Tuataras are represented by only two living species.

The archosaurs include the extant crocodilians (crocodiles and alligators), the extinct dinosaurs, and the "living dinosaurs" we know as birds (see Figure 23.50). The crocodilians—crocodiles, alligators, caimans, and ghalials—live only in tropical and warm temperate regions. They spend most of their lives in water but build nests on land. They are all carnivores, feeding on other vertebrates, including large mammals.

Dinosaurs arose among the reptiles around 215 mya and survived for about 150 million years. Both the fossil record and molecular evidence support the position of birds as a sister group of the saurischian dinosaurs.

Birds probably evolved from an ancestral theropod, a bipedal predatory dinosaur that appears to have had hollow bones, a furcula (wishbone), limbs with three digits, and a backward-thrusting pelvis.

Living birds diverged from the common flying ancestor and fall into one of two groups. The palaeognaths are flightless and include rheas, emu, kiwis, cassowaries, and the largest bird, the ostrich (see Figure 23.50B). The neognaths, most of which have retained the ability to fly, represent the remaining 9,600 species of birds.

Recent fossil discoveries have demonstrated that some dinosaurs had scales that were highly modified to form feathers. In one species, *Microraptor gui*, the feathers were structurally similar to those of modern birds (see Figure 23.51A).

*Archaeopteryx* lived 150 mya and represents the oldest known fossil of a bird. It was covered in feathers and had well-developed wings and a wishbone (see Figure 23.51B). It also had teeth, which were lost in later members of this lineage.

Feathers are complex and strong, but also lightweight. Their evolution was a major factor in diversification. The bones of dinosaurs and birds are hollow, and along with the development of feathers, they assisted in the evolution of flight.

Along with the ability to fly came a high metabolic rate needed to fuel flight. The metabolic rate of birds means that they generate a great amount of heat, and their feathers are adapted to allow for heat loss. The lungs of birds also function differently from those of other vertebrates, another adaptation to the needs of flight. Different bird groups feed on many different types of animal and plant material, including carrion. Birds that feed on fruits and seeds are major agents of plant dispersal.

Mammals radiated as nonavian dinosaurs declined in diversity. The earliest mammals lived side by side with reptiles from the first split of the two lineages early in the Mesozoic era. However, only after the large dinosaurs became extinct did mammals begin to flourish in size and number.

Four key features distinguish the mammals: (1) sweat glands in the skin that produce secretions that cool the body; (2) mammary glands that provide nutrient fluid for newborns; (3) external body hair that protects and insulates most mammals; and (4) a four-chambered heart that completely separates oxygenated from deoxygenated blood. The four-chambered heart also evolved convergently in the crocodilians and birds.

In most mammals the amniote egg became modified for growth of the embryo within the mother's uterus. There are five species of egg-laying mammals. The prototherians exist today only in Australia and New Guinea (see Figure 23.53A). They supply milk to their young as do other mammals, but they do not have nipples on their mammary glands.

Most mammals are viviparous. The remaining mammals are classed as therians. Marsupial therians give birth to tiny underdeveloped young that they nurture externally, usually in a pouch on the mother's belly (see Figure 23.53B). Marsupials were once widespread, but today the approximately 330 marsupial species are largely found in Australia and South America, with minor representatives in North America.

The 4,500 species of eutherian mammals develop in the mother's uterus. This group is widely known by the name placental mammals, from the placenta that provides nourishment for the growing embryo. However, some marsupials also have placentas.

Grazing and browsing by many herbivorous eutherian groups has transformed the terrestrial environment. In plants, this led to the evolution of spines and tough leaves. Despite such defenses, herbivores have developed adaptations to their teeth and digestive systems in order to consume these plants. This is an example of coevolution.

Eutherians exist in virtually all of Earth's environments and vary greatly in form (see Figure 23.53C–F). Several groups, most notably the cetaceans (whales and dolphins) returned to a marine lifestyle. Flight evolved in the bats (which represent the second-largest number of eutherian species, after the rodents; see Table 23.3). The 235 species of primates are the best-studied eutherians because human beings belong to this group.

**Question 12**. Amphibians are dependent on water and moist environments for their life cycle. Which stages require an aquatic environment and why is such an environment necessary? Do any amphibians have strategies that allow them to avoid the aquatic stage? If so, describe them.

**Question 13**. How have mammals and reptiles adapted to some of the challenges of a terrestrial habitat?

**Question 14.** Create a phylogenetic tree reflecting the evolutionary history of the animals listed below (review Chapter 16). Indicate the derived traits on your diagram.

Chicken
Jellyfish
Lobster
Tapeworm
Sponge
Grasshopper
Clam
Squid
Snake

## 23.7 Humans Evolved among the Primates

Two major lineages of primates split late in the Cretaceous

Bipedal locomotion evolved in human ancestors

Human brains became larger as jaws became smaller

The primate ancestor was a small, arboreal, insectivorous mammal. Primates are identifiable by the presence of opposable digits (i.e., thumbs). Early in their evolutionary history—about 65 mya—the primates split into two clades, the prosimians (lemurs, bush babies, and lorises) and the anthropoids (tarsiers, monkeys, apes, and humans; see Figure 23.54). Prosimian species were once found on all continents, but today they are restricted to Africa, especially the island of Madagascar.

Soon after the prosimian–anthropoid split, the anthropoids split further into the New World and Old World monkeys. The breakup of the African and South American continents meant that the two groups evolved in isolation. About 35 mya an Old World lineage broke off that would lead to modern apes and humans. Another split occurred around 22 mya. The Asian apes (gibbons and orangutans) descended from two groups in this lineage (see Figure 23.54). A long prehensile tail, which is unique to the New World monkeys, allows them to grasp branches. All New World monkeys are arboreal.

About 6 mya a split resulted in two primate lineages, one that led to chimpanzees and the other that led to the groups that gave rise to hominids.

Ardipithecines (the earliest protohominids) and their descendants, the australopithecines, were adapted for bipedalism, which freed up their hands for other tasks and elevated the eyes. Bipedal movement also requires less energy. *Australopithecus afarensis* is currently regarded as the ancestor of the modern humans (genus *Homo*; see Figure 23.55). There is disagreement as to how many species are represented by australopithecine fossils. It is clear that these different groups of hominids coexisted in Africa several million years ago. The genus *Homo* arose from one of the smaller lineages of australopithecines. In parallel, the genus *Parathropus* arose from larger australopithecines and later went extinct.

The oldest known member of the genus *Homo, Homo habilis,* lived about 2 mya, and tools have been found with their bones. *Homo erectus* appeared about 1.6 mya. They were as large as modern humans, used stone tools, and cooked with fire. They were the first hominid species to leave Africa.

Larger brain size appears to have evolved concurrently with a smaller, less muscular jaw structure, suggesting that the two traits may be functionally correlated.

The rapid increase in brain size may have been favored by an increasingly complex social life that thrived on ever more sophisticated communication. Any trait that allowed more effective communication would have been favored in a society based on cooperative hunting and other complex social interactions.

A number of species of *Homo* existed simultaneously over evolutionary time between 1.5 mya and 250,000 years ago. One species, *Homo neanderthalensis,* appeared about 500,000 years ago and became widespread in Europe and Asia. Although they were short, they were powerfully built. Their skulls housed a brain that is somewhat larger than that of modern humans. They manufactured tools and were able to hunt large animals.

By 200,000 years ago, *H. sapiens* was predominant in Africa, and they began to extend their range around 60,000–70,000 years ago. Around 35,000 years ago they coexisted alongside *H. neanderthalensis.* It is likely that the two species interacted, but Neanderthals abruptly became extinct 28,000 years ago, possibly exterminated by *H. sapiens.* Genetic evidence suggests that there was some interbreeding between the two species.

By about 20,000 years ago, all other *Homo* species had been supplanted by *Homo sapiens* (modern humans), the only currently existing human species. It was about this time that *H. sapiens* reached North America.

**Question 15.** What major geologic change occurred during primate evolution and how did it alter the way primate evolution proceeded?

**Question 16.** What are three traits that evolved in primates and led to evolutionary success? Which ones seem to have had the greatest impact in hominid lines?

## Test Yourself

1. Which of the following statements about echinoderms is *false?*
   a. They have a water vascular system.
   b. They have an internal skeleton.
   c. They are protostomes.
   d. The larvae have bilateral symmetry.
   e. The adult form lacks a head.
   ***Textbook Reference:*** *Concept 23.5 Deuterostomes Include Echinoderms, Hemichordates, and Chordates; Echinoderms have unique structural features*

2. Which of the following is *not* a derived trait that is shared by all animals?
   a. Hox genes
   b. The extracellular matrix molecule collagen
   c. Tight junctions, desmosomes, and gap junctions
   d. Bilateral symmetry
   e. Absence of a cell wall
   ***Textbook Reference:*** *Concept 23.2 Some Animal Groups Fall Outside the Bilateria*

3. Which of the following statements about deuterostomes is *false?*
   a. Three distinct layers of tissue are present during development.
   b. If a coelom is present, it formed within the embryonic mesoderm.
   c. Its early embryonic cleavage pattern is radial.
   d. It is diploblastic.
   e. Gastrulation occurs during development.
   ***Textbook Reference:*** *Concept 23.1 Distinct Body Plans Evolved among the Animals; Basic developmental patterns and body plans differentiate major animal groups*

4. Which of the following chordate groups evolved before the appearance of cartilaginous fishes?
   a. Ray-finned fishes and sea squirts
   b. Ascidians, lancelets, and hagfishes
   c. Ascidians, lancelets, and ray-finned fishes
   d. Lampreys and ray-finned fishes
   e. Coelacanths and ray-finned fishes
   ***Textbook Reference:*** *Concept 23.5 Deuterostomes Include Echinoderms, Hemichordates, and Chordates; There are two groups of living jawless fishes*

5. Which of the following statements about the body cavity of animals is true?
   a. The body cavity of coelomates develops from the embryonic ectoderm.
   b. The body cavity of acoelomates is filled with liquid.
   c. The pseudocoel of the pseudocoelomates has a peritoneum.
   d. The acoelomates do not have an enclosed body cavity.
   e. The coelomates have a body cavity surrounded by a peritoneum.
   ***Textbook Reference:*** *Concept 23.1 Distinct Body Plans Evolved among the Animals; The structure of the body cavity influences movement*

6. Which of the following is *not* a feature of the vertebrate body plan?
   a. A ventral spinal cord
   b. An internal skeleton
   c. A well-developed circulatory system
   d. Organs suspended in the coelom
   e. An anterior skull encasing a proportionally large brain
   ***Textbook Reference:*** *Concept 23.5 Deuterostomes Include Echinoderms, Hemichordates, and Chordates; The vertebrate body plan can support large, active animals*

7. The swim bladder of many fishes, which evolved from a lunglike sac, has the important function of

a. aiding in prey capture.
b. controlling swimming speed.
c. controlling buoyancy.
d. aiding in reproduction.
e. providing balance.

***Textbook Reference:*** *Concept 23.5 Deuterostomes Include Echinoderms, Hemichordates, and Chordates; Fins and swim bladders improved stability and control over locomotion*

8. Which of the following traits is shared by the chondrichthyans and the ray-finned fishes?
a. Gills as the major site of gas exchange
b. A skeleton composed of cartilage
c. An outer surface covered with bony plates
d. A swim bladder
e. Propulsion by means of dorsal–ventral movements of the tail

***Textbook Reference:*** *Concept 23.5 Deuterostomes Include Echinoderms, Hemichordates, and Chordates; Fins and swim bladders improved stability and control over locomotion*

9. The transition from aquatic to terrestrial lifestyles required many adaptations in the vertebrate lineage. Which of the following is *not* one of those adaptations?
a. A shift from gills to air-breathing lungs
b. Improvements in the water resistance of skin
c. An alteration in the mode of locomotion
d. Development of feathers for insulation
e. Modifications to the nitrogen elimination system

***Textbook Reference:*** *Concept 23.6 Life on Land Contributed to Vertebrate Diversification; The evolution of feathers allowed birds to fly*

10. The amniotes evolved the ability to reproduce by laying eggs that have shells. The major advantage of shelled eggs is that
a. the embryo needs only a small amount of yolk for development.
b. they do not have to be laid in a moist environment.
c. the shells increase evaporation from the egg.
d. nitrogenous wastes can be excreted across the shell.
e. a shell contributes to the efficiency of gas exchange with the environment.

***Textbook Reference:*** *Concept 23.6 Life on Land Contributed to Vertebrate Diversification; Amniotes colonized dry environments*

11. Which of the following is *not* a trait that identifies an animal as a mammal rather than an amphibian?
a. Mammary glands
b. Hair
c. Sweat glands
d. Kidneys
e. Four-chambered heart

***Textbook Reference:*** *Concept 23.6 Life on Land Contributed to Vertebrate Diversification; Mammals radiated as non-avian dinosaurs declined in diversity*

12. A key difference between Old World and New World monkeys is that the latter
a. have a prehensile tail.
b. are arboreal.
c. have a placenta.
d. are less closely related to tarsiers.
e. are omnivorous.

***Textbook Reference:*** *Concept 23.7 Humans Evolved among the Primates; Two major lineages of primates split late in the Cretaceous*

13. Which of the following is *not* part of the evidence that birds are closely related to dinosaurs?
a. The presence of feathers in representatives of both groups
b. The presence of hollow bones in representatives of both groups
c. DNA sequence data comparing birds to other living reptiles
d. A bipedal stance in fossilized therapods
e. The ability to fly

***Textbook Reference:*** *Concept 23.6 Life on Land Contributed to Vertebrate Diversification; The evolution of feathers allowed birds to fly*

14. Which of the following are *not* deuterostomes?
a. Echinoderms
b. Hemichordates
c. Cephalochordates
d. Ecdysozoans
e. Chordates

***Textbook Reference:*** *Concept 23.2 Some Animal Groups Fall Outside the Bilateria*

15. Which of the following statements about sponge structure or function is *false*?
a. Choanocytes are flagellated cells that play a role in feeding.
b. Large species are found in areas of heavy wave action, where food is most abundant.
c. Individual sponges are both male and female.
d. Water enters a sponge through pores and exits via one or more oscula.
e. Sponges have an extensive extracellular matrix holding the cells together.

***Textbook Reference:*** *Concept 23.2 Some Animal Groups Fall Outside the Bilataria; Sponges are loosely organized animals*

16. A restaurant appetizer of escargot (snails), clams on the half shell, and calamari (squid) contains which types of mollusks?
a. Chitons, bivalves, and gastropods
b. Bivalves, gastropods, and cephalopods

c. Chitons, gastropods, and cephalopods
d. Bivalves and gastropods only
e. Chitons and cephalopods only
***Textbook Reference:*** *Concept 23.3 Protostomes Have an Anterior Brain and a Ventral Nervous System; Cilia-bearing lophophores and trochophore larvae evolved among the lophotrochozoans*

17. Which of the following attributes is seen in the arthropod body plan?
a. Specialized flagellated cells for capturing food
b. Unjointed appendages
c. Muscle attachments for the appendages inside the exoskeleton
d. A soft cuticle
e. A pseudocoelom
***Textbook Reference:*** *Concept 23.4 Arthropods Are Diverse and Abundant Animals*

## Answers

### *Key Concept Review*

1. Protostomes and deuterostomes differ in the structure that becomes the anus and the structure that forms the mouth. The sequence in which these structures develop is also different. In protostomes, the mouth develops from the blastopore; in the deuterostomes, the anus develops from the blastopore. In both, the opposite structure forms later. These groups were first described according to observations of their developmental patterns, but sequence data supports this classification.
2. Hydrostatic skeletons provide for controlled movement of body parts. In terrestrial organisms, the surrounding environment does not support the body, so most organisms with hydrostatic skeletons are very small and soft bodied. Larger bodies are possible with exoskeletons or hard interior skeletons that provide structural support and protection to soft tissues.
3. Animals can either move through the environment to where the food is located or they can move the environment and the food to where they are. Sessile organisms would most likely have adaptations to bring the environment and food to them.
4. The sponge body plan consists of an aggregation of cells around a water canal system. Food particles, along with water, enter by way of small pores and are captured by choanocytes.
5.

(A) Generalized molluscan body plan

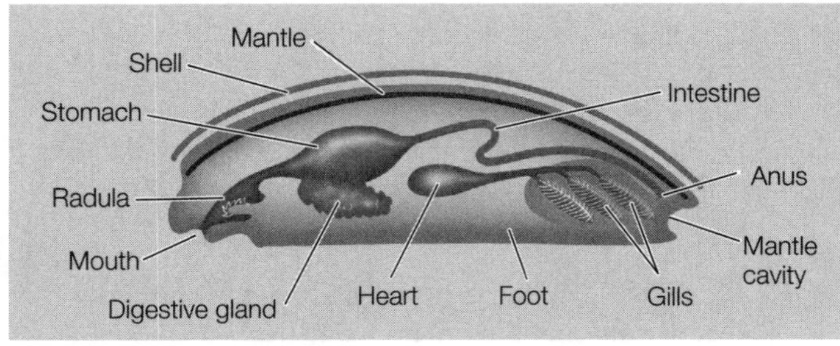


(B) Gastropods

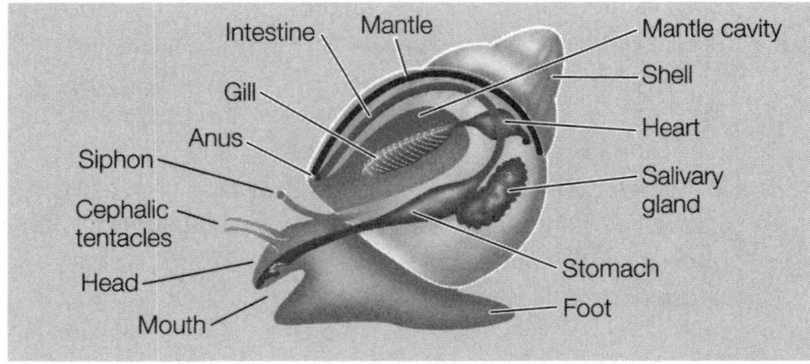


(C) Cephalopods

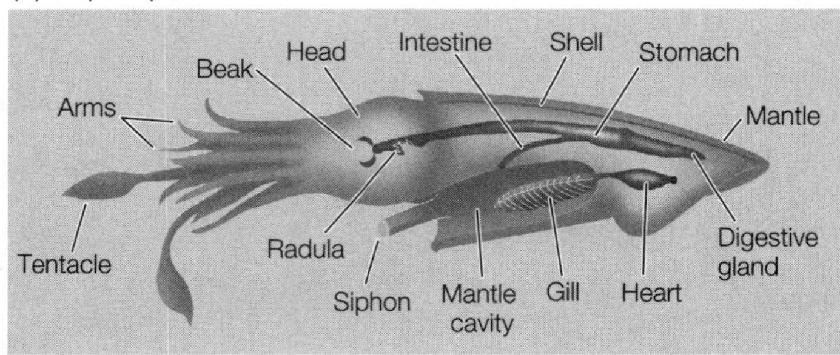


6. A cuticle reinforced with chitin provided support and protection from desiccation and predators. Its presence meant that gas exchange could not be accomplished through the body surface and that growth could not occur without molting of the rigid exoskeleton, leaving the temporarily soft-bodied animal vulnerable. Locomotion was also limited, since the animal could neither move in a wormlike fashion nor use cilia for propulsion.
7. Segmentation provides several advantages, including facilitating specialization of body regions and improved muscle coordination, as each segment can be manipulated independently.
8. Insects, like crustaceans, have three body parts: head, thorax, and abdomen. Unlike crustaceans, they have three pairs of legs attached to the thorax. They are distinguished from springtails and other hexapods by possessing external mouthparts, a pair of antennae on their heads with a motion receptor called a Johnston's organ, and tracheae.

9. Although adult ascidians and adult humans are very different animals, their embryonic stages share certain important characteristics, including the presence of a notochord (as in all deuterostomes) and a blastopore that develops into the anus. A look at embryonic grasshoppers, however, shows them to be fundamentally different. The embryonic grasshopper's blastopore develops into the mouth, placing grasshoppers squarely among the protostomes. Grasshoppers have an external rather than an internal skeleton.

10. One would look for evidence of an internal skeleton and a vascular system composed of water-filled canals. These two traits are defining characteristics of echinoderms.

11. The hinged jaw of the fishes evolved during the Devonian period. The evolution of the jaw opened up a new food source for these animals. With a jaw, fish could grasp and kill larger living prey and chew and tear body parts.

12. Amphibians generally require an aquatic environment for fertilization, larval development, and metamorphosis into a terrestrial adult. Some amphibians develop from eggs laid on land and move directly into adultlike forms, skipping the aquatic form entirely.

13. Mammals and reptiles are both amniotes. Their eggs minimize water loss while still permitting gas exchange. The skin of both groups of organisms is modified to reduce water loss, and the kidneys eliminate nitrogen while minimizing water loss. Both groups of organisms exchange gases with the atmosphere by means of lungs.

14.

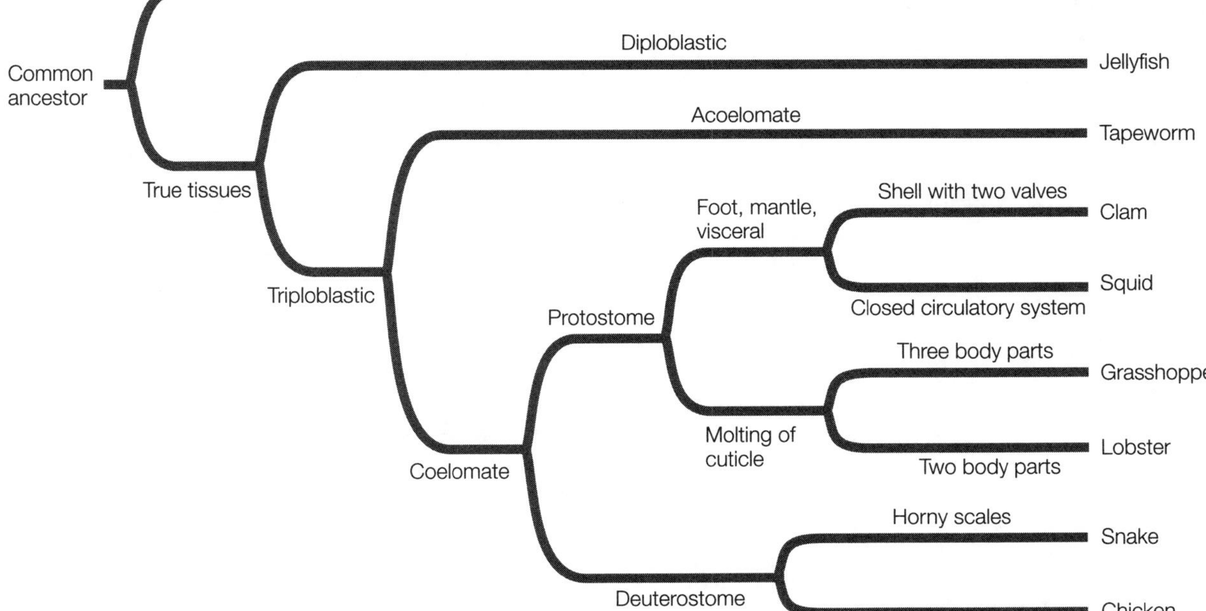

15. As the continental land masses separated 65 mya, New World monkeys were isolated from Old World monkeys and the two groups diverged.

16. Primates evolved opposable thumbs, bipedal upright locomotion, prehensile tails, and large cranium size. Opposable thumbs favored development of tool use. Bipedal locomotion increased the ability to see above vegetation and freed up hands to carry and manipulate objects. Prehensile tails increased the success of arboreal monkeys. Large cranium size permitted brain growth to accommodate language development and facilitated sophisticated social interactions.

### *Test Yourself*

1. **c.** Species from the echinoderm clade are deuterostomes.

2. **d.** Not all animals have bilateral symmetry (e.g., sponges, cnidarians, ctenophores).

3. **d.** Deuterostome embryos have three layers (the ectoderm, the mesoderm, and the endoderm), making deuterostomes triploblastic, not diploblastic.

4. **b.** The ascidians, lancelets, and hagfishes all evolved before the cartilaginous fishes.

5. **d.** The body cavity of coelomates develops from the mesoderm and contains a peritoneum. The acoelomates lack a body cavity.

6. **a.** Along with an internal skeleton, well-developed circulatory system, and organs suspended in a coelom, the vertebrates have a dorsal spinal cord.

7. **c.** The swim bladder of modern-day fishes is involved in controlling buoyancy.

8. **a.** In both the ray-finned fishes and cartilaginous fishes, the major site of gas exchange is the gills. The ray-finned fishes have a skeleton of bone and a swim bladder. These traits are not shared with the cartilaginous fishes. Neither group has bony plates in its outer surface. They all move their tails laterally, in contrast to cetaceans, which move their tails vertically.

9. **d.** The move onto land did not require the development of feathers for insulation. Amphibians and reptiles do not have an insulation layer, and many mammals have hair for insulation.

10. **b.** The shelled egg of the birds and reptiles allowed them to occupy dry terrestrial habitats because the shell decreases water loss from the egg.

11. **d.** Kidneys are present in reptiles, birds, fishes, and mammals. The presence of kidneys is not a trait that identifies an animal as a mammal.

12. **a.** Only the New World monkeys have a prehensile tail.

13. **e.** Not all birds fly and not all dinosaurs were able to fly.

14. **d.** Ecdysozoans are a type of protostome (see Table 23.1).

15. **b.** Because they are not structurally robust, large, upright sponges would be destroyed by heavy wave action.

16. **b.** Snails are in the Gastropoda, clams are in the Bivalvia, and octopuses are in the Cephalopoda.

17. **c.** Arthropods have muscles for the appendages attached inside the exoskeleton.

# 24 The Plant Body

## The Big Picture

- A plant can be thought of as having vegetative structures that carry out the major functions of day-to-day life and reproductive structures that are responsible for reproducing the plant. This chapter focuses on the vegetative plant body, consisting of the root system, which anchors the plant and absorbs water and nutrients, and the shoot system, which carries out photosynthesis and supports the plant against gravity. Modifications of the root and shoot systems lead to specialization of the plant.
- Plant cells are uniquely suited for support, transport, and carrying out cellular functions. Groups of cells form tissues that have specific roles within a plant. There are three basic tissue types in plants: dermal, ground, and vascular. Indeterminate growth and the modular organization of plants allow for regeneration of parts lost to damage and disease.
- Plant growth occurs from meristems. Apical meristems allow for elongation of the plant, whereas lateral meristems allow for secondary or woody growth. Not all plants exhibit secondary growth. All tissue types arise from the meristems and go through a process of elongation followed by differentiation.

## Study Strategies

- The best study strategy for learning this material is to study the figures, pictures, and live material. Plant anatomy is the study of structure, and since structures are three-dimensional, they are best understood visually. Review stem, root, and leaf anatomy using the figures in the textbook.
- Basic plant anatomy is not difficult, but it may be new to you. The more time spent looking at diagrams and live specimens, the easier it will be to understand the anatomy. While studying, think about the function of each structure. You will find that "form follows function."
- This chapter exposes you to many new vocabulary words. A vocabulary list will help you organize your study, but do not try merely to memorize these terms. By looking at the words and understanding their roots, you will be better able to understand their meaning.
- Go to **LaunchPad** (or use the Web addresses listed) to review the following additional resources:

  Animated Tutorial 24.1 Secondary Growth: The Vascular Cambium (PoL2e.com/at24.1)

  Activity 24.1 Eudicot Root (PoL2e.com/ac24.1)

  Activity 24.2 Monocot Root (PoL2e.com/ac24.2)

  Activity 24.3 Eudicot Stem (PoL2e.com/ac24.3)

  Activity 24.4 Monocot Stem (PoL2e.com/ac24.4)

  Activity 24.5 Eudicot Leaf (PoL2e.com/ac24.5)

## Key Concept Review

### 24.1 The Plant Body Is Organized and Constructed in a Distinctive Way

Plants develop differently than animals

The plant body has an apical–basal axis and a radial axis

The plant body is constructed from three tissue systems

Because plants are sessile (immobile), they need special adaptations to meet environmental challenges: stems, leaves, and roots allow a plant to take up resources from the plant's location; plants can continue to grow throughout their lifetime, allowing them to access additional resources. Plants are composed of root systems and shoot systems (see Figure 24.1). Root systems are responsible for mineral and water uptake and support. Shoot systems consist of leaves (and leaf derivatives) involved in photosynthesis, supportive stems, and flowers (modified shoots) for reproduction. Shoots contain repeating modules called phytomers, which comprise a node with one or more leaves, an internode, and one or more axillary buds. A bud is an undeveloped shoot that can produce a leaf, a phytomer, or a flower. Most plants are either narrow-leaved monocots or broad-leaved eudicots.

The development of plants involves determination, differentiation, morphogenesis, and growth. Meristems allow for continual growth throughout a plant's life. Plant cells differ from other eukaryotic cells in that every plant cell is bounded by cellulose-containing cell walls. The plane of cell division controls the direction of plant growth. Plant cells are pluripotent or totipotent, allowing for repair of damaged tissues.

The body plan of a plant is established in the embryo along an apical–basal axis and a radial axis (see Figure 24.4). Two unequal daughter cells are produced as the zygote goes through a mitotic division. Subsequent cell divisions result in the development of a thin suspensor and a globular embryo. The cotyledons (seed leaves) begin to form as the embryo enters the heart stage. As the cotyledons elongate, the embryo enters the torpedo stage and the internal tissues begin to differentiate. The root and shoot apical meristems develop between the cotyledons (see Figure 24.5). Meristems are undifferentiated cells found at the tips of the embryonic shoot and root that will become the organs of the plant as it grows.

Tissues are composed of cells that function together. The embryonic plant contains three tissue systems—the dermal, ground, and vascular systems (see Figure 24.6). These are arranged concentrically and will give rise to the adult tissues. The dermal tissue system makes up the outer covering of the plant and includes the epidermis and the layer of cuticle it secretes. Special epidermal cells include stomata, trichomes, and root hairs. The ground tissue system is found between the dermal and vascular tissue and is involved in storage, support, and photosynthesis. Thin-walled parenchyma cells have large central vacuoles and are frequently photosynthetic or used for storage. Collenchyma cells are support cells with special thickenings at the cell wall corners. Sclerenchyma cells have thickened secondary cell walls containing lignin and are either elongated fibers or variously shaped sclereids. Most have undergone apoptosis and provide rigid support after they die. Fibers strengthen bark and woody stems. Densely packed sclereids are found in the shells of nuts and in some seed coats and produce the gritty texture of pears and other fruit. The vascular tissue system is made of xylem and phloem, and is the conductive tissue of the plant. Xylem transports water and mineral ions from the roots to the rest of the plant. The spindle-shaped tracheids and vessel elements of the xylem are dead cells that allow movement of water (containing minerals and ions) through openings formed between the remaining cells walls. Living phloem moves carbohydrates and nutrients from sources to sinks (cells that use or store the sugars). Sources can be photosynthetic tissues or storage tissues in roots when they are tapped to provide energy for the plant. Individual phloem cells are called sieve tube elements. Plasmodesmata enlarge where sieve tube elements join, making sieve plates. Companion cells connect to the sieve tube elements via plasmodesmata and perform many of the phloem metabolic functions.

**Question 1.** Draw a typical eudicot plant. Label the shoot system and root system. Indicate on your drawing where you would find an axillary bud and where you would find a terminal bud. Indicate a phytomer unit. Label the following structures: leaf, node, internode, stem, and roots.

**Question 2.** The plant body comprises three different tissue systems. While examining a plant, you find a tissue composed of dead cells that have very thick cell walls and that contain lignin. What types of cells are you likely seeing and why?

## 24.2 Apical Meristems Build the Primary Plant Body

A hierarchy of meristems generates the plant body

The root apical meristem gives rise to the root cap and the root primary meristems

The products of the root's primary meristems become root tissues

The root system anchors the plant and takes up water and dissolved minerals

The products of the shoot's primary meristems become shoot tissues

Leaves are photosynthetic organs produced by shoot apical meristems

Plant organs can have alternative forms and functions

As opposed to animal growth, which is determinate and stops when the adult state is reached, plant growth is indeterminate and continues indefinitely. Plants use this continual growth to reach sunlight and resources in the soil. Primary growth results in lengthening of the primary plant body; it is achieved by the division of cells in the apical meristems. Secondary growth increases the thickness of the plant, and is achieved by division of cells in the lateral meristems (see Figure 24.7). Woody eudicots undergo secondary growth, but most monocots do not.

Meristems are undifferentiated cells that are able to divide and create new cells. These initial cells are similar to human stem cells. Apical meristems give rise to primary meristems that produce the primary plant body. These primary meristems are called the protoderm (dermal tissue system), the ground meristem (ground tissue system), and the procambium (vascular tissue system). All plant parts arise from division of the apical meristem cells.

Root apical meristems (protoderm, ground meristem, and procambium) produce root tissues (see Figure 24.8). At the tip of a root, the root apical meristem forms a root cap and quiescent center. The zone of cell division includes the apical and primary meristems. Above the zone of cell division is the zone of elongation. Above this is the zone of maturation where the cells differentiate and take on special functions. The root also has three primary tissue systems. The protoderm gives rise to the epidermis and root hairs, both of which are involved in water and mineral ion uptake. The ground meristem gives rise to the cortex and endodermis. The endodermal cells have a waxy coating of suberin to assist with controlling water and ion movement. The procambium gives rise to the stele, at the center of the root, which houses three tissues: the pericycle, the xylem, and the phloem. The pericycle has layers of parenchyma cells that give rise to lateral roots, create lateral root meristems for secondary growth, and transport nutrients into the xylem. In eudicots, the very center of the root is xylem, whereas in monocots the center is pith tissue (see Figure 24.9).

The roots are the main site of water and nutrient entry into the plant. In eudicots, the taproot is the primary root with lateral roots radiating out. Monocot plants have a fibrous root system that is made up of more diffuse thin roots. Monocot roots are adventitious, arising from stem tissue above the initial root. The fibrous root system helps to anchor plants in soil and prevent erosion.

Shoot growth occurs as the shoot apical meristem divides, creating the leaf primordia and axillary buds. The shoot apical meristem also gives rise to three primary meristems that produce the shoot epidermis, shoot cortex, and shoot vascular system. Shoot vascular tissues are arranged in vascular bundles containing both xylem and phloem (see Figure 24.11). In eudicots, the vascular bundles are arranged in a cylinder, allowing for woody growth. In monocots, the vascular bundles are scattered throughout the stem. Shoot apical meristems produce the leaves of the plant. Leaves arise from leaf primordia with bud primordia forming at each leaf base. Leaf growth is determinate (stops at a particular size and shape). The leaf blade is connected to the stem in a eudicot by the petiole; monocot leaves do not have petioles, but the base of the leaf wraps around the stem. Most eudicot leaves have two zones of photosynthetic cells called mesophyll (see Figure 24.12). The upper level of cylindrical mesophyll is called palisade mesophyll, and the lower level is called spongy mesophyll. Air space around mesophyll cells is necessary for $CO_2$ to reach the photosynthesizing cells. Vascular tissue extends throughout leaves as a network of veins. Veins branch throughout the leaf to carry water to cells and transport carbohydrates to sink tissues. The entire leaf is covered by a protective epidermis and is waterproofed by a waxy cuticle. Gas exchange occurs through pores called stomata.

Plant organs can be modified to perform other functions (see Figure 24.13). Taproots can be used for nutrient storage, such as in carrots and beets. Adventitious roots can be prop roots that help support a plant. Stems can be modified, such as a potato tuber and in enlarged cactus stems. Leaves can be modified for storage as in onions (for energy) or succulents (for water). The spines of cacti are modified leaves, as are the tendrils of peas that help anchor the plant as it grows upwards.

**Question 3:** What are the differences between apical and lateral meristems? Do all plants have apical meristems? Do all plants have lateral meristems?

**Question 4:** Draw a growing root. Label the primary meristems (protoderm, ground meristem, and procambium), root cap, cortex, stele, and epidermis. Discuss the function of each of these structures.

**Question 5:** Examine the leaf diagram in Figure 24.12 in the textbook. Which surface of the leaf faces the sun? How do you know? Why are the stomata opposite the palisade layer?

## 24.3 Many Eudicot Stems and Roots Undergo Secondary Growth

Secondary growth in eudicots involves the laying down of wood and bark by the two lateral meristems, vascular cambium and cork cambium. Division of cells in the vascular cambium forms new secondary xylem (wood) toward the inside of the stem and secondary phloem (inner bark) toward the outside of the stem (see Figure 24.14). The cork cambium, initially found just inside the epidermis, produces new dermal tissues (outer bark) to accommodate the increasing diameter. It inhibits water loss with the production of periderm cells. The action of the vascular cambium and cork cambium is called secondary growth. When deciduous trees begin to grow in the spring, the buds emerge as bud scales fall away. Scars show where the buds were and can be used to mark the growth each year. Buds and the newest growth from the apical meristem consist entirely of primary tissues, since secondary growth has not yet occurred in this location. Wood is made up of the layers of secondary xylem. Stretching, breaking, and flaking off of epidermis and cortex leaves the secondary phloem at risk. Cells at the surface of the phloem, the cork cambium, produce a protective layer of cork that is thickened and reinforced with waterproof suberin. New cork is produced as secondary growth proceeds. The cork cambium and the cork together constitute the periderm. The periderm and the secondary phloem (all tissues external to the vascular cambium) constitute the bark. The annual rings seen in wood are a result of climate shifts in temperate zones, particularly water availability (see Figure 24.15). Tropical trees do not undergo seasonal growth and do not lay down such visible rings. Only eudicots and nonmonocot angiosperms, as well as many gymnosperms, have vascular cambium and cork cambium and undergo secondary growth. Monocots that have thickened stems do not perform secondary growth; dead leaf bases are left around the stem, effectively widening it.

**Question 6:** Differentiate between primary and secondary xylem, and primary and secondary phloem. Where is the primary phloem in a three-year-old woody eudicot?

**Question 7:** People are discouraged from nailing objects onto trees or removing layers of bark because such practices can damage a tree. Why, then, is it possible to harvest cork year after year without harming the tree?

## 24.4 Domestication Has Altered Plant Form

From a very simple body plan, with roots, stems, leaves, and meristems, flowering plants have been able to achieve great diversity. Phenotypes that outcompete other individuals for resources will have higher fitness; natural selection acts on the phenotypes of the plant body. Humans have artificially selected plants to improve crop yield through successive breeding of individuals with the most desirable phenotypes. This is possible because of the morphological variation found within wild plant species. The domesticated forms of crops are very different from their ancestral, wild forms. Corn, which was domesticated from the grass teosinte, is a good example of the morphological differences that can exist between the parental species and later generations of a crop. Typically, crops are selected for low competition between individuals and maximal photosynthesis when planted in groups. A single species, *Brassica oleracea*, is the ancestor of many food crops, such as broccoli, brussels sprouts, and kale.

**Question 8:** How have humans domesticated plants? What are the benefits provided by domestication?

**Question 9:** The main morphological difference between corn and its ancestor grass, teosinte, is the amount of branching from the main stem. This difference is due to the activity of a single gene and its protein product, *TEOSINTE BRANCHED 1* (*TB1*). The domestication of corn occurred long before humans understood the effects of genes, or could directly manipulate them in any way. How was it possible to find and propagate individual plants with this genetic difference before modern scientific methods were available?

## Test Yourself

1. Which of the following is a *not* a component of a phytomer?
   a. Leaf
   b. Root hair
   c. Axillary buds
   d. Internode
   e. Node
   ***Textbook Reference:*** *Concept 24.1 The Plant Body Is Organized and Constructed in a Distinctive Way*

2. Suppose you are studying tropical plants in a Costa Rican cloud forest and find a tree that is a eudicot in the forest ecosystem. This plant most likely has a(n) _______ system.
   a. fibrous root
   b. taproot
   c. adventitious root
   d. rhizoid root
   e. terminal root
   ***Textbook Reference:*** *Concept 24.2 Apical Meristems Build the Primary Plant Body; The root system anchors the plant and takes up water and dissolved minerals*

3. Some plants, such as sweet peas, will attach themselves to a fence by means of tendrils, which are modifications of
   a. stems.
   b. roots.
   c. branches.
   d. leaves.
   e. seeds.
   ***Textbook Reference:*** *Concept 24.2 Apical Meristems Build the Primary Plant Body; Plant organs can have alternative forms and functions*

4. Plants are easily distinguished from animals by all of the following *except*
   a. rigid cell walls.
   b. large vacuoles.
   c. chloroplasts.
   d. apical meristems.
   e. mitochondria.
   ***Textbook Reference:*** *Concept 24.1 The Plant Body Is Organized and Constructed in a Distinctive Way; Plants develop differently than animals*

5. Plant cells that are photosynthetically active are found in the _______ layer of the leaf and are called _______ cells.
   a. mesophyll; parenchyma
   b epidermis; parenchyma
   c. mesophyll; sclerenchyma
   d. epidermis; sclerenchyma
   e. xylem; mesophyll
   ***Textbook Reference:*** *Concept 24.2 Apical Meristems Build the Primary Plant Body; Leaves are photosynthetic organs produced by shoot apical meristems*

6. Water is conducted in _______ tissue, and carbohydrates and nutrients are transported in _______ tissue.
   a. xylem; phloem
   b. phloem; xylem
   c. parenchyma; phloem
   d. parenchyma; xylem
   e. mesophyll; xylem
   ***Textbook Reference:*** *Concept 24.1 The Plant Body Is Organized and Constructed in a Distinctive Way; The plant body is constructed from three tissue systems*

7. Plants are capable of indeterminate growth because of their
   a. regions of nondividing cells.
   b. meristem tissues.
   c. epidermis.
   d. xylem.
   e. cortex.
   ***Textbook Reference:*** *Concept 24.2 Apical Meristems Build the Primary Plant Body; A hierarchy of meristems generates the plant body*

8. Which statement best describes the developmental origin of wood?
   a. Xylem cells enlarge and deposit large amounts of lignin.
   b. Primary meristems increase the amount of xylem deposited.
   c. Lateral meristems contribute to continuous increases in vascular tissue.
   d. Spongy mesophyll cells become cork cambium.
   e. Pericycle layers increase in number.
   ***Textbook Reference:*** *Concept 24.3 Many Eudicot Stems and Roots Undergo Secondary Growth*

9. Which cell type is *not* a ground tissue cell?
   a. Parenchyma
   b. Collenchyma
   c. Sclerenchyma
   d. Xylem
   e. Sclereid

   ***Textbook Reference:*** *Concept 24.1 The Plant Body Is Organized and Constructed in a Distinctive Way; The plant body is constructed from three tissue systems*

10. Which statement best describes the function of the cork cambium?
    a. It lays down a protective cork covering exposed phloem tissue.
    b. It inhibits the sloughing off of epidermal tissue.
    c. It allows for diameter shrinking in stems and roots.
    d. It supplies the secondary xylem.
    e. It supplies the secondary phloem.

    ***Textbook Reference:*** *Concept 24.3 Many Eudicot Stems and Roots Undergo Secondary Growth*

11. Sieve tube elements have sieve plates where they join other sieve tube elements. Which statement about the sieve plates is true?
    a. Sieve plate pores are enlargements of meristems.
    b. They allow conduction between sieve tube cells through plasmodesmata.
    c. They allow for the joining of cytoplasm between adjacent stomata.
    d. They contain the organelles of the cell.
    e. They filter out certain molecules to prevent their passage between cells.

    ***Textbook Reference:*** *Concept 24.1 The Plant Body Is Organized and Constructed in a Distinctive Way; The plant body is constructed from three tissue systems*

12. Plants regulate gas exchange and water loss via
    a. the cuticle.
    b. xylem.
    c. coated pits.
    d. sieve plates.
    e. stomata.

    ***Textbook Reference:*** *Concept 24.2 Apical Meristems Build the Primary Plant Body; Leaves are photosynthetic organs produced by shoot apical meristems*

13. The protoderm becomes the _______ tissue system.
    a. dermal
    b. ground
    c. vascular
    d. cortex
    e. endodermal

    ***Textbook Reference:*** *Concept 24.2 Apical Meristems Build the Primary Plant Body; A hierarchy of meristems generates the plant body*

14. Primary growth occurs at the
    a. lateral meristems.
    b. fruit.
    c. quiescent center.
    d. apical meristems.
    e. wood.

    ***Textbook Reference:*** *Concept 24.2 Apical Meristems Build the Primary Plant Body; A hierarchy of meristems generates the plant body*

15. Vascular bundles are composed of _______ and _______.
    a. root hairs; xylem
    b. cork; phloem
    c. xylem; phloem
    d. wood; cork
    e. mesophyll; xylem

    ***Textbook Reference:*** *Concept 24.2 Apical Meristems Build the Primary Plant Body; The products of the shoot's primary meristems become shoot tissues*

## Answers

### *Key Concept Review*

1.

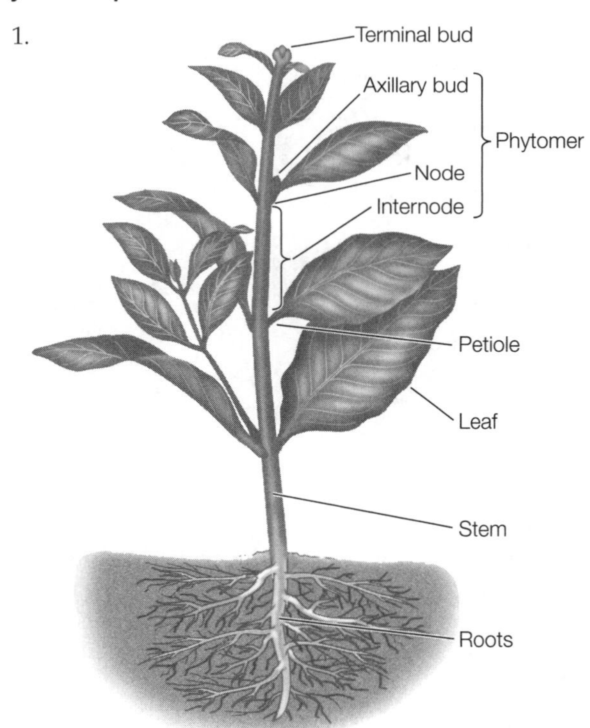


2. The three types of tissue are dermal, vascular, and ground. Dermal tissue acts as a covering for the plant and its cells are typically small. It provides protection and also functions in gas exchange and nutrient and water uptake. Vascular tissue, which is composed of cells that are either dead (xylem) or alive (phloem), is involved in transport of water and nutrients within

the plant. Certain types of ground tissue are involved in photosynthesis, and others provide support for the plant body. The cells you found are probably sclerenchyma cells, part of the ground tissue system, dead cells that have thick cell walls containing lignin that provides support.

3. Apical meristems are responsible for elongation of the plant body. Lateral meristems are responsible for an increase in girth. Lateral meristems are found only in woody eudicots and are responsible for creating wood. All plants have apical meristems.

4. 

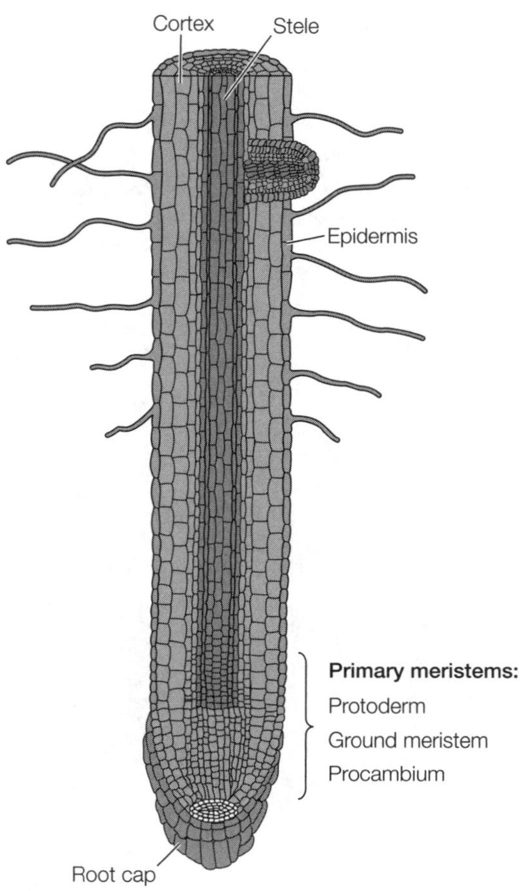


The protoderm gives rise to the epidermis for protection. The ground meristem gives rise to the cortex for storage. The procambium gives rise to the stele for transport. The root cap protects the meristem as it pushes through the soil.

5. The "top" of the leaf diagram in the textbook would be the surface that faces the sun. This is the surface where photosynthetic cells, which need maximum sun exposure, are located. The stomata are on the opposite, and cooler, side. This positioning serves to reduce water loss due to evaporation during photosynthesis.

6. Primary xylem and phloem arise from the procambium, the precursor of the vascular tissues. Divisions of the vascular cambium form secondary xylem and phloem, with xylem to the inner side and phloem to the outer side. After a year of growth, the primary phloem is crushed by the continued divisions of the vascular cambium and is no longer visible. Thus, only the latest layer of secondary phloem is ever visible in the cross-section of a woody eudicot stem.

7. Bark is defined as all tissue exterior to the vascular cambium, which includes the secondary phloem. Destruction of the secondary phloem destroys the ability of the plant to transport sugars, since there is only one layer of phloem in the tree at any given time after the first year of secondary growth. Cork, however, is the outermost layer of cells, which is reinforced by waxy suberin. Removal of only this layer leaves the cork cambium and secondary phloem intact and thus does not permanently harm the tree.

8. Humans have taken advantage of the large variation in plant shape and size within a species. They have selectively used the seeds from plants with desirable characteristics, thus ensuring that these characteristics remain in the crop. This has allowed humans to generate more productive crops, which can be easier to grow, and that give higher yields and very specific characteristics for particular applications.

9. The advent of modern genetics and advanced techniques allowed scientists to identify this very specific genomic difference between corn and teosinte, but this knowledge was not necessary for the cultivation of corn. At some point, a teosinte plant acquired a mutation in its *TB1* gene that suppressed the branching phenotype. When this plant's seeds were selected for the next generation of planting, this mutation was carried forward. Continued artificial selection pressure for this phenotype ensured that the genetic variation was maintained through successive generations. Today, scientists can target genes for variation and thus speed up the process of selection for a desired crop phenotype.

### Test Yourself

1. **b.** The phytomer is the repeating unit of a plant from node to node that includes the leaves, an internode, and one or more axillary buds.

2. **b.** A taproot would be necessary to anchor a plant of that size. Most eudicots have taproots.

3. **d.** Tendrils are modified leaves.

4. **e.** Plant cell shape is maintained by the cell walls. Animal cells do not have cell walls. Plants have vacuoles, chloroplasts, and apical meristems. Though some of these characteristics are seen in some protists, no animal cells have them.

5. **a.** Palisade and spongy mesophyll cells are photosynthetically active and derived from parenchyma cells.

6. **a.** Xylem tissue transports water from the roots throughout the plant. Phloem tissue transports carbohydrates and nutrients from source tissue to sink tissue.
7. **b.** The apical and (when present) lateral meristems are regions of continually dividing cells that can contribute to the growth of a plant throughout its life. A modular body plan also contributes to the ability of plants to grow continually.
8. **c.** Lateral meristems are responsible for the growth of new xylem and phloem. The secondary xylem gives rise to wood.
9. **d.** Parenchyma, collenchyma, and sclerenchyma are all ground tissue cells. Sclereids are a type of sclerenchyma cell. Xylem is a vascular tissue cell.
10. **a.** The cork cambium is a meristematic region outside the secondary phloem. As girth increases and splits the epidermis, causing loss of those protective layers, the cork cambium produces new cells to cover the expanding vascular tissue.
11. **b.** The sieve plates allow for the conduction of sap from one sieve tube cell to another sieve tube cell. The pores are due to enlargements of the plasmodesmata. These cells may also lose nuclei and organelles so that carbohydrates can easily pass through the sieve tubes.
12. **e.** Guard cells at the edges of stomata respond to changes in osmotic pressure (see Chapter 25). These changes lead to the opening and closing of the stomata to regulate gas exchange and water loss.
13. **a.** The protoderm becomes the dermal tissue system of the growing plant.
14. **d.** The apical meristems are the site of primary growth. Apical meristems are found at the tips of the roots, in stems, and in buds.
15. **c.** Vascular bundles are composed of the vascular tissue of the plant, which includes the xylem and phloem.

# 25 Plant Nutrition and Transport

## The Big Picture

- Plants require specific macro- and micronutrients. Deficiencies in any of these challenge the health of the plant. Essential nutrients must be available, and no substitutions will sustain the plant. Nutrients are procured from the soil solution that bathes the roots of a plant. The availability of nutrients depends on the quantity, solubility, and structure of soil. Many agricultural practices deplete soils of nutrients, and these must be replenished through fertilization.
- Some plants interact with fungi and bacteria that help them obtain needed nutrients. Some plants acquire the nitrogen essential to their growth by means of nitrogen-fixing bacteria in their root nodules, which use an enzyme called nitrogenase to convert nitrogen into ammonia. A small number of plants do not photosynthesize. These heterotrophic plants are often parasites of other plants and acquire nutrients solely through their hosts.
- In terrestrial plants, water must be acquired from the soil, transported through the plant, and used in the leaves for photosynthesis. At the same time, the nutritional products of photosynthesis must be transported throughout the plant to nonphotosynthetic tissues. This two-way transport is achieved through specialized cells that make up the vascular tissue of the plant. Water with dissolved mineral nutrients is absorbed through the roots and transported to cells in the xylem of the plant, where it is pulled up the stem to the leaves via the transpiration–cohesion–tension mechanism. Sugars and solutes are moved out of the leaves and to the rest of the plant through cells in the phloem.
- Water uptake is regulated by osmotic and water potentials in the root cells, and the rate of transport is controlled by the rate of evaporation at the leaf surface. Guard cell activity in the leaves regulates the opening and closing of stomata to match water availability, light, and drying conditions. Sucrose movement is regulated by active transport and facilitated diffusion in the phloem tissue. The rate of sucrose transport depends on the rates of loading and unloading at source and sink tissues.

## Study Strategies

- The interactions between nitrogen-fixing bacteria and plants can be confusing. Remember that it is a mutualistic relationship and that the bacteria have the enzymes necessary to fix atmospheric nitrogen.
- It is easy to fail to look at the anatomy of the structures that are used in transport. Each structure is uniquely suited to its function. Think in terms of form coupled with function to understand this process.
- Water and sucrose transport are pathways that can be understood by visually tracing a molecule of water or sucrose through the plant. Use the figures in the textbook to follow these pathways.
- In order to understand water transport, it is important to understand osmosis and the properties of water molecules. Review these concepts from earlier chapters to make this easier.
- The basis of nutritional transport in plants is cell-to-cell transport. Review the sections in the textbook on diffusion, osmosis, and active transport.
- Be sure you understand the consequences of nutrient shortages in plants.
- Go to **LaunchPad** (or use the Web addresses listed) to review the following additional resources:

  Animated Tutorial 25.1 Nitrogen and Iron Deficiencies (PoL2e.com/at25.1)

  Animated Tutorial 25.2 Water Uptake in Plants (PoL2e.com/at25.2)

  Animated Tutorial 25.3 Xylem Transport (PoL2e.com/at25.3)

  Animated Tutorial 25.4 The Pressure Flow Model (PoL2e.com/at25.4)

  Activity 25.1 Apoplast and Symplast of the Root (PoL2e.com/ac25.1)

  Analyze the Data in textbook Figure 25.2

  Apply the Concept in textbook Sections 25.2 and 25.4

## Key Concept Review

### 25.1 Plants Acquire Mineral Nutrients from the Soil

Nutrients can be defined by their deficiency

Experiments using hydroponics have identified essential elements

Soil provides nutrients for plants

Ion exchange makes nutrients available to plants

Fertilizers can be used to add nutrients to soil

The basic nutrient requirements for all living things are carbon, hydrogen, oxygen, and nitrogen. These elements are the fundamental building blocks of all macromolecules. Most plants are autotrophs and incorporate carbon through photosynthesis. Oxygen and hydrogen enter plants as water. Nitrogen's entry into plants is dependent on nitrogen-fixing bacteria in the soil. Mineral nutrients are essential to life. Sulfur, phosphorus, magnesium, and iron are all essential components of many macromolecules. Mineral nutrients enter biological organisms through soil solutions, which plants take up through their roots. Every plant requires specific essential nutrients for growth and development; these nutrients cannot be replaced by another element and must be obtained directly for the day-to-day functioning of the plant. A deficiency in any essential nutrient leads to an unhealthy plant. Macronutrients are essential elements required at a rate of 1 g per 1 kg of dry plant matter, and micronutrients are essential elements required at a rate of 100 mg per 1 kg of dry plant matter. Though many nutrient deficiencies ultimately lead to plant death, specific symptoms of deficiency are evident in a plant before it dies. Deficiency can be corrected by the addition of fertilizers to supplement the soils. Scientists determine which elements are essential by growing plants hydroponically, which allows for very precise manipulation of the nutrients that are available to the plant through alteration of the nutrient solution used (see Figure 25.2). Essential micronutrients are hard to identify because of the very low levels that are required; any small contamination or seed storage of that nutrient can confound experimental results.

Soil provides an anchor for plants, mineral nutrients, water required for growth and survival, and oxygen for the roots. The living portion of soil contains roots, protists, bacteria, fungi, and many small animals. Nonliving parts of the soil include rock fragments, water, air spaces, dead organic matter, and dissolved minerals. All soils have a soil profile consisting of several horizons (layers). Water-soluble nutrients are leached to deeper horizons through rainfall and irrigation. Three major horizons (also called zones), termed A, B, and C, can be identified in soils (see Figure 25.3).

The A horizon is topsoil that is organically rich and the most agriculturally important layer. A loam is a topsoil with an optimal mixture of sand, silt, and clay; it has high nutrient content, plentiful water, and adequate air spaces. Soils with too much sand typically do not hold nutrients or water well, and clays are too dense for the trapping of air. The B horizon is the subsoil that holds many leached nutrients. The C horizon is parent rock that gives rise to soil.

Soils form from mechanical and chemical weathering that breaks down rocks. Mechanical weathering is caused by freeze and thaw cycles, rain, and drying. Chemical weathering is caused by oxidation, water, and acids. Chemical weathering is very important in clay formation. Humus, which is formed from the breakdown of dead leaves and other organic matter, improves the quality of soil and provides air spaces.

Mineral nutrients are tied to clay particles in the soil. Because many nutrients are positively charged cations, clays with a negative charge can hold these nutrients and make them available to plants. Roots release protons into the soil that bind to clay particles, and the cation minerals are released. The $CO_2$ released from the roots can also form bicarbonate and free protons, which bind with clay. These processes are called ion exchange (see Figure 25.4). Nutrients that are negatively charged and therefore do not participate in ion exchange are rapidly leached from soil. Thus, nitrate and sulfate are often not available to plants. In the past, nutrient content of soil was replenished by shifting agriculture and leaving the land unused for long time periods. Organic fertilizers such as compost or animal manure can be used to add nutrients; the nutrients are released more slowly, limiting leaching. Inorganic fertilizers are used to quickly supply plants with nutrients, but application of too much merely leads to runoff of the excess.

Agricultural fertilizers add nitrogen, phosphorus, and potassium to soils and are rated by their "N-P-K" percentages. A 10-10-10 fertilizer, for example, contains 10 percent nitrogen (as ammonium), 10 percent phosphate, and 10 percent potash (potassium source) by weight.

**Question 1.** Differentiate between micro- and macronutrients. Where are most of these nutrients acquired?

**Question 2.** Explain how scientists determined which plant nutrients are the essential nutrients.

**Question 3.** Draw a diagram showing the different layers of the soil and describe the importance of each to a plant's survival.

### 25.2 Soil Organisms Contribute to Plant Nutrition

Plants send signals for colonization

Mycorrhizae expand the root system

Rhizobia capture nitrogen from the air and make it available to plant cells

Some plants obtain nutrients directly from other organisms

Soils are full of bacteria and fungal hyphae, some of which interact with plants to help them obtain nutrients. Plants send out signals to attract these beneficial organisms. Fungi form associations with roots to form arbuscular mycorrhizae Formation of arbuscular mycorrhizae is stimulated by the release of strigolactones by the root (see Figure 25.5A). A prepenetration apparatus helps guide the fungi as they grow into the root. Arbuscules in the cortical cells of the roots are the site of nutrient exchange. The mycorrhizae increase the surface area of the root, allowing for greater uptake of nutrients, primarily phosphorus. Bacteria known as rhizobia can form symbiotic relationships with legumes living in nodules on their roots. To establish the symbiosis, the plant releases chemical signals (including flavonoids) to attract the bacteria. In response to the flavonoids, the bacteria turn on the production of Nod factors, which cause formation of the nodule by the plant. Once housed in the nodule, the bacteria differentiate into bacteroids capable of nitrogen fixation. Both mycorrhial and rhizobial symbioses form via similar mechanisms. Mycorrhizae expand the effective surface area of the root and allow the plant to extract more nutrients from the soil. The fungus gains photosynthetic products in exchange.

Nitrogen gas is readily available in the atmosphere, but plants are unable to break the triple bonds between the two nitrogen atoms. A few bacteria species can fix nitrogen gas into biologically usable ammonia through nitrogen fixation using the enzyme nitrogenase to catalyze the reaction (see Figures 25.5B and 25.6). This reaction is energetically expensive. There are symbiotic and nonsymbiotic species that perform nitrogen fixation. Nitrogenase is inhibited by oxygen and therefore is active only under anaerobic conditions. The root nodules that house the symbiotic rhizobia maintain very low oxygen levels through the action of the protein leghemoglobin, which binds with oxygen to keep oxygen levels low.

Some plants that live in nitrogen- or phosphorus-deficient soils are carnivorous. Carnivorous plants acquire nitrogen from the proteins of trapped decaying animals. An example of a carnivorous plant is the Venus flytrap (see Figure 27.7A). Some plants have lost the ability to photosynthesize and must acquire their nutrients from other sources. Some are parasitic and acquire some or all of their nutrients from a host plant at the host plant's expense. Hemiparasites obtain water and mineral nutrients from their host, but are able to photosynthesize. Holoparasites are unable to photosynthesize.

**Question 4.** Describe how plants and bacteria interact to form nitrogen-fixing root nodules. Why do many farmers plant crops such as alfalfa and soybeans without harvesting them?

**Question 5.** Most carnivorous plants are found in boggy, wet, acidic environments. What is the effect of such an environment on nutrient availability?

## 25.3 Water and Solutes Are Transported in the Xylem by Transpiration–Cohesion–Tension

- Differences in water potential govern the direction of water movement
- Water and ions move across the root cell's cell membrane
- Water and ions pass to the xylem by way of the apoplast and symplast
- Water moves through the xylem by the transpiration–cohesion–tension mechanism
- Stomata control water loss and gas exchange

The movement of water across a semipermeable membrane is a special type of diffusion known as osmosis (see Figure 25.8). The overall tendency of a solution to take up water across a membrane (called its water potential, psi, ) is the sum of the solute potential and the pressure potential. Solute potential measures the influence of solutes on water movement (which is usually negative), while pressure potential is the pressure inside the rigid cell (turgor pressure, which is usually positive). All three parameters can be measured in megapascals (MPa). Water always moves to a region of more negative water potential. The structure of plants is maintained by osmotic phenomena. If a plant loses turgor pressure by a decrease in pressure potential, it wilts. Movement of water from cell to cell depends on the gradient of water potential. Specialized membrane channel proteins in plant cells called aquaporins can increase the rate of water movement by allowing water to cross the plasma membrane without interacting with the hydrophobic bilayer that slows water flow. Though aquaporins can increase the rate of osmosis, they cannot influence the direction of flow.

Mineral ion uptake from the soil solution requires active transport via proteins. When mineral concentrations are greater in the soil solution than in the plant, they are taken up by facilitated diffusion. If minerals are in lower concentrations outside the plant than inside the plant, or if they must be moved against an electrochemical gradient, then the plant must rely on active transport. Plants rely on a proton pump for active transport of minerals into cells. Plants actively pump protons out of cells, causing the area just outside the cell to become more positive. The result of this pumping action is that the internal environment of the plant cell becomes highly negative compared to its environment. This assists facilitated diffusion of positive ions through protein channels. It also drives the movement of negatively charged ions, such as $Cl^-$, into the cell by secondary active transport (see Figure 25.10).

Minerals and water follow two paths for reaching the vascular tissue: the rapid apoplast or the slower symplast. The apoplast is formed by cell walls and intercellular spaces. Water and minerals may move unregulated through this space without ever having to cross a membrane. The symplast is the living portion of the plant and is enclosed in plasma membranes; water and solutes move from cell to cell via the plasmodesmata. Movement of water and minerals in the symplast is highly regulated. Water and minerals can travel through the apoplast as far as the endodermis (see Figure 25.11). At the endodermis, water and minerals are stopped by the Casparian strip, which is a waxy structure surrounding the endodermal cells. Because of this, water can reach the stele only via the symplast. The transport proteins in the endodermal cell membranes determine which minerals enter the stele. Once past the endodermal barrier, water and minerals can again leave the symplast and move back to the apoplast with the aid of parenchyma cells. Ultimately, water and minerals from the soil solution end up exported into xylem cells and become the xylem sap. Before the end of the nineteenth century, upward pressure and capillary action were thought to control the movement of fluids in plants. A simple experiment in 1893, in which a sawed-off tree trunk was immersed in poison, showed that the poison continued to move upwards even as cells were killed along the way. Only when the poison reached the leaves did movement stop, showing that the leaves are critical for transport.

The mechanism that pulls water up from the roots through the plant is known as the transpiration–cohesion–tension mechanism (see Figure 25.12). This is a passive process requiring no energy input by the plant. Minerals are drawn passively along with the water column. In the process of transpiration, water evaporates from mesophyll cells of leaves, creating tension on the water associated with the mesophyll cell wall. Tension is a pulling force that drives water movement through the plant. Water molecules are cohesive, meaning that their hydrogen bonds stick together strongly enough to resist the tension, so they pull their neighboring water molecules toward the mesophyll cell walls in response to transpiration. Tension in the mesophyll in turn generates tension on the cohesive water molecules in the nearby xylem water column, which pulls water up from the roots and through the apoplast of the leaves. Transpiration also assists with temperature regulation through evaporative cooling of the leaves.

Leaf surfaces are covered with a waxy cuticle to prevent excessive water loss. However, the leaf must take up $CO_2$ for photosynthesis. Stomata, bounded by guard cells, are pores that regulate gas exchange and water loss from a leaf. In response to osmotic differences, guard cells shrink and swell to open and close the stomata. Because closed stomata prevent water loss but also exclude $CO_2$, most plants open their stomata when light intensity is sufficient to maintain photosynthesis. Specific wavelengths of light and low $CO_2$ levels stimulate a proton pump that helps regulate guard cell activity. Guard cells open when potassium ions diffuse into the cell as a result of the electrical gradient set up by the proton pump. High potassium levels cause water to move in by osmosis. Pressure potential builds in the guard cells, and they are pulled apart to reveal the stoma. For guard cells to shut, the proton pump ceases its activity, and potassium ions move back across the membrane. Water follows, and the cells go limp and seal off the stoma (see Figure 25.13). Guard cells are also regulated by water potential. If the water potential in mesophyll cells is low, the cells release the hormone abscisic acid, which causes the stomata to close. Plants also regulate the number of stomata on the leaves in order to regulate water loss and gas exchange.

**Question 6.** Create a flow chart of the path taken by a water molecule as it moves from the soil solution to the stele of a plant. Identify where the molecule is traveling through the apoplast and where it is traveling through the symplast.

**Question 7.** Under what conditions does transpiration occur most rapidly? What effect does increased transpiration have on water flow in a plant? What happens if adequate water for the plant is not available?

**Question 8.** Explain how transpiration, cohesion, and tension work together to move water in a large plant.

### 25.4 Solutes Are Transported in the Phloem by Pressure Flow

Sucrose and other solutes are carried in the phloem

The pressure flow model describes the movement of fluid in the phloem

Phloem moves materials from sources to sinks by translocation (see Figure 25.14). Sources are organs that produce more sugars than are used by metabolism, storage, and growth. Sinks are organs that do not make enough sugar for their own growth or storage needs. Sucrose, amino acids, minerals, and other substances are translocated between sources and sinks in the phloem. Translocation proceeds in both directions along the stem and stops if phloem tissue is killed. The pressure flow model explains how materials move through the phloem Sucrose and other solutes are actively transported into sieve tube companion cells, and then flow into the sieve tube elements via the plasmodesmata. This causes water to move into sieve tubes by osmosis, thereby increasing the pressure potential at the source end and pushing the contents toward the sink end. Once in the sieve tubes, sieve tube sap moves via bulk flow, which requires no energy input by the plant. Energy is required for the loading of the sieve tubes at the sources and the unloading of solutes when the sink is reached.

**Question 9.** Differentiate between source and sink tissues. What happens relative to phloem in each?

**Question 10.** Describe the pressure flow model of phloem transport. What would happen if the loading of sucrose were to be inhibited?

## Test Yourself

1. Mineral ions enter the cell due to the force of an electrochemical gradient set up by the pumping of _______ out of the cells.
   a. $K^+$
   b. $Ca^{2+}$
   c. $Na^+$
   d. $H^+$
   e. $Cl^-$
   ***Textbook Reference:*** *Concept 25.3 Water and Solutes Are Transported in the Xylem by Transpiration–Cohesion–Tension; Water and ions move across the root cell's cell membrane*

2. Compared to micronutrients, macronutrients are
   a. larger.
   b. less essential.
   c. more essential.
   d. of equal importance.
   e. needed in greater quantities.
   ***Textbook Reference:*** *Concept 25.1 Plants Acquire Mineral Nutrients from the Soil; Nutrients can be defined by their deficiency*

3. The opening and closing of the stomata is accomplished by the
   a. sieve tube.
   b. guard cells.
   c. process of translocation.
   d. aquaporins.
   e. xylem.
   ***Textbook Reference:*** *Concept 25.3 Water and Solutes Are Transported in the Xylem by Transpiration–Cohesion–Tension; Stomata control water loss and gas exchange*

4. Nitrogen gas is reduced to ammonia by which enzyme or process?
   a. Rhizobium
   b. Nitrogenase
   c. Nitrification
   d. Denitrification
   e. Rhizobenase
   ***Textbook Reference:*** *Concept 25.2 Soil Organisms Contribute to Plant Nutrition; Rhizobia capture nitrogen from the air and make it available to plant cells*

5. Clay particles in soils are important for
   a. making the soil loose and crumbly.
   b. allowing the leaching of ions.
   c. the shedding of water.
   d. weathering of the C horizon.
   e. ion exchange.
   ***Textbook Reference:*** *Concept 25.1 Plants Acquire Mineral Nutrients from the Soil*

6. Most clays form from the _______ of rock.
   a. mechanical weathering
   b. chemical weathering
   c. heaving
   d. grinding
   e. drying
   ***Textbook Reference:*** *Concept 25.1 Plants Acquire Mineral Nutrients from the Soil; Ion exchange makes nutrients available to plants*

7. Which of the following represents the correct ordering of the water potential of these root cells or regions, from most negative to least negative?
   a. Xylem, cortex apoplast, stele apoplast, soil next to root
   b. Soil next to root, xylem, stele apoplast, cortex apoplast
   c. Xylem, stele apoplast, cortex apoplast, soil next to root
   d. Stele apoplast, cortex apoplast, xylem, soil next to root
   e. Soil next to root, cortex apoplast, stele apoplast, xylem
   ***Textbook Reference:*** *Concept 25.3 Water and Solutes Are Transported in the Xylem by Transpiration–Cohesion–Tension; Water and ions pass to the xylem by way of the apoplast and symplast*

8. The relationship between rhizobium bacteria and the roots of legumes can best be described as
   a. parasitic.
   b. one-sided.
   c. mutualistic.
   d. carnivorous.
   e. detrimental.
   ***Textbook Reference:*** *Concept 25.2 Soil Organisms Contribute to Plant Nutrition*

9. Years of cotton farming in the South have stripped away much of the A horizon of the soils. The effects on agriculture have been _______ because _______.
   a. significant; the A horizon contains the most available nutrients
   b. negligible; the B horizon contains significantly more available nutrients
   c. negligible; the C horizon is most conducive to root growth
   d. significant; stripping of the A horizon leaches the C horizon of its nutrients
   e. negligible; the A horizon does not contain many nutrients
   ***Textbook Reference:*** *Concept 25.1 Plants Acquire Mineral Nutrients from the Soil; Soil provides nutrients for plants*

10. Plants are able to take up and use nitrogen in the form of
    a. ammonia.
    b. nitrogen gas.
    c. liquid nitrogen.
    d. nitrous oxide.
    e. nitric oxide.
    ***Textbook Reference:*** *Concept 25.2 Soil Organisms Contribute to Plant Nutrition; Rhizobia capture nitrogen from the air and make it available to plant cells*

11. Tension in the xylem is a result of
    a. transpiration at the leaf surface.
    b. the cohesive nature of water.
    c. the narrowness of the xylem tube.
    d. the surface area of the phloem.
    e. root pressure.
    ***Textbook Reference:*** *Concept 25.3 Water and Solutes Are Transported in the Xylem by Transpiration–Cohesion–Tension; Water moves through the xylem by the transpiration–cohesion–tension mechanism*

12. Nitrate and sulfate tend to leach from the soil because
    a. they bind with ions such as $K^+$ and $Mg^{2+}$.
    b. the $H^+$ ions released by the roots push them out.
    c. they are unable to bind with the negatively charged clay particles.
    d. they bind with the positively charged clay particles.
    e. they do not dissolve readily in water.
    ***Textbook Reference:*** *Concept 25.1 Plants Acquire Mineral Nutrients from the Soil; Soil provides nutrients for plants*

13. A plant lowers the pH of soil (makes it more acidic) by means of
    a. ion exchange.
    b. leaching.
    c. $Na^+$ pumping.
    d. proton pumping.
    e. water repulsion.
    ***Textbook Reference:*** *Concept 25.1 Plants Acquire Mineral Nutrients from the Soil; Ion exchange makes nutrients available to plants*

14. The role of leghemoglobin is to maintain _______ levels in the root nodule.
    a. high $O_2$
    b. high $CO_2$
    c. low $O_2$
    d. low $CO_2$
    e. high $N_2$
    ***Textbook Reference:*** *Concept 25.2 Soil Organisms Contribute to Plant Nutrition; Rhizobia capture nitrogen from the air and make it available to plant cells*

15. The function of the Casparian strip is to
    a. divert water and minerals around the membranes of endodermal cells.
    b. prevent water and minerals from entering the stele through the symplast.
    c. provide regulation for water and mineral movement in the plant.
    d. prevent movement of all water and minerals into the stele.
    e. absorb nutrients for even distribution throughout the root cross section.
    ***Textbook Reference:*** *Concept 25.3 Water and Solutes Are Transported in the Xylem by Transpiration–Cohesion–Tension; Water and ions pass to the xylem by way of the apoplast and symplast*

16. The primary difference between the apoplast and the symplast is that
    a. the apoplast consists of nonliving spaces and cell walls, whereas the symplast consists of living cells.
    b. apoplast movement is tightly regulated and symplast movement is not.
    c. the symplast consists of nonliving spaces and cell walls, whereas the apoplast consists of living cells.
    d. apoplast movement is slow and symplast movement is fast.
    e. the apoplast transports only ions and the symplast transports only water.
    ***Textbook Reference:*** *Concept 25.3 Water and Solutes Are Transported in the Xylem by Transpiration–Cohesion–Tension; Water and ions pass to the xylem by way of the apoplast and symplast*

17. Which statement about water transport is true?
    a. Root pressure is sufficient to drive xylem sap movement.
    b. Bulk flow is not a mechanism by which water and minerals are transported.
    c. The cohesive nature of water is central to water movement in a plant.
    d. Water transport is an active process.
    e. Leaves are not involved in water transport.
    ***Textbook Reference:*** *Concept 25.3 Water and Solutes Are Transported in the Xylem by Transpiration–Cohesion–Tension; Water moves through the xylem by the transpiration–cohesion–tension mechanism*

18. Carnivorous plants are often found in acidic and nutrient-poor environments. The main selective pressure for carnivory is
    a. lack of nitrogen and phosphorus sources.
    b. lack of iron and calcium sources.
    c. incomplete ion exchange.
    d. lack of water sources.
    e. lack of sugar sources.
    ***Textbook Reference:*** *Concept 25.2 Soil Organisms Contribute to Plant Nutrition; Some plants obtain nutrients directly from other organisms*

19. The fact that water transport continues as long as leaves are alive and active indicates that
    a. leaves pump water.
    b. leaves are necessary for transport of water.
    c. roots are active.
    d. water is not needed for leaves to remain alive.
    e. sieve tube elements are inactive.
    ***Textbook Reference:*** *Concept 25.3 Water and Solutes Are Transported in the Xylem by Transpiration–Cohesion–Tension; Water moves through the xylem by the transpiration–cohesion–tension mechanism*

20. Which statement regarding transport in phloem is true?
    a. It always moves in the direction of leaves to roots.
    b. It proceeds from source tissue to sink tissue.
    c. It requires no energy inputs from the plant.
    d. It is the same process as transport in xylem.
    e. It always moves in the direction of roots to leaves.
    ***Textbook Reference:*** *Concept 25.4 Solutes Are Transported in the Phloem by Pressure Flow*

21. If the pressure potential of a plant's cells is 0.16 megapascals (MPa) and the solute potential is –0.24 MPa, then the water potential is _______ MPa.
    a. 0.04
    b. 0.08
    c. –0.08
    d. –0.24
    e. –0.04
    ***Textbook Reference:*** *Concept 25.3 Water and Solutes Are Transported in the Xylem by Transpiration–Cohesion–Tension; Differences in water potential govern the direction of water movement*

22. The label "10-20-10" on a package of commercial fertilizer refers to the _______ the fertilizer.
    a. percentages of nitrogen, phosphorus, and potassium in
    b. percentages of nitrogen, carbon, and oxygen in
    c. percentages of phosphorus, iron, and potassium in
    d. rate at which nitrogen is released from
    e. ratio of organic to inorganic matter in
    ***Textbook Reference:*** *Concept 25.1 Plants Acquire Mineral Nutrients from the Soil; Fertilizers can be used to add nutrients to soil*

23. Which statement about xylem transport and phloem transport is true?
    a. Both are passive processes that do not require energy from the plant.
    b. Both rely on living cells only.
    c. Both rely on a water potential gradient.
    d. The direction of flow can be reversed in both.
    e. The driving force for both is in the leaves.
    ***Textbook Reference:*** *Concept 25.3 Water and Solutes Are Transported in the Xylem by Transpiration–Cohesion–Tension; Water and ions pass to the xylem by way of the apoplast and symplast*

24. Stomatal opening and closing are regulated by
    a. oxygen.
    b. sucrose.
    c. $CO_2$ concentrations.
    d. nitrogen.
    e. rhizobia.
    ***Textbook Reference:*** *Concept 25.3 Water and Solutes Are Transported in the Xylem by Transpiration–Cohesion–Tension; Stomata control water loss and gas exchange*

25. Nitrogen and potassium are acquired from
    a. micronutrients.
    b. heterotrophs.
    c. air.
    d. the soil solution.
    e. the leaves.
    ***Textbook Reference:*** *Concept 25.1 Plants Acquire Mineral Nutrients from the Soil; Soil provides nutrients for plants*

26. Regulators of stomatal opening and closing work by activating the
    a. proton pump in guard cells.
    b. proton pump in stomata.
    c. sodium–potassium pump in guard cells.
    d. sodium–potassium pump in stomata.
    e. proton pump in endodermis.
    ***Textbook Reference:*** *Concept 25.3 Water and Solutes Are Transported in the Xylem by Transpiration–Cohesion–Tension; Stomata control water loss and gas exchange*

27. Which of the following nutrients is *not* considered essential for plant growth?
    a. Cadmium
    b. Nitrogen
    c. Manganese
    d. Potassium
    e. Iron
    ***Textbook Reference:*** *Concept 25.1 Plants Acquire Mineral Nutrients from the Soil; Nutrients can be defined by their deficiency*

## Answers

### *Key Concept Review*

1. Macronutrients are needed at a rate of 1 g/1 kg of dry plant tissue. Micronutrients are needed at a rate of 100 mg/1 kg of dry plant tissue. Most of these nutrients are in soil solution and are taken up as water is drawn into roots.
2. In a series of experiments, plants were grown in cultures that lacked specific nutrients. If a plant could not complete its life cycle, the missing nutrient was said to be essential. These experiments were well-controlled hydroponic studies because nutrient content can easily be manipulated in water.
3. 

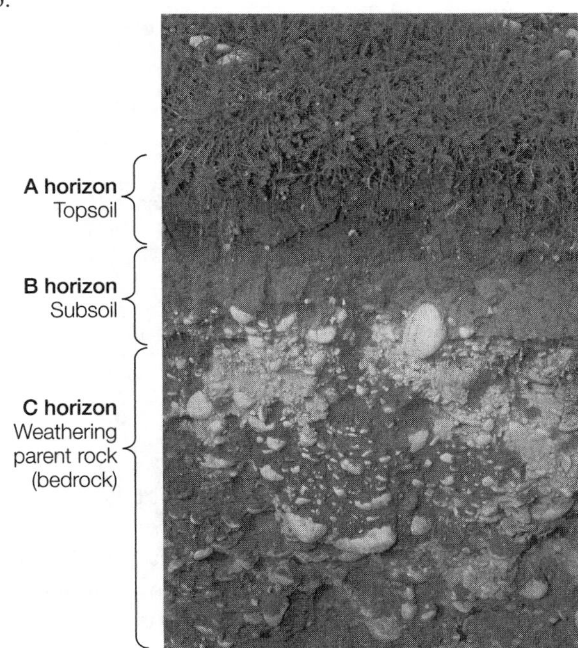


Soil is made up of three different layers: The top layer (topsoil), called the A horizon, supplies the plant's mineral nutrient needs and contains most of the soil's living and dead organic matter. The B horizon is the subsoil, which accumulates materials from the topsoil above and the parent rock below. The C horizon is the parent rock from which the soil arises.

4. Legumes can form symbioses with rhizobia, which live in nodules on the legumes' roots. The roots release flavonoids and chemical signals that attract the rhizobia. The flavonoids stimulate the bacteria to secrete Nod factors, causing the root cortex to divide and form nodules. Bacteria enter the root through an infection thread and reach the cells in the nodule. Bacteria are released into the cytoplasm of the nodule cells. The bacteria differentiate into bacteroids, which can fix nitrogen (see Figure 25.5B). By rotating crops (and especially by rotating and plowing under legume crops) farmers can organically add nitrogen to depleted soils. This results in significantly larger yields of subsequent harvested crops.
5. Acidic environments limit decomposition and thus nitrogen sources. They also limit ion exchange. Both conditions result in reduced nutrient availability that can be offset by carnivory.
6.

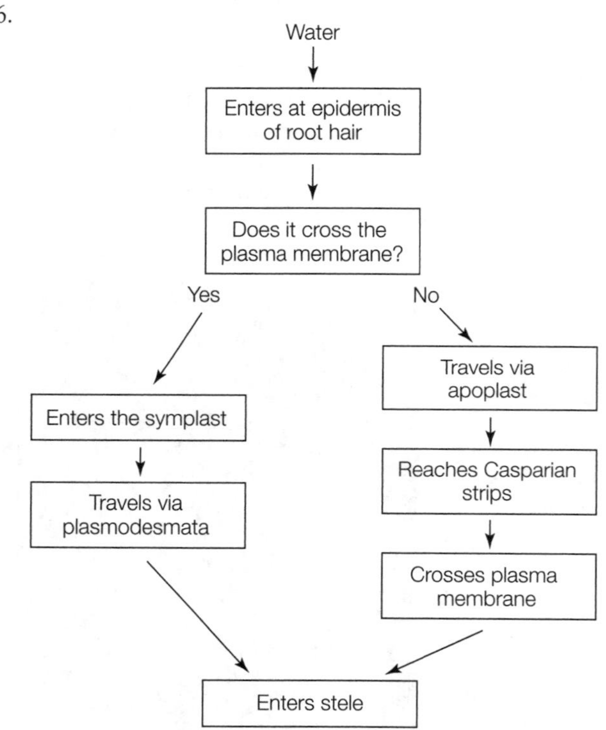


7. Transpiration occurs most rapidly in high light conditions when stomata are open, along with high wind conditions and low humidity when evaporation is greatest. This results in faster bulk flow through the xylem and increased water demands by the plant. If water is not available, plant cells lose turgor and the plant wilts.
8. The transpiration–cohesion–tension mechanism pulls water from the roots up through the plant. Water evaporates from mesophyll cells during transpiration. This puts tension on the film of water associated with the mesophyll cell wall. The tension at the mesophyll cell draws water from the xylem of the nearest vein. This creates tension in the entire xylem column, and the column is drawn upward from the roots.
9. Source tissues produce sugars in excess of what can be used and stored. Phloem loading occurs in source tissues and creates a pressure potential that results in bulk flow of sieve tube sap toward sink tissues. Sink tissues produce fewer sugars than can be stored or used and unload phloem through active transport.
10. The difference in solute concentration between sources and sinks creates a pressure potential along sieve tubes, resulting in bulk flow. For this to occur, sugars must be loaded at the source tissue and unloaded at the sink tissue through active transport, and the sieve plates must remain open and unclogged along the phloem column. If the loading of sugars is inhibited, then the plant will not be able to move fluid in the phloem and the plant will die.

### *Test Yourself*

1. **d.** Cells pump $H^+$ ions out into the soil with the help of proton pumps.
2. **e.** The main difference between micronutrients and macronutrients is in the quantity of each needed by a plant for survival.
3. **b.** Guard cells are specialized epidermal cells that regulate the opening and closing of the stomata by covering the stomata opening.
4. **b.** Nitrogenase catalyzes the reduction of nitrogen gas to ammonia. This is an energy-expensive process.
5. **e.** Clay particles are critical for ion exchange. They are also important for retaining water and for the integrity of the soil.
6. **b.** Chemical weathering leads to clay formation.
7. **c.** The xylem would be the most negative, followed by the stele, then the cortex, then the area outside the root.
8. **c.** The relationship is mutualistic, in that both the plant and the bacteria benefit from the association.
9. **a.** Topsoil, or the A horizon, is most conducive to root growth. Ample nutrients are available in the topsoil, as are air spaces and water for ease of root growth.
10. **a.** Plants can take up and use nitrogen that is in the form of ammonia (or nitrate).
11. **a.** Transpiration causes tension.
12. **c.** Nitrate and sulfate are negatively charged anions. Without any positively charged molecules in the soil with which they can interact, they leach from the soil.
13. **d.** Plants decrease the pH in the surrounding soil by pumping protons out of the roots.
14. **c.** Leghemoglobin is a protein produced by the root nodule of plants to help maintain low levels of $O_2$ so that the nitrogen-fixing bacteria are in an anaerobic environment.
15. **c.** Not all minerals that enter the apoplast of a plant's root are beneficial to the plant. The Casparian strip

prevents water and minerals from reaching the stele through the apoplast, diverting them instead through the membranes of the endodermal cells. Channel proteins in these membranes determine which minerals can enter the symplast, and from there, the stele. The Casparian strip thus contributes to the regulation of water and mineral movement in the plant.

16. **a.** The intercellular spaces and cell walls of the plant constitute the apoplast.
17. **c.** Water movement depends on the cohesive nature of water and its capacity to withstand the tension placed on the water column by transpiration.
18. **a.** Carnivory supplements insufficient nitrogen and phosphorus availability.
19. **b.** Leaves are necessary for transpiration to take place.
20. **b.** Transport in phloem does not always go from leaf to root, but it always proceeds from source tissue to sink tissue. The plant must contribute energy to create the water pressure gradient by pumping solutes into the phloem at the source and out of the phloem at the sink.
21. **c.** Water potential is equal to pressure potential plus solute potential.
22. **a.** The fertilizer is 10 percent nitrogen, 20 percent phosphorus, and 10 percent potash (a potassium source).
23. **c.** Both xylem transport and phloem transport depend on water potential.
24. **c.** Carbon dioxide levels are one of the regulators of stomatal opening and closing.
25. **d.** Nitrogen and mineral nutrients are acquired in soil solution (dissolved in water).
26. **a.** Stomatal regulators work by activating and deactivating the proton pump in guard cells.
27. **a.** Cadmium is not one of the 14 essential micro- and macronutrients.

# 26 Plant Growth and Development

## The Big Picture

- Plant growth is controlled by interactions among a plant's environment, hormones, and genetic makeup. Changes in plant growth are dependent on the information transmitted by hormones to hormone receptors, the subsequent turning on of signal transduction pathways, and the eventual alteration of gene expression. Receptors are often highly specific and respond to select hormones. Hormones are produced in a specific region of the plant body and translocated throughout the plant; therefore, their effects are often concentration-dependent. All growth in a plant is a result of changes in cell division, cell expansion, and cell differentiation.
- Seed germination, growth of the vegetative structures, reproduction, and senescence all proceed in defined patterns. Seed dormancy is broken and development begins when the seed coat is abraded, inhibitory chemicals are diluted, and the seed imbibes water. This begins a series of events that mobilize nutrients and induce growth. Once a seedling emerges from the soil, light begins to influence subsequent development under the control of hormones.
- Hormones exist in several classes, each with its own effects on growth and development. Some of these effects are antagonistic, and therefore control is maintained via relative concentration of several hormones. Hormones influence every step of development, from the breaking of seed dormancy to senescence.
- Light regulates plant processes through photoreceptors. Photoreceptors respond to very specific wavelengths and induce cascades that lead to changes in a plant.

## Study Strategies

- The way in which phytochrome shifts from the red to far-red form is a difficult concept to understand—phytochromes have puzzled researchers for many years.
- There is a tendency when studying hormones simply to memorize functions. It is important to understand the effects of different ratios of hormones in a plant, not just what one individual hormone does.
- The amount of information in this chapter may seem overwhelming. Take your time in learning how growth is regulated and the effects of the various hormones.
- Try to assimilate the big picture of how the environment, receptors, hormones, and genome interact before focusing on specific details. Make sure the broad picture is clear before you memorize functions of hormones or specific sequences of development.
- Go to **LaunchPad** (or use the Web addresses listed) to review the following additional resources:

  Animated Tutorial 26.1 Tropisms (PoL2e.com/at26.1)

  Animated Tutorial 26.2 Auxin Affects Cell Walls (PoL2e.com/at26.2)

  Activity 26.1 Monocot Shoot Development (PoL2e.com/ac26.1)

  Activity 26.2 Eudicot Shoot Development (PoL2e.com/ac26.2)

  Activity 26.3 Events of Seed Germination (PoL2e.com/ac26.3)

  Analyze the Data in textbook Figure 26.11

  Apply the Concept in textbook Section 26.2

## Key Concept Review

### 26.1 Plants Develop in Response to the Environment

The seed germinates and forms a growing seedling

Several hormones and photoreceptors help regulate plant growth

Genetic screens have increased our understanding of plant signal transduction

Plants differ from animals with their meristems, postembryonic organ formation, and differential growth. Plant development is influenced by multiple regulatory factors such as environmental cues, receptors, hormones, and the genome. Plant seeds are dormant and remain so until seed germination. Dormancy, which lasts for different periods of time depending on the plant, involves exclusion of water or oxygen from the embryo by the seed coat, mechanical restraint of the embryo by the seed coat, chemical inhibition of the embryo, and reliance on temperature and light cues. Dormancy ensures survival through adverse conditions. Germination is triggered by one or more mechanical or environmental cues. To germinate, a plant must imbibe water and draw polysaccharides, fats, and protein nutrients from the endosperm or cotyledons. Germination requires metabolic changes in the seed tissues. The germination stage is complete and the plant is called a seedling once the embryonic root, known as the radicle, emerges from the seed. Photoreceptors control the growth of the seedling as it emerges from the soil. Shoot development varies in monocots and eudicots; in monocots the shoot is protected by a sheath called the coleoptile, whereas in eudicots the cotyledons usually protect the shoot (see Figure 26.1).

There are two types of plant growth regulators: hormones and photoreceptors. Hormones are regulatory compounds that are produced in one area of a plant and translocated throughout the organism. The effects of the hormones are determined by their relative concentrations, and they play multiple regulatory roles. Photoreceptors are pigment proteins that sense light and are altered by light quality to induce changes within a plant.

Our understanding of how regulatory signaling pathways function has been enhanced through genetic screening (see Figure 26.2). Genetic screening involves mutating plants and comparing their phenotype with that of wild-type plants.

**Question 1:** Trace the basic steps that occur between the planting of a pea seed in a garden and the emergence of the pea plant.

**Question 2:** When seedlings grow in the dark they become tall and thin, a process called etiolation. Explain the advantage conferred on the plant by this response.

## 26.2 Gibberellins and Auxin Have Diverse Effects but a Similar Mechanism of Action

Gibberellins have many effects on plant growth and development

The transport of auxin mediates some of its effects

Auxin plays many roles in plant growth and development

At the molecular level, auxin and gibberellins act similarly

Gibberellins and auxin are two important plant hormones. Gibberellic acid was isolated from a fungus that infects rice plants and causes them to grow tall and spindly, and was subsequently also found in plant tissue. Charles Darwin and his son observed the function of auxin in the bending of canary grass seedlings toward the light. Asymmetrical movement of auxin in the canary grass simulates elongation of cells on the side opposite the light source. Mutant plants lacking either hormone result in short phenotypes that can be reversed by providing external sources of the hormones. Gibberellins are involved in stimulating elongation of the plant stem. Experiments with dwarf and normal plants showed that plants have innate gibberellins. Normal plants exposed to gibberellins showed no alteration in appearance, but dwarf plants exposed to gibberellins showed shoot elongation to near normal lengths (see Figure 23.3). Gibberellins play a role in fruit development. Developing seeds produce gibberellins that enhance development of fruit tissue. It is common agricultural practice to spray seedless fruits (especially grapes) with gibberellins to enhance fruit growth. In the developing seed, gibberellins stimulate the aleurone layer, a tissue layer under the seed coat, to secrete enzymes that break down the endosperm to feed the developing embryo (see Figure 26.4).

Auxin movement in plant tissues is unidirectional and polar, from apex to base. Auxin enters the cell in its nonpolar acid form by passive diffusion, and proton pumps transport $H^+$ out of the cell, causing the nonpolar auxin to become an anion and producing an electrochemical gradient (see Figure 26.5). Auxin anion efflux carriers are carrier proteins at the basal end of the cell responsible for export of auxin anions from cells, and they contribute to the unidirectional movement of auxin. Lateral redistribution of auxin is responsible for phototropism and gravitropism (see Figure 26.6). Under both conditions, higher auxin concentrations on one side of the plant cause increased rates of growth along that side and lead to bending. Auxin affects vegetative growth by initiating root growth in cuttings and promoting and maintaining the growth of a single main stem (apical dominance). Auxin inhibits abscission (the dropping of leaves) and can stimulate unfertilized fruit to form (parthenocarpy).

The effects of auxin on growth are mediated by the cell walls, which determine the rate and direction of cell growth. Cells grow by taking up water. The amount of water that can be taken up is restricted by a rigid cell wall. Cell walls must loosen, stretch, and add polysaccharides and cellulose to maintain structure as the cell expands. According to the acid growth hypothesis of cell expansion, auxin stimulates the production and insertion of proton pumps into the plasma membrane (see Figure 26.7). The subsequent decrease in pH stimulates proteins called expansins to alter polysaccharide bonding so that the cell wall becomes more pliable and readily stretched to accommodate a growing cell.

Auxin and gibberellins work through similar transduction signaling pathways (see Figure 26.8). There is a receptor in the cell that binds with auxin or gibberellins. Simulation of the pathway by these hormones promotes gene expression of growth stimulating genes. Both pathways have repressors that function by preventing transcription factors from allowing transcription of downstream genes for cell growth. Receptors for both auxin and gibberellins contain F-box regions that govern protein–protein interactions required for protein breakdown.

**Question 3:** Imagine that you are a plant physiologist interested in the hormones regulating development. Discuss an experimental approach you would use to try to determine the role of a hormone, such as gibberellins, in the regulation of development.

**Question 4:** Explain how auxin distribution regulates phototropism and gravitropism.

**Question 5:** Both gibberellins and auxin act in a similar fashion at the molecular level. Create a flow chart showing the signal transduction pathways for gibberellins and auxin. Note their similarities and their differences.

## 26.3 Other Plant Hormones Have Diverse Effects on Plant Development

Ethylene is a gaseous hormone that promotes senescence

Cytokinins are active from seed to senescence

Brassinosteroids are plant steroid hormones

Abscisic acid acts by inhibiting development

There are a number of additional hormones involved in the regulation of plant growth and development. Ethylene is a gaseous hormone that promotes leaf senescence and fruit ripening. In many instances it is given off by rotting fruit. The use of ethylene spray to promote fruit ripening is a common commercial practice. Ethylene scrubbers are used in fruit storage to prevent ethylene from accumulating and causing fruit to spoil. Other chemicals are used in the flower industry to inhibit ethylene's effects on flower senescence. Ethylene plays a role in maintaining the apical hook on emerging eudicots by inhibiting cells on the inner portion of the hook. It inhibits stem elongation, promotes lateral swelling of stems, and inhibits sensitivity to gravitropic stimulation—three responses that are known as the triple response.

Cytokinins are powerful stimulators of cell division and bud formation, aid in seed germination, inhibit stem elongation, and delay leaf senescence. Auxins and cytokinins regulate organ development based on the relative concentrations of the two. Relatively high auxin levels favor root formation and inhibit stem branching, and relatively high cytokinin levels favor shoot formation and stimulate stem branching. Cytokinins are synthesized primarily in the roots and are translocated throughout the plant. They act through a two-component system similar to that found in bacteria. A receptor (AHK) phosphorylates proteins, and a target transcription factor (ARR) acts as an effector. An intermediate protein (AHP) transfers the phosphate from the receptor to the transcription factor (see Figure 26.9).

Brassinosteroids are involved in the response to light. They have been shown to stimulate cell expansion and division in shoots and to promote xylem differentiation, pollen tube elongation, seed germination, apical dominance, and leaf senescence. They also inhibit root expansion, similarly to auxin. The brassinosteroid receptor is found on the cell membrane and initiates a transduction pathway that influences gene expression. Abscisic acid regulates the development of the seed and the production of proteins that will protect against desiccation; thus it is a stress hormone. Abscisic acid also inhibits seed germination. This condition of premature germination is called vivipary.

**Question 6:** Auxins and cytokinins appear to cancel out the effects of each other. Why is a hormone that affects bud growth produced in the roots, while the one that affects root growth is produced in the shoots? How does this relate to the polar distribution of hormones?

**Question 7:** Genetic engineering has produced fruits that are deficient in the ability to produce ethylene. How is this deficiency useful in the storage and marketing of fruits?

**Question 8:** Suppose that a scientist is experimenting in the laboratory with tissue culture methods. Pith tissue that has been isolated has grown an undifferentiated mass of cells (called a callus), and that tissue has been divided up and placed in the following culture media: (1) necessary nutrients plus auxin; (2) necessary nutrients plus zeatin (a cytokinin); (3) necessary nutrients plus equivalent concentrations of auxin and zeatin; (4) necessary nutrients plus an excess of zeatin and minimal auxin. Explain what would happen to the mass of tissue under each condition.

**Question 9:** What hormonal influences affect a pea plant from the moment it is planted until its emergence?

## 26.4 Photoreceptors Initiate Developmental Responses to Light

Phototropin, cryptochromes, and zeaxanthin are blue-light receptors

Phytochrome senses red and far-red light

Phytochrome stimulates gene transcription

Circadian rhythms are entrained by photoreceptors

Light and photoreceptors interact to stimulate a variety of events in plants, known as photomorphogenesis. Photoreceptors in plants interpret intensity, duration, and wavelength of light. Light regulates a wide variety of plant processes, including germination, flower production, and shoot elongation. Phototropin, a blue-light receptor, is a protein kinase that stimulates cell elongation by auxin. Zeaxanthin and phototropin act together to regulate light-stimulated opening of the stomata. Cryptochromes absorb blue and ultraviolet light and influence seedling development and flowering. Red light stimulates developmental and physiological events that include germination, flowering, and production of chlorophyll in seedlings.

Exposure to far-red light reverses the effects of exposure to red light, and vice-versa (see Figure 26.11). This "switching" occurs because phytochrome can be shifted from one isoform (red-absorbing $P_r$) to the other isoform (far-red-absorbing $P_{fr}$) upon absorption of light. When exposed to red light, $P_r$ is converted to $P_{fr}$. When exposed to far-red light, $P_{fr}$ is converted to $P_r$. Phytochromes are composed of a protein chain that interacts with transcription factors and the pigment chromophore. Red light changes the conformation of the protein from the $P_r$ to $P_{fr}$ form and exposes a nuclear localization sequence. The $P_{fr}$ form then moves into the nucleus, where it stimulates gene expression through a transcription factor. It can also act as a kinase and phosphorylate other proteins (see Figure 26.12).

Biological organisms exhibit a daily cycle in their functioning known as a circadian rhythm. The phytochromes are probably involved in the circadian rhythms of plants.

**Question 10:** You have produced a mutant plant that has only one isoform for the photoreceptor phytochrome. This plant flowers continuously and develops more shoots than your unmutated control plants. What would happen if you were to expose this plant to either red light or far-red light? What isoform is this plant producing and how do you know this?

**Question 11:** In what ways do circadian rhythms and photomorphogenesis overlap?

## Test Yourself

1. Plant growth is regulated by all of the following *except*
   a. environmental cues.
   b. hormones.
   c. signal transduction pathways.
   d. the expression of the plant's genome.
   e. quorum sensing.
   ***Textbook Reference:*** *Concept 26.1 Plants Develop in Response to the Environment*

2. The mechanism by which gibberellins act involves
   a. the adding of proton pumps to the plasma membrane.
   b. the phosphorylation of proteins.
   c. binding with a transcription factor.
   d. the removal of a repressor from a transcription factor.
   e. downregulation of gene expression.
   ***Textbook Reference:*** *Concept 26.2 Gibberellins and Auxin Have Diverse Effects but a Similar Mechanism of Action; At the molecular level, auxin and gibberellins act similarly*

3. Which event is the first step in germination of a seed?
   a. The imbibing of water
   b. The release of water from the fruit
   c. Chemical changes
   d. The exclusion of water
   e. Contact with soil
   ***Textbook Reference:*** *Concept 26.1 Plants Develop in Response to the Environment; The seed germinates and forms a growing seedling*

4. Which of the following events does *not* have the potential to break dormancy in seeds?
   a. Penetration of the seed coat
   b. Leaching of inhibitory compounds by water
   c. Exposure to fire
   d. Passage through an animal's digestive tract
   e. Floating long distances on wind
   ***Textbook Reference:*** *Concept 26.1 Plants Develop in Response to the Environment; The seed germinates and forms a growing seedling*

5. Which hormone is responsible for breaking winter dormancy in deciduous trees?
   a. Auxins
   b. Cytokinins
   c. Gibberellins
   d. Ethylene
   e. Brassinosteroids
   ***Textbook Reference:*** *Concept 26.2 Gibberellins and Auxin Have Diverse Effects but a Similar Mechanism of Action*

6. Which of the following processes is *not* included in the acid growth hypothesis for the regulation of cell expansion by auxin?
   a. The pumping of protons into the cell wall
   b. The pumping of protons into the cytosol
   c. Increased gene expression of the proton pump gene
   d. Increased insertion of proton pumps into the cell membrane
   e. Loosening of the cell wall due to lowered pH
   ***Textbook Reference:*** *Concept 26.2 Gibberellins and Auxin Have Diverse Effects but a Similar Mechanism of Action; Auxin plays many roles in plant growth and development*

7. Which of the following processes is *not* involved in the polar transport of auxin?
   a. Diffusion across a cell membrane
   b. Membrane protein asymmetry of auxin transport carriers
   c. Proton pumping from the cytosol
   d. Ionization of auxin as a weak acid
   e. Elimination of an electrochemical gradient
   ***Textbook Reference:*** *Concept 26.2 Gibberellins and Auxin Have Diverse Effects but a Similar Mechanism of Action; The transport of auxin mediates some of its effects*

8. A homeowner has installed an outdoor gas-burning grill on her back patio next to her favorite camellia bush. After the first few nights of using the grill, she notices that the camellia is beginning to lose its leaves. Which of the following is the best explanation for what is happening?
   a. The bush is getting too warm next to the grill.
   b. Ethylene is a by-product of the burning gas and is causing senescence in the plant.
   c. Abscisic acid is a by-product of the burning gas and is causing senescence in the plant.
   d. The plant is a biennial and is bolting.
   e. Auxin production is being inhibited.
   ***Textbook Reference:*** *Concept 26.3 Other Plant Hormones Have Diverse Effects on Plant Development; Ethylene is a gaseous hormone that promotes senescence*

9. Cytokinins interact with which of the following hormones?
   a. Ethylene

b. Abscisic acid
c. Gibberellins
d. Auxins
e. Brassinosteroids
***Textbook Reference:*** *Concept 26.3 Other Plant Hormones Have Diverse Effects on Plant Development; Cytokinins are active from seed to senescence*

10. Which of the following light receptors is responsible for absorbing both blue and ultraviolet light?
a. Phytochrome $P_r$
b. Phytochrome $P_{fr}$
c. Cryptochrome
d. Phototropin
e. Etiolatin
***Textbook Reference:*** *Concept 26.4 Photoreceptors Initiate Developmental Responses to Light; Phototropin, cryptochromes, and zeaxanthin are blue-light receptors*

11. Etiolated seedlings are produced by germinating seeds that are kept in total darkness. Plants that are kept in the dark will begin to germinate after they are given a pulse of
a. blue light.
b. red light.
c. red light followed by a pulse of far-red light.
d. far-red light.
e. ultraviolet light.
***Textbook Reference:*** *Concept 26.4 Photoreceptors Initiate Developmental Responses to Light; Phytochrome senses red and far-red light*

12. Ethylene is produced by the _______ of a plant.
a. endosperm
b. leaves
c. roots
d. pollen
e. trichomes
***Textbook Reference:*** *Concept 26.3 Other Plant Hormones Have Diverse Effects on Plant Development; Ethylene is a gaseous hormone that promotes senescence*

13. Auxin transport within a plant is said to be _______, and it is dependent on the action of _______ pumps.
a. polar; proton
b. nonpolar; potassium
c. polar; potassium
d. bidirectional; proton
e. polar; sodium
***Textbook Reference:*** *Concept 26.2 Gibberellins and Auxin Have Diverse Effects but a Similar Mechanism of Action; The transport of auxin mediates some of its effects*

14. Which action is *not* initiated by auxin?
a. Stimulation of root initiation
b. Inhibition of leaf abscission
c. Stimulation of leaf abscission
d. Maintenance of apical dominance
e. Cell expansion
***Textbook Reference:*** *Concept 26.2 Gibberellins and Auxin Have Diverse Effects but a Similar Mechanism of Action; Auxin plays many roles in plant growth and development*

15. Red light activation of phytochrome into the $P_{fr}$ state leads to
a. the inhibition of chlorophyll.
b. leaf abscission.
c. the folding of the apical hook.
d. the unfolding of the apical hook.
e. the repression of gene transcription.
***Textbook Reference:*** *Concept 26.4 Photoreceptors Initiate Developmental Responses to Light; Phytochrome stimulates gene transcription*

## Answers

### *Key Concept Review*

1. The watering of the pea seed after planting promotes the leaching of germination inhibitors. At this point the seed also begins to imbibe water, resulting in metabolic changes. DNA synthesis is halted until the radicle emerges from the seed coat. Breakdown of starch and protein reserves in the cotyledons and the endosperm begins to provide nutrients for the developing embryo. The apical hook begins to push through the soil. Upon its emergence and exposure to light, chlorophyll synthesis begins.
2. When seedlings grow thin and tall, they are searching for light in order to allow them to perform photosynthesis. This response is advantageous to a plant growing from a seed that has germinated very deep in the ground. Etiolation can help a plant reach sunlight in time to promote seedling growth.
3. To determine the role of a hormone on plant function, it helps to make a mutant plant that is unable to produce the hormone in question. In a controlled experiment, you would then grow the plant in both the absence of external application of the hormone and the presence of external application of the hormone. The differences in plant development and growth would be attributed to the action of the hormone.
4. Lateral distribution of auxin controls phototropism and gravitropism. Auxin accumulates in the shaded portions of a stem and stimulates cell growth. This uneven cell growth results in a bending of the stem toward the light. The same mechanism works in response to gravity. An accumulation of auxin occurs where the gravitational pull is the strongest. Cells grow in response to auxin, and stems bend upward, away from the gravitational force.

5.

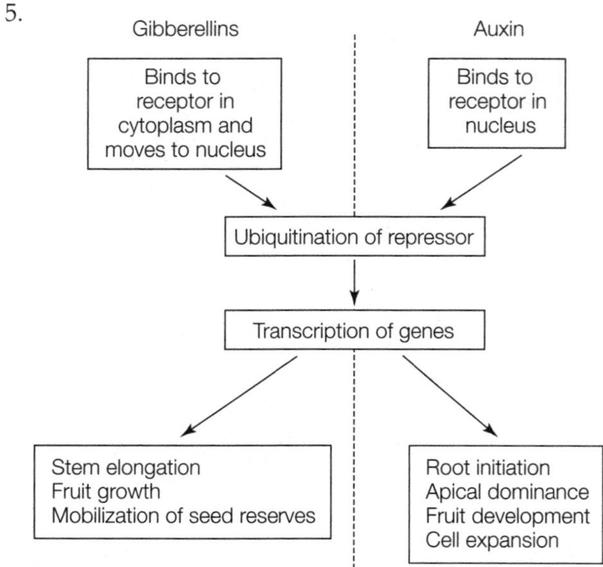


6. Roots need shoots, and vice-versa. Increases in roots require increases in shoots for photosynthesis. Regulation of the development of one by the other keeps growth in tandem. Because the distribution of the molecules is polar, a concentration gradient can be established. It is this relative gradient that controls development.
7. Fruits that are deficient in the ability to produce ethylene can be kept from ripening until they reach their destination. Once at market, they can be sprayed with ethylene to stimulate ripening. The result is fruit that can be shipped more easily and can still be ripe at any time in the market.
8. The results of the different treatments are the following: (1) roots will develop from the callus; (2) buds will develop from the callus; (3) both roots and buds will develop from the callus; (4) buds will develop from the callus.
9. Dormancy of the seed is maintained by abscisic acid until it is leached from the seed by water. Once water is imbibed, cytokinins begin to influence germination. Gibberellins assist with the mobilization of storage products to the growing embryo. Ratios of auxins and cytokinins balance root and bud formation. The apical hook is maintained by ethylene.
10. In a plant there are two isoforms of phytochrome that respond to red light and far-red light. Because your mutant plant only has one isoform for the phytochrome, exposure to one type of light or the other will have no effect. The isoform that responds to far-red light is the active isoform that stimulates flowering and shoot development. Because this plant flowers continuously and has more shoots than the control, one can conclude that its phytochrome is the active isoform.
11. Circadian rhythms operate on a daily cycle, and thus depend on the sunlight conditions through the day and night. Photomorphogenesis is the alteration of a plant based on light cues. Thus, some types of photomorphogenesis occur on a daily cycle, such as stomatal opening and the "following" of sunlight through the day by sunflowers.

### *Test Yourself*

1. **e.** The growth of a plant is regulated by environmental cues, hormones, signal transduction pathways, and genomic expression.
2. **d.** Gibberellin brings about a response in the plant cell by removing the repressor from a transcription factor, thus allowing transcription to occur.
3. **a.** The uptake of water by a seed begins the processes that lead to seed germination.
4. **e.** Mechanical abrasion, leaching of inhibitors by water, exposure to fire, and passing through a digestive tract may all trigger germination. Actual germination cannot begin until a seed imbibes water.
5. **c.** Gibberellins are responsible for bud break in deciduous trees.
6. **b.** The acid growth hypothesis involves an increase in the number of proton pumps in the cell membrane and the pumping of protons into the cell wall.
7. **e.** The proton pumping generates an electrochemical gradient across the cell membrane.
8. **b.** Ethylene gas promotes senescence and is one of the by-products of burning gas.
9. **d.** Cytokinins interact with auxins. The auxin-to-cytokinin ratio controls the bushiness of plants.
10. **c.** Cryptochromes respond to blue and ultraviolet light wavelengths.
11. **b.** The pulse of red light converts $P_r$ to $P_{fr}$, which is the active form.
12. **b.** Ethylene can be produced by the fruit, seed, or leaves in most plants.
13. **a.** Auxin transport is a polar process, meaning that it moves in only one direction with the help of proton pumps and auxin anion efflux carriers.
14. **c.** Auxin inhibits leaf abscission rather than stimulating it. Auxin is involved in all of the other processes.
15. **d.** The $P_{fr}$ phytochrome stimulates chlorophyll synthesis, hook unfolding, and leaf expansion. It activates transcription of downstream genes in these pathways.

# 27 Reproduction of Flowering Plants

## The Big Picture

- Though plants can reproduce both asexually and sexually, maintenance of genetic variability depends on sexual reproduction. The flower is the basis of sexual reproduction in plants. The flower not only produces the necessary gametes, but is also integrally involved in ensuring pollination. Eggs are produced from megaspores, and pollen is produced from microspores. Angiosperms exhibit double fertilization, which results in a diploid embryo and a triploid endosperm.
- Hormones and signaling cascades are involved in a plant's transition from the vegetative state to the reproductive state. Meristem identity genes and floral organ identity genes code for proteins necessary for flowering. The photoperiod sets the shoot apical meristem on the path to flowering. Some plants are short-day plants and flower when the day is shorter than a critical maximum, while others are long-day plants and flower once the day length reaches a critical minimum. The actual key for the photoperiod is the length of the night and not day length. The gene *FT* codes for florigen and is involved in photoperiod signaling.
- Some angiosperms are also able to reproduce asexually. This often occurs through vegetative reproduction, in which underground stems grow out from the plant. Propagation by cutting works because of the same principle. Apomixis is the asexual production of seeds without the necessity of fertilization; it produces clones of the parent plant.

## Study Strategies

- Refer to the figures in the textbook to understand flower structure, megaspore formation, microspore formation, and fertilization.
- Spend time thinking about the advantages and disadvantages of sexual and asexual reproduction, as related questions often appear in exams.
- Megaspore production is probably the most difficult concept in this chapter. It is helpful to think about the source of each cell in order to understand the whole concept.
- Remember that flowering, like all other plant processes, is a result of environmental signals, receptors, enzyme cascades mediated by hormones, and alterations in gene expression.
- Be sure you understand the particular features of double fertilization.
- Go to **LaunchPad** (or use the Web addresses listed) to review the following additional resources:

  Animated Tutorial 27.1 Double Fertilization (PoL2e.com/at27.1)

  Animated Tutorial 27.2 The Effect of Interrupted Days and Nights (PoL2e.com/at27.2)

  Activity 27.1 Angiosperm Reproduction (PoL2e.com/ac27.1)

  Analyze the Data in textbook Figure 27.9

  Apply the Concept in textbook Section 27.2

## Key Concept Review

### 27.1 Most Angiosperms Reproduce Sexually

The flower is the reproductive organ of angiosperms
Angiosperms have microscopic gametophytes
Angiosperms have mechanisms to prevent inbreeding
A pollen tube delivers sperm cells to the embryo sac
Angiosperms perform double fertilization
Embryos develop within seeds contained in fruits

Sexual reproduction is necessary for genetic recombination, and genetic variability provides plants with the ability to adapt to their environment. Plant reproduction involves alternation of diploid and haploid generations. The flower is the basis of sexual reproduction in angiosperms. The basic flower structures include carpels, stamens, petals, and sepals, all of which are modified leaves. The stamen and carpel are the male and female parts of the flower, respectively. Flowers with both a carpel and stamen are known as "perfect." "Imperfect" flowers contain either the carpel or stamen, but not both. In monoecious species, both male and female flowers are found on the same individual plant;

in dioecious species, the male and female flowers are found on separate individuals. The flower produces haploid spores that develop into gametophytes. The female gametophytes, called embryo sacs, develop in the ovule. Male gametophytes, called pollen grains, develop in the anther.

Within the ovule, a megasporocyte produces four haploid megaspores through meiosis. Only one of these megaspores survives, and it divides mitotically to produce eight nuclei within a single large cell. The nuclei migrate to either end of the cell, with two remaining in the middle. Cell walls form, isolating the three nuclei at either end into individual cells; the two nuclei in the middle remain together in one cell. At one end, the three cells become two synergid cells and one egg cell; at the other end are three antipodal cells, which eventually degenerate. In the large central cell are the two polar nuclei. This seven-celled embryo sac is the megagametophyte (see Figure 27.2). Pollen grains develop when a microsporocyte undergoes meiosis. All products of meiosis are retained and undergo mitosis to form a two-celled pollen grain composed of the tube cell and the generative cell. Further development is halted until after the pollen is transferred from the anther to the stigma during pollination.

Pollen grains are carried to female flowers via wind, animals, or other vectors. Some plants are able to self-pollinate within the same plant or flower. However, most plants prevent self-fertilization, some by physical separation in either space or time of male and female gametophytes. Plants can also prevent self-fertilization by means of genetic self-incompatibility. A gene cluster, the *S* locus, regulates self-incompatibility and prevents self-fertilization in many plants (see Figure 27.3).

Germination of a pollen grain begins when the pollen begins to take up water from the stigma. When a pollen grain germinates, a pollen tube grows down the style toward the embryo sac. Chemical signals, from synergids within the ovule, may direct the growth of the pollen tube. The pollen grain consists of two cells at the time of pollination—the tube cell and the generative cell. The tube cell controls the growth of the pollen tube. As the tube is growing, the generative cell undergoes mitosis and produces two haploid sperm cells.

Once the pollen tube enters the embryo sac, the two sperm cells are released into a synergid that disintegrates and releases the sperm nuclei. One sperm cell fuses with the egg cell, producing a diploid zygote. The other fuses with the polar nuclei, producing the triploid endosperm. All other cells disintegrate. This double fertilization resulting in a zygote and the nutritive endosperm is a characteristic feature of angiosperms (see Figure 27.4).

The embryos develop within seeds. The success of the embryo depends on its own development, as well as the development of the endosperm, the integuments, and the carpel. Ultimately, a seed coat develops from the integuments and protects the dormant embryo. As the embryo and seed are developing, the ovary begins to form the fruit (see Figure 27.5). Other parts of the flower and plant may be included in the fruit, but to be considered a fruit, only the ovary wall and the seed need be involved. The fruit disperses by various means, including by traveling on the coats of animals or by being consumed and later deposited. The fruit also protects the seed from animals and microbial diseases.

**Question 1.** In agricultural production, much effort is spent detasseling corn (removing male flowers). Why is it not possible to halt pollen production simply by spraying the corn plants with a meiosis inhibitor? (Hint: Male and female flowers occur on the same corn plant.)

**Question 2.** Describe double fertilization. What is the ploidy level of the products of double fertilization?

**Question 3.** Define a fruit. Which of the following are fruits: tomato, pear, potato, banana, cucumber, snow pea, peanut, sunflower seed?

## 27.2 Hormones and Signaling Determine the Transition from the Vegetative to the Reproductive State

Shoot apical meristems can become inflorescence meristems

A cascade of gene expression leads to flowering

Photoperiodic cues can initiate flowering

Plants vary in their responses to photoperiodic cues

Night length is the key photoperiodic cue that determines flowering

The flowering stimulus originates in the leaf

Florigen is a small protein

Flowering can be induced by temperature or gibberellins

Some plants do not require an environmental cue to flower

The beginning of flowering involves a reallocation of energy in the plant away from leaves and stem and toward flowers and gametes. Plants have one of three different life cycle patterns. Annual plants go from seed to seed set and die within one growing season. Biennial plants require one vegetative growing season before reproducing. Perennial plants flower repeatedly and live for many years.

The first transition from vegetative growth to floral production is the transition of the apical meristem into an inflorescence meristem (see Figure 27.6). The inflorescence meristem can produce bracts and floral meristems. The floral meristem differs from the apical meristem in that growth is determined. The floral meristem is programmed to produce four consecutive whorls of flower organs. A gene cascade leads to flower formation, which begins with the activation of a set of meristem identity genes. Two genes important in switching from vegetative growth to reproductive growth are *LEAFY* and *APETALA1* (or *AP1*). The expressions of floral organ identity genes specify successive whorls.

The gene cascade leading to flower development is controlled by environmental cues. Seasonal flowering is a result of changes in photoperiod, specifically the duration of continuous darkness. Experiments have shown that plants actually respond to the length of the night, rather than amount of daylight. In experiments in which daylight was interrupted, there was little effect on flowering, but in experiments in which darkness was interrupted, there were significant effects on flowering (see Figure 27.8). Each plant type has a critical day length corresponding to light availability that induces flowering. Short-day plants flower when the amount of light available is shorter than the critical maximum. Long-day plants flower only when the day is longer (more light available) than the critical minimum. Some plants require complex combinations of day lengths.

Plants, like all other organisms, have an internal mechanism for measuring the length of continuous dark periods. The duration of dark periods appears to be detected by special phytochromes that detect red light. The stimulus for flowering comes from the leaf of the plant. Florigen (FT) is thought to be the hormone responsible for flowering. The gene *FLOWERING LOCUS T* codes for the FT protein. High levels of FT induce the plant to flower. *CONSTANS* codes for a transcription factor (CO) that stimulates FT synthesis in the phloem cells. The *FLOWERING LOCUS D* gene codes for a transcription factor (FD) in the apical meristem that increases transcription of meristem identity genes (see Figure 27.10).

Vernalization is the induction of flowering by low temperatures. A transcription factor (FLC) coded by *FLOWERING LOCUS C* inhibits the FT pathway described above. Cold temperatures decrease the production of the FLC protein, leading to a functional FT pathway. Gibberellins can also stimulate a plant to flower. Those plants that do not require an environmental cue to flower rely on an "internal clock" to trigger flowering. This most likely works through FLC concentration gradients in the plant, stimulating the FT–FD pathway described above at the terminal bud.

**Question 4.** Draw a flow chart showing the mechanism of action of florigen, the plant hormone thought to be actively involved in the initiation of flower formation. Be sure to show the interaction of the three genes involved in initiating flower formation.

**Question 5.** Flowering is stimulated when light sets off a gene cascade. Explain how this may be hormonally controlled.

### 27.3 Angiosperms Can Reproduce Asexually

Angiosperms use many forms of asexual reproduction

Vegetative reproduction is important in agriculture

In certain conditions, asexual reproduction is advantageous to plants. Asexual reproduction occurs without genetic recombination. During the process of vegetative reproduction, asexual reproduction occurs through the modification of a vegetative organ. Stolons, tubers, rhizomes, and bulbs are all modifications of stems or roots that allow vegetative reproduction. Some plants, such as dandelions, reproduce asexually through seeds in a process called apomixis (see Figure 27.13). Apomictic plants skip over meiosis and fertilization and produce diploid seeds with the identical genetic makeup of the maternal plant.

Asexual reproduction is used in agriculture. Crossing two inbred homozygous genetic strains can produce plants that are superior, resulting in hybrid vigor. Cuttings are commonly used in horticulture. Grafting is widely used in fruit crops in which a shoot system from one plant (the scion) is spliced with the root system of another plant (the stock). This allows for a hardy root stock to be combined with a good but often fragile fruit producer. Meristem culture (production of plantlets from pieces of meristem) is leading to ongoing advances in agriculture.

**Question 6.** When is asexual reproduction beneficial to plants? Under what conditions is sexual reproduction beneficial?

**Question 7.** Compare and contrast asexual and sexual reproduction in plants. In which category does self-fertilization belong?

## Test Yourself

1. Suppose that you manage a greenhouse that produces roses for Valentine's Day (in February). Roses normally bloom in June. Which of the following would be the best lighting schedule for inducing flowering in the plants?
   a. 16 hours of light, followed by 8 hours of interrupted dark
   b. 16 hours of light, followed by 8 hours of uninterrupted dark
   c. 10 hours of light, followed by 14 hours of uninterrupted dark
   d. 10 hours of light, followed by 14 hours of interrupted dark
   e. None of the above schedules would induce flowering.

   ***Textbook Reference:*** *Concept 27.2 Hormones and Signaling Determine the Transition from the Vegetative to the Reproductive State; Night length is the key photo periodic cue that determines flowering*

2. After setting the correct photoperiod, the managers of a greenhouse still do not have blooming roses. Which of the following events would most likely have contributed to the problem?
   a. The heating system allowed for fluctuations in temperature between 20°C and 25°C.
   b. The furnace mechanic accidentally turned off the lights for an hour two days in a row.
   c. The cleaning crew turned the lights on for an hour three nights in a row.
   d. The staff forgot to fertilize the plants on their regular schedule.
   e. The irrigation system failed over a weekend when no staff member was around.

   ***Textbook Reference:*** *Concept 27.2 Hormones and Signaling Determine the Transition from the Vegetative to the Reproductive State; Night length is the key photo periodic cue that determines flowering*

3. You have moved into a new house. During the first summer you notice many of the plants do not bloom. During the second summer your yard is a sea of blooms. It is now spring of the third year, and there are no plants. Which of the following is the best explanation for this observation?
   a. The plants are annuals.
   b. The plants are biennials.
   c. The plants are perennials.
   d. The plants are being affected by drought.
   e. The nights are too long for the plants to thrive.
   ***Textbook Reference:*** *Concept 27.2 Hormones and Signaling Determine the Transition from the Vegetative to the Reproductive State*

4. You notice that a new houseplant sends out long stems with what look like "little plants" attached. You allow one of these to rest in a cup of water and note that roots form. What you are seeing is an example of
   a. asexual reproduction.
   b. apomixis.
   c. heterospory.
   d. parthenogenesis.
   e. vivipary.
   ***Textbook Reference:*** *Concept 27.3 Angiosperms Can Reproduce Asexually; Angiosperms use many forms of asexual reproduction*

5. Which of the following characteristics does *not* prevent self-pollination in plants?
   a. Self-incompatibility genes
   b. Physical separation of the eggs and sperm via imperfect flowers
   c. Production of pollen and eggs at different times
   d. Regular removal of pollen by symbiotic species
   e. Physical separation of pollen and stigma by long stamens
   ***Textbook Reference:*** *Concept 27.1 Most Angiosperms Reproduce Sexually; Angiosperms have mechanisms to prevent inbreeding*

6. Most grapevines that produce wine grapes are grafted onto rootstock of another species. Which of the following is *not* one of the reasons that this practice increases grape yields?
   a. A hardy rootstock can replace a weak rootstock.
   b. A high-producing vine stock can replace a low-producing vine stock.
   c. It allows vintners to select for pest resistance without losing grape quality.
   d. Grafting, as opposed to replacing whole vines, can shorten time to fruit production.
   e. Grafting can make an herbaceous plant woody.
   ***Textbook Reference:*** *Concept 27.3 Angiosperms Can Reproduce Asexually; Vegetative reproduction is important in agriculture*

7. The induction of flowering by means of exposure to low temperature is called
   a. vernalization.
   b. frigidation.
   c. apomixis.
   d. viviparity.
   e. alternation.
   ***Textbook Reference:*** *Concept 27.2 Hormones and Signaling Determine the Transition from the Vegetative to the Reproductive State; Flowering can be induced by temperature or gibberellins*

8. The *LEAFY* and *APETALA1* genes are examples of _______ genes.
   a. floral organ identity
   b. meristem identity
   c. viviparity
   d. photoperiod
   e. inflorescence
   ***Textbook Reference:*** *Concept 27.2 Hormones and Signaling Determine the Transition from the Vegetative to the Reproductive State; A cascade of gene expression leads to flowering*

9. The production of seeds without fertilization is called
   a. apomixis.
   b. parthenogenesis.
   c. conception.
   d. circadian rhythm.
   e. vernalization.
   ***Textbook Reference:*** *Concept 27.3 Angiosperms Can Reproduce Asexually; Angiosperms use many forms of asexual reproduction*

10. In the transition from vegetative growth to floral growth, the _______ must be transformed into the _______. This involves a shift from _______ growth to _______ growth.
    a. apical meristem; floral meristem; indeterminate; determinate
    b. lateral meristem; floral meristem; indeterminate; determinate
    c. apical meristem; floral meristem; determinate; indeterminate
    d. apical cambium; floral cambium; determinate; indeterminate
    e. floral meristem; apical cambium; determinate; indeterminate
    ***Textbook Reference:*** *Concept 27.2 Hormones and Signaling Determine the Transition from the Vegetative to the Reproductive State; A cascade of gene expression leads to flowering*

11. Which statement best describes the fate of the generative cell of the pollen grain?
    a. It coordinates growth of the pollen tube.
    b. It divides by meiosis to produce two sperm nuclei.
    c. It divides by mitosis to produce two sperm nuclei.

d. It forms the pollen tube.
e. It divides by meiosis to produce four sperm.
***Textbook Reference:*** *Concept 27.1 Most Angiosperms Reproduce Sexually; Angiosperms perform double fertilization*

12. Which of the following is/are *not* part of a megagametophyte?
a. Pollen grain
b. Synergids
c. Antipodal cells
d. Polar nuclei
e. Egg cell
***Textbook Reference:*** *Concept 27.1 Most Angiosperms Reproduce Sexually; Angiosperms have microscopic gametophytes*

13. Which of the following statements about double fertilization is true?
a. One sperm cell fuses with an egg cell and another sperm cell fuses with the two polar nuclei.
b. One sperm cell fuses with a generative cell and another sperm cell fuses with the two polar nuclei.
c. Two sperm cells fuse with each cell of a pollen grain.
d. One sperm cell fuses with the stigma and another sperm cell fuses with the antipodal cell.
e. Two sperm cells fuse with each synergid.
***Textbook Reference:*** *Concept 27.1 Most Angiosperms Reproduce Sexually; Angiosperms perform double fertilization*

14. Which statement about plants and photoperiodic cues is true?
a. Short-day plants flower when the day is longer than a critical maximum.
b. Short-day plants reproduce only when day and night length are equal.
c. Long-day plants flower when the night is longer than a critical maximum.
d. Long-day plants flower when the day is longer than a critical minimum.
e. Short-day plants flower when the night is shorter than a critical maximum.
***Textbook Reference:*** *Concept 27.2 Hormones and Signaling Determine the Transition from the Vegetative to the Reproductive State; Plants vary in their responses to photoperiodic cues*

15. Which of the following is *not* a part of the flower?
a. Carpels
b. Petals
c. Sieve tube
d. Stamens
e. Sepals
***Textbook Reference:*** *Concept 27.1 Most Angiosperms Reproduce Sexually; The flower is the reproductive organ of angiosperms*

## Answers

### *Key Concept Review*

1. Meiosis occurs in the megasporocyte as well as the microsporocyte. Inhibition of pollen formation via a meiosis inhibitor will also inhibit egg formation.
2. Double fertilization occurs when the two sperm nuclei produced from the generative cell of the pollen grain unite with the egg nucleus and two polar nuclei, respectively. The resulting zygote is diploid and the endosperm is triploid.
3. A fruit is the seed and the ovary wall, and it may include other structures of a flowering plant. With the exception of the potato, all of the listed structures are fruits.
4.

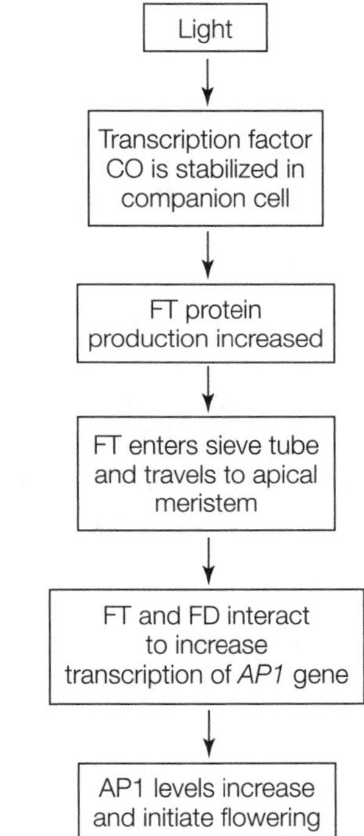


5. Light induction begins with the leaves. A single isolated leaf may stimulate floral production throughout the plant. In experiments in which leaves from one plant were grafted onto other plants, an "induced" leaf caused a plant to flower even though the plant had not been exposed to the required amount of darkness. Therefore, the leaf must send a signal (the protein florigen) that begins the gene cascade leading to flower formation.
6. Asexual reproduction is beneficial in stable environments in which many genetically identical plants are sustainable. This process can help colonize a habitat or help a population of plants to spread. The disadvantage is lack of genetic diversity, which can be detrimental if the environment changes rapidly and adaptability of the population is required.

7. Asexual reproduction does not involve meiosis, fertilization, or genetic recombination. The offspring from asexual reproduction are genetically identical to the parent plant. Sexual reproduction requires a meiotic event and fertilization. Self-fertilization is sexual reproduction, even though genetic recombination is limited. This is because a meiotic event occurs, and fertilization is necessary for reproduction to take place. Not all the offspring are identical, because of the genetic variability of the gametes.

### *Test Yourself*

1. **b.** Flowering is regulated by darkness. Interruptions in darkness can prevent flowering. A plant that blooms in the summer is typically a "long day" plant and requires that the dark period be shorter than a specified amount (and that the light period be greater than a specified amount).
2. **c.** The interruptions in the dark cycle could have affected the signals that tell the plant to begin flowering.
3. **b.** The plants are most likely biennials, which spend one season growing vegetatively and one season producing flowers before dying.
4. **a.** The runner (stolon) is propagating this plant asexually.
5. **d.** Some plants have self-incompatibility genes that prevent the growth of the pollen tube through the style. Others have mechanical barriers to their own pollen. Yet others have separate times for the development of eggs and sperm.
6. **e.** Grafting allows selection of hardy rootstock, produces well-producing vine stock, and contributes to disease resistance. By replacing grafted vine stock instead of replanting the vines, a vintner can produce a different varietal of grapes in a shorter time period.
7. **a.** The induction of flowering by means of exposure to low temperature is called vernalization. This is the practice that induces flowering in winter wheat.
8. **b.** These two genes are meristem identity genes that code for proteins involved in the initiation of flower formation.
9. **a.** Dandelions and other plants produce seeds by apomixis, without meiosis and fertilization. These seeds are genetically identical to the parent plant.
10. **a.** Apical meristems and floral meristems differ in that growth from the apical meristem is indeterminate and growth from the floral meristem is determinate, leading to four whorls of floral structures.
11. **c.** The two sperm nuclei involved in double fertilization are derived from the generative cell of the pollen grain.
12. **a.** The megagametophyte is the female gametophyte. It is initially called the embryo sac and contains three antipodal cells, two synergid cells, and two polar nuclei at the seven-cell stage.
13. **a.** Double fertilization involves the fertilization of the egg cell and the two polar nuclei by two sperm cells.
14. **d.** The cue for flowering is the length of the night, so long-day plants flower when the day is longer than a critical minimum and when the night is shorter than a critical maximum.
15. **c.** The carpels, stamens, petals, and sepals are all part of the flower. The sieve tube is part of the phloem system.

# 28 Plants in the Environment

## The Big Picture

- Plants must respond to invasion by pathogens, physical damage by natural events and herbivory, variable water supplies, temperature fluctuations, and variable soil conditions. Natural selection has made plants uniquely suited to the areas they occupy. Some thrive in very harsh environments, but others cannot.
- Interactions between plants and pathogens stimulate a series of chemical changes that ward off further infection. Plant strategies to isolate pathogens include sealing plasmodesmata, releasing proteins that interact with pathogens, and signaling other parts of the plant. A plant produces chemicals to prevent herbivory and mounts defenses around the damaged tissues.
- Specific adaptations allow plants to withstand adverse conditions. Saline environments, water loss, and high temperatures can be survived if water uptake is maximized and water loss is minimized. Sunken stomata, reduction of surface area, and increased water potential of cells are adaptations to these conditions. Plants can survive extremes in temperature by means of heat shock proteins that stabilize metabolic proteins.

## Study Strategies

- The best strategy for learning the material in this chapter is to study the various components of the chapter one at a time so that all of the defenses/modifications do not seem to run together. Focus first on pathogen interaction, then herbivory resistance, and so on.
- Think about how form follows function as you look at how plants are adapted to specific harsh environments.
- Gene-for-gene resistance can be difficult to understand, but remember that in any pathogen situation, the pathogen and the host are continually interacting; therefore, over time, interplay among the genomes has been selected for.
- The chapter provides many examples of the ways in which plants deal with pathogens and herbivory. Be sure to review them after reading the appropriate sections of the textbook.
- Go to **LaunchPad** (or use the Web addresses listed) to review the following additional resources:

  Animated Tutorial 28.1 Signaling between Plants and Pathogens (PoL2e.com/at28.1)

  Activity 28.1 Concept Matching (PoL2e.com/ac28.1)

  Analyze the Data in textbook Figure 28.4

  Apply the Concept in textbook Sections 28.1 and 28.3

## Key Concept Review

### 28.1 Plants Have Constitutive and Induced Responses to Pathogens

Physical barriers form constitutive defenses

Induced responses can be general or specific

General and specific immunity both involve multiple responses

Specific immunity is genetically determined

Specific immunity usually leads to the hypersensitive response

General and specific immunity can lead to systemic acquired resistance

The action of a pathogen signals a plant to mount a defense. In turn, the plant's defense signals the pathogen. These signals go back and forth until one or the other "wins." The defense of a plant can either always be active (constitutive) or activated in response to damage or stress (induced).

The first plant defense system is an attempt to prevent entry by pathogens. Cutin, suberin, and waxes cover the leaves and stems to exclude pathogens. Cell walls can be strengthened when induced by pathogen invasion, and particular chemicals inhibit growth of pathogens. All of these are constitutive defenses; if these fail, then the plant must use induced resistance mechanisms.

Pathogen-associated molecules called elicitors trigger induced responses in infected plants. The induced responses (the plant's "immune system") can be either general or specific. Elicitors called pathogen-associated molecular patterns (PAMPs), which are recognized in generic classes such as bacterial flagellins or chitins, induce general immunity. Pattern recognition receptors on the plant cells activate signal transduction pathways in response to PAMPs that activate general responses to infection. Effectors are specific elicitors that generate specific responses in the infected plant. Binding of effectors to cytoplasmic receptors called R proteins triggers specific immunity. The types of cellular responses are similar in general and specific immunity: generation of NO and reactive oxygen species, polymer deposition, hormone signaling, and altered gene expression. Phytoalexins are nonspecific antifungal and antibacterial compounds produced by the plant. Pathogenesis-related proteins (PR proteins) are enzymes that function to break down the walls of pathogens or serve as signaling molecules to pathogen-free cells in the plant.

Gene-for-gene resistance is a highly specific type of resistance that depends on the presence of a plant allele for a gene that matches an allele in the pathogen. *R* genes code for receptors in plants and *Avr* genes code for elicitors in the pathogen. If the receptor from a plant's *R* gene and the elicitor from a pathogen's *Avr* gene match, a defensive response is elicited in the plant (see Figure 28.2). Though this mechanism is not completely understood, it is thought to be mediated by nitric oxide and peroxide.

In the hypersensitive response, pathogen-containing tissue and surrounding tissues undergo apoptosis upon infection and become necrotic lesions. This serves to contain and isolate the pathogen to the infected tissues. As these cells are dying, they release phytoalexins. Upon recognition of a pathogen, polysaccharides are deposited in the cell wall and seal off the plasmodesmata to form a barrier between cells.

Systemic acquired resistance protects the whole plant against further infection. Salicylic acid produced at the site of an infection stimulates PR protein production. Salicylic acid is also transported to other parts of the plant to signal the presence of an invader. Infected plants release methyl salicylate, which can become airborne and serve as a signal to other plant parts or to neighboring plants to mount a defense via PR proteins. Plants can use RNA interference (RNAi, also known as posttranscriptional gene silencing) to respond to attack by RNA viruses. A plant produces small pieces called small interfering RNA (siRNA) from interactions with the viral RNA and these siRNAs help to degrade the viral mRNA.

**Question 1.** It has long been a practice in rural areas to nail fencing to trees. Diagram how the act of nailing a fence to a tree trunk can introduce pathogens into the tree. Be sure to show what measures the tree takes to ward off disease.

**Question 2.** Explain the difference between general and specific immune responses. What is the benefit in having both types of responses instead of one or the other?

## 28.2 Plants Have Mechanical and Chemical Defenses against Herbivores

Constitutive defenses are physical and chemical

Plants respond to herbivory with induced defenses

Why don't plants poison themselves?

Plants don't always mount a successful defense

Plants use a number of physical and chemical defenses against herbivores. In some plants, protection is provided by constitutive anatomical defenses, including morphological features such as thorns, spines, and hairs. Plants also produce secondary metabolites, which serve no basic cellular function but act as defenses against grazers. The effects of these chemicals are diverse, ranging from neurotoxins to hormone mimics (see Table 28.1). Canavanine is a defensive secondary metabolite that plays multiple roles in a plant, including assisting with nitrogen storage in the seed. Because it is similar to arginine, it is incorporated into proteins after consumption by an herbivore. Canavanine alters the tertiary structure of proteins in an herbivore and can be lethal. Nicotine kills insects by acting as an inhibitor of nervous system function (see Figure 28.4).

Plants perceive the damage caused by herbivores through membrane signaling and chemical signaling. When an herbivore begins to eat the plant, changes in the plasma membrane electrical potential pass along to every cell within the plant. Chemical signaling occurs when the herbivore's saliva combines with plant fatty acids to elicit local and systemic responses in the plant, or send messages to other individuals to stimulate their defenses. The response of a plant to perceived damage stimulates a signal transduction pathway involving jasmonic acid (jasmonate). Wounding causes the release of an elicitor, which travels to other parts of the plant and induces hydrolysis of a membrane lipid. A by-product of this breakdown is jasmonate, which enters the nucleus and activates production of a protease inhibitor that inhibits digestion in the digestive system of the herbivore (see Figure 28.5). Jasmonate also triggers the formation of volatile compounds that attract potential predators of the herbivorous insect.

Plants use a variety of strategies to prevent harming themselves with their own defensive mechanisms. One is to isolate toxic substances in compartments. Water-soluble toxins are stored in vacuoles, hydrophobic poisons are stored in latex, and other toxins are dissolved in the waxes covering the epidermis. Precursors of toxic material can be stored separately from the enzymes that convert them into active poison, combining only after the cell is damaged. Some plants have evolved proteins that do not react with their toxins, thus protecting against accidental toxicity. Insects sometimes find ways to get around the mechanisms used by plants to deter them. One example is a beetle that feeds on the leaves of milkweed. The leaves would normally release large amounts of latex from specialized tubes called laticifers; certain species strategically cut the laticifers around the site at which they would like to feed, thus eliminating the possibility of being drowned by the latex.

**Question 3.** Describe some of the strategies by which plants can defend themselves against herbivory.

**Question 4.** Many plants produce toxins that damage eukaryotic predators. Considering that the cell structures of the herbivore and the plant are similar, why is the plant not affected by its own toxins?

### 28.3 Plants Adapt to Environmental Stresses

Some plants have special adaptations to live in very dry conditions

Some plants grow in saturated soils

Plants can respond to drought stress

Plants can cope with temperature extremes

Some plants can tolerate soils with high salt concentrations

Some plants can tolerate heavy metals

Plants have evolved the ability to live in harsh environments. In dry conditions such as deserts, some annuals are drought avoiders. They survive in these areas by completing their entire life cycle in the short period that water is available. Xerophytes are plants specifically adapted for year-round life in dry areas. They may have special modifications such as thickened cuticles, epidermal hairs, and stomatal crypts (see Figure 28.7) to prevent water loss. Trichomes help to decrease the intensity of light hitting the surface of the plant. Succulents have water-storing leaves or stems. Roots can also be adapted for drought conditions. Long taproots can reach water supplies far underground, and shallow fibrous root systems that grow only during rainy seasons are also adapted to desert growth. Xerophytic plants are able to extract more water when it is available because they store proline or secondary metabolites in their vacuoles, thus raising the water potential.

Too much water can be as dangerous to plant survival as too little. Roots submerged in water do not get adequate oxygen to sustain respiration. Some plants that grow in standing water overcome this by producing pneumatophores, which are root extensions that grow above water and provide oxygen to the entire root system. Others have leaf modifications called aerenchyma for buoyancy and oxygen storage (see Figure 28.10B).

Inadequate water supply results in changes in membrane integrity and in the three-dimensional structure of proteins. Drought conditions stimulate the roots to release abscisic acid, which travels to the leaves. Abscisic acid closes the stomata and starts gene expression of late embryogenesis abundant (LEA) proteins. The LEA proteins stabilize membranes and other proteins (see Figure 28.11).

Temperature extremes pose yet another stress to plant survival. High temperatures denature proteins and destabilize membranes. Cold temperatures decrease membrane fluidity, and freezing causes rupturing of membranes if ice crystals form. Plants have evolved a number of adaptations to deal with heat, including hairs and spines to dissipate heat. Plants produce heat shock proteins that act as chaperonins and help stabilize protein structure against denaturation. Plants can adjust to cold through a process of cold-hardening, which involves repeated exposure to cool, non-damaging temperatures over many days. This alters the saturated and unsaturated fatty acid composition of membranes, allowing them to remain fluid at cooler temperatures and making them more resistant to rupture. Some plants have antifreeze compounds that prevent ice crystal formation.

Some plants are adapted to survive in saline environments. Halophytes (salt-loving plants) are adapted to living in saline environments and are exposed to an osmotic challenge. They are the only plants that accumulate sodium and chloride ions, which they store in leaf vacuoles. The increased salt concentration of the halophyte means it has a water potential that is more negative than that of the soil solution, allowing it to take up water from the saline environment. Some plants are able to excrete salt so that it does not reach toxic proportions. Salt glands move salt to the leaf surface, where it can be lost to wind or rain. Salt glands on the leaf assist with water procurement from the roots and with reduction of water loss to evaporation.

Heavy metals in soils, such as chromium, mercury, lead, and cadmium, are poisonous to most plants. Some geographic areas are rich in these metals as a result of normal geological processes, and others have been contaminated by human activity. Plants living in these areas have adaptations that allow them to accumulate large quantities of heavy metals. These plants, known as hyperaccumulators, increase ion transport into the roots, increase translocation, accumulate ions in shoot vacuoles, and resist the toxin. These plants are important in bioremediation cleanup efforts.

**Question 5.** At a T-intersection in a small northern town, a local business plants several ornamental shrubs. During the winter, snowplows push snow from the intersection around the shrubs. In an attempt to save the shrubs, the business covers the shrubs with burlap. Will this save the shrubs? What conditions in the intersection are likely to cause the greatest damage to the shrubs?

**Question 6.** Pansies are planted in the winter in the southeastern United States. To ensure that the plants are ready to enter cool ground, nurseries cold-harden the plants for several days. Describe this process and explain why it is done.

**Question 7.** Imagine that you are put in charge of the bioremediation of a mine and ore refinery site in your city. How can plants help with the cleanup of sites contaminated with heavy metals?

## Test Yourself

1. Defensive strategies in plants that are always turned on are called
   a. induced defenses.
   b. heat shock proteins.
   c. alternating defenses.
   d. constitutive defenses.
   e. constitutional defenses.
   ***Textbook Reference:*** *Concept 28.1 Plants Have Constitutive and Induced Responses to Pathogens; Physical barriers form constitutive defenses*

2. Plants acquire systemic resistance in much the same way that people acquire resistance to pathogens; however, the mechanism of systemic acquired resistance is quite different in plants. Which of the following does *not* have a role in acquired resistance in plants?
   a. Salicylic acid
   b. *R* genes
   c. Methyl salicylate
   d. PR proteins
   e. siRNA

   ***Textbook Reference:*** *Concept 28.1 Plants Have Constitutive and Induced Responses to Pathogens; General and specific immunity can lead to systemic acquired resistance*

3. Halophytes are different from all other types of plants in that they
   a. can accumulate sodium and chloride ions.
   b. have a positive water potential.
   c. contain no sodium or chloride ions.
   d. contain no stomata.
   e. break down sodium atoms.

   ***Textbook Reference:*** *Concept 28.3 Plants Adapt to Environmental Stresses; Some plants can tolerate soils with high salt concentrations*

4. Upon infection by a pathogen, plant cells increase synthesis of polysaccharides. The function of these polysaccharides is to
   a. destabilize the cell walls.
   b. synthesize antibodies to the pathogen.
   c. isolate the pathogen in the invaded tissue.
   d. break down the wax barrier.
   e. communicate the existence of the pathogen to the leaves.

   ***Textbook Reference:*** *Concept 28.1 Plants Have Constitutive and Induced Responses to Pathogens; General and specific immunity both involve multiple responses*

5. Canavanine is toxic to many herbivores but not to plants. Which statement regarding this differential toxicity is true?
   a. Canavanine is confused with arginine in plants but not in animals.
   b. Canavanine is toxic to herbivores but is incorporated into plant cells and causes them to fold properly.
   c. Plants are able to differentiate between arginine and canavanine, whereas animals cannot.
   d. Plants store canavanine in vacuoles, whereas animals metabolize it.
   e. Plants metabolize canavanine, whereas animals store it.

   ***Textbook Reference:*** *Concept 28.2 Plants Have Mechanical and Chemical Defenses against Herbivores; Constitutive defenses are physical and chemical*

6. Which of the following best describes how plants produce their own insecticide?
   a. Wounded cells release an elicitor that directly induces synthesis of protease inhibitors, and protease inhibitors act as insecticides.
   b. Wounded cells release jasmonates, jasmonates stimulate elicitor synthesis, elicitor causes the production of protease inhibitors, and protease inhibitors act as insecticides.
   c. Wounded cells release an elicitor that causes membrane breakdown, membrane breakdown releases jasmonates, and jasmonates act as insecticides.
   d. Wounded cells release an elicitor that causes membrane breakdown, membrane breakdown releases jasmonates, jasmonates induce synthesis of protease inhibitors, and protease inhibitors act as insecticides.
   e. Wounded cells release protease inhibitors, protease inhibitors cause release of elicitors, and elicitors act as insecticides.

   ***Textbook Reference:*** *Concept 28.2 Plants Have Mechanical and Chemical Defenses against Herbivores; Plants respond to herbivory with induced defenses*

7. Which of the following is *not* an adaptation to drought conditions?
   a. Water-storing leaves
   b. Leaf loss
   c. Sunken stomata
   d. Increased stomata number
   e. Water-storing stems

   ***Textbook Reference:*** *Concept 28.3 Plants Adapt to Environmental Stresses; Some plants have special adaptations to live in very dry conditions*

8. Gene-for-gene resistances depend on which of the following?
   a. Compatible alleles in plant and pathogen
   b. Incompatible alleles in plant and pathogen
   c. Recessive *Avr* genes only
   d. Recessive *R* genes only
   e. PR proteins

   ***Textbook Reference:*** *Concept 28.1 Plants Have Constitutive and Induced Responses to Pathogens; Specific immunity is genetically determined*

9. Which of the following conditions stimulates the production of heat shock proteins?
   a. Abnormally high temperatures
   b. Abnormally low temperatures
   c. Both abnormally high and abnormally low temperatures
   d. Slightly lowered temperature
   e. Heat shock proteins are continually available in a plant, no matter what the temperature.

   ***Textbook Reference:*** *Concept 28.3 Plants Adapt to Environmental Stresses; Plants can cope with temperature extremes*

10. The main function of heat shock proteins is to
    a. stabilize proteins necessary to a cell's survival.
    b. reinforce membranes that lose fluidity.
    c. cause the plant to enter dormancy.
    d. act as an antifreeze compound.
    e. stimulate bolting.
    ***Textbook Reference:*** *Concept 28.3 Plants Adapt to Environmental Stresses; Plants can cope with temperature extremes*

11. Which of the following is *not* involved in the process of cold-hardening?
    a. Production of antifreeze proteins
    b. Increased production of unsaturated fatty acids
    c. Production of phytoalexins
    d. Production of heat shock proteins
    e. Repeated exposure to cool temperatures over several days
    ***Textbook Reference:*** *Concept 28.3 Plants Adapt to Environmental Stresses; Plants can cope with temperature extremes*

12. Plants that produce toxins for defense avoid poisoning themselves by a number of strategies, including by storing the toxic substances in
    a. roots.
    b. latex.
    c. soil.
    d. the same place as enzymes that convert them to the active form.
    e. leaf palisade cells.
    ***Textbook Reference:*** *Concept 28.2 Plants Have Mechanical and Chemical Defenses against Herbivores; Why don't plants poison themselves?*

13. Which of the following is *not* a secondary plant metabolite?
    a. Phenolics
    b. Alkaloids
    c. Terpenes
    d. Glucosinolates
    e. PR proteins
    ***Textbook Reference:*** *Concept 28.2 Plants Have Mechanical and Chemical Defenses against Herbivores; Constitutive responses are physical and chemical*

14. The hypersensitive response involves the release of
    a. lignin.
    b. PR proteins.
    c. *R* genes.
    d. phytoalexins.
    e. secondary metabolites.
    ***Textbook Reference:*** *Concept 28.1 Plants Have Constitutive and Induced Responses to Pathogens; Specific immunity usually leads to the hypersensitive response*

15. A plant's immune response to RNA viruses involves
    a. heat shock proteins.
    b. secondary metabolites.
    c. small interfering RNA.
    d. PR proteins.
    e. *R* genes
    ***Textbook Reference:*** *Concept 28.1 Plants Have Constitutive and Induced Responses to Pathogens; General and specific immunity can lead to systemic acquired resistance*

## Answers

### *Key Concept Review*

1.

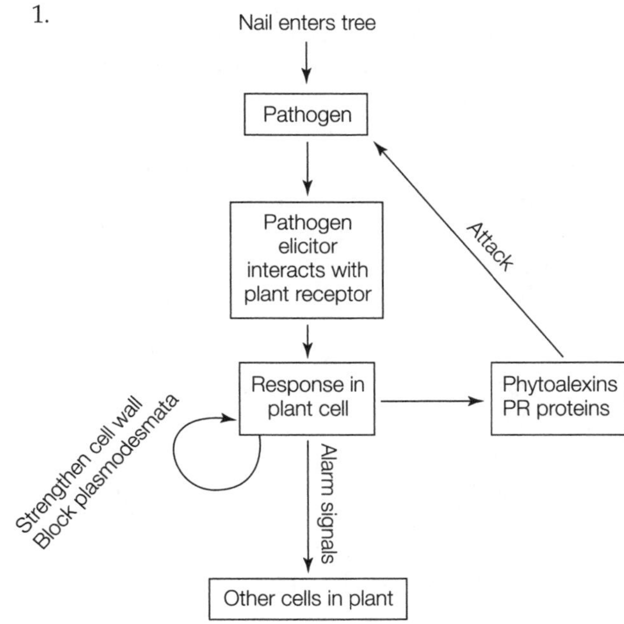


2. General immunity is stimulated by elicitors called pathogen associated molecular patterns (PAMPs), whereas specific immunity is triggered by effectors. The differences between the two pathways are in the downstream proteins that are activated, and the fact that PAMPs react with membrane receptors whereas effectors enter the cell and react with intracellular receptors. The benefit to having general responses is that they can be activated quickly, with a reasonably good expectation of fighting off the invading organism. This can be followed by specific attacks that will be able to effectively kill the pathogen.

3. Some plants have physical characteristics that deter herbivory, such as crystals of calcium oxalate, thick outer surfaces, or thorns. Plants can also prevent herbivory by producing secondary metabolites that affect potential herbivores. These compounds may act as neurotoxins, inhibit digestion, or have a number of other consequences (see Table 28.1). Induced defenses can also be employed to defend against herbivory.

4. Plants may isolate the toxins in vacuoles within their cells so that the toxins do not come into contact with the active machinery of the cells. Plants may restrict toxin production to already-damaged tissue. Plants

may store toxins as precursor molecules that need to be exposed to particular enzymes in order to become activated. Plants may also have modified proteins that are not affected by the toxin.

5. The burlap will most likely do little for the shrubs; the roots are in most danger. Because streets are heavily salted in the winter, the snow around them creates a saline environment. If the shrubs are not adapted to a saline environment, little will help them survive the salt accumulation around their roots.
6. Cold-hardening is gradual exposure to adverse temperatures. This allows for structural changes in cell membranes to help them cope with extreme temperatures. It also allows for the production of heat shock proteins and other compounds necessary for survival at temperature extremes.
7. Only plants that have evolved a strategy for taking up and coping with heavy metals can grow in these contaminated sites. Deliberately planting them at these sites initiates the slow process of removing the metals from the soil.

### *Test Yourself*

1. **d.** If something is always turned on, it is said to be constitutive.
2. **b.** Plants that have been exposed to pathogens wall off invaded tissue. This tissue releases salicylic acid to the rest of the plant, which stimulates PR proteins. The tissue also releases airborne methyl salicylate, which causes release of PR proteins in distant regions of the plant and even in neighboring plants. Another type of specific immunity uses RNA interference through small interfering RNA (siRNA).
3. **a.** They can accumulate sodium and chloride ions. This makes their water potential more negative, and they are able to extract more water from their surroundings than plants that do not store the ions.
4. **c.** Polysaccharides seal plasmodesmata and reinforce cell walls. The purpose of this is to contain the pathogen in a minimal number of cells.
5. **c.** Canavanine's mechanism of action is to be taken up by the herbivore and incorporated into its proteins. Because the structure of canavanine is different from that of the arginine it replaces, the herbivore's proteins take up a structure that is lethal to the herbivore. The plant itself and some herbivores have enzyme mechanisms to distinguish canavanine from arginine and prevent its incorporation into proteins.
6. **d.** See Figure 28.5 in the textbook.
7. **d.** Water storage tissues allow the plant to take advantage of any water that is available and hold on to it for lean times. Leaf loss, when photosynthesis cannot be sustained, prevents loss of resources. Sunken stomata prevent excessive water loss to evaporation.
8. **a.** The plant must have an *R* allele, and the pathogen must have an *Avr* allele. Compatibility between these alleles leads to resistance.
9. **a.** Heat shock proteins are produced in response to extreme high temperatures.
10. **a.** Heat shock proteins are chaperonin proteins. They function to stabilize and prevent denaturation (unfolding) of essential proteins.
11. **c.** Cold-hardening involves production of antifreeze proteins, heat shock proteins, and an increase in unsaturated fatty acids in the membrane. The production of phytoalexins is involved in chemical defenses against pathogens.
12. **b.** Plants that produce toxins as a defense typically store them in special locations, such as latex, where they will not damage tissue.
13. **e.** Alkaloids, phenolics, terpenes, and glucosinolates are all examples of secondary plant metabolites. PR proteins are proteins that function in defense, and as such are primary products of cell metabolism.
14. **d.** The hypersensitive response involves the release of antibiotic phytoalexins.
15. **c.** Plants that are invaded by RNA viruses use the RNA of the invading virus to interfere with and block viral replication.

# 29 Fundamentals of Animal Function

## The Big Picture

- One of the most important roles of tissues, organs, and organ systems is to help maintain homeostasis within the body's cells. Only by maintaining relatively constant intracellular conditions can basic metabolic reactions continue. Food and the energy it provides is essential for creating and maintaining the organization that allows animals to function.
- Animals either generate their own body heat (homeotherms) or rely on the environment to determine body temperature (poikilotherms). Control of body temperature can occur by behavioral and physiological mechanisms. Homeotherms typically have more insulation and higher metabolic rates than poikilotherms. The metabolic rates of homeotherms and poikilotherms respond differently to changes in environmental temperature. Homeotherms increase metabolism as environmental temperature decreases; they also increase metabolic rate at high temperatures when energy is needed for sweating and panting. The metabolic rate of poikilotherms is temperature-dependent, fluctuating with environmental temperature.
- Metabolic rates are affected by the amount of energy needed to maintain the body as well as any physical activity performed by the animal. Metabolic rate is also proportional to body size, but there are scaling relationships that make this relationship nonlinear.
- Animal functions are governed by control mechanisms. The phenotype of an animal can change, within limits, based on changes in the external conditions.

## Study Strategies

- Remember that homeostasis does not necessarily mean holding every body variable absolutely constant, but rather making sure that body variables are held at the correct level for the circumstances. Even body temperature in humans, which we think of as being a constant 37°C, drops slightly in the early morning hours and rises considerably during extended vigorous exercise.
- Understanding the difference between poikilotherms and homeotherms can be difficult. These terms refer to the relative stability of an animal's temperature. Whereas body temperatures of poikilotherms are determined primarily by external sources of heat and thus can fluctuate, those of homeotherms are determined by heat generated metabolically (i.e., within their bodies) and are maintained at a relatively stable level.
- Refer to Figures 29.16 and 29.17 to understand how homeostasis is maintained through negative- and positive-feedback systems.
- Review the organization of physiological systems, from cells to tissues to organs to organ systems. This will help you when you study each of the organ systems in later chapters.
- Go to **LaunchPad** (or use the Web addresses listed) to review the following additional resources:

  Animated Tutorial 29.1 The Hypothalamus (PoL2e.com/at29.1)

  Animated Tutorial 29.2 Circadian Rhythms (PoL2e.com/at29.2)

  Activity 29.1 Thermoregulation in a Homeotherm (PoL2e.com/ac29.1)

  Activity 29.2 Tissues and Cell Types (PoL2e.com/ac29.2)

  Analyze the Data in textbook Figure 29.7

  Apply the Concept in textbook Sections 29.2 and 29.6

## Key Concept Review

### 29.1 Animals Eat to Obtain Energy and Chemical Building Blocks

#### Animals need chemical building blocks to grow and to replace chemical constituents throughout life

#### Animals need inputs of chemical-bond energy to maintain their organized state throughout life

Animals are heterotrophs, and need to obtain energy from the chemical bonds in organic compounds from other living organisms. Autotrophs, such as plants, can obtain energy from the chemical bonds in inorganic molecules and can generate organic compounds from inorganic sources. In most ecosystems, the energy input is in the form of sunlight and is incorporated into organic molecules by plants and algae. Animals ingest these compounds as sources of energy and chemical building blocks.

Animals need chemical building blocks for growth and because the molecules in their bodies are constantly exchanged as they are recycled when they are no longer useful. Nearly all cells in the body, such as red blood cells, are constantly broken down and replaced. During this process molecules are lost and must be replaced from food sources.

Energy is required to maintain the organization of the body. Organization is present from the molecular level up to the organismal level. The second law of thermodynamics tells us that any system tends toward disorganization; thus, energy inputs are required to fight this tendency and maintain organization. Energy can be defined, in biological systems, as the capacity to create or maintain this organization. Only some forms of energy are available for this purpose (can perform work); heat energy is not available to perform work. Every time an organism converts energy into another form, some is lost to heat energy. Animals need a constant input of energy from organic molecules.

**Question 1.** Explain the difference between heterotrophs and autotrophs. To which group do animals belong?

**Question 2.** Why do animals need to eat?

### 29.2 An Animal's Energy Needs Depend on Physical Activity and Body Size

- We quantify an animal's metabolic rate by measuring heat production or $O_2$ consumption
- Physical activity increases an animal's metabolic rate
- Among related animals, metabolic rate usually varies in a regular way with body size

The amount of energy an individual animal needs depends on many different variables. Energy is used for different purposes in the body: to maintain organization, create force, and synthesize molecules, among other functions.

When an animal consumes energy, the energy is converted to heat, which cannot be used to perform work. An animal's metabolic rate is defined as its rate of energy consumption. The rate is important because this energy needs to be replaced by eating; the amount of food needed will depend on the energy usage. Since there is a direct relationship between oxidation of the matter and heat produced, metabolic rate is quantified by the amount of oxygen needed for the aerobic metabolism of the ingested organic matter. The animal "consumes" oxygen in the metabolism of the food, meaning that the oxygen is combined with other atoms.

The amount of physical activity will affect the amount of energy needed. This is characterized by the basal metabolic rate (BMR). Sometimes the relationship between speed and energy usage is not linear; when animals swim or fly, there can be an exponential relationship between speed and energy usage. Among related animals, the metabolic rate will vary with body size, but not in a linear fashion. The BMR per gram will differ based on the size of the animal; a small animal consumes much more energy per gram of weight than a larger animal. This is called a scaling relationship (see Figure 29.3).

**Question 3.** Define metabolic rate and explain how it is quantified in animals.

**Question 4.** Explain why a bird in flight does not consume as much energy at an intermediate air speed as it does at a low or high air speed.

### 29.3 Metabolic Rates Are Affected by Homeostasis and by Regulation and Conformity

- Animals are classed as regulators and conformers
- Regulation is more expensive than conformity
- Homeostasis is a key organizing concept
- Animals are classed as homeotherms or poikilotherms based on their thermal relationships with their external environment
- Homeothermy is far more costly than poikilothermy
- Homeotherms have evolved thermoregulatory mechanisms
- Hibernation allows mammals to reap the benefits of both regulation and conformity

Cells in an animal's body are bathed with tissue fluids (also called interstitial fluids), which constitute the cell's environment. An animal's internal environment consists of this tissue fluid, and the external environment is the environment outside the body. Some animals attempt to maintain a constant internal environment regardless of changes in the external environment; these animals are called regulators. Animals that do not tightly control their internal environment are called conformers (see Figure 29.4). Regulation (maintenance of a constant internal environment) and conformity (variability of the internal environment) can be applied separately to each characteristic; fish are conformers in terms of body temperature but regulators in terms of ion concentrations.

It is always more difficult to regulate than to conform; regulation requires more energy because of the extra work required. The benefit of regulation is stability. The benefit of conformity is the low energy required, but a drawback is the lack of stability. Regulation increases the metabolic rate of an organism.

Physiologists in the nineteenth century noticed that the characteristics of blood plasma in humans and other animals is consistent over time. This led to the concept of homeostasis, the idea that there is a stable internal environment and that mechanisms exist to maintain this stability. This is a very important concept in physiology and related biological fields of study and is similar to the concept of regulation. Animals that exhibit thermoregulation (regulate their internal temperature) are called homeotherms. In most homeotherms, if a resting individual is exposed to a variety of external temperatures, its body temperature remains constant but its metabolic rate varies (see Figure 29.5). The animal's metabolic rate is fairly consistent within a range of temperatures called the thermoneutral zone (TNZ). As the external temperature falls below the TNZ, the metabolic rate of the animal increases to offset the heat loss; as the external temperature rises, metabolic rate increases because the body works harder to cool itself. Most animals are not homeotherms, but instead allow their body temperature to match that of the external environment; they are called poikilotherms (or ectotherms). Poikilotherms have wide ranges of permissible body temperatures. The metabolic rate of a poikilotherm varies exponentially as the external temperature changes. All physiological processes in organisms are sensitive to temperature. Increases in temperature usually increase the rate of a given process. The $Q_{10}$ is a measure of change in a physiological process as the temperature increases or decreases by 10°C (see Figure 29.6). Most biological $Q_{10}$ values range from 2 to 3. A $Q_{10}$ of 2 means that the reaction rate doubles when temperature increases by 10°C. Poikilotherms can affect their body temperature through behavior by moving into cooler or warmer areas, if possible (see Figure 29.8). This will also affect their net metabolism. For any given external temperature, a homeotherm will have a higher metabolic rate than a poikilotherm of the same body size (see Figure 29.9). At colder temperatures this difference becomes magnified because the homeotherm must generate heat to stay warm, whereas the poikilotherm experiences slower metabolism due to temperature effects on reactions.

Mammals, birds, and insects are the major homeotherm groups. Not all insects exhibit homeothermy, but those that do only do at specific times, such as in flight.

All homeothermic animals have mechanisms for increasing metabolic rate in order to generate body heat when the environment is cold. Sometimes ATP is required, such as in the skeletal muscle contractions that constitute shivering. Some animals perform nonshivering thermogenesis, which occurs in brown adipose tissue (BAT) (see Figure 29.10). This process uses uncoupling of oxidative phosphorylation, so that no ATP is generated, but heat is made in the mitochondria due to a short circuit of the proton gradient. Insulation, such as fur and feathers, and specialized blood flow patterns also allow homeotherms to maintain high body temperatures. When the environment becomes too hot, homeotherms need to get rid of excess body heat. Sweating and panting are two methods that utilize evaporation to cool the body.

Hibernation is a method by which mammals can take advantage of regulation and conformity. During the warm parts of the year the animal performs regulation, but during the cold months it hibernates, and thus exhibits conformity. The hibernation allows the animal to conserve energy because of the low metabolic costs of remaining relatively dormant, and the animal shifts between two metabolism-temperature curves (see Figure 29.12).

**Question 5.** Why is it important for an organism to maintain homeostasis?

**Question 6.** Design a study to demonstrate that the vertebrate thermostat is located in the hypothalamus.

**Question 7.** In response to the smell of food, the stomach produces hydrochloric acid in anticipation. Is this an example of negative feedback, positive feedback, or another type of control? Explain your answer.

**Question 8.** Why would a homeothermic animal's metabolic rate be higher at 40°C than at 30°C? What physiological phenomenon would explain why its metabolic rate rises when the environmental temperature is above the upper critical limit?

**Question 9.** What types of insulation do mammals have, and what physiological controls exist to increase insulation?

## 29.4 Animals Exhibit Division of Labor, but Each Cell Must Make Its Own ATP

- Fluid compartments are separated from one another by physiologically active epithelia and cell membranes
- Animals exhibit a high degree of division of labor
- Division of labor requires a rapid transport system
- Each cell must make its own ATP
- Animal cells have aerobic and anaerobic processes for making ATP

Animals are mostly (about 60%) water, which is divided into various compartments. Most of the fluid is within cells, as intracellular fluid. The water outside of cells is the extracellular fluid. There are two categories of extracellular fluid: the blood plasma and the interstitial fluid. These compartments are kept separate by epithelia and cell membranes. An epithelium is a sheet of cells that covers a surface or organ or lines a body cavity (see Figure 29.13). A simple epithelium is a single layer of cells on top of a basement membrane (basal lamina). Epithelia play additional roles, such as pumping ions, secreting substances like digestive enzymes and milk, absorbing molecules, and acting as sensors (for example, for smell and taste). Cell membranes separate fluid compartments on a much smaller scale. Again, this barrier serves other functions, such as secreting ions and receiving signals.

Multicellular organisms coordinate the activities of different types of cells in order to perform all the necessary functions to sustain life. A tissue is a group of cells of similar type. An organ is two or more types of tissue with a defined relationship. A multi-organ system is a group of organs working together (see Figure 29.14). Each organ may have different tissue types: muscle, nervous, connective, and epithelial. Each tissue type consists mostly of one type of cell. Each organ is specialized to perform a certain set of tasks, and is reliant on other organs to perform other vital functions. Because of this networking of the organ systems, a complex multicellular organism requires a method to deliver molecules to target cells quickly. In animals, this is often achieved with a circulatory system. Because of the oxygen demands of a high metabolic rate, animals with higher metabolic rates require high rates of oxygen transport and thus have well-developed and highly efficient circulatory systems.

Despite the ability of the multicellular organism to divide its functions among different organs, each cell must make its own ATP, which is not transported from cell to cell like other molecules. Each cell uses food compounds, such as fatty acids or glucose, to generate ATP. Mostly this process is aerobic, requiring oxygen, and takes place in the mitochondria. All types of food molecules—sugars, lipids, and proteins—can be used for ATP synthesis by aerobic metabolism. If oxygen is not present in sufficient amounts, then some cells can activate alternative processes to generate ATP that are anaerobic. This usually refers to anaerobic glycolysis, or glycolysis without subsequent oxidative phosphorylation. Only sugars can be used as fuels, since only sugars can enter the glycolysis pathway. Anaerobic glycolysis is highly inefficient and cannot be maintained long-term due to the accumulation of lactic acid.

**Question 10.** Label the organs and four main tissue types in the diagram below.

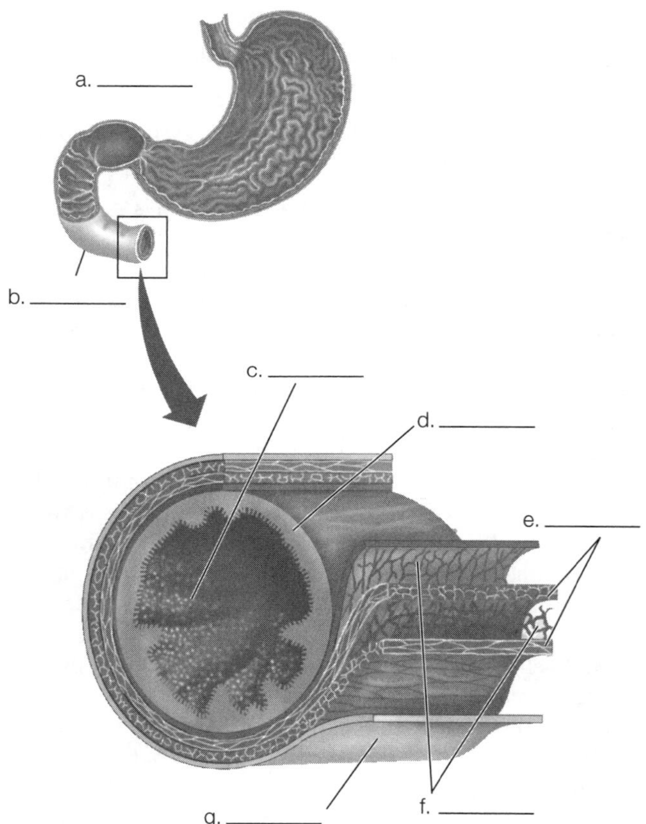


**Question 11.** Explain why each cell must produce its own ATP and does not depend on other cells to deliver energy in this form.

## 29.5 The Phenotypes of Individual Animals Can Change during Their Lifetimes

Phenotypic plasticity is common at the biochemical level

Phenotypic plasticity also occurs at the scales of tissues and organs

Phenotypic plasticity is under genetic control

The genotype of an individual is constant, but observable phenotypes can change; this is known as phenotypic plasticity. When phenotype changes due to long-term exposure to a particular environment, the individual has acclimated (or acclimatized) to that environment. Phenotypic plasticity occurs often at the biochemical level through inducible enzymes, enzymes whose levels vary depending on environment or experiences. Fish and invertebrates will adapt to different seasons through acclimation and changes in enzyme levels. Liver enzymes are another set of inducible enzymes that exhibit phenotypic plasticity; levels of detoxification enzymes depend on previous exposures to potential toxins. Tissues and organs can also exhibit phenotypic plasticity, such as the growth of brown adipose tissue used for thermogenesis when the environment becomes colder. The phenotypic plasticity is driven by changes in gene expression based on external conditions. Natural selection acts on programs of responses, which in turn depend on the external conditions.

**Question 12.** Animals have the ability to undergo acclimatization. Describe this process in relation to temperature regulation.

**Question 13.** Explain why phenotypic plasticity is intimately related to genetics and evolution.

## 29.6 Animal Function Requires Control Mechanisms

Homeothermy exemplifies negative-feedback control

Positive feedback occurs in some cases

Biological clocks make important contributions to control

The nervous and endocrine systems control the ways tissues and organs interact, and there are control mechanisms at the cellular level that govern cellular processes. When control mechanisms fail, evidence exists in the form of a disease or disorder such as cancer or a deadly fever.

In order to have a control mechanism for a characteristic, or controlled variable, there has to be a sensor that detects the current value of that variable. For instance, body temperature is sensed at different places in the body such as the brain and skin. Information about the state of the variable is sent to the brain. Effectors are tissues or organs that are able to change the controlled variable, such as shivering muscles. The control mechanism decides when and where to activate the effectors in order to keep the controlled variable within a prescribed range. For mammalian body temperature, this mechanism is located in the hypothalamus of the brain. The control mechanism compares the information about the current status of the controlled variable with a desired set point, and then activates different effectors to minimize the difference between the actual variable level and the set point. This is referred to as a negative-feedback system, since the goal is to reduce the difference between variable and set point (see Figure 29.16).

Positive feedback control mechanisms are not as common in animals but they do occur. Positive feedback control has a destabilizing influence on the controlled variable. This can be advantageous in particular situations as long as it is eventually resolved. One example of positive feedback control is mammalian childbirth and the uterine contractions that deliver the fetus. The control center is the hypothalamus, and the effector is the pituitary gland (see Figure 29.17).

Biological clocks are another control mechanism in animals. These do not require outside cues in order to maintain a regular cycle. Free-run biological clocks, operating without external timing cues such as sunrise and sunset, do not operate on a perfect 24-hour cycle. Animals require entrainment in order to maintain a cycle that matches a daylight cycle. Biological clocks with free-run cycles of about 24 hours are circadian clocks. There are also clocks that match years and tidal cycles in certain animals. In animals, a master clock controls all the cycles in the body such as sleep, digestion, and reproductive cycles; in mammals this is located in the suprachiasmatic nuclei of the brain.

**Question 14.** During childbirth, the pressure exerted on the mother's cervix by the emerging infant leads to increased contraction of the uterus. What type of feedback—negative or positive—is involved in this situation?

**Question 15.** Complete the table below, comparing some general traits of homeotherms and poikilotherms.

| | Homeotherm | Poikilotherm |
|---|---|---|
| Heat source | | |
| Efficiency of energy usage | | |
| Resting metabolic rate | | |
| Temperature control center | | |
| Insulation | | |

**Question 16.** You measure the metabolism of two animals of similar size at 15°C and 25°C. You find that the metabolic rate of Animal 1 is 115 ml of oxygen per hour at 15°C and 55 ml of oxygen per hour at 25°C. The metabolic rate of Animal 2 is 5.5 ml of oxygen per hour at 15°C and 11.5 ml of oxygen per hour at 25°C. Calculate the $Q_{10}$ for the metabolic rate in both animals, and determine if they are homeotherms or poikilotherms.

**Question 17.** In the figure below, graph the lizard's predicted body temperature based on your knowledge of how poikilotherms regulate their temperature with behavior.

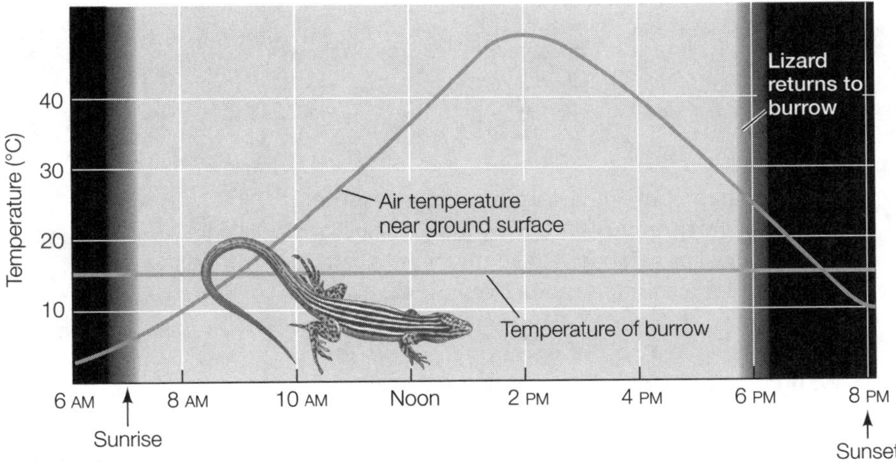

## Test Yourself

1. Which statement about interstitial fluid is true?
   a. It is found within cells.
   b. It represents most of the water in an animal's body.
   c. It makes up blood plasma.
   d. It does not exchange molecules with intracellular fluid.
   e. It is found between the cells of the body.
   ***Textbook Reference:*** *Concept 29.3 Metabolic Rates Are Affected by Homeostasis and by Regulation and Conformity*

2. Which situation is *not* paired correctly with the likely phenotypic plasticity that would occur?
   a. An animal that usually lives close to sea level is relocated to a high elevation; hemoglobin production increases.
   b. A lanky person becomes a bodybuilder; his or her muscles become extremely large.
   c. An animal finds that the easiest food source is high in the trees; its neck grows in length to reach this food source.
   d. A small mammal moves from a warm environment to a cold environment; it generates a larger brown adipose tissue deposit to increase thermogenesis.
   e. An individual becomes a heavy drinker and experiences an increase in the liver enzymes that are responsible for detoxifying alcohol.
   ***Textbook Reference:*** *Concept 29.5 The Phenotypes of Individual Animals Can Change during Their Lifetimes; Phenotypic plasticity also occurs at the scales of tissues and organs*

3. In an environment with an ambient temperature lower than an animal's core body temperature, which physiological or behavioral response would be *inappropriate* if the animal needed to eliminate excess body heat?
   a. Wallowing in a pool of water
   b. Inhibiting blood flow to peripheral vessels
   c. Sweating
   d. Decreasing physical activity
   e. Finding a cool, shady spot
   ***Textbook Reference:*** *Concept 29.3 Metabolic Rates Are Affected by Homeostasis and by Regulation and Conformity; Homeotherms have evolved thermoregulatory mechanisms*

4. A lizard lives in a desert environment where the temperature is low at night and high during the day. What might it do to maintain the most stable body temperature?
   a. Stay in a burrow during the night and shuttle between the sun and shade on the surface during the day
   b. Stay on the surface during the night and move to a burrow during the day
   c. Increase metabolism and heat production during the night and decrease it during the day
   d. Decrease metabolism and heat production during the night and increase it during the day
   e. Move into a state of hypothermia during the night and a state of torpor during the day
   ***Textbook Reference:*** *Concept 29.3 Metabolic Rates Are Affected by Homeostasis and by Regulation and Conformity; Animals are classed as homeotherms or poikilotherms based on their thermal relationships with their external environment*

5. What is the difference between negative feedback mechanisms and positive feedback mechanisms?
   a. Negative feedback mechanisms exist only in the circulatory system.
   b. Negative feedback mechanisms return a system to a set point, whereas positive feedback mechanisms amplify a response.
   c. Negative feedback mechanisms move a system away from a set point, whereas positive feedback mechanisms stabilize a system toward a set point.
   d. Negative feedback mechanisms stabilize a system toward a set point, whereas positive feedback mechanisms reset the set point.
   e. There is essentially no difference between the two feedback systems.
   ***Textbook Reference:*** *Concept 29.6 Animal Function Requires Control Mechanisms; Homeothermy exemplifies negative-feedback control*

6. On a hot day, an active dog will pant heavily and visit his water bowl frequently. Which statement about the dog's condition or behavior is *false*?
   a. He is lowering his core body temperature by means of evaporative cooling.
   b. He is operating at a basal metabolic rate.
   c. He is at or above the upper end of the thermoneutral zone.
   d. He is sending more blood to his skin.
   e. He is adjusting his behavior in order to thermoregulate.
   ***Textbook Reference:*** *Concept 29.3 Metabolic Rates Are Affected by Homeostasis and by Regulation and Conformity; Homeotherms have evolved thermoregulatory mechanisms*

7. Evaporative cooling is an effective way to increase heat loss, but it carries the physiological drawback of
   a. an increased use of ATP and substantial water loss.
   b. the lowering of the hypothalamic thermal set point.
   c. a decreased use of ATP and substantial water gain.
   d. exhausting the supply of brown fat.
   e. a lowered basal metabolic rate.
   ***Textbook Reference:*** *Concept 29.3 Metabolic Rates Are Affected by Homeostasis and by Regulation and Conformity; Homeotherms have evolved thermoregulatory mechanisms*

8. Animals need to eat to obtain
   a. energy for maintaining organization.
   b. amino acids to rebuild proteins.
   c. energy for maintaining body temperature.
   d. iron and other minerals.
   e. all of the above.
   ***Textbook Reference:*** *29.1 Animals Eat to Obtain Energy and Chemical Building Blocks*

9. Which statement about metabolic rate is *false*?
   a. A small mammal needs much more food per gram of body weight than a similar large mammal.
   b. Metabolic rate is inversely related to an animal's activity.
   c. Basal metabolic rate is expressed in rates per unit of body weight.
   d. Relative basal metabolic rates have ecological implications.
   e. Metabolic rate is measured indirectly by assessing oxygen consumption.
   ***Textbook Reference:*** *Concept 29.2 An Animal's Energy Needs Depend on Physical Activity and Body Size; Among related animals, metabolic rate usually varies in a regular way with body size*

10. Which of the following is *not* a barrier between fluid compartments in animals?
    a. Lining of kidney tubules
    b. Cell membranes
    c. Intestinal lining
    d. The skin of a frog sitting in a pond
    e. Heart valve
    ***Textbook Reference:*** *Concept 29.4 Animals Exhibit Division of Labor, but Each Cell Must Make Its Own ATP; Fluid compartments are separated from one another by physiologically active epithelia and cell membranes*

11. Which type of molecule is used for energy during anaerobic ATP production?
    a. Sugars
    b. Lipids
    c. Proteins
    d. Nucleic acids
    e. Fatty acids
    ***Textbook Reference:*** *Concept 29.4 Animals Exhibit Division of Labor, but Each Cell Must Make Its Own ATP; Animal cells have aerobic and anaerobic processes for making ATP*

12. A number of physiological processes can undergo acclimatization. Which statement about acclimatization is *false*?
    a. It occurs in response to seasonal temperature changes.
    b. Acclimatization of metabolic rate occurs because enzyme expression changes.
    c. It involves the changing of a set point.
    d. All multicellular animals are capable of acclimatizing to all the environmental changes they face.
    e. Acclimatization is primarily seen in organisms with body temperatures tightly coupled to environmental temperatures.
    ***Textbook Reference:*** *Concept 29.5 The Phenotypes of Individual Animals Can Change during Their Lifetimes; Phenotypic plasticity is common at the biochemical level*

13. Which of the following is an example of negative feedback?
    a. Blood clotting platelet activation releases chemicals that cause more platelet activation
    b. Contractions during childbirth: a uterine contraction stimulates the release of oxytocin, which increases uterine contractions
    c. Lactation: a suckling baby stimulates the release of prolactin, which increases milk production
    d. Regulation of blood calcium levels: a person's blood calcium level rises, signaling the bone cells to deposit more calcium in the bone tissue, thus lowering blood calcium
    e. Nerve impulse/action potential: leakage of sodium ions through membrane channels changes membrane potential, which causes opening of more membrane channels
    ***Textbook Reference:*** *Concept 29.6 Animal Function Requires Control Mechanisms; Positive feedback occurs in some cases*

14. Some animals use brown adipose tissue (BAT) as a source of heat generation. Which statement about BAT is *false*?
    a. BAT is involved in nonshivering thermogenesis.
    b. BAT burns fuel without producing ATP.
    c. BAT is found in some mammals.
    d. BAT is highly vascularized.
    e. BAT cells contain few mitochondria.
    ***Textbook Reference:*** *Concept 29.3 Metabolic Rates Are Affected by Homeostasis and by Regulation and Conformity; Homeotherms have evolved thermoregulatory mechanisms*

15. The metabolic rate of an animal is
    a. the rate at which it absorbs chemical energy from heat.
    b. the rate at which it digests its food.
    c. the amount of food it eats.
    d. its normal level of physical activity.
    e. its rate of energy consumption.
    ***Textbook Reference:*** *Concept 29.2 An Animal's Energy Needs Depend on Physical Activity and Body Size; We quantify an animal's metabolic rate by measuring heat production or $O_2$ consumption*

## Answers

### Key Concept Review

1. An autotroph can use inorganic molecules as sources of energy, and can create organic molecules from inorganic starting material. A heterotroph needs to find a source of organic molecules from another organism, both for energy and for chemical building blocks. Animals are heterotrophs.
2. Animals require energy to do work; this work is necessary to maintain structure and organization in the body. Energy is lost every time it is converted to another form, and thus must be replenished. Energy is found in the chemical bonds of organic matter, which is internalized through eating. Animals need a contant source of organic matter.
3. The metabolic rate is the rate of energy consumption, or the rate at which chemical energy is converted to heat. The most direct way to determine metabolic rate is to measure the amount of heat generated. Because this is very difficult to do, we make use of the nearly linear relationship between the amount of oxygen used during aerobic metabolism and the amount of heat generated. Thus, metabolic rate is usually calculated by measuring the amount of oxygen needed by an organism.
4. While in flight, the bird makes use of air flow over the wings to maintain speed and altitude. It takes more energy to take off and accelerate to an intermediate speed than it does to maintain this speed. Similarly, it takes excess energy to increase the speed above that which is easily maintained with air currents.
5. Homeostasis is the maintenance of a constant internal environment. The body functions with the help of proteins and enzymes. Changes in pH, temperature, glucose level, and oxygen and carbon dioxide levels (among other things) of the internal environment can affect cellular function. Loss of homeostasis can lead to improper functioning of proteins and cell membranes and, ultimately, to cell death.
6. Two possible approaches would be to conduct a lesion experiment and a thermal stimulation experiment. In a lesion experiment, the region of the hypothalamus thought to function as the thermostat would be destroyed in order to observe the lesioned animal's ability to regulate body temperature. In a thermal stimulation experiment, the hypothalamus could be cooled or warmed and physiological and behavioral changes associated with thermoregulation (such as constriction or dilation of blood vessels in the skin, and sweating or panting) would be monitored.
7. This is neither negative nor positive feedback, but instead an example of feed-forward information. The stomach does not continually produce hydrochloric acid as would be expected in positive feedback, but just needs to prepare for the presence of food. Feed-forward information is predictive of a change in the internal environment before that change occurs. Feed-forward information functions to change the set point.
8. A homeotherm in a hot environment will attempt to keep its core body temperature from increasing. The heat-loss mechanisms available to a homeotherm—sweating and panting—require the use of ATP and result in an increased metabolic rate. Homeotherms will expend energy to protect body temperature and lose heat to the environment.
9. Fur and fat are effective insulators in mammals. Fur is an effective insulator because it traps still, warm air from the body; the longer the fur, the more effective it is as an insulator. Fat works in a similar manner because it has a low thermal conductance. Humans do not have sufficient hair on their bodies for it to provide adequate insulation; therefore, clothes are used as insulation. Physiologically, mammals can change their blood flow patterns to increase or decrease the amount of blood that is received by the skin. When they are active, they get rid of excess heat by transporting heat to hairless skin surfaces.
10.
    a. Stomach
    b. Small intestine
    c. Epithelial cells
    d. Mucosa
    e. Smooth muscle
    f. Nervous tissue
    g. Epithelial cells and connective tissue
11. ATP is the energy currency of the cell, but it is not transported between cells (it is not stable enough). Therefore, the body transports energy in another form (usually sugars, like glucose) which is used by the cell's own processes to supply ATP for cellular work on an as-needed basis.
12. Animals undergo acclimatization in response to changes in seasonal conditions, such as temperature. Often these changes are brought about by the production of enzymes that function better at the new temperature. Because of acclimatization, metabolic functions are less sensitive to long-term changes in temperature than to short-term changes.
13. Phenotypic plasticity refers to changes in phenotype in response to environmental conditions. A particular phenotype is a direct result of the expression of particular genes in the organism's genome, and thus are effects of the proteins that are encoded by these genes. Natural selection does not only exert pressure favoring a particular gene and phenotype, but also can favor the ability of the organism to adapt to its surroundings. This pattern of response can generate a new phenotype based on the external conditions.
14. Increased uterine contraction in response to pressure on the cervix is an example of a positive feedback.

15.

| | **Homeotherm** | **Poikilotherm** |
|---|---|---|
| Heat source | Behavioral and internal metabolism | Behavioral and environmental |
| Efficiency of energy usage | Inefficient | More efficient |
| Resting metabolic rate | Higher | Lower |
| Temperature control center | Hypothalamus | Hypothalamus |
| Insulation | Fur, feathers, fat layer beneath skin | Little to none |

16. Metabolic rate can be calculated using the equation $Q_{10} = (R_T/R_{T-10})$. The second animal shows a direct correlation between temperature and metabolism, and compared to the first animal, it has a lower overall metabolic rate at both temperatures. This information can be organized into a table such as the following:

| | Metabolic rate at 15°C | Metabolic rate at 25°C | $Q_{10}$ | Response of metabolism to decreasing temperature | Type of animal |
|---|---|---|---|---|---|
| Animal 1 | 115 ml $O_2$/hour | 55 ml $O_2$/hour | 0.48 (=55/115) | Metabolism increases | Endotherm |
| Animal 2 | 5.5 ml $O_2$/hour | 11.5 ml $O_2$/hour | 2.1 (=11.5/5.5) | Metabolism decreases | Ectotherm |

17.

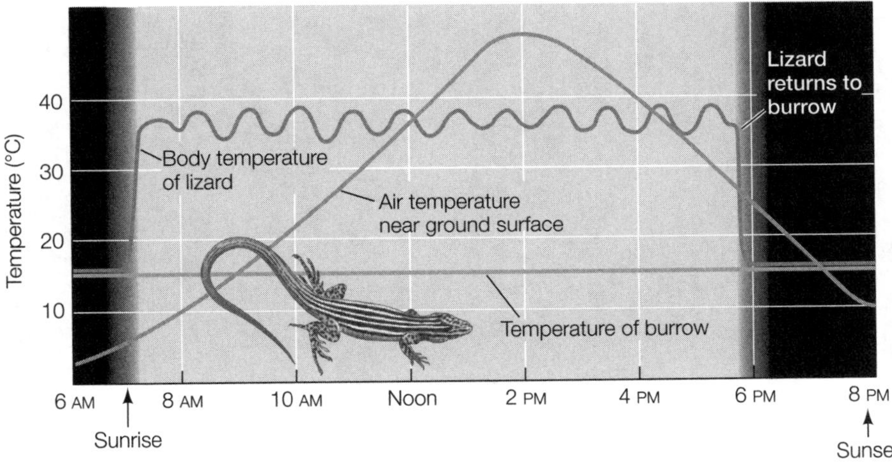


## *Test Yourself*

1. **e.** Interstitial fluid is found between the cells of the body.
2. **c.** The other answers detail changes at the molecular, tissue, and organ level that occur when environment changes. Answer c should seem familiar as the Lamarckian concept of evolution, which was ultimately discarded in favor of Darwin's theory of natural selection.
3. **b.** If the environment is cooler than the animal, the response would be to lose heat to the environment across the skin, which would mean increasing blood flow to peripheral vessels, not inhibiting it.
4. **a.** For a poikilothermic lizard in the desert to maintain the most stable body temperature, it should remain in a burrow during the night, where temperatures do not drop very much, and then shuttle between the sun and shade during the day.
5. **b.** Negative feedback loops use information about the state of a system to bring the internal environment back toward the set point values. Positive feedback loops amplify the response of a system, moving the system away from the set point.
6. **b.** After exercise, a dog will have a high metabolic rate and will pant in an attempt to cool down. Its body temperature will most likely be above the upper thermoneutral zone.

7. **a.** Evaporative cooling is a costly means to lower body temperature. ATP must be used during either sweating or panting. Water loss can also be high during evaporative cooling.
8. **e.** Food is a source of chemical building blocks and energy. It is also used in the regulation of body temperature.
9. **b.** Metabolic rate is positively correlated with an animal's activity; energy usage goes up as an animal becomes more active. Scaling relationships exist for energy usage among similar animals based on size. These scaling relationships mean that smaller animals require more food per individual than larger animals do, so that prey individuals typically eat more food compared to predator individuals.
10. **e.** A heart valve is a barrier that temporarily stops blood flow through the heart during its beating. Since there is blood on either side, it does not separate different fluid compartments.
11. **a.** Sugars, lipids (fatty acids), and proteins can all be used to produce ATP during aerobic respiration, but only sugars are used in anaerobic ATP production. Anaerobic ATP production makes use only of the glycolysis pathway, which uses glucose as its input.
12. **d.** Physiological processes acclimatize in response to a seasonal change in the environment. Acclimatization may involve changes in enzyme expression and a resetting of a physiological set point. Every organism has limits to its ability to acclimatize.
13. **d.** In this example, the effectors that are activated by the elevated calcium serve to bring the calcium levels back down to normal, thus reducing the difference between the actual level and the set point.
14. **e.** Brown fat, which is found in some mammals, burns fuel without producing ATP. This process of heat production is termed nonshivering thermogenesis. Brown fat is highly vascularized and rich in mitochondria.
15. **e.** The metabolic rate is a measure of all the energy used by an individual for all reasons, including intracellular reactions and physical activity.

# 30 Nutrition, Feeding, and Digestion

## The Big Picture

- All animals are heterotrophs—they can acquire energy and nutrients by eating plants, or animals, or both. Animals' digestive systems are specialized for efficient digestion of this food. Food passing through the gut is first broken down mechanically into small pieces and then broken down enzymatically into small molecules that can be absorbed across the gut. As nutrients and water are absorbed from ingested material, the undigested material forms the feces, which are eliminated. Numerous hormones control the movements of food through the gut and the secretion of enzymes and other chemicals involved in digestion. Hormones are also involved in storage of nutrients, breakdown of stored nutrients, and signaling of hunger and satiety.
- There are specific nutrient requirements for any organism. Carbohydrates, proteins, and lipids are broken down into their components and used as building blocks. Some of these components are essential when an organism cannot make them from other subunits. There are certain essential vitamins, minerals, amino acids, and several fatty acids required by humans.

## Study Strategies

- The study of digestive function can give you insights into your own health. The list of essential vitamins, minerals, amino acids, and fatty acids, for example, may show you that certain important elements are missing from your diet.
- The complexity of the digestive process may surprise you. Pay attention to the differences in how proteins, carbohydrates, and fats are digested, absorbed, and transported, and which parts of the digestive tract are involved in these processes.
- Though the guts of vertebrates are structurally complex, they can all be divided into three functional regions: a foregut, midgut, and hindgut. If you break down the structure of the gut into these regions and determine which physiological processes occur in each region, you will have divided the task into smaller pieces.
- Because different types of food are digested and absorbed by different mechanisms, learn how a particular type of food is digested and match the appropriate enzymes to that process. For example, it will be easier to remember how proteins are digested than to recount a long list of digestive enzymes and try to pick the one that might break down protein. In short, learn the process, not the list!
- Go to **LaunchPad** (or use the Web addresses listed) to review the following additional resources:

  Animated Tutorial 30.1 Parabiotic Mice Simulation (Pol2e.com/at30.1)

  Animated Tutorial 30.2 Insulin and Glucose Regulation (Pol2e.com/at30.2)

  Activity 30.1 Vitamins in the Human Diet (Pol2e.com/ac30.1)

  Activity 30.2 Mineral Elements Required by Animals (Pol2e.com/ac30.2)

  Activity 30.3 The Human Digestive System (Pol2e.com/ac30.3)

  Analyze the Data in textbook Figure 30.15

  Apply the Concept in textbook Section 30.1

## Key Concept Review

### 30.1 Food Provides Energy and Chemical Building Blocks

Food provides energy

Chemical energy from food is sometimes stored for future use

Food provides chemical building blocks

Some nutrients in foods are essential

Nutrient deficiencies result in diseases

Animals are heterotrophs, meaning that they must get their energy and nutrients from the lipids, carbohydrates, and proteins of plants and other animals. Energy is the ability to do work. The energy in food is in the form of chemical energy. The calorie (cal) and kilocalorie (kcal) are measures of heat energy released when food is burned completely. A single calorie is the amount of heat needed to raise the temperature of 1 gram of water by 1°C. (The "calories" we often count and that are reported on food labels are actually kilocalories.)

Metabolic rate is a measure of the total energy used by an animal. Basal metabolic rate is the resting metabolic rate or the energy consumption needed for all essential physiological functions. Energy needed for metabolism can be obtained from food taken in or from stored food. Lipids, carbohydrates, and proteins are the components of food that provide energy. Animals store energy in the form of lipids and glycogen. Lipids have far more energy per gram than glycogen, so lipids are a more compact way of storing energy. Glycogen, however, is broken down directly into glucose. Some tissues such as the brain need glucose almost exclusively to meet their energy needs, so glycogen stores, found in the liver and muscles, can serve as ready sources of energy for these tissues.

A nutrient that is required but cannot be synthesized by the body is called an essential nutrient. Essential nutrients must be supplied by the food an organism eats or by symbiotic microbes within the organism. The essential nutrients differ among different organisms.

Amino acids are the building blocks of proteins. Some amino acids can be synthesized, but the essential amino acids can be obtained only from outside sources. In adult humans, there are eight essential amino acids: isoleucine, leucine, lysine, methionine, phenylalanine, threonine, tryptophan, and valine. All eight essential amino acids can be found in foods such as meat, eggs, and milk. Combinations of grains and legumes, such as corn and beans, also provide all eight essential amino acids (see Figure 30.3). Human infants require an additional amino acid, histidine. The essential fatty acids for humans are alpha-linolenic acid, an omega-3 fatty acid, and linoleic acid, an omega-6 fatty acid.

Vitamins are carbon compounds that are required in small amounts and often function as coenzymes. Lipid-soluble vitamins can accumulate in the liver, whereas excess water-soluble vitamins are eliminated in urine. Humans require a number of vitamins (see Table 30.1). Vitamin D is needed to help in the uptake of calcium into the body. Your body synthesizes vitamin D when your skin is exposed to sunlight. This is the major source of vitamin D, though there are a few food sources as well. Diseases that can result from deficiencies in vitamins are numerous (see Table 30.1).

An essential mineral is a chemical element required in the diet; examples include calcium and iron (see Table 30.2). Calcium is needed in large amounts for bones, teeth, proper nerve function, and muscle contraction. Iron is needed in small amounts and is an important component of hemoglobin, myoglobin, and certain enzymes in the respiratory chain.

Malnutrition is caused by the lack of any essential nutrient in the diet. Chronic malnutrition leads to deficiency diseases. For example, a chronic lack of vitamin C in the diet leads to scurvy, which initially causes bleeding gums but eventually is fatal. Citrus fruits are a rich source of vitamin C, and in the eighteenth century the simple act of stocking sailing ships with limes helped put an end to the high incidence of scurvy among sailors.

**Question 1.** From a nutritional standpoint, what is the best way to cook a vegetable containing folic acid—boiling it in water or steaming it?

**Question 2.** If a person suffers from anemia (low number of red blood cells or low levels of hemoglobin) due to malnourishment, what are some of the likely vitamins and minerals that may be missing from the diet? If the person is a vegan, and therefore does not eat meat, eggs, or dairy, which vitamin deficiency is the most likely cause of the anemia?

## 30.2 Animals Get Food in Three Major Ways

Some animals feed by targeting easily visible, individual food items

Suspension feeders collect tiny food particles in great numbers

Many animals live symbiotically with microbes of nutritional importance

There are three major ways animals can acquire food. Many target easily visible food, some filter their food out of the surrounding environment, and some rely on microbes to obtain necessary nutrients.

Animals that feed by targeting choose their specific food items. Examples include animals that graze, such as cattle, birds that fish, or crabs that grab their prey.

A number of animals are suspension feeders. They filter suspended material from their environment using a sieve-like structure in their bodies. Baleen whales, for example, have no teeth and sift out small organisms from seawater using plates of material called baleen. Many aquatic invertebrates, including mussels and clams, filter small particles from the water.

Many animals live symbiotically with microbes and gain nutrients from the microbes. The corals that build coral reefs contain photosynthetic algae in their tissues that provide the corals with glucose and other food molecules. If the corals are stressed, they lose their symbiotic algae and die if they cannot regain them.

Ruminants, such as cattle, sheep, and deer, are herbivorous mammals that rely on large populations of microbes for nutrition. These animals eat mostly grasses, but they are unable to use their own enzymes to digest the cellulose in the plants. Instead, a large community of microbes in two chambers (the rumen and reticulum) of their four-chambered stomach breaks down the cellulose for the animal (see Figure 30.7). The contents of the rumen are periodically regurgitated into the mouth for rechewing. From the rumen, food passes into the omasum, where water absorption occurs, and then into the abomasum, or true stomach. In the abomasum, hydrochloric acid and proteases kill the microorganisms and partly digest them; digestion then continues in the small intestine. Thus, ruminants acquire glucose from the breakdown of cellulose and they acquire protein from digesting the microorganisms. The microbes also supply

B vitamins and absorb nitrogen wastes, used to make microbial protein. The animal is then able to harvest this protein when it digests the microbes.

Like most animals, humans have populations of microbes in their guts, and these play important roles in our nutrition and health.

**Question 3.** If you were snorkeling over a coral reef and saw huge swaths of coral that looked white, what would you be able to conclude?

**Question 4.** In the United States, beef cattle are often fed corn and other grains to fatten them. Do you think these cattle can remain healthy on a diet of grains containing negligible amounts of cellulose? Explain the reasoning behind your answer.

## 30.3 The Digestive System Plays a Key Role in Determining the Nutritional Value of Foods

- Digestive abilities determine which foods have nutritional value
- Animals are diverse in the foods they can digest
- Digestive abilities sometimes evolve rapidly
- Digestive abilities are phenotypically plastic

Animals typically have a tubular digestive tract lined by an epithelium. Nutrients from the food digested within the lumen of the gut pass through the epithelial lining and are absorbed by capillaries in close proximity to the epithelium. The digestive tract therefore has two major functions: digestion and absorption. Digestion breaks down the food within the lumen of the gut into smaller molecules using enzymes; absorption transports the molecules from the lumen of the gut into the blood.

The nutritional value of a food is based on an animal's ability to digest and absorb it. Humans, for example, are unable to digest cellulose, so it holds no nutritional value for us.

The food an animal can digest depends upon the enzymes it can secrete or those provided by microbes in its gut. Enzymes are specific for the chemical bonds they break. For example, carboxypeptidase B, chymotrypsin, and trypsin are enzymes that are specific for different types of bonds between amino acids in a protein. An enzyme that breaks down trehalose, a sugar found in insects, is found in insectivorous mammals. Mammals that do not produce this enzyme cannot get the same nutrition from eating insects as an insectivorous mammal can.

Evolutionary changes that alter what an organism can digest can occur relatively rapidly. In humans, for example, the ability to digest lactose, a milk sugar, is present in infants but often disappears in adults because they lose the ability to produce lactase, the enzyme that breaks down lactose. Several populations of humans, however, have independently evolved the ability to retain lactase into adulthood, and the mutations that resulted in this change appeared as recently as 5,000 years ago.

Individual animals adjust their digestion based on their diet, and these adjustments can occur within a short time scale. Rats fed on a low-protein diet that are switched to a high-protein diet, for example, increase their production of protein-digesting enzymes within 24 hours. If the rats are switched from a high-protein to a low-protein diet, their production of protein-digesting enzymes diminishes. Absorption rates can be similarly affected. A person who starts eating a lot of sugar, for example, increases his or her efficiency in absorbing the products of sugar digestion.

**Question 5.** Can you use the number of calories in a specific food as a measure of its nutritional value?

**Question 6.** If a person who normally eats high amounts of sugary foods and a person who normally eats little sugar both drink a liter of soda containing 120 g of sugar, will they absorb the same amount of monosaccharides from the soda?

## 30.4 The Vertebrate Digestive System Is a Tubular Gut with Accessory Glands

- Several classes of digestive enzymes take part in digestion
- Processing of food starts in the foregut
- Food processing continues in the midgut and hindgut
- The midgut is the principal site of digestion and absorption
- The hindgut reabsorbs water and salts

The vertebrate gut is a tube from the mouth to the anus. Its organization of tissue layers is similar throughout its length (see Figures 30.11 and 30.12). Its innermost layer is an epithelium supported by an underlying connective tissue. Together they make up the mucosa. Outside of this is the submucosa, a layer of tissue containing blood vessels, lymph vessels, and a network of neurons that is part of the enteric nervous system. (The enteric nervous system is part of the autonomic nervous system.) Two layers of smooth muscle lie outside of the submucosa: an inner circular layer and an outer longitudinal layer. Between these muscle layers is a second network of neurons which coordinates peristalsis to push food within the gut. This network is also part of the enteric nervous system. Sphincter muscles control the entry of food into the stomach from the esophagus and the exit of stomach contents into the small intestine. The gut lumen is anoxic, meaning it does not contain $O_2$. A parasite entering the gut, therefore, can survive only if it can live in an anoxic environment.

Hydrolytic enzymes found in the gut are used to break down proteins, carbohydrates, lipids, and nucleic acids into their component monomers. During hydrolysis, water is added to break a chemical bond. Digestive enzymes are classified according to the substances they hydrolyze: carbohydrases hydrolyze carbohydrates; proteases hydrolyze proteins; peptidases hydrolyze peptides; lipases hydrolyze lipids; and nucleases hydrolyze nucleic acids.

The vertebrate gut can be divided into three sections: the foregut, consisting of the mouth, pharynx, esophagus, and stomach; the midgut, consisting of the small intestines;

and the hindgut, consisting of the large intestines and ending at the anus. In the foregut, food is physically broken up into smaller pieces in preparation for digestion by enzymes. Food enters the mouth, and in some animals it is broken up by grinding mechanisms that can include teeth and jaw muscles. In mammals, saliva in the mouth containing the enzyme amylase starts breaking down carbohydrates. The food then moves through the pharynx and down the esophagus to the stomach; in some animals this is preceded by a storage chamber called a crop or a grinding chamber called a gizzard. The stomach serves as a storage chamber and a mixing chamber. It also secretes HCl, which lowers the pH of the stomach contents and helps to kill microbes. The HCl also activates pepsinogen into its active form, pepsin, which is an enzyme that breaks down proteins. Pepsinogen and HCl are secreted into the stomach lumen by stomach epithelial cells.

Stomach contents enter the midgut (small intestine) through the pyloric sphincter, which allows only small amounts of material to pass through at a time. This process is controlled by hormones and the enteric nervous system. The midgut is the primary site of digestion and absorption. In many vertebrates, the parts of the midgut that absorb nutrients have increased surface area through folds as well as fingerlike projections known as villi, which in turn have microscopic projections called microvilli (see Figure 30.13).

In the small intestine, carbohydrate and protein digestion continue, the digestion of lipids begins, and nutrients are absorbed. Lipid digestion begins with the help of secretions and enzymes from the liver and pancreas. The liver secretes bile through a side branch of the hepatic duct into the gallbladder, where the bile is stored. When lipids are present, bile is secreted by the gallbladder into the small intestine through the common bile duct (see Figure 30.11). Bile molecules increase the exposed surface area of the lipids for digestion by fat-digesting enzymes, the lipases.

The pancreas secretes enzymes that break down proteins, carbohydrates, and lipids. These are brought to the small intestine through the pancreatic duct. Protein-digesting enzymes secreted by the pancreas are released as proenzymes and are converted into their active form in the midgut. The pancreas also helps maintain a slightly alkaline pH in the small intestine by secreting bicarbonate ions. This neutralizes the acidic contents coming from the stomach.

Final digestion of proteins, carbohydrates, and lipids occurs in the small intestine. Proteins are broken down into amino acids, carbohydrates are broken down into monosaccharides, and lipids are broken down into fatty acids and glycerol. These small molecules are then absorbed.

Material entering the large intestine (colon) has had most of the nutrients removed, but still contains important ions and water. The water and ions are absorbed in the large intestine, leaving behind semisolid feces, which are stored in the final region of the large intestine, the rectum, until they are eliminated. Diseases such as cholera that prevent reabsorption of water and ions produce severe diarrhea that can lead to severe dehydration and death.

Many regions of the gut house symbiotic bacteria, which supply important nutrients to the animal.

**Question 7.** Create a flow chart of the following organs, indicating the order in which they occur from the start to the end of the human gut: anus, esophagus, large intestine, mouth, small intestine, and stomach. Indicate in which of these organs enzymatic digestion occurs, and the particular nutrient (e.g., lipids, protein) being broken down. Also indicate in which of the organs absorption occurs, and the particular nutrient or substance absorbed.

**Question 8.** In ruminants, such as cows and bison, microorganisms that break down cellulose are found in two chambers of the greatly enlarged stomach. In other herbivores, including rabbits, the microorganisms that break down cellulose are found in the cecum, a chamber off the large intestine. In which type of herbivore is the absorption of nutrients from cellulose more efficient?

## 30.5 The Processing of Meals Is Regulated

### Hormones help regulate appetite and the processing of a meal

### Insulin and glucagon regulate processing of absorbed food materials from meal to meal

The absorptive state is the period after a meal, when food is in the gut. The postabsorptive state is the period when the gut is empty and the body uses its energy reserves.

Hormones control many digestive functions (see Figure 30.14). When food arrives in the stomach, cells in its wall secrete gastrin, which stimulates the stomach to secrete HCl and pepsinogen. When the acidic contents of the stomach arrive in the small intestine, the small intestine releases secretin, which stimulates the pancreas to release bicarbonate ions into the small intestine to neutralize the pH of the contents. Lipids and proteins stimulate the mucosa of the small intestine to secrete cholecystokinin, which stimulates the release of bile and pancreatic digestive enzymes. Both secretin and cholecystokinin act to inhibit HCl secretion and muscle contraction in the stomach, which helps to control the timing of contents being released from the stomach into the small intestine.

Ghrelin, a hormone produced by the stomach, is secreted when the stomach is empty and stimulates appetite. Leptin, a hormone produced by fat cells, provides the brain with information about the body's fat reserves. High leptin levels indicate that fat reserves are plentiful and there is no need to eat, so appetite is suppressed.

Insulin and glucagon are endocrine hormones produced by the pancreas that control metabolic fuel use (see Figure 30.16). In response to high blood glucose levels, the pancreas releases insulin. Insulin facilitates the diffusion of glucose into cells, where it can be used for energy and the synthesis of fuel storage molecules—glycogen and lipids. In the postabsorptive state, if blood glucose levels fall below normal, the pancreas releases glucagon, which stimulates the liver to break down glycogen into glucose and release it into the blood.

**Question 9.** The mice shown in the diagram below are a parabiotic pair, meaning their circulatory systems have been surgically joined. One is a normal wild type mouse (WT). The other has two recessive *ob* alleles (*ob*/*ob*), causing loss of function of the *Ob* gene, which codes for leptin. Explain why both the wild-type and *ob*/*ob* mice are of normal weight.

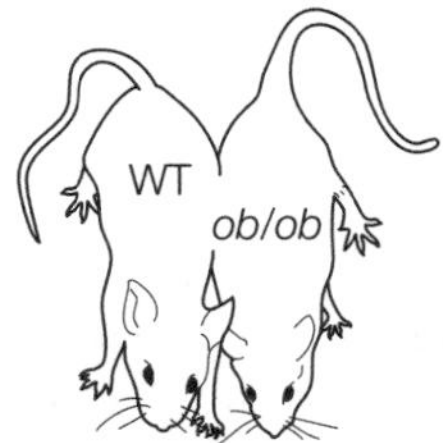


**Question 10.** The diagram below illustrates how glucose concentration in the blood is regulated. Fill in the missing words in the diagram in each of the numbered boxes.

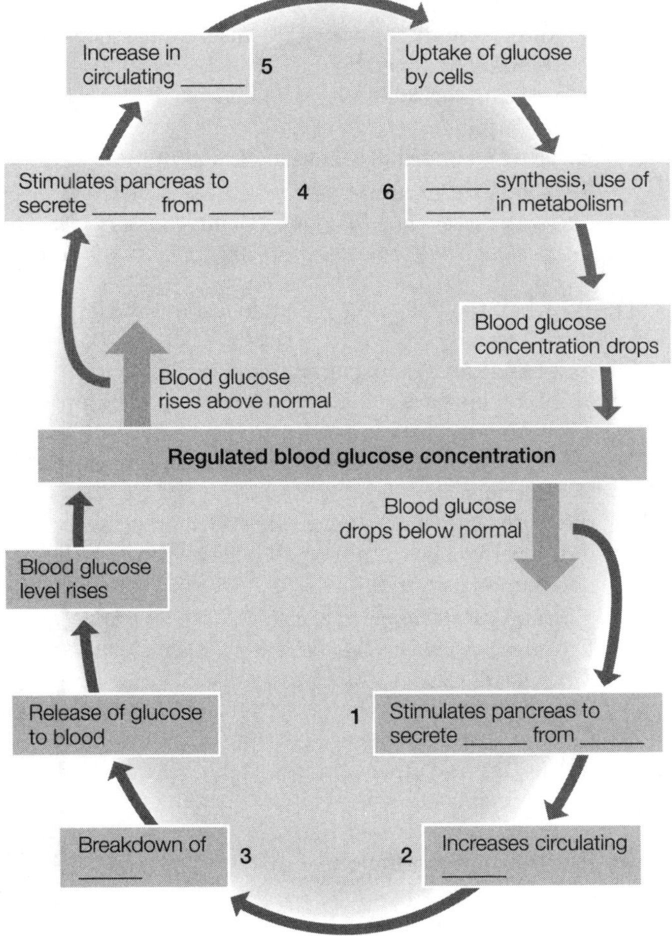


## Test Yourself

1. Certain amino acids are essential to the diet of animals because they
   a. prevent overnourishment.
   b. are cofactors and coenzymes that are required for normal physiological function.
   c. cannot be synthesized directly.
   d. are needed to make stored lipids that are used during hibernation and migration.
   e. are the most important source of stored energy.

   ***Textbook Reference:*** *Concept 30.1 Food Provides Energy and Chemical Building Blocks; Some nutrients in foods are essential*

2. In the figure below, photo(s) _______ show(s) suspension feeders; photo(s) _______ show(s) target feeders.

1. 2.

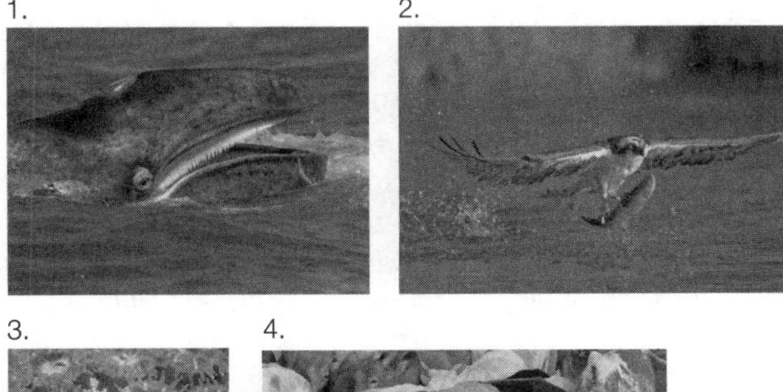

3. 4.

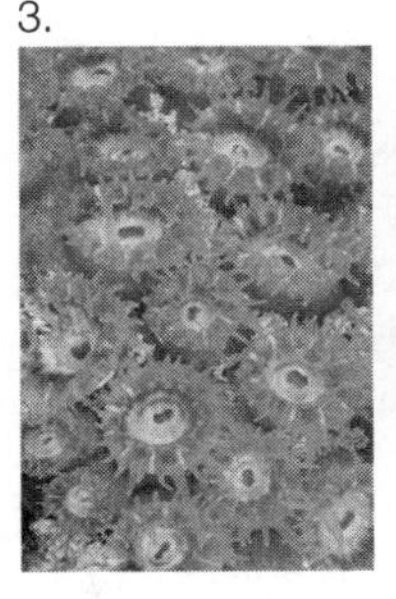

   a. 4; 1 and 2
   b. 1 and 4; 2 and 3
   c. 4; 1, 2 and 3
   d. 1; 2, 3 and 4
   e. 1 and 3; 2 and 4

   ***Textbook Reference:*** *Concept 30.2 Animals Get Food in Three Major Ways; Suspension feeders collect tiny food particles in great numbers*

3. If you switch to a low-carbohydrate/high-protein diet by cutting out starchy foods and adding large quantities of meat to your diet, what change can you expect in your body?
   a. Increased production of enzymes that break down meat proteins.
   b. Loss of the ability to digest starch.
   c. Genes coding for protein enzymes will mutate.
   d. Evolution of the ability to digest meat.
   e. Genetic changes such that your offspring will not be able to digest starch.

   ***Textbook Reference:*** *Concept 30.3 The Digestive System Plays a Key Role in Determining the Nutritional Value of Foods; Digestive abilities are phenotypically plastic*

4. The gallbladder
   a. produces bile.
   b. is part of the liver.
   c. stores bile produced by the liver.
   d. produces cholecystokinin.

e. is vestigial in humans.
***Textbook Reference:*** *Concept 30.4 The Vertebrate Digestive System Is a Tubular Gut with Accessory Glands; The midgut is the principal site of digestion and absorption*

5. The pancreas
   a. is exclusively an endocrine gland responsible for the production and release of insulin and glucagon.
   b. is exclusively an endocrine gland that produces salivary amylase.
   c. contains villi to increase surface area of the small intestine.
   d. produces bile for digestion of lipids.
   e. produces exocrine products involved in digestion of products from the stomach.
   ***Textbook Reference:*** *Concept 30.4 The Vertebrate Digestive System Is a Tubular Gut with Accessory Glands; The midgut is the principal site of digestion and absorption*

6. Hydrochloric acid
   a. is secreted by the gastric glands of the liver.
   b. is secreted by epithelial cells of the stomach.
   c. increases the pH of the contents in the small intestine.
   d. promotes the growth of microorganisms in the stomach.
   e. is secreted by salivary glands.
   ***Textbook Reference:*** *Concept 30.4 The Vertebrate Digestive System Is a Tubular Gut with Accessory Glands; Processing of food starts in the foregut*

7. Bile produced in the liver is associated with which of the following?
   a. Emulsification of lipids into tiny droplets in the small intestine
   b. Digestive action of pancreatic amylase
   c. Emulsification of lipids into tiny droplets in the stomach
   d. Digestion of proteins into amino acids
   e. Emulsification of lipids into tiny droplets in the large intestine
   ***Textbook Reference:*** *Concept 30.4 The Vertebrate Digestive System Is a Tubular Gut with Accessory Glands; The midgut is the principal site of digestion and absorption*

8. Most of the enzymatic digestion of food in humans is completed in the
   a. mouth.
   b. large intestine.
   c. pancreas.
   d. stomach.
   e. small intestine.
   ***Textbook Reference:*** *Concept 30.4 The Vertebrate Digestive System Is a Tubular Gut with Accessory Glands; The midgut is the principal site of digestion and absorption*

9. Which of the following does *not* contribute to the large surface area available for nutrient absorption in the small intestines?
   a. Villi
   b. Intestinal length
   c. Microvilli
   d. Bile duct
   e. Surface folds
   ***Textbook Reference:*** *Concept 30.4 The Vertebrate Digestive System Is a Tubular Gut with Accessory Glands; The midgut is the principal site of digestion and absorption*

10. Cystic fibrosis causes the production of unusually thick mucus, which sometimes blocks the pancreatic duct. As a result, individuals with cystic fibrosis commonly
    a. experience heartburn caused by food backing up into the esophagus.
    b. suffer from disruption of the mechanical digestion of food in the stomach.
    c. are malnourished.
    d. are obese.
    e. have an abnormal secretion of insulin.
    ***Textbook Reference:*** *Concept 30.4 The Vertebrate Digestive System Is a Tubular Gut with Accessory Glands; The midgut is the principal site of digestion and absorption*

11. Which of the following statements about the large intestine is true?
    a. It has almost no bacterial populations.
    b. Its contents are at a low pH due to high amounts of HCl.
    c. It absorbs much of the water remaining in waste materials.
    d. It is the site of most of the digestive processes.
    e. It receives bile from the gall bladder through the common bile duct.
    ***Textbook Reference:*** *Concept 30.4 The Vertebrate Digestive System Is a Tubular Gut with Accessory Glands; The hindgut reabsorbs water and salts*

12. Certain laxatives increase peristalsis in the large intestine, leaving other portions of the gut unaffected. Use of such laxatives
    a. results in loss of body fat.
    b. results in dehydration.
    c. results in constipation.
    d. increases the absorption of water.
    e. increases the absorption of ions.
    ***Textbook Reference:*** *Concept 30.4 The Vertebrate Digestive System Is a Tubular Gut with Accessory Glands; The hindgut reabsorbs water and salts*

13. If you fast for a day, which of the following hormones is most likely to be secreted in response to your fasting?
    a. Glucagon
    b. Insulin

c. Leptin
d. Secretin
e. Gastrin

***Textbook Reference:*** *Concept 30.5 The Processing of Meals Is Regulated; Insulin and glucagon regulate processing of absorbed food materials from meal to meal*

14. Taking mega doses of vitamin A is potentially more of a health risk than taking mega doses of niacin because
    a. vitamin A is water-soluble, so excess amounts are reabsorbed in the kidneys.
    b. vitamin A is lipid-soluble, so excess amounts are absorbed in the stomach.
    c. niacin is lipid-soluble, so excess amounts are emulsified by bile in the small intestines.
    d. niacin is water-soluble, so excess amounts can be eliminated in the urine.
    e. vitamin A is much larger than niacin, so vitamin A remains in the stomach for longer periods.

    ***Textbook Reference:*** *Concept 30.1 Food Provides Energy and Chemical Building Blocks; Some nutrients in foods are essential*

15. If you develop gallstones and your condition is serious enough that you have your gallbladder removed, what advice is your physician likely to give you about your diet?
    a. Cut out all fats, because you are no longer able to secrete bile.
    b. Eat fats in moderation, because your intestines will not receive large amounts of bile at any one time.
    c. Eat a low-carbohydrate diet, because your intestines will now be receiving lower amounts of carbohydrases.
    d. Cut out all high-protein foods, because your stomach will be receiving less HCl.
    e. Eat a low-protein diet, because your pancreas will have a reduced ability to secrete proteases.

    ***Textbook Reference:*** *Concept 30.1 Food Provides Energy and Chemical Building Blocks; Some nutrients in foods are essential*

16. The diagram below is of the (part of the digestive system) _______; from the diagram it can be concluded that if you take antacids, then your ability to _______.

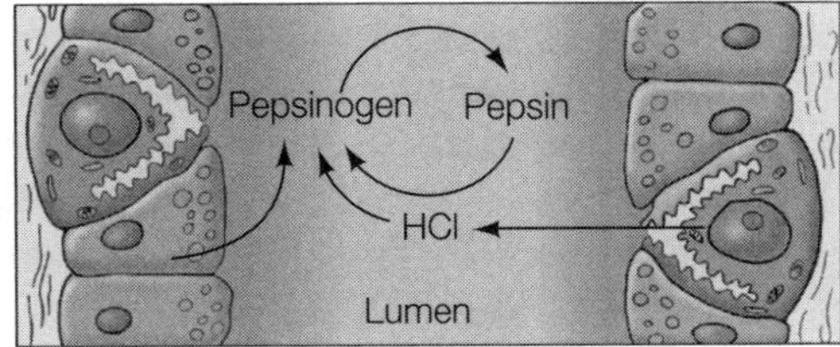


    a. mouth; begin the digestion of starch will decrease
    b. small intestine; convert pepsinogen into pepsin will increase
    c. small intestine; convert pepsin into pepsinogen will decrease
    d. stomach; digest protein will increase
    e. stomach; digest protein will decrease

    ***Textbook Reference:*** *Concept 30.4 The Vertebrate Digestive System Is a Tubular Gut with Accessory Glands; Processing of food starts in the foregut*

## Answers

### *Key Concept Review*

1. Folic acid is a water-soluble vitamin found in vegetables. When vegetables are boiled, the folic acid is lost. Therefore, the best cooking method to retain folic acid in the vegetables is steaming.
2. Deficiencies in any of the following vitamins or minerals can lead to anemia: vitamins $B_6$, $B_{12}$, folic acid, vitamin E, iron, and copper. The vitamin deficiency most likely to be causing anemia in a person who is a vegan would be a vitamin $B_{12}$ deficiency.
3. You would be able to conclude that the coral are stressed because they have lost their symbiotic algae (a process called bleaching). These coral are deteriorating and will die if they are unable to obtain new symbiotic algae.
4. Beef cattle fed grains instead of grass cannot remain healthy, because their digestive tracts are designed for digesting grass, not grain. Beef cattle are ruminants and rely on large communities of microbes in their rumen for many of their nutritional needs. These microbes digest the cellulose in the grass that is eaten and provide the animal with B vitamins and short-chain fatty acids such as acetic acid; the microbes also take up waste nitrogen from the animal, converting it into microbial protein, which the animal then harvests for protein. If the cattle are switched from a diet of grass to a diet of grains, the microbes cannot perform these functions for the animal.
5. The nutritional value of food depends on what an organism can digest and absorb from that food, not on the number of calories it contains. For example, humans have no way of digesting cellulose, so the nutritional value of grass for humans is negligible, regardless of the caloric content of the grass.
6. No. The person who normally eats high amounts of sugar will absorb more monosaccharides from the soda than the person who normally eats little sugar. Digestive and absorptive processes are able to adapt to different types of diet. The absorption mechanisms for taking up the products of sugar digestion become more effective in a person who eats lots of sugar than a person who eats little sugar.

7.

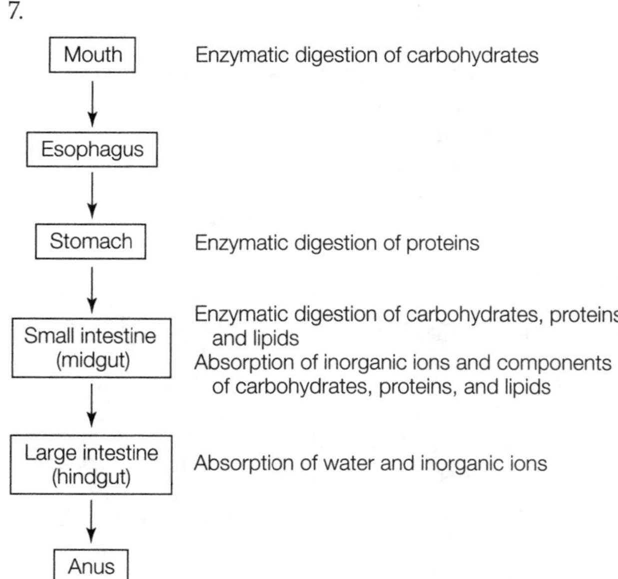


8. In vertebrates, most nutrients are absorbed in the small intestine. Because ruminants break down cellulose in chambers of their stomach (which comes before the small intestine in position along the gut), the nutrients from cellulose are available for absorption in the small intestine, and absorption is relatively efficient. In nonruminant herbivores, however, cellulose is broken down in the cecum, after it has passed through the small intestine, so nutrient absorption is less efficient. (Note of interest: Some nonruminant species produce two types of feces and eat the one containing cecal material to gain greater access to nutrients.)

9. Leptin is a hormone produced by fat cells; a high level of leptin signals the brain to suppress appetite. The WT mouse can produce leptin, but the *ob/ob* mouse cannot. If the *ob/ob* mouse were not attached to the circulatory system of the WT mouse, it would be obese due to the lack of leptin. But because the *ob/ob* mouse is receiving leptin through the circulatory system of the WT mouse, its brain responds to that leptin and suppresses the animal's appetite when leptin levels rise. Therefore, the animal does not overeat.

10. The blanks in the numbered boxes on the diagram should be filled in as follows:
    1: glucagon; alpha cells
    2: glucagon
    3: glycogen
    4: insulin: beta cells
    5: insulin
    6: glycogen; glucose

### *Test Yourself*

1. **c.** Essential amino acids must be acquired through the diet because animals cannot directly synthesize all the amino acids needed for protein production.

2. **e.** The whale shown in photograph 1 is a baleen whale and is a suspension feeder, filtering out small organisms from the seawater. The corals shown in photograph 4 are also suspension feeders. The bird in photograph 2 is a target feeder, as are the cattle in photograph 3. They feed by targeting easily visible food items.

3. **a.** By adding much more of meat to your diet, the production of enzymes for breaking down meat proteins will increase. You will not lose your ability to digest starch, but you will become less effective at absorbing the products of starch digestion. Your gut is able to adapt to changes in your diet by producing more or less enzymes as needed.

4. **c.** Bile is produced by the liver, stored in the gallbladder, and released into the small intestine to aid in lipid digestion.

5. **e.** The pancreas has an endocrine function, producing and releasing insulin and glucagon, but it is not exclusively an endocrine gland. It also produces exocrine products such as lipases, nucleases, amylases, and trypsin, all of which are involved in digestion of products from the stomach.

6. **b.** Hydrochloric acid is a strong acid secreted by epithelial cells in the lining of the stomach. It lowers the pH of the stomach fluid.

7. **a.** Bile aids in the digestion of lipids in the small intestine by acting as a detergent, emulsifying fat into tiny lipid droplets.

8. **e.** The small intestine is the main site for enzymatic digestion in humans.

9. **d.** The sheer length, surface folds, and the presence of microvilli and villi increase the surface area of the small intestine.

10. **c.** People with cystic fibrosis often suffer from malnourishment because pancreatic enzymes cannot reach the small intestine to promote the breakdown of complex food molecules. As a result, nutrients are not absorbed in adequate amounts.

11. **c.** The large intestine is the site of water and ion absorption. The large populations of bacteria found in the large intestine contribute useful vitamins to their hosts.

12. **b.** Laxatives that enhance peristalsis in the large intestine cause rapid passage of material through the large intestine. This interferes with the reabsorption of water and ions, a major function of the large intestine, and can lead to dehydration.

13. **a.** When a person is fasting, blood glucose levels will drop. This will stimulate the pancreas to secrete glucagon, which will cause the breakdown of glycogen and the release of glucose into the blood to keep blood sugar levels within a normal range.

14. **d.** Niacin is a water-soluble vitamin, so excesses are eliminated in the urine. Vitamin A, however, is lipid-soluble and when taken in excess, it can accumulate in body fat and reach toxic levels in the liver.

15. **b.** If your gallbladder is removed, you will not be able to store bile and deliver it to the intestines in large amounts when needed. Instead, bile will be secreted by the liver and carried in small amounts directly to the small intestine. Bile helps in the digestion of fats by preventing the formation of large lipid droplets and promoting tiny droplets. This allows lipases to break down lipids. If the gallbladder is removed, therefore, fats can still be digested, but they should be eaten in moderation because large amounts of bile will not be available after a single meal to help digest the fats.

16. **e.** HCl is secreted by epithelial cells in the stomach. HCl turns pepsinogen into its active form, pepsin. Pepsin digests protein. Antacids lower the levels of HCl in the stomach, which will lower the conversion of pepsinogen to pepsin. This, in turn, decreases a person's ability to digest protein. (Note of interest: The lower amounts of HCl in the stomach can also allow microbes to survive that otherwise would have been killed in the extremely low pH of the stomach.)

# 31 Breathing

## The Big Picture

- Cells exchange gases with their environment—they pick up $O_2$ for creating large amounts of ATP from the breakdown of food subunits in their mitochondria and unload $CO_2$ as a waste product of this metabolism. Animals have evolved diverse structures and mechanisms for exchanging these gases with the environment. Tracheal systems in insects, gills in fishes, and lungs in terrestrial vertebrates are three examples of gas exchange organs. Short diffusion distances and large surface areas are adaptations designed to maximize gas exchange and characterize all gas exchange organs.
- If distances between gas exchange organs and systemic cells are great, bulk flow methods that allow gases to travel quickly are needed. In the human body, the blood vascular system serves this purpose. The blood picks up waste $CO_2$ from systemic tissues and quickly brings it to the lungs, where the $CO_2$ is unloaded and $O_2$ is picked up. The blood then carries $O_2$ back to the systemic tissues.

## Study Strategies

- Refer to Fick's law of diffusion as a guideline for understanding gas exchange. The various components of the law—distance for diffusion, difference in the partial pressures of the gas at two locations, surface area over which gas exchange occurs—provide a good framework for understanding why gas exchange organs have evolved with certain characteristics in common. For example, gas exchange organs tend to have large surface areas and very short diffusion distances, with a large difference in the partial pressure for $O_2$ across their surfaces. All of this makes sense in the context of Fick's law of diffusion. Fick's law of diffusion also governs the movement of $O_2$ and $CO_2$ in the body, and it can help you remember where and why these gases are picked up and released in the body.
- The structure of the avian and fish respiratory systems is complex and very different from the mammalian pattern. Study Figures 31.7, 31.8, and 31.9. Follow the countercurrent exchange of gases in fish and be sure to remember that avian air sacs are not sites of gas exchange.
- When you breathe, you use a number of muscles to bring air into and out of your lungs. Focus on the muscles and ribs of your thorax as you inhale and exhale; put your hands on your chest to determine how your thoracic cavity changes as you breathe. This will help you to more clearly understand which elements are used in a breathing cycle.
- Your own lungs contain more air than you can breathe in or out. Study Figure 31.12—there is always residual air left in the lungs. This residual air is important for helping to keep the alveoli of the lungs from collapsing.
- Although you might expect that the $O_2$ content of the blood would regulate respiratory rate in mammals, including humans, it turns out that respiratory rate is primarily regulated by the $CO_2$ content of the blood.
- Go to **LaunchPad** (or use the Web addresses listed) to review the following additional resources:

  Animated Tutorial 31.1 Airflow in Birds (PoL2e.com/at31.1)

  Animated Tutorial 31.2 Airflow in Mammals (PoL2e.com/at31.2)

  Activity 31.1 The Human Respiratory System (PoL2e.com/ac31.1)

  Activity 31.2 Concept Matching (PoL2e.com/ac31.2)

  Analyze the Data in textbook Figure 31.15

  Apply the Concept in textbook Section 31.1

## Key Concept Review

### 31.1 Respiratory Gas Exchange Depends on Diffusion and Bulk Flow

The diffusion of gases depends on their partial pressures

Diffusion can be very effective but only over short distances

Gas transport in animals often occurs by alternating diffusion and bulk flow

### Breathing is the transport of $O_2$ and $CO_2$ between the outside environment and gas exchange membranes

### Air and water are very different respiratory environments

The respiratory gas oxygen ($O_2$) is required by mitochondria to produce energy in the form of ATP. The respiratory gas carbon dioxide ($CO_2$) is one of the waste by-products of this ATP production and must be eliminated from an animal's body. The transfer of these gases occurs by simple diffusion in the respiratory systems of animals.

Partial pressures are used to express the concentrations of gases in a mixture. The difference between the partial pressures of $O_2$ and $CO_2$ between the two sides of a respiratory surface drives the movement of $O_2$ into the body and $CO_2$ out of the body.

The tendency for a gas to move by diffusion across gills, body surfaces, or lungs depends on its partial pressure. The sum of all the gases' partial pressures in a gas mixture is the total partial pressure, which in the atmosphere equals the atmospheric pressure. $O_2$ makes up about 21 percent of the atmospheric pressure. As elevation increases and atmospheric pressure decreases, the total amount of $O_2$ in air decreases.

Rate of diffusion, described by Fick's law of diffusion, depends on the cross-sectional area over which the gas is diffusing and the difference in the partial pressure of the gas in the location the gas is diffusing from and the location it is diffusing to. It also depends in part on a diffusion coefficient that varies according to temperature, the diffusion medium, and the particular gas that is diffusing.

When distances between the respiratory exchange surface and the systemic cells are too large for diffusion to be an effective method of bringing $O_2$ to the tissues and removing $CO_2$, bulk flow methods are used to move the respiratory gases in a medium such as blood. If you were to rely on diffusion to get $O_2$ from your lungs to cells in your lower leg, for example, it would take over 30 years for enough $O_2$ to get to these cells.

A useful rule of thumb, calculated by August Krogh, is that in an animal, tissue can only be about 0.5 mm thick for $O_2$ to diffuse through it at a rate that adequately supports cellular respiration. This means that when blood delivers $O_2$ to your cells, the thickness between the red blood cells carrying the $O_2$ and the cells that will be using it can only be about 0.5 mm. This suggests why capillary walls are very thin and why capillaries are so numerous that they come within close proximity to every cell in the human body.

Breathing is defined as the process by which $O_2$ and $CO_2$ are exchanged between the environment and the respiratory surfaces. Breathing organs such as lungs and gills have evolved in air-breathing and water-breathing animals for bulk flow of air or water to and from the respiratory surfaces (ventilation). Lungs usually involve tidal ventilation, in which airflow moves first in one direction and then in the opposite direction. Ventilation in gills, by contrast, is unidirectional, since water is pumped in a one-way stream over the gills.

Some aquatic amphibians have external gills for exchange of respiratory gases with water (see Figure 31.4A). Internal gills, such as those in crayfish or fishes, have a large surface area and are protected from the environment within a cavity inside the animal's body (see Figure 31.4A). Many vertebrates have lungs, and air-breathing insects have a system of tracheae (see Figure 31.4B).

There are major differences in the $O_2$ carrying capacity of air and water. Water contains far less $O_2$ compared to air, and $O_2$ also diffuses far more slowly in water than in air. In water-breathing animals, temperature affects respiration because there is less $O_2$ in warm water than in cold, so in air-breathing animals such as fish, the need for $O_2$ increases as the temperature of the water increases.

**Question 1.** The equation for Fick's law of diffusion appears below. Based on this equation, what modifications in a breathing system would increase the diffusion rate of gases at a respiratory surface?

$$\text{Rate of diffusion between two locations} = DA\,\frac{P_1 - P_2}{L}$$

**Question 2.** Why would a tropical fish in a fish tank face severe respiratory problems (and possibly death) if the heater in the tank malfunctioned and caused extremely high water temperatures?

## 31.2 Animals Have Evolved Diverse Types of Breathing Organs

### Specialized breathing organs have large surface areas of thin membranes

### The directions of ventilation and perfusion can greatly affect the efficiency of gas exchange

### Many aquatic animals with gills use countercurrent exchange

### Most terrestrial vertebrates have tidally ventilated lungs

### Birds have rigid lungs ventilated unidirectionally by air sacs

### Insects have airways throughout their bodies

Animals maximize respiration by decreasing the distances over which respiratory gases must diffuse. This is accomplished by very thin respiratory surface membranes. In animals such as flatworms, the animals themselves are so thin that they can use their body surface for gas exchange. Animals such as sponges have extensive internal passageways that bring every cell into close proximity to the external environment. Most animals, however, use ventilation to actively move the external medium over the respiratory surface and perfuse the internal side of the respiratory organ with blood that carries the respiratory gases.

Fish have internal gills with a large surface area that they ventilate with the unidirectional flow of water to maximize the uptake of oxygen by the blood. Each gill has hundreds of gill filaments with folds, or lamellae, that act as respiratory gas exchange surfaces. Countercurrent flow (blood in the lamellae flowing in the opposite direction to that of water flowing over the lamellae) maximizes the $O_2$ difference between the partial pressure of $O_2$ in the water and the blood (see Figures 31.6 and 31.7). As blood flows through the lamellae, it is always in contact with water that has a higher $O_2$ partial pressure, resulting in a continuous difference in partial pressure of $O_2$ between the blood and the water, which maximizes the uptake of $O_2$ into the blood.

Lungs are found in amphibians, reptiles (including birds), and mammals. They are also found in air-breathing fish, such as lungfish. The inside surfaces of lungs are coated with surfactant, a lipid-protein substance that reduces the surface tension of lung epithelium and helps to keep the epithelial surfaces from sticking together.

In mammals, lungs are dead-end sacs in which ventilation is tidal. In tidal ventilation, fresh air flows in and exhaled gases flow out by the same route. Tidal breathing limits the difference in the partial pressure of $O_2$ between the air and the blood for driving the diffusion of $O_2$ from air into blood. Fresh air is not moving into the lungs during some of the breathing cycle, and when it does enter the lungs, it mixes with stale air.

Birds have a unidirectional flow of air through their lungs. Gas exchange occurs in air capillaries that branch off the parabronchi of the lungs; gas exchange does not occur in the associated air sacs (see Figure 31.8). During inhalation, the posterior air sacs receive the incoming air, and the anterior air sacs receive the air that was in the lungs. During exhalation, the air in the posterior air sacs flows into the lungs, and the air in the anterior air sacs leaves the bird.

Insects have a tracheal system—a series of air tubes that open to the environment through spiracles and end in the tissues as air capillaries. Gases diffuse through the tracheae into the air capillaries, but can also be moved by movements of the insect's body (see Figure 31.10).

**Question 3.** Why would the internal gills of fishes fail to work as respiratory organs on land?

**Question 4.** On the diagram of the avian respiratory system below, label the following structures: anterior air sacs, posterior air sacs, trachea, lung, and bronchus. When the bird is exhaling air, the stale air is in which structures before it moves into the trachea?

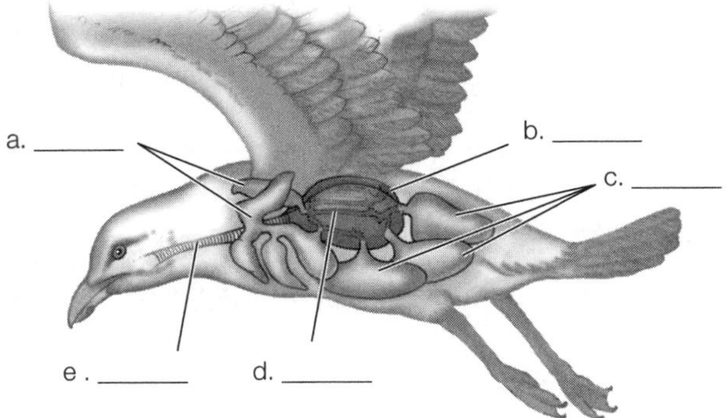


### 31.3 The Mammalian Breathing System Is Anatomically and Functionally Elaborate

At rest, only a small portion of the lung volume is exchanged

The lungs are ventilated by expansion and contraction of the thoracic cavity

The breathing rhythm depends on nervous stimulation of the breathing muscles

Breathing is under negative-feedback control by $CO_2$

Breathing is also under control of factors in addition to $CO_2$

Mammalian lungs have a very large surface area and a very short gas diffusion pathway. Air enters at the oral cavity or nasal passage, travels down the pharynx through the larynx, and then through the trachea, which branches into two bronchi, one leading to each lung. In the lung, more branching occurs to produce bronchioles and finally the primary site of gas exchange, the small air sacs called the alveoli (see Figure 31.11). $O_2$ diffuses across the thin-walled alveoli into many surrounding capillaries—a distance of only about 0.5 μm. There is also some gas exchange in the last few branches of the bronchioles, called respiratory bronchioles.

The tidal breathing of mammals can be described in terms of a series of lung volumes measured with a spirometer (see Figure 31.12). The volume of air that is moved during one cycle is known as the tidal volume. A person's resting tidal volume is less than the exercising tidal volume. The maximum tidal volume is called the vital capacity. This is the maximum amount of air that moves in and out per breath when a person is inhaling and exhaling as much as possible. Not all of the air can be forced out of the lungs. The air that remains inside is called the residual volume. The respiratory minute volume is the amount of air that is inhaled and exhaled each minute. This increases when a person goes from resting to exercising.

Lungs are inflated by increasing the volume of the thoracic cavity. This is done by contraction of the diaphragm and some intercostal muscles that pull the ribs outward. During exhalation, the diaphragm relaxes, the contracted intercostal muscles relax, and a second set of intercostal muscles contracts, causing the ribs to be pulled inward and the volume of the thoracic cavity to be reduced (see Figure 31.13). When a person is at rest, inhalation is an active process requiring muscle contraction, but exhalation is a passive process, requiring only the elastic recoil of the thoracic cavity and lungs to decrease the volume of the thoracic cavity. At times of strenuous exercise, however, intercostal muscles and abdominal muscles help reduce the volume of the thoracic cavity and exhalation becomes an active process.

A person's breathing rhythm is controlled by neurons in the region of the brain called the medulla oblongata (see Figure 31.14). Two different areas in the medulla are involved in controlling the muscles involved in breathing: the paired hypoglossal nuclei and the paired pre-Bötzinger complexes.

Tissues produce $CO_2$ as a by-product of metabolism, and this must be moved from the tissues and excreted into the environment. Most of this $CO_2$ reacts with $H_2O$ in the blood, which forms carbonic acid ($H_2CO_3$) and dissociates into hydrogen ions ($H^+$) and bicarbonate ions ($HCO_3^-$). In humans and other mammals, chemosensitive neural centers on the ventral surface of the medulla monitor the partial pressure of $CO_2$ and $H^+$ concentrations in the blood. These breathing control centers increase ventilation rates when blood levels of $CO_2$ and $H^+$ increase. When increased ventilation brings the levels of $CO_2$ and $H^+$ back to normal, the breathing control centers no longer stimulate increased ventilation rates. Thus, these breathing control centers work in a negative-feedback manner.

Control of breathing is relatively insensitive to blood $O_2$ levels, but chemoreceptors in the arteries leaving the heart (carotid and aorta) can signal the breathing control centers in the medulla oblongata in response to low levels of $O_2$, thereby increasing ventilation rates. In addition, information from receptors detecting muscle and joint movements during exercise can also cause increased ventilation rates.

**Question 5.** Are inhalation and exhalation active processes (requiring muscle contraction) or passive processes (not requiring muscle contraction)?

**Question 6.** During an episode of hyperventilation, a person breathes very deeply and rapidly, and large amounts of $CO_2$ are eliminated from the blood. What happens next and why?

## Test Yourself

1. The figure below shows a graph of the $O_2$ supplied to turtle eggs buried in the sand near the edge of the water. Which of the following is likely to have happened between days 50 and 60 and was this event likely to have sped up or slowed down development and possibly killed the embryos within the eggs?

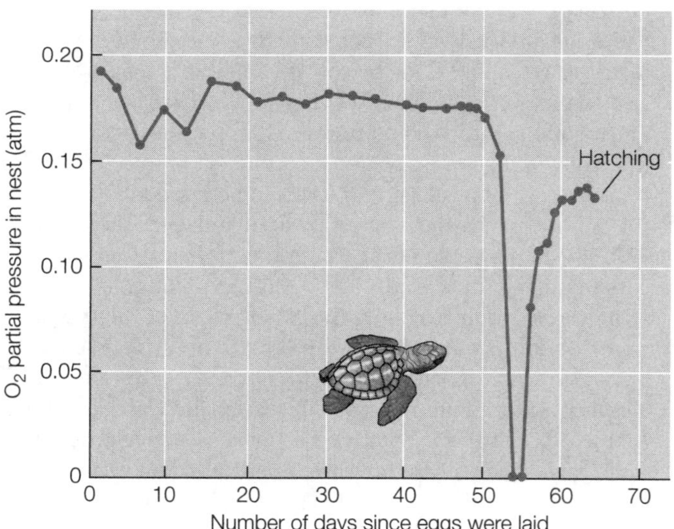


   a. Water temperatures decreased, likely slowing down development and possibly killing the embryos.
   b. Water inundated the nest, likely slowing down development and possibly killing the embryos.
   c. Air temperatures decreased, likely slowing down development and possibly killing the embryos.
   d. Air temperatures increased, likely speeding up development.
   e. Water temperatures increased, likely speeding up development.

   ***Textbook Reference:*** *Concept 31.1 Respiratory Gas Exchange Depends on Diffusion and Bulk Flow; Air and water are very different respiratory environments*

2. Can an aquatic worm with no specialized respiratory surfaces get enough $O_2$ to all of its cells to survive?
   a. No, breathing organs are required in all animals.
   b. Yes, but only if it is able to breathe air.
   c. Yes, but only if it is in warm rather than cold water.
   d. Yes, but only if it is no more than about 1 mm thick.
   e. Yes, but only if it is spherical in cross section, with a diameter of 1 cm or less.

   ***Textbook Reference:*** *Concept 31.1 Respiratory Gas Exchange Depends on Diffusion and Bulk Flow*

3. Which of the following statements is *false*?
   a. Compared to a given volume of water, the same volume of air is moved more easily across a respiratory surface.
   b. Compared to water, air holds more $O_2$ per unit volume.
   c. Global warming is likely to have a positive effect on fish.
   d. Increases in temperature decrease the $O_2$ content of water.
   e. $O_2$ diffuses more rapidly in air than in water.

   ***Textbook Reference:*** *Concept 31.1 Respiratory Gas Exchange Depends on Diffusion and Bulk Flow; The diffusion of gases depends on their partial pressures*

4. External gills, tracheal systems, and lungs share which of the following sets of characteristics?
   a. They are parts of gas exchange systems, they exchange both $CO_2$ and $O_2$, and they increase surface area for diffusion.
   b. They are used by water breathers, their functioning is based on countercurrent exchange, and they make use of negative pressure for breathing.
   c. They exchange only $O_2$, they are associated with a circulatory system, and they are found in vertebrates.
   d. They are found in insects, they employ positive-pressure pumping, and their functioning is based on crosscurrent flow.
   e. They are parts of gas exchange systems, they exchange both $CO_2$ and $O_2$, and they decrease surface area for diffusion.

   ***Textbook Reference:*** *Concept 31.1 Respiratory Gas Exchange Depends on Diffusion and Bulk Flow; Breathing is the transport of $O_2$ and $CO_2$ between the outside environment and gas exchange membranes*

5. All of the following represent a larger volume of air than is normally found in the resting tidal volume of a human lung *except*
   a. residual volume.
   b. maximum exhalation.
   c. maximum inhalation.
   d. vital capacity.
   e. capacity area.

   ***Textbook Reference:*** *Concept 31.3 The Mammalian Breathing System Is Anatomically and Functionally Elaborate; At rest, only a small portion of the lung volume is exchanged*

6. Both bird and mammal lungs
   a. have tidal ventilation.
   b. contain alveoli at the terminal ends.
   c. have residual air that cannot be exhaled.
   d. exchange $O_2$ and $CO_2$ with blood in capillaries.
   e. have a unidirectional flow of air moving through them.

   ***Textbook Reference:*** *Concept 31.2 Animals Have Evolved Diverse Types of Breathing Organs; Birds have rigid lungs ventilated unidirectionally by air sacs*

7. Which of the following does *not* play a role in the diffusion of $O_2$ across a membrane (according to Fick's law of diffusion)?
   a. Surface area
   b. Volume of $O_2$
   c. Difference in concentration or partial pressure
   d. Diffusion distance
   e. Diffusion coefficient

   ***Textbook Reference:*** *Concept 31.1 Respiratory Gas Exchange Depends on Diffusion and Bulk Flow; The diffusion of gases depends on their partial pressures*

8. Because of the relatively high altitude of Antonito, Colorado, the town has a normal barometric pressure of about 600 mm Hg rather than 760 mm Hg as at sea level. The partial pressure of $O_2$ in Antonito's air is approximately _______ mm Hg. (Hint: Air is approximately 21 percent oxygen.)
   a. 75
   b. 126
   c. 160
   d. 76
   e. 21

   ***Textbook Reference:*** *Concept 31.1 Respiratory Gas Exchange Depends on Diffusion and Bulk Flow; The diffusion of gases depends on their partial pressures*

9. Air flows into the lungs of mammals during inhalation because the
   a. pressure in the lungs falls below atmospheric pressure.
   b. volume of the lungs decreases.
   c. pressure in the lungs rises above atmospheric pressure.
   d. diaphragm moves upward toward the lungs.
   e. internal intercostal muscles contract.

   ***Textbook Reference:*** *Concept 31.3 The Mammalian Breathing System Is Anatomically and Functionally Elaborate; The lungs are ventilated by expansion and contraction of the thoracic cavity*

10. The movement of $O_2$ and $CO_2$ between the blood in the tissue capillaries and the cells in tissues depends most directly upon
    a. active transport of $O_2$ and $CO_2$.
    b. total atmospheric (barometric) pressure differences across the cell membranes.
    c. diffusion of $O_2$ and $CO_2$ from an area of lower concentration of the gas to an area of higher concentration of the gas.
    d. diffusion of $O_2$ and $CO_2$ from an area of higher partial pressure of the gas to an area of lower partial pressure of the gas.
    e. osmosis across cell membranes.

    ***Textbook Reference:*** *Concept 31.1 Respiratory Gas Exchange Depends on Diffusion and Bulk Flow; The diffusion of gases depends on their partial pressures*

11. Although 21 percent of air is oxygen, the alveoli of the lungs do not contain air with this much oxygen because
    a. we normally do not ventilate our lungs at a high enough rate to meet this level.
    b. the lungs have too many alveoli to ventilate them all at this level.
    c. there is residual stale air in the lungs that cannot be exhaled.
    d. the trachea and bronchi are too small to contain this much oxygen.
    e. most $O_2$ has been exchanged before reaching the alveoli.

    ***Textbook Reference:*** *Concept 31.3 The Mammalian Breathing System Is Anatomically and Functionally Elaborate; At rest, only a small portion of the lung volume is exchanged*

12. The presence of $CO_2$ in blood will lower the pH of blood because $CO_2$ combines with
    a. $H_2O$ to form $H^+$ and $HCO_3^-$, increasing the acidity of the blood
    b. $H_2O$ to form only $HCO_3^-$, increasing the alkalinity of the blood
    c. $H_2O$ to form only $H^+$, increasing the acidity of the blood
    d. $H^+$ to form $HCO_3^-$, making the blood more alkaline
    e. $H_2O$ to form $H^+$ and $HCO_3^-$, decreasing the acidity of the blood

    ***Textbook Reference:*** *Concept 31.3 The Mammalian Breathing System Is Anatomically and Functionally Elaborate; Breathing is under negative-feedback control by $CO_2$*

13. All of the following are involved in the control of breathing *except*
    a. neurons in the medulla.

b. chemosensitive regions on the surface of the medulla.
c. chemosensitive regions along the aorta.
d. chemosensitive regions along the carotid arteries.
e. neurons in muscle.

***Textbook Reference:*** *Concept 31.3 The Mammalian Breathing System Is Anatomically and Functionally Elaborate; Breathing is also under control of factors in addition to* $CO_2$

14. The diagram below shows two possible arrangements for blood flowing through the gas exchange region of a gill. What is a likely sequence of values for partial pressures of $O_2$ in the blood in figure 2, going from left to right? (Hint: Use information in figure 1 to help you figure out the likely values.)

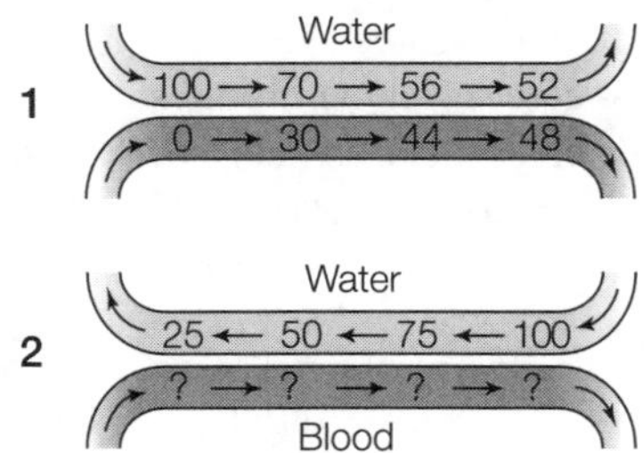


a. 0, 30, 44, 48
b. 100, 70, 56, 52
c. 25, 50, 75, 100
d. 0, 25, 50, 75
e. 30, 48, 78, 100

***Textbook Reference:*** *Concept 31.2 Animals Have Evolved Diverse Types of Breathing Organs; The directions of ventilation and perfusion can greatly affect the efficiency of gas exchange*

15. Why are you more likely to find birds than mammals at high altitudes?
a. Birds use tidal ventilation, whereas mammals use unidirectional ventilation.
b. Birds have a lower metabolism than mammals.
c. Birds require less $O_2$ for energy production than mammals.
d. Birds have more alveoli in their lungs than mammals.
e. Birds have more efficient respiratory systems than mammals.

***Textbook Reference:*** *Concept 31.2 Animals Have Evolved Diverse Types of Breathing Organs; Birds have rigid lungs ventilated unidirectionally by air sacs*

# Answers

## Key Concept Review

1. A number of modifications would increase the diffusion rate of gases at a respiratory surface. An increase in the number of alveoli in a lung would increase the surface area for gas exchange. A decrease in the thickness of capillary walls would decrease the distance needed by gas to diffuse. A countercurrent exchange arrangement or unidirectional flow would increase the difference between the partial pressures of the gas across the respiratory surface.

2. The tropical fish would be caught in the double bind of needing more $O_2$ as the water temperature increased (due to its increasing metabolic rate with increasing water temperature), and yet facing lower levels of $O_2$ in the warmer water (since warm water holds less dissolved $O_2$ than cold water does).

3. The internal gills of fishes consist of thin, delicate tissue. On land, such gills would collapse and clump together, significantly reducing the surface area for gas exchange.

4. 
a. Anterior air sacs
b. Lung
c. Posterior air sacs
d. Bronchus
e. Trachea

The stale air being exhaled is leaving the anterior air sacs and entering the trachea.

5. Inhalation is always an active process because it involves contraction of the diaphragm and certain intercostal muscles. Exhalation is normally a passive process (the diaphragm relaxes), except during strenuous exercise when a second set of intercostal muscles and abdominal muscles contract.

6. Following hyperventilation, breathing temporarily stops because so much $CO_2$ has been removed from the blood, dropping the partial pressure of $CO_2$ and $H^+$ concentration enough that the respiratory control centers temporarily stop sending signals to the diaphragm and intercostal muscles. Breathing begins again when the partial pressure of $CO_2$ and the $H^+$ concentration build to sufficient levels in the blood to prompt the respiratory control centers to send their signals.

## Test Yourself

1. **b.** The eggs buried in sand will get adequate $O_2$ from the air as long as the sand is dry. When the sand is inundated with water, the amount of $O_2$ available to the developing embryos is greatly diminished because the amount of $O_2$ water can contain is far less than the amount in air. The decreased access to $O_2$ will slow the development of the embryos and possibly kill them.

2. **d.** If an aquatic worm has no respiratory surfaces, it must rely on simple diffusion from the environment into its cells. For simple diffusion to be an effective means of $O_2$ transport, an animal must be very thin, with all of its cells no more than about 0.5 mm from the environment, meaning the worm could be no more than 1 mm thick.

3. **c.** Water breathing is more difficult than air breathing because the higher density of water makes it more expensive to move across the respiratory surfaces than air, it has less $O_2$ than air, and the $O_2$ content is very dependent on the temperature of the water. As water temperatures rise, the amount of $O_2$ in the water decreases. As global warming increases the temperatures of water, fish will be faced with decreased amounts of $O_2$, and this will have a detrimental effect on their ability to survive.

4. **a.** External gills, tracheal systems, and lungs are all examples of gas exchange systems used by various animals to exchange both $CO_2$ and $O_2$. One of the special properties of all respiratory systems is that they increase surface area for diffusion.

5. **e.** Answers a, b, c, and d have a volume that is larger than the resting tidal volume.

6. **d.** The lungs of birds and mammals are both sites for the exchange of $O_2$ and $CO_2$ with blood in capillaries. The mammalian lung is the only one that ends in alveoli, where the air is motionless. Flow of air is unidirectional in birds, but tidal in mammals.

7. **b.** The volume of the gas is not part of Fick's law of diffusion. All the other parameters are important in determining the rate of diffusion of gases in the respiratory system of animals.

8. **b.** Air at sea level with an atmospheric pressure of 760 mm Hg has a partial $O_2$ pressure of 160 mm Hg (21 percent of 760). Therefore, the partial pressure of $O_2$ at Antonito, Colorado, is 126 mm Hg (21 percent of 600).

9. **a.** Inhalation in the mammalian lung occurs by means of negative pressure produced by contraction of the diaphragm. Thus, the pressure in the lungs falls below atmospheric pressure.

10. **d.** Movement of $O_2$ and $CO_2$ from the blood to the tissues always occurs by diffusion of $O_2$ and $CO_2$ from a region of higher partial pressure of the gas to a region of lower partial pressure of the gas.

11. **c.** The alveoli do not contain air with 21 percent $O_2$ because incoming air is mixed with residual air left in the lungs, which has had some of the $O_2$ removed by the lungs.

12. **a.** $CO_2$ combines with $H_2O$ in the plasma to form $H^+$ and $HCO_3^-$. The increase in $H^+$ lowers the pH of the blood, thereby increasing the acidity of the blood.

13. **e.** Answers a, b, c, and d are involved in the control of breathing.

14. **d.** Of the values listed, the most likely sequence of partial pressures of $O_2$ in the blood in figure 2, going from left to right, is 0, 25, 50, 75. In figure 2, blood flows in a direction opposite to that of water; it is therefore in a countercurrent exchange arrangement. The blood is able to pick up more $O_2$ from the water than in the cocurrent arrangement shown in figure 1. In the countercurrent exchange arrangement, the difference in partial pressures of $O_2$ between the blood and water is kept at a maximum as the water and blood flow through the exchange surface. Notice that as blood is leaving the exchange surface, it is meeting fresh water from the environment.

15. **e.** Birds have an extremely efficient respiratory system, characterized by the continuous unidirectional flow of air through the lungs. In contrast, mammals use tidal ventilation, a less efficient method, in which air flows in and exhaled gases flow out the same route. Tidal ventilation results in the mixing of fresh air with stale air left in the lungs. $O_2$ availability decreases with altitude. Because of the increased efficiency of their lungs, birds are more likely than mammals to be able to tolerate lower $O_2$ levels and are therefore more likely to be found at very high elevations.

# 32 Circulation

## The Big Picture

- Cardiovascular systems are required when animals grow too big to acquire nutrients and eliminate waste by diffusion alone. Animals have evolved diverse cardiovascular systems, which can be classified as open or closed. An open system has relatively low pressure and large sinuses that bring blood directly to tissues and organs. Closed systems have an interconnected series of vessels that take blood to tissues while keeping the blood contained within these vessels.
- During the evolution of vertebrates, the heart has changed from the simple two-chambered heart of fishes to the complex four-chambered hearts of birds and mammals. The four-chambered heart, such as your own, is really two pumps in one. The left pump provides oxygenated blood under high pressure to the systemic tissues, and the right pump sends deoxygenated blood under lower pressure to the lungs. A cardiac pacemaker in the heart sets the cardiac rhythm. In the blood, $O_2$ is carried primarily by respiratory pigments such as hemoglobin, increasing the amount of $O_2$ the blood can carry.

## Study Strategies

- It is frequently thought that the hearts of amphibians and certain reptiles such as turtles and lizards are primitive structures, awaiting "repair through evolution" to the more highly evolved bird and mammal heart. In fact, the hearts of amphibians and these reptiles can separate the oxygenated and deoxygenated blood coming into their single ventricle by its particular architecture. In addition, reptiles such as turtles are able to shunt blood from pulmonary to systemic circuits when not breathing, such as when diving under water, thereby saving energy that would otherwise be wasted if blood were being sent to the temporarily non-ventilated lungs.
- People often fail to appreciate that the human heart is two distinct pumps—the right heart pumping blood to the lungs and the left heart pumping blood to the body. The two pumps just happen to be packaged in the same structure. In fact, cardiologists refer to the "left heart" and "right heart," emphasizing the separate nature of these two pumps. Understanding the flow of blood through the human heart becomes easier when it is considered from this perspective. Also, determining the pattern of blood flow through the human heart will be easier if you mentally "unfold" the circulation into a great circle in which there are two pumps at opposite sides, with vascular beds for the lungs and systemic circulation intervening between the pumps. Once you have mastered this concept, you will see that to place the two pumps in proximity in the same structure (the heart) you need to twist the circle, but the connections remain the same—blood from the right heart is going to the lungs and coming back to the left heart; blood from the left heart is going to the systemic circulation and coming back to the right heart, and the circle is completed.
- Following from the preceding, learn the various chambers and valves in the context of a sequence of structures in each of the right and left pumps of the heart, rather than as a list of terms printed on a complex, anatomically accurate diagram of the heart.
- Go to **LaunchPad** (or use the Web addresses listed) to review the following additional resources:

  Animated Tutorial 32.1 The Cardiac Cycle (PoL2e.com/at32.1)

  Animated Tutorial 32.2 Hemoglobin Loading and Unloading Simulation (PoL2e.com/at32.2)

  Activity 32.1 Vertebrate Circulatory Systems (PoL2e.com/ac32.1)

  Activity 32.2 The Human Heart (PoL2e.com/ac32.2)

  Activity 32.3 Structure of a Blood Vessel (PoL2e.com/ac32.3)

  Activity 32.4 Oxygen-Binding Curves (PoL2e.com/ac32.4)

  Apply the Concept in textbook Sections 32.1 and 32.3

## Key Concept Review

### 32.1 Circulatory Systems Can Be Closed or Open

Closed circulatory systems move blood through blood vessels

In open circulatory systems, blood leaves blood vessels

An animal's circulatory system transports nutrients and wastes to and from tissues and cells; of all these substances, $O_2$ must be transported the most quickly. Animals with high metabolic rates, such as mammals and birds, certain fish such as tuna, and fast-swimming mollusks such as squid need a high-performance circulatory system, whereas animals with low metabolic rates, such as snails and clams, can function with low-performance circulatory systems. An exception is insects, which have high demands for $O_2$ but low-performance circulatory systems. Their blood does not need to carry $O_2$, since $O_2$ is supplied to each cell by the network of air-filled tubes of their tracheal system.

Circulatory systems may be classified as open or closed. In an open circulatory system, blood and interstitial fluids are not separated; the blood containing interstitial fluids is therefore often referred to as hemolymph. A pump helps to move the blood (hemolymph) through large lacunae and sinuses. A lobster, for example, has an open circulatory system (see Figure 32.4). Blood in the lobster leaves the heart by going through arteries that extend both anteriorly and posteriorly from the dorsal heart. It then pours into lacunae and sinuses before it returns to the heart.

In a closed circulatory system, blood is pumped by the heart through a series of interconnected closed vessels: arteries, arterioles, capillaries, venules, and veins. The blood is kept within the vessels, separate from the interstitial fluid. The exchange of nutrients, respiratory gases, and waste occurs through the walls of the smallest vessels, the capillaries. There are numerous advantages to closed systems compared to open ones: more rapid transport of blood in the closed vessels, greater control over diversion of blood flow between different tissues, and greater control over hormone and nutrient transport to the tissues. Smooth muscle cells around arterioles dilate or constrict these vessels, thereby restricting or increasing blood flow to different areas. When you exercise, for example, arterioles in your skin dilate, helping to cool you off (see Figure 32.3).

**Question 1.** You are in the Arctic and discover a new species of animal that maintains a body temperature of 37°C and actively hunts down its food in the ocean. Without dissecting the animal, you must explain to others what type of circulatory system this animal most likely has. What will you tell them?

**Question 2.** When you go out on a very cold day, your fingers are usually the first part of the body that feels cold, even if you are wearing gloves. How is your circulatory system involved in causing your fingers to feel cold before your core?

### 32.2 The Breathing Organs and Systemic Tissues Are Usually, but Not Always, in Series

Most fish have the systemic circuit and gill circuit connected in series

Mammals and birds have the systemic circuit and lung circuit connected in series

Amphibians and most non-avian reptiles have circulatory plans that do not anatomically guarantee series flow

Breathing organs and systemic tissues are usually organized in series, whether it is a gill circuit or a lung circuit. Fish have a gill circuit. Their heart is a single tube containing one atrium leading into one ventricle. Blood pumped through the heart then enters the gills, where gas exchange occurs. Blood leaving the gills travels to the systemic organs and tissues and then returns to the heart (see Figure 32.5A).

Birds and mammals have a four-chambered heart that allows for complete separation and no mixing of systemic and pulmonary blood, and their systemic and lung circuits are connected in series (see Figure 32.5B). This cardiovascular arrangement maximizes $O_2$ transport by preventing mixing between pulmonary and systemic circulations. The pulmonary circulation operates at low pressure, protecting the delicate lung membranes, whereas the systemic circuit operates at higher pressure, allowing the perfusion of tissues and organs distant from the heart.

Adult amphibians and most non-avian reptiles have a heart with two atria and a single ventricle. One atrium receives deoxygenated blood from the tissues and the other receives oxygenated blood from the lungs (see Figure 32.5C). Blood from both atria enter the single ventricle where oxygenated and deoxygenated blood can mix, but anatomical features of the ventricle help direct deoxygenated blood preferentially to the lungs and oxygenated blood to the systemic body tissues. In certain reptiles such as turtles, which stop breathing while they are diving, vessel constriction in the lungs during the dives causes blood to largely bypass the pulmonary circulation and flow to the systemic circuit.

**Question 3.** What are the important differences between the circulatory system of a fish and that of a turtle?

**Question 4.** Add arrows to the diagram below to show the direction of blood flow through this circulation plan. This circulation pattern is found in what group of animals?

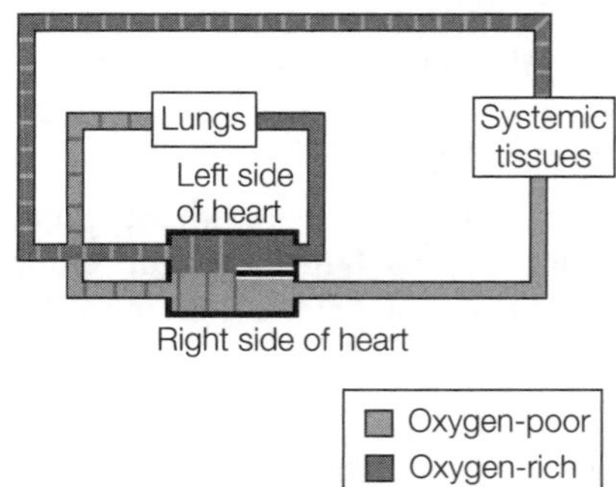

## 32.3 A Beating Heart Propels the Blood

Vertebrate hearts are myogenic and multi-chambered

The myocardium must receive $O_2$

An electrocardiogram records the electrical activity of the heart

Crustacean hearts are neurogenic and single-chambered

A heart is a pumping structure that propels blood through the circulatory system by contraction of its muscle cells, the myocardium. The cardiac output of a heart is equal to its stroke volume times the number of times it beats per minute; in your own heart, this is about 5 L/min when you are at rest.

The human heart, like that of other vertebrates, is multi-chambered and myogenic, meaning that it is able to contract without innervation from the nervous system. Cycles of contractions occur in two phases: systole when the two ventricles contract, and diastole, when the ventricles relax (see Figure 32.7). At the very end of diastole, the two atria contract. The contraction of the ventricles pressurizes the blood within them. Over a single cardiac cycle the pressure is highest during ventricle contraction (systole) and lowest during relaxation (diastole). This pressure can be measured in an artery using a sphygmomanometer and stethoscope; in a healthy adult, systole is about 120 mm Hg and diastole is about 80 mm Hg, typically reported as a ratio of "120 over 80."

The human heart, like that of mammals in general, has two atria and two ventricles (see Figure 32.6). Between each atrium and ventricle an atrioventricular valve prevents the backflow of blood from the ventricle into the atrium. The pulmonary valve resides between the right ventricle and the pulmonary artery and the aortic valve resides between the left ventricle and the aorta; these two valves help maintain one-way flow through the heart.

Deoxygenated blood flows from the systemic tissues through the superior vena cava and inferior vena cava into the right atrium and then into the right ventricle. Blood flows from the right ventricle to the pulmonary artery and on to the lungs, where it becomes oxygenated (see Figure 32.6). Oxygenated blood from the lungs flows back to the heart by way of the pulmonary veins into the left atrium, then into the left ventricle, and finally out through the aorta to the systemic tissues. Ventricular contraction pushes blood out of the heart to the lungs and systemic tissues. The left ventricle is thicker and more muscular than the right ventricle and develops higher pressures for the systemic circuit than are produced by the right ventricle for the lung circuit.

The human heartbeat is generated in the heart's pacemaker. The pacemaker, or sinoatrial (S-A) node, is a small section of highly modified cardiac muscle tissue located in the wall of the right atrium where the superior vena cava enters the atrium. The S-A node starts the cycle of contraction by sending out a wave of electrical depolarization that spreads through the muscle cells of the atria. The signal is able to spread because the cardiac muscle cells are electrically coupled with their neighboring cells. The wave of contraction moves from the atria to the ventricles through the atrioventricular node (A-V node) and down through conducting fibers to the base of the ventricles, where it fans out into the ventricular muscle (see Figure 32.8). The spread of the contraction signal allows the ventricles to contract slightly after the atria contract.

The rhythm of the human heartbeat is modified by the autonomic nervous system (sympathetic and parasympathetic) and endocrine system, as you have no doubt experienced whenever you feel a burst of adrenaline—your heart beats faster and harder. Norepinephrine from the sympathetic nervous system increases heart rate; acetylcholine from the parasympathetic nervous system decreases heart rate (see Figure 32.9).

The muscle cells of hard-working, high-performance hearts, such as the human heart, need arteries to bring them $O_2$-rich blood. This is accomplished by means of coronary arteries. If these arteries are blocked by a build-up of fatty deposits, a person can have a heart attack.

An electrocardiogram (ECG, also abbreviated EKG) is a measure of the electrical activity of the heart as a complex wave pattern (see Figure 32.10). The wave has five parts: P (depolarization and contraction of the atria), Q, R, and S (depolarization and contraction of the ventricles), and T (relaxation and repolarization of the ventricles).

The heart of crustaceans, such as lobsters and crayfish, is neurogenic rather than myogenic. This means that it requires nervous stimulation in order to beat. In a lobster, the heart pushes blood out to body sinuses through anterior and posterior arteries. Blood is then drawn back into the heart through openings in the heart wall, called ostia, as stretched ligaments attached to the heart pull the heart wall outward.

**Question 5.** Trace a drop of blood through your own circulatory system, from the time it leaves the right ventricle to when it returns to its starting place in the right ventricle.

**Question 6.** If a person's heart stopped beating, but he or she was still able to breathe, what would be the most immediate consequence? Explain the reasoning behind your answer?

## 32.4 Many Key Processes Occur in the Vascular System

Pressure and linear velocity vary greatly as blood flows through the vascular system

Animals have evolved arrangements of blood vessels that help them conserve heat

Blood flow leaves behind fluid that the lymph system picks up

Arteries have thicker walls compared to veins and contain elastic fibers in addition to layers of smooth muscle and connective tissue. This allows them to stretch when the heart beats and squeeze the blood when they return to their unstretched state, thereby preventing large pressure changes during a cardiac cycle. Veins have one-way valves that prevent blood from flowing backward. The contractions of skeletal muscles around a vein squeeze it and help move the blood forward (see Figure 32.14). Capillaries, the smallest vessels, are the site of exchange of gases, nutrients, and wastes between the blood and tissues.

The linear velocity of blood is highest in the arteries, falls precipitously in the capillaries, and rises again in the veins; blood pressure is highest in arteries, drops in the capillaries, and remains low in the veins (see Figure 32.13).

In many types of mammals and birds, various arrangements of blood vessels establish a countercurrent heat exchanger, retaining heat within the organism rather than losing it to the environment (see Figure 32.16). This can be seen, for example, in the legs of the arctic fox. It is also seen in some fish, such as in the red swimming muscles of tuna (see Figure 32.17).

When blood moves through capillaries, blood pressure from the arterial side forces fluids, ions, and small molecules out of the capillaries into the interstitial fluids. Near the venule end of the capillaries, the difference in osmotic potential between the plasma inside the capillaries and the surrounding interstitial fluids creates an osmotic pressure resulting in fluid recovery back into the capillaries. Thus the net flow of water between the plasma and tissue fluid is determined by the difference in blood pressure and osmotic pressure (see Figure 38.18).

There would be a net loss of fluid from your blood into your interstitial fluids if there weren't a system to return this fluid to the blood. The system that serves this retrieval function is the lymphatic system, a network of vessels with one-way valves that collects excess interstitial fluid and returns it to veins in the neck. Lymph nodes associated with the lymphatic vessels filter foreign materials and microorganisms from the blood and are part of your immune system.

**Question 7.** Blood moves through the arteries and veins at a faster velocity than the blood moving through the capillaries. What is the cause and effect of blood slowing down as it moves through the capillaries?

**Question 8.** If you found a bird in the Arctic walking across the arctic ice, what adaptations to the circulation going into its legs and feet would you expect it to have?

## 32.5 The Blood Transports $O_2$ and $CO_2$

### Hemoglobin and hemocyanin are the two principal respiratory pigments

### Respiratory pigments combine with $O_2$ reversibly

One of the most important functions of blood is to deliver $O_2$ to tissues and remove $CO_2$. $O_2$ is not very soluble in liquids, so blood cannot carry large amounts of $O_2$ unless it also contains a respiratory pigment that is able to bind to $O_2$. The principal respiratory pigments found in animals are hemoglobin and hemocyanin.

Hemoglobin is found in almost all vertebrates as well as in annelids, such as earthworms, and a number of other invertebrate groups. Hemoglobin molecules have four protein subunits, each of which contains one iron atom and can bind one molecule of $O_2$. In vertebrates, hemoglobin is found only in red blood cells; it is not found in the blood plasma. Hemocyanin is found in mollusks such as squid, in arthropods such as lobsters, and in spiders. Instead of containing iron, it contains copper and it is never found inside blood cells but instead is found dissolved in the blood plasma. Hemoglobin is bright red when oxygenated and purple-red when deoxygenated. Hemocyanin is blue when it is oxygenated and colorless when it is deoxygenated.

Respiratory pigments can combine reversibly with $O_2$. This process is summarized by an oxygen equilibrium curve (see Figure 32.20). Human hemoglobin picks up oxygen in the lungs, where the partial pressure of $O_2$ is high; it releases $O_2$ in systemic tissues, where the partial pressure of $O_2$ is low. The lower the partial pressure of $O_2$ in the tissues, the more oxygen the hemoglobin molecules release.

**Question 9.** You find a crab along a rocky shore and observe that its gills are bright blue. You then find a similar type of crab in a tide pool whose gills are pale. What can you conclude about the two crabs?

**Question 10.** In the diagram below, which numbered point is associated with hemoglobin found in red blood cells of capillaries in the lungs? Which point is associated with hemoglobin found in red blood cells of capillaries in systemic tissues of a person at rest?

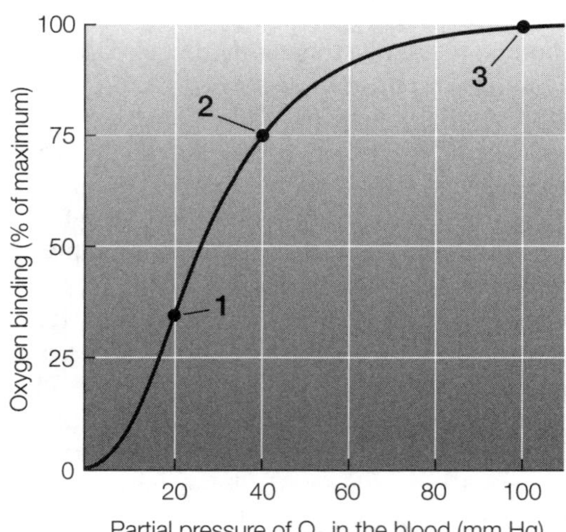


## Test Yourself

1. Which of the following is *not* one of the reasons that closed circulatory systems are more efficient than open circulatory systems?
   a. Closed systems rely exclusively on simple diffusion for transport, whereas open systems rely on pumping mechanisms.
   b. Transport within closed systems is more rapid than in open systems.
   c. Blood can easily be directed to specific areas in closed systems, but not in open systems.

d. Compared to open systems, closed systems operate better under higher pressure.
e. In closed systems molecules and cells that transport hormones and nutrients can be kept in vessels until they unload their goods at specific tissues.
***Textbook Reference:*** *Concept 32.1 Circulatory Systems Can Be Closed or Open; In open circulatory systems, blood leaves blood vessels*

2. The left ventricle exceeds the right ventricle in
a. the amount of blood that enters during heart contraction.
b. the volume expelled during contraction.
c. the pressure developed during contraction.
d. the speed with which it contracts.
e. its elasticity.
***Textbook Reference:*** *Concept 32.3 A Beating Heart Propels the Blood; Vertebrate hearts are myogenic and multi-chambered*

3. When a turtle stops breathing during a dive,
a. blood is shunted from the systemic circuit to the pulmonary circuit.
b. blood is shunted specifically to the brain.
c. blood is shunted from the pulmonary circuit to the systemic circuit.
d. blood is shunted to the skin for gas exchange.
e. its pattern of breathing and nonbreathing periods does not change.
***Textbook Reference:*** *Concept 32.2 The Breathing Organs and Systemic Tissues Are Usually, but Not Always, in Series; Amphibians and most non-avian reptiles have circulatory plans that do not anatomically guarantee series flow*

4. Which of the following structures of the human lymphatic system acts primarily as a filter for detecting and destroying microorganisms in lymph traveling through major lymph vessels?
a. Lymph nodes
b. Thymus
c. Lymph capillaries
d. Thyroid
e. Thoracic ducts
***Textbook Reference:*** *Concept 32.4 Many Key Processes Occur in the Vascular System; Blood flow leaves behind fluid that the lymph system picks up*

5. Which of the following statements about vertebrate circulatory systems is *false*?
a. Birds have a four-chambered heart.
b. In birds and mammals, pressures in the pulmonary circuit are lower than those in the systemic circuit.
c. The ventricles of turtles and lizards are functionally divided.
d. Amphibians have two atria and one ventricle.
e. In fishes, blood passes through the gills, returns to the heart, and then is pumped to the systemic tissues.
***Textbook Reference:*** *Concept 32.2 The Breathing Organs and Systemic Tissues Are Usually, but Not Always, in Series; Mammals and birds have the systemic circuit and lung circuit connected in series*

6. Which of the following represents the correct sequence in which cardiac action potentials pass through the heart?
a. Conducting system between ventricles, A-V node, S-A node, atrial fibers
b. A-V node, atrial fibers, S-A node, conducting system between ventricles
c. S-A node, conducting system between ventricles, atrial fibers, A-V node
d. Conducting system between ventricles, A-V node, atrial fibers, S-A node
e. S-A node, atrial fibers, A-V node, conducting system between ventricles
***Textbook Reference:*** *Concept 32.3 A Beating Heart Propels the Blood; Vertebrate hearts are myogenic and multi-chambered*

7. The atrial walls are _______ the ventricular walls, and pressure generated in the atrial chambers is _______ the pressure in the ventricles.
a. thinner than; higher than
b. thinner than; lower than
c. thicker than; higher than
d. thicker than; lower than
e. the same thickness as; the same as
***Textbook Reference:*** *Concept 32.3 A Beating Heart Propels the Blood; Vertebrate hearts are myogenic and multi-chambered*

8. Which of the following regions of the vascular bed is the actual site of gas exchange with surrounding tissue?
a. Arteries
b. Capillaries
c. Lymphatic vessels
d. Veins
e. Venules
***Textbook Reference:*** *Concept 32.1 Circulatory Systems Can Be Closed or Open; Closed circulatory systems move blood through blood vessels*

9. Which of the following statements about the control of circulation in humans is *false*?
a. Sympathetic nerve input to skeletal muscle would most likely cause the arterioles in the muscle to dilate.
b. Blood flow is increased when the S-A node is stimulated by the sympathetic nervous system.
c. When a person is exercising on a hot day, arterioles in the skin usually constrict.

d. Hormones such as adrenalin increase the rate of depolarization of heart muscle cells.
e. The S-A node is the pacemaker and consists of modified cardiac muscle cells.

***Textbook Reference:*** *Concept 32.3 A Beating Heart Propels the Blood; Vertebrate hearts are myogenic and multi-chambered*

10. Red blood cells
a. contain hemoglobin and are part of the immune system.
b. are spherical cells that are slightly larger than the diameter of a capillary.
c. are biconcave cells containing platelets.
d. are bright red in the lungs and purple-red in venules.
e. secrete a respiratory pigment into the plasma.

***Textbook Reference:*** *Concept 32.5 The Blood Transports $O_2$ and $CO_2$; Hemoglobin and hemocyanin are the two principal respiratory pigments*

11. In which of the following would the highest blood pressure be recorded?
a. The ventricle supplying blood to the gills of a fish
b. The anterior dorsal artery of an ant
c. The pulmonary vein of a frog
d. The ventricle supplying blood to the systemic circuit of a bird
e. The vessels leaving the gills of a mollusk

***Textbook Reference:*** *Concept 32.2 The Breathing Organs and Systemic Tissues Are Usually, but Not Always, in Series; Mammals and birds have the systemic circuit and lung circuit connected in series*

12. The net loss of fluid from blood capillaries increases if
a. plasma filtration decreases.
b. the osmotic pressure of plasma increases.
c. blood pressure increases in the capillaries.
d. the osmotic pressure of interstitial fluid decreases.
e. blood pressure drops below osmotic pressure.

***Textbook Reference:*** *Concept 32.4 Many Key Processes Occur in the Vascular System; Blood flow leaves behind fluid that the lymph system picks up*

13. Blood consists of a fluid fraction of _______ and a cellular fraction of _______.
a. plasma; water, red blood cells, and platelets
b. erythrocytes; leukocytes, red blood cells, and platelets
c. plasma; red blood cells, platelets, and leukocytes
d. leukocytes; erythrocytes and platelets
e. interstitial fluid; white blood cells, leukocytes, and platelets

***Textbook Reference:*** *Concept 32.5 The Blood Transports $O_2$ and $CO_2$*

14. Cigarette smoke contains nicotine, which increases heart rate and causes blood vessels to constrict. These changes would most likely
a. increase diastole but lower systole.
b. lower diastole but increase systole.
c. result in blood pressure readings higher than 120 over 80 mm Hg.
d. lower both diastole and systole.
e. cause hypotension.

***Textbook Reference:*** *Concept 32.3 A Beating Heart Propels the Blood; Vertebrate hearts are myogenic and multi-chambered*

15. In which region of the human body would hemoglobin most likely be fully deoxygenated?
a. Capillaries of lungs
b. Veins of muscles when a person is at rest
c. Arterioles of leg muscles when a person is running
d. Carotid artery
e. Venules of leg muscles when a person is walking

***Textbook Reference:*** *Concept 32.5 The Blood Transports $O_2$ and $CO_2$; Hemoglobin and hemocyanin are the two principal respiratory pigments*

## Answers

### *Key Concept Review*

1. Due to the high metabolic needs of the animal, it most likely has a closed circulatory system with lung and systemic circuits in a high-performance series arrangement similar to that of a mammal or bird.
2. When you are cold, arterioles in your extremities constrict to maintain heat in your core. Your feet are likely to remain warm longer than your hands and fingers because you are using your feet to walk, and this requires more widely open arterioles to deliver $O_2$ to the working foot muscles.
3. Fish have a two-chambered heart (one atrium and one ventricle), with blood flowing from the heart to the gills and then directly to the systemic tissues. The blood loses pressure going through the gills and does not return to the heart before going to the systemic tissues; it therefore perfuses the tissues under low pressure. Turtles have a more complex three-chambered heart (two atria and one ventricle), with a pulmonary and systemic circuit. The single ventricle allows for some mixing of oxygenated and deoxygenated blood in the ventricle. However, the ventricle is functionally divided, so outflow can be directed to the pulmonary circuit and to the systemic circuit. After going through the lungs, blood returns to the heart before being pumped out to the systemic tissues, thereby receiving an extra boost in pressure. This means that blood flowing to the systemic tissues is under higher pressure in the turtle than it is in the fish.

4.

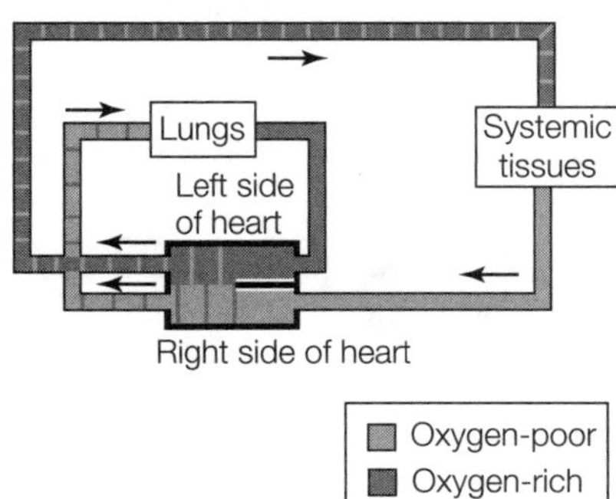


5. Blood from the right ventricle exits through the pulmonary valve into the pulmonary artery and then goes to the lungs, where it is oxygenated. From the lungs, the oxygenated blood returns to the left atrium via the pulmonary veins and passes through the atrioventricular valve into the left ventricle. Blood from the left ventricle is pumped to the systemic circuit through the aortic valve and into the aorta. From the aorta, blood moves to the tissues in arteries and arterioles, passes through tissues in the capillaries, and returns via the venules, veins, and ultimately the superior vena cava and inferior vena cava to the right atrium of the heart. From the right atrium, blood is pumped back where it started—into the right ventricle through the atrioventricular valve.

6. If a person's heart stopped beating, even if breathing continued, the person would quickly go unconscious. Blood would not be moving, so it would not be picking up $O_2$ from the lungs and delivering it to any of the systemic tissues, including the brain. If the brain is deprived of $O_2$ for just 20 seconds, the result is loss of consciousness.

7. Blood is pumped from the heart into the arteries, where it has a relatively fast velocity. Upon entering the capillaries, much of the velocity is lost. The capillaries have a much higher total cross-sectional area than the arteries, and the velocity of blood flow is inversely proportional to the cross-sectional area; thus, the more area through which the blood flows, the more slowly it will flow. Just as water in a stream that widens to become a river flows more quickly in the stream than it does in the river, blood moves slowly through the capillaries and more quickly through the arteries and veins. The slow velocity of blood through the capillaries allows for greater exchange of materials with the tissues.

8. It is likely that the arctic bird has a countercurrent heat exchange arrangement of blood vessels going into its legs and feet. If the arteries lie in close proximity to the veins in the legs, then heat can move out of the warm arterial blood into the cooler venous blood as the blood circulates. This will help to retain heat in the bird's core, while keeping its feet much cooler than its core.

9. Crabs have hemocyanin in their blood. This is a respiratory pigment that is blue when it is oxygenated and colorless when it is deoxygenated. The crab whose gills are blue must have been in highly oxygenated water, whereas the crab whose gills are pale must have been in poorly oxygenated water.

10. In the diagram, point 3 is associated with hemoglobin in red blood cells of capillaries in the lungs, where the hemoglobin becomes fully oxygenated; point 2 is associated with hemoglobin in red blood cells of capillaries in systemic tissues of a person at rest, where the hemoglobin is becoming partially deoxygenated. Hemoglobin doesn't become fully deoxygenated until it reaches tissues where the partial pressure of $O_2$ is lower, as at point 1.

### *Test Yourself*

1. **a.** Both closed systems and open systems rely on pumping mechanisms (and not just simple diffusion) to distribute fluid throughout the body.
2. **c.** The pressure generated by the left ventricle in the blood flowing to the systemic circuit is greater than the pressure generated by the right ventricle in the blood flowing to the pulmonary circuit.
3. **c.** During non-breathing periods, the blood is shunted from the pulmonary circuit to the systemic circuit.
4. **a.** The lymph nodes filter and destroy microorganisms that are traveling through the lymphatic system.
5. **e.** In fishes, blood passes through the gills and directly out to the systemic tissues; it does not return to the heart for additional pumping.
6. **e.** The cardiac action potential passes through the S-A node, atrial fibers, A-V node, and down the conducting system between the ventricles to the base of the ventricles.
7. **b.** The atrium has thinner walls and generates lower pressure than the ventricles.
8. **b.** Gas exchange at the tissues occurs across the capillaries.
9. **c.** When a person is exercising on a hot day, the arterioles in the skin dilate and the skin flushes. This allows for the loss of more heat to the environment, which helps prevent overheating.
10. **d.** Red blood cells are bright red when the hemoglobin they contain is fully oxygenated; hence they would be bright red where they pick up $O_2$ in the lungs. Hemoglobin is purple-red when deoxygenated, so the red blood cells would be this color after they release $O_2$ in capillaries and enter the venules of systemic tissues.
11. **d.** Of the vertebrate groups, mammals and birds tend to have the highest blood pressure. Therefore, the

ventricle supplying blood to the systemic circuit of a bird has the highest pressure.

12. **c.** Changes in blood pressure will change the amount of fluid that moves out of the capillaries into the interstitial fluid. Increases in blood pressure will result in greater rates of fluid moving out of the capillaries.

13. **c.** The fluid portion of blood is the plasma; three components of the solid fraction are red blood cells (erythrocytes), platelets, and leukocytes (white blood cells).

14. **c.** Increased heart rate and constriction of blood vessels would cause blood pressure to rise. Thus, cigarette smoking is associated with high blood pressure, which in a resting adult is a reading above 120 over 80 mm Hg.

15. **e.** Hemoglobin releases oxygen in the capillaries of systemic tissues. Since the blood moves from capillaries into venules, the hemoglobin in venules will carry less $O_2$ than the other structures listed.

# 33 Muscle and Movement

## The Big Picture

- Actin and myosin are the "universal" proteins for motion. Whether they are located in a unicellular animal or the leg muscle of a human, the molecular interactions of actin and myosin produce movement in the structures in which they reside.
- The sliding filament contractile mechanism of muscle contraction consists of actin and myosin filaments sliding past each other, resulting in the shortening (contraction) or lengthening (relaxation) of muscle cells. The whole process of actin and myosin interaction is tightly regulated by the movement of calcium ions into and out of the intracellular spaces of muscle cells, which in turn is activated by the arrival of action potentials in motor neurons. Muscle contraction requires energy in the form of ATP.
- In vertebrates, muscles act in concert with an internal skeleton made of bone. Bone is living tissue that is constantly remodeled. Bones are articulated, forming joints that provide for specialized directional movements of the tissues supported by the bones. Muscles controlling joint movement are often located in pairs acting antagonistically, with one set of muscles causing bending of the joint and the other causing straightening of the joint. Some invertebrates have hydrostatic skeletons (a fluid-containing body cavity surrounded by muscles), while others have exoskeletons (rigid outer coverings).
- Not all muscle cells are identical; many different types of muscle have evolved in many different phyla. The power output of a muscle cell varies with the ATP supply. In order to have sustained power, a muscle cell requires oxygen. If a short burst of power is needed, then the cell can rely upon its ATP stockpile or use anaerobic glycolysis.

## Study Strategies

- The interactions of myosin, actin, troponin, tropomyosin, and calcium and the role of action potentials in stimulating muscle contraction constitute a complex, multistep process. First, break the process down into its constituents and learn their locations and general structures. Second, determine how actin and myosin move relative to each other through a series of power strokes. Finally, understand how calcium ions released from the sarcomeres by action potentials initiate and maintain the whole process of muscle contraction.
- The sarcomere is the functional unit of the muscle cell, and until you understand its fine structure—Z lines, H lines, etc.—it will be difficult to appreciate how the sarcomere shortens through the actions of actin and myosin.
- Many people think of bone as tissue that is not living. In fact, bone is a living tissue that is constantly remodeled.
- Go to **LaunchPad** (or use the Web addresses listed) to review the following additional resources:

  Animated Tutorial 33.1 Molecular Mechanisms of Muscle Contraction (PoL2e.com/at33.1)

  Animated Tutorial 33.2 Smooth Muscle Action (PoL2e.com/at33.2)

  Activity 33.1 The Structure of a Sarcomere (PoL2e.com/ac33.1)

  Activity 33.2 The Neuromuscular Junction (PoL2e.com/ac33.2)

  Analyze the Data in textbook Figure 33.16

  Apply the Concept in textbook Section 33.1

## Key Concept Review

### 33.1 Muscle Cells Develop Forces by Means of Cycles of Protein–Protein Interaction

Contraction occurs by a sliding-filament mechanism

Actin and myosin filaments slide in relation to each other during muscle contraction

ATP-requiring actin–myosin interactions are responsible for contraction

Excitation leads to contraction, mediated by calcium ions

Muscle tissue makes up a large portion of the body in many animal species. Muscles are important because they are the basis of most animal behavior; they are necessary for movement and the operation of some internal organs. Skeletal muscle is attached to the skeleton and responsible for movements of the body. Muscles create mechanical forces through contraction (though not necessarily shortening). The sliding-filament theory describes the molecular mechanism of muscle contraction.

A muscle cell is sometimes called a muscle fiber. Vertebrate muscle cells are very large and multinucleate; they form through fusion of embryonic cells called myoblasts. Each muscle contains hundreds or thousands of muscle fibers organized into bundles and surrounded by connective tissue (see Figure 33.2A). Contraction occurs due to the interaction of actin and myosin proteins, arranged in filaments, in the muscle fiber. The two types of filaments overlap in parallel arrays (see Figure 33.2B). Inside each muscle fiber are separate collections of actin and myosin filaments called myofibrils. The myofibrils are divided along their length into contractile units called sarcomeres; a Z line at either end defines each sarcomere. This organization gives skeletal muscle its striated appearance under the microscope. The center of each sarcomere is the M band. The myosin filaments extend symmetrically from the M bands. The region that contains the myosin filaments is the A band. The actin filaments extend from the Z lines. There are three light bands in a sarcomere: the H zone, centered on the M band, and two I band halves at each end. The protein titin runs from Z line to Z line and provides resistance to stretch in relaxed skeletal muscle. During muscle contraction, the Z lines move toward each other, and the H zone and I band shrink in size due to the sliding of actin filaments along the myosin filaments. This is the sliding filament contractile mechanism of muscle contraction (see Figure 33.3).

The proteins myosin and actin are the key to muscle contraction. Myosin molecules are made of two polypeptide chains wrapped around each other, each with a globular head at one end, much like two twisted golf clubs. A myosin filament is composed of many myosin molecules (see Figure 33.4). Actin filaments are composed of two monomer chains in a helical arrangement and look like two linear strings of pearls wrapped around each other. The proteins tropomyosin and troponin are associated with actin (see Figure 33.4).

Myosin heads change conformation when they bind to actin filaments at myosin binding sites, forming a cross-bridge connection. The conformational change in the myosin pulls the actin in toward the middle of the sarcomere. ATP then binds to an ATP binding site on myosin, resulting in the release of actin from the myosin and the return of myosin to its original conformation. Many myosin molecules cycle through binding with actin to shorten a sarcomere.

Muscle cells are excitable, as are nerve cells, meaning that the cell membrane can generate and conduct electrical impulses (action potentials). Action potentials (impulses) are regions of reversed polarity, also referred to as depolarizations. Action potentials initiate at one point in the cell membrane and travel down its length. Each muscle fiber has contact with the axon of a nerve cell, at a point called the neuromuscular junction. Excitation is the event that occurs when a nerve impulse reaches the neuromuscular junction and excites the muscle fiber. Excitation of a muscle fiber leads to contraction of a muscle by excitation–contraction coupling. Calcium ions play an important role in muscle contraction. Action potentials spread deep into the sarcoplasm (cytoplasm) of the muscle through transverse tubules (T tubules) that are in contact with the sarcoplasmic reticulum (endoplasmic reticulum) throughout the sarcoplasm (cytoplasm) (see Figure 33.6). When stimulated by a nerve impulse, the sarcoplasmic reticulum releases calcium ions. Calcium regulates contraction by binding with troponin, causing tropomyosin to twist and expose the actin–myosin binding sites on the actin filaments, in turn allowing myosin to form cross-bridges with the actin (see Figure 33.7). The release of myosin from actin requires ATP, accounting for the high energy needs of muscle contraction. After an excitation ends, calcium is pumped back into the sarcoplasmic reticulum, and tropomyosin and troponin again cover the myosin binding sites on actin filaments, preventing muscle contraction until another impulse arrives and the cycle begins again.

**Question 1.** Label the following structures in the diagram of skeletal muscle below: muscle, tendon, single muscle fiber, single myofibril, sarcomere, actin filament, myosin filament, Z line, A band, H zone, I band, M band. Also label the terms on the enlargement of the sarcomere.

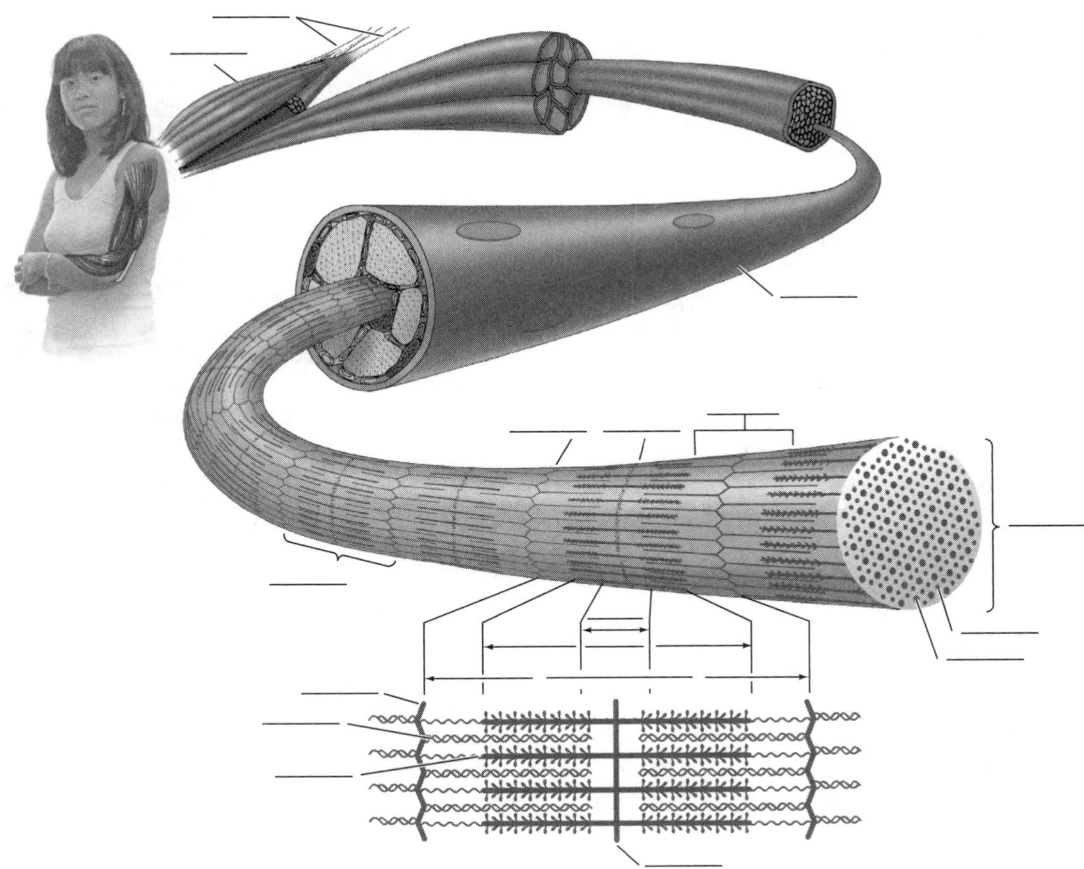

**Question 2.** What is the cause of rigor mortis, the stiffening of muscles after death?

### 33.2 Skeletal Muscles Pull on Skeletal Elements to Produce Useful Movements

In vertebrates, muscles pull on the bones of the endoskeleton

In arthropods, muscles pull on interior extensions of the exoskeleton

Hydrostatic skeletons have important relationships with muscle

Skeletal systems provide solid supports for the body, and structure against which the skeletal muscles pull to generate movement. Vertebrates have an endoskeleton, surrounded by the other tissues of the body. Arthropods have exoskeletons, with all of the tissue inside.

Endoskeletons are mainly made of bone (see Figure 33.8). Bone is a living, dynamic tissue that is a matrix of collagen fibers impregnated with calcium phosphate crystals. The several types of cells in bone are responsible for its maintenance and repair; bone is constantly altered throughout an individual's life. Bone is a reservoir of calcium, under the control of hormones and vitamin D. Cartilage in the endoskeleton is a flexible extracellular matrix. Joints are structures where two bones meet; muscles work at joints to produce movement. Flexible connective tissue called tendon attaches muscles to bones. A particular muscle can work only in one direction; there must be an antagonistic pair of muscles to create opposite movements. For instance, the biceps and triceps work to flex and extend the forearm, respectively. Sets of muscles work together to create complex movements under the control of the nervous system (see Figure 33.10).

Arthropod exoskeletons are make of chitin, and sometimes hardened with calcium deposits. The exoskeleton protects the animal, but must be shed when the animal grows. At joints, the muscles attach to inward projections of the exoskeleton called apodemes. Muscles work in antagonistic pairs to move parts of the exoskeleton with respect to each other.

Many soft-bodied invertebrates have a fluid-filled body cavity that acts as a hydrostatic skeleton, with high fluid pressure. Earthworms have circular muscles and longitudinal muscles that oppose each other to burrow through the soil (see Figure 33.12).

**Question 3.** Describe how the earthworm's hydrostatic skeleton is used to move the worm through the soil.

**Question 4.** Osteoporosis is a decrease in bone density that results when the destruction of bone outpaces the formation of new bone, leading to thin, brittle bones. Why does weight-bearing exercise help prevent osteoporosis?

**Question 5.** Hydrostatic skeletons are found in annelids, such as earthworms, and cnidarians, such as hydras. Would a hydrostatic skeleton work for a terrestrial animal that moves by walking? Why or why not?

### 33.3 Skeletal Muscle Performance Depends on ATP Supply, Cell Type, and Training

Muscle power output depends on a muscle's current rate of ATP supply

Muscle cell types affect power output and endurance

Training modifies muscle performance

Muscles use the immediate, glycolytic, and oxidative systems to obtain ATP needed for contraction (see Figure 33.13). The immediate system utilizes preformed ATP and creatine phosphate. The glycolytic system anaerobically metabolizes carbohydrates to generate ATP. The oxidative system requires oxygen and completely metabolizes carbohydrates and fats to water and carbon dioxide. The three systems vary in how rapidly they produce ATP and how long they can continue to generate ATP. Depending on which system the cell is using, the power output and endurance of the muscle cell will change. The immediate system can supply ATP rapidly, but it can only function briefly. The glycolytic system can synthesize ATP at a slightly lower rate than the immediate system can supply, but it can also operate for only a short period of time (though longer than the immediate system). The oxidative system generates ATP more slowly than the other two, but it can operate indefinitely with sufficient oxygen supply. A muscle cell varies its performance depending on which system it uses. This explains the inverse relationship between speed and endurance.

Not all muscle cells have the same capacity to use the glycolytic and oxidative systems. Slow oxidative (slow-twitch) cells and fast glycolytic (fast-twitch) cells both occur in any given species of animal. They differ in three main ways: their principal mechanism for making ATP, their mechanism of oxygen uptake from the blood, and their form of myosin.

Slow oxidative cells use the oxidative system and thus need a lot of oxygen. In vertebrates, these cells have high levels of myoglobin, which increases the speed of oxygen entry, making the cells red. They also have large numbers of mitochondria to support oxidative phosphorylation. Their form of myosin has relatively low ATPase activity, meaning they have slow turnover of actin–myosin cross-bridges. They develop tension slowly and are best suited for sustained, low-power output.

Fast glycolytic cells have high levels of the enzymes needed to perform glycolysis, and have myosin with high ATPase activity. These cells recycle actin–myosin cross-bridges rapidly. They contract and develop tension quickly, but fatigue rapidly as well. They have relatively few mitochondria and no myoglobin and so do not appear red.

Individual people vary in their proportions of red and white muscle cells, which makes a difference in high-level physical competition. Endurance athletes tend to have mostly slow oxidative fibers, and other elite-level athletes, who need short bursts of high energy, tend to have higher numbers of fast glycolytic cells (see Figure 33.15). Training can alter the proportion of red and white cells in muscle, and is a good example of phenotypic plasticity (see Section 29.5). Lifting weights increases muscle size due to increase of actin and myosin content of the cells, not an increase in the number of cells. Endurance exercise is steady and continuous, such as cross-country running and distance cycling. Resistance exercise features relatively few repetitions over a short period of time, such as lifting weights and wrestling. Endurance exercise causes muscle cells to increase numbers of mitochondria, causes some fast glycolytic cells to become slow oxidative cells, and stimulates growth of blood capillaries. These effects are due to alteration in expression of over 100 different genes. Resistance exercise causes muscles to build larger contractile apparatus, improving the force delivered by contraction, and causes some cells to convert from slow oxidative to fast glycolytic.

**Question 6.** White muscle and red muscle are found in different parts of the body and are used for different types of movement. What are the physiological and morphological characteristics that distinguish the two types of muscle?

**Question 7.** In turkeys, the breast (flight) muscles are light in color while in mallard ducks they are dark. What does this tell you about the locomotor capabilities of these two birds?

**Question 8.** Some athletes hoping to improve their performance take creatine as a dietary supplement. Some reports suggest that creatine supplements boost performance in activities that require short bursts of energy, such as sprinting, but not in those that require endurance. Why might this be the case?

### 33.4 Many Distinctive Types of Muscle Have Evolved

Vertebrate cardiac muscle is both similar to and different from skeletal muscle

Vertebrate smooth muscle powers slow contractions of many internal organs

Some insect flight muscle has evolved unique excitation–contraction coupling

Catch muscle in clams and scallops stays contracted with little ATP use

Fish electric organs are composed of modified muscle

Vertebrate cardiac (heart) muscle shares some characteristics with vertebrate skeletal muscle: they both appear striated because of the regular pattern of sarcomeres. However, cardiac muscle cells are much smaller than skeletal muscle cells and have only one nucleus. Gap junctions join two adjacent cardiac muscle cells where the two sarcoplasms (cytoplasms) are continuous through pores. If one cell is excited, its neighbors will also be excited due to this continuity. This guarantees that all the cardiac muscle cells are excited at the same time, allowing synchronous contraction and pumping of the blood through the heart. The cardiac cells themselves generate the heart rhythm in a vertebrate, which is unlike the breathing rhythm (controlled by skeletal muscle). Thus, a heart can continue beating in the absence of a nerve impulse.

Vertebrate smooth muscle cells are spindle-shaped with a single nucleus. They do not have regular arrangements of the contractile filaments and thus do not have the striated appearance of skeletal and cardiac muscle cells. Smooth muscles are found in the internal organs, such as the blood vessels and digestive tract, and are controlled by the nervous system. Some smooth muscle tissue is arranged in sheets in which gap junctions join the cells. Excitation of one cell spreads, creating coordinated contraction of the cells; this can create peristalsis, a wave of contraction from one end of the gut to the other. Calcium influx is required for excitation–contraction coupling in cardiac and smooth muscle, just as in skeletal muscle, but the details are slightly different.

There are many different types of invertebrate muscle, two of which are asynchronous flight muscle and catch muscle. In 25 percent of insect species, each contraction requires a separate excitation. Higher frequencies of contraction require excitations that occur at higher frequencies, but contraction becomes compromised. Some insects have evolved asynchronous muscle in which each excitation causes many contractions. At high contraction frequency, asynchronous muscle maintains greater power and higher efficiency than synchronous muscle. Asynchronous muscle is found in about 75 percent of insect species.

Bivalves, like scallops and clams, close their shells by contraction of adductor muscles, which can be used for defense against predators. The adductors can enter a state called "catch" in which they maintain contraction for protracted periods with almost no use of ATP. The way this works is not well understood. The catch state can be quickly ended by a signal from the nervous system. It is adaptive for animals that need to use muscular contraction to make their body armor effective.

Electric eels and similar fish produce high voltages or currents in the water around them (up to 700 volts or 20 amperes). The electric organs are evolved skeletal muscle structures and consist of modified muscle cells. The contractile apparatus is poorly developed and so they are not able to contract; however, the voltage differences across each cell membrane are added together when the cells are excited. Each cell has a voltage of about 0.1 V, but when several cells are added together significant voltages can be achieved.

**Question 9.** Curare is a poison from South America that is applied to the tips of poison arrow darts. Mammals hit by the darts die by asphyxiation because their respiratory muscles cannot contract. Suggest a mechanism by which curare might work.

**Question 10.** A sample of electric eel tissue has 2,000 cells. How large a voltage would you expect this electric organ to generate?

## Test Yourself

1. Which statement is *false*?
   a. Peristalsis is a random set of smooth muscle cell contractions.
   b. Glycolysis is the source of anaerobic ATP in vertebrate muscle.
   c. Myoglobin is responsible for maintaining large reserves of oxygen in muscle tissue.
   d. A hydrostatic skeleton maintains its shape with high fluid pressure.
   e. Parathyroid hormone and vitamin D regulate the deposition of calcium in bone.

   ***Textbook Reference:*** *Concept 33.4 Many Distinctive Types of Muscle Have Evolved; Vertebrate smooth muscle powers slow contractions of many internal organs*

2. Endoskeletons
   a. are characteristic of arthropods.
   b. are located on the inside of the body.
   c. lack joints.
   d. require molting as the animal grows.
   e. provide support to earthworms, along with their hydrostatic skeleton.

   ***Textbook Reference:*** *Concept 33.2 Skeletal Muscles Pull on Skeletal Elements to Cause Useful Movements*

3. $Ca^{2+}$ binds to _______ in skeletal muscle and leads to exposure of the binding site for _______ on the _______ filament.
   a. troponin; myosin; actin
   b. troponin; actin; myosin
   c. actin; myosin; troponin
   d. tropomyosin; myosin; actin
   e. myosin; actin; troponin

   ***Textbook Reference:*** *Concept 33.1 Muscle Cells Develop Forces by Means of Cycles of Protein–Protein Interaction; Excitation leads to contraction, mediated by calcium ions*

4. Which of the following would be the source of ATP during the last mile of a three-mile walk?
   a. ATP stored in muscle
   b. Glycolysis
   c. Creatine phosphate
   d. Oxidative metabolism
   e. Calcium ions

   ***Textbook Reference:*** *Concept 33.3 Skeletal Muscle Performance Depends on ATP Supply, Cell Type, and Training; Muscle power output depends on a muscle's current rate of ATP supply*

5. You place a sample of muscle tissue under a light microscope and see that most of the cells are dark-colored and filled with mitochondria. The cell extract has relatively low ATPase activity. This muscle tissue most likely came from which of the following sources?
   a. The leg muscle of a marathon runner
   b. The leg muscle of a sprinter

c. A biceps muscle from a weightlifter
d. A biceps muscle from a pole vaulter
e. There is too little information to make a hypothesis.
***Textbook Reference:*** *Concept 33.3 Skeletal Muscle Performance Depends of ATP Supply, Cell Type, and Training; Training modifies muscle performance*

6. Which statement is *false*?
a. Cardiac muscle is striated.
b. Actin is absent from smooth muscle.
c. Skeletal muscle is considered voluntary.
d. Smooth muscle is found in the digestive tract and the walls of the bladder.
e. A single skeletal muscle cell has many nuclei.
***Textbook Reference:*** *Concept 33.4 Many Distinctive Types of Muscle Have Evolved; Vertebrate smooth muscle powers slow contractions of many internal organs*

7. The oxygen-binding molecule in skeletal muscle is
a. myoglobin.
b. hemoglobin.
c. ATP.
d. myokinase.
e. creatine phosphate.
***Textbook Reference:*** *Concept 33.3 Skeletal Muscle Performance Depends on ATP Supply, Cell Type, and Training; Muscle cell types affect power output and endurance*

8. The action potential that triggers a muscle contraction travels deep within the muscle cell by means of
a. the sarcoplasmic reticulum.
b. transverse (or T) tubules.
c. synapses.
d. motor end plates.
e. neuromuscular junctions.
***Textbook Reference:*** *Concept 33.1 Muscle Cells Develop Forces by Means of Cycles of Protein–Protein Interaction; Excitation leads to contraction, mediated by calcium ions*

9. A sarcomere is best described as a
a. structural unit within a myofibril bounded by H zones.
b. structural unit within a myofibril bounded by I bands.
c. structural unit within a myofibril bounded by A bands.
d. structural unit within a myofibril bounded by Z lines.
e. collection of myofibrils.
***Textbook Reference:*** *Concept 33.1 Muscle Cells Develop Forces by Means of Cycles of Protein–Protein Interaction; Actin and myosin filaments slide in relation to each other during muscle contraction*

10. Which protein moves tropomyosin?
a. Calmodulin
b. Acetylcholine
c. Actin
d. Troponin
e. Titin
***Textbook Reference:*** *Concept 33.1 Muscle Cells Develop Forces by Means of Cycles of Protein–Protein Interaction; Excitation leads to contraction, mediated by calcium ions*

11. Rank the three ATP generation systems of muscle tissue based on power output, from highest to lowest.
a. Oxidative, glycolytic, immediate
b. Glycolytic, oxidative, immediate
c. Immediate, oxidative, glycolytic
d. Immediate, glycolytic, oxidative
e. Oxidative, immediate, glycolytic
***Textbook Reference:*** *Concept 33.3 Skeletal Muscle Performance Depends on ATP Supply, Cell Type, and Training; Muscle power output depends on a muscle's current rate of ATP supply*

12. Which type of muscle has evolved in insects to allow high frequencies of contraction while maintaining power output?
a. Cardiac muscle
b. Synchronous muscle
c. Asynchronous muscle
d. Catch muscle
e. Myoblasts
***Textbook Reference:*** *Concept 33.4 Many Distinctive Types of Muscle Have Evolved; Some insect flight muscle has evolved unique excitation–contraction coupling*

13. A soccer player who has suffered a knee injury that damages the tissue holding his anterior thigh muscle to his knee bone (patella) has most likely damaged _______ tissue.
a. muscle
b. tendon
c. ligament
d. cartilage
e. membrane
***Textbook Reference:*** *Concept 33.2 Skeletal Muscles Pull on Skeletal Elements to Cause Useful Movements; In vertebrates, muscles pull on the bones of the endoskeleton*

14. ATP provides the energy for muscle contraction by allowing for the
a. formation of an action potential in the muscle cell.
b. breaking of actin–myosin bonds.
c. formation of actin–myosin bonds.
d. release of calcium by the sarcoplasmic reticulum.
e. formation of T tubules.
***Textbook Reference:*** *Concept 33.1 Muscle Cells Develop Forces by Means of Cycles of Protein–Protein Interaction; ATP-requiring actin–myosin interactions are responsible for contraction*

15. Which of the following people would likely have the densest bones?
a. A 14-year-old female who swims daily
b. A 20-year-old female who bicycles daily

c. An astronaut who has just returned from six weeks in space
d. A 30-year-old male who eats calcium-rich foods and jogs regularly
e. A 60-year-old male who does sit-ups daily

***Textbook Reference:*** *Concept 33.2 Skeletal Muscles Pull on Skeletal Elements to Cause Useful Movements; In vertebrates, muscles pull on the bones of the endoskeleton*

# Answers

## *Key Concept Review*

1.

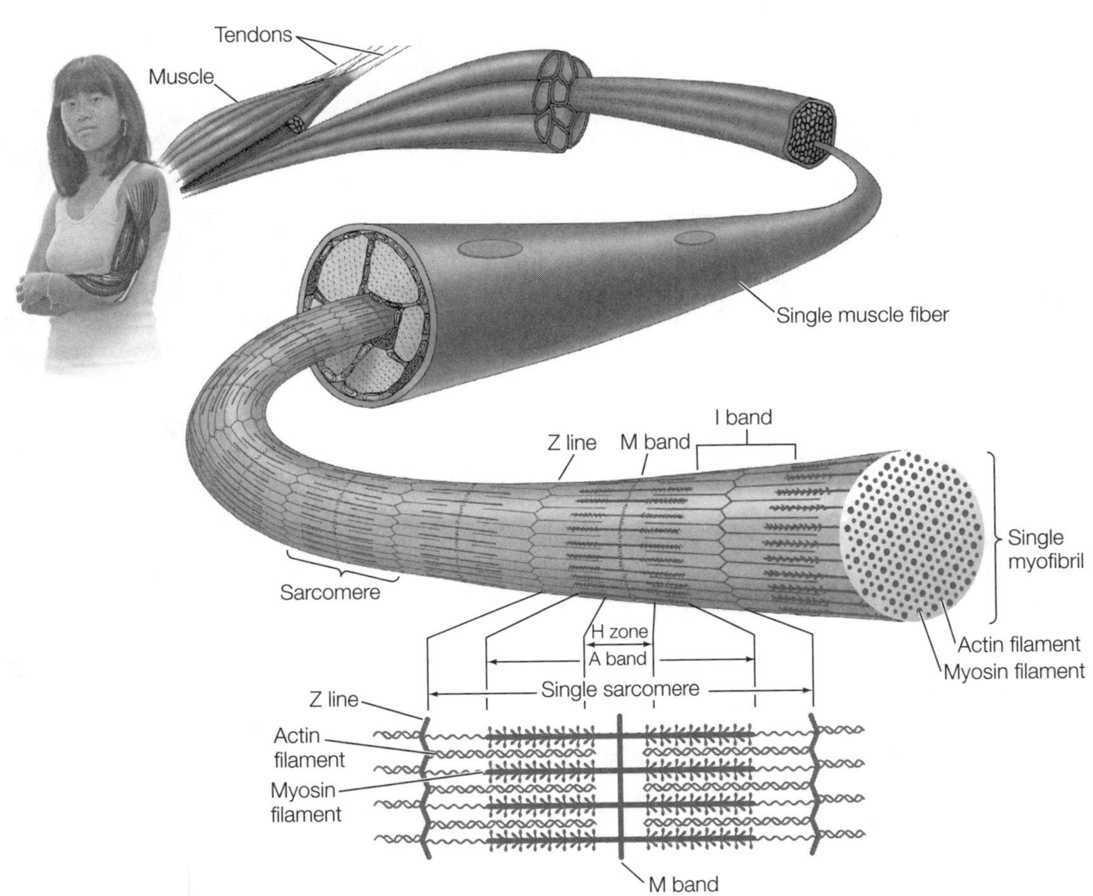


2. ATP is needed to break the actin–myosin bonds. ATP production stops at death, so the actin–myosin bonds cannot be broken and muscles stiffen. Eventually, the proteins deteriorate and the muscles soften.

3. The hydrostatic skeleton of the earthworm is an incompressible fluid-filled cavity surrounded by longitudinal and circular muscles. Contractions of particular muscles raise the fluid pressure in the worm high enough to expand forward through the soil. The worm then anchors itself, and contraction of different muscles shortens the length of the worm and brings the back end closer to the front. The process repeats to create net movement.

4. Weight-bearing exercises place stress on bone, ultimately altering the activity of bone cells to induce thickening of bone.

5. Hydrostatic skeletons work well for aquatic animals, such as hydras, and for terrestrial animals, such as earthworms that move by crawling through a substrate. However, a hydrostatic skeleton would not work for most terrestrial animals. Such skeletons provide little or no protection against drying out (dehydration is always a danger on land), and they do not provide sufficient support for a large animal that walks by holding its body off the ground.

6. Fast glycolytic fibers are known as white muscle and have few mitochondria, small amounts of myoglobin, and few blood vessels. Muscles with many fast glycolytic fibers are good for short-term work that requires maximum strength. Slow oxidative fibers are known as red muscle. Red muscle has many mitochondria, large amounts of myoglobin, and many blood vessels. Muscles with many slow oxidative fibers function well in endurance activities.
7. The breast muscles of turkeys are fast glycolytic, or white, muscle (= "white meat"), and their leg muscles are slow oxidative, or red, muscle (= "dark meat"). This tells us that turkeys fly only in short bursts and typically walk or run. In contrast, the breast muscles of mallard ducks are slow oxidative, or red, muscle, indicating that they are capable of prolonged flight.
8. Taking creatine supplements may increase stores of creatine phosphate in muscle. Creatine phosphate stores energy in a phosphate bond, which it can transfer to ADP to form ATP. However, the energy is available immediately and the supply is quickly exhausted (creatine phosphate is part of the immediate system for supplying ATP to muscle). Thus, we would predict that creatine supplements would be most useful for activities such as sprinting, in which fast glycolytic fibers generate a lot of force quickly.
9. Skeletal muscles are used in breathing. Curare acts by preventing the neuromuscular junction from operating properly and sending a nerve impulse from the neuron to the muscle fiber.
10. Each cell generates a membrane potential of 0.1 V (100 mV). When all the cells depolarize together, the total voltage generated would be 200 V (200,000 mV).

### *Test Yourself*

1. **a.** Peristalsis is a wave of contractions moving along the gut tube from one end to the other.
2. **b.** Endoskeletons (such as those of mammals) are found inside the body, and exoskeletons (such as those of insects) are found outside the body.
3. **a.** Calcium is released from the sarcoplasmic reticulum and binds with troponin, resulting in exposure of the myosin-binding site on actin.
4. **d.** Oxidative metabolism would supply ATP needed by muscles in the last mile of a three-mile walk.
5. **a.** This muscle tissue is mostly slow oxidative cells, and thus comes from the distance athlete who needs to maintain effort over a relatively long period.
6. **b.** Although contraction of smooth muscle is controlled differently from that of skeletal muscle, smooth muscle does contain actin and myosin.
7. **a.** Myoglobin is the main oxygen-carrying molecule in skeletal muscle.
8. **b.** The action potential arriving to the muscle travels into the muscle through the transverse (or T) tubules.
9. **d.** A sarcomere is a structural unit within a myofibril bounded by Z lines; it contains actin and myosin.
10. **d.** The binding of calcium with troponin causes a conformational change in tropomyosin.
11. **d.** The immediate system supplies very little ATP, but allows muscle cells to reach the highest power output. The oxidative system can generate ATP indefinitely, but does not generate much power in the muscle contractions.
12. **c.** Asynchronous muscle does not have a 1:1 ratio of excitations to contractions, and thus allows one excitation to stimulate multiple contractions.
13. **b.** Tendons connect muscles to bone.
14. **b.** ATP provides energy that is used to break actin–myosin bonds.
15. **d.** Placing stress on bones, which would occur through jogging, keeps them dense and healthy. Swimming, cycling, and a zero-gravity environment, such as space, do not stress bones.

# Neurons, Sense Organs, and Nervous Systems

## The Big Picture

- The neuron is the functional unit of the nervous system. Neurons generate membrane potentials, which are differences in charge across their membranes. Nerve impulses pass down the axon of a neuron as action potentials—disruptions in the resting neuron membrane due to the opening and closing of sodium and potassium voltage-gated channels.
- Axon terminals form synapses with target cells. Neurotransmitters transmit a nerve impulse from the presynaptic cell to the postsynaptic cell. A neurotransmitter is specific to the receptor to which it binds.
- Sensory structures convert stimuli into action potentials in the nervous system, which then are interpreted by the central nervous system.
- The anatomical divisions of the mammalian nervous system are the central nervous system (CNS) and peripheral nervous system (PNS). The functional divisions are the sympathetic and parasympathetic nervous systems. Afferent nerves carry information from the PNS to the CNS, and efferent nerves carry information from the CNS to the PNS.

## Study Strategies

- One of the most difficult challenges when learning about the nervous system is understanding how the resting membrane potential and action potential are produced. Remember that the neuron at rest is like a battery, and that the charge of the membrane is dependent on the ions on either side. Learn what each important anion and cation does at rest and during the action potential.
- Make sure you understand the concept of "pre-" and "postsynaptic" neurons. Try to remember the sequence of events and relate them to the function of the neuron transmitting a stimulus from one cell to another.
- Examine the function of a synapse and how the action potential is transmitted across the synapse.
- It is easy to confuse "afferent" and "efferent," and it is important in your understanding of the nervous system to differentiate between them. Think of *a*fferent as *a*rriving, and *e*fferent as *e*xiting the reference point—usually the CNS.
- Several key terms in this chapter occur repeatedly: parasympathetic, sympathetic, afferent, efferent, preganglionic, and postganglionic. Until you master this vocabulary, it will be difficult to put together a comprehensive picture of the nervous system. Create your own list of the terms you see repeatedly, and make sure that you understand their meanings.
- All of the senses seem to have different mechanisms for the transmission of information to the brain. Remember that there are only a few types of receptors that respond to stimuli and that they all generate action potentials. The steps in sensory transduction are also similar in all the different sensory systems.
- The neurons of the ear, eye, knee, stomach, or any other part of the body all fire action potentials. However, the action potentials are interpreted differently (e.g., action potentials coming from the eye are interpreted as light) because of the region of the brain that receives and analyzes them.
- The route by which sound travels in the ear can be very confusing. View the cochlea in the uncoiled form as shown in Figure 34.16A. This will help you visualize how pressure waves of different wavelengths produce different sounds.
- Go to **LaunchPad** (or use the Web addresses listed) to review the following additional resources:

  Animated Tutorial 34.1 The Resting Membrane Potential (PoL2e.com/at34.1)

  Animated Tutorial 34.2 The Action Potential (PoL2e.com/at34.2)

  Animated Tutorial 34.3 Synaptic Transmission (PoL2e.com/at34.3)

  Animated Tutorial 34.4 Neurons and Synapses (PoL2e.com/at34.4)

  Animated Tutorial 34.5 Mechanoreceptor Simulation (PoL2e.com/at34.5)

  Animated Tutorial 34.6 Sound Transduction in the Human Ear (PoL2e.com/at34.6)

Animated Tutorial 34.7 Photosensitivity (PoL2e.com/at34.7)

Animated Tutorial 34.8 Information Processing in the Spinal Cord (PoL2e.com/at34.8)

Activity 34.1 The Nernst Equation (PoL2e.com/ac34.1)

Activity 34.2 Structures of the Human Ear (PoL2e.com/ac34.2)

Activity 34.3 Structure of the Human Eye (PoL2e.com/ac34.3)

Activity 34.4 Structure of the Human Retina (PoL2e.com/ac34.4)

Activity 34.5 Structures of the Human Brain (PoL2e.com/ac34.5)

Apply the Concept in textbook Section 34.3

## Key Concept Review

### 34.1 Nervous Systems Are Composed of Neurons and Glial Cells

Neurons are specialized to produce electric signals

Glial cells support, nourish, and insulate neurons

The nervous system is made up of two major types of cells: neurons and glia (also called glial cells). A neuron is composed of four main parts: the cell body, dendrites, axon, and axon terminals (see Figure 34.2). The cell body contains the nucleus and other organelles. Dendrites extend from the cell body and receive information from other neurons. The axon conducts action potentials away from the cell body. Axon terminals interact with the neuron's target cells to form a synapse, a tiny gap where information is passed from one neuron to another neuron. The first neuron is the presynaptic neuron, and the second is the postsynaptic neuron. Synapses can be either chemical or electrical.

Glia serve many functions in the nervous system, such as supplying nutrients to neurons, removing wastes, and helping neurons make the proper connections during development. Glia that insulate axons in the central nervous system are oligodendrocytes and those that insulate axons in the peripheral nervous system are Schwann cells (see Figure 34.3). The covering produced by Schwann cells and oligodendrocytes is called myelin; myelinated axons conduct action potentials more rapidly than do axons lacking myelin. Star-shaped glia, called astrocytes, contribute to the blood–brain barrier, which protects the brain from some toxins in the blood.

**Question 1.** Why do certain substances, such as anesthetics and alcohol, have rapid effects on the brain, whereas others cannot reach the brain?

**Question 2.** In Lou Gehrig's disease, motor neurons die and muscles no longer receive neural messages. Affected individuals gradually lose control over their limbs and body, and the eventual cause of death is respiratory failure due to deterioration of the muscles that control breathing. Sensory neurons and interneurons are not affected by the disease, and the individuals do not have loss of cognitive function. Explain in terms of the neurological processes why cognition is unimpaired.

### 34.2 Neurons Generate Electric Signals by Controlling Ion Distributions

Only small shifts of ions are required for rapid changes in membrane potential

The sodium–potassium pump sets up concentration gradients of $Na^+$ and $K^+$

The resting potential is mainly a consequence of $K^+$ leak channels

The Nernst equation predicts an ion's equilibrium potential

Gated ion channels can alter membrane potential

Changes in membrane potential can be graded or all-or-none, depending on whether a threshold is crossed

An action potential is a large depolarization that propagates with no loss of size

Action potentials travel particularly fast in large axons and in myelinated axons

Current is a flow of electric charges from one place to another. Voltage (electric potential difference) exists if positive charges are concentrated in one place and negative charges are concentrated in another place. Voltage differences exist only across cell membranes; they do not exist in open solutions. Because of the nature of plasma membranes (phospholipid bilayer), ions do not readily cross the plasma membrane. Bulk solutions are not immediately in contact with a membrane and are electrically neutral.

The sodium–potassium pump (also known as sodium–potassium ATPase) transports $Na^+$ out of the cell and $K^+$ into it, thereby maintaining higher concentrations of $Na^+$ ions outside the cell and higher concentrations of $K^+$ ions inside it. At rest, neurons have a specific charge due to $K^+$ movement to the outside of the cell, resulting in a resting potential. $K^+$ channels are the most commonly open (sometimes called $K^+$ leak channels), allowing $K^+$ to diffuse out of the cell down the concentration gradient that has been set up by the $Na^+$–$K^+$ pump. $K^+$ leak channels are largely responsible for the membrane's resting potential.

Resting neurons have a negative charge inside and a positive charge outside, resulting in a difference in electrical charge across the membrane known as the membrane potential. In an unstimulated neuron, this voltage difference is called a resting potential. Electrodes can be used to measure resting potentials; such measurements show that the resting potential is typically between –60 and –70 mV (see Figure 34.4). The electrical charge across the membrane at rest is due to differences in concentrations of the charged ions sodium ($Na^+$), chloride ($Cl^-$), potassium ($K^+$), and calcium ($Ca^{2+}$). The lipid bilayer of the plasma membrane is impermeable to ions. Ions move across the plasma membrane through ion transporters and channels.

Ion channels are selective pores in the plasma membrane that allow specific ions to diffuse across the membrane. Whereas some are always open (such as $K^+$ channels), others are gated (open under certain conditions and closed under other conditions). Ion channels can be voltage-gated (responding to changes in the voltage across the membrane), stretch-gated (responding to mechanical force applied to the membrane) or ligand-gated (responding to the binding of a specific molecule) (see Figure 34.5).

Membranes can be depolarized or hyperpolarized (see Figure 34.6). Depolarization occurs when the inside of a neuron becomes less negative compared to the resting potential. Hyperpolarization occurs when the inside of a neuron becomes more negative compared to the resting potential.

Small, local changes in membrane potential are called graded membrane potentials. Action potentials are large but short-lived changes in membrane potential (see Figure 34.7). Action potentials are generated when the membrane reaches a threshold potential, $Na^+$ voltage-gated channels open, and $Na^+$ enters the cell to make the inside of the axon positive. Voltage-gated $K^+$ channels then open, allowing $K^+$ to leave the axons to help return the membrane potential back to the resting level. As the $K^+$ channels open, the $Na^+$ channels close and cannot be opened for a few milliseconds, which is called the refractory period. The $Na^+$–$K^+$ pump helps return the concentration of ions back to the resting levels.

Action potentials travel down an axon by a positive feedback mechanism that stimulates adjacent regions of an axon to generate the action potential (see Figure 34.8). The $Na^+$ ions that enter during an action potential flow to adjoining regions of the axon, stimulating depolarization and the movement of the action potential along the axon. The refractory period, during which the $Na^+$ channels cannot act, can be explained by the presence of two gates in the channel, an activation gate and an inactivation gate. The refractory period keeps an action potential moving in one direction, away from the cell body.

An action potential is an all-or-none response: the depolarization must reach a threshold level for an action potential to occur. An action potential is also a self-regenerating response; once an action potential occurs at one location on an axon, it stimulates the adjacent area to generate an action potential.

In the nervous systems of invertebrates, the conduction velocity of axons increases with the increasing diameter of axons. In the nervous systems of vertebrates, the conduction velocity of axons is increased by myelination, or the concentrated layers of myelin formed by glia that wrap themselves around the axons. The glia leave regularly spaced gaps called nodes of Ranvier, which are the sites where depolarization can occur. As action potentials jump from node to node, the speed of transmission increases in a process known as saltatory conduction (see In-Text Art, p. 708 [1]).

**Question 3.** In order to determine the role of the potassium channels in a neuron, a researcher has knocked out all the functional potassium channels and depolarized the membrane potential. What will happen to the membrane potential after depolarization?

**Question 4.** Label the following structures on the unmyelinated motor neuron shown below: axon, axon hillock, axon terminals, cell body, and dendrites. What is the direction and manner in which an action potential is conducted along the neuron?

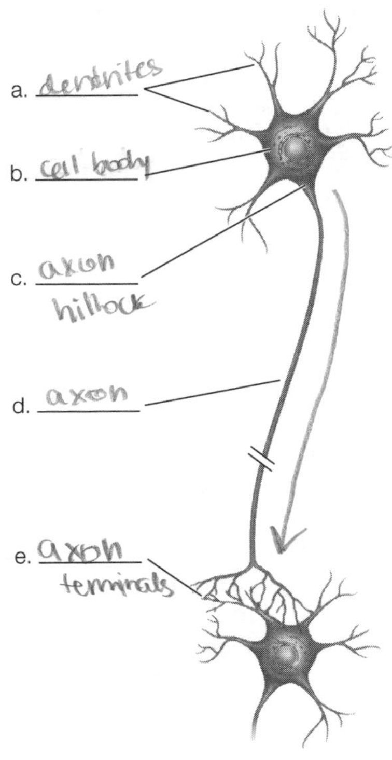


**Question 5.** Explain how an action potential travels more quickly down an axon wrapped in myelin than it does down an unmyelinated axon.

## 34.3 Neurons Communicate with Other Cells at Synapses

- Chemical synapses are most common, but electrical synapses also exist
- The vertebrate neuromuscular junction is a model chemical synapse
- Many neurotransmitters are known
- Synapses can be fast or slow depending on the nature of receptors
- Fast synapses produce postsynaptic potentials that sum to determine action potential production
- Synaptic plasticity is a mechanism of learning and memory

Chemical synapses are more common than electrical synapses. At a chemical synapse, an action potential arriving at the axon terminals of a presynaptic neuron causes the release of neurotransmitter, which travels across the synaptic cleft to bind with receptors on the postsynaptic neuron. Electrical synapses are formed by direct contact between adjacent neurons; these synapses contain numerous gap junctions. Electrical synapses allow for rapid communication.

Neuromuscular junctions are chemical synapses between motor neurons and skeletal muscle cells. Each muscle fiber in vertebrate skeletal muscle is typically innervated by one motor neuron. At the junction, the axon of the neuron branches into axon terminals that are in synaptic contact with the muscle cells. When an action potential reaches the axon terminal, calcium channels on the presynaptic membrane open, causing calcium to diffuse into the cell. This results in the release of the neurotransmitter acetylcholine (ACh) into the synaptic cleft. Acetylcholine binds to ligand-gated receptors on the muscle cell, causing ACh-gated receptors on the postsynaptic membrane to open, allowing $Na^+$ to diffuse into the cell. This causes a large, graded depolarization. If depolarization is great enough to exceed threshold, an action potential is propagated through the muscle cell, activating contraction.

Neurotransmitters fall into three major categories: amino acids, biogenic amines, and peptides. Presynaptic neurons produce and release specific neurotransmitters. In the brain, a postsynaptic neuron may have hundreds or thousands of presynaptic neurons. Each may produce different neurotransmitters, so many types of neurotransmitters may be released.

Two general categories of neurotransmitter receptors are ionotropic and metabotropic. Ionotropic receptors are ion channels on the postsynaptic membrane that are activated by binding of the neurotransmitter. They allow fast, short-lived responses. Metabotropic receptors are not ion channels. They act by initiating signaling cascades, which eventually cause changes in ion channels. When mediated by metabotropic receptors, postsynaptic cell responses are usually slower and longer-lived than those generated by ionotropic receptors.

Excitatory synapses between motor neurons and muscle cells in vertebrates depolarize the postsynaptic membrane, and inhibitory synapses between neurons hyperpolarize the postsynaptic membrane. Neurons may receive synaptic inputs from many neurons.

Excitatory and inhibitory postsynaptic potentials are summed by spatial summation (adding up of simultaneous potentials at different sites; see Figure 34.11) or by temporal summation (adding up of the postsynaptic potentials generated at the same site in rapid sequence). The region of the cell body at the base of the axon, called the axon hillock, is the "decision-making" area of a neuron. If the axon hillock is depolarized, the axon will fire an action potential.

Synapses of animals can undergo long-term changes in their functional properties and physical shape over the individual's lifetime, a phenomenon known as synaptic plasticity. This is proposed as one of the mechanisms of learning and memory. Learning refers to the ability of an individual animal to change its behavior in response to earlier experiences. Sea hares (marine mollusks), for example, become sensitized to noxious agents and withdraw their gills. The synapses between sensory neurons and motor neurons become functionally strengthened over time, leading to synaptic plasticity. The mammalian hippocampus shows synaptic plasticity as well. In living hippocampus tissue, repeated stimulation of neuron circuits causes them to grow physically and strengthen functionally.

**Question 6.** The active ingredients in many nerve gases belong to a class of chemicals called anticholinesterases (chemicals that block acetylcholinesterase). Explain a possible synaptic mechanism by which these chemicals can damage an animal's nervous system.

**Question 7.** Clinical depression is thought to be due, in part, to insufficient levels of the neurotransmitter serotonin. Drugs known as selective serotonin reuptake inhibitors (SSRIs) can be used to treat depression. Taking the name of this class of drugs as a clue, propose a mechanism by which they may act.

## 34.4 Sensory Processes Provide Information on an Animal's External Environment and Internal Status

- Sensory receptor cells transform stimuli into electrical signals
- Sensory receptor cells depend on specific receptor proteins and are ionotropic or metabotropic
- Sensation depends on which neurons in the brain receive action potentials from sensory cells
- Sensation of stretch and smell exemplify ionotropic and metabotropic reception
- Auditory systems use mechanoreception to sense sound pressure waves
- The photoreceptors involved in vision detect light using rhodopsins
- The vertebrate retina is a developmental outgrowth of the brain and consists of specialized neurons
- Some retinal ganglion cells are photoreceptive and interact with the circadian clock
- Arthropods have compound eyes
- Animals have evolved a remarkable diversity of sensory abilities

Sensory receptor cells are specialized cells (typically neurons) that transform stimulus energy into electrical signals. These cells carry out sensory transduction, which is the process of changing one type of energy into another. The signal produced then causes action potentials that convey the information to the brain or other parts of the nervous system. Sensory receptors are highly specific in the stimuli to which they respond. Heat, light, mechanical motion, and chemicals are the major types of stimulus energy detected by receptors.

Sensory receptor proteins in sensory receptor cells initially detect the stimulus, producing graded changes called receptor potentials. Receptor cells often have cell membranes with exceptionally large surface areas, making cells highly sensitive to stimuli. Ionotropic receptor cells typically have stimulus-gated $Na^+$ channels. Metabotropic receptor cells typically have receptor proteins that activate G proteins, leading to changes in concentrations of second messengers in the receptor cell.

Sensory systems convey information from specific stimuli due to physical separation of the brain and sensory axons innervating those separate parts. Olfactory receptors in mammals are found in the olfactory epithelium that lines the internal nasal cavity. Odorants dissolve in mucus covering the epithelium. G proteins are activated when odorant molecules bind to receptor proteins of the epithelium. Second messengers are produced that cause a graded depolarization. Each receptor cell produces one type of G protein–linked receptor protein (GPLR) which detects only the molecules that bind to it. Collectively, the GPLRs can detect a spectrum of odorants.

The hearing organs of many animals contain a thin membrane that moves inward when high-pressure sound waves strike it and moves outward when low-pressure sound waves strike it. Mechanoreceptors linked to the membrane detect these movements. Insects have simple membranes in which a single area of the membrane responds to all sound frequencies. Such a system simply detects sound pressure or the absence of it. In the mammalian ear (see Figure 34.15), the tympanic membrane (ear drum) initially responds to pressure waves. Three interconnected bones—the incus, malleus, and stapes—of the air-filled middle ear act as a lever system that turns vibration of the tympanic membrane into stronger vibrations on the oval window (part of the fluid-filled inner ear). The inner ear transmits vibrations from the oval window to the cochlea, a coiled, fluid-filled tube. This is where sound energy is transformed into electrical signals. The basilar membrane, which is found within the cochlea, varies in width and stiffness. Near the oval window, it is thin and stiff; at its far end, it is thick and less stiff. Low-frequency vibrations affect the wide end farthest from the oval window more than other areas of the membrane (see Figures 34.15C and 34.16B). The organ of Corti sits on the basilar membrane and houses the mechanoreceptors called hair cells. When oscillations reach the hair cells, they bend in response. This bending gets transduced into electrical signals. The cells synapse with neurons, releasing neurotransmitters that produce action potentials (see Figure 34.16A).

Many animals have visual systems that sense and respond to light. Simple systems allow animals to orient to the sun and sky. Complex systems provide animals with fast and detailed information about their environment. Photoreceptors are the sensory receptor cells. They are sensitive to light and act as metabotropic receptor cells. Rhodopsins are the sensory receptor proteins. They act as G protein–linked receptors and are made of two parts: the protein opsin and the nonprotein 11-*cis*-retinal. When rhodopsins absorb photons of light, they change conformation, with 11-*cis*-retinal twisting, changing the conformation of opsin (see In-Text Art, p. 718). This activates a G protein–mediated signal, with a second messenger affecting voltage-gated $NA^{+}$ channels in the photoreceptor membrane.

Vertebrate eyes form images. The cornea is transparent connective tissue at the front of the eye. In the center of the colored iris is the pupil, through which light enters the light-sensing part of the eye. The iris regulates the size of the pupil, and thus the amount of light entering the eye. The lens lies behind the pupil. In mammals, muscles change the shape of the lens to focus light on the retina, which is the photosensitive layer. The retina contains rod cells and cone cells, both of which have large numbers of rhodopsin molecules. In humans, rods outnumber cones. Rods are more sensitive to light and are important in dim light. Cones provide high-acuity color vision. Rods and cones produce graded membrane potentials and become hyperpolarized in response to light. Both synapse with other neurons in the retina and send signals to them by graded neurotransmitter release. These integrating neurons (bipolar cells, horizontal cells, amacrine cells, and ganglion cells; see Figure 34.19) are arranged in the retina such that light passes through them to reach the rods and cones. Horizontal and amacrine cells detect relative signal strengths of neighboring photoreceptor and bipolar cells. Horizontal cells also sharpen contrast between light and dark. Amacrine cells detect motion. All this information converges onto ganglion cells, which produce action potentials. The axons of ganglion cells form the optic nerve and carry action potentials to the visual cortex of the brain. Up to 2 percent of ganglion cells are photoreceptive. They have low light sensitivity and do not produce images. They provide information about the presence and intensity of light, which is used to entrain circadian biological clocks and in the regulation of pupil size.

Arthropods have compound eyes made up of optical units called ommatidia (see Figure 34.20). Each ommatidium has one lens. Light entering the ommatidium strikes the photoreceptor (retinula) cells, generating action potentials. Each ommatidium of a compound eye is directed at a slightly different part of the visual world. The more ommatidia in the eye, the higher the resolution of the image compiled in the brain.

**Question 8.** In the diagram of the vertebrate eye below, label each of the following structures: sclera, cornea, iris, pupil, lens, retina, fovea, vitreous humor, and optic nerve.

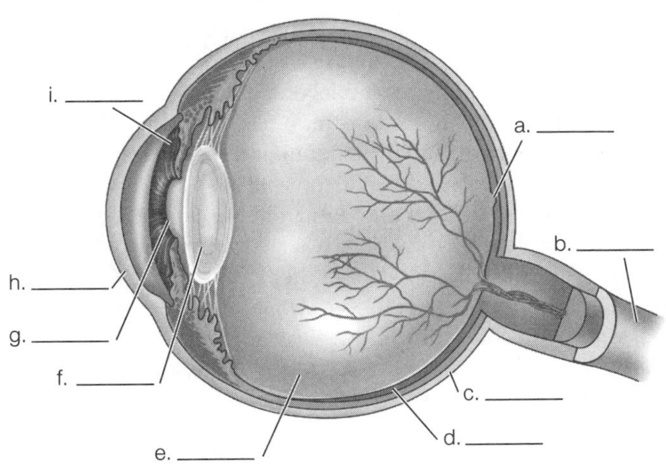


**Question 9.** You have just given a presentation in your biology class that was illustrated by many elaborate red- and green-colored slides. Afterward, a male friend tells you that

he could not see any of the differences you were reporting. Why could your friend not see the differences?

**Question 10.** Design a study to determine the lowest threshold of hearing for a mammal other than a human.

### 34.5 Neurons Are Organized into Nervous Systems

The autonomic nervous system controls involuntary functions

Spinal reflexes represent a simple type of skeletal muscle control

The most dramatic changes in vertebrate brain evolution have been in the forebrain

Location specificity is an important property of the mammalian cerebral hemispheres

Nervous systems consist of many neurons, estimated up to 86 billion in the human brain. The simplest nervous systems consist of a nerve net that is widely and randomly dispersed throughout the body. Cnidarians possess this type of system. With the evolution of increased complexity in nervous systems, centralization (the clustering of neurons) and cephalization (the concentration of integrating areas toward the head region) occurred. As animals move forward through their environment, the anterior head makes first contact with the environment.

The central nervous system (CNS) consists of large structures composed of integrating neurons and associated glial cells. The brain is the largest part of the CNS. The CNS interacts with sensory cells and sense organs, which provide information about the external environment and internal status of the organism. Effectors are cells or tissues that "carry out orders." These include muscles and glands. The peripheral nervous system (PNS) is made up of neurons and parts of neurons outside of the CNS. These neurons bring sensory information to the CNS and carry orders from the CNS to the effectors.

Neurons are classified as interneurons (neurons confined to the CNS that function in integration and information storage), sensory neurons (neurons that act as sensory receptor cells or that carry signals to the CNS from sensory cells or organs), and motor neurons (neurons that convey signals from the CNS to effectors).

The vertebrate CNS is positioned in the dorsal part of the body and is mostly hollow. It consists of the brain and spinal cord and is not differentiated into subparts. In arthropods, the CNS is ventral and solid, and it is composed of subparts, specifically ganglia, at each segment.

The autonomic nervous system (ANS) exists in all groups of vertebrates and controls effectors (autonomic effectors) other than skeletal muscle. It has components in the CNS and PNS. There is no ability to consciously or voluntarily control these effectors. The ANS controls smooth muscle, exocrine glands, and some endocrine glands such as the adrenal medulla. The three divisions of the ANS are the enteric, sympathetic, and parasympathetic divisions. The sympathetic and parasympathetic divisions consist mainly of efferent neurons (see Figure 34.22). Each efferent pathway consists of two neurons that synapse at a ganglion. These ganglion are discrete, anatomically clustered neuron cell bodies in the PNS. Preganglionic neurons (first of the two in the pathway) of the parasympathetic division exit the CNS from the cranial region of the brain and sacral region of the spinal cord. They extend almost to the target cells before synapsing with postganglionic neurons. Preganglionic neurons of the sympathetic division exit the CNS from the thoracic and lumbar regions of the spinal cord. Ganglia of the sympathetic division lie close to the spinal cord, and postganglionic neurons span greater distances to the effector organs. Sympathetic neurons release the neurotransmitter norepinephrine, while parasympathetic neurons release acetylcholine. In organs that have innervation of both sympathetic and parasympathetic neurons, target cells respond in opposite ways to the neurotransmitters. The divisions often work in opposition to each other. The flight-or-fight response is under sympathetic control. It greatly increases heart rate, force of heart contraction, cardiac output, and glucose release, and it dilates the passageways of the lungs.

Skeletal muscles are under voluntary control and are innervated by motor neurons and sensory neurons. Spinal reflexes are neuron-mediated responses that do not involve the brain. These neuron interactions occur in the spinal cord and require no conscious attention. The knee-jerk reflex is a good example of a spinal reflex. In the spinal cord, sensory axons synapse with motor neurons, generating an action potential. Limb muscles are organized into opposing pairs. When muscles contract, sensory neurons from opposing muscles relax due to inhibitory synapses.

The vertebrate brain consists of three parts: forebrain, midbrain, and hindbrain. All information traveling between the brain and spinal cord passes through the medulla oblongata, the posterior part of the hindbrain. Cerebral hemispheres carry out high-order sensory, motor, and integrative function, and are highly developed in birds and mammals. In humans, they are responsible for language and reasoning. Mammals and birds have evolved large cerebral hemispheres, which are important in higher brain function.

The cerebral cortex, the outermost layer of the cerebral hemispheres, is only 4 mm thick but is folded into ridges that increase its size. In the mammalian nervous system, the left side of the body is served mostly by the right side of the brain, and the right side of the body is served by the left brain. Location specificity is used by the brain to interpret action potentials. In each hemisphere specific regions are specialized for specific sensory and motor function. Functional magnetic resonance imaging (fMRI) can be used to indicate brain regions where there is increased neuronal activity. Brain region maps show that parts of the brain that serve various anatomical regions are physically related to each other in ways that mirror physical relationships of the rest of the body. The somatosensory part of the cerebral cortex is a well-known example (see Figure 34.29A). Language processing in humans occurs in one hemisphere, which for most people is the left hemisphere.

**Question 11.** Draw a diagram comparing the neural pathways for both the sympathetic and parasympathetic nervous systems. In your diagram, indicate relative locations of structures and lengths of preganglionic and postganglionic neurons.

**Question 12.** Species that can recognize themselves in a mirror, including humans, great apes, dolphins, and elephants, have expanded insulas (a section of the cerebral cortex) and are considered by some researchers to be self-aware. What do these species have in common, and how might the validity of the mirror test be challenged? What factors need to be considered when one is designing an experiment to test self-awareness in another species?

## Test Yourself

1. The extensions of postsynaptic neurons that provide the main receptive surface for presynaptic neurons are the
   a. nuclei.
   b. somas.
   c. axons.
   d. dendrites.
   e. glial cells.
   ***Textbook Reference:*** *Concept 34.1 Nervous Systems Are Composed of Neurons and Glial Cells; Neurons are specialized to produce electric signals*

2. In the human visual system, _______ send information directly to the brain.
   a. amacrine cells
   b. bipolar cells
   c. ganglion cells
   d. rods and cones
   e. horizontal cells
   ***Textbook Reference:*** *Concept 34.4 Sensory Processes Provide Information on an Animal's External Environment and Internal Status; The vertebrate retina is a developmental outgrowth of the brain and consists of specialized neurons*

3. The long extension from the cell body of a neuron that provides the pathway for action potentials to the synapse is the
   a. dendrite.
   b. Schwann cell.
   c. axon.
   d. presynaptic membrane.
   e. nerve net.
   ***Textbook Reference:*** *Concept 34.1 Nervous Systems Are Composed of Neurons and Glial Cells; Neurons are specialized to produce electric signals*

4. The threshold of a neuron is the
   a. amount of inhibitory neurotransmitter required to inhibit an action potential.
   b. membrane voltage at which an axon potential will be suppressed.
   c. amount of excitatory neurotransmitter required to elicit an action potential.
   d. membrane voltage at which the membrane potential develops into an action potential.
   e. closing of numerous sodium channels.
   ***Textbook Reference:*** *Concept 34.2 Neurons Generate Electric Signals by Controlling Ion Distributions; Changes in membrane potential can be graded or all-or-none, depending on whether a threshold is crossed*

5. When a membrane is at the resting potential, the concentration of
   a. sodium and potassium ions is higher on the inside of its membrane than on the outside.
   b. sodium and potassium ions is higher on the outside of its membrane than on the inside.
   c. sodium ions is higher on the inside of its membrane and of potassium ions is higher on the outside.
   d. sodium ions is higher on the outside of its membrane and of potassium ions is higher on the inside.
   e. sodium ions inside the cell equals the concentration of potassium ions inside the cell.
   ***Textbook Reference:*** *Concept 34.2 Neurons Generate Electric Signals by Controlling Ion Distributions; The resting potential is mainly a consequence of $K^+$ leak channels*

6. Glial cells are specialized to do all of the following *except*
   a. generate neural impulses.
   b. insulate axons.
   c. supply neurons with nutrients.
   d. help maintain a proper ionic environment for the neuron.
   e. guide neurons to make proper contacts during development.
   ***Textbook Reference:*** *Concept 34.1 Nervous Systems Are Composed of Neurons and Glial Cells; Glial cells support, nourish, and insulate neurons*

7. Which of the following structures is *not* found in the inner ear?
   a. Reissner's membrane
   b. Tectorial membrane
   c. Tympanic membrane
   d. Basilar membrane
   e. Semicircular canal
   ***Textbook Reference:*** *Concept 34.4 Sensory Processes Provide Information on an Animal's External Environment and Internal Status; Auditory systems use mechanoreceptors to sense sound pressure waves*

8. _______ are neurotransmitter receptors that are commonly G protein–linked and control production of second messengers.
   a. Ionotropic receptors
   b. Voltage-gated channels

c. Chemoreceptors
d. Biogenic receptors
e. Metabotropic receptors
***Textbook Reference:*** *Concept 34.3 Neurons Communicate with Other Cells at Synapses; Synapses can be fast or slow depending on the nature of receptors*

9. Which of the following statements about the sympathetic division of the autonomic nervous system is *false*?
a. It increases heart rate.
b. It relaxes the urinary bladder.
c. It stimulates digestion.
d. It increases blood pressure.
e. It relaxes airways.
***Textbook Reference:*** *Concept 34.5 Neurons Are Organized into Nervous Systems; The autonomic nervous system controls involuntary functions*

10. Which of the following statements about neurotransmitter receptors is *false*?
a. Ionotropic receptors are ion channels.
b. The acetylcholine receptor of the motor end plate is a metabotropic receptor.
c. Metabotropic receptors are not ion channels.
d. Metabotropic receptors induce signaling cascades in the postsynaptic cell.
e. Responses in the postsynaptic cell mediated by metabotropic receptors are usually slower than those mediated by ionotropic receptors.
***Textbook Reference:*** *Concept 34.3 Neurons Communicate with Other Cells at Synapses; Synapses can be fast or slow depending on the nature of receptors*

11. The rapid depolarization of a neuron during the first half of an action potential is due to the
a. exit of $K^+$ ions from the cell through gated potassium channels.
b. rapid reversal of ion concentration caused by the action of the sodium–potassium pump.
c. entry of $Na^+$ ions into the cell through gated sodium channels.
d. movement of both $Na^+$ and $K^+$ ions through appropriate open channels.
e. closing of sodium channels.
***Textbook Reference:*** *Concept 34.2 Neurons Generate Electric Signals by Controlling Ion Distributions; An action potential is a large depolarization that propagates with no loss of size*

12. In _______ synapses, cell membranes of the pre- and postsynaptic cells are joined by gap junctions at which cytoplasms of the two cells are continuous.
a. chemical
b. gated
c. electrical
d. biogenic
e. receptive
***Textbook Reference:*** *Concept 34.3 Neurons Communicate with Other Cells at Synapses; Chemical synapses are most common, but electrical synapses also exist*

13. In mammalian cerebral hemispheres, the major mechanism the brain uses to interpret action potentials is _______, whereby specific regions in adults are specialized to carry out specific sensory and motor functions.
a. autonomic function
b. location specificity
c. somatosensory mapping
d. neuromuscular sensitivity
e. synaptic plasticity
***Textbook Reference:*** *Concept 34.5 Neurons Are Organized into Nervous Systems; Location specificity is an important property of the mammalian cerebral hemispheres*

14. Which of the following is an example of a signal sent to skeletal muscles that does *not* involve participation of the brain?
a. Chemoreception
b. Vision
c. Sympathetic stimulation of adrenal glands
d. Parasympathetic stimulation of the stomach
e. Spinal reflex
***Textbook Reference:*** *Concept 34.5 Neurons Are Organized into Nervous Systems; Spinal reflexes represent a simple type of skeletal muscle control*

15. The substance that wraps around the axon of many neurons and provides for increased conduction speed is
a. dendrase.
b. histamine.
c. acetylcholine.
d. myelin.
e. microglia.
***Textbook Reference:*** *Concept 34.1 Nervous Systems Are Composed of Neurons and Glial Cells; Glial cells support, nourish, and insulate neurons*

16. All sensory systems, no matter which type of stimulus they detect, convey information
a. from the CNS to the peripheral nervous system.
b. in the form of action potentials.
c. to the visual cortex.
d. to the CNS as receptor potentials.
e. to the CNS as graded membrane potentials.
***Textbook Reference:*** *Concept 34.4 Sensory Processes Provide Information on an Animal's External Environment and Internal Status; The vertebrate retina is a developmental outgrowth of the brain and consists of specialized neurons*

# Answers

## *Key Concept Review*

1. Astrocytes are glia that help form the blood–brain barrier by surrounding tiny, very permeable blood vessels in the brain. The barrier prevents most water-soluble substances and large molecules from reaching the brain. However, because the barrier is made of plasma membranes, fat-soluble substances such as anesthetics and alcohol can pass through it.

2. A person with Lou Gehrig's disease would have no cognitive deficits because only the motor neurons (concerned with output) are affected by the disease; sensory neurons (concerned with input) and interneurons (concerned with integration) are unaffected.

3. The potassium voltage-gated channels are responsible for setting up the resting potential of a membrane. Potassium ions have a tendency to diffuse out of the cell, leaving a negative charge inside. Knocking out the function of the potassium voltage-gated channels would result in the cell's being unable to maintain resting potential. If the cell was depolarized by the opening of sodium voltage-gated channels, then it might not repolarize because the potassium channels that help repolarize the membrane would not be functioning.

4.
   a. Dendrites
   b. Cell body
   c. Axon hillock
   d. Axon
   e. Axon terminals

   The action potential travels down the axon, away from the axon hillock.

5. The conduction of an action potential down a myelinated axon is called saltatory conduction. The myelin acts to insulate areas of the axon, preventing depolarization. The areas of the axon between the myelin sheaths are known as nodes of Ranvier. Depolarization can occur only at these nodes. As the action potential moves down a myelinated axon, the influx of sodium ions at one node diffuses down the axon. This results in the depolarization of the next node of Ranvier. Depolarization can occur only in the downstream nodes because the upstream nodes are in a refractory period. As a result, the action potential moves quickly down the axon to the synapse.

6. Acetylcholine is the neurotransmitter used by all neuromuscular synapses in vertebrates. It transmits the action potential from a presynaptic cell to a postsynaptic cell. The enzyme acetylcholinesterase is found in the synaptic cleft, and it cleaves acetylcholine to help remove it from the synaptic cleft after an action potential. A nerve gas with components that block the action of acetylcholinesterase would cause acetylcholine to build up in the synaptic cleft. This buildup would mean that the receptors on the postsynaptic cell would remain bound with acetylcholine, resulting in prolonged muscle contraction.

7. Selective serotonin reuptake inhibitors, such as Prozac, increase the level of serotonin at the synapse by reducing its rate of removal.

8.
   a. Fovea
   b. Optic nerve
   c. Sclera
   d. Retina
   e. Vitreous humor
   f. Lens
   g. Pupil
   h. Cornea
   i. Iris

9. Your friend has red–green color blindness. The cones in our eyes allow us to see color. We have cones for red, green, and blue. A lack of one of type of cone, or a reduced number, can cause color blindness, as can problems in the functioning of one type of cone. Red–green color blindness is a sex-linked trait that is more common in men than in women.

10. The lowest threshold of human hearing can be determined by presenting auditory stimuli to subjects and asking them to respond with a yes or no answer as to whether a particular stimulus can be heard. In other mammals, the electrical activity of the auditory nerve and auditory regions of the brain can be monitored directly. The lowest threshold of hearing can be determined by presenting sounds that stimulate the ear, auditory nerve, and parts of the brain involved with hearing and then recording, by means of electrodes placed strategically on particular locations of the head, the differences in electrical potentials elicited by the different sounds. Subjects are typically anesthetized during the procedure.

11.

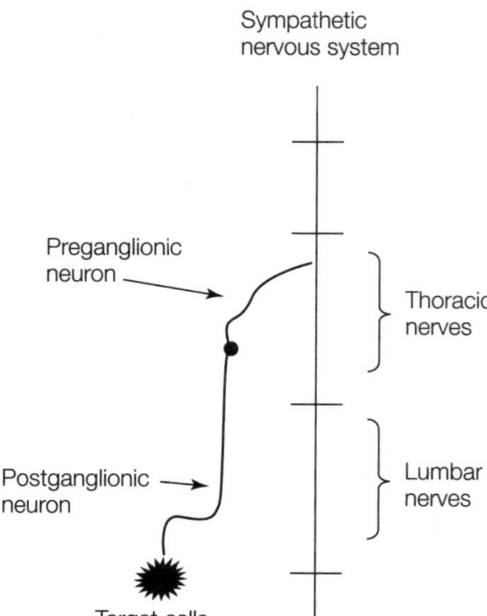


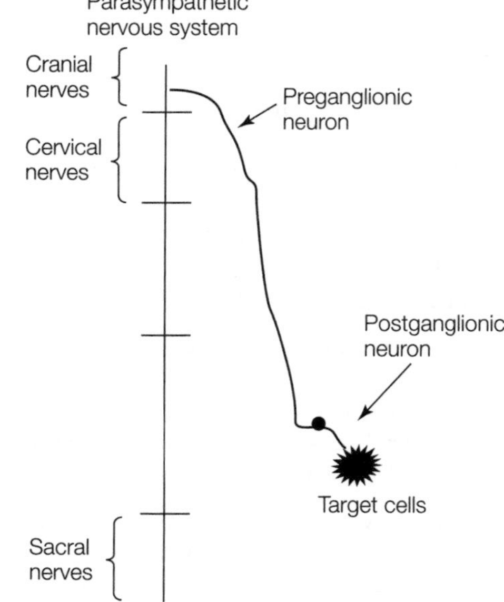


Nerves in the sympathetic nervous system originate from the thoracic and lumbar regions of the spinal cord. The preganglionic neurons are short, postganglionic neurons are long. In the parasympathetic nervous system, nerves originate in the cranial and sacral regions. The preganglionic neurons are long, postganglionic neurons are short.

12. Species such as humans, great apes, dolphins, and elephants are visually oriented animals. This raises the question of whether the mirror test is valid when applied to animals whose primary sensory modality is not vision, but perhaps olfaction or hearing. In designing an experiment to test self-awareness, a researcher should consider whether the mirror test is a valid test for the study species or whether the mirror test should be adapted to better match the sensory modality of the study species.

### *Test Yourself*

1. **d.** The neuron is composed of a cell body, an axon, and dendrites. The dendrites form synapses with presynaptic cells to create the junction where information from one neuron is transferred to another neuron.
2. **c.** The ganglion cells transmit information from the bipolar cells to the brain. The axons of the ganglion cells connect with the optic nerve.
3. **c.** The neuron is composed of the cell body, the dendrite, and the axon. The axon carries action potentials away from the cell body to the synapses.
4. **d.** For an action potential to occur in an axon, the membrane must be depolarized above the level known as the threshold.
5. **d.** The resting potential of a neuron membrane occurs when the sodium ion concentration is higher on the outside and the potassium ion concentration is higher on the inside.
6. **a.** Glia perform many functions in the nervous system, but they do not generate neural impulses.
7. **c.** Although the tympanic membrane is found in the human ear, it is not found in the inner ear. The tympanic membrane is the membrane that transmits sounds from the auditory canal to the middle ear.
8. **e.** Metabotropic receptors are commonly G protein–linked receptors.
9. **c.** The sympathetic division of the autonomic nervous system inhibits digestion rather than stimulating it.
10. **b.** The acetylcholine receptor of the motor end plate is an ionotropic receptor.
11. **c.** The first step in an action potential is the influx of $Na^+$, leading to a depolarization of the axon membrane. $Na^+$ rushes into the cell due to the higher concentration outside of the cell and the negative membrane potential.
12. **c.** Electrical synapses occur in cells joined by gap junctions.
13. **b.** Location specificity refers to specific regions of the cerebral hemispheres carrying out specific sensory and motor functions.
14. **e.** In spinal reflexes, all the neuron interactions occur in the spinal cord.
15. **d.** The glia that coat the axon of some neurons form myelin.
16. **b.** All sensory systems, no matter which type of stimulus they detect, convey information in the form of action potentials.

# Control by the Endocrine and Nervous Systems

## The Big Picture

- Nerve and endocrine cells work together to coordinate animal functions. Neural control is fast acting and delivered to highly defined target cells. Endocrine control is slow and broadcast; all cells are potentially exposed the released hormone. Only cells with receptors will respond to hormones.
- Hormones travel via the bloodstream to distant sites within the body to affect target tissues. They can act locally (paracrine) or on the cells that secrete them (autocrine). Target cells respond to specific endocrine signals.
- Endocrine glands are ductless and secrete their products into the blood. Neurosecretory cells propagate action potentials, and neuroendocrine cells respond to other hormones. Hormones are classified by their type and effect on target cells.
- Hormones regulate functions ranging from growth to sexual maturity. Hormone action is controlled by several different mechanisms. In negative feedback, a released hormone inhibits the tissue that may have stimulated its release in the first place. Tropic hormones influence other endocrine glands.

## Study Strategies

- The pituitary gland produces numerous regulatory hormones. The functions of the anterior and posterior pituitary are quite distinct, and keeping them separate is important for understanding the endocrine system.
- Remember that male sex steroids (androgens) and female sex steroids (estrogens and progesterone) are made and used by both sexes, but the relative levels vary in the two sexes.
- The many hormones may seem to constitute a long list of random compounds. Try to learn the function along with the name of each hormone, as this will help you remember the target tissues.
- Some hormones have extremely specific actions (e.g., follicle-stimulating hormone), whereas others have broad effects on many target tissues (e.g., epinephrine and thyroxine). Learning which hormones are "specialists" and which are "generalists" will help you better understand the endocrine system.
- Biologists studying the endocrine system frequently use abbreviations for the names of hormones. Learn these abbreviations along with the full names.
- Go to **LaunchPad** (or use the Web addresses listed) to review the following additional resources:

  Animated Tutorial 35.1 The Hypothalamus and Negative Feedback (PoL2e.com/at35.1)

  Animated Tutorial 35.2 Complete Metamorphosis (PoL2e.com/at35.2)

  Activity 35.1 The Human Endocrine Glands (PoL2e.com/ac35.1)

  Activity 35.2 Concept Matching: Vertebrate Hormones (PoL2e.com/ac35.2)

  Analyze the Data in textbook Figure 35.9

  Apply the Concept in textbook Section 35.2

## Key Concept Review

### 35.1 The Endocrine and Nervous Systems Play Distinct, Interacting Roles

The nervous and endocrine systems work in different ways

Nervous systems and endocrine systems tend to control different processes

The nervous and endocrine systems work together

Chemical signaling operates over a broad range of distances

Endocrine cells, like nerve cells, are specialized for control and coordination. Nerve and endocrine cells work together to ensure that an animal functions in coordinated, harmonious ways. Most of this communication occurs when neural or endocrine chemical signals travel to target cells (see Figure 35.1). Signals bind to receptors on the target cell and trigger a response. Neurons are specially adapted to generate action potentials. These neural impulses travel down the axon and cause release of neurotransmitters at the axon terminals. Neurotransmitters travel across synapses and bind to target cells. Neural signals are fast and addressed, meaning they are delivered to highly defined target cells.

Endocrine cells release chemical signals (hormones) into the blood. Endocrine control has two essential features: it is slow, taking minutes to days to have its effect, and it is broadcast, meaning that hormones released in the blood are potentially exposed to all cells. Only cells with receptors for the hormones will respond to them.

Nervous systems are capable of much finer control than endocrine systems. The two systems tend to be used to control different functions in the body. Nervous systems tend to control skeletal muscle, while endocrine systems typically control prolonged activities like developmental or metabolic changes. Most body tissues are under control of both the nervous and endocrine systems, which do not operate in mutually exclusive ways. The nervous system exerts control over the endocrine system by initiating endocrine function, such as gonadal or thyroid hormones throughout life. The endocrine system sometimes exerts control over the nervous system. Sex hormones affect brain development during puberty in mammals.

Chemical signaling occurs over a broad range of spatial scales (see Figure 35.2). Besides nervous and endocrine signaling, many chemicals diffuse from cell to cell without entering the blood. Paracrines are chemicals secreted by one cell and affect functions of neighboring cells. Autocrines are chemicals that are secreted by cells into intercellular fluid and diffuse to receptors on that same cell and affect its functions. Neurotransmitters resemble paracrines in that they move short distances from cell to cell by diffusion. They differ in that their secretion is controlled from farther away by electrical signals that travel the length of the presynaptic cell. Pheromones are chemical signals that an individual animal releases into its external environment and have their effects on individuals of the same species.

**Question 1.** Compare the signals of the nervous system and endocrine system. How are they similar? How do they differ?

**Question 2.** How are paracrine hormones similar to neurotransmitters? How do they differ?

## 35.2 Hormones Are Chemical Messengers Distributed by the Blood

Endocrine cells are neurosecretory or nonneural

Most hormones belong to one of three chemical groups

Receptor proteins can be on the cell surface or inside a cell

Hormone action depends on the nature of the target cells and their receptors

A hormonal signal is initiated, has its effect, and is terminated

Animals have two types of glands that produce and secrete materials. Exocrine glands have ducts that carry secretions away from the gland. Endocrine glands do not have ducts. By definition, their products are secreted into blood flowing through nearby vessels. In some cases, endocrine cells exist as single cells scattered within tissues composed of other cells, or they may be aggregated to form discrete tissues or organs, such as the thyroid and adrenal glands. The vertebrate testes, ovaries, and pancreas are mixes of endocrine and non-endocrine tissues.

A hormone is a chemical substance that is secreted into the blood by endocrine cells that regulates the function of other cells that it reaches by blood circulation. They act at very low blood concentrations, and their effect on target cells is to initiate noncovalent binding of the hormone to the receptor proteins on the target cell. Growth, development, reproductive cycles, water balance and long-term stress responses are typically under hormonal control in animals.

Two broad classes of endocrine cells exist (see Figure 35.3). Neurosecretory cells (neuroendocrine cells) are excitable cells that propagate action potentials. Nonneural endocrine cells (epithelial endocrine cells) are not excitable. Cell bodies of neurosecretory cells are located in the central nervous system (CNS). They have axons that typically extend outside of the CNS. They release hormones into the blood at their axon terminals. They often have axon terminals positioned in neurohemal organs, specialized parts of the circulatory system where terminals are closely located to the blood. Neurosecretory cells provide a direct interface between the nervous and endocrine systems. They respond by secreting hormones into the blood. Nonneural endocrine cells do not generate action potentials. They secrete their hormones in response to other hormones. The typical order of events is that a cell receives a hormonal signal from another gland and that hormone binds to receptor proteins, which initiate secretion by the nonneural endocrine cell. The anterior pituitary gland and thyroid gland are examples of this.

Three classes of hormones include peptides or protein hormones, steroid hormones, and amine hormones (see Figure 35.4). Most hormones are peptides. These hormones are water-soluble and easily transported in the blood. Before release, peptide and protein hormones are packaged in vesicles and released by exocytosis. Because they do not pass easily through the cell membrane, their receptors are on the exterior of the target cell. Steroid hormones are derived from cholesterol and are lipid-soluble. Steroid hormones easily pass through membranes of the cells that make them, but they require carrier proteins to transport them in the blood. They also readily pass through the membranes of target cells, so receptors for steroid hormones are often inside target cells. Amine hormones are small molecules made from single amino acids. Some are water-soluble and some are lipid-soluble.

Many hormone receptors, such as peptide and some amine hormones, are positioned in the cell membranes of target cells. Many of these receptors are G protein–linked receptors that, after binding, initiate second messenger cascades in the target cell. Other receptors are located in the cytoplasm or nucleus of target cells. Receptors of steroid hormones and some amine hormones fall under this category. Upon binding, steroid hormones alter gene expression, resulting in protein synthesis. Testosterone functions this way in skeletal muscle, leading to synthesis of contractile proteins (actin and myosin). The number of receptor molecules in a target cell can vary, depending on hormone concentration. Positive or negative feedback affects the number of receptor molecules a target cell expresses. In some cases, high concentrations of circulating hormones may cause target cells to decrease the number of receptors.

Hormone action depends on characteristics of target cells. Hormones can have dramatically different effects on different cells. As animals evolved, hormone function has changed over evolutionary time because receptors have changed. Prolactin, the hormone that stimulates milk synthesis in mammals), can be found in all other vertebrate phyla. In salmon, it is involved in stabilizing blood ion concentration during migration. In birds, it can stimulate nest building and parental care. Within individual animals, target cells can have different receptor proteins and systems. When the fight-or-flight response is activated during emergencies, the adrenal medullary glands secrete epinephrine and norepinephrine at the same time the sympathetic nervous system exerts its effects. There are at least five types of receptors for the hormones that bind with different degrees of ease.

Hormone release varies. Peptide and amine hormones are usually synthesized prior to use and can be secreted rapidly. Steroid hormones are typically synthesized on demand, so initiation is relatively slow. Removal of hormones from the blood can be through enzymatic degradation by the target cells or certain organs, or they can be excreted. The half-life of a hormone in the blood is the time required for half of a group of simultaneously secreted hormone molecules to be removed from the blood. Half-life can be minutes (epinephrine) to days (thyroxine). Hormone removal tends to take place at a steady rate. Some hormones are bound to carrier proteins while circulating in the blood. Carrier proteins often extend the half-lives of hormones. Peripheral activation occurs when hormones are converted to more active forms after secretion. Thyroxine that is secreted will be enzymatically modified in peripheral tissues to form triiodothyronine.

**Question 3.** Describe the modes of action of lipid-soluble and water-soluble hormones. Which type of hormone would likely produce more rapid effects?

**Question 4.** Compare and contrast the two types of endocrine cells.

## 35.3 The Vertebrate Hypothalamus and Pituitary Gland Link the Nervous and Endocrine Systems

Hypothalamic neurosecretory cells produce the posterior pituitary hormones

Hormones from hypothalamic neurosecretory cells control production of the anterior pituitary hormones

Endocrine cells are organized into control axes

Hypothalamic and anterior pituitary hormones are often secreted in pulses

The pituitary gland, which is attached to the hypothalamus, is a key meeting point for nervous and endocrine systems. The hypothalamus produces two hormones (antidiuretic hormone and oxytocin) that are stored and then secreted by the pituitary gland. Other hormones released by the hypothalamus influence release of pituitary hormones. Hormones released by neurons are called neurohormones.

The pituitary gland produces hormones that control many of the other endocrine glands in the body. It is made up of the anterior pituitary and the posterior pituitary. Each part uses different control mechanisms and releases different hormones.

Long axons from the hypothalamus extend into the posterior pituitary (see Figure 35.6). The posterior pituitary releases antidiuretic hormone and oxytocin, both of which are produced by neurons in the hypothalamus. Antidiuretic hormone (ADH, also called vasopressin) acts on the kidneys to stimulate reabsorption of water when blood pressure drops or salt increases in the blood. Oxytocin stimulates uterine contractions during childbirth and ejection of milk from mammary glands. It can be secreted simply in response to the sight and sound of a baby—a good example of how an external stimulus received by the nervous system controls a hormonal process. Oxytocin can also promote pair bonding and maternal bonding. The anterior pituitary gland releases four tropic hormones: thyroid-stimulating hormone (TSH), which controls thyroid function; luteinizing hormone (LH), which controls the function of the testes; follicle-stimulating hormone (FSH), which controls ovarian function; and adrenocorticotropic hormone (ACTH), which controls the adrenal cortex.

A different cell type produces each of the tropic hormones. The anterior pituitary also produces several other peptide hormones, including growth hormone and prolactin. Growth hormone (GH) stimulates the liver to release chemical signals called somatomedins, or insulin-like growth factors (IGFs), which stimulate bone and cartilage growth. GH also stimulates amino acid uptake by other cells. Over- or underproduction of GH can cause gigantism or pituitary dwarfism, respectively.

The anterior pituitary is composed of endocrine cells that also respond to neurohormones that are secreted by the hypothalamus and transported to the anterior pituitary by portal blood vessels (see Figure 35.7). One of these hypothalamic hormones is thyrotropin-releasing hormone (TRH), which causes the release of thyroid-stimulating hormone (TSH) by the anterior pituitary. TSH, in turn, stimulates the thyroid gland to release thyroxine. As with cortisol, these steps are controlled by a negative feedback loop.

The hypothalamo-hypophysial portal system consists of two capillary beds connected so that blood flows through the hypothalamus first, then through the anterior pituitary gland. Neurosecretory cells in the hypothalamus secrete hormones into the first capillary bed, which travel to the capillary bed in the anterior pituitary gland. There they control secretion of the nonneural endocrine cells. Hormones added to the blood by the neurosecretory cells of the hypothalamus are called releasing hormones (RHs) and inhibiting hormones (IHs). Releasing hormones cause target cells to release their hormone, while IHs inhibit release. A hormone axis describes a sequence of hormonal release. For example, neurosecretory cells of the brain control secretion of tropic hormones by the anterior pituitary cells, which causes hormone secretion by cells in a gland somewhere else in the body. The hypothalamus–pituitary–adrenal cortex (HPA) axis is an example of this. When an animal is under stress, the hypothalamus secretes corticotropin-releasing hormone, which travels through the portal system to the anterior pituitary, which releases adrenocorticotropic hormone (ACTH). ACTH travels through the bloodstream to the adrenal cortex, which responds by increasing secretion of glucocorticoids into the blood. The hypothalamus secretes hormones in pulses, which seems to be necessary for certain hormones to act. Continuous secretion can lead to reduction in receptor molecules of target cells, thus reducing sensitivity to the hormones.

**Question 5.** Alcohol inhibits release of antidiuretic hormone (ADH). What effect would alcohol consumption have on urination?

**Question 6.** Why does breast-feeding an infant soon after birth help a mother's uterus return to its prepregnancy size?

## 35.4 Hormones Regulate Mammalian Physiological Systems

The thyroid gland is essential for normal development and provides an example of hormone deficiency disease

Sex steroids control reproductive development

Figure 35.10 shows the principal mammalian endocrine glands and summarizes their major hormones and actions.

The thyroid gland wraps around the front of the trachea, expanding into a lobe on each side. It produces the hormones thyroxine and calcitonin. Thyroxine, also known as $T_4$ because it has four iodine atoms, is an amine hormone. The thyroid also produces the hormone $T_3$, which is nearly identical to thyroxine except that it has only three iodine atoms and is more active. Target cells have enzymes that can convert $T_4$ to $T_3$, thereby setting their own sensitivity to thyroid hormones. The thyroid hormones are is especially important during development and growth because they promote uptake of amino acids and synthesis of proteins. Insufficient thyroxine during prenatal or early postnatal development can cause cretinism, a condition characterized by delayed physical and mental development. Thyroid hormones elevate metabolic rate in mammals and birds.

In males, testes produce androgens (testosterone); in females, the ovaries produce estrogens and progesterone. Both sexes make and use androgens and estrogens, but the relative levels differ between the sexes.

Early in development, the sex organs of human embryos are similar. Later, the presence of the Y chromosome in male embryos causes the undifferentiated gonads to begin producing androgens. In response to androgens, the reproductive system develops into a male system. In the absence of androgens, the reproductive system develops into a female system (see Figure 35.13).

Sex steroids increase in concentration at the time of puberty. Both luteinizing hormone (LH) and follicle-stimulating hormone (FSH) from the anterior pituitary gland (collectively called gonadotropins) control the production of sex steroids. The production of these tropic hormones by the anterior pituitary is controlled by gonadotropin-releasing hormones (GnRH). At the onset of puberty, the GnRH-producing cells are no longer under negative feedback, so the level of GnRH increases, which stimulates the production of LH and FSH. Increased levels of LH and FSH in females and LH in males stimulate the gonads to produce more sex hormones. These sex steroids promote development of secondary sexual characteristics.

**Question 7.** In the diagram below, identify and match the gland with its location in the body. In the case of gonads, you may use the same location as long as you identify the sex of the individual.

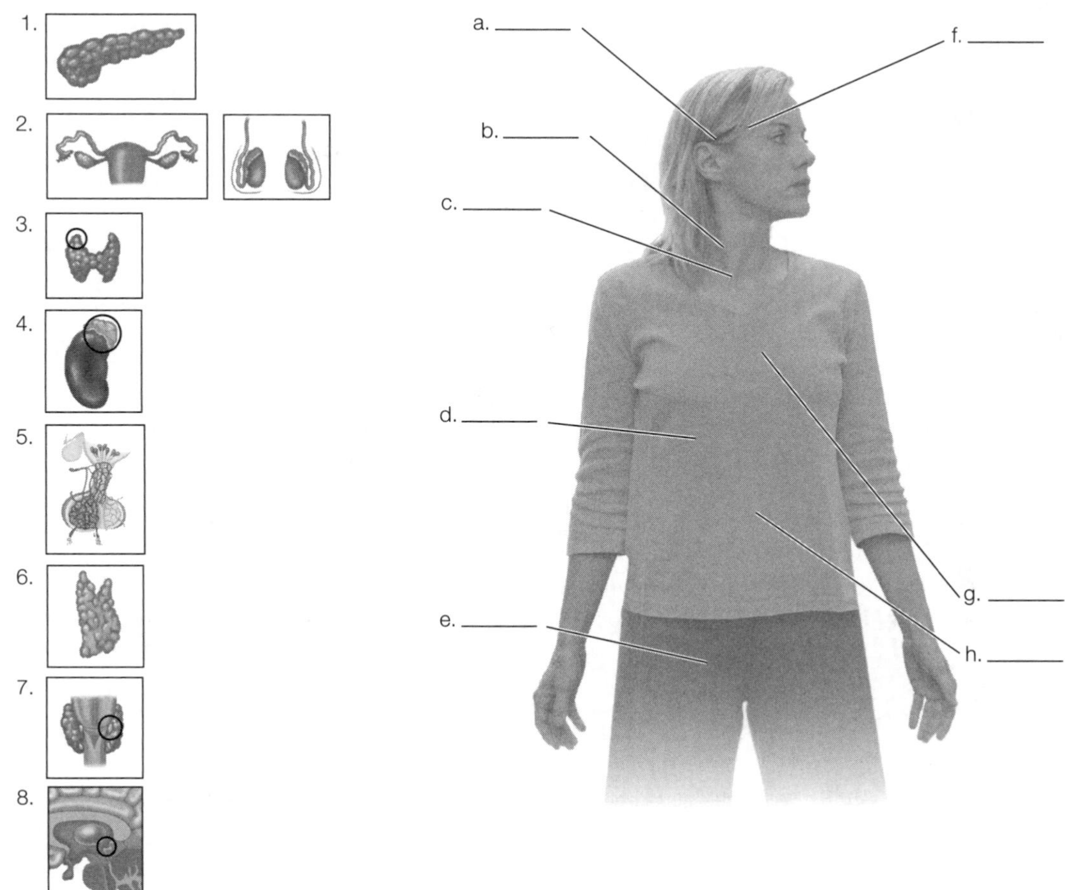


**Question 8.** The hormone testosterone is produced and released mainly by the testes. Design a study to determine the effects of testosterone on the behavior of male rats.

**Question 9.** Would castration (removal of the testes) lead to an immediate and complete absence of testosterone in a person's bloodstream? Why or why not?

## 35.5 The Insect Endocrine System Is Crucial for Development

Insects and other arthropods have elaborate endocrine systems. Some blood-sucking insects secrete diuretic hormones after a meal, promoting excretion of water in the blood and concentrating the blood proteins in the gut.

In insects, hormones control the processes of molting (shedding of the exoskeleton) and metamorphosis (transformation to the adult stage). Each growth stage between molts is called an instar (see Figure 35.15).

Insects produce prothoracicotropic hormone (PTTH). PTTH diffuses in extracellular fluid to the prothoracic gland. The prothoracic gland releases a steroid hormone called ecdysone, which diffuses to the target tissues to stimulate molting (see Figure 35.14). The control of insect molting by PTTH and ecdysone illustrates that the nervous and endocrine systems often act in concert to control growth and development; this link is also common in many vertebrates.

Juvenile hormone, which is secreted by the corpora allata, prevents premature maturation. In insects with incomplete metamorphosis, high levels of juvenile hormone result in the molting bug's becoming a slightly larger juvenile. When levels of juvenile hormone drop, the insect becomes an adult. In insects such as butterflies or moths, which undergo complete metamorphosis, the levels of juvenile hormone decrease as the larva molts a fixed number of times. When juvenile hormone falls below a certain level, the larva spins into a cocoon and molts into a pupa. No juvenile hormone is secreted during the pupal stage, so the pupa metamorphoses into the adult (see Figure 35.15).

**Question 10.** Describe what would likely happen to a fourth-instar silkworm moth *(Hyalophora)* if it were either partially or fully decapitated.

**Question 11.** Describe the hormonal pathways involved in molting in insects. What is the interrelationship of these hormones?

## Test Yourself

1. A paracrine signal
   a. circulates in the bloodstream and affects distant cells.
   b. always acts on a wide variety of target tissues.
   c. acts on nearby cells.
   d. acts on neuronal cells only.
   e. acts on glands only.
   ***Textbook Reference:*** *Concept 35.1 The Endocrine and Nervous Systems Play Distinct, Interacting Roles; Chemical signaling operates over a broad range of distances*

2. A target cell's response to a hormone depends on all of the following *except*
   a. the amount of hormone released.
   b. the number of receptors in or on that target cell.
   c. how well the hormone binds with the receptor.
   d. the presence of the proper receptor in or on the target cell.
   e. the neural connectivity of the target cell.
   ***Textbook Reference:*** *Concept 35.2 Hormones Are Chemical Messengers Distributed by the Blood; Hormone action depends on the nature of the target cells and their receptors*

3. Pheromones are
   a. chemical signals released by animals into the environment that have their effect on the same species.
   b. hormone receptors found on neurosecretory cells.
   c. hormone receptors found on nonneural endocrine cells.
   d. G protein–linked receptors found within cells.
   e. hormones that have their effect on other endocrine cells.
   ***Textbook Reference:*** *Concept 35.1 The Endocrine and Nervous Systems Play Distinct, Interacting Roles; Chemical signaling operates over a broad range of distances*

4. Which statement about hormonal functioning is *false*?
   a. Hormones act in very low concentrations.
   b. Many hormones act at sites distant from where they are produced.
   c. Hormones are transported in the blood.
   d. The same hormone can influence activities of several different target cells.
   e. Hormones exert their effects by diffusing into cells.
   ***Textbook Reference:*** *Concept 35.2 Hormones Are Chemical Messengers Distributed by the Blood; A hormone signal is initiated, has its effect, and is terminated*

5. In insects, juvenile hormone
   a. is produced by the corpora cardiaca.
   b. is also known as brain hormone.
   c. prevents maturation.
   d. is produced only by those with complete metamorphosis.
   e. is stored in the prothoracic gland.
   ***Textbook Reference:*** *Concept 35.5 The Insect Endocrine System Is Crucial for Development*

6. Thyroxine is important during development because it
   a. directs prenatal development of human sex organs.
   b. promotes uptake of amino acids and synthesis of proteins.
   c. is necessary for formation of the pituitary gland.
   d. stimulates formation of the thyroid gland.
   e. affects hormone release in other endocrine organs.
   ***Textbook Reference:*** *Concept 35.4 Hormones Regulate Mammalian Physiological Systems; The thyroid gland is essential for normal development and provides an example of hormone deficiency disease*

7. Which of the following has one part that is nonneural and a second part that is a neurohemal organ?
   a. Adrenal gland
   b. Thyroid gland
   c. Hypothalamus
   d. Pituitary gland
   e. Parathyroid gland
   ***Textbook Reference:*** *Concept 35.3 The Vertebrate Hypothalamus and Pituitary Gland Link the Nervous and Endocrine Systems; Hormones from hypothalamic neurosecretory cells control production of the anterior pituitary hormones*

8. Which of the following regulates daily rhythms?
   a. Pineal gland
   b. Thyroid gland
   c. Adrenal glands
   d. Ovaries
   e. Pituitary gland
   ***Textbook Reference:*** *Concept 35.4 Hormones Regulate Mammalian Physiological Systems*

9. Which of the following represents the correct sequence of steps leading to the release of glucocorticoids by the adrenal gland?
   a. Thyroid-stimulating hormone, thyrotropin-releasing hormone, glucocorticoids
   b. Adrenocorticotropic hormone, thyrotropin-releasing hormone, glucocorticoids
   c. Corticotropin-releasing hormone, adrenocorticotropic hormone, glucocorticoids
   d. Adrenocorticotropic hormone, corticotropin-releasing hormone, glucocorticoids
   e. Prolactin, thyroid-stimulating hormone, glucocorticoids
   ***Textbook Reference:*** *Concept 35.3 The Vertebrate Hypothalamus and Pituitary Gland Link the Nervous and Endocrine Systems; Endocrine cells are organized into control axes*

10. The target tissues of hormones are those tissues that
    a. can be penetrated by the particular hormones.
    b. have specific enzymes with which the hormones interact directly.
    c. have high concentrations of the "second messenger."
    d. have receptors for the particular hormones.
    e. have the particular genes that the hormones can express.

    ***Textbook Reference:*** *Concept 35.1 The Endocrine and Nervous Systems Play Distinct, Interacting Roles; The nervous and endocrine systems work in different ways*

11. Which of the following vertebrate hormones are produced in the anterior pituitary gland?
    a. Somatostatin, antidiuretic hormone, and insulin
    b. Prolactin, growth hormone, and enkephalins
    c. Oxytocin, prolactin, and adrenocorticotropin
    d. Estrogen, progesterone, and testosterone
    e. Growth hormone, gonadotropin-releasing hormone, and thyroid-releasing hormone

    ***Textbook Reference:*** *Concept 35.3 The Vertebrate Hypothalamus and Pituitary Gland Link the Nervous and Endocrine Systems; Hormones from hypothalamic neurosecretory cells control production of the anterior pituitary hormones*

12. Hormones that are secreted by one endocrine gland and control the activities of another endocrine gland are called _______ hormones.
    a. growth
    b. obstructive
    c. tropic
    d. selective
    e. paracrine

    ***Textbook Reference:*** *Concept 35.3 The Vertebrate Hypothalamus and Pituitary Gland Link the Nervous and Endocrine Systems; Hormones from hypothalamic neurosecretory cells control production of the anterior pituitary hormones*

13. Which statement about parathyroid hormone (PTH) is true?
    a. It increases calcium in the blood.
    b. It decreases calcium in the blood.
    c. It alters metabolic rate.
    d. It increases blood sodium.
    e. It initiates the fight-or-flight response.

    ***Textbook Reference:*** *Concept 35.4 Hormones Regulate Mammalian Physiological Systems*

14. Which statement about sexual differentiation in human embryos is *false*?
    a. Estrogens are needed for development of female sex organs.
    b. Androgens promote the development of male sex organs.
    c. In the absence of androgens, female sex organs develop.
    d. Sexual differentiation begins at about the seventh week of development.
    e. Presence of a Y chromosome prompts production of testosterone by embryonic gonads.

    ***Textbook Reference:*** *Concept 35.4 Hormones Regulate Mammalian Physiological Systems; Sex steroids control reproductive development*

15. Which statement about steroid hormones is true?
    a. They are water-soluble.
    b. They are produced by the thyroid gland.
    c. They are lipid-soluble.
    d. They are derived from the amino acid tyrosine.
    e. They are associated with second messenger cascades.

    ***Textbook Reference:*** *Concept 35.2 Hormones Are Chemical Messengers Distributed by the Blood; Most hormones belong to one of three chemical groups*

## Answers

### *Key Concept Review*

1. The signals of the nervous system are rapid and addressed, traveling to specific target cells, while signals of the endocrine system are slow and broadcast, transmitted to all cells but only affecting those cells with receptors for the hormone. The nervous system tends to control skeletal muscle, while the endocrine system controls prolonged activity such as developmental and metabolic changes. Both deliver signals to target cells, and work together to coordinate an animal's functions.
2. Neurotransmitters are similar to paracrine hormones in that they move short distances between cells through diffusion. They differ in that secretion of neurotransmitters is regulated by electrical impulses that travel along the presynaptic cell from farther away.
3. The plasma membrane of a cell is hydrophobic. Lipid-soluble hormones, which (as their name suggests) dissolve in lipids (fats), move through the plasma membrane and act on receptors inside the target cells. Water-soluble hormones cannot cross the lipid-based plasma membrane and act on receptors on the outside of target cells. Such binding by water-soluble hormones initiates changes in the cell by activating enzymes. Because lipid-soluble hormones cross the plasma membrane and enter the cell, they tend to take longer to act, but they also remain active for a longer time period than do water-soluble hormones.
4. Neurosecretory cells are excitable cells that propagate action potentials. They are similar to neurons. Cell bodies of neurosecretory cells are found in the central nervous system, with axons extending outside of the

CNS. The axon terminals of these cells are located close to blood vessels. Nonneural endocrine cells do not generate action potentials. They release their hormones into the bloodstream in response to other hormones. Both types of cells release their product into the bloodstream and have their effect on distant target cells. They differ in that neurosecretory cells release products in response to action potentials, while nonneural endocrine cells release products in response to hormones.

5. Because alcohol temporarily inhibits secretion of antidiuretic hormone (ADH) by the posterior pituitary, it will increase the amount of urine produced. The main function of ADH is to conserve water by decreasing urine output.

6. Suckling by a baby stimulates the release of oxytocin from the posterior pituitary, and this leads to milk ejection from the mammary glands. Oxytocin also causes uterine contractions, and after birth this helps the mother's uterus return to its prepregnancy size.

7.
   a. 8. Pineal gland
   b. 3. Thyroid gland
   c. 7. Parathyroid gland
   d. 4. Adrenal gland
   e. 2. Gonads (ovaries; testes)
   f. 5. Hypothalamus/Posterior pituitary/ Anterior pituitary
   g. 6. Thymus
   h. 1. Pancreas

8. One method would be to remove the testes, the main glands that produce testosterone, and record the behavior of the castrated rats once they have recovered from surgery. Later, testosterone could be replaced through injections or implants and the behavior of the rats could be examined to see if it returned to presurgery levels. Another method would be to monitor levels of testosterone during one year, while simultaneously monitoring the behavior of the rats to see if there is a correlation between high levels of testosterone and certain behaviors, such as aggression. A third method would be to administer a drug that temporarily and reversibly blocks the effects of testosterone to see if it changed the behavior of the rats.

9. Castration would not lead to an immediate absence of testosterone, a lipid-soluble hormone, in the bloodstream. Like many hormones, testosterone has a half-life of days to weeks. In addition, small amounts of the sex steroids are produced by the adrenal glands in both males and females.

10. If the *Hyalophora* larva is partially decapitated, leaving the corpora allata, it will molt into a fifth-instar juvenile because the corpora allata produces juvenile hormone, which prevents maturation into an adult. An insect that is fully decapitated will molt into an adult rather than another juvenile instar. The fully decapitated animal has enough circulating prothoracicotropic hormone (brain hormone) to molt, but in the absence of juvenile hormone, it molts into an adult.

11. Prothoracicotropic hormone (PTTH) is released from neurosecretory cells of the brain and diffuses in extracellular fluid to the prothoracic gland, which releases ecdysone, a steroid hormone. Ecdysone diffuses to epidermal cells that secrete enzymes that loosen connections of the epidermal cells with the exoskeleton, allowing it to be shed. During development of insects that undergo complete metamorphosis, nonneural endocrine cells of the corpora allata release juvenile hormone (JH). This hormone regulates body form, and when present in high concentrations, retains the larval form. As larvae molt and mature, JH concentrations drop, and the insect body form changes into a pupa. PTTH regulates molting, but JH regulates body form.

### Test Yourself

1. **c.** Paracrine signals act on nearby cells. Autocrine signals act on the cells that secrete them. Circulating hormones enter the bloodstream and affect distant cells.

2. **e.** The response of target cells to a hormone depends on how much of the hormone is present and acting on the target, how many receptors are present on the surface of the cell that the hormone can act on, and how well the hormone binds with the receptor.

3. **a.** Pheromones are chemicals released by individual animals that have their effect on other animals of the same species.

4. **e.** Lipid-soluble hormones do diffuse into cells, but peptide and some amine hormones are water-soluble and cannot cross the plasma membrane.

5. **c.** Juvenile hormone is secreted continuously by the corpora allata of insects and prevents maturation. Prothoracicotropic hormone (brain hormone) is secreted by the corpora cardiaca and is involved with ecdysone in molting.

6. **b.** Thyroxine promotes uptake of amino acids and synthesis of proteins. Insufficient thyroxine during prenatal or early postnatal development can lead to cretinism.

7. **d.** The pituitary gland has an anterior lobe that is nonneural and a posterior lobe that is a neurohemal organ.

8. **a.** The pineal gland secretes melatonin, a hormone that controls daily rhythms.

9. **c.** Under stress, the hypothalamus releases corticotropin-releasing hormone, which targets cells of the anterior pituitary to release adrenocorticotropic hormone (ACTH). ACTH targets cells of the adrenal cortex, which release glucocorticoids.

10. **d.** For a hormone to act on a target cell, the target cell must have the receptors for the specific hormone to bind, and trigger the hormonal action.
11. **b.** Prolactin, growth hormone, and enkephalins are produced in the anterior pituitary.
12. **c.** Hormones that control endocrine gland function are known as tropic hormones.
13. **a.** Parathyroid hormone increases the concentration of calcium in the blood.
14. **a.** Estrogens are not needed for the development of female sex organs. Female sex organs form in the absence of androgens. Male sex organs form when androgens are present.
15. **c.** Steroid hormones are lipid-soluble and are produced by the gonads and adrenal glands. They do not act through second messenger systems.

# 36 Water and Salt Balance

## The Big Picture

- Different animals experience different types of stresses related to water and salt balance. Marine animals have problems with water loss due to osmosis, whereas freshwater animals lose salts, so both must be conserved. Regardless of their environment, all animals produce metabolic wastes as a by-product of metabolism. The kidney is the excretory organ responsible for removing metabolic wastes from blood. The nephron is the "functional unit" of the vertebrate kidney.
- Kidneys work by differentially processing nitrogenous wastes, ions, and water. In the kidney, large volumes of plasma are filtered from the blood into the interior of the nephron. As the fluid passes through the various specialized regions of the nephron, the desirable components to be retained are pulled back into the tissues, leaving behind the waste products in the more concentrated form of urine. As urine leaves the nephron, water content is adjusted in the collecting ducts through hormonal alteration of their water permeability.

## Study Strategies

- Depending on whether they are living in fresh water or salt water, fishes may excrete copious amounts of dilute urine or small amounts of more concentrated urine. Because both environments create osmoregulatory problems, think about whether the tissues have higher or lower solute concentration relative to the surrounding water, and predict water and ion movements on that basis.
- Understanding how the nephron functions, and in particular how the loop of Henle works as a "countercurrent multiplier system," is one of the greater challenges in this chapter. This topic can be understood more easily if you imagine a single $Na^+$ ion in the blood as it enters the nephron. Mentally follow its pathway—and its potential pathways—as it traverses the nephron, eventually ending up in the urine. Note especially that it may take the ion some time to pass through the loop of Henle as it leaves the ascending limb, only to recycle back into the descending limb—a process that can be repeated many times before it finally escapes from the countercurrent multiplier. Repeat this imaginary journey from the perspective of a water molecule.
- Go to **LaunchPad** (or use the Web addresses listed) to review the following additional resources:

  Animated Tutorial 36.1 The Mammalian Kidney (PoL2e.com/at36.1)

  Animated Tutorial 36.2 Kidney Regulation Simulation (PoL2e.com/at36.2)

  Activity 36.1 The Human Excretory System (PoL2e.com/ac36.1)

  Activity 36.2 The Vertebrate Nephron (PoL2e.com/ac36.2)

  Analyze the Data in textbook Figure 36.10

  Apply the Concept in textbook Section 36.5

## Key Concept Review

### 36.1 Kidneys Regulate the Composition of the Body Fluids

Kidneys make urine from the blood plasma

Kidneys regulate the composition and volume of the blood plasma

Urine/plasma (U/P) ratios are essential tools for understanding kidney function

Our day-to-day urine concentrations illustrate these principles

The range of action of the kidneys varies from one animal group to another

Extrarenal salt excretion sometimes provides abilities the kidneys cannot provide

Regulating body fluids is crucial to an animal's survival. Three interrelated properties are critical for maintaining body fluid homeostasis: osmotic pressure (a measure of the total concentration of solutes), ionic composition, and volume. When even one of these three properties is compromised, bodily functions are disrupted.

An animal's kidneys (excretory organs) produce urine from blood plasma. While overall kidney complexity varies, the basic structure consists of tubules that discharge urine. The simplest kidney is a single tubule, whereas the more complex kidney has microscopic tubules (nephrons).

Urine is made from the water and solute in plasma. Thus, the kidney can regulate the composition of plasma by changing the composition of the urine. This relationship is expressed as the urine/plasma (U/P) ratio. Both osmotic pressure and ion concentration can be described using U/P ratios.

Kidneys will adjust the U/P ratio in response to bodily conditions. For example, if you drink a glass of water, your plasma becomes more dilute. This change is detected by sensory structures. In response, the kidneys will remove more water than solute from the plasma, creating dilute urine. The opposite happens if plasma concentrations increase. Animals differ greatly in their ability to concentrate urine. In fact, only mammals, birds, and insects can concentrate their urine at levels higher than their plasma level.

Animals that are either at a high risk of dehydration or that gain excess salt during their daily lives can perform extrarenal salt excretion. Ocean- and desert-dwelling animals have salt glands to excrete excess salt. Saltwater fish have salt-secreting cells. However, mammals do not have extrarenal salt excretion even if they live in the ocean.

**Question 1.** If a terrestrial animal is prevented from taking in water, what can its kidneys do to compensate?

**Question 2.** Describe a mechanism other than the kidneys that can be used by animals living in a salty environment to excrete excess salt.

## 36.2 Nitrogenous Wastes Need to Be Excreted

- Most water-breathing aquatic animals excrete ammonia
- Most terrestrial animals excrete urea, uric acid, or compounds related to uric acid

Osmolarity is the number of osmoles of solute particles per liter of solvent. Regulation of the osmolarity of extracellular fluid is critical. Differences in osmolarity between the extracellular fluid and the fluid inside cells can lead to changes in cell volume, ultimately causing cells to expand and burst or to shrink and die.

The breakdown of carbohydrates and fats produces carbon dioxide and water; these two end products can be eliminated easily. Proteins and nucleic acids contain nitrogen, so their breakdown produces carbon dioxide, water, and nitrogenous wastes, which can be toxic.

Ammonia ($NH_3$) is a highly toxic nitrogenous waste that either must be excreted or converted to the less toxic urea and uric acid (see Figure 36.4). Ammonotelic animals excrete ammonia to the aquatic environment, usually across gill membranes. Examples include most bony fishes and aquatic invertebrates. Ureotelic animals excrete nitrogenous waste as urea. Examples include mammals, cartilaginous fishes, and most amphibians. Uricotelic animals, such as birds and other reptiles, excrete nitrogenous waste as uric acid. Insects also are uricotelic. Most species produce more than one nitrogenous waste.

**Question 3.** When there is too much uric acid in your blood, it can no longer dissolve and it crystallizes. Your body then attacks these crystals, causing the tissues to become inflamed, resulting in gout. Discuss two treatment strategies that could be pursued by a pharmacology lab working on the development of a new medication for gout.

**Question 4.** Why do water-breathing animals typically excrete ammonia as their primary nitrogenous waste?

## 36.3 Aquatic Animals Display a Wide Diversity of Relationships to Their Environment

- Most invertebrates in the ocean are isosmotic with seawater
- Ocean bony fish are strongly hyposmotic to seawater
- All freshwater animals are hyperosmotic to fresh water
- Some aquatic animals face varying environmental salinities

Aquatic organisms have developed multiple mechanisms for survival within their environment. Animals whose body fluids have the same osmotic pressure as the water in which they live are said to be isosmotic (equal) to their environment. Those whose body fluids have a higher osmotic pressure than the environment are termed hyperosmotic (higher), and those with a lower pressure are hyposmotic (lower).

Animals that are isosmotic allow the ionic composition of their extracellular fluid to match that of the environment. Thus, they do not actively regulate the osmolarity of their tissues. Most marine invertebrates are isosmotic.

Ocean bony fish are hyposmotic to their environment and thus are hyposmotic regulators. They tend to lose water by osmosis and gain ions by diffusion from the ocean. Since they are unable to make their urine more dilute than the seawater, their gills have specialized cells that pump salt into the water.

Freshwater animals are hyperosmotic to their environment and thus are hyperosmotic regulators. They tend to gain water and lose ions. The water is excreted as urine. To contend with ion loss, freshwater animals have specialized cells that use ATP to pump chloride and sodium ions from the environment into their blood. These cells are in the gills of fish, crayfish, and clams.

Some aquatic species live in environments with extreme osmotic fluctuations. Osmotic regulators (osmoregulators) actively regulate the osmolarity of their tissues, even as environmental osmolarity changes. An example is salmon, which live part of their lives in fresh water and part in the ocean. They transition from hyperosmotic regulators to hyposmotic regulators. This is in contrast to osmotic conformers (osmoconformers) such as marine invertebrates, which do not regulate their osmolarity.

**Question 5.** Why are terrestrial animals always osmoregulators?

**Question 6.** Osmoconformers occur in marine environments but not in freshwater environments. Why?

## 36.4 Dehydration Is the Principal Challenge for Terrestrial Animals

- Humidic terrestrial animals have rapid rates of water loss that limit their behavioral options
- Xeric terrestrial animals have low rates of water loss, giving them enhanced freedom of action
- Some xeric animals are adapted to live in deserts

Animals whose outer body coverings are highly permeable to water are classed as humidic. Since they rapidly lose water to their environment, they have adapted behaviors to compensate. As an example, a frog will hop into a sunlit area to eat, but within a few minutes must return to a pond or other watery environment.

Animals that have evolved body coverings that are waxy or greasy are protected from dehydration and can spend indefinite periods of time in the open air. These organisms are classed as xeric and include mammals, birds, insects, and spiders. There are some xeric animals that are so well adapted to life in an arid environment they can survive without drinking water. They subsist on metabolic water which is produced by the oxidation of organic molecules during metabolism.

**Question 7.** Some xeric animals are adapted to live in deserts. What are some of the adaptations they have in order to survive in the dry environment of a desert?

**Question 8.** What are some of the adaptations that humidic animals have that allow them to live on land?

## 36.5 Kidneys Adjust Water Excretion to Help Animals Maintain Homeostasis

- Fluid enters a nephron by ultrafiltration driven by blood pressure
- The processing of the primary urine in amphibians reveals fundamental principles of nephron function
- Mammalian kidneys produce exceptionally high urine concentrations
- The Malpighian tubules of insects employ a secretory mechanism of producing primary urine

Kidney tubules make urine in two steps that result in the production of primary and definitive urine. The first structure in a vertebrate nephron is the Bowman's capsule, which encloses the glomerulus (see Figure 36.8). The glomerulus, a dense cluster of capillaries, is the location of blood filtration. The walls of these capillaries do not allow the components of the blood to pass through into the lumen of the capsule. Water, salts, sugars, and other small organic molecules can pass through the walls of the glomerular blood capillaries. Ultrafiltration is the process of forming primary urine and the rate of formation is called the glomerular filtration rate (GFR). Definitive urine is produced by the reabsorption of water and small molecules from the primary urine.

Nephrons are divided into sections. Those that are nearest to the Bowman's capsule are called early or proximal, while those that are distant are called late or distal. The entire length of the nephron is lined with a single layer of epithelial cells, but the structure and protein expression varies considerably along its length. In proximal segments, the nephron wall is highly permeable to water due to a high concentration of aquaporins in the epithelial cells lining this region. In addition, $Na^+$ and $Cl^-$ are removed by active transport. This creates tubular fluid within the lumen of the nephron. The distal tubule is similar to the proximal tubule except that the aquaporins are inserted or removed as needed. Ions are still actively pumped out of the urine.

Antidiuretic hormone (ADH, also called vasopressin) controls the permeability of the proximal tubules to water by stimulating the insertion of aquaporins, specifically AQP-2, into the plasma membranes of the cells in this region of the tubule. The hypothalamus stimulates the release of ADH in response to increased serum osmolarity, resulting in increased water reabsorption and dilution of the blood, and the production of small amounts of concentrated urine.

A critical difference between amphibian and mammalian kidneys is that the distal tubules (collecting ducts or loops of Henle) of mammalian kidneys are surrounded by tissue fluids that are much more concentrated than plasma. The loop of Henle produces a concentration gradient in the medulla by acting as a countercurrent multiplication system (see Figure 36.11). This concentration gradient moves water across fluid compartments by osmosis.

The concentration of the fluids surrounding the loops of Henle is raised by the active transport of $Cl^-$ and $Na^+$ into the tissue from the loop of Henle that ascends out of the medulla. The ascending limb is impermeable to water, so only solutes leave. The loop of Henle that descends into the medulla is permeable to water, and water moves out into the surrounding tissues due to the osmotic pressure deep in the medulla. In the ascending loop, the filtrate is more concentrated than the fluid in the surrounding tissues. Water cannot move out because of the low permeability of the ascending loop.

The fluid reaching the distal convoluted tubule is less concentrated than blood plasma. The distal convoluted tubule fine-tunes the ionic composition of the urine. As the fluid leaves the distal convoluted tubule for the collecting duct, its major solute is urea.

Malpighian tubules are the excretory organ of insects. Malpighian tubules actively transport uric acid and KCl (or other salts) from the tissue into the tubules, with water following passively (see Figure 36.12). The tubules lead into the hindgut, where uric acid precipitates and water is reabsorbed. Insects eliminate semisolid matter containing uric acid and other wastes.

**Question 9.** In the diagram below, label the following structures: medullary blood vessels, nephron, proximal convoluted tubule, Bowman's capsule, collecting duct, loops of Henle, and distal convoluted tubule.

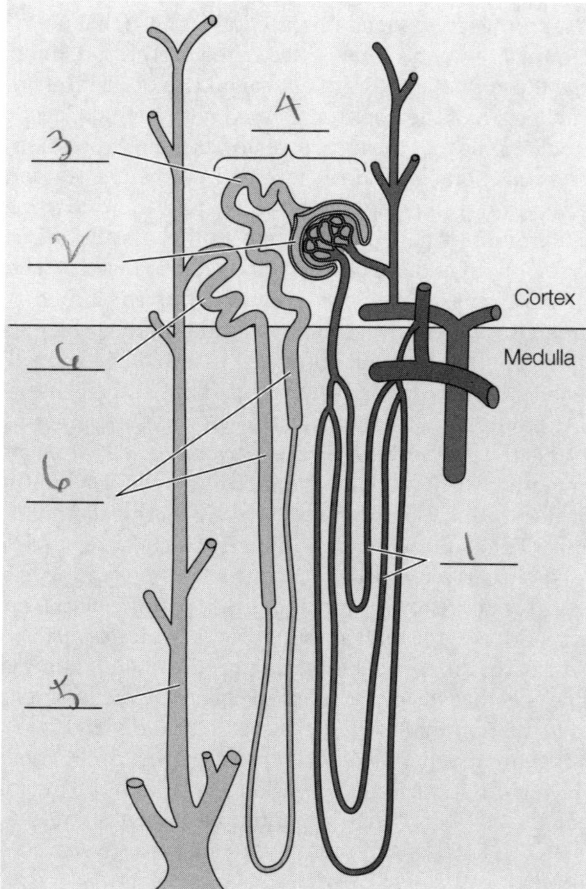


**Question 10.** The drug urizadole inhibits antidiuretic hormone secretion. How does this inhibition affect urine volume and glomerular filtration rate in patients taking this medication?

**Question 11.** You have just eaten a large number of very salty potato chips, and the osmolarity of your blood has increased. What response will your body have in order to bring your blood osmolarity back to homeostasis?

## Test Yourself

1. Which of the following is *not* a normal constituent of the glomerular filtrate?
   a. Red blood cells
   b. Urea
   c. Sodium ions
   d. Glucose
   e. Amino acids
   ***Textbook Reference:*** *Concept 36.5 Kidneys Adjust Water Excretion to Help Animals Maintain Homeostasis; Fluid enters a nephron by ultrafiltration driven by blood pressure*

2. Which end product of catabolism is the most toxic?
   a. Water
   b. Ammonia
   c. Carbon dioxide
   d. Urea
   e. Uric acid
   ***Textbook Reference:*** *Concept 36.2 Nitrogenous Wastes Need to Be Excreted; Most water-breathing aquatic animals excrete ammonia*

3. The sole mechanism for water reabsorption by the renal tubules is
   a. active transport.
   b. osmosis.
   c. cotransport with sodium ions.
   d. cotransport with bicarbonate ions.
   e. diffusion.
   ***Textbook Reference:*** *Concept 36.5 Kidneys Adjust Water Excretion to Help Animals Maintain Homeostasis; Mammalian kidneys produce exceptionally high urine concentrations*

4. In order to maintain homeostasis, a marine bony fish
   a. excretes only small amounts of water and pumps sodium out of its body at the gills.
   b. excretes large amounts of water and pumps sodium into its body.
   c. converts nitrogenous wastes to urea.
   d. excretes uric acid.
   e. excretes small amounts of water but does not excrete sodium.
   ***Textbook Reference:*** *Concept 36.3 Aquatic Animals Display a Wide Diversity of Relationships to Their Environment; Ocean bony fish are strongly hyposmotic to seawater*

5. Why can xeric animals stay out all day in the sun but humidic animals cannot?
   a. Xeric animals have a thin outer covering that is so permeable to water that they can absorb moisture from the air.
   b. Humidic animals have exposed gills that dry out rapidly, whereas xeric animals do not.
   c. Xeric animals have body coverings embedded with layers of lipids to prevent dehydration.

d. Unlike humidic animals, xeric animals do not sweat, and thus they do not lose moisture to the environment.
e. Humidic animals secrete slime from their skin, which causes them to dehydrate rapidly.
***Textbook Reference:*** *Concept 36.4 Dehydration is the Principal Challenge for Terrestrial Animals; Humidic terrestrial animals have rapid rates of water loss that limit their behavioral options*

6. Which of the following statements about the excretory system of insects is *false*?
a. Active transport moves materials from the coelomic fluid into the Malpighian tubules.
b. The Malpighian tubules can produce a highly concentrated waste product, allowing insects to inhabit some of Earth's driest habitats.
c. Reabsorption of salts takes place mostly in the gut.
d. Water reabsorption takes place by osmotic movement only.
e. Insects excrete urea.
***Textbook Reference:*** *Concept 36.5 Kidneys Adjust Water Excretion to Help Animals Maintain Homeostasis; The Malpighian tubules of insects employ a secretory mechanism of producing primary urine*

7. Which of the following represents the correct pathway of water and solutes traveling through a nephron?
a. Glomerulus, Bowman's capsule, renal tubule, collecting ducts
b. Bowman's capsule, glomerulus, renal tubule, collecting ducts
c. Renal tubule, glomerulus, Bowman's capsule, collecting ducts
d. Collecting ducts, glomerulus, Bowman's capsule, renal tubule
e. Glomerulus, Bowman's capsule, collecting ducts, renal tubule
***Textbook Reference:*** *Concept 36.5 Kidneys Adjust Water Excretion to Help Animals Maintain Homeostasis; Mammalian kidneys produce exceptionally high urine concentrations*

8. In the loops of Henle there is countercurrent multiplication. The countercurrent multiplication works because the osmotic pressure in the medulla varies from _______ m*Osm* at the surface to _______ m*Osm* deep in the medulla.
a. 300; 1,200
b. 1,200; 300
c. 0; 300
d. 1,200; 0
e. 0; 300
***Textbook Reference:*** *Concept 36.5 Kidneys Adjust Water Excretion to Help Animals Maintain Homeostasis; Mammalian kidneys produce exceptionally high urine concentrations*

9.–13. Refer to the diagram below of the mammalian nephron to answer the questions that follow. Some can have more than one answer.

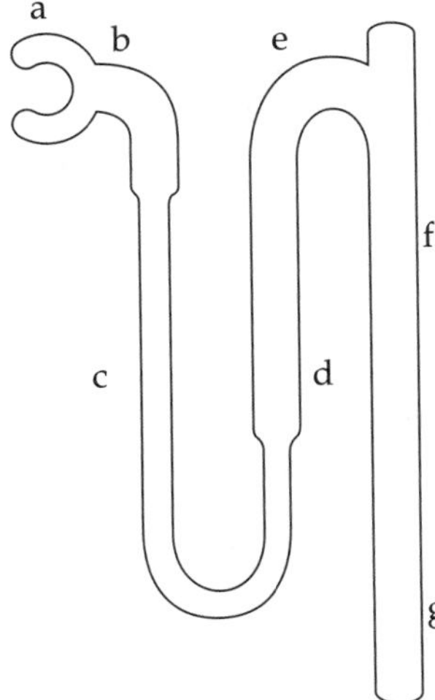


9. The composition of the filtrate would be most like plasma in the tubule next to letter(s) _______.
10. The NaCl concentration in the extracellular fluid would be greatest in the area of letter(s) _______.
11. The osmolarity of the filtrate next to letter(s) _______ is similar to the osmolarity of blood plasma.
12. The urine would be most concentrated in the collecting duct next to letter(s) _______.
13. Most of the glomerular filtrate is reabsorbed into the blood in capillaries next to letter(s) _______.
***Textbook Reference:*** *Concept 36.5 Kidneys Adjust Water Excretion to Help Animals Maintain Homeostasis; Mammalian kidneys produce exceptionally high urine concentrations*

14. Which statement about excretory systems is *false*?
a. All marine invertebrates conform to the osmotic pressure of the environment.
b. Osmotic conformers are found only in marine environments.
c. Terrestrial animals are always osmotic regulators.
d. Some marine vertebrates are osmotic conformers.
e. Most marine invertebrates are osmotic conformers.
***Textbook Reference:*** *Concept 36.3 Aquatic Animals Display a Wide Diversity of Relationships to Their Environment; Some aquatic animals face varying environmental salinities*

15. Several hormones help regulate water and solute uptake and release in the nephron. Antidiuretic hormone (ADH) promotes _______ in response to _______.
a. active transport of $Cl^-$; increased solute concentration
b. active transport of $Na^+$; increased blood pressure
c. increased permeability of the collecting duct to water; increased plasma concentration

d. decreased permeability of the collecting duct to water; increased solute concentration
e. decreased permeability of the collecting duct to water; decreased blood pressure

***Textbook Reference:*** *Concept 36.5 Kidneys Adjust Water Excretion to Help Animals Maintain Homeostasis; The processing of the primary urine in amphibians reveals fundamental principles of nephron function*

16. Osmotic pressure is
 a. a measure of the principal solutes in most biological solutions.
 b. a measure of the total concentration of solutes (dissolved matter).
 c. a measure of the partial pressures of solutes in a medium.
 d. unimportant to kidney function.
 e. the only important factor in kidney function.

 ***Textbook Reference:*** *Concept 36.1 Kidneys Regulate the Composition of the Body Fluids*

17. If the incoming arteriole that supplies blood to the glomerulus becomes dilated, then
 a. the protein concentration of the filtrate decreases.
 b. hydrostatic pressure in the glomerulus decreases.
 c. the glomerular filtration rate increases.
 d. the glomerular filtration rate decreases.
 e. the water concentration of the filtrate decreases.

 ***Textbook Reference:*** *Concept 36.5 Kidneys Adjust Water Excretion to Help Animals Maintain Homeostasis; Fluid enters a nephron by ultrafiltration driven by blood pressure*

## Answers

### *Key Concept Review*

1. As the body becomes dehydrated, kidneys can filter less water from the blood, thus attempting to preserve homeostasis.
2. Animals living in a salty environment have salt glands in their heads that excrete salt.
3. The symptoms of gout are caused by the precipitation of uric acid crystals in joints due to high levels of uric acid in the extracellular fluid. Two potential approaches for new medications would involve blocking the production of uric acid by the body or improving the removal of uric acid from the body.
4. Water-breathing animals typically excrete ammonia as their primary nitrogenous waste because ammonia is highly soluble in water and diffuses rapidly. Ammonia excretion occurs continuously across the gills.
5. On land, water and salts are usually in short supply, so terrestrial animals are osmoregulators, actively regulating the osmolarity of their extracellular fluid.
6. Osmoconformers allow their extracellular fluid to equilibrate with their surroundings. This can work in marine environments because there are many solutes in saltwater, some of which are necessary to support life. In addition, many marine osmoconformers have the ability to regulate the concentrations of certain ions in their extracellular fluids. In fresh water, however, there are few solutes, so the body fluids of an osmotic conformer would be too dilute to support life.
7. They minimize water loss by staying in deep, cool underground burrows during the heat of the day, they produce dry feces, and they have highly efficient kidneys that produce extremely concentrated urine
8. They live in highly humid environments, such as under logs or in dense vegetation, or they leave their humid environments for only brief periods of time. If they remain in the open air and/or sunlight for too long, they risk dehydration.
9. 

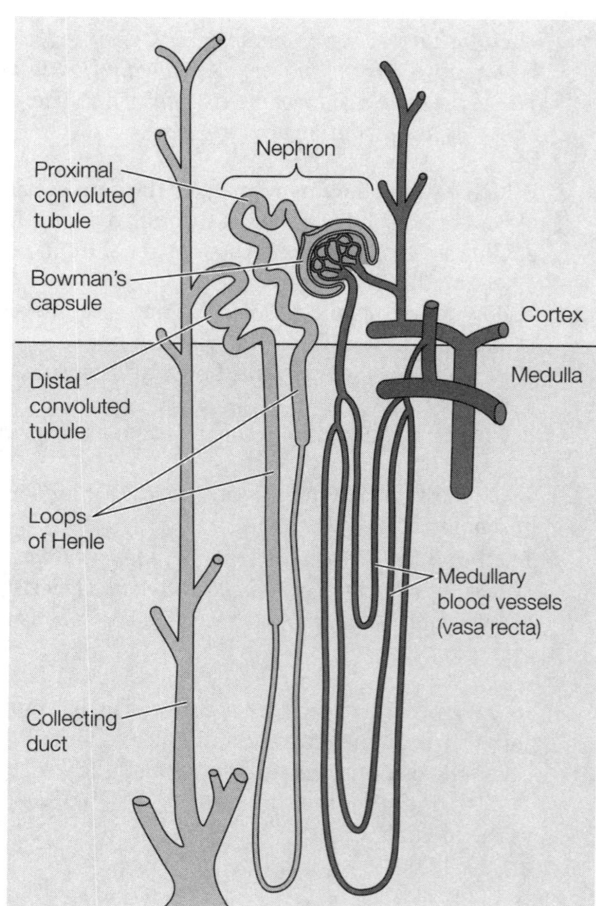


10. Inhibition of antidiuretic hormone by urizadole affects the permeability of the collecting ducts to water. This results in increased urine volume because water will not be reabsorbed across the collecting ducts. Blockage of ADH has no effect on the glomerular filtration rate.

11. In response to increased blood osmolarity, the hypothalamus stimulate the release of ADH. ADH acts to increase the permeability of the collecting ducts to water so that increased amounts of water can be reabsorbed to bring down the blood osmolarity. The hypothalamus will also stimulate thirst, causing you to increase your water intake.

### *Test Yourself*

1. **a.** Red blood cells are too large to be filtered out of the blood at the glomerulus and thus will not be found in the filtrate.
2. **b.** Ammonia is the most toxic end product of catabolism.
3. **b.** The sole mechanism for water reabsorption in the renal tubules is by osmosis.
4. **a.** Marine bony fishes live in an environment in which salts tend to be drawn into the body and water tends to be drawn out of the body. To counter these effects, these fishes excrete small amounts of water and actively pump sodium out of the body across the gills.
5. **c.** Xeric animals have layers of lipids (waxy or greasy) in their body coverings to prevent water in their body fluids from freely evaporating into the air. Humidic animals (e.g., millipedes, centipedes, and amphibians) have outer body coverings that are highly permeable to water.
6. **e.** Insects excrete uric acid, not urea.
7. **a.** The route of water and solutes through the nephron is from the glomerulus, to Bowman's capsule, to the renal tubule, to collecting ducts.
8. **a.** Countercurrent multiplication is achieved a flow of tubular fluid in opposite directions through a gradually increasing osmotic concentration. This pulls water from the fluid in the tubules and results in a more concentrated urine, since the concentration gradient starts at 300 m*Osm* at the surface of the medulla and is 1,200 m*Osm* deep in the medulla.
9. **a.** At Bowman's capsule the filtrate is most similar to plasma.
10. **g.** The sodium concentration is the highest in the extracellular fluid near the middle of the medulla.
11. **a, b, and e.** The osmolarity of the filtrate is similar to that of plasma in the cortex of the kidney (including the glomerulus and Bowman's capsule), the proximal convoluted tubule, and the distal convoluted tubule.
12. **g.** The highest concentration of the filtrate in the collecting ducts will be near their ends, deep in the medulla.
13. **b.** The bulk of the water and solute is reabsorbed at the proximal convoluted tubule.
14. **e.** Marine invertebrates are primarily osmotic conformers, but some can regulate their osmolarity (osmotic regulators).
15. **c.** Antidiuretic hormone acts on the collecting ducts by increasing permeability to water. Antidiuretic hormone secretion is stimulated by rises in plasma concentration.
16. **b.** Osmotic pressure is, by definition, a measure of the total concentration of solutes (dissolved matter).
17. **c.** Changes in incoming arteriole affect glomerular filtration rate. Dilation of the incoming arteriole increases pressure, which will increase filtration rate.

# 37 Animal Reproduction

## The Big Picture

- Animals reproduce both asexually and sexually. Asexual reproduction typically produces genetically identical individuals. Sexual reproduction has the advantage of increasing genetic diversity by creating gametes that have gone through meiosis (gametogenesis). Haploid gametes (eggs and sperm) fuse at fertilization, restoring the diploid number of chromosomes to the zygote. Fertilization can be external, typically by gametes that are spawned into the environment, or internal, occurring within the female's body.
- In women, gametogenesis involves an ovarian cycle, during which an egg is produced and released, and a uterine cycle that prepares the uterus for implantation and pregnancy. These are coordinated by complex changes in hormone levels. In men, gametogenesis produces motile sperm that exit the body through ducts, along with fluid that is added by glands along the route. Among animals, a variety of methods integrate reproduction with the life cycle, with events often timed so young emerge when food is plentiful.

## Study Strategies

- Knowledge of the ploidy level of the developing egg and sperm is essential in understanding how genetic inheritance works. Study the diagrams in the chapter that summarize the essential aspects of meiosis.
- Figures will teach you a great deal. Many of the details of the chapter are contained within the figures. Having a picture in your mind of an event greatly increases your understanding and ability to remember what you have read.
- Following the hormonal changes associated with human ovarian and menstrual cycles can be difficult. Study the diagrams of the ovarian and menstrual cycles in parallel to see how they are coordinated.
- When you are trying to understand a general phenomenon, study the specific examples described. Often you will remember the examples best, and you can work back from there to remember the general phenomenon they represent.
- Go to **LaunchPad** (or use the Web addresses listed) to review the following additional resources:

  Animated Tutorial 37.1 Fertilization in a Sea Urchin Egg (PoL2e.com/at37.1)

  Animated Tutorial 37.2 The Menstrual Cycle (PoL2e.com/at37.2)

  Activity 37.1 Spermatogenesis (poL2e.com/ac37.1)

  Activity 37.2 The Human Female Reproductive Tract (poL2e.com/ac37.2)

  Activity 37.3 The Human Male Reproductive Tract (poL2e.com/ac37.3)

  Analyze the Data in textbook Figure 37.9

  Apply the Concept in textbook Section 37.2

## Key Concept Review

### 37.1 Sexual Reproduction Depends on Gamete Formation and Fertilization

Most animals reproduce sexually

Gametogenesis in the gonads produces the haploid gametes

Fertilization may be external or internal

The sex of an offspring is sometimes determined at fertilization

Some animals undergo sex change

A variety of animals, mostly among the invertebrates, reproduce asexually, producing offspring that are genetically identical to one another and to the parent. Asexual reproduction has several advantages: time and energy are not wasted on mating, and every member of the population can produce offspring. The main disadvantage of asexual reproduction is that it does not generate genetic diversity, which can be essential if the environment changes.

One method of asexual reproduction is budding, in which outgrowths from the parent produce new individuals (see Figure 37.2). Another method is fission, whereby an individual splits into two or more pieces and each piece grows into a new individual.

Most animals reproduce sexually, which involves the joining of two haploid gametes to form a diploid cell. Sexual reproduction has three major stages: gametogenesis (production of gametes), mating or spawning (bringing gametes together), and fertilization (fusing of gametes). Costs of sexual reproduction include the need to search for mates and the requirements of complex and expensive courtship behavior. It also breaks up favorable combinations of genes. The main advantage to sexual reproduction is that it generates genetic diversity. A great deal of genetic diversity is generated by the formation of gametes through meiotic cell divisions. During meiosis, crossing over between homologous chromosomes and independent assortment of chromosomes contribute to genetic diversity. Male and female haploid gametes then fuse at fertilization, creating a new individual that is genetically distinct from both parents.

In sexually reproducing animals, the sexes differ from each other. The primary reproductive organs in males are the testes and in females are the ovaries. The secondary reproductive organs in human males include the ducts and associated glands for carrying sperm out of the body as well as the penis. In human females secondary reproductive organs include the oviduct and uterus. Secondary sexual characteristics are nonreproductive structures such as facial hair and a deep voice in men.

Sexual reproduction is based on gametogenesis (oogenesis and spermatogenesis), which occurs in an animal's gonads. Female gametes—eggs or ova—are produced by oogenesis in the ovaries, and male gametes—sperm—are produced by spermatogenesis in the testes.

Oogenesis begins with a germ cell ($2n$) that proliferates through mitosis (see Figure 37.4B). The first meiotic division produces two haploid cells, but the division is unequal and results in one large oocyte and a tiny polar body. When the oocyte goes through the second meiotic division, two haploid cells are produced, but again the division is unequal, resulting in one large ovum and a tiny polar body. The two meiotic divisions in a developing ovum, therefore, maintain the bulk of the cytoplasm and the nutrients in a single large cell, while the other cells produced as tiny polar bodies are discarded and later degenerate. It is thought that mitosis in the germ cells of a female placental mammal stops when the female is still a fetus, and the germ cells start meiosis but are arrested in the midst of their first meiotic division. In human females these cells remain arrested for years. At puberty, one oocyte a month matures and ovulates, while the rest remain arrested.

Spermatogenesis begins with a germ cell ($2n$) that proliferates through mitosis (see Figure 37.4A). When a germ cell in a testis enters meiosis, the first meiotic division divides the cell into two equal-sized haploid cells ($n$), and the second meiotic division divides each of these cells into two equal-sized haploid cells ($n$). From one germ cell, therefore, four haploid cells are produced that go on to differentiate into sperm cells. Sperm cells are streamlined and have a flagellum.

In male mammals, sperm cells are produced in the seminiferous tubules (see Figure 37.5). Sertoli cells surround the developing germ cells within the seminiferous tubules. Sertoli cells nourish and protect the germ cells as they go through meiosis and develop into sperm. Interstitial cells (Leydig cells) outside the seminiferous tubules secrete testosterone, which is necessary for sperm production.

In female mammals, the germ cells in the ovary are surrounded by follicle cells that support development of the ova. The structure consisting of follicle cells surrounding the ovum is called a follicle. The follicle cells increase in number as the ovum matures; the maturing follicle eventually forms a large blister on the surface of the ovary just before the ovum is released (ovulated).

Fertilization can be external or internal. Many aquatic animals have external fertilization. Eggs and sperm are released into the water (in a process called spawning) and must find each other before fertilization can take place. Many animals employ internal fertilization, in which sperm and egg fuse within the female reproductive tract rather than in the external environment. Organisms that nurture the embryos within the female's body, such as placental mammals, have internal fertilization. Organisms in which the female lays eggs encased in an impenetrable shell also must have internal fertilization, because fertilization must take place before the shell is laid down around the eggs. Examples are birds, turtles, and some sharks.

For successful fertilization to occur, the sperm and egg must recognize each other. Specific recognition molecules on the gametes mediate recognition and prevent eggs from being fertilized by sperm from a different species. Sea urchins are a well-studied example. Sea urchin gametes are spawned into the seawater, and molecules on the sperm recognize species-specific molecules in the jelly of the eggs (see Figure 37.7). The acrosome then breaks open, releasing enzymes that digest a path for the sperm through the protective jelly surrounding the egg. While this is happening, actin within the head of the sperm assembles and pushes the acrosomal sac out into an acrosomal process. The acrosomal process contains molecules that recognize species-specific molecules on the vitelline envelope, which lies just outside the egg cell membrane. Fertilization occurs when the egg and sperm cell membranes fuse, allowing the nucleus of the sperm to enter the egg cytoplasm. This fusion of sperm and egg cell membranes triggers blocks to polyspermy, which prevents more than one sperm from entering the egg. Entry by more than one sperm usually means death of the embryo. Sea urchins have a fast block to polyspermy, which takes less than a second to complete, and a slow block to polyspermy, which takes about a minute to complete.

The sex of the offspring is often determined at fertilization. In placental mammals (including humans), sex is determined by the presence or absence of a Y chromosome: an XX individual is female, and an XY individual is male. This is due primarily to a gene called *SRY* (*s*ex-determining *r*egion of the *Y* chromosome). A mutation that removes the *SRY* gene from the Y chromosome can result in an XY individual being female; a mutation that inserts an *SRY* gene into an X chromosome can result in an XX male. Sex determination is some organisms is determined by environmental factors. In crocodilians and turtles, for example, the temperature at which embryos are incubated determines the sex of the embryos.

Some organisms are not determined as a single sex but are able to change sex—these animals are sequential hermaphrodites. Anemonefish, for example, live in groups that include a number of nonbreeding males, one breeding male, and one breeding female. If the female dies or is removed, the breeding male turns into a breeding female.

**Question 1.** The diagram below shows meiosis in what type of gonad? If this were meiosis in a human ($2n = 46$), how many chromosomes would you find in the cell marked 1? How many chromosomes would you find in the cell marked 2? What is the fate of the cell marked 2?

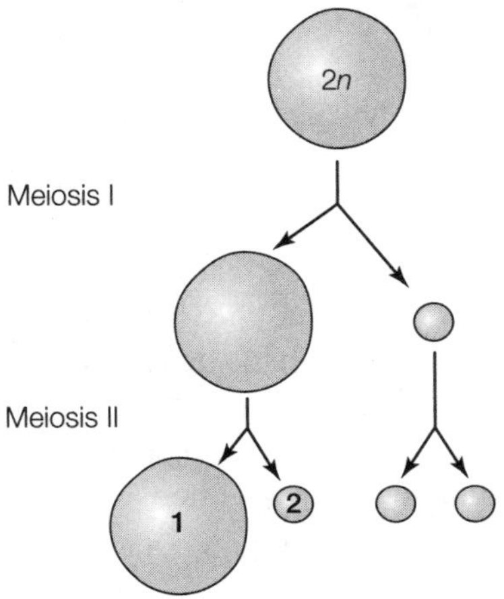


**Question 2.** Why is it problematic if more than one sperm enters an ovum at fertilization?

### 37.2 The Mammalian Reproductive System Is Hormonally Controlled

- Ova mature in the ovaries and move to the uterus
- Ovulation is either induced or spontaneous
- Pregnancy is a specialized hormonal state
- Male sex organs produce and deliver semen
- Many contraceptive methods are available

In mammals, the ovaries release eggs into the body cavity close to the opening of the oviducts. Cilia guide the ovum to the oviduct and sweep it down the length of the oviduct to the uterus (see Figure 37.10). At the bottom of the uterus is the cervix, which leads into the vagina. Sperm deposited in the vagina during copulation move through the cervix and uterus, into the oviduct, and up to the upper end of the oviduct. If there is an ovulated egg at the upper end of the oviduct, then fertilization can occur. If this happens, then the fertilized egg begins development. The endometrium of the uterus is thickened during this stage of the uterine cycle, and the embryo can then implant in the endometrium.

What induces an egg to be ovulated depends upon the species. Rabbits, for example, are induced ovulators. Copulation itself induces ovulation by causing gonadotropin-releasing hormone (GnRH) to be released from the hypothalamus, which causes a surge in secretion of luteinizing hormone from the anterior pituitary gland, which in turn induces ovulation. Most mammals, however, are spontaneous ovulators, and ovulation occurs according to a cycle. In humans, for example, a mature egg is released about every 28 days. Female primates menstruate during each cycle. Other female mammals show specific mating behavior around the time they are ovulating; this is called entering estrus, or "heat," and their cycles are called estrous cycles.

In female placental mammals, the ovarian and uterine cycles are coordinated by hormones. Humans are a good example. Mature eggs are produced during a 28-day ovarian cycle (see Figure 37.11B). A follicle containing an oocyte matures, and the oocyte is ovulated along with a halo of follicle cells at the midpoint of the cycle, around day 14. The remaining follicle cells become the corpus luteum, a glandular structure that produces progesterone and estrogen (see Figure 37.11B,C). These hormones cause the uterine endometrium to thicken and become highly vascularized, preparing it for implantation of an embryo (see Figure 37.11D). If the embryo does not implant, the corpus luteum degenerates, estrogen and progesterone levels decrease, and the endometrium breaks down and is expelled—a process called menstruation. The first day of menstruation is counted as the beginning of the cycle.

Luteinizing hormone (LH) and follicle-stimulating hormone (FSH) secreted by the anterior pituitary gland control the maturation of the follicles within the ovary (see Figure 37.11A). These hormones are secreted under the control of estrogen and of gonadotropin-releasing hormone (GnRH) from the hypothalamus of the brain. Just prior to the midpoint of the cycle, a surge in LH triggers ovulation.

If the ovulated egg meets sperm in the upper end of the oviduct, fertilization can occur. The zygote begins dividing, and this development continues as the embryo is moved down the oviduct into the uterus, a trip that takes about 4 days. The embryo arrives in the uterus as a blastocyst (blastula). After several more days the embryo implants by burrowing into the endometrium. The embryo secretes human chorionic gonadotropin (hCG), which keeps the corpus luteum functional. Continued production of estrogen and progesterone by the corpus luteum supports development and maintenance of the endometrium, which is needed for pregnancy. A woman who wants to know if she is pregnant can use a pregnancy test that detects the presence of hCG.

During pregnancy, growth of the embryo is supported by the placenta, a structure in which embryonic and maternal blood vessels are in close proximity. Fetal blood and maternal blood do not mix in the placenta, but the barrier between them is extremely thin, allowing for efficient exchange of nutrients, respiratory gases, urinary wastes, and other components carried in the blood.

At the end of pregnancy, oxytocin, a hormone from the posterior pituitary and under the control of the hypothalamus, stimulates the uterus to contract in ever-increasing force until the baby is pushed out of the uterus. After birth, oxytocin, along with prolactin from the anterior pituitary, stimulates the mother's breasts to produce and eject milk.

Sperm are produced in the testes, which in most mammals are held in the scrotum (see Figure 37.12). This pouch of skin holds the testes outside the body cavity, where temperatures are about 2°C lower than normal body temperature and optimal for spermatogenesis. Male mammals produce semen, which contains sperm and fluids that support the sperm and facilitate fertilization. Immature sperm move from the seminiferous tubules of the testis to the epididymis, where they mature and are stored. The epididymis connects to the vas deferens, which connects to the urethra, which opens to the outside of the body at the tip of the penis.

The fluids in semen are contributed by accessory glands along the route of the sperm (see Figure 37.12). The seminal vesicles and prostate gland secrete fluids that provide a protective environment and contain fructose as an energy source for the sperm. The bulbourethral glands secrete a clear, lubricating mucus.

During sexual arousal, the penis becomes engorged with blood, and contraction of smooth muscle in the ducts and at the base of the penis results in ejaculation of semen. The dilation of blood vessels in the penis during an erection is initiated by the neurotransmitter nitric oxide (NO) from parasympathetic nerve endings. NO increases levels of the second messenger cGMP. Erectile dysfunction (or impotence), the inability to achieve or sustain an erection, can be treated with medications that prevent the breakdown of cGMP.

Testosterone, produced in the Leydig cells of the testes, is necessary for sperm production. In human males testosterone is secreted during fetal development but then remains low until puberty. After puberty it stays high, enabling sperm production for the rest of the man's life. Other hormones are involved in sperm production in humans as well. The hypothalamus secretes GnRH, which prompts the anterior pituitary to release LH and FSH. LH stimulates the Leydig cells to secrete testosterone, and FSH and testosterone stimulate the Sertoli cells to support spermatogenesis.

There are many contraceptive methods available to humans to prevent pregnancy. Non-hormonal methods include abstinence, the rhythm method (avoidance of intercourse around the time of ovulation), and coitus interruptus (withdrawing the penis before ejaculation). They also include barrier methods—the male and female condoms and female diaphragm (that caps the cervix), and spermicides—spermicidal jellies, and the sponge containing spermicides. Hormone-based methods include oral hormones ("the pill") and ovulation-preventing hormones that are administered non-orally (through a patch or injections). Intrauterine devices (IUDs) prevent implantation. Sterilization methods are vasectomy (for males) and tubal ligation (for females). A vasectomy involves cutting and tying off the cut ends of the vasa deferentia, thereby preventing sperm from leaving the body. A tubal ligation in a woman ties off the oviducts, preventing an egg from descending into the uterus and therefore preventing sperm from reaching the ovulated egg.

The failure rates of contraceptive methods vary (see Table 37.1), with the rhythm method resulting in the second highest pregnancy rate (i.e, compared to unprotected sex). Abstinence has the lowest failure rate, with sterilization methods producing the next lowest rate. The "pill" has very low failure rates; combinations of barrier methods, such as a diaphragm coated with spermicidal jelly used in combination with a male condom, also have very low failure rates.

**Question 3.** The diagram below shows the ovarian cycle. Label the following: mature follicle, oocyte, corpus luteum, ovulation. Add arrows and labels to indicate when progesterone levels are highest and when LH and FSH levels are highest.

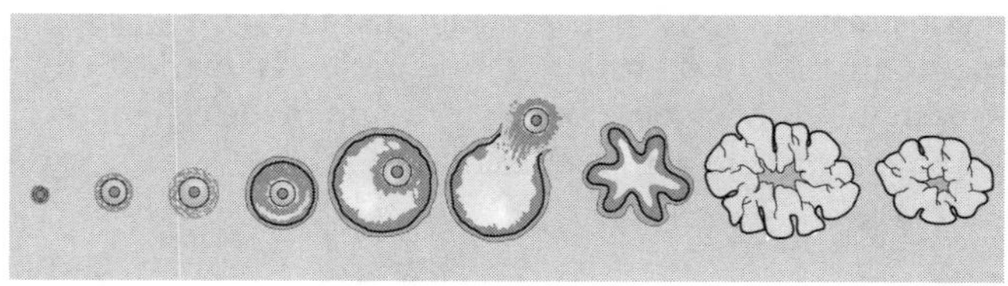

**Question 4.** Why might a physician recommend that men who want children but who have been diagnosed with low sperm counts avoid wearing tight-fitting underpants or pants?

### 37.3 Reproduction Is Integrated with the Life Cycle

Animals often gain flexibility by having mechanisms to decouple the steps in reproduction

Some animals can reproduce only once, but most can reproduce more than once

Seasonal reproductive cycles are common

Reproduction is an important part of the life cycle, and various methods have evolved that fit reproduction into the life cycle in ways that increase success. Being able to store sperm for long periods of time within the female reproductive tract, for example, is a way of decoupling the timing of copulation and fertilization. Examples of organisms that do this include the female blue crab and queen honey bees. In some organisms embryonic development can be arrested in order to delay the time when the young emerge. This is called embryonic diapause, and it is used by silkworm moths to ensure that the larvae emerge in the spring when food is plentiful rather than in the winter months. Delayed implantation is a mechanism found in some placental mammals. In the Antarctic seal, for example, the female mates soon after giving birth in the early summer. The embryo does not implant, however, until early autumn, which allows the birth of the offspring to occur in the early summer when the offspring will not be endangered by winter storms and will have enough time to grow before the next winter arrives.

In some animals, reproduction is a once-in-a-lifetime affair. Organisms that are programmed physiologically for this are called semelparous. The common octopus (*Octopus vulgaris*) and the Pacific salmon (*Oncorhynchus nerka*) are examples. The female common octopus guards her eggs until they hatch, and then both the male and female die. The Pacific salmon swim from the ocean upstream in rivers and streams to spawn, and both females and males die after spawning. Animals that are capable of reproducing more than once are called iteroparous.

Many animals have reproductive cycles that are tied to the seasons. Some of these breed only once in a reproductive season. The red fox (*Vulpes vulpes*) is an example. The female ovulates only once per year. If she loses her kits in that year, she is unable to breed again until the next year. Cues that seasonal breeders use to time their breeding often involve photoperiods, such as the long days of late spring and early summer. Some animals can time events based on the entire year—these animals are using circannual biological clocks.

**Question 5.** The Antarctic fur seal exhibits delayed implantation. Mark the diagram below to show when the pregnant fur seal will give birth, when she will mate again, when fertilization will occur, when the embryo will implant, and when the next pup will be born.

Pregnant female

| SPRING | SUMMER | AUTUMN | WINTER | SPRING | |
|---|---|---|---|---|---|

**Question 6.** Pacific salmon live in the ocean for 1–4 years and then swim up rivers and streams in the Pacific Northwest to spawn. After spawning, both the male and female salmon die, filling the streams with their carcasses. What are some of the advantages to this type of reproduction?

## Test Yourself

1. You have a prize female rabbit that you want to breed with a prize male rabbit. The best time to schedule a short visit between your rabbit and the male rabbit to ensure offspring from this mating would be
   a. after the female rabbit enters estrus.
   b. just after the female rabbit ovulates.
   c. while the female rabbit is pregnant.
   d. after the female rabbit has given birth.
   e. in the middle of the female rabbit's estrus cycle.
   ***Textbook Reference:*** *Introduction*

2. Asexual reproduction is an effective strategy in stable environments because
   a. gametogenesis is most efficient under these conditions.
   b. the resulting offspring, typically genetically identical to their parents, are preadapted to their environment.
   c. asexual reproduction produces a large amount of genetic diversity.
   d. animal cells tend to be more totipotent under stable conditions.
   e. sessile animals and sparse populations are more common under stable conditions.
   ***Textbook Reference:*** *Concept 37.1 Sexual Reproduction Depends on Gamete Formation and Fertilization*

3. In mammals, the egg and sperm differ from each other in
   a. the number of chromosomes they contribute to the zygote.
   b. the amount of cytoplasm they contain.
   c. the number of meiotic divisions they go through.
   d. the number of chromosomes in the germ cell that produces them.
   e. their dependency on FSH and LH for development.
   ***Textbook Reference:*** *Concept 37.1 Sexual Reproduction Depends on Gamete Formation and Fertilization; Gametogenesis in the gonads produces the haploid gametes*

4. External fertilization
   a. is typical in birds.
   b. requires a penis.
   c. does not require species-specific recognition molecules of sperm and ova.
   d. occurs in some, but not all, aquatic animals.
   e. occurs before spawning.
   ***Textbook Reference:*** *Concept 37.1 Sexual Reproduction Depends on Gamete Formation and Fertilization; Fertilization may be external or internal*

5. Which of the following best represents the normal path of a mammalian sperm cell as it makes its way from the point of entry into a female's reproductive tract to the location where fertilization typically occurs?
   a. Cervix, vagina, ovary, oviduct
   b. Vagina, cervix, uterus, oviduct
   c. Uterus, cervix, vagina, oviduct
   d. Vagina, uterus, cervix, oviduct
   e. Cervix, vagina, oviduct, ovary

   ***Textbook Reference:*** *Concept 37.2 The Mammalian Reproductive System Is Hormonally Controlled; Ova mature in the ovaries and move to the uterus*

6. In the diagram of the human male reproductive system, sperm are stored in the _______ and clear mucus is added to the semen by the _______.

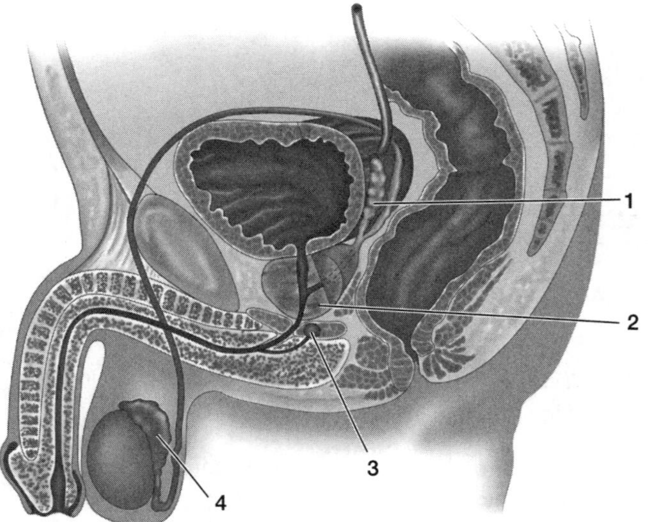


   a. seminal vesicle (1); prostate gland (2)
   b. prostate (2); seminal vesicle (4)
   c. epididymis (4); seminal vesicle (2)
   d. bulbourethral gland (4); epididymis (1)
   e. epididymis (4); bulbourethral gland (3)

   ***Textbook Reference:*** *Concept 37.2 The Mammalian Reproductive System Is Hormonally Controlled; Male sex organs produce and deliver semen*

7. When meiosis occurs in germ cells within an ovary,
   a. polar bodies contain half the number of chromosomes as the mature ovum.
   b. polar bodies contain no chromosomes, while the ovum retains all the chromosomes.
   c. meiotic divisions divide the cytoplasm evenly.
   d. the oocyte is haploid after the first meiotic division.
   e. the oocyte becomes haploid, while the polar body remains diploid.

   ***Textbook Reference:*** *Concept 37.1 Sexual Reproduction Depends on Gamete Formation and Fertilization; Gametogenesis in the gonads produces the haploid gametes*

8. Which of the following is the correct order of events during fertilization in a sea urchin?
   a. Acrosomal process forms; acrosome reaction; sperm penetrates vitelline envelope; sperm fuses with egg cell membrane
   b. Sperm penetrates vitelline envelope; acrosomal process forms; acrosome reaction; sperm fuses with egg cell membrane
   c. Acrosome reaction; acrosomal process forms; sperm penetrates vitelline envelope; sperm fuses with egg cell membrane
   d. Sperm fuses with egg cell membrane; sperm penetrates vitelline envelope; acrosome reaction; acrosomal process forms
   e. Sperm penetrates vitelline envelope; sperm fuses with egg cell membrane; acrosomal process forms; acrosome reaction

   ***Textbook Reference:*** *Concept 37.1 Sexual Reproduction Depends on Gamete Formation and Fertilization; Fertilization may be external or internal*

9. In the human female's menstrual cycle, when are progesterone concentrations high enough to maintain the uterus in a proper condition for pregnancy?
   a. For the entire duration of the cycle
   b. During the first day of the cycle
   c. During the first half of the cycle
   d. During the second half of the cycle
   e. During a 5-day window in the middle of the cycle

   ***Textbook Reference:*** *Concept 37.2 The Mammalian Reproductive System Is Hormonally Controlled; Ovulation is either induced or spontaneous*

10. Antarctic fur seals display _______, whereas the common octopus displays _______.
    a. circannual biological clocks; embryonic diapause
    b. a seasonal reproductive cycle; iteroparity
    c. delayed implantation; semelparity
    d. embryonic diapause; a seasonal reproductive cycle
    e. semelparity; iteroparity

    ***Textbook Reference:*** *Concept 37.3 Reproduction Is Integrated with the Life Cycle; Some animals can reproduce only once, but most can reproduce more than once*

11. Which of the following contraceptive methods has the lowest failure rate?
    a. Intrauterine device
    b. Rhythm method
    c. Diaphragm
    d. Coitus interruptus
    e. Condom

    ***Textbook Reference:*** *Concept 37.2 The Mammalian Reproductive System Is Hormonally Controlled; Many contraceptive methods are available*

12. If you compared the genetic makeup of an animal produced by budding with that of its parent, which of the following would you expect to discover?
    a. About 100 percent genetic similarity
    b. About 50 percent genetic similarity
    c. About 33 percent genetic similarity
    d. About 25 percent genetic similarity
    e. No genetic similarity

    ***Textbook Reference:*** *Concept 37.1 Sexual Reproduction Depends on Gamete Formation and Fertilization*

13. Spermatogenesis
    a. results in two sperm cells from each germ cell.
    b. involves a period of arrest during the first meiotic division.
    c. results in diploid gametes.
    d. apportions cytoplasm equally among resulting cells.
    e. apportions cytoplasm unequally among resulting cells.

    ***Textbook Reference:*** *Concept 37.1 Sexual Reproduction Depends on Gamete Formation and Fertilization; Gametogenesis in the gonads produces the haploid gametes*

14. During fertilization in the sea urchin,
    a. the egg permits several sperm to enter it.
    b. the vitelline envelope and cell membrane of the sperm fuse.
    c. the egg has more than one method for blocking polyspermy.
    d. species-specific recognition occurs when the egg and sperm fuse.
    e. the acrosomal reaction occurs just before the sperm makes contact with the egg jelly.

    ***Textbook Reference:*** *Concept 37.1 Sexual Reproduction Depends on Gamete Formation and Fertilization; Fertilization may be external or internal*

15. If a woman thinks she may be pregnant, she can use a pregnancy test from a pharmacy that will detect the presence of
    a. LH.
    b. FSH.
    c. chorionic gonadotropin.
    d. progesterone.
    e. estrogen.

    ***Textbook Reference:*** *Concept 37.2 The Mammalian Reproductive System Is Hormonally Controlled; Pregnancy is a specialized hormonal state*

16. Oral contraceptives that are hormonally based and referred to as "the pill"
    a. work by preventing implantation.
    b. work by preventing ovulation.
    c. contain low doses of LH and FSH.
    d. contain low doses of testosterone.
    e. have a high failure rate.

    ***Textbook Reference:*** *Concept 37.2 The Mammalian Reproductive System Is Hormonally Controlled; Many contraceptive methods are available*

## Answers

### *Key Concept Review*

1. The diagram shows meiosis in an ovary. If this were in a human, the cells marked 1 and 2 would both contain 23 chromosomes. The cell marked 2 will degenerate.
2. More than one sperm entering an ovum would result in an abnormal number of chromosomes in the zygote, which would make it incapable of normal development. One sperm entering the egg ensures equal genetic contributions by each parent.
3. The diagram should be marked as shown below.

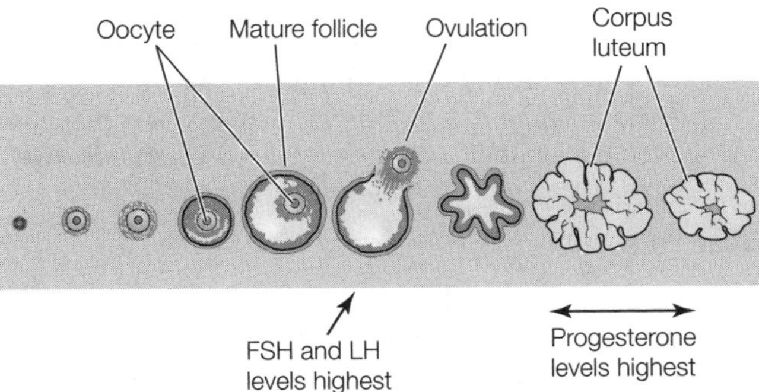


4. The optimal temperature for spermatogenesis is slightly below normal body temperature, which explains why the testes are located outside the body cavity in the scrotum. Tight clothing pulls the testes close to the body, exposing them to higher than optimal temperatures for spermatogenesis.

5.

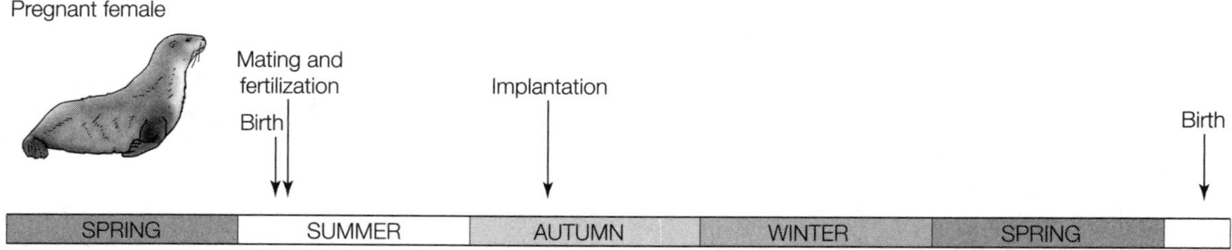


6. It may be a reproductive advantage for the salmon to delay reproduction until they are large in size, able to produce enormous numbers of gametes. They live and feed in the ocean where food is more plentiful than in the streams. By returning to the streams for reproduction, they are most likely returning to an environment that is safer for the embryonic fish with less predation. Meanwhile, they are maximizing their reproductive capacity by being so large for this one-time spawning.

### *Test Yourself*

1. **d.** The female rabbit can be mated after she has given birth. Rabbits exhibit postpartum estrus; that is, they go into "heat" after giving birth, and the act of copulation induces ovulation.

2. **b.** The parents that have survived to reproduce asexually are able to survive in the current stable environment. Therefore, the offspring should be preadapted for this stable environment.

3. **b.** There are many differences between eggs and sperm. One major difference is that eggs are larger and contain more cytoplasm and nutrients than sperm. They do not differ in the number of chromosomes they are carrying, and in mammals, both depend upon LH and FSH for their development.

4. **d.** External fertilization occurs in some, but not all, aquatic animals. It does not occur in birds, and it does not require a penis. It occurs after spawning and does involve species-specific recognition molecules.

5. **b.** In mammals, a sperm cell is ejected by the male into the vagina. From the vagina the sperm moves through the cervix into the uterus and finally into the upper end of the oviduct, where fertilization occurs.

6. **e.** In the human male, sperm are stored in the epididymis (4) and clear mucus is added to the semen by the bulbourethral gland (3).

7. **d.** During oogenesis, the oocyte becomes haploid after the first meiotic division. The polar bodies are created by an unequal division at the first and second meiotic divisions. The polar body and large egg cell are haploid after the first meiotic division and continue to be haploid after the second meiotic division.

8. **c.** During fertilization in the sea urchin, the sperm undergoes the acrosome reaction when it reaches the jelly surrounding the egg; this causes the acrosomal process to form. After the sperm penetrates the vitelline envelope it fuses with egg cell membrane.

9. **d.** High levels of progesterone are needed to maintain the uterus in the proper condition for pregnancy. The levels of progesterone are high only during the second half of the uterine cycle.

10. **c.** Antarctic fur seals display delayed implantation. The common octopus displays semelparity.

11. **a.** Of the methods listed, the intrauterine device (IUD) has the lowest failure rate (<1%).

12. **a.** Budding is a form of asexual reproduction in which new individuals form from outgrowths of the parent organism. These buds are therefore genetically identical to the parent organism.

13. **d.** During spermatogenesis, cytoplasm is apportioned equally among resulting cells. This differs from oogenesis, in which cytoplasm is apportioned unequally among daughter cells.

14. **c.** Similar to fertilization in other animals, during fertilization in the sea urchin there are two blocks to polyspermy, a fast block followed by a slow block. This prevents any further sperm from entering the egg. The acrosomal reaction in the sperm is induced by the sperm making contact with the jelly surrounding the egg. Species-specific recognition occurs when the sperm meets the jelly and when it meets the vitelline envelope. Fertilization occurs when the sperm and egg cell membranes fuse.

15. **c.** These pregnancy tests detect the presence of human chorionic gonadotropin, which is secreted by the embryo.

16. **b.** The birth control pill works by blocking ovulation, not implantation. It typically contains estrogen and progesterone, or just progesterone. It does not contain FSH, LH, or testosterone.

# Animal Development

## The Big Picture

- Animal development begins at fertilization, when an egg and sperm fuse to form a diploid zygote. This is followed by a period of cell division called cleavage, and the result is typically a ball of undifferentiated cells called the blastula. Gastrulation is the next stage, when the cells of the embryo rearrange themselves to form germ layers. In most animals there are three germ layers: ectoderm, mesoderm, and endoderm. Ectoderm forms the outer epidermis and nervous system of an animal; endoderm forms the inner lining of the gut and organs derived from the gut; and mesoderm forms the layers of muscle, connective tissue, skeleton, and organs lying between the outer epidermis and inner lining of the gut.
- In reptiles, including birds, and mammals, embryos develop a set of extraembryonic membranes that protect and nourish the embryo. Development does not end at birth or hatching but continues throughout life.

## Study Strategies

- There are a variety of cleavage patterns among different groups of animals. Learning these cleavage patterns becomes easier if you remember that yolk is an impediment to cell division. In an amphibian egg, for example, there are moderate amounts of yolk that are concentrated in the lower half of the egg. The yolk slows down cleavage, so cells in the lower half of the egg divide more slowly than cells in the upper half of the egg. In a bird egg, where there are very large amounts of yolk, the yolk is such an impediment to cleavage that the yolky cytoplasm remains undivided and only the non-yolky cytoplasm divides to form the cells of the blastula.
- Gastrulation can be difficult to picture and understand, especially given the different ways in which gastrulation occurs in different organisms. Study the figures in the textbook. Remember that the basic pattern is that the endoderm and mesoderm move inside the embryo, where the endoderm forms a gut tube and the mesoderm surrounds the endoderm. On the outside of the embryo the ectoderm spreads to cover the outside of the embryo. This results in a three-layered embryo, with an innermost endoderm that is surrounded by mesoderm, which in turn is surrounded by ectoderm.
- A key point in understanding animal development is remembering what the three germ layers give rise to. This is much easier if you recognize that the endoderm forms the innermost tube and becomes the inner lining of the gut. The ectoderm forms the outermost covering of the animal, which is the epidermis, along with the nervous system. Everything in between these two layers is formed primarily by mesoderm. Think of a cross section through your own body. Innermost there is your gut tube, and the inner epithelium lining of that tube was formed by endoderm. Your outermost covering, your epidermis, as well as your nerve cord, was formed by ectoderm. Everything in between was formed primarily by mesoderm; that includes the connective tissue and muscles of your gut wall, your circulatory system, skeleton, kidneys, reproductive organs, the muscles of your body wall, and the dermis of your skin. If you understand this overall plan, it will be easy for you to predict germ-layer origins of tissues and organs of any animal. For example, you can predict from this that skin (made up of epidermis and dermis) is derived from both the ectoderm (forming the epidermis) and mesoderm (forming the dermis).
- This chapter contains a great deal of terminology, but many terms are common to many animals. Create a list of new terms from the chapter. Then determine which are the more general terms, applicable to numerous animals (e.g., "somites" and "gastrulation"), and learn these first. Then tackle those referring to development in specific groups of animals (e.g., "dorsal lip of the blastopore" and "primitive streak").
- Go to **LaunchPad** (or use the Web addresses listed) to review the following additional resources:

  Animated Tutorial 38.1 Gastrulation (PoL2e.com/at38.1)

  Animated Tutorial 38.2 Tissue Transplants Reveal the Process of Determination (PoL2e.com/at38.2)

Activity 38.1 Extraembryonic Membranes (PoL2e.com/ac38.1)

Analyze the Data in textbook Figure 38.15

Apply the Concept in textbook Section 38.4

## Key Concept Review

### 38.1 Fertilization Activates Development

Egg and sperm make different contributions to the zygote

Polarity is established early in development

Development begins when a haploid sperm fuses with a haploid egg. However, fertilization does more than produce a diploid zygote; fertilization activates development. The egg contributes most of the cytoplasm and organelles (including a haploid nucleus) to the zygote. The sperm contributes its haploid nucleus and a centriole that becomes the centrosome, which plays an essential role in microtubule organization and setting up the first cleavage spindle for cell division.

In an unfertilized amphibian egg, nutrients are concentrated in the lower half of the egg, called the vegetal hemisphere. The upper half of the egg, known as the animal hemisphere, has heavy pigmentation in its outer cytoplasm. Upon fertilization, the cytoplasm undergoes rearrangement (see Figure 38.1). A sperm enters in the animal hemisphere at what will become the ventral side of the animal. This causes rotation of the cortical (outer) cytoplasm, which uncovers a lighter region of cytoplasm on the side opposite sperm entry. This lighter cytoplasmic region is called the gray crescent, which will become the dorsal region of the animal. After sperm entry, therefore, the zygote has already established the embryo's anterior–posterior and dorsal–ventral axes. The shifting of the cytoplasm that uncovers the gray crescent is orchestrated by the sperm centriole that entered the egg at fertilization. It organizes microtubules into an array that extends into the vegetal hemisphere and guides the shifting of the cytoplasm.

An experiment by the biologist Hans Spemann revealed that cytoplasmic determinants important to development are found in the gray crescent. By tying a thin hair around a salamander zygote, he was able to cut the zygote in half. He discovered that when the gray crescent was bisected so that each half of the zygote received half of the gray crescent, each half developed into a normal larva. If the entire gray crescent was on one side of the constricting hair, that half of the zygote developed into a normal larva and the other half developed into an abnormal growth that Spemann called a "belly piece" (see Figure 38.2). These results showed that information necessary for normal development is not distributed evenly throughout the amphibian zygote's cytoplasm, and that cytoplasmic determinants necessary for establishing polarity of the embryo reside in the gray crescent.

**Question 1.** Which two organelles are contributed by the sperm at fertilization, and why is each organelle important?

**Question 2.** The diagram below shows an experiment performed on a salamander zygote. In the first part of the experiment, a hair is used to divide the zygote into two separate halves. In the second part of the experiment, each half is allowed to develop. Label the gray crescent in each picture. Draw the results of the second part of the experiment. What can you conclude from this experiment?

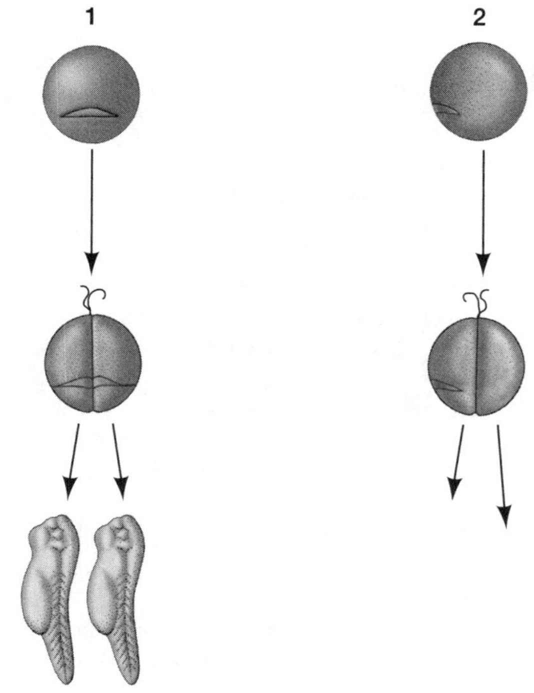


### 38.2 Cleavage Creates Building Blocks and Produces a Blastula

Specific blastomeres generate specific tissues and organs

The amount of yolk affects cleavage

Cleavage in placental mammals is unique

Following fertilization, the zygote begins a stage called cleavage, when cells divide rapidly but do not grow. The cells of the embryo, therefore, become progressively smaller as cleavage proceeds. By the end of the cleavage stage, the embryo is called a blastula. Often this is a ball of cells with a central fluid-filled cavity called a blastocoel. The cells of the blastula are called blastomeres. By labeling these cells and observing what they become later on, scientists have been able to create blastula fate maps. For example, a fate map of the frog blastula shows the positions of cells that will form each of the germ layers (see Figure 38.3).

The amount of yolk in the egg determines the pattern of cleavage, since cleavage furrows are impeded by yolk. An egg with small amounts of yolk can divide completely (complete cleavage), as seen in the sea urchin embryo (see

In-Text Art, p. 790). In eggs with large amounts of yolk, such as bird eggs, there is incomplete cleavage. Amphibian eggs, with moderate amounts of yolk, also have complete cleavage, but the yolk that is concentrated in the vegetal half of the embryo slows down the process, resulting in fewer, larger cells in the vegetal half than the animal half of the embryo (see Figures 38.3, 38.6, and 38.7). The eggs of fishes and reptiles, including birds, and of egg-laying mammals (monotremes), have large amounts of yolk, which prevent the eggs from dividing completely. In these organisms cleavage is confined to a region of non-yolky cytoplasm on top of the large sphere of yolky cytoplasm and forms a disc of cells (known as the blastodisc) on top of the yolk mass (see Figure 38.7). Insects also have yolky eggs, but the yolk is concentrated in the center of the egg. The nucleus divides many times in the yolky cytoplasm without any cytoplasmic division, forming a syncytium. Eventually the nuclei migrate to the peripheral cytoplasm, where cell membranes develop, separating the nuclei into separate cells (see Figure 38.7).

Among organisms with complete cleavage, there are two main patterns of cleavage—radial and spiral. These patterns are determined by the orientation of the mitotic spindles at each cleavage. Sea urchins have radial cleavage. At the 8-cell stage, the four daughter cells sit directly above the four parent cells (see In-Text art, p. 790). Annelids and mollusks have spiral cleavage. In spiral cleavage, the mitotic spindles align so that at the 8-cell stage the four daughter cells sit in the grooves between the four parent cells (see In-Text art, p. 791).

Blastomeres become determined (committed to a specific fate) at different times in development. In some embryos, even at the 2-cell stage, each blastomere has a specific fate such that if a blastomere is removed, the embryo that develops will be missing a specific region of the body. This is called mosaic development. A dramatic example of this is seen in the spirally cleaving eggs of a snail egg, in which an "out-pouching" called a polar lobe forms during early cleavage stages. The polar lobe is attached to one of the blastomeres, ensuring that specific cytoplasm is isolated to that blastomere. If this polar lobe is experimentally removed, the larva that develops is incomplete. In sea urchins, by contrast, the cells of the 2-cell and even the 4-cell stage can be separated from one another, and each cell will form a normal larva. An embryo that can compensate for the loss of a cell and develop normally is said to have regulative development. Regulated development is typical of vertebrates, including humans.

The early cell divisions of placental mammals are unique. Though the egg divides completely, the pattern is neither radial nor spiral. Mitotic spindles align so that at the second division, the cells divide in different planes, with one cell dividing vertically and the other dividing horizontally (see In-Text art, p. 793). In addition, the embryo's genes are transcribed very early in cleavage; in most other animals, transcription of the embryo's own genome does not take place until toward the end of the blastula stage. Also unique to placental mammals is the separation of cells during cleavage into the inner cell mass (which will become the embryo proper) and the trophoblast (which attaches to the uterus during implantation and becomes the embryonic side of the placenta). At this stage, the mass of cells is called a blastocyst and consists of an inner cell mass attached to a fluid-filled blastocoel (see Figure 38.8).

**Question 3.** How does cleavage in birds differ from cleavage in placental mammals?

**Question 4.** In the figure below, predict the outcome of the experiment by explaining the likely developmental fate of cells 2–5. Compare each of these results to the result labeled 1.

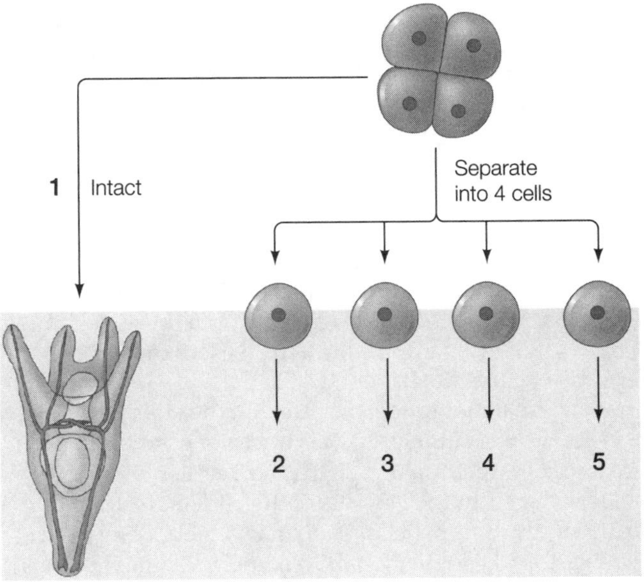


## 38.3 Gastrulation Produces a Second, then a Third Germ Layer

Following the blastula stage is a stage called gastrulation. This is a period when the cells of the embryo undergo major rearrangements, with many cells moving to the inside of the embryo while others spread to cover the outside of the embryo, setting the stage for development of the organs. In triploblastic organisms, three germ layers are formed by gastrulation: ectoderm, mesoderm, and endoderm. The endoderm is a tube inside the embryo, the mesoderm surrounds the endoderm, and the ectoderm surrounds the mesoderm and forms the outside of the embryo.

Just as the yolk affects how cells divide during the cleavage stage, it also affects how gastrulation proceeds. In embryos with little yolk, as in the sea urchin, the ventral wall of the embryo invaginates to form the archenteron ("first gut") consisting of endoderm and mesoderm cells. This leaves ectoderm on the outside of the embryo (see Figure 38.9). The archenteron is elongated by thin, contractile extensions of cells at the top. These attach to the inside of the blastula wall and contract, pulling the archenteron toward the region that will later become the mouth. The blastopore, where invagination began, will become the anus. The sea urchin is thus a deuterostome: an animal in which the anus forms from the blastopore and the mouth forms secondarily. Animals in which the blastopore becomes the mouth are called protostomes.

In amphibians with moderate amounts of yolk, cells move inward starting at the dorsal lip of the blastopore, which forms in the region of the gray crescent. These cells are destined to become the notochord, which directs much of the later development of the embryo. Cells continue to move inward along the lateral sides and finally the ventral lip of the blastopore, completing a circle of inward movement around a yolk plug (see Figure 38.10). In amphibians, the blastopore becomes the anus.

In yolky eggs such as those of birds, gastrulation movements are restricted to the flat plate of cells lying on top of the yolk. First the disc splits into two layers, an upper epiblast and a lower hypoblast. Cells of the epiblast move toward the midline of the disc, forming the primitive streak, and then move inward along this line. This inward movement creates a groove called the primitive groove. The anterior end of the groove is Hensen's node, where cells that will form the notochord move inward and than anteriorly (see Figure 38.11). Hensen's node is therefore comparable to the dorsal lip of the blastopore in the amphibian embryo. The cells left on the outside of the epiblast form the ectoderm. The cells that move inward form the mesoderm and endoderm (see Figure 38.11).

In placental mammals, even though the embryo develops from an egg with little yolk, the pattern of gastrulation still reflects their evolutionary origins from earlier forms that had yolky eggs. In the gastrula of the placental mammal, the inner cell mass splits into an upper epiblast and a lower hypoblast, and gastrulation movements are similar to those in birds.

Among other functions, gastrulation sets up the position of germ layers for the future formation of the body coelom, a cavity that is lined with mesoderm. The inner layer of mesoderm surrounds the gut and other organs, and the outer layer of mesoderm lies up against the body wall. The coelom provides space for the organs to move independently of the outer body wall. In deuterostomes, the coelom often forms from an out-pouching of the archenteron; in protostomes, it forms as a split in the mesoderm (see Figure 38.12).

**Question 5.** In the diagram below of a sea urchin gastrula, label the ectoderm, endoderm, mesenchyme, blastocoel, archenteron, and thin, contractile extensions. What will the structure labeled "A" become in the larva?

**Question 6.** Although details of gastrulation differ among sea urchins, frogs, and chickens, what is the common result in all of these animals?

## 38.4 Gastrulation Sets the Stage for Organogenesis and Neurulation in Chordates

### The notochord induces formation of the neural tube

### Mesoderm forms tissues of the middle layer

During gastrulation, three germ layers form and take up positions that allow organogenesis to proceed (see Figures 38.13 and 38.16). Gastrulation makes possible the inductive interactions between cells, which trigger differentiation and organ formation. The inner layer, or endoderm, will become the lining of the digestive tract, the lining of the respiratory tract, and the lining of some other organs, such as the pancreas and liver. The outer layer, or ectoderm, will become the epidermis (outer layer of the skin), derivatives of the epidermis (such as hair, feathers, and sweat glands), and the nervous system. The middle layer, or mesoderm, will become the heart, blood vessels, urinary and reproductive systems, and tissues such as muscle, bone, and dermis of the skin.

In chordates, an important tissue that forms during gastrulation is the notochord, a structure that lies ventral to the neural ectoderm. Numerous experiments have shown that the notochord is responsible for inducing the neural ectoderm to invaginate and form the neural tube. Anteriorly the neural tube becomes the forebrain, midbrain, and hindbrain. Posteriorly it becomes the spinal cord. If the neural tube doesn't close, birth defects such as spina bifida in the spinal region or anencephaly in the brain region can result.

Experiments by Hans Spemann and Hilde Mangold on salamander embryos showed that the gray crescent (which becomes the notochord) is able to organize the rest of the tissues of the embryo. They found that when the dorsal lip of the blastopore from one embryo was transplanted to another embryo on the side opposite to the embryo's own dorsal lip, the result was a second site of gastrulation and eventually two embryos attached belly-to-belly. They concluded that the dorsal lip acts as the primary embryonic organizer (see Figure 38.14).

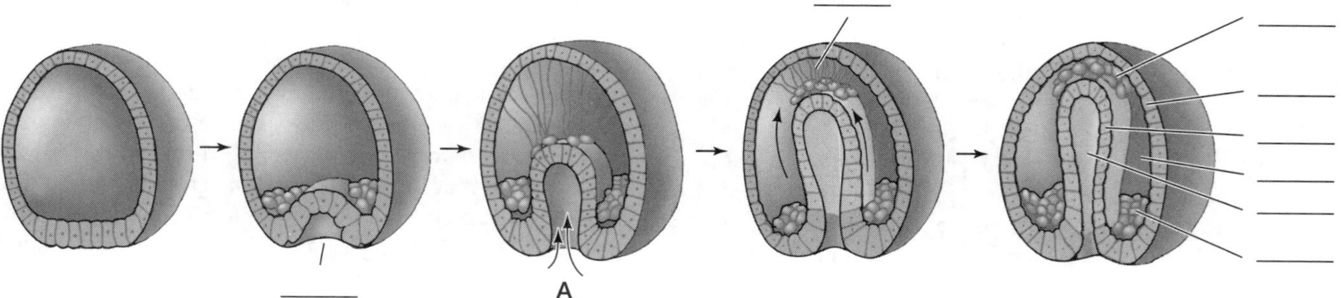

Further studies showed that β-catenin, a transcription factor, is both necessary and sufficient for creating the organizer (see Apply the Concept, p. 798). Further research on amphibian embryos showed that polypeptides secreted by the notochord control the induction of the neural ectoderm. Noggin and Chordin, released by the notochord, block the activity of BMP4, which normally induces the ectoderm to form epidermis. When this action is blocked, the ectoderm forms neural ectoderm (see Figure 38.15). The transcription factor Sonic hedgehog (Shh) is also secreted from the notochord and helps to induce the ventral region of the neural tube to differentiate.

Once the neural tube invaginates and closes (a process called neurulation), neural crest cells that had been associated with the neural ectoderm separate from the neural tube and migrate to many places in the body, developing into numerous diverse structures such as skull bones, pigment cells, and sensory neurons (see Figure 38.16). It is because of the neural crest cells that vertebrates are able to have heads.

In vertebrates, the mesoderm becomes subdivided into several regions. The somites are a repeating series of paired blocks of tissue on either side of the neural tube and running down the length of the body of the embryo. These will form bones, cartilage, muscle, and dermis. The intermediate mesoderm sits next to the somites and will form the urinary and reproductive systems. The lateral plate mesoderm lying next to the intermediate mesoderm will split into two layers to line the body coelom. The inner lateral plate mesoderm will form muscles of the gut, most of the circulatory system, including the heart, and the peritoneum that lines the inner side of the coelom. The outer lateral plate mesoderm will form many of the muscles of the body wall and the peritoneum that lines the outer side of the coelom.

**Question 7.** The diagram shows a transplant experiment on a salamander embryo. Diagram the results of this experiment. What do these results tell you about the region that was transplanted?

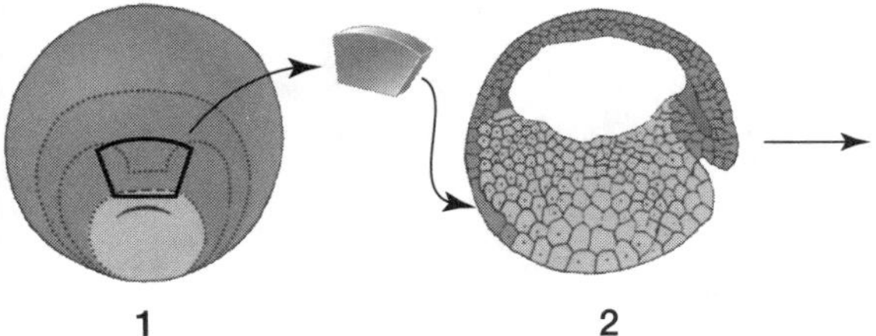


**Question 8.** Refer to the diagram below. What are the structures labeled 1–5. What will each of these form in the organism?

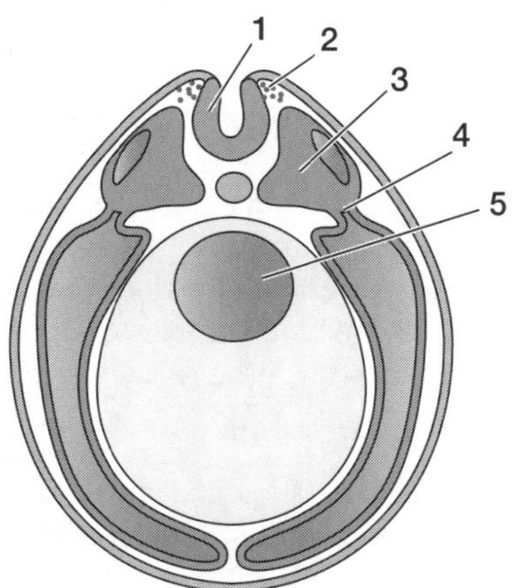


## 38.5 Extraembryonic Membranes Protect and Nourish the Embryo

Extraembryonic membranes form with contributions from all germ layers

Extraembryonic membranes in mammals form the placenta

Fish also make yolk sacs

Extraembryonic membranes surround the embryos of reptiles, birds, and mammals. In birds, the yolk sac is derived from the endoderm of the hypoblast and the inner layer of the lateral plate mesoderm. The yolk sac surrounds the yolk and becomes vascularized with the vitelline arteries, veins, and capillaries. Digestion of nutrients from the yolk occurs in the yolk sac, and these nutrients are then delivered to the embryo by way of the vitelline veins (see Figure 38.18). The allantois, derived from endoderm and the inner layer of the lateral plate mesoderm, also becomes vascularized. It forms a sac that stores metabolic wastes. The amnion, derived from ectoderm and the outer layer of the lateral plate mesoderm, is a sac that is not vascularized. It forms around the embryo, creating a sac filled with amniotic fluid that protects the embryo. The chorion, derived from ectoderm and the outer layer of the lateral plate mesoderm, lies outside of the amnion. It is also unvascularized and helps to prevent water loss from the embryo. It eventually fuses with the allantois to form the chorioallantoic membrane, which is pushed up against the shell. The chorioallantoic membrane provides for exchange of oxygen and carbon dioxide between the embryo and the outside environment.

In placental mammals, such as humans, the embryo arrives in the uterus as a blastocyst, and the trophoblast layer of the embryo attaches to the endometrium of the uterine wall and begins implantation (see Figures 38.8 and 38.19). The chorion and allantois combine to form the chorioallantoic placenta. The yolk sac is an early site of blood formation, as in the other amniotes. The amnion surrounds the embryo and fills with amniotic fluid that bathes the embryo. When a pregnant woman's "water breaks" near the beginning of labor, the amnion has burst, releasing the amniotic fluid.

In humans, gestation can be divided into three trimesters, each about 12 weeks in length. During the first trimester, cell division and tissue differentiation are rapid, and organ development begins. At this stage, the developing human is most sensitive to environmental disrupters. By the end of the first trimester, the embryo is considered a fetus and is about 8 cm long.

Many fish have heavily yolky eggs. Though they are not amniotes, they still form a yolk sac that surrounds the yolk, is vascularized, and brings yolk nutrients back to the embryo through its vitelline blood vessels (see Figure 38.20). Unlike in amniotes, however, the fish yolk sac is derived from all three germ layers.

**Question 9.** The diagram below shows a 9-day chick embryo with the shell removed. Label each of the extraembryonic membranes. Put an "A" in the region where amniotic fluid is found, "G" where the gut cavity of the embryo is, and "W" where metabolic waste is stored.

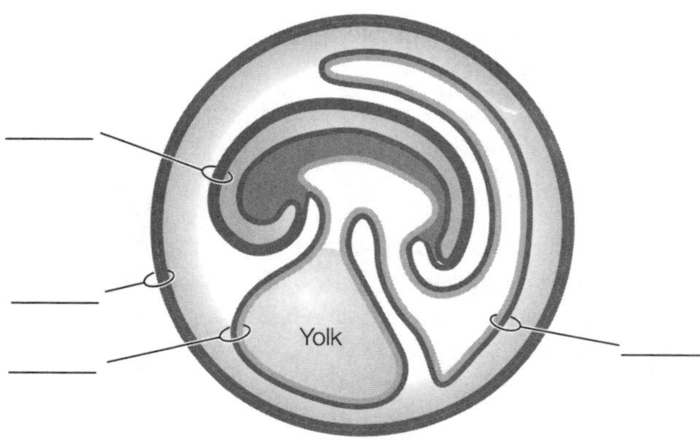


**Question 10.** Why would exposure to a harmful chemical during the first trimester of a woman's pregnancy be more likely to produce a major birth defect than exposure to the same chemical during the last trimester?

### 38.6 Development Continues throughout Life

Development does not stop at hatching or birth, but continues throughout the life of an animal. Many animals go through direct development to reach full size, meaning that they hatch or are born looking similar to their parents and then grow from there. Many animals go through indirect development, meaning that they spend part of their lives in a form that is different from that of their parents. Indirect development can involve dramatic changes in morphology. For example, a monarch butterfly produces an embryo that hatches as a caterpillar, a feeding stage that eventually metamorphoses into the winged adult. Species that are sessile, such as corals, typically produce swimming larva that can disperse. The metamorphosis can be so dramatic that scientists sometimes have had a hard time identifying the correct species of an organism from its larval stage. In the tunicate (sea squirt), for example, the larva contains a notochord, dorsal hollow nerve cord, and post-anal tail, identifying it as a chordate—yet none of these characteristics remain in the adult following metamorphosis.

Growth in an organism can be isometric, in which all parts grow larger but maintain their original proportions, or allometric, in which the parts of the body do not maintain the same proportions as they grow. Humans provide an example of allometric growth. The head of a human baby, for example, is much larger relative to the rest of its body than the head of an adult is.

**Question 11.** What is the advantage to a monarch butterfly of indirect development and going through a larval caterpillar stage?

**Question 12.** Which of the following terms apply to the sea urchin?

Protostome
Deuterostome
Vertebrate
Invertebrate
Chordate
Direct development
Indirect development

## Test Yourself

1. During its development, the human embryo is contained within a fluid-filled chamber enclosed by the extraembryonic membrane called the
   a. yolk sac.
   b. amnion.
   c. chorion.
   d. allantois.
   e. trophoblast.
   ***Textbook Reference:*** *Concept 38.5 Extraembryonic Membranes Protect and Nourish the Embryo; Extraembryonic membranes in mammals form the placenta*

2. The dorsal lip of the blastopore organizes embryo formation in salamanders. The equivalent structure in chickens is
   a. the epiblast.
   b. the hypoblast.
   c. Hensen's node.
   d. the notochord.
   e. the gray crescent.
   ***Textbook Reference:*** *Concept 38.4 Gastrulation Sets the Stage for Organogenesis and Neurulation in Chordates; The notochord induces formation of the neural tube*

3. When a sperm fuses with the egg cell membrane of an amphibian egg,
   a. the sperm enters in the ventral region of the egg where it is least pigmented.
   b. the site of sperm entry defines the anterior–posterior but not the dorsal–ventral polarity of the embryo.
   c. entry of the sperm centriole results in a shift in the pigmented outer cytoplasm of the egg.

d. the site of sperm entry defines the dorsal–ventral but not the anterior–posterior polarity of the embryo.
e. the sperm enters in the region of the gray crescent.
***Textbook Reference:*** *Concept 38.1 Fertilization Activates Development; Polarity is established in development*

4. The location of the _______ determines the anterior–posterior axis of the embryo.
a. primitive streak
b. blastopore
c. vegetal hemisphere
d. hypoblast
e. mesenchyme
***Textbook Reference:*** *Concept 38.3 Gastrulation Produces a Second, then a Third Germ Layer*

5. In which animal does division of the nucleus precede the division of the cytoplasm during the cleavage stage?
a. Wood frog
b. Fruit fly
c. Mouse
d. Zebrafish
e. Chicken
***Textbook Reference:*** *Concept 38.2 Cleavage Creates Building Blocks and Produces a Blastula; The amount of yolk affects cleavage*

6. In the diagram below, the embryo _______; the blastocoel of the embryo is labeled _______; the endoderm of the embryo is labeled _______.

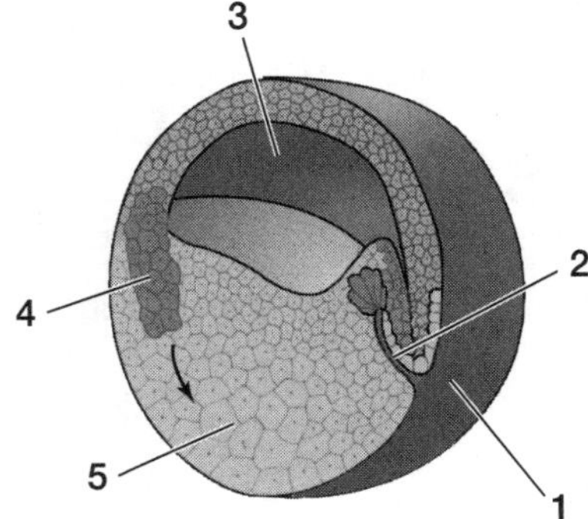


a. is from an amphibian; 3; 5
b. is from a sea urchin; 2; 4
c. has undergone incomplete cleavage; 3; 5
d. has undergone complete cleavage; 2; 1
e. has undergone spiral cleavage; 3; 1
***Textbook Reference:*** *Concept 38.3 Gastrulation Produces a Second, then a Third Germ Layer*

7. Which sequence represents the correct order of the germ layers, from the inside to the outside of a gastrula?
a. Mesoderm, ectoderm, endoderm
b. Endoderm, ectoderm, mesoderm
c. Ectoderm, mesoderm, endoderm
d. Endoderm, mesoderm, ectoderm
e. Mesoderm, endoderm, ectoderm
***Textbook Reference:*** *Concept 38.4 Gastrulation Sets the Stage for Organogenesis and Neurulation in Chordates*

8. The _______ eventually develop into vertebrae, ribs, and trunk muscles, as well as the dermis of the skin, and are found along the sides of the _______.
a. somites; neural tube
b. neural tube cells; notochord
c. blastopore cells; dorsal lip
d. neural crest cells; dorsal lip
e. neural plate cells; archenteron
***Textbook Reference:*** *Concept 38.4 Gastrulation Sets the Stage for Organogenesis and Neurulation in Chordates; Mesoderm forms tissues of the middle layer*

9. The tissues and organs that eventually will develop from specific germ layers can be depicted visually by means of a fate map. On a fate map of a salamander embryo, ectoderm will form the
a. lining of the gut.
b. nervous system and epidermal layer of the skin.
c. kidneys.
d. heart.
e. lining of the respiratory tract.
***Textbook Reference:*** *Concept 38.2 Cleavage Creates Building Blocks and Produces a Blastula*

10. The primary embryonic organizer is most likely initiated by
a. the yolk.
b. Noggin.
c. β-catenin.
d. cAMP.
e. Chordin
***Textbook Reference:*** *Concept 38.4 Gastrulation Sets the Stage for Organogenesis and Neurulation in Chordates; The notochord induces formation of the neural tube*

11. Birds develop extraembryonic membranes during development. Which statement about avian extraembryonic membranes is *false*?
a. The yolk sac surrounds the yolk and provides nutrients.
b. The amnion and chorion are derived from ectoderm and the outer layer of the lateral plate mesoderm.
c. The allantois stores nutrients.
d. The chorioallantoic membrane exchanges gases between the embryo and the environment.
e. The allantois is derived from endoderm and the inner layer of the lateral plate mesoderm.
***Textbook Reference:*** *Concept 38.5 Extraembryonic Membranes Protect and Nourish the Embryo; Extraembryonic membranes form with contributions from all germ layers*

12. Based on what you know about fertilization, what can you surmise about the inheritance of mitochondria in an individual?
    a. Mitochondria are maternally inherited.
    b. Mitochondria are paternally inherited.
    c. Half the mitochondria come from the mother and half from the father.
    d. There is no way to determine which parent contributes mitochondria to the zygote.
    e. Mitochondria arise anew in the zygote.
    ***Textbook Reference:*** *Concept 38.1 Fertilization Activates Development; Egg and sperm make different contributions to the zygote*

13. Which statement about the human blastocyst is true?
    a. The trophoblast gives rise to the embryo proper.
    b. The embryo arrives in the uterus in the early stages of cleavage.
    c. The blastocyst implants itself by first attaching the inner cell mass to the endometrium.
    d. The fetal blood mixes with the maternal blood in the chorionic villi.
    e. The trophoblast forms the fetal side of the placenta.
    ***Textbook Reference:*** *Concept 38.5 Extraembryonic Membranes Protect and Nourish the Embryo; Extraembryonic membranes in mammals form the placenta*

14. Hans Spemann called the dorsal lip of the blastopore the embryonic organizer because it
    a. is the point where gastrulation begins.
    b. becomes part of the nervous system.
    c. becomes part of the notochord.
    d. leads to the establishment of the embryonic axes.
    e. is the location at which the sperm enters the egg.
    ***Textbook Reference:*** *Concept 38.4 Gastrulation Sets the Stage for Organogenesis and Neurulation in Chordates; The notochord induces formation of the neural tube*

15. Which statement about complete cleavage versus incomplete cleavage is true?
    a. Incomplete cleavage occurs in species whose eggs have small volumes of cytoplasm.
    b. Complete cleavage is found in all mammals and is the more evolved condition.
    c. Incomplete cleavage occurs in species whose eggs have large amounts of yolk.
    d. Complete cleavage occurs only in eggs that have been fertilized by two sperm.
    e. Incomplete cleavage occurs in species whose eggs have small amounts of yolk.
    ***Textbook Reference:*** *Concept 38.2 Cleavage Creates Building Blocks and Produces a Blastula*

16. Refer to the diagram below. In a newborn baby, the ratio of head size to body size is about 1:5. Based on the diagram, you would expect the ratio in an adult to be

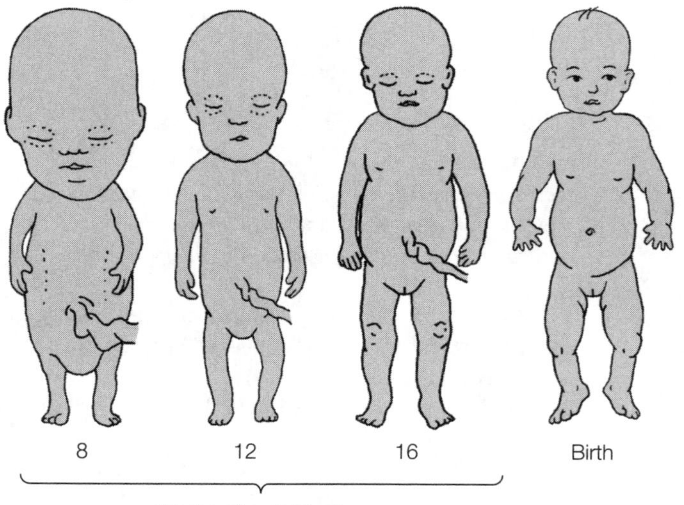


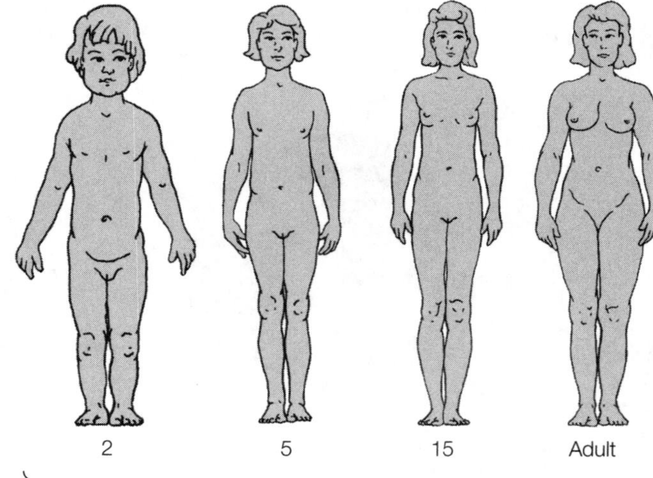


    a. the same (about 1:5).
    b. much smaller (about 1:20).
    c. smaller (about 1:10).
    d. larger (about 1:4).
    e. much larger (about 1:3).
    ***Textbook Reference:*** *Concept 38.6 Development Continues throughout Life*

17. Refer to the figure below. If you found an organism floating in the ocean that looked similar to the organism in the photograph, you would conclude that it is most likely a(n)

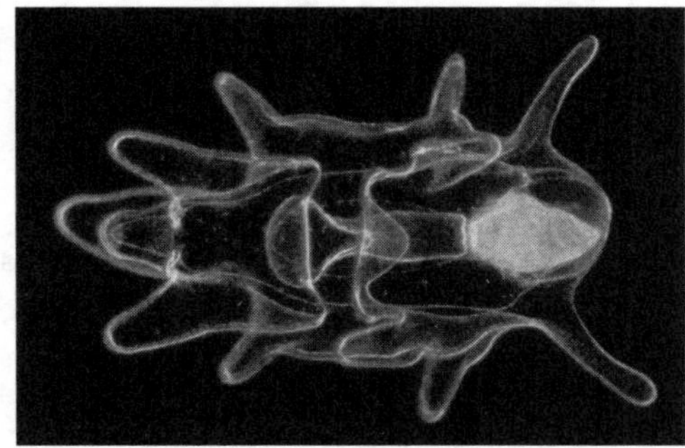

a. organism with direct development.
b. organism in the larval stage for dispersal.
c. adult sea star.
d. organism that will metamorphose into a winged adult.
e. chordate in the larval stage.

***Textbook Reference:*** *Concept 38.6 Development Continues throughout Life*

## Answers

### *Key Concept Review*

1. The sperm contributes a haploid nucleus, which, when combined with the egg's haploid nucleus, produces a diploid zygote. The sperm also contributes a centriole, which becomes the centrosome of the zygote. The centrosome functions as a microtubule-organizing center. The centriole of the sperm organizes the microtubules of the first mitotic spindle as the zygote starts cleavage.

2. See the diagram below. One can conclude from this experiment that cytoplasmic determinants are located within the gray crescent and that these are necessary for normal development of the salamander embryo. In addition, one can conclude that a normal larva can develop from half of the zygote as long as it receives at least half of the gray crescent.

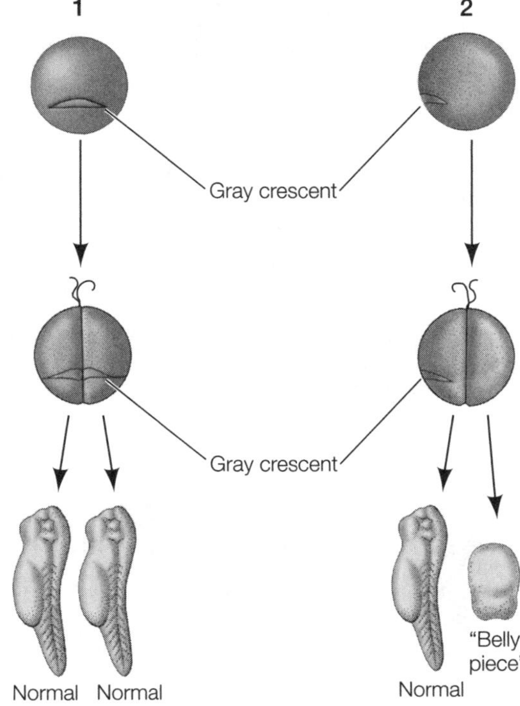


3. Due to the large amount of yolk in the avian egg, cleavage in birds is incomplete, and the result is a blastodisc that sits on top of the yolk (see Figure 38.7B; although the egg pictured is from a zebrafish, bird eggs show a similar cleavage pattern). Embryos of a placental mammal undergo complete cleavage, resulting in a structure called a blastocyst (see Figure 38.8A). Genes are transcribed in the early cleavage stages of a placental mammal and are transcribed much later in the chick embryo. Finally, cleavage is unique in a placental mammal in that the cells divide in a pattern called rotational cleavage (see In-Text art, p. 793) and the cells separate into an inner cell mass and trophoblast.

4. The cells marked 2–5 will all develop into normal larvae, and each will be a quarter of the size of the larva labeled 1.

5. See the diagram below. The structure labeled "A" will become the anus of the larva.

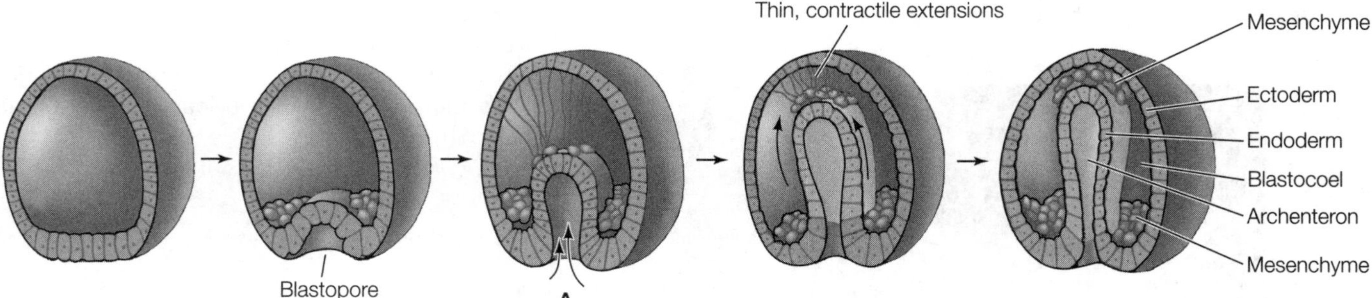


6. Although the details of gastrulation differ among sea urchins, frogs, and chickens, the common result is the formation of germ layers—ectoderm, mesoderm, and endoderm—from which all tissues and organs will form.
7. The results of the experiment are shown on the diagram below. These results tell us that the transplanted tissue is capable of organizing tissues around it to form a second embryo.

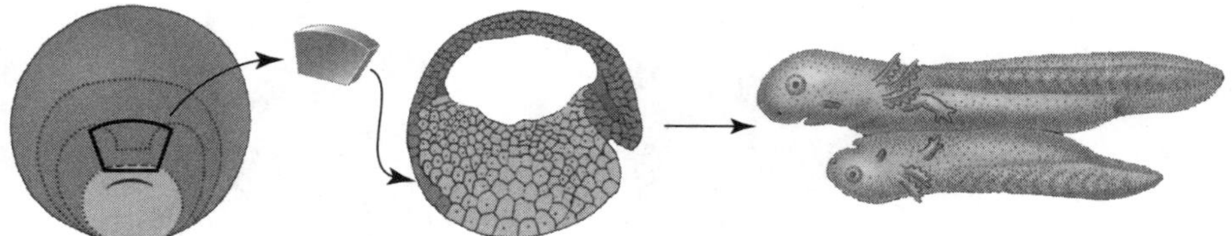

8. The structures in the diagram and their fates are as follows:

   1: Neural ectoderm; invaginates to form a neural tube, which will form the brain and spinal cord

   2: Neural crest; will form many structures, including sensory neurons, skull bones, and pigment cells

   3: Somite; will form bones, cartilage, skeletal muscle, and the dermis of the skin

   4: Intermediate mesoderm; will form the urinary and reproductive systems

   5: Archenteron; will form the lining of the gut and respiratory system as well as the lining of structures such as the liver and pancreas
9. See the diagram below.

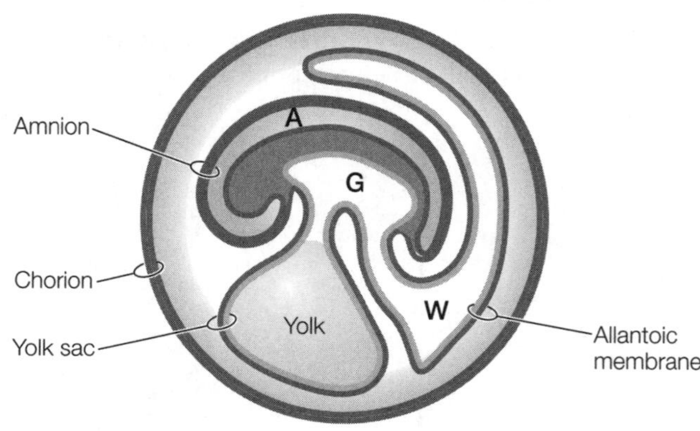


10. Exposure to a harmful chemical during the first trimester would be more likely to produce a major birth defect than exposure to the same chemical during the last trimester because tissues and organs are forming during the first trimester. The last trimester involves mostly rapid growth.
11. The caterpillar stage of the monarch butterfly is a voracious feeding stage that allows the animal to increase dramatically in size before it metamorphoses into the winged adult.
12. The terms from the list that apply to sea urchin are as follows: deuterostome, invertebrate, and indirect development (the organism has a larval stage).

### *Test Yourself*

1. **b.** The amnion is the membrane that surrounds the developing human fetus in its fluid-filled amniotic cavity.
2. **c.** In chickens, Hensen's node is the equivalent of the dorsal lip of the salamander blastopore.
3. **c.** When a sperm fuses with the egg cell membrane of an amphibian egg, entry of the sperm centriole results in a shift in the pigmented outer cytoplasm of the egg. The centriole organizes microtubules within the cytoplasm that cause the shift.

4. **b.** The blastopore, which will eventually become the anus, marks the posterior of the embryo.
5. **b.** In the fruit fly embryo, as in other insect embryos, the nuclei initially divide in the absence of cytoplasmic division. Only after the nuclei have migrated to the peripheral cytoplasm are cell membranes put into place to separate the nuclei into separate cells.
6. **a.** The embryo in the diagram is from an amphibian; the blastocoel of the embryo is labeled 3; the endoderm of the embryo is labeled 5.
7. **d.** The germ layers that are formed during gastrulation are the inner layer of endoderm, the middle layer of mesoderm, and the outer layer of ectoderm.
8. **a.** Somites are located along the neural tube in the developing vertebrate. These blocks of mesoderm cells will develop into the vertebrae, ribs, and trunk muscles as well as the dermis of the skin.
9. **b.** Ectoderm will form the epidermal layer of the skin and the nervous system.
10. **c.** β-catenin is thought to be the initiator of organizer activity during early development.
11. **c.** During development, an avian embryo produces wastes, but because birds develop in a shell, the wastes must be stored in the egg. The allantois forms a sac that is used for the storage of metabolic wastes.
12. **a.** Mitochondria are maternally inherited. The sperm does not contribute mitochondria to the zygote. All the mitochondria within the zygote, therefore, come from the oocyte that was created within the mother.
13. **e.** In the human blastocyst, the trophoblast forms the fetal side of the placenta. It does not give rise to the embryo proper, and it arrives in the uterus as a blastocyst, not in the early stages of cleavage. It is the trophoblast that attaches to the endometrium of the uterus. When the placenta is established, the blood of the mother and fetus do not mix, though the tissue barrier between them is thin.
14. **d.** The dorsal lip leads to the establishment of the embryonic axes.
15. **c.** Incomplete cleavage occurs because the cleavage furrows cannot completely penetrate the yolk. Incomplete cleavage occurs in animals, such as reptiles, including birds, whose eggs have large amounts of yolk.
16. **c.** When the baby grows up to be an adult, you would expect that the ratio of head size to the rest of its body will be smaller. If you make the measurements on the diagram shown, you will see that the ratio in the adult is about 1:10.
17. **b.** The organism in the photograph is a larval stage of a sea star and is used for dispersal.

# 39 Immunology: Animal Defense Systems

## The Big Picture

- Animals have two main ways to defend against pathogens: innate defenses and adaptive defenses. Innate defenses are nonspecific and include barriers, phagocytic cells, specialized proteins, and inflammation. Adaptive defenses are specific, slow to develop, and long-lasting. All animals have innate defenses, whereas only vertebrates have developed adaptive immune defenses. The adaptive defenses are coordinated by humoral (B cells and antibodies) and cellular (T cells) responses.

## Study Strategies

- The adaptive immune response may seem complex initially, but there are only two major components: B cells, which secrete antibodies, and T cells, which regulate the activities of other white blood cells and thereby help direct the immune response.
- MHC proteins may also seem complicated at first, but they have only a few essential functions in the immune system. These proteins are specific for each individual and are essential for presenting antigens to the two different types of T lymphocytes: Class I MHC proteins (found on all nucleated cells in mammals) present antigens to $T_C$ cells, while Class II MHC proteins (found on macrophages, B cells, and dendritic cells) present antigens to $T_H$ cells.
- The specificity of the immune response is complex, but it can be understood by reviewing cellular communication. Antibodies of B cells bind with antigen on the surface of pathogens; T cell receptors bind antigen presented on MHC I and MHC II proteins on the surface of cells. Once these membrane proteins interact, cytokines are secreted, and cells begin to divide and proliferate. B cells secrete antibodies.
- The rearrangement of DNA in the nucleus of the B cell may seem foreign to students who think the integrity of the genetic information must be preserved. But this change in the genetic makeup of activated B and T cells is essential to providing the organism with a diverse and specific set of antibodies and cell receptors.
- List the innate (nonspecific) responses that a human body can make to an invading pathogen. Next, list cells that are part of the adaptive (specific) responses. Include proteins that are crucial to generating that specificity. Review how genetic rearrangements generate a diverse set of antibodies for B cells.
- Diagram an antibody and label the important components. Highlight those areas where mutations and alterations change the amino acid sequence of the heavy and light chains. Note which regions remain the same when a B cell switches immunoglobulin classes (e.g., IgM to IgG) and which regions change.
- Review the expression and function of MHC I and MHC II proteins.
- Make a list of the functions of $T_H$ cells and $T_C$ cells.
- Draw a diagram of a chromosome and label it with 4 *V*s (*V*1–*V*4), 2 *D*s (*D*1 and *D*2), 3 *J*s (*J*1–*J*3), and 8 *C*s (*C*1–*C*8). Imagine that a B cell is maturing to produce antibody on its surface. Select one *V*, one *D*, one *J*, and one *C* segment to make an antibody. Repeat the process using different segments. Predict how many different antibodies can be generated from this set of gene segments.
- List some disorders of the immune system and their probable causes.
- Go to **LaunchPad** (or use the Web addresses listed) to review the following additional resources:

  Animated Tutorial 39.1 Cells of the Immune System (PoL2e.com/at39.1)

  Animated Tutorial 39.2 Humoral Immune Response (PoL2e.com/at39.2)

  Animated Tutorial 39.3 A B Cell Builds an Antibody (PoL2e.com/at39.3)

  Animated Tutorial 39.4 Cellular Immune Response (PoL2e.com/at39.4)

  Activity 39.1 Inflammation Response (PoL2e.com/ac39.1)

  Activity 39.2 Immunoglobulin Structure (PoL2e.com/ac39.2)

Activity 39.3 The Major Organ Systems (PoL2e.com/ac39.3)

Analyze the Data in textbook Figure 39.5

Apply the Concept in textbook Sections 39.3 and 39.4

## Key Concept Review

### 39.1 Animals Use Innate and Adaptive Mechanisms to Defend Themselves against Pathogens

Innate defenses evolved before adaptive defenses

Mammals have both innate and adaptive defenses

Pathogens are harmful organisms and viruses that cause disease. Animal responses to these pathogens involve recognition that the pathogen is nonself (the recognition phase), activation of cells and molecules against the invader (the activation phase), and the attack of the invader (the effector phase).

Once an organism has been exposed to a pathogen, it can develop immunity, which is the ability to produce a rapid and strong adaptive immune response when later exposed to the same pathogen, thereby preventing disease.

Innate, nonspecific defenses are inherited mechanisms that act rapidly to protect the body from pathogens. Innate defenses include physical barriers and cellular and chemical defenses (see Table 39.1). All animals have innate responses.

An innate defense shared by diverse animals, including humans and flies, are Toll-like receptors. Toll-like receptors recognize and bind specific conserved molecules on pathogens (see Figure 39.1). Activation of Toll-like receptors initiates signaling pathways, which induces the expression of genes that encode anti-pathogen molecules.

Adaptive, specific defenses are mechanisms aimed at particular targets. The specific defenses involve humoral (B cells and antibodies) and cellular (T cells) responses (see Table 39.1). These responses develop slowly and last a long time. Adaptive defenses evolved later than innate defenses and are found only in vertebrate animals. There are two main types of white blood cells (leukocytes): phagocytes and lymphocytes (see Figure 39.2). Phagocytes, such as macrophages and dendritic cells, engulf pathogens and are part of innate and adaptive responses. Lymphocytes include B cells and T cells, which are involved in the adaptive response, and natural killer cells, which exhibit both innate and adaptive responses.

**Question 1.** Human macrophages recognize lipopolysaccharide, a component of the cell wall of Gram-negative bacteria. How do macrophages recognize these components and how do they respond?

**Question 2.** Deer ticks transmit the bacterium that causes Lyme disease. The test for Lyme disease involves drawing blood and looking for antibodies to the bacterium. Why would a doctor refuse to test for Lyme disease in a patient who found a deer tick attached to her body and asked to be tested immediately?

### 39.2 Innate Defenses Are Nonspecific

Barriers and local agents defend the body against invaders

Cell signaling pathways stimulate additional innate defenses

Inflammation is a coordinated response to infection or injury

Inflammation can cause medical problems

The innate defenses encountered first by a pathogen on the skin include the physical barrier of the skin, the saltiness of the skin, and the presence of normal flora (bacteria and fungi) with which the pathogen must compete. Within the skin, innate defenses include phagocytosis, complement proteins, interferons, inflammation, and fever (see Figure 39.3).

Innate defenses encountered by pathogens that enter the nose or other internal organs include mucus (a secretion that traps microorganisms), lysozyme (an enzyme that causes bacteria to burst), defensins (peptides that are toxic to many bacteria and enveloped viruses), and harsh conditions (for example, acidic gastric juice in the stomach).

Some animal and plant cells activate pattern recognition receptors (PRRs), such as Toll-like receptors that recognize pathogen associated molecular patterns (PAMPs). PAMPs include bacteria-expressed flagellin and lipopolysaccharide, fungal chitin, and nucleic acid variants that are expressed by viruses, such as double-stranded RNA. Activation of PRRs leads to anti-pathogen responses that include phagocytosis of invading organisms, activation of natural killer cells, activation of the complement system, and production of cytokines.

Phagocytes ingest cells and viruses, which are then killed by either hydrolysis or defensins inside the phagocyte. Natural killer cells are white blood cells that can recognize virus-infected or cancerous cells and induce them to undergo apoptosis. They also interact with the adaptive defense mechanisms by lysing antibody-labeled target cells.

About 20 complement proteins act as antimicrobial proteins in vertebrate blood. They function in the following characteristic sequence: one protein binds to an invading cell so that phagocytes can recognize and destroy it; another protein activates the inflammatory response; still other proteins lyse the invading cell. They are an important part of innate (nonspecific) and adaptive (specific) defense responses.

Cytokines are signaling molecules released by many types of cells and include inflammatory cytokines and interferons. Interferons are small signaling proteins produced in response to viral infection. They bind to uninfected cells and stimulate signaling pathways that inhibit viral reproduction should the cells later become infected. They also stimulate cells to digest bacterial and viral proteins into smaller peptides; this is an important first step in adaptive immunity.

Inflammation, a response induced by infection or injury, is responsible for the symptoms of redness, swelling, and heat at infected or injured sites. Mast cells, one of the first cells to respond to infection and/or injury, release chemical signals, including tumor necrosis factor, prostaglandins, and histamine, that induce inflammation. Important inflammation-associated defenses include isolation of the affected area to contain the damage, and dilation of local blood vessels, which allows complement proteins and phagocytic cells to enter the inflamed area. Phagocytes engulf pathogens and dead cells and promote healing through the expression of cytokines. Inflammation can also induce fever, which can have anti-pathogen effects. Platelets are activated by injuries and release growth factors that induce neighboring skin cells to divide and heal the wound (see Figure 39.4).

An inflammation response that is inappropriately strong can result in allergic responses (histamine release and inflammation prompted by a nonself molecule that is typically harmless), autoimmune diseases (in which the immune system attacks tissues in the organism's own body), or sepsis (a severe bacterial infection that extends throughout the body). In sepsis, blood vessels throughout the body dilate, causing a severe drop in blood pressure.

**Question 3.** Pathogenic bacteria in the digestive tract often cause diarrhea. Why might diarrhea be considered a protective response by the body? Also, what other characteristic of the large intestines might help protect against invaders?

**Question 4.** How might urine function as an innate defense?

## 39.3 The Adaptive Immune Response Is Specific

### Adaptive immunity has four key features

### Macrophages and dendritic cells play a key role in activating the adaptive immune system

### Two types of adaptive immune responses interact

Adaptive immunity develops in response to a pathogen; after exposure, an individual develops antibodies against that pathogen. Passive immunity refers to the acquisition of immunity from antibodies received from another individual (see Figure 39.5).

One feature of adaptive immunity is its specificity, which allows it to focus on pathogens that are present. T cell receptors and antibodies produced by B cells recognize specific antigens, which are nonself proteins or polysaccharides. More specifically, the immune cells recognize antigenic determinants, or epitopes (small portions of antigens), on invading pathogens. There can be many antigenic determinants on a pathogen surface.

A second feature of adaptive immunity is its capacity to distinguish self from nonself. Animals are able to tolerate their own antigens due to clonal deletion, a process that removes any B and T cells that show the potential to mount a strong response against self-antigens. Such cells undergo programmed cell death (apoptosis). Failure of clonal deletion leads to autoimmunity (an immune response within an individual to self-antigens, which may result in disease).

A third feature of adaptive immunity is its diversity, meaning that it is able to respond to an enormous range of different pathogens. In order to do so, the body must generate diverse lymphocytes that are specific for particular antigens. The genetic changes resulting in diversity are produced by DNA changes, including chromosomal rearrangements and other mutations that occur as B and T cells are produced in the bone marrow. Each B cell can make only one type of antibody, and the T cells have a specific receptor to recognize an antigen. There are millions of different kinds of B cells and T cells with specific receptors. Essentially, the immune system has the machinery in place to recognize antigens before they are encountered. Also, when an antigen that fits the surface antibody binds to the B cell, that cell is activated and divides to produce clonal B cells, which secrete antibodies (see Figure 39.6). T cells are clonally selected in a similar manner (clonal selection).

A fourth feature of adaptive immunity is its immunological memory (i.e., it "remembers" a pathogen), which allows it to respond even more effectively to subsequent encounters with the same pathogen. In the primary immune response, activated B and T cells produce effector cells and memory cells. Effector cells attack a pathogen by producing specific antibodies (in the case of B cells) or cytokines (in the case of T cells). Effector B cells are called plasma cells. This primary immune response occurs during a first encounter with an antigen. Effector cells generally live only a few days. Following the primary immune response, long-lived memory cells remain and divide at a slow rate. During subsequent encounters with the same pathogen, memory B and T cells divide rapidly to produce effector cells and more memory cells to serve as a more powerful immune response. This rapid response, the secondary immune response, provides a natural immunity to diseases caused by those pathogens. Vaccination triggers a primary response, so the body will be prepared to mount a stronger secondary response if it encounters the pathogen again.

There are two interactive immune responses against invaders. The humoral immune response relies on B cells that make antibodies. The cellular immune response relies on cytotoxic T cells ($T_C$) that bind to and destroy self-cells that are mutated or infected by pathogens. These systems act in concert and share some of the same mechanisms (see Figure 39.7).

There are three phases to the adaptive immune response. In the recognition phase, the organism detects a pathogen by distinguishing self and nonself. Presentation of the antigen to the immune system is a key event in both the humoral and cellular responses. In the humoral immune response, presentation occurs when an antigen binds to a B cell that has an antibody on its surface that is specific for that antigen. In the cellular immune response, a T-helper cell ($T_H$) with a specific receptor binds to an antigen-presenting cell, which has a particular antigen inserted into its plasma membrane. In both responses, binding initiates the activation phase. In the activation phase, cells and molecules mobilize to fight the invader. During the effector phase, B cells produce antibodies that bind to the pathogen and/or infected cells. The bound antibodies attract phagocytes that engulf the pathogen or infected cell, and complement proteins aid in the destruction of the pathogen or cell, as well. In cellular immunity to viruses, $T_C$ cells and molecules are mobilized to destroy the invader during the effector phase.

**Question 5.** Imagine that a particular bacterium has surface molecules that resemble membrane proteins found on heart valves. What would happen to an individual infected with this bacterium if the infection were not treated?

**Question 6.** In the diagram below, label each structure indicated by the letters a–f. At the bottom of the diagram (letters g and h), identify which represents the cellular immune response and which represents the humoral immune response.

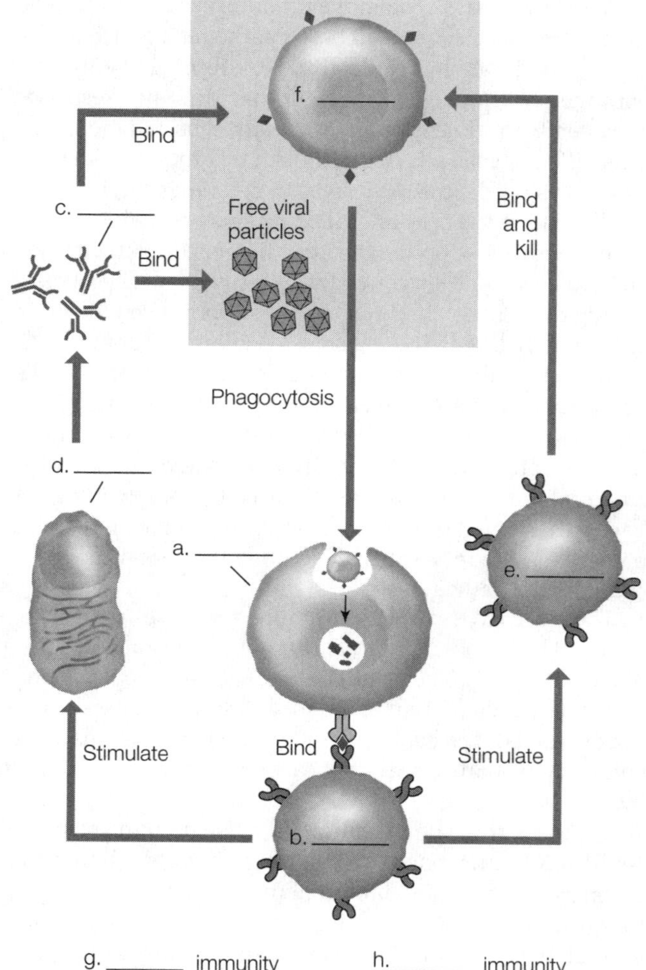


## 39.4 The Adaptive Humoral Immune Response Involves Specific Antibodies

Plasma cells produce antibodies that share a common overall structure

Antibody diversity results from DNA rearrangements and other mutations

Antibodies bind to antigens and activate defense mechanisms

Activation of B cells results from (1) the binding of an antibody to a particular antigenic determinant, and (2) the arrival of a signal from a T-helper ($T_H$) cell that has recognized a fragment of the same antigen, which is presented by the B cell on its cell surface. Once this occurs, the plasma cell (effector B cell) is able to secrete antibodies. As the effector B cells proliferate, they produce antibodies that are specific for the antigen that bound to the parent B cell.

Antibodies (also called immunoglobulins) are proteins that can be grouped into five different classes, although they are all similar in structure. Antibody molecules consist of two identical heavy polypeptide chains and two identical light polypeptide chains held together by disulfide bonds. Each polypeptide consists of a constant region (which is similar in amino acid sequence from one immunoglobulin to another within a class) and a variable region that forms the antigen-binding site (see Figure 39.8). The variable region produces the specificity of the millions of antibodies.

The antigen-binding sites on each immunoglobulin are identical, so antibodies are bivalent. Each antibody can recognize two antigen molecules, and because most antigens have multiple epitopes, antibodies can form large complexes with antigens. These complexes are easy targets for phagocytes.

The five classes of immunoglobulins expressed by B cells are IgM, IgD, IgG, IgA, and IgE. For all classes expressed by the same B cell clone, the variable region of the immunoglobulin (which binds the antigen) remains the same. The different classes are based on differences in the constant region of the heavy chain. All classes are secreted by B cells, but IgG is the most abundant circulating immunoglobulin.

The genome of the B cell undergoes genetic rearrangement during differentiation, and the combination of different alleles allows a B cell's genome to encode for a unique immunoglobulin. DNA fragments are rearranged and joined during B cell development to generate antibody supergenes (see Figures 39.9 and 39.10). The variable region of the light chain originates from two families of genes (*V* and *J*), and the variable region of the heavy chain originates from three families of genes (*V*, *D*, and *J*). For the heavy chain in mice, there are multiple genes coding for each of the three parts of the variable region: 100 *V*, 30 *D*, and 6 *J* genes (see Figure 39.9). Each B cell randomly selects one gene from each of these multiple genes to make the final coding sequence, *VDJ*, for the heavy-chain variable region. Light chains are similarly constructed from DNA segments. Light and heavy chains are combined to create billions of possible antibodies. Additional diversity is achieved by imprecise recombination during DNA rearrangements, by the addition of extra nucleotides to cut DNA fragments as insertion mutations by the enzyme terminal transferase, and by an increased spontaneous mutation rate in immunoglobulin genes.

Antibodies have two roles in B cells. First, because they are expressed on the cell surface, they can act as a receptor for an antigen in the recognition phase of the humoral response. Second, during the effector phase of the humoral response, antibodies are produced in large amounts by activated B cells. These antibodies enter the bloodstream, where they may bind to an antigen on the surface of a pathogen, stimulating macrophages or natural killer cells to dispense with the pathogen. Alternatively, if the antigen is free in the bloodstream, the antibodies may join to form large insoluble antibody–antigen complexes that are destroyed by phagocytic cells.

**Question 7.** What might happen to a person receiving a blood transfusion if the donor blood cells express foreign antigens on their cell surfaces?

**Question 8.** The enzymes RAG1 and RAG2 initiate DNA recombination in B cells and T cells. What would happen if a person did not express these enzymes?

### 39.5 The Adaptive Cellular Immune Response Involves T Cells and Their Receptors

T cell receptors specifically bind to antigens on cell surfaces

MHC proteins present antigen to T cells and result in recognition

$T_H$ cells contribute to the humoral and cellular immune responses

Activation of the cellular response results in death of the targeted cell

Regulatory T cells suppress the humoral and cellular immune responses

AIDS is an immune deficiency disorder

T cells have specific glycoprotein surface receptors that are made up of two different polypeptide chains, both with a variable and a constant region (see Figure 39.11). The variable region provides the specificity of the T cell receptor. T cell receptors bind antigen fragments that are displayed on antigen-presenting cells in the presence of an MHC protein.

MHC proteins are plasma membrane glycoproteins that present antigens to $T_H$ and $T_C$ cells. Class I MHC proteins are present on the surface of every nucleated cell in the mammalian body. They present antigens (fragments of virus proteins in virus-infected cells or abnormal proteins made by cancer cells) to $T_C$ cells. Class II MHC proteins are found on the surfaces of B cells, macrophages, and dendritic cells. They present antigens to $T_H$ cells (see Figure 39.12).

Activation of a $T_H$ cell results in the production of clones with the specific T cell receptor. $T_H$ cells also produce cytokines that can induce B cells and $T_C$ cells to propagate.

B cells internalize antigens recognized and bound by their B cell receptors. Antigens are degraded into fragments, and these fragments are presented on MHC class II molecules. When a $T_H$ cell recognizes and binds the antigen–MHC complex, it stimulates the B cell to undergo clonal propagation and secrete antibodies.

Activation of a $T_C$ cell results in the production of clones with the specific T cell receptor. These $T_C$ cells bind to cells carrying the antigen and the Class I MHC protein. Once bound, the $T_C$ cell kills the antigen-carrying cell; it accomplishes this either by producing perforins that lyse the cell or by stimulating the cell to undergo apoptosis.

Like $T_H$ and $T_C$ cells, regulatory T cells (Tregs) develop in the thymus, express the T cell receptor, and are activated if they bind to antigen–MHC complexes. However, Tregs recognize self-antigens. When Tregs are activated, they secrete a cytokine called interleukin-10. This blocks the activation of $T_C$ and $T_H$ cells and causes them to undergo apoptosis (see Figure 39.13).

AIDS is caused by the human immunodeficiency virus (HIV), a virus that eventually destroys $T_H$ cells. HIV is transmitted through body fluids, including blood and semen. HIV initially infects macrophages, $T_H$ cells, and antigen-presenting dendritic cells. There is an immune response to the viral infection and some $T_H$ cells are activated. However $T_H$ cells eventually decline: HIV itself kills them and $T_C$ cells lyse infected $T_H$ cells. Extensive production of HIV by infected cells activates the humoral response, antibodies bind HIV, and the level of HIV in the blood decreases (see Figure 39.14). Nevertheless, the infection remains at a low level because of the depletion of $T_H$ cells. A person may remain at this "set point" for many years. Eventually, the $T_H$ cells are destroyed and the patient becomes susceptible to opportunistic infections, including Kaposi's sarcoma and pneumonia caused by the fungus *Pneumocystis jirovecii*.

Drug treatments generally focus on inhibiting processes associated with viral entry, assembly, and replication. Combinations of treatments have resulted in long-term survival of HIV-infected people.

**Question 9.** If cytotoxic T cells were eliminated from a person's array of immune defenses, what kinds of disease would the person be susceptible to?

**Question 10.** Why are organ transplants more successful when the donor is a relative of the recipient?

**Question 11.** The polio vaccine comes in two forms: an injected dose of inactivated (dead) virus and an oral vaccine of attenuated virus. Which would be the safer vaccine for a child with an immunodeficiency disorder? Explain your reasoning.

**Question 12.** Draw a flowchart depicting the body's three lines of defense: first line of innate mechanisms, second line of innate mechanisms, and third line of adaptive mechanisms. Within each of the three lines of defense, include the component mechanisms.

## Test Yourself

1. Which statement about phagocytes is true?
   a. They are derived from T and B cells.
   b. Only vertebrates have them.
   c. They engulf nonself materials.
   d. They are a type of erythrocyte.
   e. They secrete antibodies.
   ***Textbook Reference:*** *Concept 39.1 Animals Use Innate and Adaptive Mechanisms to Defend Themselves against Pathogens; Innate defenses evolved before adaptive defenses*

2. Which statement about $T_H$ cells is true?
   a. They secrete antibodies.
   b. They present antigen on MHC II.
   c. They ingest antigens.
   d. They regulate both the cellular and the humoral response.
   e. They nonspecifically activate B cells.

   ***Textbook Reference:*** *Concept 39.5 The Adaptive Cellular Immune Response Involves T Cells and Their Receptors; $T_H$ cells contribute to the humoral and cellular immune responses*

3. Which of the following represents an adaptive, specific defense of the immune system?
   a. Macrophages
   b. Defensins
   c. Lysozyme
   d. Mucus
   e. Antibodies

   ***Textbook Reference:*** *Concept 39.2 Innate Defenses Are Nonspecific; Barriers and local agents defend the body against invaders*

4. When the receptor of a $T_H$ cell binds to a pathogen presented on a macrophage, it
   a. inactivates itself.
   b. secretes cytokines.
   c. inactivates B cells.
   d. inactivates the macrophage.
   e. becomes a $T_C$ cell.

   ***Textbook Reference:*** *Concept 39.3 The Adaptive Immune Response Is Specific; Two types of adaptive immune responses interact*

5. Part of the normal adaptive immune response includes
   a. the production of B memory cells.
   b. the production of memory macrophages.
   c. antibody secretion by eosinophils.
   d. the production of B cells that attack the individual's own cells.
   e. the production of complement proteins as a specified immune response.

   ***Textbook Reference:*** *Concept 39.3 The Adaptive Immune Response Is Specific; Adaptive immunity has four key features*

6. Antibody molecules are
   a. produced by B cells and have only a constant region.
   b. secreted by B cells once a signal (a cytokine) is received from a T cell.
   c. produced by T cells and have a variable and a constant region.
   d. proteins that induce B cells to form memory cells.
   e. proteins with antibody binding sites.

   ***Textbook Reference:*** *Concept 39.4 The Adaptive Humoral Immune Response Involves Specific Antibodies; Plasma cells produce antibodies that share a common overall structure*

7. Which statement about cytotoxic T cells is true?
   a. They release cytokines that activate B cells.
   b. They attack pathogens by binding to cell surface antigens on those pathogens.
   c. They destroy pathogens by engulfing them.
   d. They destroy host cells that are infected with virus.
   e. They stimulate the classical complement pathway.

   ***Textbook Reference:*** *Concept 39.5 The Adaptive Cellular Immune Response Involves T Cells and Their Receptors; Activation of the cellular response results in death of the targeted cell*

8. Which statement about the humoral response is true?
   a. It involves the formation of antibodies.
   b. It occurs when T cells lyse target cells.
   c. It is due to T cells' secretion of their receptors.
   d. It occurs when natural killer cells engulf cancer cells.
   e. It is initiated when macrophages engulf bacteria.

   ***Textbook Reference:*** *Concept 39.4 The Adaptive Humoral Immune Response Involves Specific Antibodies; Plasma cells produce antibodies that share a common overall structure*

9. An autoimmune disease is
   a. active when organ transplantation is successful.
   b. caused by viruses.
   c. a response in which the immune cells attack the body's own tissues.
   d. a result of the destruction of the immune system.
   e. inflammation that is always targeted to a specific location in the body.

   ***Textbook Reference:*** *Concept 39.2 Innate Defenses Are Nonspecific; Inflammation can cause medical problems*

10. Patients with HIV are susceptible to a variety of infections because
    a. the virus produces cell surface receptors that bind to pathogens, making it easier for those pathogens to be infective.
    b. the synthesis of a DNA copy of the viral genome makes a person susceptible to infection.
    c. HIV attacks and destroys the $T_H$ cells, which are central to mounting an effective immune response.
    d. HIV destroys B cells, so antibodies cannot be made in response to invading pathogens.
    e. HIV mutates the B cells, so they cannot make the huge array of antibodies needed for an effective immune response.

    ***Textbook Reference:*** *Concept 39.5 The Adaptive Cellular Immune Response Involves T Cells and Their Receptors; AIDS is an immune deficiency disorder*

11. DNA rearrangements in the B cell
    a. are responsible for generating single B cells that can express many different antibodies.
    b. lead to mutations in T cells, resulting in the elimination of essential T cell genes.
    c. occur only in B memory cells.

d. are responsible for generating many different antibodies, with each B cell only expressing antibodies with identical antigen specificity.
e. occur only in fully mature B cells.
***Textbook Reference:*** *Concept 39.4 The Adaptive Humoral Immune Response Involves Specific Antibodies; Antibody diversity results from DNA rearrangements and other mutations*

12. Major histocompatibility proteins function in the immune system by
a. engulfing pathogens.
b. presenting antigens to $T_C$ and $T_H$ cells.
c. generating antibodies to different pathogens.
d. presenting antigen fragments to B cells.
e. presenting macrophages to $T_H$ cells.
***Textbook Reference:*** *Concept 39.5 The Adaptive Cellular Immune Response Involves T Cells and Their Receptors; MHC proteins present antigen to T cells and result in recognition*

13. In a process called _______, immature B or T cells with the potential to respond strongly to self-antigens undergo apoptosis
a. clonal selection
b. immunological memory
c. clonal deletion
d. the primary immune response
e. the secondary immune response
***Textbook Reference:*** *Concept 39.3 The Adaptive Immune Response Is Specific; Adaptive immunity has four key features*

14. Inflammation occurs when _______ release _______.
a. mast cells; histamine
b. neutrophils; toxins
c. B cells; histamine
d. mast cells; toxins
e. neutrophils; antihistamine
***Textbook Reference:*** *Concept 39.2 Innate Defenses Are Nonspecific; Inflammation is a coordinated response to infection or injury*

15. The process by which an antigen binds to a specific B cell and the B cell begins to divide is called
a. a nonspecific defense.
b. clonal selection.
c. immune activation.
d. meiosis.
e. a secondary immune response.
***Textbook Reference:*** *Concept 39.3 The Adaptive Immune Response Is Specific; Adaptive immunity has four key features*

## Answers

### *Key Concept Review*

1. Macrophages recognize lipopolysaccharide via a Toll-like receptor. Upon binding lipopolysaccharide, the Toll-like receptor activates signal transduction pathways that induce the expression of genes that encode anti-bacterial molecules.
2. Antibodies are part of the adaptive, specific immune response, and such responses typically take at least 96 hours to develop (see Table 39.1). Thus, it would not make sense to test immediately for antibodies to the bacterium that causes Lyme disease. Ideally, the test would be performed at least 96 hours after the deer tick was found.
3. Diarrhea helps move the pathogen out of the digestive tract quickly. The large intestine is home to many bacteria that keep invaders in check.
4. Urine flushes pathogens from the urinary tract. The chemical composition of urine (e.g., its acidity) makes the urinary tract inhospitable to some pathogens.
5. If the bacterial infection were not treated, the individual's immune response to this pathogen would include the secretion of antibodies. Such antibodies could cross-react with the individual's own membrane proteins on the cells of the heart valve, leading to an autoimmune response and possibly heart disease. This is an example of molecular mimicry inducing autoimmunity.
6. 
a. Antigen-presenting cell
b. T-helper cell ($T_H$)
c. Antibodies
d. B cell
e. Cytotoxic T cell ($T_C$)
f. Virus-infected cell
g. Humoral
h. Cellular
7. If a person receives a blood transfusion and the donor blood cells express foreign antigens, then antibodies in the recipient's blood will bind to these antigens and cause the donor cells to clump or agglutinate, which can be damaging and possibly fatal. This is an example of antibodies forming large, insoluble antibody-antigen complexes when confronted with an antigen free in the bloodstream.
8. If a person did not express RAG1 and RAG2, their B cells and T cells could not undergo DNA recombination during their development. Consequently, the person would lack mature B cells and mature T cells that exhibit diverse antigen specificities. The individual would be extremely susceptible to bacterial and viral infections.

9. Cytotoxic T cells target virally infected cells and some cancer cells. If cytotoxic T cells were eliminated, the individual would be much more susceptible to viral infections and cancer.

10. Organ transplants are successful if the MHC proteins and other cell surface markers are similar in the donor and recipient, making the immune system of the recipient more tolerant of the foreign tissue. Since close relatives have more genes in common than unrelated people do, they are more likely to have similar surface antigens on their cells. Organ transplants from unrelated individuals require the administration of immunosuppressive drugs to the recipient so that the patient's own immune system does not reject the foreign tissue. The drug cyclosporin, which inhibits T cell development, is often used.

11. The injected dose of inactivated poliovirus would be safer for a patient with a compromised immune system because an inactivated pathogen cannot cause disease. In contrast, the attenuated vaccine contains a mutant form of the virus that does reproduce itself in the patient, although at a very slow rate. However, it is possible for the attenuated virus to mutate back to a virulent form (which does happen occasionally) and cause disease.

12.

**First line of defense (innate)**
Physical barriers (e.g., skin, mucous membranes)
Chemical barriers (e.g., acid conditions, lysozyme)

**Second line of defense (innate)**
Defensive cells (e.g., phagocytic cells, natural killer cells)
Defensive proteins (e.g., complement system, interferons)
Inflammation
Fever

**Third line of defense (adaptive)**
Humoral immune response (B cells make antibodies)
Cellular immune response (T-helper cells, cytotoxic T cells)

### *Test Yourself*

1. **c.** Phagocytes (not B cells, T cells, or erythrocytes) are nonspecific cells that engulf nonself materials. All animals have them. They do not produce antibodies.

2. **d.** T-helper ($T_H$) cells specifically activate B cells and $T_C$ cells that recognize the same antigen. Thus, they regulate both the cellular and the humoral systems. B cells are antigen-presenting cells; they bind antigen, digest it, present fragments of that antigen on their MHC II proteins to $T_H$ cells, and secrete antibodies as part of the humoral response.

3. **e.** Macrophages, defensins, lysozyme, and cilia on mucous membranes are all part of the nonspecific response of the immune system. Antibodies represent the specific response of the immune system.

4. **b.** When the $T_H$ cell binds antigen being presented on a macrophage, it secretes cytokines, which stimulate other immune cells to divide.

5. **a.** In a normal adaptive immune response, B memory cells are produced, allowing the organism to mount a faster and more effective response to any subsequent encounter with the pathogen. Memory macrophages do not exist. Eosinophils do not secrete antibodies. In an abnormal immune response, B cells that attack the individual's own cells can be activated, as seen in autoimmune diseases. Complement proteins are part of the innate, nonspecific immune response.

6. **b.** T cells have cell surface receptors with a variable and a constant region and do not produce antibody molecules. Antibodies are not produced by macrophages. Only B cells produce antibodies in response to cytokines released from the $T_H$ cell. Each antibody molecule has two identical heavy chains and two identical light chains, and each of these chains has a variable and a constant region. Antigens have antibody binding sites called epitopes.

7. **d.** Cytotoxic T cells bind to virus antigen presented on MHC I protein and destroy those cells. $T_H$ cells release cytokines to activate B cells. The antibodies of B cells bind cell surface antigens on pathogens. Cytotoxic T cells do not engulf pathogens. T cells have no involvement in the classical complement pathway.

8. **a.** The humoral response refers to that part of the specific response that releases antibodies (B cells) in the lymph and blood (the "humors" of the body). One of the cellular responses includes activated cytotoxic T cells lysing target cells. T cells do not secrete their receptors. Macrophages are part of the nonspecific response and can activate $T_H$ cells, but not B cells.

9. **c.** Autoimmune diseases occur when the immune cells attack the body's own cells. Transplant rejection can occur if the transplanted tissue is recognized by the body as nonself, so this is not an autoimmune response. The immune system can be destroyed by an immune deficiency disorder.

10. **c.** An HIV-infected individual is more susceptible to a variety of infections because the virus destroys $T_H$ cells, which are essential for mounting an effective immune response. HIV does not bind to pathogens and does not destroy B cells or cause mutations in their DNA that alter antibody production.

11. **d.** DNA rearrangements occur in B cell precursors and result in the expression of antibodies with the

same unique antigen specificity by each mature B cell. DNA rearrangements also occur in T cells to generate T cell receptors. This rearrangement does not destroy essential T cell genes. Memory B cells are cells in which DNA rearrangement has already occurred.

12. **b.** MHC proteins present antigens to $T_C$ and $T_H$ cells. The cytokines of T cells (and not MHC proteins) activate B cells. MHC proteins do not generate antibodies. Macrophages are antigen-presenting cells, but they are not themselves presented as antigens to $T_H$ cells.

13. **c.** Clonal deletion is the process by which immature B cells or T cells that show the potential to mount a strong immune response to self-antigens undergo apoptosis.

14. **a.** Inflammation occurs in response to an infection or tissue injury. Once an infection or injury occurs, the local mast cells release histamine. The histamine, in turn, makes the capillaries leak, and fluid accumulating in the area produces the inflammation.

15. **b.** When an antigen binds to a B cell, the B cell begins to divide and make clone cells, which results in more B cells that recognize the antigen. This process is known as clonal selection. A secondary immune response occurs when memory cells encounter an antigen and begin to divide actively.

# 40 Animal Behavior

## The Big Picture

- Behavior has a neural basis and it evolves. Biological determinism is the belief that behavior is entirely determined by genes. We know this to be inaccurate because even genetically similar individuals can exhibit dramatic differences in behavior as a result of learning and epigenetic effects. Behavior is closely integrated with physiology, growth, and body size.
- Navigation is movement toward a particular destination. Orientation is the adoption of a particular path with respect to an environmental cue. Animals typically have multiple ways of determining direction. Migration is the periodic movement by animals from one location to another, where they remain for a significant period of time before returning to their previous location.
- Living in groups provides benefits, such as enhanced thermoregulation, increased ability to detect predators, and more efficient discovery of preferred environments. Costs include greater conspicuousness to predators, increased risk of disease transmission, and increased competition for resources. Behavior structures ecological communities.

## Study Strategies

- Be sure to understand how genes, learning, and epigenetic effects influence behavior.
- When thinking about a particular behavior, consider how the behavior is integrated with other characteristics of the animal such as its physiology, development, and body size. Also consider how that behavior might help to structure ecological communities.
- We often read about a particular mechanism of orientation used by a particular species (e.g., the time-compensated sun compass by homing pigeons or magnetic cues by loggerhead turtles). It is important to remember, however, that a species typically has multiple mechanisms for determining direction.
- Go to **LaunchPad** (or use the Web addresses listed) to review the following additional resources:

  Animated Tutorial 40.1 Time-Compensated Solar Compass (PoL2e.com/at40.1)

  Animated Tutorial 40.2 Homing Simulation (PoL2e.com/at40.2)

  Animated Tutorial 40.3 The Costs of Defending a Territory (PoL2e.com/at40.3)

  Animated Tutorial 40.4 Foraging Behavior (PoL2e.com/at40.4)

  Activity 40.1 Honey Bee Dance Communication (PoL2e.com/ac40.1)

  Activity 40.2 Concept Matching (PoL2e.com/ac40.2)

  Analyze the Data in textbook Figure 40.9

  Apply the Concept in textbook Section 40.3

## Key Concept Review

### 40.1 Behavior Is Controlled by the Nervous System but Is Not Necessarily Deterministic

Many types of evidence point to the neural basis of behavior

Behaviors evolve

Despite its neural basis, behavior is not necessarily simplistically deterministic

Behaviors are performed and ephemeral; they are not material objects. Nevertheless, behaviors have their basis in the nervous system. Evidence that behaviors have a neural basis can be negative (e.g., what results when a particular brain region is damaged or destroyed?), positive (e.g., which brain region is active during the performance of a specific behavior?), or come in the form of fixed action patterns. Fixed action patterns are highly stereotyped behaviors that are displayed without prior learning; their performance simply requires a functioning nervous system. Examples of fixed action patterns include web building by spiders and begging by gull chicks (see Figure 40.2).

Genes encode the properties of neural tissue. If particular alleles result in a nervous system that produces more adaptive behaviors, then those alleles will be favored by natural selection. Thus, behavior evolves. Experiments using artificial selection (when humans choose which animals breed) show that behavior can evolve rapidly. The belief that behavior is hard-wired by genetics is called biological determinism. However, even genetically similar individuals can exhibit dramatic differences in behavior as a result of learning and epigenetic effects (nongenomic effects that occur as animals interact in unique ways with their environment).

**Question 1.** Design an experiment to show that song learning in birds has a neural basis.

**Question 2.** Respond to the statement, "Human behavior is hard-wired by genetics."

## 40.2 Behavior Is Influenced by Development and Learning

Specific information of critical survival value is often learned during early postnatal development

Early experience also has other, more global effects on behavior

Learning is the ability to modify behavior based on previous experience. Behavioral imprinting is a form of learning that occurs during a brief window of time early in life. Preferences established during this brief window are difficult to change. Soon after hatching, geese develop a preference for their parents and follow them everywhere. Young geese also learn the characteristics of a suitable mate during a brief window of time. Other examples of learning in young animals include indigo buntings' learning the North Star, which is critical for later navigation, and the learning of species-specific song by males of other bird species.

Experience early in life can have multiple long-term effects. Rat mothers can be categorized as "high-caring" and "low-caring" based on the time they spend licking their young and adopting a favorable nursing posture. When tested in adulthood, offspring of "low-caring" mothers are more fearful than offspring of "high-caring" mothers. Epigenetic effects are responsible, in part, for the observed differences between the two types of offspring. Specifically, early experience differentially affects key regulatory genes in stress-response pathways in the two types of offspring and such differences continue throughout life.

**Question 3.** You have just been asked to develop a protocol for raising young California condors from hatching to eventual release to the wild. Given what you know about the effects of early experience on subsequent behavior, what recommendations would you make?

**Question 4.** A researcher interested in studying the effects on adult behavior of malnourishment early in life establishes two groups of young rats. One group has free access to food and the other has limited access to food. In adulthood, both groups have free access to food and individuals from each group are scored on several behavioral tests. Respond to the statement, "Because rats in both groups have free access to food in adulthood, their behavior on tests will be indistinguishable."

## 40.3 Behavior Is Integrated with the Rest of Function

Toads and frogs have evolved contrasting behavioral specializations that depend on their biochemistry of ATP synthesis

Behaviors are often integrated with body size and growth

Animal behavior is closely integrated with physiology. The escape behavior of sustained running by pronghorn is possible because these animals have exceptionally efficient delivery of oxygen to their muscles, use the oxygen at very high rates to make ATP in their muscle cells, and use the ATP at very high rates for muscle contraction. Pronghorn also have very large lungs and muscles for their size.

Toads and frogs further illustrate how escape behavior is tied to muscle physiology. Toads escape predators by hopping at moderate speeds, which they can sustain for a long period of time. In contrast, frogs escape predators by hopping very quickly, and they can keep this up for only a short period of time before becoming fatigued. Whereas aerobic ATP production takes time and is resistant to fatigue, anaerobic ATP production is fast and susceptible to fatigue. The leg muscles of toads are high in enzymes needed for aerobic ATP production. In contrast, the leg muscles of frogs are high in enzymes needed for anaerobic ATP production.

Animal behavior is also closely integrated with body size. The tonal frequencies of insect and frog calls vary with body size, with low frequencies being associated with large body size. Additionally, successful reproduction often requires sufficient body size, as in the case of male elk, which have to be large enough to dominate other males in order to breed.

**Question 5.** Turkeys flee from predators either by running, which they can sustain for a long period of time, or by bursting into flight, which they can sustain for a very short period of time. Would you predict the leg muscles or breast muscles of turkeys to specialize in aerobic ATP production?

**Question 6.** A researcher studying male frogs in their night chorus wishes to get an estimate of male body size without catching and measuring individual males. What could the researcher monitor at a distance that would provide such an estimate?

## 40.4 Moving through Space Presents Distinctive Challenges

Trail following and path integration are two mechanisms of navigation

Animals have multiple ways of determining direction

Honey bee workers communicate distance and direction by a waggle dance

Migration: Many animals have evolved periodic movements between locations

Navigation is the act of moving toward a particular destination. Orientation is the adoption of a particular path with respect to an environmental cue. Trail following is one form of navigation used by ants. Worker ants that have found food mark their path home with a pheromone. Pheromones are chemicals released by individuals into the environment that prompt responses from other individuals of that species. Once the trail is marked with pheromone, other worker ants can follow the trail from the nest back to the food. Path integration is another form of navigation used by ants (see Figure 40.9). Although worker ants may wander quite a bit while foraging, they can return straight back to their nest (i.e., they don't retrace their steps). In this case, a worker ant integrates information about the lengths of trail segments (measured by number of steps) and direction (using landmarks, and sun and polarized light compasses) to return straight home. Homing pigeons return home using a sun compass to determine north, south, east, and west, together with their circadian clock to tell time of day (see Figure 40.10). Animals typically use several mechanisms of orientation. In addition to their time-compensated sun compass, homing pigeons orient using the Earth's magnetic field, landmarks, odors, and low-frequency sounds. Pigeons also learn routes from other pigeons.

Honey bees dance to communicate the location of a food source in the environment (see Figure 40.11). The waggle dance is used to communicate distance and direction to food. Duration of the straight run of the waggle dance indicates distance to the food source. Direction of the straight run of the waggle dance (specifically its angle from vertical) indicates the direction of the food source relative to the sun.

Some animals migrate, which means they periodically move from one location to another, stay for a significant period of time, and then return to their previous location. Hatchling loggerhead turtles that enter the coastal waters off Florida migrate to African waters where they remain for a while and then return to coastal Florida. Loggerhead turtles use the Earth's magnetic field to orient along their journey (see Figure 40.12).

**Question 7.** To test whether mice use landmarks when returning home to their nests, a researcher introduced a mouse to an experimental room and gave it a few days to make a nest. Then the researcher placed obvious landmarks, such as logs and rocks, near the nest. At the start of the second week, the researcher sees the mouse leave its nest and quickly moves the logs and rocks to a new location in the room. This test is repeated with several different mice. Each mouse always returned to its nest without difficulty; none of the mice headed toward the new location of the rocks and logs. Should the researcher conclude that landmarks are unimportant to mice when orienting in their environment? Why or why not?

**Question 8.** How does a bee that has found a patch of flowers measure the distance to the patch and the direction of the patch so it can inform other members of the hive of the patch's location?

## 40.5 Social Behavior Is Widespread

### Some societies consist of individuals of equal status

### Some societies are composed of individuals of differing status

### Eusociality represents an extreme type of differing status

Many animals live in groups. A society is a group of individuals of the same species that cooperate to some degree. Group living has costs and benefits. Costs include greater conspicuousness to predators, increased risk of disease transmission, and increased competition for resources among group members. When groups consist of individuals of similar status, benefits include enhanced thermoregulation by huddling in cold temperatures, increased ability to detect predators because of more eyes, ears, and noses (see Figure 40.13), and more efficient discovery of preferred environments (see Figure 40.14). When groups consist of individuals of differing status, additional benefits include increased access to females for dominant males and, for females that mate with dominant males, genetically well-endowed offspring

Eusociality is an extreme form of group living in which some members of a group are nonreproductive. These nonreproductive individuals help fertile group members, usually their mother, to reproduce. Eusocial species include honey bees and naked mole-rats. Altruism is any act by an individual that imposes a cost to that individual while helping another individual.

**Question 9.** For each of the two *y*-axes in the following graph, draw a labeled curve that correctly summarizes observations made on goshawks attacking wood pigeons, as described in the textbook.

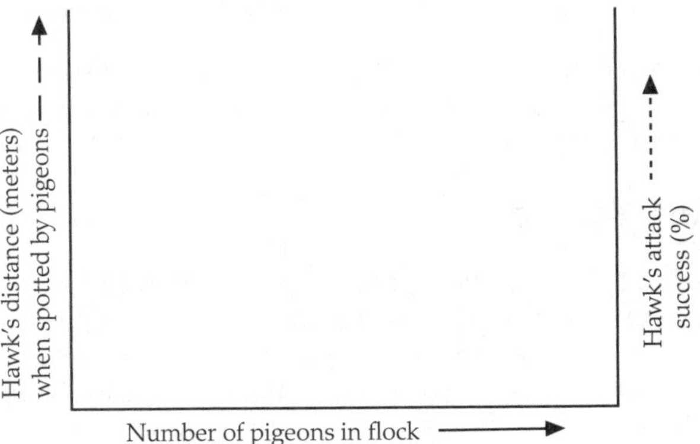


**Question 10.** While studying a colony of vampire bats you notice that individuals that have successfully obtained a blood meal on a given night often share some of their meal with other colony members that have been unsuccessful in obtaining a meal. Develop two hypotheses to explain this apparently altruistic act.

## 40.6 Behavior Helps Structure Ecological Communities and Processes

### Behavior helps maintain species

### Animals often behaviorally partition space into territories or home ranges

### Behavior helps structure relationships among species

Animal behavior influences the use of time within ecological communities, and as a result behavior influences interactions between species. For example, two species that are active during daylight hours are likely to interact. In contrast, a species that is active during daylight hours is unlikely to interact with a species that is active at night. Behavior also helps to maintain species as distinct entities. Individuals of two closely related species may have different habitat preferences, so they rarely encounter each other under natural conditions. Should individuals of the two species meet, subtle differences in their behavior will prompt them to separate before mating.

The behavior of animals also provides structure to the use of space within ecological communities. The territory of an animal is an area from which other individuals of its own species (and sometimes individuals of other species) are excluded. The home range of an animal is the area that it occupies without excluding other individuals. Home ranges often overlap. Both territories and home ranges provide familiarity, but only territories provide exclusive use of resources.

Cost–benefit analysis can be used to examine the interactions between species within ecosystems. This approach can be applied to foraging behavior (i.e., what food an animal selects and when and where it searches for it). More specifically, foraging behavior can be examined in terms of the energy and time expended (costs) as compared to the energy obtained (benefits).

**Question 11.** How do territories differ from home ranges? Name one cost and one benefit of maintaining a territory.

**Question 12.** Why are matings between individuals of different species less common under natural conditions than under captive conditions?

## Test Yourself

1. Some birds give a species-specific vocalization called an "alarm call" when they see a predator. Other members of their species respond to these calls by taking cover. This would be an example of altruistic behavior if
   a. giving the alarm call posed no cost to the caller.
   b. receivers of the call did not benefit.
   c. the call imposed costs on the caller and the receivers.
   d. the caller incurred a cost and the receivers benefited.
   e. the call benefited both the caller and the receivers.
   ***Textbook Reference:*** *Concept 40.5 Social Behavior Is Widespread; Some societies are composed of individuals of differing status*

2. In Galápagos finches, species-specific songs are
   a. inherited.
   b. learned once a male reaches adulthood.
   c. learned during a brief window of time early in development.
   d. unaffected by songs produced by fathers.
   e. learned when young males listen to the songs of their mothers.
   ***Textbook Reference:*** *Concept 40.2 Behavior Is Influenced by Development and Learning; Specific information of critical survival value is often learned during early postnatal development*

3. Toads escape predators by hopping at moderate speeds, and they have good endurance. The leg muscles of toads therefore
   a. rely entirely on glycolysis.
   b. have enzyme profiles similar to those of frogs, which hop quickly and fatigue rapidly.
   c. reflect their escape behavior with respect to size but not function.
   d. rely on aerobic respiration.
   e. have numerous muscle fibers that need oxygen immediately.
   ***Textbook Reference:*** *Concept 40.3 Behavior Is Integrated with the Rest of Function; Toads and frogs have evolved contrasting behavioral specializations that depend on their biochemistry of ATP synthesis*

4. Which statement about homing pigeons is *false*?
   a. Homing pigeons use olfactory cues to orient.
   b. Homing pigeons sometimes make use of landmarks to return home.
   c. On cloudy days, homing pigeons switch from using the sun to orient to using the Earth's magnetic field.
   d. Homing pigeons use low-frequency sound to orient.
   e. Experimental shifts in the circadian clocks of homing pigeons do not disrupt their use of the sun as a compass.
   ***Textbook Reference:*** *Concept 40.4 Moving through Space Presents Distinctive Challenges; Animals have multiple ways of determining direction*

5. Which statement about biological determinism is *false*?
   a. Belief in human biological determinism has at different times increased and decreased in popularity over the years.
   b. Proponents of biological determinism once believed, incorrectly, that inherited brain size determines a person's intelligence.
   c. Biological determinism emphasizes the role of learning in behavior.
   d. Epigenetic effects can result in substantial behavioral differences among genetically similar individuals; this argues against biological determinism.
   e. According to biological determinism, behavior is inflexible.
   ***Textbook Reference:*** *Concept 40.1 Behavior Is Controlled by the Nervous System but Is Not Necessarily Deterministic; Despite its neural basis, behavior is not necessarily simplistically deterministic*

6. Which of the following is a cost of living in groups?
   a. Compromised thermoregulation
   b. Increased risk of disease transmission
   c. Reduced competition
   d. Decreased ability to detect predators
   e. Decreased conspicuousness to predators
   ***Textbook Reference:*** *Concept 40.5 Social Behavior Is Widespread; Some societies consist of individuals of equal status*

7. Pheromones
   a. are used to communicate between individuals of different species.
   b. can be used to mark trails.
   c. are used by pigeons to find their way home.
   d. are used by animals for navigation but not for mate attraction.
   e. produce their effects by circulating within the animal's body.
   ***Textbook Reference:*** *Concept 40.4 Moving through Space Presents Distinctive Challenges; Trail following and path integration are two mechanisms of navigation*

8. Which statement about learning is true?
   a. Learning plays no role in the escape behavior of white-footed mice.
   b. Learning allows animals to modify their behavior based on previous experiences.
   c. Indigo buntings learn to recognize the true North Star during their first migration.
   d. The learning of species-specific songs in Galápagos finches requires particular brain regions, which are larger in females than in males.
   e. Learning that occurs early in life has minimal effects on survival.
   ***Textbook Reference:*** *Concept 40.2 Behavior Is Influenced by Development and Learning; Specific information of critical survival value is often learned during early postnatal development*

9. Which statement about honey bees is *false*?
   a. Honey bees communicate the distance and direction of a food source to other colony members by performing a waggle dance inside the hive.
   b. Because the sun is important for determining the direction of the food source from the hive, bees cannot find food on cloudy days.
   c. The duration of the straight run of the waggle dance conveys distance to a food source.
   d. During a waggle dance, the worker bee moves in a repeating figure-eight pattern.
   e. The angle of the straight run of the waggle dance conveys direction of the food source.
   ***Textbook Reference:*** *Concept 40.4 Moving through Space Presents Distinctive Challenges; Honey bee workers communicate distance and direction by a waggle dance*

10. By occupying a home range, an animal
    a. gains exclusive access to mates.
    b. benefits from being familiar with the environment.
    c. gains exclusive access to food resources.
    d. incurs costs associated with keeping neighbors out.
    e. gains exclusive access to shelter.
    ***Textbook Reference:*** *Concept 40.6 Behavior Helps Structure Ecological Communities and Processes; Animals often behaviorally partition space into territories or home ranges*

11. Which statement about migration is *false*?
    a. Migration is a one-way trip.
    b. Loggerhead turtles migrate long distances using magnetic fields to orient along the way.
    c. Migration occurs periodically in an animal's lifetime.
    d. Some shorebirds migrate long distances over the open ocean.
    e. Miniaturized electronic devices can be used to monitor animal movements, including migration.
    ***Textbook Reference:*** *Concept 40.4 Moving through Space Presents Distinctive Challenges; Migration: Many animals have evolved periodic movements between locations*

12. Which statement about eusociality is *false*?
    a. Most known cases of eusociality occur among birds.
    b. In eusocial groups, nonreproductive members help fertile members reproduce.
    c. Naked mole-rats are eusocial.
    d. Eusociality occurs in ants and wasps.
    e. In a colony of honey bees, only one female is reproductive.
    ***Textbook Reference:*** *Concept 40.5 Social Behavior Is Widespread; Eusociality represents an extreme type of differing status*

13. The escape behavior of pronghorn illustrates that
    a. behavior is integrated with physiology and structure.
    b. growth influences behavior.
    c. antipredator behavior is hard-wired.
    d. hormones influence behavior.
    e. antipredator behavior is learned during a brief window of time early in development.
    ***Textbook Reference:*** *Concept 40.3 Behavior Is Integrated with the Rest of Function*

14. A study of golden shiners in aquaria revealed that
    a. small schools are more conspicuous to predators than are large schools.
    b. costs of thermoregulation decrease as size of school increases.
    c. as size of school increases, so too does the efficiency with which preferred environments are found.
    d. as size of school decreases, so too does risk of disease transmission.

e. small schools are better than large schools at detecting predators.

*Textbook Reference:* Concept 40.5 Social Behavior Is Widespread; Some societies consist of individuals of equal status

15. Fixed action patterns
   a. typically change with experience.
   b. are displayed by invertebrates but not by vertebrates.
   c. demonstrate the importance of endocrine control of behavior.
   d. demonstrate the importance of epigenetic effects.
   e. are highly stereotyped and performed without prior learning.

   *Textbook Reference:* Concept 40.1 Behavior Is Controlled by the Nervous System but Is Not Necessarily Deterministic; Many types of evidence point to the neural basis of behavior

## Answers

### Key Concept Review

1. To show that song learning in birds has a neural basis, you could damage or destroy a region of the brain thought to be important in song learning and then look at the effects; this is considered "negative evidence." In addition, you could use brain-imaging studies to see whether a particular region of the brain is active during song learning; this is considered "positive evidence."
2. The statement is only partially true. Genetically similar humans can display very different behavior due to learning and epigenetic effects that occur as each person develops and interacts uniquely with the environment.
3. Because young birds learn the characteristics of an appropriate mate soon after hatching, it would be important to provide the condor chicks with cues appropriate to their species when caring for them. When feeding and interacting with the young, caretakers should use puppets that resemble adult condors; in this way, the young will learn the characteristics of their species, and not those of the human caretakers.
4. Malnourishment early in life affects epigenetic marking in rats, and these marks persist into adulthood. Thus, rats malnourished when young will likely differ behaviorally from rats that had free access to food when young, even if both groups of rats have free access to food in adulthood.
5. Because aerobic respiration takes time and is resistant to fatigue, the leg muscles of turkeys would specialize in aerobic ATP production. The breast muscles would specialize in anaerobic respiration, which is fast but susceptible to fatigue.
6. In frogs, the tonal frequency of their call is closely associated with male body size, with large males producing calls of lower frequency than small males. Thus, the researcher could estimate body size of individual male frogs by monitoring call frequencies.
7. The researcher should not conclude that landmarks are unimportant in mouse orientation. Most animals have redundancy in their mechanisms of orientation, so if they are prevented from using a particular mechanism, they can switch to another. Mice may use landmarks when they are available, but switch to another mechanism when they are unavailable. Also, although we usually think of landmarks as visual, the guideposts can be in any sensory modality. It is possible that the mice used landmarks other than the logs and rocks placed around the nest by the researcher.
8. Bees that have found a patch of flowers measure the distance to the patch by monitoring the rates at which they fly past local landmarks. Bees measure the direction of the patch by monitoring the angles at which they fly relative to the sun's compass position.
9. Your curves should show a positive relationship between the hawk's distance when spotted and pigeon flock size and a negative relationship between the hawk's attack success and pigeon flock size (see below).

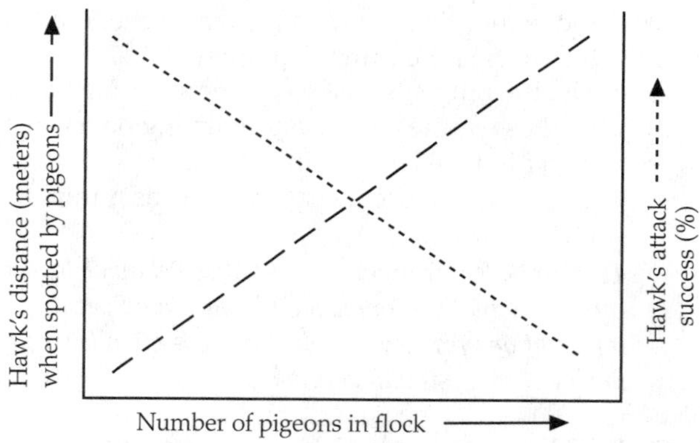


10. It would be important to make sure that the act of sharing meals indeed fits the definition of altruism (i.e., that the sharing bat suffers a cost and the receiving bat benefits). If that is indeed the case, there are different possible explanations. Vampire bats that share some of their blood meal may be sharing it with relatives. By helping relatives, an individual can increase the representation of some of its own alleles in the population. An alternative hypothesis is that by helping a certain group member on one night, that individual may, in turn, help the altruist on another night. This alternative hypothesis applies to unrelated individuals.
11. An individual's territory is an area from which it excludes other individuals. An individual's home range is an area that it occupies without excluding other

individuals. Home ranges often overlap, whereas territories do not. One benefit of a territory is exclusive access to resources on the territory. One cost of maintaining a territory is the time and energy required to exclude other individuals.

12. Matings between individuals of different species typically do not occur under natural conditions because behavioral differences, such as differing habitat preferences or activity periods, usually keep such individuals apart. In addition, should individuals of different species meet, subtle differences in their behavioral interactions cause them to separate before mating.

### *Test Yourself*

1. **d.** For giving an alarm call to be considered an altruistic act, the caller must incur a cost and the receivers of the call must benefit.
2. **c.** In Galápagos finches, species-specific songs are not inherited. Instead, they are learned during a relatively brief window of time (about 30 days) early in life (from 10–40 days after hatching). Young males learn by listening to the songs of their fathers. Females do not sing, so young males cannot learn their species-specific song by listening to their mothers.
3. **d.** The leg muscles of toads would rely on aerobic respiration, which takes time and is fatigue-resistant.
4. **e.** Experimental shifting of the circadian clock of homing pigeons disrupts their use of the sun as a compass. Homing pigeons use multiple cues to orient, including the sun, odor, sound, landmarks, and Earth's magnetic field. They also learn from other pigeons.
5. **c.** Biological determinism does not emphasize the role of learning in behavior. Instead, it emphasizes that behavior is predictable, inflexible, and determined entirely by genes.
6. **b.** Increased risk of disease transmission is a cost of group living. Other costs include greater conspicuousness to predators and increased competition for resources among group members.
7. **b.** Pheromones can be used to mark trails. Pheromones are used in communication between individuals of the same species and they are released to the external environment.
8. **b.** Learning is the ability to modify behavior based on previous experience. Learning shapes escape behavior of mice, as shown by the experiments of Lee Metzgar. Information critical to survival is often learned early in life. In Galápagos finches, learning species-specific song requires particular brain regions that are larger in males than in females.
9. **b.** On cloudy days bees can use atmospheric polarization patterns to infer the position of the sun.
10. **b.** Animals that occupy home ranges benefit from being familiar with the environment. They can avoid predators and find food more readily as a result of this familiarity. Territories, not home ranges, provide exclusive access to resources, such as food, shelter, and mates.
11. **a.** An animal that migrates leaves one location, stays for a substantial amount of time at another location, and then returns. Thus, migration entails a "round trip." Migration occurs periodically during an animal's lifetime.
12. **a.** Most known cases of eusociality occur among insects, including species of ants, bees, wasps, and termites.
13. **a.** The escape behavior of pronghorn illustrates that behavior is integrated with physiology (e.g., pronghorn have efficient delivery of oxygen to muscles and use of oxygen to make ATP) and structure (e.g., pronghorn have large lungs and muscles).
14. **c.** The study of golden shiners revealed that as size of school increases, so too does the efficiency with which fish find preferred environments
15. **e.** Fixed action patterns are highly stereotyped and performed without prior learning. They occur in diverse animals, including vertebrates such as gulls.

# 41 The Distribution of Earth's Ecological Systems

## The Big Picture

- Organisms and their environments constitute ecological systems. The components of ecological systems, listed in order of increasing scales of space and time, include individual organisms, populations, communities, landscapes, ecosystems, and the biosphere. Much evidence supporting the theory of continental drift is based on the distributions of organisms on Earth. The acceptance by geologists and biologists of the reality of continental drift revolutionized the field of biogeography.
- The distribution of Earth's physical environments shapes the distribution of organisms. Biomes are distinct physical environments inhabited by ecologically similar organisms. Terrestrial biomes are distinguished largely by differences in vegetation, while aquatic biomes are distinguished by abiotic characteristics such as salinity, depth, and temperature. Human activities result in ecosystems that have fewer species and a more uniform physical structure than they would have in the absence of human activities.

## Study Strategies

- For an overview of terrestrial biomes, study Figure 41.10, which shows how terrestrial biomes are determined by average annual temperature and precipitation. These physical variables largely determine the type of vegetation in each biome.
- Study Figure 41.11 to get a sense of the geographical location and extent of each terrestrial biome. A useful exercise to help you learn the relative locations of terrestrial biomes is to imagine a journey (e.g., from equatorial South America to Alaska, or from Maine to California), naming the biomes that you would pass through. Next study Figure 41.13, which shows the water-depth zones of aquatic biomes. To learn these zones, imagine moving from the shallowest area to the deepest area in first a lake and then an ocean, and see if you can name the zones.
- For an overview of Earth's terrestrial biogeographic regions, study Figure 41.14. Then, without looking at Figure 41.14, try to recreate the map from the figure on a separate sheet of paper. On your map, indicate the seven biogeographic regions, and wherever possible, indicate the barriers to movement that separate adjacent biogeographic regions.
- Go to **LaunchPad** (or use the Web addresses listed) to review the following additional resources:

  Animated Tutorial 41.1 Rain Shadow (PoL2e.com/at41.1)

  Animated Tutorial 41.2 Terrestrial Biomes (PoL2e.com/at41.2)

  Activity 41.1 Major Biogeographic Regions (PoL2e.com/ac41.1)

  Apply the Concept in textbook Sections 41.2 and 41.4

## Key Concept Review

### 41.1 Ecological Systems Vary over Space and Time

Organisms and their environments are ecological systems

Ecological systems can be small or large

Ecological systems vary, but in ways that can be understood with scientific methods

Physical geography concerns the distribution of Earth's climates and surface features. Biogeography is the study of the distributions of species. Ecology is the study of interactions among organisms and between organisms and their physical environment.

Ecological systems include organisms and the environment with which they interact. The environment has both abiotic (nonliving) components and biotic (living) components. Ecological systems can be studied from the following perspectives, listed in order of increasing scales of space and time: individual organism; population (a group of organisms of the same species living in a particular area at the same time); community (an assemblage of interacting populations of different species within a particular geographic area); landscape (several communities); ecosystem (communities plus their physical environment); and biosphere (all the organisms and environments of Earth).

Although large ecosystems tend to be more complex than small ones, even small ecosystems can be very complex.

For example, the microbial community of the human gut contains hundreds of species and is one of the most densely populated ecosystems on Earth. In addition, this community varies from one person to the next and with diet (see Figure 41.1).

**Question 1.** Distinguish between a community and an ecosystem.

**Question 2.** Describe the human gut and its microbial inhabitants in terms of the following categories: organism, population, community, and ecosystem.

### 41.2 Solar Energy Input and Topography Shape Earth's Physical Environments

Variation in solar energy input drives patterns of weather and climate

The circulation of Earth's atmosphere redistributes heat energy

Ocean circulation also influences climate

Topography produces additional environmental heterogeneity

Climate diagrams summarize climates in an ecologically relevant way

The climate of a region is the average of the atmospheric conditions found in that region over the long run. Weather is the short-term state of those conditions.

Solar energy varies with latitude. Regions near the poles receive less energy per unit of ground area than regions near the equator because the angle of the sun at high latitudes is shallow rather than steep (see Figure 41.3). The tilt of Earth's axis of rotation causes seasonality (see Figure 41.4).

Rising air expands and cools, releasing moisture, whereas descending air is compressed and warmed, taking up moisture. Unequal heating of the atmosphere at low and high latitudes produces vertical and latitudinal movements of air masses. These movements lead to very moist climates at the equator and at 60°N and 60°S latitudes (where air rises) and to arid climates at about 30°N and 30°S latitudes and near the poles (where air descends). Two Hadley cells, one north and one south of the equator, produce these latitudinal patterns of precipitation (see Figure 41.5).

The spinning of Earth on its axis causes air masses moving latitudinally to be deflected to the right in the Northern Hemisphere and to the left in the Southern Hemisphere. Thus, winds blowing toward the equator at low latitudes veer to become the northeast and southeast trade winds, whereas winds blowing away from the equator at mid-latitudes are deflected to become the prevailing westerlies (see Figure 41.6).

Ocean currents are driven primarily by prevailing winds but are deflected by continents. The poleward movement of ocean water warmed in the tropics is a major mechanism of heat transfer to high latitudes. Winds can also cause upwellings, which are areas where colder water from deep below the water's surface rises to mix with and replace warmer surface water. Water has a high heat capacity, so oceans moderate terrestrial climates.

Topography is variation in the elevation of Earth's surface; it influences physical conditions in terrestrial and aquatic environments. In terrestrial environments, for example, precipitation tends to be greater on the windward side of mountains (where rising air cools and releases moisture) than on the leeward side (where air descends, warms, and holds moisture) (see Figure 41.8).

Heinrich Walter developed climate diagrams that plot average monthly temperature and precipitation throughout the year. These diagrams summarize climate in a particular location and illustrate when conditions allow for terrestrial plant growth (see Figure 41.9).

**Question 3.** Evaluate the following remark: "Temperatures have been so cold over the last few days that global climate change—in particular global warming—cannot possibly be happening."

**Question 4.** If you were going to plant a crop that required a lot of water, would you be better off planting it on the windward or leeward side of a mountain? Explain your choice.

### 41.3 Biogeography Reflects Physical Geography

Similarities in terrestrial vegetation led to the biome concept

Climate is not the only factor that molds terrestrial biomes

The biome concept can be extended to aquatic environments

A biome is a distinct physical environment inhabited by ecologically similar organisms. In biomes that occur in several widely separated areas of the globe, species occurring in different locations are unlikely to be closely related phylogenetically, but they are likely to share many adaptations to their environment as a result of convergent evolution.

Differences in vegetation distinguish one terrestrial biome from another. These differences in vegetation reflect gradients in annual patterns of temperature and precipitation (see Figure 41.10). On a global scale, the distribution of terrestrial biomes reflects latitudinal and elevational gradients in temperature and precipitation. A particular biome can occur in widely separated regions (see Figure 41.11). Fire and soil fertility also influence vegetation, and therefore biomes.

Biomes also occur in aquatic environments (see Table 41.1). However, unlike terrestrial biomes that are distinguished by their vegetation, aquatic biomes are not characterized by a structurally dominant group of organisms. Instead, aspects of the physical environment, such as the water depth, temperature, and salinity distinguish aquatic biomes.

Salinity is the prime characteristic used to distinguish aquatic environments because it is the most important determinant of the organisms that can live in a particular aquatic habitat. Freshwater biomes include streams, ponds, and lakes, and saltwater biomes include oceans and salt lakes. Estuarine biomes occur where freshwater and saltwater mix at river mouths.

Freshwater environments can be classified as running water environments (streams and rivers) and standing water environments (lakes and ponds). Bodies of standing fresh water and oceans can be divided into zones based on depth and light penetration (see Figure 41.13).

**Question 5.** How does the basis for distinguishing terrestrial biomes differ from the criteria used to distinguish aquatic biomes?

**Question 6.** Construct a concept map whose theme is "Oceans." Include in your map the following terms: oceans, aphotic zone, currents, depth, direction of prevailing winds, distance from shore, intertidal zone, latitudinal differences in solar energy input, location of continents, pelagic zone, photic zone, photosynthetic organisms, rotation of Earth, and zones. Connect these concepts by verbs, phrases, or comparative terms to indicate the relationships among them.

## 41.4 Biogeography Also Reflects Geological History

Barriers to dispersal affect biogeography

The movements of continents account for biogeographic regions

Phylogenetic methods contribute to our understanding of biogeography

Earth can be divided into several biogeographic regions, each containing characteristic assemblages of species. The biotas of the biogeographic regions differ because barriers, such as oceans or mountains, restrict the dispersal of organisms (see Figure 41.14).

Two scientific advances changed the field of biogeography: the acceptance of the theory of continental drift and the development of phylogenetic taxonomy. Continental drift has influenced the evolution and mixing of species throughout the history of life on Earth. It explains some discontinuous distributions that include several biogeographic regions (see Figure 41.15). During biotic interchange, two different biota merge following the fusion of two formerly separated land masses. For example, following the formation of the Central American land bridge connecting North America (the Nearctic region) and South America (the Neotropical region), many species of mammals that had evolved on one continent colonized the other, in what became known as the Great American Interchange.

Biogeographers use phylogenetic information, together with the fossil record and geological history, to study modern distributions of organisms. For example, they can compare the sequence and timing of splits in a phylogenetic tree with the sequence and timing of movements and splits of geographic areas.

**Question 7.** South America is a center of diversification for many groups of freshwater fishes, including characins, the group that includes piranhas. Characins also occur in Africa. How would you explain the modern distribution of characins, given that these fishes cannot cross open salt water?

**Question 8.** As a biogeographer, you notice that a phylogenetic split in your study organisms coincided in time with the formation of a mountain range. What connection can be made between these simultaneous occurrences?

## 41.5 Human Activities Affect Ecological Systems on a Global Scale

We are altering natural ecosystems as we use them

We are replacing natural ecosystems with human-dominated ones

We are blurring biogeographic boundaries

Science provides tools for conserving and restoring ecological systems

When humans use natural ecosystems, such as when hunting and fishing, we change the abundance of individuals within species, which in turn alters patterns of interaction between species and ecosystem function. Humans have converted much of Earth's land into croplands, pasturelands, and urban areas. These human-dominated ecosystems are less complex than the natural ecosystems they have replaced. For example, when compared with natural ecosystems, agricultural areas have lower species diversity and are more spatially and physically uniform.

Human activities also influence surrounding natural ecosystems that remain. Examples of such activities include water control measures, the release of pollutants, and the introduction of new species. The deliberate or inadvertent introduction of nonnative species to areas is blurring biogeographic boundaries and homogenizing Earth's biota.

Two subdisciplines of ecology, conservation ecology (the study of extinction and its prevention) and restoration ecology (the study of how to restore damaged ecosystems), have emerged to help scientists better understand the impacts of humans on ecological systems. Ecologists often use the tools of natural history and mathematical modeling. Natural history is the informal observation of nature. Such observations provide critical information at each step of a study, including hypothesis development, experimental design, and interpretation of results. Lack of natural history information limits our ability to answer questions. Models can be built on the basis of natural history information.

**Question 9.** At the end of their day of fishing at a local pond, some sportsmen have released a pail of leftover bait fish that are not native to the area. What potential repercussions could result from such an action?

**Question 10.** In your ecology class, you have been given an assignment to conduct a field study concerning the effects on vegetation of excluding white-tailed deer from a particular area. Although your instructor has suggested a period of informal observation of deer feeding in the study area, your lab partner suggests skipping that phase and going straight to developing an hypothesis and designing the study. Would you agree with your lab partner's plan? Why or why not?

## Test Yourself

1. The region of the ocean that lies close enough to shore to be affected by wave action is the _______ zone.
   a. abyssal
   b. pelagic
   c. benthic
   d. aphotic
   e. intertidal
   ***Textbook Reference:*** *Concept 41.3 Biogeography Reflects Physical Geography; The biome concept can be extended to aquatic environments*

2. Which statement about the human gut is *false*?
   a. It is one of Earth's most densely populated ecosystems.
   b. Its biotic components include bacteria, archaea, and yeasts.
   c. The gut microbial communities of relatives are more similar than those of unrelated individuals.
   d. The microbial community of an individual's gut remains unchanged when diet changes.
   e. The gut microbial communities of obese people are enriched in bacteria that are efficient at extracting energy from food.
   ***Textbook Reference:*** *Concept 41.1 Ecological Systems Vary over Space and Time; Ecological systems vary, but in ways that can be understood with scientific methods*

3. Multiple communities form
   a. populations.
   b. biospheres.
   c. landscapes.
   d. abiotic components of an ecosystem.
   e. abiotic components of a region's physical geography.
   ***Textbook Reference:*** *Concept 41.1 Ecological Systems Vary over Space and Time; Ecological systems can be small or large*

4. Which statement about biogeography is *false*?
   a. North America and South America have always been connected by a land bridge.
   b. Gondwana was the southern supercontinent.
   c. The Nearctic and Palearctic are northern biogeographic regions.
   d. The Himalayas separate the Oriental and Palearctic regions.
   e. Deserts separate the Palearctic and Ethiopian regions.
   ***Textbook Reference:*** *Concept 41.4 Biogeography Also Reflects Geological History; The movements of continents account for biogeographic regions*

5. If Earth did not spin on its axis, the northeast trade winds would blow from the
   a. northeast.
   b. south.
   c. north.
   d. east.
   e. southwest.
   ***Textbook Reference:*** *Concept 41.2 Solar Energy Input and Topography Shape Earth's Physical Environments; The circulation of Earth's atmosphere redistributes heat energy*

6. Which of the following statements about oceans is *false*?
   a. Prevailing winds drive ocean currents.
   b. Ocean currents are deflected by land masses.
   c. Oceans intensify Earth's terrestrial climates.
   d. Water circulation in oceans is three-dimensional.
   e. The Gulf Stream brings warm water from the tropical Atlantic Ocean and Gulf of Mexico northward.
   ***Textbook Reference:*** *Concept 41.2 Solar Energy Input and Topography Shape Earth's Physical Environments; Ocean circulation also influences climate*

7.–8. The diagram below shows a mountain with a sea breeze blowing in the direction indicated by the arrow.

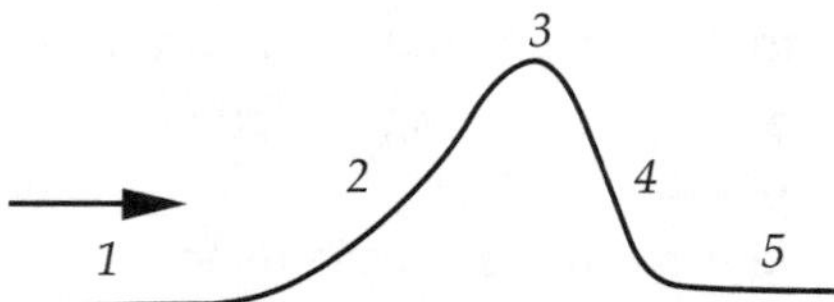


7. Which number corresponds to the area in which air would be *both* relatively warm and relatively dry?
   a. 1
   b. 2
   c. 3
   d. 4
   e. 5
   ***Textbook Reference:*** *Concept 41.2 Solar Energy Input and Topography Shape Earth's Physical Environments; Topography produces additional environmental heterogeneity*

8. In which area of the diagram is a process occurring that is similar to the process that occurs in the region near the equator?
   a. 1
   b. 2
   c. 3
   d. 4
   e. 5
   ***Textbook Reference:*** *Concept 41.2 Solar Energy Input and Topography Shape Earth's Physical Environments; Topography produces additional environmental heterogeneity*

9. In comparison to natural ecosystems, human-dominated ecosystems
   a. contain more interacting species.
   b. are structurally more complex.
   c. are more spatially uniform.

d. have greater overall plant diversity.
e. are less likely to include monocultures.
***Textbook Reference:*** *Concept 41.5 Human Activities Affect Ecological Systems on a Global Scale; We are replacing natural ecosystems with human-dominated ones*

10. Kangaroo rats of the deserts of the southwestern United States closely resemble jerboas, rodents that inhabit deserts in Asia and Africa. The two groups of rodents are not closely related, yet they both have small forelimbs, large hind limbs, and a long tail. These similarities are an example of
a. competition.
b. convergent evolution.
c. dispersal.
d. divergent evolution.
e. biotic interchange.
***Textbook Reference:*** *Concept 41.3 Biogeography Reflects Physical Geography; Similarities in terrestrial vegetation led to the biome concept*

11. Which term represents the largest scale in space and time?
a. Landscape
b. Biosphere
c. Ecosystem
d. Population
e. Community
***Textbook Reference:*** *Concept 41.1 Ecological Systems Vary over Space and Time; Ecological systems can be small or large*

12. Which statement about aquatic biomes is *false*?
a. In freshwater biomes, slow-flowing water is associated with a soft bottom.
b. Salinity is the primary characteristic that distinguishes aquatic biomes.
c. Photosynthetic organisms are confined to the photic zone in both freshwater and saltwater biomes.
d. Organisms of the abyssal zone of oceans experience high pressure.
e. The limnetic zone of lakes and the pelagic zone of oceans are near-shore environments.
***Textbook Reference:*** *Concept 41.3 Biogeography Reflects Physical Geography; The biome concept can be extended to aquatic environments*

13. An aquatic biome in which fresh water and salt water mix is called a(n)
a. estuary.
b. coral reef.
c. seagrass bed.
d. pelagic zone.
e. hydrothermal vent.
***Textbook Reference:*** *Concept 41.3 Biogeography Reflects Physical Geography; The biome concept can be extended to aquatic environments*

14. An acre of land in Colombia and an acre of land in Michigan would be exactly the same in terms of
a. the angle of the sun reaching the ground in the month of July.
b. solar energy input in the month of July.
c. annual solar energy input.
d. total hours of daylight per year.
e. the extent of temperature fluctuations over the course of a year.
***Textbook Reference:*** *Concept 41.2 Solar Energy Input and Topography Shape Earth's Physical Environments; Variation in solar energy input drives patterns of weather and climate*

15. Wallace's line separates the _______ and _______ biogeographic regions.
a. Nearctic; Palearctic
b. Oriental; Australasian
c. Neotropical; Nearctic
d. Oriental; Palearctic
e. Ethiopian; Antarctic
***Textbook Reference:*** *Concept 41.4 Biogeography Also Reflects Geological History; The movements of continents account for biogeographic regions*

16. Which terrestrial biome has the highest average temperatures and the most precipitation?
a. Boreal forest
b. Tropical savanna
c. Temperate seasonal forest
d. Tundra
e. Tropical rainforest
***Textbook Reference:*** *Concept 41.3 Biogeography Reflects Physical Geography; Similarities in terrestrial vegetation led to the biome concept*

17. Which statement about natural history observations is *false*?
a. They lead to new questions.
b. They allow for the formulation of appropriate hypotheses.
c. They help in the designing of appropriate experiments.
d. They provide needed context for the interpretation of laboratory results.
e. They typically include observations of nature during a formal hypothesis-testing experiment.
***Textbook Reference:*** *Concept 41.5 Human Activities Affect Ecological Systems on a Global Scale; Science provides tools for conserving and restoring ecological systems*

## Answers

### *Key Concept Review*

1. A community is an assemblage of interacting populations of different species in a particular geographic

area. An ecosystem includes communities plus their physical environment.

2. A bacterium within the phylum Firmicutes is one example of an individual organism inhabiting the human gut. At a particular time, the gut of a person might have several hundred of these particular bacteria at a specific location within the gut, which would constitute a population (i.e., individuals of one species living in the same area, at the same time, and interacting). There would also be many other species of microbes in this person's gut, and these interacting populations would constitute a community. This community plus the gut environment would make up the ecosystem.
3. The remark demonstrates confusion about what is meant by weather (conditions of the atmosphere over a short period of time such as days) and what is meant by *climate* (conditions of the atmosphere over relatively long periods of time).
4. It would be better to plant on the windward side of the mountain because rising air cools and releases moisture, which would be better for the crop. On the leeward side, air descends, warms, and little precipitation falls.
5. Terrestrial biomes are distinguished on the basis of vegetation. Aquatic biomes are not distinguished on the basis of a structurally dominant group of organisms, but instead on the basis of abiotic factors such as water salinity, movement, and depth.
6.

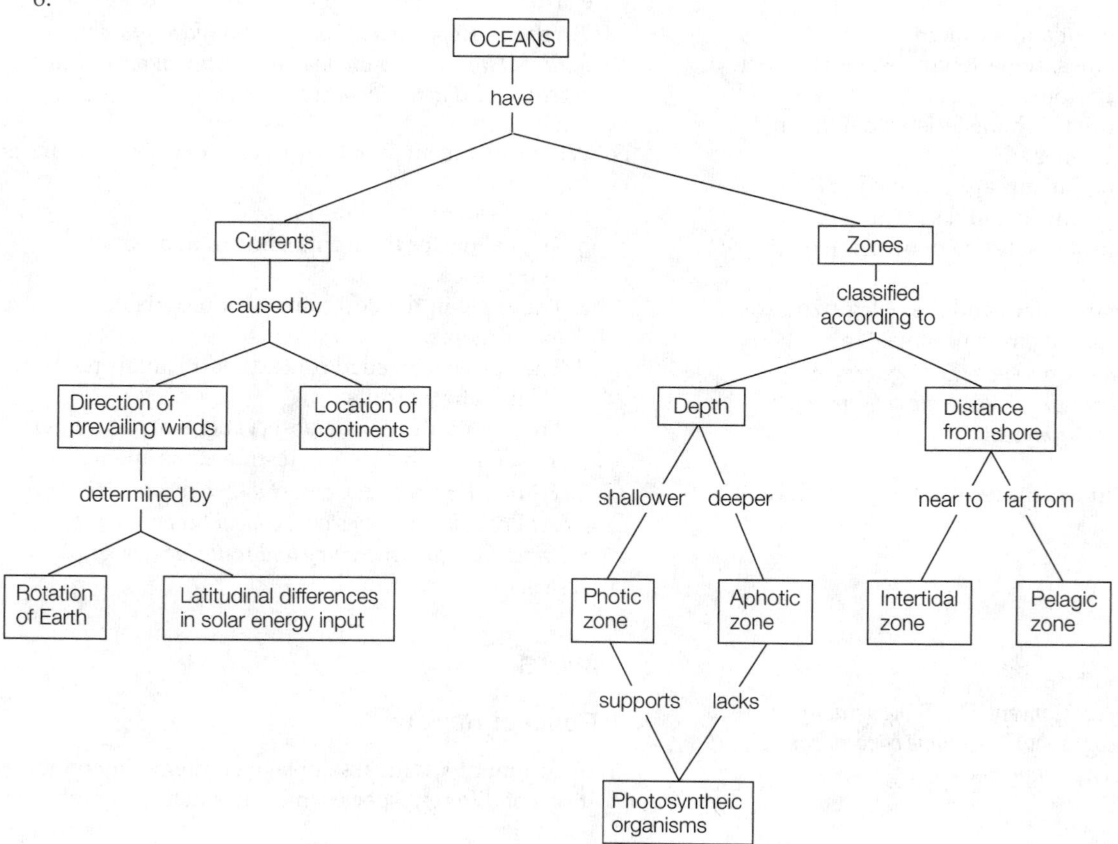


7. The characins existed before the continents split and drifted to their current locations.
8. If a phylogenetic split coincides in time with the formation of a barrier to dispersal (such as the uplift of a mountain range), it is reasonable to conclude that the barrier caused the phylogenetic split by subdividing the original range of the ancestral species.
9. Releasing a nonnative species of fish is a bad idea because the released individuals may survive and thrive in their new environment. If male and female individuals are released, they may breed and produce offspring, thereby increasing the number of nonnative individuals in the pond. Individuals of a nonnative species may outcompete members of a similar native species or directly eliminate native species through predation. Sometimes nonnative species introduce diseases to which native species are susceptible. Introductions of nonnative species blur biogeographic boundaries and lessen spatial heterogeneity in Earth's species composition.
10. The lab partner's plan is a bad idea. Informal natural history observations are critical to every stage of an ecological study. Such observations help in the generation of questions, development of hypotheses, design of experiments, and interpretation of results. Also, natural history information should be used when building mathematical models of any phenomenon under study.

## *Test Yourself*

1. **e.** The intertidal zone is affected by wave action.
2. **d.** Microbial communities change as diet changes. For example, a calorie-restricted diet prompts the microbial community of obese people to change and approach that of lean people.
3. **c.** Multiple communities form landscapes. Communities represent interacting populations of different species in a particular area, so they are biotic (living) components of the environment.
4. **a.** North and South America have not always been connected by a land bridge. The land bridge that formed between these two continents resulted in the Great American Interchange.
5. **c.** If Earth did not spin on its axis, air flowing south toward the equator would not be deflected to the right. Therefore, the northeast trade winds would blow directly from the north instead of from the northeast.
6. **c.** Oceans do not intensify terrestrial climates; instead, they moderate them.
7. **e.** The air would have lost most of its moisture while rising on the windward side of the mountain, and in descending to area *e* it would have warmed again.
8. **b.** In regions surrounding the equator, warm air rises and loses much of its moisture. These same events occur when moist air rises over a mountain.
9. **c.** When compared with natural ecosystems, human-dominated ecosystems are more spatially uniform.
10. **b.** The resemblance between kangaroo rats and jerboas is an example of convergent evolution.
11. **b.** The biosphere includes all of Earth's organisms and their environments, and represents the largest scale in space and time.
12. **e.** The limnetic zone of lakes and the pelagic zone of oceans are open-water, off-shore environments.
13. **a.** An estuary is the body of water found at the mouth of a river, where salt water mixes with fresh water.
14. **d.** All areas on Earth receive equal hours of daylight per year. The seasonal distribution of those hours, the solar energy input, the sun angle, and the extent of seasonal fluctuations in temperature do vary latitudinally.
15. **b.** Wallace's line separates the Oriental and Australasian regions.
16. **e.** Tropical rainforests have the highest average temperatures and the most precipitation.
17. **e.** Natural history observations are critical to all stages of an ecological study, including the generation of new questions, hypothesis development, experimental design, and the interpretation of results. However, these observations are performed outside a formal hypothesis-testing investigation.

# 42 Populations

## The Big Picture

- Concepts of population ecology such as multiplicative growth and carrying capacity are fundamental to the study of human population growth. They also contribute to the preservation of biodiversity and the control of undesirable species. Darwin's realization that all populations have the inherent capacity for multiplicative growth was crucial to the development of his theory of natural selection.
- In many species, individuals occur in subpopulations that occupy suitable habitat patches separated by unsuitable habitat. The set of subpopulations in a particular region is a metapopulation. Sometimes dispersal is possible between subpopulations. The BIDE model of population size differs from the BD model because in addition to considering births (B) and deaths (D), it includes the number of individuals moving into a subpopulation (immigrants; I) and the number moving out (emigrants; E). Thus, whereas the BD model considers populations to be closed systems, the BIDE model considers subpopulations to be open systems.

## Study Strategies

- It can be difficult to interpret graphs depicting concepts of population ecology. Look carefully at how the axes of a graph are labeled. Ask yourself questions such as: Are the scales arithmetic or logarithmic? Does the graph trace the fate of a cohort of the population or the growth pattern of the entire population? Redrawing graphs from the textbook is a good way to reinforce your understanding of the concepts being presented.
- It is important to understand how an ever-faster per capita *growth* rate of the human population can be accompanied by a steadily *decreasing* doubling time. Review the definition of doubling time in the text (p. 11) and study Figure 42.9A.
- Review how the BD and BIDE models of population size differ.
- Take advantage of tutorials and laboratory activities involving computer simulations of population growth or other aspects of population ecology.
- Go to **LaunchPad** (or use the Web addresses listed) to review the following additional resources:

  Animated Tutorial 42.1 Multiplicative Population Growth Simulation (PoL2e.com/at42.1)

  Animated Tutorial 42.2 Density-Dependent Population Growth Simulation (PoL2e.com/at42.2)

  Animated Tutorial 42.3 Metapopulation Simulation (PoL2e.com/at42.3)

  Animated Tutorial 42.4 Habitat Corridors (PoL2e.com/at42.4)

  Activity 42.1 Population Growth (PoL2e.com/ac42.1)

  Analyze the Data in textbook Figures 42.6 and 42.11

  Apply the Concept in textbook Sections 42.3, 42.4, and 42.5

## Key Concept Review

### 42.1 Populations Are Patchy in Space and Dynamic over Time

Population density and population size are two measures of abundance

Abundance varies in space and over time

A population consists of the individuals of a species that interact with one another within a given area at a particular time. A species' role in a particular community is determined by characteristics of individuals and by the relative abundance of the species, which is a population-level characteristic. Abundance can be measured as population density (the number of individuals per unit of area or volume) or as population size (the total number of individuals in the population). In particular areas, population densities change over time. The region in which a species is found is called its geographic range. Within the geographic range, a species occupies particular habitats. These may exist as habitat patches (also called habitat "islands") —areas of suitable habitat surrounded by unsuitable habitat.

**Question 1.** You have been live-trapping, marking, and releasing voles in a 1-hectare plot for several months. Based on your field work, you calculate the population density of the study plot to be 150 voles per hectare. The total area occupied by the population is 4 hectares. What is the total population size?

**Question 2.** Timber rattlesnakes are found in the eastern half of the United States: north to southern Maine, south to northern Florida, and west to central Texas. This species of snake prefers forested areas, particularly those containing wooded rocky ledges with southern exposures for basking. Areas with high rodent densities are ideal. Roads pose a great risk and agricultural fields are avoided. Explain the following concepts and how they apply to this species: geographic range, habitat, and habitat patch (or habitat "island").

## 42.2 Births Increase and Deaths Decrease Population Size

Births add individuals to populations and deaths remove them. The "birth–death," or BD, model of population change states that the number of individuals in a population at some time in the future equals the number now, plus the number that are born, minus the number that die ($N_{t+1} = N_t + B - D$, where *N* is the population size, *B* is the number of births in the time interval from time *t* to time $t + 1$, and *D* is the number of deaths in that same time interval; equation 42.1). The growth rate of a population refers to the change in population size in a certain period of time ($\Delta N/\Delta T = B - D$, where the Greek symbol Δ means "change in"; equation 42.2).

It is often difficult to directly measure change in size of a total population of individuals, so ecologists keep track of a sample of individuals, called a cohort. Using this sample, ecologists can calculate the per capita birth rate (*b*), which is the number of offspring produced by an average individual, and the per capita death rate (*d*), which is the average individual's chance of dying. The per capita birth rate minus the per capita death rate represents the average individual's contribution to the total population growth rate; this value is called the per capita growth rate and is symbolized by *r*. The following equation allows us to predict changes in population size: $\Delta N/\Delta T = rN$ (equation 42.4). If the per capita birth rate is greater than the per capita death rate, then *r* will be greater than zero and the population will grow. If the per capita birth rate is less than the per capita death rate, then *r* will be less than zero and the population will shrink. If the per capita birth rate equals the per capita death rate, then *r* will equal zero and the size of the population will not change.

**Question 3**. An ecologist who has been monitoring a cohort of Couch's spadefoot toads in a locality in California describes the total population size as declining. What does this tell you about the per capita birth rates and death rates of the study cohort?

**Question 4.** Why is it unusual for population densities to remain unchanged over time?

## 42.3 Life Histories Determine Population Growth Rates

Life histories are diverse

Resources and physical conditions shape life histories

Species' distributions reflect the effects of environment on per capita growth rates

The life history of a species includes information on the time course of an average individual's growth, development, reproduction, and death. This information can be summarized in a table (see Table 42.1). Survivorship is the fraction of individuals that survive to different ages; it can also be expressed as mortality (which equals 1 – survivorship). Fecundity is the average number of offspring each individual produces at each age. Survivorship and fecundity influence the per capita growth rate (*r*). Life histories vary considerably, both across species and within species.

Organisms need resources (materials and energy), and the time to acquire them. They also require physical conditions they can tolerate. The distinction between resources and conditions is that resources can be used up whereas conditions are experienced. The rate of resource acquisition increases with resource availability (see Figure 42.4). The principle of allocations states that a unit of an obtained resource can be used for only one function at a time. The resources obtained by organisms must be divided among competing functions, which include body maintenance (usually the first priority), growth, reproduction, and defense (see Figure 42.5). Resource allocation to specific functions changes in relation to the abundance of resources and whether conditions are typical or stressful. Life-history tradeoffs are the negative relationships among growth, reproduction, and survival. For example, species that invest in reproduction have high fecundity but low survival. Species distributions can be predicted once we know how resource availability and physical conditions affect survivorship and fecundity (see Figures 42.6 and 42.7).

**Question 5.** Predict how increased survivorship and increased fecundity would influence per capita growth rate.

**Question 6.** Many small mammals produce large litters every few weeks during the breeding season. Given this aspect of their life history, what would you expect regarding adult survivorship? This type of adult survivorship is an example of what phenomenon?

## 42.4 Populations Grow Multiplicatively, but the Multiplier Can Change

Multiplicative growth with constant *r* can generate large numbers very quickly

Populations growing multiplicatively with constant *r* have a constant doubling time

Density dependence prevents populations from growing indefinitely

Changing environmental conditions cause the carrying capacity to change

Technology has increased Earth's carrying capacity for humans

For a period of time, populations can grow multiplicatively, which means that a constant multiple of the population size (*N*) is added during each time period. In contrast, when populations exhibit additive growth, a constant number of individuals is added during each time period. Multiplicative growth generates large numbers very quickly (the growth pattern is J-shaped) and has a constant doubling time, providing $r$ does not change.

Populations cannot exhibit multiplicative growth indefinitely. Typically, population growth slows and levels off at the carrying capacity ($K$), which is the number of individuals the environment can support indefinitely. Populations stop growing because $r$ is density-dependent; $r$ is highest when population densities are low, and decreases as population densities increase (see Figure 42.8). At high population densities, each individual has access to fewer resources, and as a result, birth rates decrease and death rates increase, causing decreases in $r$.

The carrying capacity varies over space because resource availability varies over space, as do physical conditions that affect the costs of body maintenance. The carrying capacity also varies over time because resource availability and physical conditions vary over time, such as over years or across seasons.

Among populations of large animals, the human population stands out as one that has continued to grow at an ever-faster per capita rate, with its doubling time decreasing dramatically from the 1500s to the 1900s (see Figure 42.9A). The fast rate of growth of the human population has been made possible by technological advances that increase food production or improve health and therefore increase the carrying capacity (see Figure 42.9B). It is possible, however, that the human population has now exceeded its carrying capacity; the evidence for this includes the predicted decline in our finite fuel resources upon which technological advances are based, and environmental problems such as climate change and degradation of ecosystems. Change in size of the human population can occur through voluntary reductions in per capita birth rate or by the less appealing prospect of increasing mortality.

**Question 7.** Why is the per capita growth rate described as density-dependent?

**Question 8.** Just as the per capita growth rate of a population can be described as density-dependent, so too can other factors that influence population size. Of the following three factors, which would you describe as density-dependent: (1) disease; (2) weather; and (3) starvation?

### 42.5 Immigration and Emigration Affect Population Dynamics

Most populations are divided into geographically separated subpopulations that live in habitat patches—areas of suitable habitat that are separated from other patches by unsuitable environments. Individuals sometimes can disperse among the patches. The larger population to which the subpopulations belong is called the metapopulation.

Each subpopulation changes through births and deaths. Subpopulations also gain individuals through immigration and lose individuals through emigration. The BIDE model states that the number of individuals in a population at some time in the future equals the number now, plus the number that are born, plus the number that immigrate, minus the number that die, minus the number that emigrate.

Because individual subpopulations are much smaller than the metapopulation, they are prone to fluctuations leading to extinction. As long as subpopulations are connected by dispersal, however, they can be rescued from extinction by immigration.

**Question 9.** How does the division of some populations into discrete subpopulations relate to the possible "rescue" of a subpopulation from extinction? How would barriers to immigration influence the potential for rescue?

**Question 10.** Metapopulation A of crustaceans has eight subpopulations, each with more than 50 individuals. Metapopulation B of crustaceans has four subpopulations, each with fewer than 30 individuals. In the face of environmental disturbance, which metapopulation would likely go extinct sooner and why?

### 42.6 Ecology Provides Tools for Conserving and Managing Populations

Knowledge of life histories helps us manage populations

Knowledge of metapopulation dynamics helps us conserve species

Life history information can identify particular life stages that are most important for reproduction, survival, and hence the population growth rate. This information, in turn, can be used to manage populations of species considered desirable or undesirable.

Conservation strategies also have been informed by knowledge of metapopulations. For example, once any remaining habitat has been identified and the risks to habitat patches have been evaluated, conservation efforts often focus on protecting as many habitat patches as possible, with priority given to the largest patches. Because opportunities for dispersal among patches are critical, corridors for dispersal may be protected or created.

**Question 11.** What are dispersal corridors, and how have experiments demonstrated that they help species persist in patchy environments?

**Question 12.** Construct a concept map whose theme is "Metapopulation." Include in your map the following terms: metapopulation, subpopulations, births, deaths, immigrants, emigrants, recolonization, and dispersal corridors. Connect these concepts by verbs, phrases, or comparative terms to indicate the relationships among them.

## Test Yourself

1. There are 150 golden shiners living in a pond. Which characteristic would you need to know in order to calculate their population density?
   a. Their birth rate
   b. Their growth rate

c. Their death rate
d. The volume of water in which they live
e. Their fecundity
***Textbook Reference:*** *Concept 42.1 Populations Are Patchy in Space and Dynamic over Time; Population density and population size are two measures of abundance*

2. *Per capita* means per
a. population.
b. cohort.
c. individual.
d. family.
e. community.
***Textbook Reference:*** *Concept 42.2 Births Increase and Deaths Decrease Population Size*

3. Of the following animals, which would likely need a continuous corridor of habitat for successful dispersal to a new patch of suitable habitat?
a. Bat
b. Bird
c. Butterfly
d. Salamander
e. Dragonfly
***Textbook Reference:*** *Concept 42.6 Ecology Provides Tools for Conserving and Managing Populations; Knowledge of metapopulation dynamics helps us conserve species*

4. A decreasing doubling time indicates
a. an increasing per capita growth rate.
b. a decreasing per capita growth rate.
c. higher per capita death rates than per capita birth rates.
d. that the population has reached its carrying capacity.
e. that per capita death rates equal per capita birth rates.
***Textbook Reference:*** *Concept 42.4 Populations Grow Multiplicatively, but the Multiplier Can Change; Technology has increased Earth's carrying capacity for humans*

5. When resources such as food and roost sites are available at normal levels but environmental temperatures drop to stressful levels, a crow would likely
a. increase allocation of resources to body maintenance.
b. increase allocation of resources to reproduction.
c. increase allocation of resources to growth.
d. increase allocation of resources to defense.
e. keep its resource allocation the same as when conditions were typical.
***Textbook Reference:*** *Concept 42.3 Life Histories Determine Population Growth Rates; Resources and physical conditions shape life histories*

6. A metapopulation
a. includes a cluster of distinct subpopulations linked by dispersal.
b. has no carrying capacity.
c. occurs when births equal deaths.
d. has a constant population density.
e. is an isolated subpopulation.
***Textbook Reference:*** *Concept 42.5 Immigration and Emigration Affect Population Dynamics*

7.–8. Refer to the graph below.

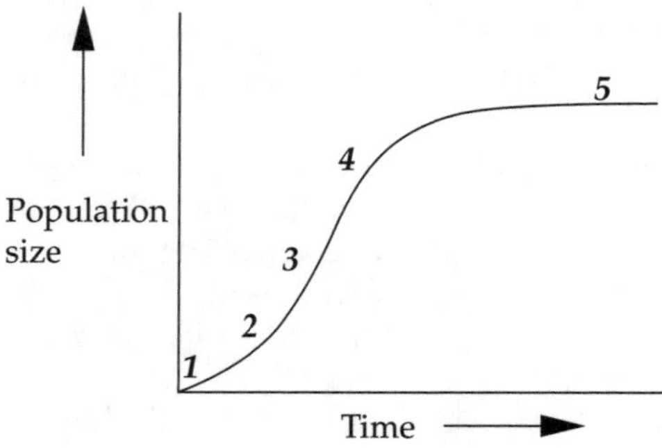


7. In the population growth curve shown in the graph, the rate of growth is greatest at which point?
a. 1
b. 2
c. 3
d. 4
e. 5
***Textbook Reference:*** *Concept 42.4 Populations Grow Multiplicatively, but the Multiplier Can Change; Density dependence prevents populations from growing indefinitely*

8. At which point in the graph would there be zero population growth ($r = 0$)?
a. 1
b. 2
c. 3
d. 4
e. 5
***Textbook Reference:*** *Concept 42.4 Populations Grow Multiplicatively, but the Multiplier Can Change; Density dependence prevents populations from growing indefinitely*

9. Which statement about life histories is *false*?
a. Life histories can be either qualitative or quantitative descriptions.
b. A species' life history reflects the timing of growth, development, reproduction, and death of an average individual.
c. Life histories vary across species.
d. Within species, life histories are invariant.
e. A species' life history can provide information that is useful to understanding its demography.
***Textbook Reference:*** *Concept 42.3 Life Histories Determine Population Growth Rates; Life histories are diverse*

10. Which statement about carrying capacity is *false*?
    a. Technology has increased Earth's carrying capacity for humans.
    b. At carrying capacity, $r = 0$ and populations stop growing.
    c. Carrying capacity is the number of individuals that the environment can support indefinitely.
    d. The carrying capacity of a particular environment can fluctuate over time.
    e. Per capita growth rate is negative when the population is below the carrying capacity.

    ***Textbook Reference:*** *Concept 42.4 Populations Grow Multiplicatively, but the Multiplier Can Change; Technology has increased Earth's carrying capacity for humans*

11. The BIDE model of population growth differs from the BD model in that it
    a. includes immigration but not emigration.
    b. includes emigration but not immigration.
    c. includes both immigration and emigration.
    d. focuses only on births and deaths.
    e. does not account for the possibility of extinction.

    ***Textbook Reference:*** *Concept 42.5 Immigration and Emigration Affect Population Dynamics*

12. Which population would grow most rapidly?
    a. A population exhibiting additive growth
    b. A population at its carrying capacity
    c. A population approaching its carrying capacity
    d. A population well below its carrying capacity and exhibiting multiplicative growth
    e. All of the populations would grow at the same rate.

    ***Textbook Reference:*** *Concept 42.4 Populations Grow Multiplicatively, but the Multiplier Can Change; Density dependence prevents populations from growing indefinitely*

13. Which characteristic of a subpopulation is associated with longer persistence?
    a. Small size
    b. Isolation
    c. High numbers of immigrants
    d. High numbers of emigrants
    e. Few surrounding subpopulations

    ***Textbook Reference:*** *Concept 42.5 Immigration and Emigration Affect Population Dynamics*

14. A positive per capita growth rate
    a. occurs when the per capita birth rate is greater than the per capita death rate.
    b. is symbolized by $N$.
    c. occurs when the per capita birth rate is less than the per capita death rate.
    d. is expected in populations that do not persist over time.
    e. occurs when the per capita birth rate equals the per capita death rate.

    ***Textbook Reference:*** *Concept 42.2 Births Increase and Deaths Decrease Population Size*

15. Which statement concerning the management and conservation of populations is *false*?
    a. For black-legged ticks, a successful blood meal for larvae is the prime determinant of subsequent nymph abundance.
    b. Availability of larval food plants is key to maintaining populations of Edith's checkerspot butterfly.
    c. Successful black rockfish populations rely on the protection of some large females.
    d. Conservation planners often use population density to estimate the carrying capacity of a particular habitat patch.
    e. The most effective way to reduce disease risk from black-legged ticks is to control deer populations.

    ***Textbook Reference:*** *Concept 42.6 Ecology Provides Tools for Conserving and Managing Populations; Knowledge of metapopulation dynamics helps us conserve species*

## Answers

### *Key Concept Review*

1. It is usually impossible to count all of the individuals in a population, so researchers typically measure population density and then multiply it by the area occupied by the population to calculate total population size. In this example, 150 voles per hectare × 4 hectares = 600 voles for the total population size.
2. A species' geographic range is the region where it is found. For timber rattlesnakes, the range is the eastern half of the United States, north to southern Maine, south to northern Florida, and west to central Texas. Within its geographic range, a species may be restricted to a particular type of environment, which is called its habitat. The habitat of timber rattlesnakes is forested areas with rocky ledges and high rodent densities. Sometimes suitable habitat occurs in patches surrounded by unsuitable habitat. Roads and agricultural areas may fragment the forested habitat of timber rattlesnakes into patches (or "islands") of suitable habitat.
3. If the total population size is described as declining, then the per capita death rate must exceed the per capita birth rate of the cohort under study.
4. It is unusual for population densities to remain unchanged over time because this situation would occur only when the number of births exactly equals the number of deaths.
5. Higher survivorship and higher fecundity would typically result in a higher per capita growth rate.
6. Species of small mammals that invest heavily in reproduction often do so at the expense of survivorship, so we would predict short life expectancies in these species. The negative relationship between reproduction and survival is an example of a life-history tradeoff.

7. As population density increases and resources become scarcer, birth rates decline and death rates increase. In other words, *r* decreases as the population becomes more crowded. Thus, the per capita growth rate is said to be density-dependent.
8. Disease and starvation would be considered density-dependent factors because they have a greater impact as conditions become more crowded (i.e., as population density increases). Weather is not a density-dependent factor.
9. Rescue occurs if a subpopulation that has undergone a large decline is saved from extinction by the immigration of individuals from other subpopulations. As one would predict, barriers to immigration reduce the potential for rescue.
10. Metapopulation B will likely go extinct sooner because the time to extinction is shorter for the metapopulation with fewer and smaller subpopulations.
11. Dispersal (habitat) corridors are relatively thin strips of habitat of a particular type that connect larger patches of the same type of habitat. Their importance in permitting individuals to disperse from one patch to another was demonstrated in experiments described in Figure 42.11 of the textbook.
12.

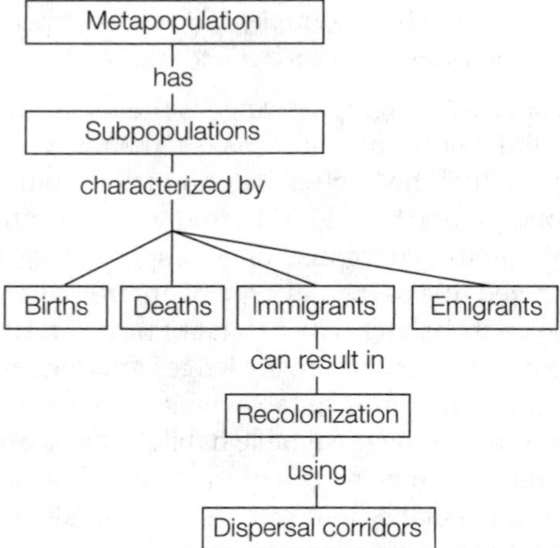


### Test Yourself

1. **d.** For organisms living in water, population density is the number of individuals per unit volume.
2. **c.** Per capita means per individual.
3. **d.** A salamander is a terrestrial animal with limited dispersal abilities, so it would need a continuous corridor of habitat. Animals that fly can more readily reach patches separated by unsuitable habitat.
4. **a.** A decreasing doubling time indicates an increasing per capita growth rate. This was the case for the human population from the 1500s to the 1900s (see Figure 42.9A).
5. **a.** When resources are available at normal levels but environmental temperatures drop to stressful levels, a crow would likely allocate more resources to body maintenance.
6. **a.** A metapopulation has many subpopulations linked by dispersal.
7. **c.** The curve showing the relationship between population size and time is steepest at 3, so the growth rate would be greatest at that point.
8. **e.** The growth rate at 5 would be zero. This is the point called the environmental carrying capacity at which the birth and death rates are equal.
9. **d.** Life histories vary across species and within species, and can be either qualitative or quantitative descriptions.
10. **e.** Per capita growth rate is positive when the population is below the carrying capacity.
11. **c.** The BIDE model of population growth differs from the BD model by including immigration and emigration. In the BIDE model, immigration can replace an extinct population.
12. **d.** A population well below its carrying capacity and exhibiting multiplicative growth would grow the most rapidly.
13. **c.** A subpopulation with a large number of immigrants will have longer persistence than those that are isolated, small in size, or have many emigrants. Also, the presence of many surrounding subpopulations is associated with longer persistence.
14. **a.** A positive per capita growth rate occurs when the per capita birth rate is greater than the per capita death rate, and this would be expected to characterize populations that persist over time.
15. **e.** Scientists have found that the most effective way to reduce disease risk from black-legged ticks is to control small mammal populations (which are larval hosts) rather than deer populations (which are adult hosts).

# Ecological and Evolutionary Consequences of Interactions within and among Species

## The Big Picture

- An individual's fitness is affected by both intraspecific interactions and interspecific interactions. The five categories of interspecific interactions are competition, consumer–resource, mutualism, commensalism, and amensalism. Interspecific interactions affect per capita growth rates and species distributions. Additionally, whereas resource partitioning and coexistence can result from interspecific competition, evolutionary arms races can result from consumer–resource interactions.
- Consumer–resource interactions include predation, herbivory, and parasitism. Energy passes from primary producers (also called autotrophs or photosynthesizers) to heterotrophs, beginning with primary consumers (herbivores), followed by secondary consumers, which feed on herbivores, and so on. Decomposers feed on the wastes and dead bodies of other organisms, and omnivores feed at more than one trophic level. Food webs are diagrams of trophic interactions. When species are added to or removed from communities, effects extend across trophic levels. These trophic cascades illustrate the complex web of interspecific interactions that characterizes ecological communities.

## Study Strategies

- This chapter describes many different categories of interspecific interactions and introduces terms relating to these categories that may be new to you. To learn the basic types of interaction, study the chart presented in Figure 43.1A. Then create your own charts or concept maps to organize the material in more detail and to master the terminology. Focus on how each type of interspecific interaction influences fitness and per capita growth rate. Study the word equation concerning interspecific competition (p. 886) and develop similar word equations for the other interspecific interactions.
- Review the major trophic levels described in Table 43.1. Understanding trophic levels is necessary to understanding food webs.
- Go to **LaunchPad** (or use the Web addresses listed) to review the following additional resources:

  Animated Tutorial 43.1 Species Interactions and Population Dynamics (PoL2e.com/at43.1)

  Animated Tutorial 43.2 An Ant–Plant Mutualism (PoL2e.com/at43.2)

  Activity 43.1 Ecological Interactions (PoL2e.com/ac43.1)

  Activity 43.2 The Major Trophic Levels (PoL2e.com/ac43.2)

  Analyze the Data in textbook Figure 43.8

  Apply the Concept in textbook Sections 43.2 and 43.4

## Key Concept Review

### 43.1 Interactions between Species May Increase, Decrease, or Have No Effect on Fitness

Interspecific interactions are classified by their effects on fitness

The effects of many interactions are contingent on the environment

Interactions between individuals of different species are called interspecific interactions. Such interactions affect the survival and reproduction of the individuals involved and therefore their fitness, which in turn contributes to total population growth rate. Interspecific interactions influence population densities, species distributions, and lead to evolutionary change in one or both species.

Interactions between species in a community fall into five general categories that reflect whether the interaction's affect on fitness is positive (+), negative (–), or neutral (0) (see Figure 43.1A). This system yields five broad categories of interspecific interactions: competition (–/–), consumer–resource (+/–), mutualism (+/+), commensalism (+/0), and amensalism (–/0). Interspecific competition occurs when two or more species use the same resource, such as food, shelter, water, space, and light (see Figure 43.1B). In consumer–resource interactions, the consumer benefits and the consumed organism (the resource) loses (see Figure 43.1C). These interactions can be further classified as (1) predation; (2) herbivory; and (3) parasitism. Mutualism is a type of interaction between species that benefits both (see Figure 43.1D). Examples include leaf-cutter ants and their fungi, plants and the animals that pollinate them or disperse their seeds, and humans and their beneficial gut bacteria. Commensalism is an interaction that benefits one participant while leaving the other unaffected. The interactions between cowbirds and ungulates are commensal because the cowbirds benefit from insects flushed from the grass as cows and bison walk, but the ungulates are unaffected by cowbird presence. Amensalism is harmful to one participant while leaving the other unaffected. These interactions are often accidental, such as when members of a herd of elephants crush insects and plants as they walk. These five types of interactions are not always easy to distinguish, and their outcomes depend on ecological circumstances (see Figure 43.2).

**Question 1.** Two of the main characters in the animated film *Finding Nemo* are anemonefish (also known as clown fish), which live inside of sea anemones and are unaffected by their stings. How is this relationship an example of the ways in which interactions between species are not always easy to classify?

**Question 2.** You have discovered two individuals of different species living in close association. Design an experiment to determine if the relationship is mutualistic.

## 43.2 Interactions within and among Species Affect Population Dynamics and Species Distributions

Interspecific interactions can modify per capita growth rates

Interspecific interactions affect population dynamics and can lead to extinction

Interspecific interactions can affect species distributions

Rarity advantage promotes species coexistence

Recall from Chapter 42 that per capita growth rates typically decrease with increasing population density; this reflects intraspecific interactions, and likely *intraspecific* competition for resources. The word equation on page 874 of Chapter 42 can be extended to take into account *interspecific* competition: per capita growth rate ($r$) of species A = {maximum possible $r$ for species A in uncrowded conditions – an amount that is a function of A's own population density} – {an amount that is a function of the population density of competing species B}. Interspecific interactions other than competition can be similarly described, with the effects of the other species either subtracted or added, depending on the nature of the interaction. For example, in a mutualistic interaction, the effect of each species would be added to the equation for the other because both species benefit.

Interspecific interactions have the following general consequences: (1) the per capita growth rate of each species is modified by the presence of the other species; (2) the average population densities of each species differ in the presence and absence of the other species; and (3) in certain interactions, such as consumer–resource, extinction of one or both of the interacting species can occur. These consequences are illustrated in Figure 43.3 for competing species of *Paramecium*. Interspecific interactions also influence the distributions of species. For example, competitive interactions can restrict distributions (see Figure 43.4). Mathematical models of interspecific competition reveal that two competing species can coexist when intraspecific competition is stronger than interspecific competition. When species A is at low density and competing species B is at high density, species A gains the rarity advantage. This advantage prevents species A from decreasing to zero, and coexistence with species B results. Interspecific competition may be weaker than intraspecific competition due to resource partitioning, which occurs when competing species differ somewhat in their resource use (see Figure 43.5).

**Question 3.** How might the rarity advantage apply to a consumer–resource interaction, such as that between predators and their prey?

**Question 4.** Two species, A and B, have a commensal relationship in which species A benefits and species B is unaffected. Develop a word equation for species A that describes the effect of species B on its growth rate. As an example, see the word equation for interspecific competition on page 886 of the chapter. Do you need a reciprocal term in the word equation for species B in the commensal relationship?

## 43.3 Species Are Embedded in Complex Interaction Webs

Consumer–resource interactions form the core of interaction webs

Losses or additions of species can cascade through communities

The cascading effects of ecological interactions have implications for conservation

Webs of interspecific interactions are organized around consumer–resource interactions, also called feeding or trophic interactions. The organisms in a community use diverse sources of energy, and are grouped into trophic levels according to the number of steps through which energy passes to reach them. Energy passes, in sequence, from autotrophic primary producers (photosynthesizers) to heterotrophs, beginning with primary consumers (herbivores), followed by secondary consumers (which feed on herbivores), and so on. Decomposers feed on the dead bodies and waste products of other organisms. Omnivores obtain their food from more than one trophic level (see Table 43.1). A food web is a diagram depicting the linkages among interacting species, with these linkages organized vertically by trophic level (see Figure 43.6).

Recall that interspecific interactions influence per capita growth rates and population densities of the species involved. The elaborate web of interactions within a community means that change in one species—for example, the arrival of a new species or an increase or decrease in the population density of a resident species—can affect the entire community. In a trophic cascade, changes at one trophic level cause changes at other trophic levels (see Figure 43.7). Introduced species that reproduce rapidly and spread widely are described as invasive. Invasive species usually affect native species negatively by outcompeting or eating them. Some introductions are accidental while others are deliberate (for example, species are sometimes introduced to control pests). Invasive species alter interactions among native species. Some invasive species alter ecological interactions by causing extinctions. Because species are parts of a web of interactions; impacts on one species can influence interspecific interactions throughout a community.

**Question 5.** After the European rabbit (an herbivore) was introduced into Australia in 1859, it quickly became a devastating invasive species, causing the extinction or decline of many native plants, major damage to crops, and soil erosion. Rabbits also had a negative impact on some of Australia's native mammals, such as bettongs, because the rabbits competed with them for food. Is it likely that the impacts of the rabbits would be confined to native herbivores? To control the exponential growth of the rabbit population (which had grown to an estimated 600 million individuals from an original 24), biologists infected rabbits with the myxoma virus in 1950. It is estimated that in the first years after introduction of the virus, as many as 99 percent of infected rabbits died. Over a period of years, however, biologists observed an increase both in the average life span of infected rabbits and in the percentage of rabbits surviving infection. How would you explain these later changes, and what do they suggest about the effectiveness of this control measure?

**Question 6.** The Great Lakes of North America have been subjected to many waves of introduced species as a result of canal construction in the eighteenth and nineteenth centuries. This led to colonization by species from eastern North America, such as the sea lamprey. Subsequently, increases in international shipping traffic to the Great Lakes introduced species from other continents via discharge of ballast water, including the zebra mussel from Europe. Other species, such as coho salmon, a species native to the Pacific coast of North America, were deliberately introduced to enhance recreational fishing. Most recently, species of carp native to Asia have escaped from fish farms and now threaten to invade the Great Lakes via canals in the Chicago area that connect rivers of the Mississippi system to Lake Michigan. Draw a diagram indicating these four invasive pathways to the Great Lakes, specifying whether the species introductions were accidental or intentional. Since there are different sources and pathways for these invasive species, and many invasive species now are well established in the Great Lakes, can you envision any way that the negative impacts of such species can be minimized without altering current Great Lakes ecosystems or the human infrastructure of canals, shipping pathways, or fish farms?

### 43.4 Interactions within and among Species Can Result in Evolution

Intraspecific competition can increase the carrying capacity

Interspecific competition can lead to resource partitioning and coexistence

Consumer–resource interactions can lead to an evolutionary arms race

Mutualisms involve conflict of interest

Intraspecific competition can increase the carrying capacity of a population. This occurs when competition among individuals in the population results in selection for traits that allow individuals to be more efficient in their use of resources. Interspecific competition can lead to resource partitioning, which in turn can allow competing species to coexist. Resource partitioning is thus an evolutionary response to interspecific competition (see Figure 43.9).

Consumer–resource interactions are those involving predators and their prey, parasites and their hosts, and herbivores and the plants on which they feed. In each of these three interactions, the species have opposing interests, and this can lead to an evolutionary arms race. For example, prey species continually evolve better defenses and predators continually evolve better ways to find, capture, and consume their prey. In an evolutionary arms race, neither prey nor predator has any lasting advantage. In mutualisms, both species benefit. When considering mutualisms, it is important to recognize that species do not display traits that have evolved solely to help another species. Instead, species A displays traits that have evolved because they benefited individuals of species A; these traits may happen to benefit individuals of species B as well. For example, pollinators visit flowers to get food and happen to pollinate the flowers when doing so. Mutualisms involve the exchange of resources and services. The fitness effects of mutualistic relationships can vary with environmental conditions. For example, mycorrhizal fungi are beneficial to plants in nutrient-poor soils, but they are costly to plants in nutrient-rich soils.

**Question 7.** Many bats prey on moths. Foraging bats use echolocation to detect moths; the bats emit calls and then listen to the echoes of calls returning from objects in the environment. Moths, in turn, have evolved ears capable of detecting bat echolocation calls; this allows the moths to avoid foraging bats. In response, some bats emit calls outside the hearing range of moths. Evaluate the statement, "Bats have clearly won the evolutionary arms race with moths."

**Question 8.** Interspecific cleaning interactions occur in many natural systems. In a familiar example, large predatory fish come to cleaning stations because their gill cavities are infested with crustacean ectoparasites that interfere with respiration. Small cleaner fish remove the ectoparasites. Evaluate the statement, "The cleaning behavior of the small fish evolved to help the larger fish."

## Test Yourself

1. Species in a(n) _______ interaction increase each other's per capita growth rate.
   a. commensal
   b. competitive

c. mutualistic
d. amensal
e. consumer–resource
***Textbook Reference:*** *Concept 43.2 Interactions within and among Species Affect Population Dynamics and Species Distributions; Interspecific interactions can modify per capita growth rates*

2. Certain birds follow swarms of foraging army ants and prey upon the insects that the ants flush out. Which term best describes the relationship between the birds and the ants?
a. Competition
b. Predation
c. Mutualism
d. Amensalism
e. Parasitism
***Textbook Reference:*** *Concept 43.1 Interactions between Species May Increase, Decrease, or Have No Effect on Fitness; Interspecific interactions are classified by their effects on fitness*

3. Which statement about parasitism is true?
a. It is a form of competition.
b. It differs from predation in that parasites often do not kill their hosts, while predators do kill their prey.
c. It is a form of predation.
d. It does not affect the fitness of the host.
e. It does not affect the per capita growth rate of the host.
***Textbook Reference:*** *Concept 43.1 Interactions between Species May Increase, Decrease, or Have No Effect on Fitness; Interspecific interactions are classified by their effects on fitness*

4. Rarity advantage
a. results in coexistence of competing species.
b. allows predators to drive their prey to extinction.
c. reflects the situation in which per capita growth rates of a species involved in an interspecific interaction may rebound at high density.
d. promotes the extinction of an interacting species.
e. occurs when interspecific competition is stronger than intraspecific competition.
***Textbook Reference:*** *Concept 43.2 Interactions within and among Species Affect Population Dynamics and Species Distributions; Rarity advantage promotes species coexistence*

5. Evolutionary arms races are *not* associated with
a. predator–prey interactions.
b. parasite–host interactions.
c. herbivore–plant interactions.
d. consumer–resource interactions.
e. mutualisms.
***Textbook Reference:*** *Concept 43.4 Interactions within and among Species Can Result in Evolution; Consumer–resource interactions can lead to an evolutionary arms race*

6. Which statement about competition is *false?*
a. Competitors can coexist but they achieve lower densities than either would alone.
b. Presence of a competitor always reduces population growth rate.
c. Resource use is more similar among interspecific competitors than among intraspecific competitors.
d. One competitor can drive another extinct.
e. Interspecific competition can restrict species distributions.
***Textbook Reference:*** *Concept 43.2 Interactions within and among Species Affect Population Dynamics and Species Distributions; Rarity advantage promotes species coexistence*

7. In which relationship is the fitness of both participants negatively affected?
a. Mutualism
b. Amensalism
c. Commensalism
d. Consumer–resource
e. Competition
***Textbook Reference:*** *Concept 43.1 Interactions between Species May Increase, Decrease, or Have No Effect on Fitness; Interspecific interactions are classified by their effects on fitness*

8. Which of the following organisms is *incorrectly* matched with its trophic level?
a. Bison – primary consumer
b. Earthworm – decomposer
c. Grass – primary producer
d. Parasite – secondary or tertiary consumer
e. Fungus – primary producer
***Textbook Reference:*** *Concept 43.3 Species Are Embedded in Complex Interaction Webs; Consumer–resource interactions form the core of interaction webs*

9. If a species of plant that is trampled by an animal eventually evolves sharp spines that prevent trampling, this means that its association with the animal has changed from _______ to _______.
a. amensalism; competition
b. amensalism; commensalism
c. commensalism; competition
d. commensalism; mutualism
e. commensalism; amensalism
***Textbook Reference:*** *Concept 43.1 Interactions between Species May Increase, Decrease, or Have No Effect on Fitness; Interspecific interactions are classified by their effects on fitness*

10. In which pair of relationships is the fitness of one participant unaffected?
a. Mutualism and amensalism
b. Predation and commensalism
c. Competition and parasitism
d. Herbivory and predation

e. Commensalism and amensalism
***Textbook Reference:*** *Concept 43.1 Interactions between Species May Increase, Decrease, or Have No Effect on Fitness; Interspecific interactions are classified by their effects on fitness*

11. The phenomenon whereby two competing species differ somewhat in their use of resources is called
a. resource partitioning.
b. rarity advantage.
c. extinction.
d. a consumer–resource interaction.
e. heterotrophy.
***Textbook Reference:*** *Concept 43.2 Interactions within and among Species Affect Population Dynamics and Species Distributions; Rarity advantage promotes species coexistence*

12.–13. Refer to the food web below to answer the questions that follow.

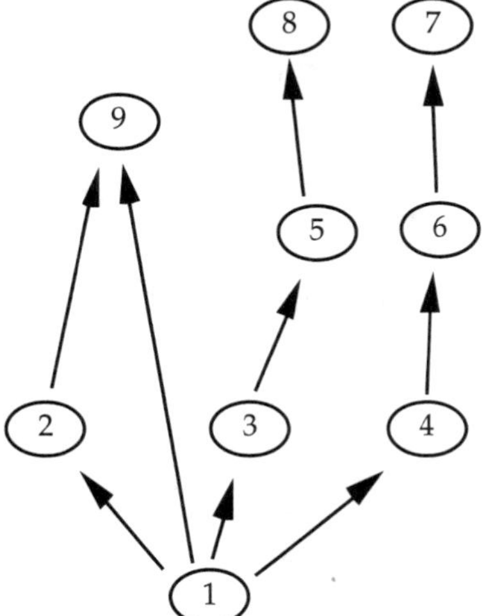


12. Organism 9 is a(n)
a. herbivore.
b. primary carnivore.
c. secondary carnivore.
d. primary producer.
e. omnivore.
***Textbook Reference:*** *Concept 43.3 Species Are Embedded in Complex Interaction Webs; Consumer–resource interactions form the core of interaction webs*

13. The food web has _______ trophic levels.
a. two
b. three
c. four
d. five
e. six
***Textbook Reference:*** *Concept 43.3 Species Are Embedded in Complex Interaction Webs; Consumer–resource interactions form the core of interaction webs*

14. A bird eats the fruit of a plant species. The seeds are not digested and germinate in the bird's excrement at some distance from the parent plant. This is an example of
a. predation.
b. competition.
c. commensalism.
d. mutualism.
e. amensalism.
***Textbook Reference:*** *Concept 43.1 Interactions between Species May Increase, Decrease, or Have No Effect on Fitness; Interspecific interactions are classified by their effects on fitness*

15. Which statement about ecological interactions is *false*?
a. Intraspecific competition has no effect on carrying capacity.
b. Interspecific competition can lead to evolution.
c. In mutualisms, each species acts in its own self-interest.
d. Resource partitioning can lead to the coexistence of two competing species.
e. Intraspecific competition can lead to evolution.
***Textbook Reference:*** *Concept 43.4 Interactions within and among Species Can Result in Evolution; Mutualisms involve conflict of interest*

## Answers

### *Key Concept Review*

1. The anemonefish clearly benefit from the interaction with their host by gaining protection from their enemies. If the sea anemones were unaffected, the relationship would be one of commensalism, but it is not clear that this is so. The anemones may gain nutrients (especially nitrogen) from the feces of the fish, which would make the interaction mutualistic. The fish may occasionally steal prey from the anemones, which would make the interaction competitive. The net effect of the relationship may depend on environmental conditions, such as the availability of nitrogen-rich food to anemones.
2. Both species benefit in a mutualistic relationship. One way to test for a mutualistic relationship is to establish two groups, one in which the two species are associated and another in which the species have been experimentally separated. Each species in the two groups should be monitored over time and the per capita growth rate determined. If both species display higher per capita growth rates when together than when apart, then a mutualistic relationship is suggested. Further experiments would likely be necessary to reveal the details of the relationship (i.e., what critical resource each species is obtaining from the interaction).
3. Once prey become rare, they may be harder to find because high-quality shelters are more available

than when the prey population was at higher densities. Rare prey also may be in better physical shape (and therefore better able to escape or defend themselves) because their food resources are now shared among fewer conspecifics. Also, if finding rare prey becomes too time-consuming and difficult, predators may switch to another prey species. For all of these reasons, the per capita growth rates of the prey may increase rather than decline to zero.

4. The following word equation describes the effect of species B on species A in this commensal relationship:
   per capita growth rate ($r$) of species A = {maximum possible $r$ for species A in uncrowded conditions – an amount that is a function of A's own population density} + {an amount that is a function of the population density of species B}.

   According to the definition of a commensal interaction, species B is unaffected by the interaction, so the equation for species B does not have a reciprocal term (i.e., a term concerning the effect of species A on species B).

5. The introduced rabbits likely affected species at trophic levels other than just the primary consumer (herbivore) level. For example, if rabbits caused decreases in populations of native herbivores, this could affect the primary producers eaten by native herbivores as well as the secondary consumers (carnivores) that fed on the native herbivores. Also, rabbits might compete with native mammals—potentially with omnivores or carnivores—for resources other than food, such as burrows. Introductions of nonnative species can cause a cascade of effects across trophic levels. Over time, the rabbits evolved resistance to the myxoma virus. Thus, the effectiveness of this control measure was limited over time. It did not solve the long term problems resulting from the introduction of rabbits, and it did not restore the affected communities to their original state.

6.

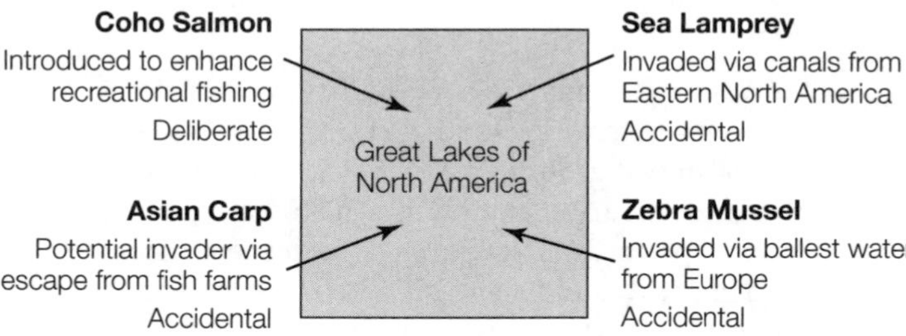


   There are no possible solutions that could return the Great Lakes to the previous uninvaded state, and it is likely that solutions for one invasive species would have impacts on many other species in the Great Lakes ecosystem.

7. The statement is inaccurate, because in an evolutionary arms race neither species gains a lasting victory. Bats will continually evolve better ways to detect moths and moths will continually evolve better ways to avoid detection.

8. The statement is inaccurate, because species do not display traits that have evolved solely to help another species. Instead, individuals of one species display traits that have evolved because they benefited individuals of that species; these traits may happen to benefit individuals of another species as well. Thus, cleaning behavior evolved to help individual small cleaner fish, and such behavior just happens to benefit the large predatory fish.

### *Test Yourself*

1. **c.** Species in a mutualistic relationship will increase each other's per capita growth rates.
2. **a.** Competition, because the ants and birds are competing for the same food supply.
3. **b.** Parasitism is a consumer–resource interaction in which the parasite gains at the host's expense (both in terms of fitness and per capita growth rate). However, parasites often do not kill their hosts; predators do kill their prey.
4. **a.** Rarity advantage describes the rebound of per capita growth rates at low density; this can result in the coexistence of competing species.
5. **e.** Evolutionary arms races are associated with consumer–resource interactions, which include predator–prey interactions, parasite–host interactions, and herbivore–plant interactions.
6. **c.** Individuals of the same species use very similar resources. In contrast, individuals of competing species typically differ to a greater extent in their resource use.
7. **e.** Competition negatively affects the fitness of both participating species.
8. **e.** Fungi are decomposers; they do not perform photosynthesis, so they cannot be primary producers.
9. **a.** Initially the plant is harmed and the animal is unaffected (amensalism). After the evolution of sharp spines, the two species are competitors for space.
10. **e.** In commensalism and amensalism the fitness of one participant is unaffected by the interaction.
11. **a.** The phenomenon whereby two competing species differ somewhat in their use of resources is called resource partitioning.
12. **e.** Because organism 9 eats from both the primary producer level (1) and the herbivore level (2), it is an omnivore.
13. **c.** Four trophic levels are depicted. The levels are primary producer (1), primary consumer (also called herbivore: 2, 3, 4), secondary consumer (those that eat

herbivores, also called carnivore: 5, 6), and tertiary consumer (those that feed on secondary consumers: 7, 8). The omnivore (9) occupies either the herbivore or secondary consumer (carnivore) level, depending on whether it is feeding on a primary producer or on an herbivore.

14. **d.** It is an example of mutualism, because both species have benefited; the bird dispersed and provided fertilizer for the plant's seed, and the plant provided food for the bird.

15. **a.** Intraspecific competition can increase carrying capacity. Both intraspecific and interspecific competition can lead to evolution. Additionally, interspecific competition can lead to resource partitioning and the coexistence of two species. Mutualisms are based on the self-interest of each species.

# 44 Ecological Communities

## The Big Picture

- Community function relies on species diversity, which has two components: species richness and species evenness. Human-induced disturbance and intensive management alter complex community interactions and community function.
- The theory of island biogeography, which relates the area of an island and the extent of isolation of the island to the equilibrium number of species inhabiting the island, has applications in conservation biology. For example, habitat destruction often results in habitat islands which, if small, inevitably lose some of the species that existed in the previously more extensive habitat. Also, if a habitat island is located far from an area of more extensive habitat, it will have lower colonization rates and lower equilibrium species richness. Ecological communities provide goods and services to humans. The ability of a community to provide goods and services often depends on its diversity, which in turn is influenced by habitat complexity and rates of colonization and extinction.

## Study Strategies

- The niche of a species is often confused with its habitat. Habitat refers to the preferred physical environment of a species (e.g., pond versus stream), whereas niche refers to the entire set of physical and biological conditions a species needs in order to survive and reproduce. Niche also includes the role of the species in a community. Thus the habitat of a species is just one of many attributes of its niche.
- The graphical representations of the concepts of island biogeography theory are sometimes hard to follow (see Figure 44.11). Try redrawing the graphs for yourself, paying particular attention to how the axes are labeled.
- This chapter contains descriptions of a number of studies of community structure, including several that relate species richness to other community characteristics. As you read about these studies, focus on the particular question that each study was designed to answer and how the results support the conclusions drawn from the study.
- Go to **LaunchPad** (or use the Web addresses listed) to review the following additional resources:

  Animated Tutorial 44.1 Succession after Glacial Retreat (PoL2e.com/at44.1)

  Animated Tutorial 44.2 Island Biogeography Simulation (PoL2e.com/at44.2)

  Animated Tutorial 44.3 Fragmentation Effects (PoL2e.com/at44.3)

  Activity 44.1 Energy Flow through an Ecological Community (PoL2e.com/ac44.1)

  Activity 44.2 Measures of Species Diversity (PoL2e.com/ac44.2)

  Analyze the Data in textbook Figures 44.15

  Apply the Concept in textbook Section 44.5

## Key Concept Review

### 44.1 Communities Contain Species That Colonize and Persist

A community consists of all the species that live and interact in a particular area. Thus, a community includes all the species that have colonized it minus those that have gone locally extinct, perhaps due to inappropriate environmental conditions, lack of a critical resource, or competition with resident species. Community structure can be characterized by the particular collection of species present (species composition), number of species, and the relative abundance of each species. The species composition of communities changes over time and space.

**Question 1.** A small lake contains vertebrates, such as snapping turtles, painted turtles, bluegill sunfish, and largemouth bass. What additional information is needed to characterize this vertebrate community?

**Question 2.** A landowner in Minnesota, hoping to develop a bass population for fishing, introduces six juvenile largemouth bass of unknown sex, each about 10 cm in length, into a small and shallow pond. Largemouth bass, a predatory species, do not reproduce until reaching about 25 cm in length. The pond already contains herbivorous minnows and tadpoles and carnivorous salamanders. The population of largemouth bass fails to persist in the pond. What factors might have influenced its extinction?

### 44.2 Communities Change over Space and Time

Species composition varies along environmental gradients

Several processes cause communities to change over time

Species composition changes along environmental gradients. When ecologists establish a transect (a straight line used when conducting surveys) along an environmental gradient and sample the species present, they find that certain species drop out and others appear (see Figure 44.2). Thus, species turnover occurs when there is spatial variation in environmental conditions. Habitat structure reflects the horizontal and vertical distribution of objects in the environment. Plants contribute to habitat structure and often determine the animals present in a particular community.

Extinction and colonization, disturbance, and climate change prompt changes in the species composition of communities through time. In a particular community, resident species may go extinct and new species may arrive. Sudden environmental change, whether caused by human activities or natural phenomena such as volcanic eruptions or hurricanes, causes turnover of species. Following a disturbance, species often replace one another in a fairly predictable sequence called succession (see Figure 44.4). The precise sequence reflects species differences in colonizing ability and environmental tolerance. Early arriving species tend to have superior dispersal abilities. Environmental conditions change over time as a result of physical processes and the activities of the early colonists, creating conditions that later-arriving species can better tolerate. When a community is destroyed by disturbance, succession can lead to establishment of a community similar to the original one. Sometimes, however, an ecological transition to a different community occurs. Climate change causes temporal variation in the species composition of communities by influencing the geographic ranges of species (see Figure 44.5).

**Question 3.** White-tailed deer are often associated with wooded areas that contain a mix of trees, shrubs, forbs, grasses, and sedges. What might form the basis of this animal–vegetation association?

**Question 4.** Forensic entomologists can estimate how many days a corpse has been dead by identifying the particular insects present on the body. This estimate is possible because different species of insects are found on corpses in a predictable sequence that reflects changes in physical conditions of the body over time (e.g., decreases in moisture). What ecological process forms the basis for the technique used by forensic entomologists to determine date of death?

### 44.3 Community Structure Affects Community Function

Energy flux is a critical aspect of community function

Community function is affected by species diversity

Community function can be assessed by the amount of energy or matter that moves into and out of the community during a particular period of time. Gross primary productivity (GPP) is the total amount of energy that primary producers capture and convert to chemical energy during a particular time interval. Net primary productivity (NPP) is the energy contained in the tissues that primary producers have made during that time interval; in other words, this is the portion of GPP that becomes available to consumers during the time period. The ecological efficiency of energy transfer from one trophic level to the next is about 10 percent. Three factors account for such low ecological efficiency, and these also are the reasons that most communities have fewer than four trophic levels. First, organisms use most of the energy they accumulate for respiration and other metabolic processes; this energy is ultimately dissipated as heat. Second, consumers do not ingest all of the biomass available to them (for example, herbivores may avoid plants with well-developed chemical defenses). Third, consumers cannot assimilate (digest and absorb) all of the biomass they ingest.

The niche of a species is the set of physical and biological conditions necessary for its persistence. Within a community, each species has a unique niche. Species diversity has two main components, species richness and species evenness, both of which affect community function (see Figure 44.7). Species richness is the number of species in the community. Communities with many species are more diverse than those with fewer species, all else being equal. Species evenness considers the distribution of species' abundances within a community. The more even the distribution, the more diverse the community. Community function can be measured by the outputs of species interactions. Such outputs increase with species diversity. For example, within a particular type of community, NPP is typically greater and more stable over time as species diversity increases. This association between species diversity and community function may reflect sampling (just by chance, communities with many species may be more likely to have some species with a strong influence on community output) or niche complementarity (communities with many species are more likely to have species with complementary niches, with the result that all available resources are used).

**Question 5.** Under what circumstances would a forest with many species have relatively low diversity?

**Question 6.** Growing crops as monocultures is standard practice in modern agriculture. Why are monocultures unstable? What agricultural practices might result in more stable and more productive ecological communities?

### 44.4 Diversity Patterns Provide Clues to What Determines Diversity

Species richness varies with latitude

Species richness varies with the size and isolation of oceanic islands

For many taxa, species diversity is greatest in the tropics and decreases with increasing latitude (see Figure 44.9). Ecologists have suggested several explanations for latitudinal gradients in diversity. One possibility is that organisms in tropical regions have not been subjected to large-scale disturbance, such as the glacial cycles that affected temperate regions, and as a result the tropics have retained more species. Another possibility is that the greater productivity of the tropics (made possible by primary producers thriving in the warm, wet conditions) allows species with narrow, specialized niches to coexist. Greater niche specialization could interact with greater habitat complexity to amplify tropical diversity.

According to the theory of island biogeography, equilibrium species richness (the number of species living in an area) on islands is determined by the rate of arrival of new species (= rate of colonization) and the rate of extinction of species already present. The theory further predicts that the equilibrium number of species should increase with island size and decrease with distance from the mainland (see Figure 44.10). Scientists have confirmed the predictions of the theory both by observation and experiments (see Figure 44.11).

**Question 7.** You are asked to lead a tour highlighting breeding bird species diversity, and you can select among the following three sites: Alaska, Mexico, or Ecuador. Which site would you choose and why?

**Question 8.** How might the theory of island biogeography be applied to conservation efforts of mainland species?

### 44.5 Community Ecology Suggests Strategies for Conserving Community Function

Ecological communities provide humans with goods and services

Ecosystem services have economic value

Island biogeography suggests strategies for conserving community diversity

Trophic cascades suggest the importance of conserving certain species

The relationship of diversity to community function suggests strategies for restoring degraded habitats

Ecological communities provide humans with goods (for example, wood from trees and food from plants and animals) and services (for example, flood control, soil stabilization, and climate regulation) (see Table 44.1). These goods and services are called ecosystem services, because they often result from interactions between communities and the physical environment. The explicit acknowledgement that ecological systems have economic value is a relatively recent phenomenon.

Habitat fragmentation creates habitat islands—isolated patches of suitable habitat surrounded by areas of unsuitable habitat. Fragmentation is predicted to result in the loss of species, because as the total amount of habitat decreases, the average size of a habitat patch decreases, and patches become more isolated from one another. The theory of island biogeography can be applied to habitat patches. Steps that might enhance colonization and reduce extinction in fragmented habitat include making sure that patches are close to one another, providing dispersal corridors among patches, and retaining large patches of the original habitat. Community function can be improved by targeting species with particularly important roles in their community and by restoring the original species diversity of communities.

**Question 9.** List five ecosystem goods or services.

**Question 10.** In the diagram below, the oval shape represents a large unbroken tract of forest. The four circles to the right represent habitat fragments: two large ones (signifying large area) and two smaller ones (signifying smaller area). The two filled-in rectangles represent dispersal corridors. Select two of the four habitat fragments, along with the two dispersal corridors and the unbroken tract, and design an arrangement to achieve the goal of habitat fragments with the highest species richness.

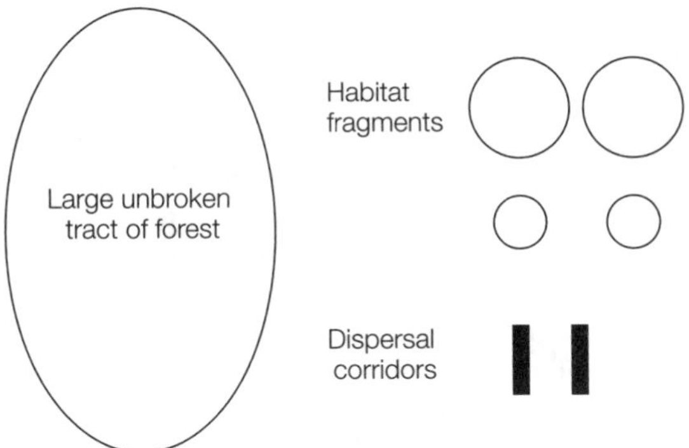


## Test Yourself

1. The rate at which new species arrive on an island
   a. increases with closeness to the mainland.
   b. decreases with closeness to larger islands.
   c. increases as more species occupy the island.
   d. is lower than the extinction rate at the equilibrium species richness.
   e. is greater than zero when all species on the mainland are present on the island.

   ***Textbook Reference:*** *Concept 44.4 Diversity Patterns Provide Clues to What Determines Diversity; Species richness varies with the size and isolation of oceanic islands*

2. Which of the following communities of 24 individuals each is the most diverse?
   a. A community with three species, each of which has eight individuals
   b. A community with four species, each of which has six individuals
   c. A community with four species; one species has 18 individuals and the remaining three species each have two individuals
   d. All three communities are equally diverse.
   e. It is impossible to determine from the information given.

   ***Textbook Reference:*** *Concept 44.3 Community Structure Affects Community Function; Community function is affected by species diversity*

3. Which of the following organisms would be the *least* likely to be one of the first colonists at an oceanic island where disturbance has wiped out all species that were members of the original community?
   a. Bird
   b. Bat
   c. Plant with wind-dispersed seeds
   d. Flying insect
   e. Salamander
   ***Textbook Reference:*** *Concept 44.2 Communities Change over Space and Time; Several processes cause communities to change over time*

4. Which statement about succession is *false*?
   a. Ecological transition describes the phenomenon whereby disturbance wipes out a community and succession leads to the eventual establishment of a community similar to the original one.
   b. Species that are early colonists may either facilitate or inhibit colonization by later-arriving species.
   c. Succession often leads to the establishment of a community that resembles the original one.
   d. Succession is the relatively predictable sequence of species replacements following a disturbance.
   e. Because environmental conditions at a site change over time, the successional sequence often reflects species differences in environmental tolerance.
   ***Textbook Reference:*** *Concept 44.2 Communities Change over Space and Time; Several processes cause communities to change over time*

5. If the net primary productivity of a community is 2,000 kcal/m$^2$/yr, what would be the best estimate of the productivity (in kcal/m$^2$/yr) of the secondary consumers in the community?
   a. 4,000
   b. 2,000
   c. 200
   d. 20
   e. 2
   ***Textbook Reference:*** *Concept 44.3 Community Structure Affects Community Function; Energy flux is a critical aspect of community function*

6. Ecosystems provide goods and services that
   a. typically cost more than artificial solutions.
   b. depend on species diversity.
   c. increase with environmental disturbance.
   d. benefit from habitat fragmentation.
   e. rely on intense management by humans.
   ***Textbook Reference:*** *Concept 44.5 Community Ecology Suggests Strategies for Conserving Community Function; Ecosystem services have economic value*

7. Which phenomenon does *not* cause changes in species composition within communities over time?
   a. Climate change
   b. Local extinction of a species
   c. Disturbance
   d. Colonization
   e. An environmental gradient in heavy metals
   ***Textbook Reference:*** *Concept 44.2 Communities Change over Space and Time; Several processes cause communities to change over time*

8. Experiments on wetland restoration have demonstrated that planting a single species in experimental plots, as opposed to several species, is associated with
   a. greater percentage of ground cover.
   b. more complex vegetation structure.
   c. slower progress toward restoring the community's original function.
   d. faster accumulation of nitrogen in roots.
   e. more rapid development of vegetation cover.
   ***Textbook Reference:*** *Concept 44.5 Community Ecology Suggests Strategies for Conserving Community Function; The relationship of diversity to community function suggests strategies for restoring degraded habitats*

9. Species richness
   a. is lowest in the tropics.
   b. decreases with increasing latitude.
   c. usually decreases when climates are more stable.
   d. typically decreases with greater habitat complexity.
   e. is linked to lower NPP.
   ***Textbook Reference:*** *Concept 44.4 Diversity Patterns Provide Clues to What Determines Diversity; Species richness varies with latitude*

10. Which statement about ecological communities is *false*?
    a. The ecological community of a pond includes all the species living in the pond and the abiotic environment (water, sunlight, etc.) that supports them.
    b. The ecological community of a pond includes all the species living in the pond.
    c. Communities can be characterized by their species composition.
    d. Communities gain and lose species.
    e. Community structure consists of three components: species composition, number of species, and abundance of each species.
    ***Textbook Reference:*** *Concept 44.1 Communities Contain Species That Colonize and Persist*

11. In the figure below, the species numbers of many islands of different sizes are plotted against their distance from the mainland. Data points for islands of similar size have been connected to form the four curves. Which curve corresponds to the group of islands with the smallest size?

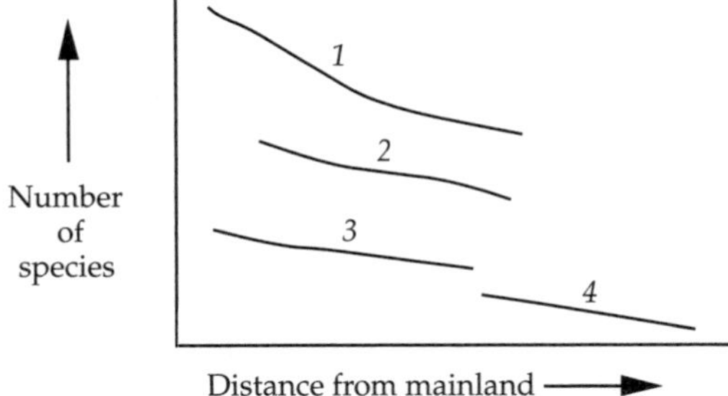


a. 1
b. 2
c. 3
d. 4

***Textbook Reference:*** *Concept 44.4 Diversity Patterns Provide Clues to What Determines Diversity; Species richness varies with the size and isolation of oceanic islands*

12. The theory of island biogeography makes predictions about the effects of species number on the rate of extinction. Refer to the figure below, in which the solid curve shows this relationship for a large island. Which of the dashed curves shows the expected relationship for a small island?

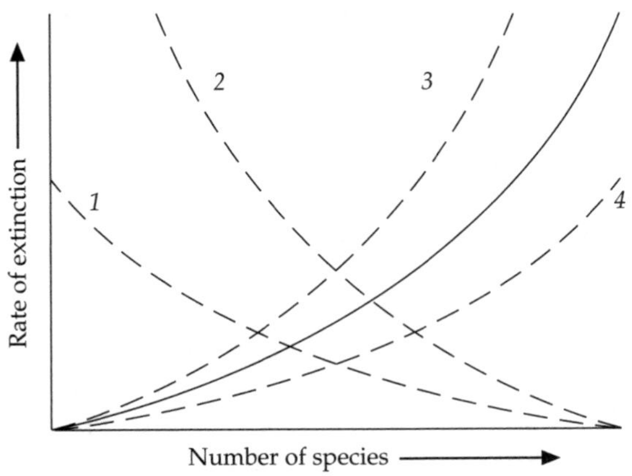


a. 1
b. 2
c. 3
d. 4

***Textbook Reference:*** *Concept 44.4 Diversity Patterns Provide Clues to What Determines Diversity; Species richness varies with the size and isolation of oceanic islands*

13. Which of the following statements about habitat fragmentation is *false*?
a. Small, isolated patches lose species more rapidly than larger, isolated patches.
b. Isolated patches lose species more rapidly than patches of similar size that are near other patches.
c. Habitat fragmentation results in lower species richness in the fragments than in the original habitat.
d. Human-dominated habitat surrounding patches increases the colonization rate of patches.
e. Connecting fragments with dispersal corridors enhances colonization.

***Textbook Reference:*** *Concept 44.5 Community Ecology Suggests Strategies for Conserving Community Function; Island biogeography suggests strategies for conserving community diversity*

14. Which of the following statements about community structure is *false*?
a. Species richness is the number of species in a community.
b. Species evenness reflects the similarity among species in regard to their abundance.
c. NPP typically decreases with increases in species richness.
d. Species richness and species evenness determine species diversity.
e. A species' niche includes its functional role in a community.

***Textbook Reference:*** *Concept 44.3 Community Structure Affects Community Function; Community function is affected by species diversity*

15. Which of the following trophic levels can support the most individuals?
a. Secondary consumers
b. Primary consumers
c. Secondary producers
d. Tertiary consumers
e. Omnivores

***Textbook Reference:*** *Concept 44.3 Community Structure Affects Community Function; Energy flux is a critical aspect of community function*

## Answers

### *Key Concept Review*

1. In order to characterize the vertebrate community of the lake, we would also need to know the total number and the relative abundance of each vertebrate species present.
2. Extinction of the introduced bass population could have occurred for many reasons. For example, the bass might not have been able to tolerate the environmental conditions of the pond. More specifically, because the pond is small and shallow and located in Minnesota, it is subjected to temperature extremes that the fish might have been unable to tolerate. Related aspects of the physical environment, such as low oxygen

levels in warm water during summer, could also have been problematic. Although food resources were initially present in the form of minnows, tadpoles, and salamanders, the prey supply might have decreased to levels that could not support a population of bass. Predatory birds, such as herons, might also have decreased the number of bass. Finally, the introduced population might simply have been too small in size to successfully colonize the pond, and there is no guarantee that both males and females were part of the original introduction.

3. A certain species will be found in the plant communities that provide its preferred food. Animal–plant associations also may occur because plants modify physical conditions, such as temperature and humidity, and thereby make the environment more suitable for a particular animal. Plants also determine habitat structure (the horizontal and vertical distribution of objects in a particular habitat), which influences an animal's ability to avoid detection by predators and to evade them once detected. In the case of white-tailed deer, their association with a habitat that contains this mix of vegetation centers on the species' need for year-round access to high-quality food. Deer also use the vegetation for shelter and concealment.
4. The forensic entomologists are using the process of ecological succession to determine date of death. Ecological succession is the phenomenon whereby species often replace one another in a predictable sequence following a disturbance.
5. A forest with many species could have relatively low diversity if it contained one or a few species that were especially abundant and the other species in the forest community had only a small number of individuals.
6. Community outputs are generally greater and more stable over time as species diversity increases. Monocultures are unstable because they are subject to attack by insect pests and pathogens that destroy or damage crops. More diverse systems in which two or more crops are grown on the same plot are less subject to pest outbreaks, and NPP would be expected to be greater in these systems than in single crop systems.
7. The best site for breeding bird diversity would likely be Ecuador because it is closest to the equator. Species richness tends to increase toward the equator.
8. Human activities have fragmented mainland areas, effectively establishing islands of suitable habitat surrounded by unsuitable habitat (for example, agricultural land). The theory of island biogeography predicts that we will lose species from habitat patches as the size of patches declines and as patches become more isolated from one another, making colonization more difficult.
9. Ecosystem goods and services may include: (1) food supplied by plants and animals; (2) wood from trees; (3) maintenance of fertile soil through decomposition; (4) waste treatment through decomposition; and (5) flood control. (See Table 44.1 for a more complete list.)
10. The general arrangement that would achieve the highest species richness in the habitat fragments is shown below.

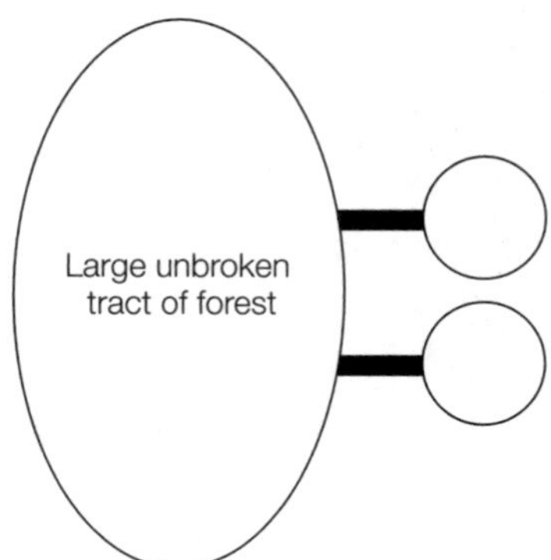


The essential elements are: (1) selecting the two largest habitat fragments from the four provided, because larger areas support larger populations, which are less prone to extinction; (2) arranging the two fragments and the large tract of unbroken forest as close together as possible to allow for the greatest dispersal among them, which increases colonization rates; and (3) placing the dispersal corridors between each fragment and the large unbroken tract of forest, because this too would maximize colonization rates.

### *Test Yourself*

1. **a.** The rate at which new species arrive on an island increases with closeness to the mainland and declines as the island increases in number of species. The equilibrium species richness occurs when the colonization rate equals the extinction rate. The rate at which new species arrive on an island also increases with closeness to larger islands.
2. **b.** The community with four species, each of which has six individuals, is the most diverse because it has the largest number of equally abundant species.
3. **e.** Following environmental disturbance, the first species to arrive on a site are those with superior dispersal abilities. Animals that can fly and plants with wind-dispersed seeds are all good dispersers. Salamanders cannot cross salt water and have limited dispersal abilities, so they would be unlikely to colonize an oceanic island.
4. **a.** Ecological transition occurs when a distinctly different community emerges following disturbance, rather than a community similar to the original one. For example, following intensive grazing by cattle, certain grasslands in the United States/Mexico Borderlands changed to shrublands.
5. **d.** Net primary productivity is the rate at which energy is incorporated into the bodies of primary producers through growth and reproduction. As a rule,

only about 10 percent of the energy at one trophic level is transferred to the next. Because the secondary consumers are two trophic levels above the primary producers, their productivity would be expected to be only 1 percent of net primary productivity, which in the example equals 20.

6. **b.** Ecosystem goods and services depend on species diversity and are harmed by disturbance, intensive management, and habitat fragmentation. Ecosystems goods and services often are less expensive than artificial solutions.

7. **e.** Temporal variation in species composition can result from climate change, disturbance, extinction, and colonization. In contrast, spatial variation in species composition results from environmental gradients.

8. **c.** Planting a single species is associated with slower progress toward restoring the wetland's original function. Planting a mixture of species resulted in more rapid development of vegetative cover, more extensive ground cover, more complex vegetative structure, greater accumulation of nitrogen in the roots of plants, and more rapid progress in restoring the wetland (see Figure 44.15).

9. **b.** Species richness is greatest in the tropics and decreases with increasing latitude. Species richness may be enhanced by greater NPP, climatic stability, and habitat complexity.

10. **a.** An ecological community comprises all of the species living in a defined area but excludes its nonliving components.

11. **d.** Curve *4*. Each curve corresponds to a group of similar-sized islands whose species number is plotted against their distance from the mainland. For each curve, species number decreases with distance from the mainland, and the curve that is lowest relative to the vertical axis would be the group of smallest islands.

12. **c.** Curve *3*. With less space available, populations of the different species would be smaller. This would subject them to higher extinction rates than would be expected on larger islands.

13. **d.** Human-dominated habitat surrounding patches acts as a barrier to dispersal, so it decreases the colonization rate of patches.

14. **c.** NPP typically increases with increases in species richness.

15. **b.** The level of primary consumer can support the most individuals because this is the lowest trophic level listed; with each move to a higher trophic level, fewer and fewer individuals can be supported. This occurs because only about 10 percent of energy is transferred from one trophic level to the next.

# 45 The Global Ecosystem

## The Big Picture

- Net primary productivity (NPP) of terrestrial ecosystems varies with temperature and precipitation. NPP of aquatic systems varies with availability of light and nutrients. NPP is a measure of ecosystem function.
- Earth is an open system with respect to energy and a closed system with respect to matter. Photosynthetic organisms capture energy from the sun. Energy flows from the primary producers to other organisms and ultimately is dissipated as heat. In contrast, chemical elements, which are present in fixed amounts, come from within the system of Earth itself and cycle through ecosystems. Knowledge of how materials move through biogeochemical cycles is crucial for understanding and predicting the effects of human alterations of these cycles.
- Human activities have increased greenhouse gases in Earth's atmosphere and increases in global temperatures have followed. Climate change is altering the distributions, abundances, and interspecific interactions of species. International cooperation is needed to address global climate change.

## Study Strategies

- Eutrophication—the "enrichment" of a body of water with nutrients—often results in a "dead zone" because extra nutrients stimulate rapid growth of phytoplankton populations to a point at which consumers cannot eat them all. Respiration by the phytoplankton and by the decomposers that process their dead bodies depletes the water of oxygen, making it difficult for other aquatic organisms to survive. The idea that excess nutrients can lead to death of aquatic organisms can be confusing, so remember that the link between excess nutrients and death is oxygen availability.
- When studying the material on biogeochemical cycles, focus on the following basic questions: Where are the abiotic reserves of an element located? How does the element leave the reserve and enter living organisms? Why do living organisms need the element? How does the element return to its abiotic reserve?
- Go to **LaunchPad** (or use the Web addresses listed) to review the following additional resources:

  Animated Tutorial 45.1 The Global Water Cycle (PoL2e.com/at45.1)

  Animated Tutorial 45.2 The Global Nitrogen Cycle (PoL2e.com/at45.2)

  Animated Tutorial 45.3 The Global Carbon Cycle (PoL2e.com/at45.3)

  Animated Tutorial 45.4 Earth's Radiation Budget (PoL2e.com/at45.4)

  Activity 45.1 The Phosphorus and Sulfur Cycles (PoL2e.com/ac45.1)

  Activity 45.2 The Benefits of Cooperation (PoL2e.com/ac45.2)

  Activity 45.3 Concept Matching (PoL2e.com/ac45.3)

  Analyze the Data in textbook Figure 45.8

  Apply the Concept in textbook Sections 45.3 and 45.5

## Key Concept Review

### 45.1 Climate and Nutrients Affect Ecosystem Function

NPP is a measure of ecosystem function

NPP varies predictably with temperature, precipitation, and nutrients

An ecosystem is an ecological community plus the abiotic environment with which the organisms interact. Large-scale movements of organisms (e.g., migrations) and materials (e.g., by means of flowing water or circulating air) link ecosystems. Energy flow in ecosystems typically originates with photosynthesis. As discussed in Chapter 44, net primary productivity (NPP) is the portion of assimilated energy left over after the energy used by primary producers for their own metabolism is subtracted; in other words, this is the portion of gross primary productivity (GPP) that becomes available to consumers. All of the other organisms in an ecosystem derive their energy directly or indirectly from net primary productivity. NPP is essentially a measure of the influx of energy and materials into a community, and can serve as a measure of ecosystem function.

NPP varies with type of ecosystem (see Figure 45.1). Within terrestrial ecosystems, tropical forests are the most productive area per unit of area, and tundra and deserts are the least productive. NPP varies with latitude, with productive areas in the tropics and less productive areas at high latitudes and in the dry regions around 30°N and S (see Figure 45.2). Terrestrial NPP generally increases with temperature, and increases to a certain level with precipitation and then declines (see Figure 45.3). Within aquatic ecosystems, swamps, marshes, and coral reefs are the most productive and open ocean is the least productive. Nutrient availability is the main determinant of aquatic NPP. Nutrients are most abundant where land runoff brings nutrients into shallow coastal waters and where upwellings bring nutrients from the sediments.

**Question 1.** On the graph in part A, draw the general relationship between NPP and mean annual temperature for a terrestrial ecosystem. On the graph in part B, draw the general relationship between NPP and mean annual precipitation for a terrestrial ecosystem.

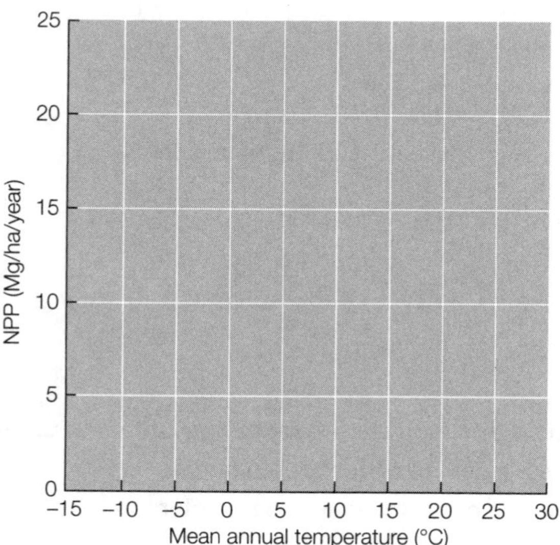


(B) Precipitation

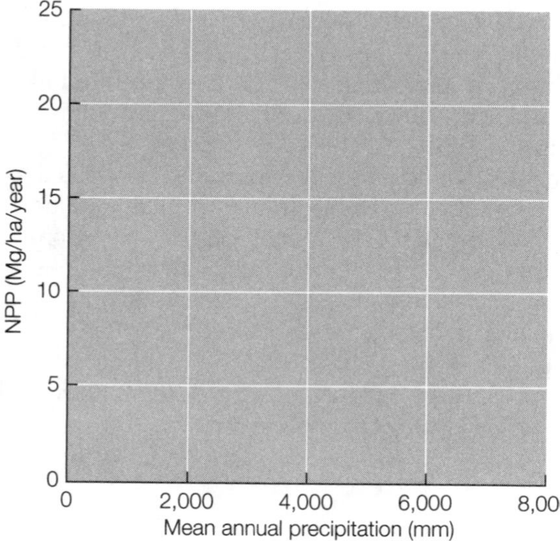


**Question 2.** Explain why marine NPP is highest in coastal areas.

## 45.2 Biological, Geological, and Chemical Processes Move Materials through Ecosystems

The forms and locations of elements determine their accessibility to organisms

Movement of matter is driven by biogeochemical processes

Earth is essentially an open system with respect to radiant energy from the sun, but it is a closed system with respect to matter. Nevertheless, the distribution of matter is not static because energy from the sun and heat from Earth's interior drive the processes that move materials around the planet.

Elements exist in different chemical forms and occur in different locations; some of these forms and locations are accessible to life while others are not (see Figure 45.5). These different forms and locations can be described as compartments, and biological, geological, and chemical processes cycle matter among compartments. The biosphere is the part of Earth where life exists; it is where atmosphere, land, and water come into contact. The pattern of movements of elements through organisms and compartments of the physical environment are called biogeochemical cycles. An accumulation of an element in a compartment is a pool; movement of elements into and out of a compartment is called flux.

**Question 3.** By what processes do inorganic compounds in rocks become accessible to organisms?

**Question 4.** Through what processes do organic and inorganic compounds enter compartments where they are inaccessible to organisms?

## 45.3 Certain Biogeochemical Cycles Are Especially Critical for Ecosystems

Water transports materials among compartments

Within-ecosystem recycling dominates the global nitrogen cycle

Movement of carbon is linked to energy flow through ecosystems

Biogeochemical cycles are not independent

Most of Earth's water (96.5%) is in the oceans, with the remainder divided among other compartments, such as ice and snow, groundwater, the atmosphere, and fresh water. Water moves between these compartments through gravity-driven flow when in its liquid state (e.g., when precipitation falls from the atmosphere to Earth's surface) or with changes in its physical state (e.g., when changing from liquid to gas during evaporation and transpiration). The sun powers the water cycle by causing evaporation, most of it from ocean surfaces (see Figure 45.6). Human activities that reduce vegetative cover disrupt the water cycle by increasing runoff and drying out local ecosystems.

Nitrogen is often in limited supply in biological communities. Although nitrogen gas ($N_2$) makes up 78 percent of Earth's atmosphere, most organisms cannot use this form of nitrogen. Some microorganisms, however, can convert nitrogen gas to ammonium ($NH_4^+$) in a process called nitrogen fixation. Other microorganisms convert ammonium into nitrate ($NO_3^-$). Primary producers can take up ammonium and nitrate, thereby making nitrogen available to consumers. Microorganisms also perform denitrification, which returns nitrogen gas to the atmosphere (see Figure 45.7). Human activities have affected the nitrogen cycle. Industrial and agricultural nitrogen fixation now rival natural terrestrial nitrogen fixation. When nitrogen applied as fertilizer to cropland enters bodies of water, it can lead to eutrophication, a process in which high nutrient levels promote excessive algal growth, depleting oxygen in the water and creating a "dead zone" (see Figure 45.9). Human-caused perturbations of the nitrogen cycle have other adverse effects, including increased air pollution and acid rain.

The largest pools of carbon occur in fossil fuels and carbonate rocks; other pools include biomass and carbon dioxide ($CO_2$) and methane ($CH_4$) in the atmosphere (see Figure 45.10). Energy flow and the flux of carbon through biological communities are linked. Photosynthesis moves carbon from inorganic compartments in the atmosphere and water into the organic compartment, and respiration reverses this. When carbon moves from the atmosphere into surface waters, some of it is used by producers, some precipitates as carbonate (which forms rocks such as limestone), and some falls to the sea floor as detritus (which may eventually form fossil fuels). Burning fossil fuels increases the atmospheric pool of carbon dioxide. Livestock production increases the atmospheric pool of methane. Biogeochemical cycles interact, and perturbations in one cycle can affect others in ways that are difficult to predict.

**Question 5.** Analyze the changes to the water cycle in an ecosystem that has undergone deforestation.

**Question 6.** In what ways are farming methods that require the large-scale use of manufactured fertilizers contributing to eutrophication and other adverse environmental effects?

## 45.4 Biogeochemical Cycles Affect Global Climate

Earth's surface is warm because of the atmosphere

Recent increases in greenhouse gases are warming Earth's surface

Human activities are contributing to changes in Earth's radiation budget

About 49 percent of incoming solar energy is absorbed by Earth's surface. The rest is absorbed by the atmosphere or reflected back into space (see the left side of Figure 45.11). Earth's surface re-radiates some of the absorbed solar radiation in infrared wavelengths, and some of this is absorbed by molecules of gas in the atmosphere and radiated back to Earth's surface.

The greenhouse effect describes the warming of Earth as a result of its atmosphere trapping heat and then radiating it back to the surface. Greenhouse gases (such as carbon dioxide, water vapor, methane, and nitrous oxide) absorb strongly in the infrared wavelengths. Thus, these gases effectively trap outgoing infrared radiation emitted by Earth and radiate it back to the surface, raising surface temperatures (see the right side of Figure 45.11).

Greenhouse gases are increasing in the atmosphere (see Figure 45.12), and global temperatures also are increasing (see Figure 45.13). Higher global temperatures will affect climate, with wet regions expected to get wetter and dry regions drier. Storm intensity is predicted to increase. Human activities, including the burning of fossil fuels, deforestation, and livestock production, are adding greenhouse gases to the atmosphere.

**Question 7.** If you were asked to develop a computer model to predict climate change, what factors would you include in your model?

**Question 8.** Analyze how global warming could influence the spread of human disease.

## 45.5 Rapid Climate Change Affects Species and Communities

Rapid climate change presents ecological challenges

Changes in seasonal timing can disrupt interspecific interactions

Climate change can alter community composition by several mechanisms

Extreme weather events also have an impact

The fast pace at which Earth's climate is changing is challenging the ability of species to evolve in response. Climate change is altering the timing of some environmental cues but not others, so the temporal relationships among cues are shifting. Such mismatches in timing can disrupt interactions among community members who use different cues.

Climate change can alter the species composition of communities. For example, local populations unable to respond to changing environmental conditions may go extinct. Alternatively, changing conditions may favor one species over another and thereby alter the relative abundances of species. Climate change also can lead to shifts in the geographic ranges of species, which may lead to the formation of new communities. Finally, the geographic ranges of species and the composition of communities can be affected by the increased frequency of the extreme weather events that are predicted to accompany global warming.

**Question 9.** Many mammals exhibit seasonal reproductive rhythms that are largely determined by day length. Day length is an environmental cue that is independent of climate. Does this mean that mammals with seasonal reproduction will not be affected by climate change? Explain your answer.

**Question 10.** Speculate about the potential influence of climate change on bird migration and the impact this will have on communities.

### 45.6 Ecological Challenges Can Be Addressed through Science and International Cooperation

Climate change has occurred in Earth's past. However, current climate change is unique because it has been precipitated by the diverse activities of a single species, *Homo sapiens*. Through science and cooperative interactions, humans are uniquely equipped to devise solutions and meet the challenges of developing sustainable economies and curbing the multiplicative growth of our population.

**Question 11.** Evaluate the following statement: "The current period of climate change is unprecedented."

**Question 12.** Assess whether or not you think humans are capable of addressing the environmental problems they have created.

## Test Yourself

1. Areas of upwelling in the ocean have high primary productivity because such areas
   a. bring nutrients from the seafloor up to the surface.
   b. are characterized by clear water, which permits light to penetrate to an unusually great depth.
   c. bring warm water to the surface.
   d. trap nutrients washed into the ocean from nearby land masses.
   e. have water of lower salinity than other oceanic regions.

   ***Textbook Reference:*** *Concept 45.1 Climate and Nutrients Affect Ecosystem Function; NPP varies predictably with temperature, precipitation, and nutrients*

2. Which of the following statements about the history of climate change on Earth is *false*?
   a. Asteroid impacts, continental drift, and volcanic activity have caused changes in temperature in the past.
   b. Organisms have previously induced changes in the composition of the atmosphere.
   c. Current climate change is unique because it has been caused by a single species, *Homo sapiens*.
   d. Mass extinctions have never been associated with changes in temperature.
   e. Sunspots have caused changes in the Earth's climates.

   ***Textbook Reference:*** *Concept 45.6 Ecological Challenges Can Be Addressed through Science and International Cooperation*

3. The process by which nitrogen gas is converted to ammonium by microorganisms is called
   a. denitrification.
   b. eutrophication.
   c. nitrification.
   d. nitrogen fixation.
   e. oxidation.

   ***Textbook Reference:*** *Concept 45.3 Certain Biogeochemical Cycles Are Especially Critical for Ecosystems; Within-ecosystem recycling dominates the global nitrogen cycle*

4. Global warming is expected to
   a. decrease storm intensity.
   b. cause wet regions to become drier.
   c. produce less intense fluxes of water.
   d. prompt Hadley cells to expand in the direction of the equator.
   e. increase weather variability.

   ***Textbook Reference:*** *Concept 45.4 Biogeochemical Cycles Affect Global Climate; Recent increases in greenhouse gases are warming Earth's surface*

5. Which statement about climate change and environmental cues is *false*?
   a. Temporal relationships among cues are shifting due to climate change.
   b. Climate change is influencing day length.
   c. Mismatches in timing caused by climate change disrupt species interactions within communities.
   d. Not all environmental cues respond to the same degree to climate change.
   e. Not all environmental cues respond in the same direction to climate change.

   ***Textbook Reference:*** *Concept 45.5 Rapid Climate Change Affects Species and Communities; Changes in seasonal timing can disrupt interspecific interactions*

6. The largest pool of nitrogen is located in
   a. benthic sediments.
   b. ocean waters.
   c. soil.
   d. vegetation.
   e. the atmosphere.

   ***Textbook Reference:*** *Concept 45.3 Certain Biogeochemical Cycles Are Especially Critical for Ecosystems; Within-ecosystem recycling dominates the global nitrogen cycle*

7. Terrestrial NPP
   a. generally increases with temperature.
   b. is high when soils are low in nutrients.
   c. increases with latitude.
   d. does not vary with precipitation.
   e. is high in the dry regions around 30°N and S.

   ***Textbook Reference:*** *Concept 45.1 Climate and Nutrients Affect Ecosystem Function; NPP varies predictably with temperature, precipitation, and nutrients*

8. Which statement about biogeochemical cycles is *false*?
   a. Decomposition returns elements within macromolecules to organic compartments.

b. Carbon and nitrogen cycles are linked because both elements have water-soluble forms.
c. Biogeochemical cycles are linked when biomass is produced.
d. Biogeochemical cycles all include both organismal and nonliving components.
e. Perturbations in one biogeochemical cycle can have major effects on other cycles.

***Textbook Reference:*** *Concept 45.3 Certain Biogeochemical Cycles Are Especially Critical for Ecosystems; Biogeochemical cycles are not independent*

9. The largest pool of carbon is located in
a. soil.
b. plant biomass.
c. the atmosphere.
d. ocean waters.
e. fossil fuels.

***Textbook Reference:*** *Concept 45.3 Certain Biogeochemical Cycles Are Especially Critical for Ecosystems; Movement of carbon is linked to energy flow through ecosystems*

10. Which statement about the greenhouse effect is *false*?
a. Carbon dioxide and methane absorb strongly in infrared wavelengths.
b. Perturbations to the nitrogen cycle have no effect on Earth's radiation balance.
c. Earth's surface re-radiates some of the heat that is absorbed from the atmosphere.
d. The pools of greenhouse gases in the atmosphere determine the amount of energy kept in the lower atmosphere.
e. Earth's surface reemits photons at infrared wavelengths.

***Textbook Reference:*** *Concept 45.4 Biogeochemical Cycles Affect Global Climate; Earth's surface is warm because of the atmosphere*

11. About _______ percent of incoming solar radiation is absorbed by Earth's surface.
a. 20
b. 31
c. 49
d. 80
e. 90

***Textbook Reference:*** *Concept 45.4 Biogeochemical Cycles Affect Global Climate; Earth's surface is warm because of the atmosphere*

12. The biosphere
a. includes land and water, but not the atmosphere.
b. is thick relative to Earth's radius.
c. is where life exists on Earth.
d. includes biotic but not abiotic components.
e. includes the interior of Earth.

***Textbook Reference:*** *Concept 45.2 Biological, Geological, and Chemical Processes Move Materials through Ecosystems; The forms and locations of elements determine their accessibility to organisms*

13. Which statement about water and its cycling is *false*?
a. Most of Earth's water is in the oceans.
b. Runoff is the gravity-driven flow of liquid water from land to oceans.
c. Transpiration is the process by which water changes from gas to liquid.
d. Water vapor is a greenhouse gas.
e. Evaporation is the process by which water changes from liquid to gas.

***Textbook Reference:*** *Concept 45.3 Certain Biogeochemical Cycles Are Especially Critical for Ecosystems; Water transports materials among compartments*

14. Which statement about the effects of climate change on community composition is *false*?
a. Extinction of local populations due to climate change causes loss of diversity, which leads to loss of community function.
b. Climate change has caused species to move to lower latitudes.
c. Climate change can alter the species compositions of communities by causing shifts in geographic ranges.
d. Climate change has caused species to move up mountains.
e. Climate change can alter the species compositions of communities by altering the relative abundances of species.

***Textbook Reference:*** *Concept 45.5 Rapid Climate Change Affects Species and Communities; Climate change can alter community composition by several mechanisms*

15. Which terrestrial ecosystem has moderate NPP?
a. Tropical forest
b. Cultivated land
c. Desert
d. Tundra
e. Temperate forest

***Textbook Reference:*** *Concept 45.1 Climate and Nutrients Affect Ecosystem Function; NPP varies predictably with temperature, precipitation, and nutrients*

## Answers

### *Key Concept Review*

1.

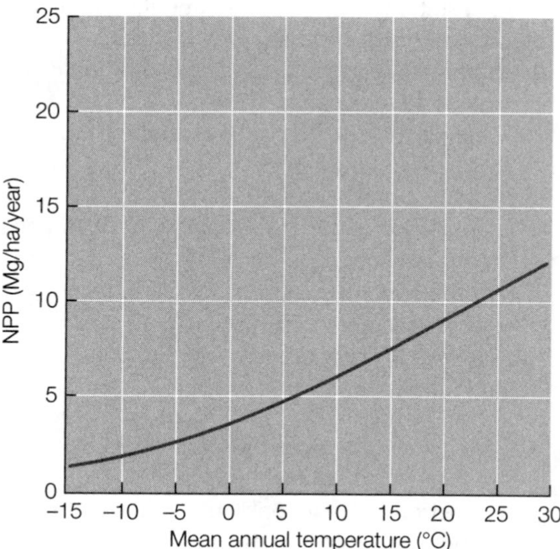


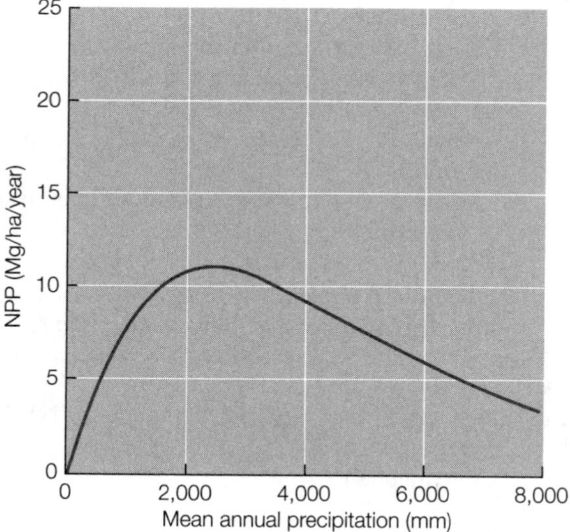


2. Marine NPP is highest in coastal areas because runoff from land brings nutrients into shallow waters. In these waters, photosynthesis occurs in the photic zone, where light can penetrate.
3. Inorganic compounds in rocks become accessible to living organisms when uplift, weathering, and erosion release them to soil and water, where they can be taken up by autotrophs (see Figure 45.5).
4. Via deposition and compaction, organic and inorganic compounds enter compartments where they are inaccessible to organisms (see Figure 45.5).
5. Trees take up water through their roots, and this water is released to the atmosphere via transpiration (the evaporation of water from plants). Transpiration also increases local humidity. Removal of trees disrupts the water cycle by decreasing transpiration, which can decrease precipitation and local humidity. Tree removal also lessens the ability of the land to retain water, so surface runoff increases. Overall, deforestation tends to dry out local ecosystems.
6. If farmers apply fertilizers to croplands in quantities that exceed the plants' abilities to absorb them, some of the excess nitrates leave the soil and enter bodies of water. This process contributes to eutrophication of lakes and the creation of "dead zones" in the sea, such as the one in the Gulf of Mexico around the mouth of the Mississippi River. In nutrient-poor terrestrial ecosystems, deposition of excess nitrogen can change the species composition of plant communities and convert species-rich communities to species-poor communities.
7. Factors that would be important to incorporate in a computer model of global climate change include incoming and outgoing radiation, patterns of circulation in the atmosphere and oceans, and interactions with Earth's surface (including those associated with human activities).
8. Global warming can affect human disease by increasing the population sizes and geographic ranges of vectors (such as insects and rodents) and pathogens (such as disease-causing viruses and bacteria) and the length of the transmission season. Cold temperatures in winter kill many pathogens. The milder winters associated with global warming will allow increasing numbers of pathogens to survive; hence the diseases they cause will become more common. Also, disease vectors such as mosquitoes, ticks, and rodents may thrive in a warmer climate and possibly extend their geographic ranges; this could result in the spread of certain diseases to new areas.
9. Mammals whose reproduction is cued to day length can still be affected by climate change because the species with which they interact may be influenced by climate change. Different species make use of different cues, and so they respond in diverse ways to climate change. This makes possible the development of timing mismatches among species in a community, which could disrupt interactions between mammals and the organisms on which they feed (e.g., trees may leaf out earlier or later, and prey species could go extinct locally). Timing mismatches could also disrupt interactions between mammals and their predators (e.g., a new species of predator could arrive through extension of its geographic range).
10. Climate change could affect bird migration in several ways. Species that typically migrate long distances

could migrate shorter distances, leading to geographic redistribution and changes in the species composition of communities. Species that typically migrate short distances could become sedentary. Species could continue to travel their usual distance but arrive earlier on their breeding grounds. This could be problematic if the timing of their reproduction failed to coincide with an adequate supply of prey to feed their nestlings, leading to a decline in their population relative to resident bird species. Changes in their abundance would in turn affect other species in the community (e.g., predators, competitors, and parasites).

11. The current period of climate change is not unprecedented from the standpoints of temperature change and change in atmospheric composition. Over the course of Earth's history, temperatures have changed and extinctions have occurred as a result of asteroids, continental drift, and volcanic activity. Organisms have also changed the composition of Earth's atmosphere, such as when the first photosynthetic microbes increased atmospheric oxygen. What is unprecedented about current climate change is that it is being caused by the activities of a single species, *Homo sapiens*.
12. Humans are capable of solving environmental problems because we can use knowledge gained through scientific study to develop solutions. Also, compared with other species, we have well-developed cooperative interactions, with regard to interactions among unrelated individuals. Nevertheless, we face the challenges of changing to more sustainable economies and curbing our multiplicative population growth.

### Test Yourself

1. **a.** Low concentrations of nutrients limit the primary productivity of most ocean waters. In zones of upwelling, nutrient-rich water is brought from the ocean bottom to the surface, thereby enhancing the growth of photosynthesizing plankton.
2. **d.** Temperature change has been linked to five mass extinctions in Earth's past.
3. **d.** Nitrogen fixation is the process by which nitrogen gas is converted to ammonium by microorganisms. It is performed, for example, by certain bacteria in the soil and by symbiotic bacteria associated with plant roots.
4. **e.** Global warming is predicted to increase weather variability and storm intensity, cause wet regions to become wetter, produce more intense fluxes of water, and cause Hadley cells to expand poleward.
5. **b.** Day length is an environmental cue that is independent of climate.
6. **e.** The largest pool of nitrogen is nitrogen gas in the atmosphere.
7. **a.** Terrestrial NPP generally increases with temperature. It is highest in the tropics and declines at high latitudes. Terrestrial NPP increases with increasing precipitation to a point and then declines. Terrestrial NPP is low in the dry regions around 30°N and S and in areas where soils are low in nutrients.
8. **a.** Decomposition returns elements within macromolecules to inorganic compartments.
9. **e.** The largest pool of carbon is found in fossil fuels.
10. **b.** Nitrous oxide is a greenhouse gas, so perturbations to the nitrogen cycle can affect Earth's radiation balance. Burning fossil fuels releases oxides of nitrogen.
11. **c.** About 49 percent of incoming solar radiation is absorbed by Earth's surface. Of the remaining amount, about 20 percent is absorbed by the atmosphere and 31 percent is reflected.
12. **c.** The biosphere is where life exists on Earth. It includes land, water, atmosphere, and organisms, and is thin relative to Earth's radius. It does not include the interior of Earth, just the thin skin at the surface.
13. **c.** Transpiration is the process by which water from plants evaporates (changes from a liquid to a gas). Plants move water from the ground to the atmosphere via transpiration.
14. **b.** Climate change has caused species to move to higher latitudes.
15. **b.** Cultivated land has moderate NPP. Terrestrial NPP is highest in tropical and temperate forests and lowest in deserts and tundra.

Notes

Notes

Notes